# LIMIT ANALYSIS *of* SOLIDS *and* STRUCTURES

JACOV A. KAMENJARZH

CRC Press
Boca Raton New York London Tokyo

**Library of Congress Cataloging-in-Publication Data**

Catalog record available from the Libarary of Congress.



Direct all inquiries to CRC Press, Inc., 2000 Corporate Blvd., N.W., Boca Raton, Florida 33431.



International Standard Book Number 0-8493-2873-x
Printed in the United States of America 1 2 3 4 5 6 7 8 9 0
Printed on acid-free paper

# CONTENTS


**Chapter I. Rigid perfectly plastic body** 1

1. Plastic deformation 1
   - 1.1. Elastic and residual strain 1
   - 1.2. Yield stress 2
   - 1.3. Elasticity domain 3
   - 1.4. Rate insensitivity 3
   - 1.5. Basic properties of plastic materials 4
2. Admissible stresses. Yield surface 5
   - 2.1. Stress space 5
   - 2.2. Admissible stresses 6
   - 2.3. Yield surface 6
3. Constitutive relations 8
   - 3.1. Normality flow rule 9
   - 3.2. Constitutive maximum principle 12
   - 3.3. Rigid perfectly plastic body 14
4. Dissipation 15
   - 4.1. Constitutive maximum principle and dissipation 15
   - 4.2. Properties of dissipation 16
   - 4.3. Yield surface determined by dissipation 18
5. Rigid–plastic problem 19
   - 5.1. Quasistatic problem 19
   - 5.2. First example of the limit load 21
   - 5.3. Rigid–plastic problem 23
   - 5.4. On the admissibility and equilibrium conditions 24
6. Beams and trusses 25
   - 6.1. Beam. Kirchhoff hypothesis 25
   - 6.2. Internal forces in a beam 27
   - 6.3. Constitutive relations for rigid perfectly plastic beam 28
   - 6.4. Rigid–plastic problem for a beam 29
   - 6.5. Truss 30
   - 6.6. Internal forces in a truss 32
   - 6.7. Rigid–plastic problem for a truss 33

Appendix A. Stress and strain 34
   - A.1. Stress 34
   - A.2. Deformation 35
   - A.3. Elasticity law 37

Appendix B. Convex sets and convex functions 37
B.1. Definitions and examples 37
B.2. Separation theorem 38
Appendix C. Extremums 39
C.1. Minimum and maximum 39
C.2. Extremum problems 40
C.3. Conditions for the minimum of a convex function 41
Comments 42

**Chapter II. Virtual work principle** 45
1. Bodies under standard loading 45
1.1. Virtual velocity fields and power 45
1.2. Virtual work principle 46
1.3. Generalized equilibrium conditions 50
2. Bodies under mixed boundary conditions 51
2.1. Tangent load and surface slip 51
2.2. Rigid punch loading 52
3. Beams and trusses 54
3.1. Virtual velocity fields and power for a beam 54
3.2. Virtual work principle for a beam 55
3.3. Generalized conditions for equilibrium of a beam 55
3.4. Virtual work principle for a truss 56
Comments 57

**Chapter III. Fundamentals of limit analysis** 59
1. Rigid–plastic problem 59
1.1. Stress fields and equilibrium conditions 59
1.2. Admissible stress fields 60
1.3. Velocity fields 61
1.4. Formulation of the problem 63
2. Safety factor 63
2.1. Admissible and inadmissible loads 64
2.2. Safety factor 65
3. Safe and limit loads 67
3.1. Safe loads 67
3.2. Safety criterion 69
3.3. Limit load and failure mechanism 70
3.4. On the failure mechanism existence 71
4. Problems of limit analysis 72
4.1. Limit stress state principle 72
4.2. Safety of a loading 73
4.3. Limit surface 74
4.4. Limit analysis in presence of a permanent load 76
5. Basic statements and methods of limit analysis 77
5.1. Lower bound for safety factor: static multiplier 78
5.2. Upper bound for safety factor: kinematic multiplier 79
5.3. Criterion for static and kinematic multipliers equality 82

5.4. Rigid-plastic solution method 84
5.5. On the kinematic method 84
6. Examples of limit analysis 85
6.1. Formulation of the axially symmetric plane strain problem 85
6.2. A pipe under internal pressure 87
6.3. A pipe under internal torsion 88
Comments 91

**Chapter IV. Limit analysis: general theory** 95
1. Rigid–plastic problem: general formulation 95
1.1. Equilibrium conditions 96
1.2. Kinematic relations 98
1.3. Constitutive relations 99
1.4. General formulation of the problem 103
1.5. Local description of material properties 105
2. Examples 106
2.1. Rigid-plastic problem for a beam 106
2.2. Rigid–plastic problem for a discrete system 108
3. Safe, limit and inadmissible loads 110
3.1. Safety factor 111
3.2. Safe stress fields 112
3.3. Safe loads 113
3.4. Safety criterion 113
3.5. Limit analysis 114
4. Static and kinematic multipliers 115
4.1. Static multiplier 115
4.2. Kinematic multiplier 115
4.3. Criterion for static and kinematic multipliers equality 117
4.4. On methods for limit analysis 119
5. Integral formulation of constitutive maximum principle 120
5.1. Integral formulation 120
5.2. Extremum property and computation of dissipation 121
5.3. Admissible stresses: a set-valued mapping 123
5.4. Conditions for equivalence of constitutive principle formulations 125
6. Extremum of integral functional 126
6.1. Integral functional 127
6.2. Evaluating the extremum 130
Appendix A. Linear spaces 136
A.1. Definitions and examples 136
A.2. Subspace 137
A.3. Linear operator 137
A.4. Pairing between linear spaces 138
Appendix B. Measurable sets and measurable functions 139
B.1. Measure 139

B.2. Measurable functions 140
Comments 140

**Chapter V. Extremum problems of limit analysis** 143
1. Static and kinematic extremum problems 143
1.1. Limit static and kinematic multipliers 143
1.2. Main results 145
2. Static and kinematic extremum problems: standard formulation 146
2.1. Minkowski function 147
2.2. Static extremum problem standard form 149
2.3. Kinematic extremum problem standard form 151
3. Dual extremum problem 153
3.1. Fenhel transformation 153
3.2. Constructing the dual problem 157
3.3. Applying the dual problem 158
3.4. Conditions for extremums equality 159
4. Conditions for equality of limit multipliers – I 159
4.1. Static extremum problem dual of the kinematic problem 160
4.2. Repeated Fenhel transformation 161
4.3. Limit multipliers equality 164
5. Bodies with bounded yield surfaces 166
5.1. Set of admissible stress fields 166
5.2. Spaces of stress and strain rate fields 170
5.3. Equality of limit multipliers 170
6. Conditions for equality of limit multipliers – II 172
6.1. Kinematic extremum problem dual of the static problem 173
6.2. Continuity of convex functions 174
6.3. Equality of limit multipliers 175
7. Bodies with cylindrical yield surfaces 177
7.1. Cylindrical yield surfaces 177
7.2. Spaces of stress and strain rate fields 178
7.3. Equality of limit multipliers 180
7.4. Another case of limit multipliers equality 182
8. Counterexamples 183
8.1. Unequality of limit multipliers 183
8.2. Unattainability of extremums over smooth fields 186
Appendix A. Normed spaces 190
A.1. Definitions and examples 190
A.2. Space of essentially bounded functions 192
A.3. Convergency. Closure. Continuity. 193
A.4. Conjugate space 194
Appendix B. Duality theorem 195
Comments 199

**Chapter VI. Reduction of limit analysis extremum problems** 201
1. Reduction of static and kinematic extremum problems 201
   1.1. Static problem reduction 201
   1.2. Kinematic problem reduction 203
   1.3. Reduced extremum problems: main results 204
   1.4. Safety factor as extremum in the reduced problems 205
2. Pressure field restoration 207
   2.1. Regularity of body boundary 207
   2.2. Distribution restoration 209
   2.3. Pressure field in a body with fixed boundary 210
   2.4. Pressure field in a body with fixed part of boundary 211
3. Approximations to vector fields 215
   3.1. Regularity of the free part of the body boundary 215
   3.2. Conditions for approximation 216
4. Approximations to solenoidal vector fields 221
   4.1. Approximation in case of fixed boundary 221
   4.2. Vector fields with a given divergence 223
   4.3. Approximation conditions – I 225
   4.4. Approximation conditions – II 228
Appendix A. Distributions 231
Appendix B. Sobolev spaces 235
   B.1. Definition and main properties 235
   B.2. Spaces of traces 238
Comments 240

**Chapter VII. Limit state** 243
1. Stress field 243
   1.1. Limit state problem 243
   1.2. Stress field 244
2. Failure mechanism 245
   2.1. Strain rate field 245
   2.2. Extension scheme 246
   2.3. Rigid-plastic problem weak formulation 248
   2.4. Limit state 250
Comments 253

**Chapter VIII. Discontinuous fields in limit analysis** 255
1. Kinematic multiplier for discontinuous velocity field 255
   1.1. On definition of kinematic multiplier 255
   1.2. Dissipation at discontinuity surface 256
   1.3. Surface slip 258
   1.4. Main property of kinematic multiplier 260
2. Methods for limit analysis 262
   2.1. Kinematic method 263
   2.2. Criterion for static and kinematic multipliers equality 263
   2.3. Rigid-plastic solutions method 267

3. Discontinuity relations 269
3.1. Normality law 269
3.2. Maximum principle 270
3.3. Normality law for velocity jump 271
3.4. On the possibility of velocity jump 272
4. Bodies with jump inhomogeneity 274
4.1. Jump inhomogeneity 274
4.2. Dissipation at a discontinuity surface 275
4.3. Kinematic multiplier and kinematic method 280
4.4. Rigid-plastic solutions method 280
4.5. Discontinuity relations 281
5. Examples of limit analysis 283
5.1. Lateral stretching of strip 283
5.2. Stretching of a strip with a hole 285
5.3. Limit surface for biaxial stretching of the plane with holes 289
5.4. A pipe under internal torsion 290
5.5. Shear of a parallelepiped with jump inhomogeneity 291
6. Derivation of the formula for kinematic multiplier 293
6.1. Formula for kinematic multiplier 293
6.2. Smoothing the jump 295
6.3. Smoothing the jump on standard domain boundary 298
6.4. Smoothing with a given trace 301
6.5. Derivation of the formula for kinematic multiplier 305
Comments 309

**Chapter IX. Numerical methods for limit analysis** 313
1. Approximations for the kinematic extremum problem 313
1.1. Formulation of the problem 313
1.2. Approximations 316
2. Discretization: finite element method 319
2.1. Idea of the method 319
2.2. Approximation for velocity fields space 321
2.3. Approximation for solenoidal velocity fields space 323
2.4. Discretized problem of limit analysis 331
3. Minimization: separating plane method 335
3.1. Subgradients 335
3.2. Infimum and $\epsilon$-subdifferentials 336
3.3. Separating plane method 337
3.4. Algorithm and convergence of iterations 341
3.5. Finding a subgradient 345
Comments 348

**Chapter X. Shakedown theory** 349
1. Elastic–plastic problem 349
1.1. Elastic perfectly plastic body 349
1.2. Elastic–plastic problem: strong formulation 350
1.3. A way to generalize formulation: examples 351

1.4. General formulation of the problem 355
1.5. Formulation in stresses 358
2. Elastic–plastic body under variable load: examples 359
2.1. Residual stresses and shakedown 360
2.2. Nonshakedown at bounded plastic strain 364
2.3. Nonshakedown at unbounded plastic strain 369
2.4. Shakedown at nonstop plastic flow 374
3. Conditions for shakedown. Safety factor 380
3.1. Definitions of shakedown and nonshakedown 381
3.2. Elastic reference body 382
3.3. Shakedown conditions and safety factor: main results 383
4. Shakedown and nonshakedown theorems 388
4.1. Shakedown theorem 388
4.2. Lower bound for plastic work 391
4.3. Damaging cyclic loading 393
4.4. Nonshakedown theorem 394
4.5. Reduction of nonshakedown theorem assumptions 397
5. Problems of shakedown analysis 398
5.1. Shakedown to a set of loads 399
5.2. One-parametric problems of shakedown analysis 401
5.3. Shakedown under thermomechanical loading 408
6. Extremum problems of shakedown analysis 410
6.1. Static extremum problem 410
6.2. Kinematic extremum problem 413
6.3. Conditions for equality of the extremums – I 418
6.4. Conditions for equality of the extremums – II 419
7. Kinematic method for safety factor evaluation 422
7.1. Formulation of the method 422
7.2. Modified kinematic problem 424
7.3. Formula for the safety factor upper bound 426
7.4. Possibility of safety factor evaluation 427
7.5. Finite element method 430
Comments 432
**Bibliography** 435

# PREFACE

Solids normally change their elastic behavior to inelastic under the action of a sufficiently large load. In particular, they undergo plastic strain that does not vanish after unloading. Plastic deformation is sometimes so intensive that it resembles fluid flow and thus becomes one of the main causes of bearing capacity loss (along with instability, fracture and fatigue). The limit analysis problem is to find out whether a given loading is safe against such plastic failure.

Engineers to a certain extent anticipated the idea of limit analysis for a long time. A systematic approach was developed by A.A.Gvozdev in the thirties and independently established by D.C.Drucker, W.Prager and H.J.Greenberg a decade later. Since the thirties, methods for limit analysis were first worked out for discrete systems and shortly after for continual ones. The last twenty five years have seen considerable development both in theory and methods of limit analysis following the recognition of convex analysis as its adequate mathematical basis. Limit analysis is now an almost complete branch of solid mechanics and applied mathematics with powerful methods and an expanding field of application.

The first aim of this book is to comprehensively and clearly present state of the art limit analysis including:

- concepts, basic ideas, and setting of problems arising from solid mechanics and mechanical engineering,
- mathematical foundations,
- analytical and numerical methods,
- examples of concrete problems solutions.

The other purpose is to present mathematical means useful in limit analysis as well as in numerous domains of science and technology.

Comprehension of the main results and their applications does not presume any knowledge beyond the regular courses in linear algebra and calculus. All necessary concepts are introduced gradually with appropriate explanations, some basic concepts and information are recalled in the Appendices. Fairly simple descriptions of ideas and main results always precede rigorous reasoning. The only exception to the rule is the chapter dealing with existence theorems, on which the rest of the chapters do not depend. Hopefully, the book will be found intelligible not only by experts but also by students specializing in applied mathematics, solid mechanics and mechanical engineering.

## Brief survey of contents

**Chapter I** presents the concept of *plastic deformation* and introduces a model of the *rigid perfectly plastic body.* A simple example illustrates specific features of perfectly plastic behavior: existence of the *limit load* and *plastic failure mechanism*, which are the main subjects of the book. Three-dimensional bodies are considered as well as beams and trusses which are examples of one-dimensional continual systems and discrete systems.

Appendices to Chapter I present some basic concepts of the mechanics and convex analysis we use throughout the book.

**Chapter II** deals with the virtual work principle, which gives a unified formulation of equilibrium conditions for mechanical systems. This formulation makes possible the development of a unified limit analysis theory for various types of loadings and for mechanical systems of all types: solids, beams, trusses, etc.

**Chapter III** gives the reader an idea of the limit analysis. Here is its concise formulation.

• The effect of the load $\mathbf{l}$ on a rigid perfectly plastic body is characterized by a nonnegative number $\alpha_l$ called the *safety factor* of the load $\mathbf{l}$:

- the body does not deform if $\alpha_l > 1$,
- the body deforms at an arbitrarily large strain rate if $\alpha_l = 1$,
- the body cannot bear the load if $\alpha_l < 1$.

• Any stress field equilibrating the load $\mathbf{l}$ determines *a lower bound* for the safety factor. In principle, it is possible to evaluate the safety factor by the static method, that is, as the best estimation obtainable this way.

• Any kinematically admissible velocity field determines *an upper bound* for the safety factor. The kinematic method suggests computing the best estimation of this type. Examples show that either the best estimation equals the safety factor or there is a gap between them. In this connection the problem arises of establishing conditions under which there is no such gap, that is, the kinematic method gives not only the upper bound for the safety factor but also its value.

• The best lower (static) and the best upper (kinematic) bounds are equal and hence are also equal to the safety factor under certain conditions. These conditions lead to formulation of a rigid-plastic problem whose solution determines the value of the safety factor.

Some simple examples in Chapter III illustrate applications of the above-mentioned methods to the limit analysis. Further development of the methods and refinement of the basic propositions of limit analysis are the subject of subsequent chapters.

**Chapter IV** establishes basic propositions of the limit analysis in the unified form for all rigid perfectly plastic systems and without smoothness hypotheses. Ideas of proofs remain the same as in the case of three-dimensional bodies (Chapter III); the proofs become more distinct since they get rid of specifically three-dimensional features.

**Chapter V** formulates extremum problems of limit analysis. These problems determine the best lower (static) and the best upper (kinematic) bounds for the safety factor. Conditions under which the safety factor can be evaluated, not only by the static method but also by the more effective kinematic method, are established.

The source of the conditions is extremum problems duality theory. Sufficiently complete presentation of the duality approach and a procedure for constructing the dual extremum problem in Chapter V enables the reader to master the widely applicable duality method.

**Chapter VI** proceeds with studying the extremum problems of limit analysis. It focuses on the important class of plastic bodies with cylindrical yield surfaces; such surfaces are normally used for constitutive modelling of metals and alloys. The original formulations of limit analysis extremum problems admit substantial reduction in the case of such bodies. The simplification arises from a theorem on restoration of the pressure field previously eliminated from the original formulation of the problem. The theorem is established in a sufficiently general case and is applicable to a wide class of problems for incompressible media.

Another important proposition established in Chapter VI is an approximation theorem for solenoidal vector fields. The theorem has some useful applications, in particular, for formulating numerical methods.

**Chapter VII** considers behavior of rigid perfectly plastic bodies under the action of limit loads, that is, of loads with the safety factor $\alpha_l = 1$. A plastic failure mechanism exists in this case: the rigid-plastic problem has solutions with an arbitrarily large strain rate.

**Chapter VIII** deals with the application of discontinuous fields to limit analysis. Making use of such fields extends the capabilities of the kinematic and rigid plastic solutions methods. For example, discontinuous solutions may be available while applying the latter method in case no smooth solution exists. A key to the extension is an explicit formula for the safety factor upper bound determined by a discontinuous velocity field. The formula takes into account the possibility of material properties jump in composite bodies. Discontinuity relations are presented as well as several examples illustrating the applications of discontinuous fields to limit analysis.

**Chapter IX** presents numerical methods for limit analysis: finite element method and separating plane method. The former allows, in particular, descretization of the kinematic extremum problem; the latter provides tools for solving the discrete problem. The separating plane algorithm applicable to a wide class of convex optimization problems is described. An important feature of the method is that it does not require differentiability of the goal function. The separating plane algorithm results in successive approximations converging to the safety factor.

**Chapter X** deals with the elastic-plastic shakedown theory which estimates the elastic-plastic body response to all possible loadings within a given set of loads. For example, analysis of structures subjected to cyclic mechanical or thermal loadings often calls for such estimating. Mechanical and mathematical foundations of the shakedown theory are presented as well as a number of examples illustrating the behavior of an elastic-plastic body under a variable loading.

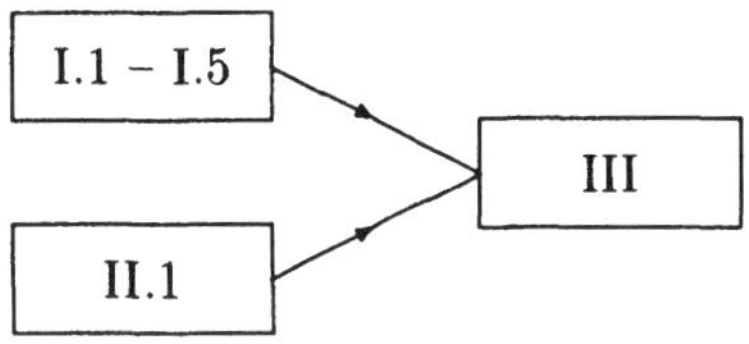


**Figure 0.1**

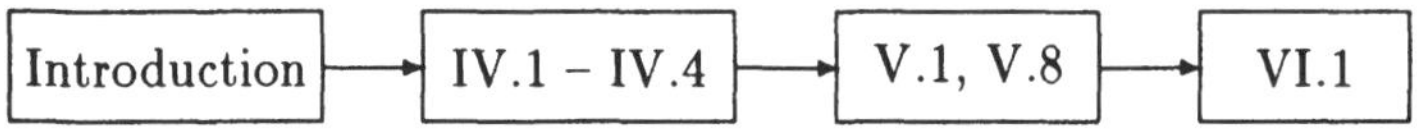


**Figure 0.2**

Despite considerable difference between the original formulations, the problem of shakedown analysis turns out to be an immediate generalization of the limit analysis problem. The experience with the limit analysis is extended to develop the shakedown theory: from establishing basic propositions to formulating effective numerical methods.

Every chapter includes **Comments** as its last section. The Comments give some additional remarks and a bibliography (not claiming to be complete).

## Guide to the book

The structure of this book enables the reader to choose his own way through the text according to his interests. Some possible routes are proposed below. Each of them corresponds approximately to the program of one of the courses delivered by the author at Moscow State University and University Lille I.

Roman and Arabic numerals stand for chapters and sections, respectively; for example, IV.2 denotes Section 2 of Chapter IV.

**1. Introduction to limit analysis.** Figure 0.1 shows the reading plan. The indicated sections contain information that could be found in many textbooks on plasticity. It is sufficient to read these sections to get an idea of limit analysis and is necessary to be ready for the further studying of the subject. The reader who already has a notion of the limit analysis may just gloss over these sections to accustom himself to the notations we use throughout this book.

The indicated sections deal with three-dimensional bodies under standard loading. It is useful to add three more sections: Sections I.6 and II.3 add other rigid-plastic systems to the picture; Section II.2 considers more general types of loading.

In what follows "Introduction" means the above-described Introduction to Limit Analysis.

**2. Limit analysis: general theory.** Figure 0.2 shows the reading plan. The program covers all the basic propositions of limit analysis. Their for-

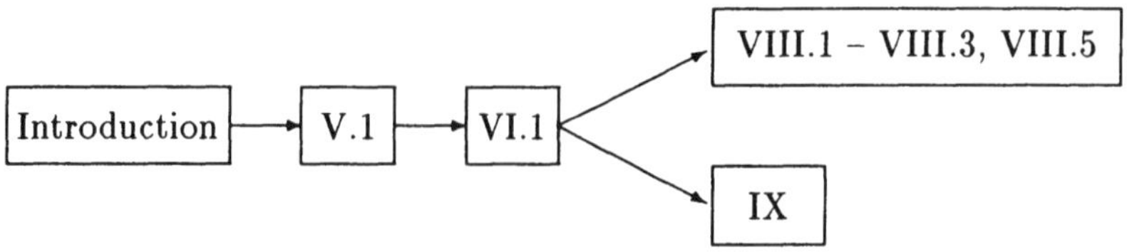


**Figure 0.3**

mulations are simple; the greater part of the proofs is not included in the program. Thus, the main results are considered rather than the methods to obtain them. The indicated sections are intelligible for any reader and give him a state-of-the-art notion of limit analysis theory.

The basic propositions of limit analysis are established in a unified form for various rigid perfectly plastic systems: solids, beams, frames, etc. Formulations of the static and kinematic extremum problems are given for evaluating the safety factor. The formulations are substantially reduced for bodies with cylindrical yield surfaces.

**3. Methods for limit analysis.** Figure 0.3 shows the reading plan. The program addresses the reader interested in the applications of limit analysis, that is, in practical calculation of the safety factor. The program includes formulations of the extremum problems for evaluating the safety factor, their modifications, and methods for solving these problems. The modified formulations extend capabilities of the limit analysis by involving discontinuous fields, whose application to solving concrete problems is illustrated in Section VIII.5.

The finite element method and the separating plane method for computing the safety factor are described in Chapter IX.

We do not consider the well known slip lines method in this book. The reader interested in limit analysis and not familiar with this method should definitely learn about it. Formulation of this method and examples of its applications to limit analysis can be found in almost every textbook on plasticity.

**4. Mathematical foundations of limit analysis.** Figure 0.4 shows the reading plan. The dotted arrows indicate optional reading paths. The program covers formulations and proofs of the main propositions of limit analysis as well as formulations of the limit analysis methods.

The program also includes descriptions of adequate mathematical tools and a scheme for applying them to various mechanical problems. In this book we make use of these tools twice: in limit analysis and in the shakedown theory. They are also useful in many other domains of science and technology. The main tools are: a theorem on extremum of integral functional (Section IV.6), a procedure for constructing dual extremum problems and conditions for equality of the extremums in the dual and original problems (Section V.3). For problems of incompressible media mechanics, these tools are substantially supplemented by theorems on pressure field restoration (Section VI.2) and an approximation theorem for solenoidal vector fields (Section VI.4).

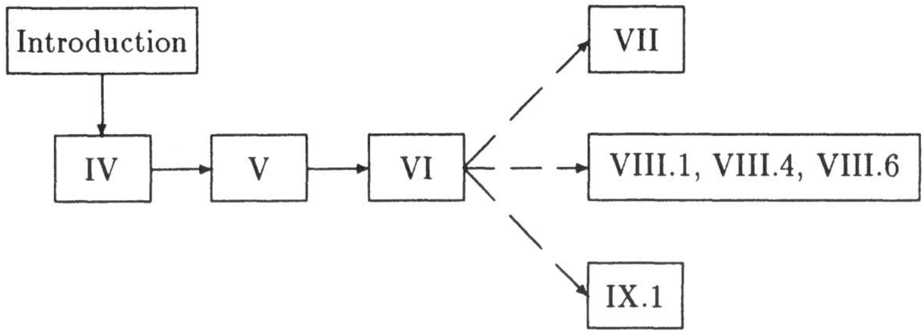


**Figure 0.4**

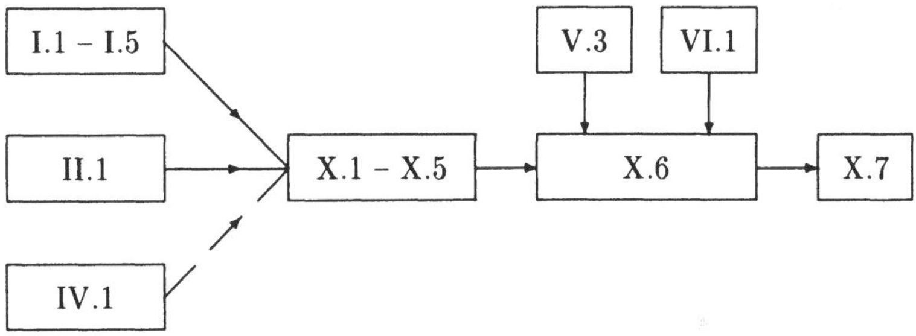


**Figure 0.5**

The optional reading paths lead to the formula for the safety factor upper bound as determined by a discontinuous velocity field (sections of Chapter VIII), convergence of finite element approximations to the safety factor (Sections IX.1, IX.2), and the existence theorem (Chapter VII).

**5. Foundations of shakedown theory.** Figure 0.5 shows the reading plan. Sections I.1 to I.5 and II.1 play the part of an Introduction, and it is worthwhile to supplement the latter with Section IV.1. Sections X.1 to X.5 deal with the mechanical foundations of shakedown theory: from formulation of the elastic-plastic problem to the shakedown criterion. The criterion results in the concept of (shakedown) safety factor. Mathematical foundations of the shakedown theory are considered in Sections X.6 and X.7 following the same scheme as in limit analysis. The reader interested in the mechanical aspect of the theory may skip Section X.6 and read the part of Section X.7 presenting a kinematic method for evaluating the safety factor.

## Notations

The main variables limit analysis deals with are velocity $\mathbf{v}$, strain rate and stress tensors $\mathbf{e}$ and $\boldsymbol{\sigma}$. We use bold characters for them as well as for other vectors and tensors. Default coordinates $x_1$, $x_2$, $x_3$ are Cartesian. Components of vectors and tensors, for example, $v_i$, $e_{ij}$, $\sigma_{ij}$ are considered in the Cartesian frame everywhere in this book apart from solving some concrete problems. Since the coordinate frame is fixed, the tensor nature of variables does not come into the picture; therefore, the reader may identify, for example, the second rank tensor with the matrix of its components.

We use the expression $\mathbf{e} \cdot \boldsymbol{\sigma}$ to denote the "complete" contraction of second rank tensors $\mathbf{e}$ and $\boldsymbol{\sigma}$: $\mathbf{e} \cdot \boldsymbol{\sigma} = e_{ij}\sigma_{ij}$. We adopt the standard summation over repeated indices convention.

We use decimal section numbering, for example, "Subsection IV.1.3" refers to the third subsection of the first section in Chapter IV. We omit the chapter number when referring to a section within the chapter. Analogously, decimal numbering is used for formulas and figures.

* *
*

I am very grateful to Prof. M.E. Eglit for her discussion and important remarks on some chapters. I would also like to thank Mr. G.Reiss–Souto for discussing the numerical methods for limit analysis. Ms. A.Makhlin substantially improved my translation of the manuscript which was a great help in the preparation of this book.

I am thankful to my colleagues from the Fluid Mechanics Department of Moscow State University for a friendly and stimulating atmosphere.

*May 21, 1995.* *J.Kamenjarzh*

Some sections in this book are based on research which was made possible in part by Grant No ND6000 from the International Science Foundation.

CHAPTER I

# RIGID PERFECTLY PLASTIC BODY

This chapter presents the rigid perfectly plastic model of a solid. A complete system of equations and boundary conditions describing behavior of such a body is formulated. A simple example illustrates specific features of the perfectly plastic behavior, in particular, existence of the limit load and failure mechanism, which are the main subjects of this book. Beside three-dimensional bodies, beams and trusses are considered as examples of one-dimensional continual systems and discrete systems.

## 1. Plastic deformation

A steel spring subjected to a small load behaves elastically; it deforms during loading and regains its initial length after unloading. However, if the load is sufficiently large, the strain to the spring does not vanish after unloading. In this section, we consider some characteristic features of such behavior, which is called plastic.

**1.1. Elastic and residual strain.** The plastic response to a loading studied in detail in numerous experimental works is typical for metals, alloys, and many other materials. Figure 1.1 shows, for example, test data for three rod specimens made of aluminum of two different brands ( curves 1 and 2) and of steel (curve 3) subjected to quasistatic, isothermal, axial loading. The diagram shows the strain (relative elongation) $\varepsilon$ versus the tensile stress $\sigma$. During the first loading while the stress is sufficiently small, the strain is proportional to the stress in accordance with the linear elasticity law: $\varepsilon = \sigma/E$, E=const. Specimens also obey this law under any loading no matter if $\sigma$ increases or decreases provided the stress remains sufficiently small. In particular, the strain $\varepsilon$ vanished after unloading when $\sigma = 0$.

However, the specimens' behavior changes as soon as the stress reaches a certain critical value, for example, $\sigma_A$ which corresponds to the state $A$ of specimen 1 (see Figure 1.1). The diagram shows that the stress–strain relation is not linear, and moreover, is not elastic any longer. Consider the test data for specimen 1. The rod was subjected to an increasing load. During the loading the point representing the current state of the specimen moved along the curve $OAB$. After the loading stopped and the stress reached the value $\sigma_B$ corresponding to the state $B$, the unloading started. If the deformation was elastic, the current state point would move along the same curve $OAB$ at both loading and unloading (in the opposite directions). As far the

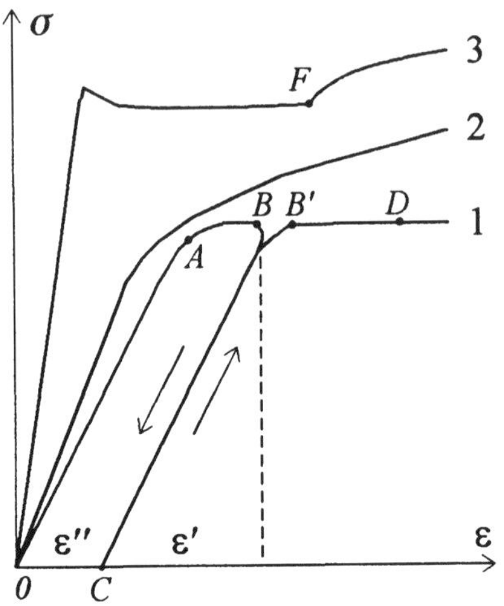


**Figure 1.1**

(Figure 1.1 is from Drucker, D.C., *Introduction to Mechanics of Deformable Solids*, McGraw-Hill, New York, 1967. With permission.)

curve $BC$ along which the point moved at the unloading, it can be concluded that the loading–unloading process corresponding to the curve $OABC$ is not elastic.

The linear elasticity law, nevertheless, holds for the strain and stress *increments* during unloading. The increments are proportional with the same elasticity coefficient that relates the strain and stress during the initial loading $OA$ (see Figure 1.1):

$$\varepsilon - \varepsilon_B = \frac{1}{\mathrm{E}} (\sigma - \sigma_B).$$

There remains the nonzero strain $\varepsilon = \varepsilon_B - \sigma_B/\mathrm{E}$ in the rod after the unloading ended at $\sigma = 0$. The elastic part $\varepsilon' = \sigma_B/\mathrm{E}$ of the strain $\varepsilon_B$ was removed during the unloading, and the remaining part $\varepsilon'' = \varepsilon_B - \varepsilon'$ of the strain is *a residual* or *permanent* strain. Thus, we can consider the strain $\varepsilon$ at any current state as the sum $\varepsilon = \varepsilon' + \varepsilon''$ of the elastic and residual parts. One can measure both parts of the strain using the unloading test, that is, unloading the specimen from the current state. The elastic part can also be evaluated using the elasticity law.

**1.2. Yield stress.** The residual strain appears in specimen 1 if it is loaded, for example, up to the state $B$. On the other hand, a process in which the stress does not reach a certain critical value generates no residual strain. We refer to this critical stress value as the *elasticity limit* (elasticity limit in tension in the case under consideration), which in the case of specimen 1 corresponds to the state $A$ in Figure 1.1. Elasticity limit draws the boundary between two types of material response to a loading. Accumulation of residual strain becomes possible when the stress reaches the elasticity limit, while the residual strain rate may be rather high at almost constant stress equal to the elasticity limit. Therefore, we also refer to the latter as *yield stress.*

Elasticity limit may vary with strain. As far as the uniaxial tension–compression of a rod is concerned, the initial value of the elasticity limit is replaced by the largest value the stress takes during the process. Suppose, for example, that specimen 1 is loaded up to the state $B$, then unloaded and reloaded again. The two latter steps correspond to the curves $BC$ and $CB'D$ in Figure 1.1. There is only a negligible difference between the segments $BC$ and $C'B'$ of the curves, which, in particular, allows us to consider the stresses $\sigma_B$ and $\sigma_{B'}$ practically equal. During the loading, the stress $\sigma$ varies from 0 up to $\sigma_B \approx \sigma_{B'}$, and the linear elasticity law relates the strain and stress increments: $\varepsilon - \varepsilon_B = (\sigma - \sigma_B)/E$. The specimen shows the same behavior for any process in which the stress varies between 0 and $\sigma_B$ as the residual strain does not vary. As soon as the stress exceeds the value $\sigma_B$, the behavior becomes inelastic, which is represented by the segment $BD$ in Figure 1.1. Residual strain increases along this path and it is easy to detect its increment by unloading the specimen starting, for example, from the state $D$. The critical stress value, which separates the domains of the elastic and inelastic responses, is $\sigma_B$. Thus, $\sigma_B$ is the new value of the elasticity limit.

The elasticity limit (the yield stress) of a certain material practically may not vary for a wide range of strain values. The residual strain is accumulated at almost constant stress in this case. A horizontal segment of the stress–strain diagram referred to as a *yield plateau* corresponds to such a process. The diagram for specimen 1 in Figure 1.1 contains the yield plateau, and the edge of the plateau was not reached in the experiment which we discuss here. The yield plateau for specimen 3 ends in the state $F$, and there is no yield plateau for specimen 2.

It is worth noting, that while deforming along the yield plateau, the elastic strain remains constant together with the stress, and considerable residual strain may be accumulated in such a process.

**1.3. Elasticity domain.** As we see from the test data, stress and strain increments are related by the elasticity law for tensile stresses ranging from 0 to the elasticity limit in tension. The elasticity law also holds in the case of compression as long as the compressive stress stays lower than its critical value, that is, the elasticity limit in compression. Variation of the stress between the elasticity limits in tension and compression produces no variation of the residual strain. The elasticity law relates stress and strain increments during such processes. Therefore, we refer to the domain between the yield stresses in tension and compression as *an elasticity domain* (in tension–compression).

**1.4. Rate insensitivity.** Test data for identical specimens subjected to the same load may differ substantially if the loadings are carried out at different rates. It is quite possible, for example, to obtain large residual strain in a specimen stressed to a level less than its elastic limit.

Consider, for example, a rod specimen under a constant load resulting in a stress much lower than the elasticity limit. As we see from the test data, no residual strain appears in the specimen. However, this is true provided that the time the specimen remains loaded is not too long; otherwise, considerable residual strain may be accumulated in the specimen (it is called creep strain). On the other hand, when loaded at a high rate, the specimen behaves similarly to the patterns shown in Figure 1.1. At the same time, the yield stress turns out to be much greater in this case than in the case of a loading at a moderate rate. For many materials, there exists a large interval between the above mentioned extremes, within which the response of a specimen to the loading does not depend on the rate. We refer to this property as *rate insensitivity.* Many metals and alloys are rate insensitive within the strain rate interval $10^{-5}$ to $10^{-2}$ 1/sec. The test data in Figure 1.1 were obtained within the interval of rate intensity.

**1.5. Basic properties of plastic materials.** In the preceding section, some characteristic features of plastic behavior in the case of uniaxial tension–compression were discussed. These are the most important properties of plastic materials. We take them as the basis of the plastic material definition and will now summarize and formulate them more accurately.

In what follows we restrict ourselves to isothermal processes, in which the temperature $T$ is the same at every point of the body and does not vary with time. This assumption is satisfied for a body with sufficiently high thermoconductivity if the body is surrounded by a medium with the constant temperature $T_0$ and is subjected to a loading at a moderate rate. In such a case, the intensive heat conduction equalizes the temperature everywhere in the body, which takes a much shorter than characteristic time of the load variation. Normally, many metal structures work under such conditions.

The properties we discuss below relate to states and processes at a point of the body. One may equivalently say that the properties relate to *uniform* states and processes in the whole body. We refer to a state of a body as uniform if the states at all points of the body are the same. A process is uniform if all the states of the body are uniform during the process. When studying material properties experimentally, one normally arranges testing so as to make the process in the specimen as close to a uniform process as possible.

Formulations of the plastic material properties generalize test data obtained in numerous experiments, and idealize these data. Therefore, formulating the properties concerning the response of a plastic body to various loadings is a first step towards constructing a mathematical model of the plastic behavior. Here, *loading* is a process of continuous variation of the stress. We refer to the state with zero stress as *unloaded.*

Material is referred to as *plastic material* if it possesses the following properties:

- Starting from any state, a loading which (i) results in stress and strain increments related by the elasticity law, and (ii) ends in the unloaded state can be carried out. This process is referred to as *complete unloading.*

  Normally, for a given state there are many processes which start

at this state and are complete unloadings. The strain remaining after complete unloading is referred to as *residual strain*. The residual strain at a given state does not depend on the choice of complete unloading: strain variation obeys the elasticity law and therefore is only determined by the initial and final stresses.
- There exist loadings which result in the variation of residual strain.
- Material is rate intensive. More precisely, two loadings along the same path in the same direction (possibly at different rates) result in the same state.

Residual strain of bodies which possess these properties is referred to as *plastic strain*.

Many plastic materials also possess the following additional properties:
- For every state, there exists the maximally extended stress domain such that the stress and strain increments are related by the elasticity law for any loading within the domain. This domain is referred to as *elasticity domain*.
- Stresses lie either in the elasticity domain that corresponds to the current state or on its boundary.
- Elasticity domain is convex and contains the zero stress.
- Elasticity domain can vary provided plastic strain varies.

We only consider materials possessing all of the above properties. These are the most general properties of plastic materials. To construct a concrete model of a plastic body one specifies its elasticity domain and evolution laws for the elasticity domain and plastic strain.

## 2. Admissible stresses. Yield surface

Plastic behavior of solids as observed experimentally is rather complicated. Fortunately, it is not necessary to go into details to solve many important problems, and it is sufficient to make use of simplified models. The rigid perfectly plastic model is the simplest one; however, it possesses all of the basic properties listed in the previous section. The main concept of this model is a set of admissible stresses (or, equivalently, the yield surface, which is the boundary of this set). In this section we introduce these concepts and some examples of yield surfaces.

**2.1. Stress space.** Stresses in solids are represented by symmetric second rank tensors. Such tensors also represent strain and strain rate. A second rate tensor may be identified with the matrix composed of the tensor's components as long as a certain coordinate system is chosen. The matrix is symmetric in the case of a symmetric tensor. The sum of two second rate tensors is a tensor with a matrix of components equal to the sum of the matrices of the tensors-summands' components. Similarly, the product of a second rank tensor and a number is a second rate tensor with a matrix of components equal to the product of the matrix of tensor-multiplier's components and the number. Symmetric second rank tensors form a linear space with respect to these operations. The space is six-dimensional if tensors are considered in a three-dimensional space. We denote by *Sym* the six-dimensional space of symmetric second rank tensors.

The complete contraction $\mathbf{a} \cdot \mathbf{b} = a_{ij} b_{ij}$ of tensors $\mathbf{a}$, $\mathbf{b}$ defines a scalar multiplication in *Sym*, which makes the space Euclidean. This allows interpreting and depicting symmetric second rank tensors as vectors. For example, one can speak of their orthogonality. Such interpretation makes some structures visual, and we will often use it in what follows. The symbol $|\mathbf{a}| = (\mathbf{a} \cdot \mathbf{a})^{1/2}$ denotes the norm in *Sym*.

We denote by $\mathbf{I}$ the identity tensor (metric tensor of the three-dimensional Euclidean space). Every tensor $\mathbf{t}$ in *Sym* is decomposed into *deviatoric* and *spheric* parts $\mathbf{t}^d$ and $\mathbf{t}^s$:

$$\mathbf{t} = \mathbf{t}^d + \mathbf{t}^s, \quad \mathbf{t}^s = \frac{1}{3} t_{ii} \mathbf{I}, \quad \mathbf{t}^d = \mathbf{t} - \frac{1}{3} t_{ii} \mathbf{I}.$$

We denote the subspace of deviatoric tensors, that is, tensors with the zero spheric part by $Sym^d$,

$$Sym^d = \{\mathbf{t} \in Sym : \ \mathbf{t}^s = 0\}.$$

This subspace is often called a *deviatoric plane.*

**2.2. Admissible stresses.** To introduce a model of a plastic body, we start by describing the elasticity domain and adopting a constitutive law for its evaluation. We consider materials with a large yield plateau, that is, with an almost constant elasticity limit. We will assume the elasticity domain to be constant, thus restricting ourselves to the simplest case of its evolution law. Models making use of this assumption are referred to as *perfectly plastic* and are the only ones considered throughout this book.

Stresses in a plastic body lie either inside the elasticity domain or on its boundary, see Subsection 1.5. Hence, admissible stress in a perfectly plastic body belongs to a fixed set consisting of the elasticity domain and its boundary.

REMARK 1. Restricting the range of possible stresses is not an original idea of plasticity. The first model of continuum mechanics, the perfect fluid, imposes a well known limitation: stresses are reduced to pressure, that is, the deviatoric part of the stress tensor is zero.

We consider a perfectly plastic body and denote by $C_x$ the set of stresses admissible at its point $x$. It will always be assumed that the set $C_x$ is convex, contains the zero stress, and is closed, that is, $C_x$ includes its boundary. We admit dependence of $C_x$ upon the point $x$; this means that, generally speaking, we consider inhomogeneous bodies.

**2.3. Yield surface.** The boundary of the set $C_x$ of admissible stresses is referred to as the *yield surface.* As long as the stress varies within the interior of $C_x$, the plastic strain $\boldsymbol{\varepsilon}^p$ remains constant: the interior of $C_x$ is the elasticity domain. The plastic flow, that is, variation of $\boldsymbol{\varepsilon}^p$ can occur provided the stresses are on the yield surface (this is where its name originated from). Apparently, the set $C_x$ uniquely determines the yield surface and vice versa.

Normally, the yield surface is specified by the equation $F_x(\boldsymbol{\sigma}) = 0$, where $F_x$ is a convex continuous function determined on the space *Sym* of symmetric second rank tensors and is referred to as a *yield function.* The inequality $F_x(\boldsymbol{\sigma}) \le 0$ specifies the set of admissible stresses:

$$C_x = \{\boldsymbol{\sigma} \in Sym : \ F_x(\boldsymbol{\sigma}) \le 0\}.$$

Convexity of the yield function implies convexity of the set $C_x$.

EXAMPLE 1. *Mises yield surface.* Consider the function $F_x(\boldsymbol{\sigma}) = |\boldsymbol{\sigma}^d| - \sqrt{2}k_x$, $k_x > 0$ on *Sym*. It is convex and continuous, and it determines a set $C_x$ of admissible stresses:

$$C_x = \left\{\boldsymbol{\sigma} \in Sym : \ |\boldsymbol{\sigma}^d| \le \sqrt{2}k_x\right\}.$$

The set $C_x$ is convex and closed, and it contains the zero stress. The boundary of $C_x$ is determined by the equation $|\boldsymbol{\sigma}^d| = \sqrt{2}k_x$ and is called *Mises yield surface.* The parameter $k_x$ has a clear mechanical meaning. Namely, consider pure shear stress $\boldsymbol{\sigma}$ with the only nonzero components $\sigma_{12} = \sigma_{21}$. Then $\boldsymbol{\sigma}$ is on the yield surface provided $|\sigma_{12}| = k_x$. Thus $k_x$ turns out to be the value of the shear stress for which plastic flow is possible. In other words, $k_x$ is the yield stress or the elasticity limit in shear.

EXAMPLE 2. *Tresca yield surface.* Consider a point $x$ of a plastic body. Let $\boldsymbol{\sigma}$ be the stress at $x$ and $\mathbf{p}_{\boldsymbol{\nu}}$ be the traction exerted by $\boldsymbol{\sigma}$ on a surface element at $x$ with the unit normal $\boldsymbol{\nu}$ (see Appendix A).

Let $\tau(\boldsymbol{\sigma}, \boldsymbol{\nu})$ be the tangent component of $\mathbf{p}_{\boldsymbol{\nu}}$ and $\tau_{\max}$ be the maximum value of $\tau(\boldsymbol{\sigma}, \boldsymbol{\nu})$ over all the normals of surface elements at $x$:

$$\tau_{\max} = \max\left\{\tau(\boldsymbol{\sigma}, \boldsymbol{\nu}) : \ \boldsymbol{\nu} \in \mathbf{R}^3\right\}.$$

A simple computation results in an expression for $\tau_{\max}$ through principal stresses $\sigma_1(\boldsymbol{\sigma})$, $\sigma_2(\boldsymbol{\sigma})$, $\sigma_3(\boldsymbol{\sigma})$ determined as the roots of the equation

$$\det \|\sigma_{ij} - \lambda\delta_{ij}\| = 0.$$

Namely, we have

$$\tau_{\max} = \frac{1}{2}\max\{\, |\sigma_1(\boldsymbol{\sigma}) - \sigma_2(\boldsymbol{\sigma})|\,,\ |\sigma_2(\boldsymbol{\sigma}) - \sigma_3(\boldsymbol{\sigma})|\,,\ |\sigma_3(\boldsymbol{\sigma}) - \sigma_1(\boldsymbol{\sigma})|\,\}.$$

Making use of this expression, it is easy to show that $\tau_{\max}$ is a convex and continuous function of $\boldsymbol{\sigma}$. We introduce the yield function $F_x(\boldsymbol{\sigma}) = \tau_{\max}(\boldsymbol{\sigma}) - k_x$, $k_x > 0$, which determines the set $C_x$ of admissible stresses in *Sym*:

$$C_x = \{\boldsymbol{\sigma} \in Sym : \ \tau_{\max}(\boldsymbol{\sigma}) \le k_x\}.$$

The set $C_x$ is convex and closed, and it contains the zero stress. The boundary of $C_x$ is determined by the equation $\tau_{\max}(\boldsymbol{\sigma}) = k_x$ and is called *Tresca yield*

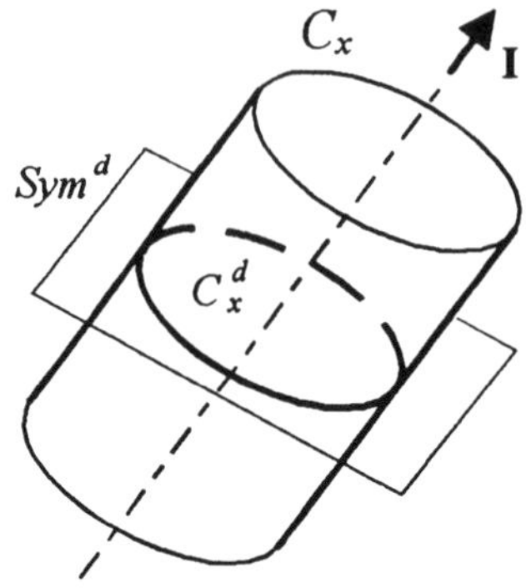


**Figure 2.1**

*surface*. Unlike the smooth Mises yield surface, the Tresca surface is piecewise smooth with the corners formed by its smooth parts.

EXAMPLE 3. *Cylindrical yield surfaces.* The Mises and Tresca yield functions possess a common feature: they do not depend on the spheric part of the stress. This means that the yield function $F_x$ satisfies the equality $F_x(\boldsymbol{\sigma}) = F_x(\boldsymbol{\sigma} - p\mathbf{I})$ for every stress $\boldsymbol{\sigma}$ and every number $p$. If the yield function $F_x$ satisfies this condition, then adding any spheric tensor $-p\mathbf{I}$ to an admissible stress $\boldsymbol{\sigma}$ results in the admissible stress $-p\mathbf{I} + \boldsymbol{\sigma}$. In other words, the set $C_x$ of admissible stresses is a cylinder in the space $Sym$ with its axis directed along the identity tensor $\mathbf{I}$. We refer to such sets of admissible stresses as *cylindrical*; the axis direction is always presumed to be the $\mathbf{I}$-direction. The corresponding yield surface is also called *cylindrical.* Special attention will be paid to cylindrical yield surfaces since they are typical for metals and alloys.

The cylinder $C_x$ is uniquely determined by its cross-section $C_x^d$, that is, by the intersection of $C_x$ and the deviatoric plane $Sym^d$ orthogonal to $\mathbf{I}$ (see Figure 2.1):

$$C_x^d = \{\boldsymbol{\sigma} \in C_x : \ \boldsymbol{\sigma}^s = 0\}.$$

The cylinder $C_x$ is convex and closed, and it contains the zero stress if and only if the cross-section $C_x^d$ possesses these properties.

## 3. Constitutive relations

Admissible stresses in a perfectly plastic body lie either inside or on the yield surface. Although the plastic strain does not vary in the first case, the second case allows plastic flow, thus requiring certain relations to describe the process. Two equivalent formulations of plastic constitutive relations will be introduced in this section. The first one, the normality flow rule, is more convenient for solving concrete problems. The second one, the constitutive maximum principle, is preferable for studying general properties of plastic bodies.

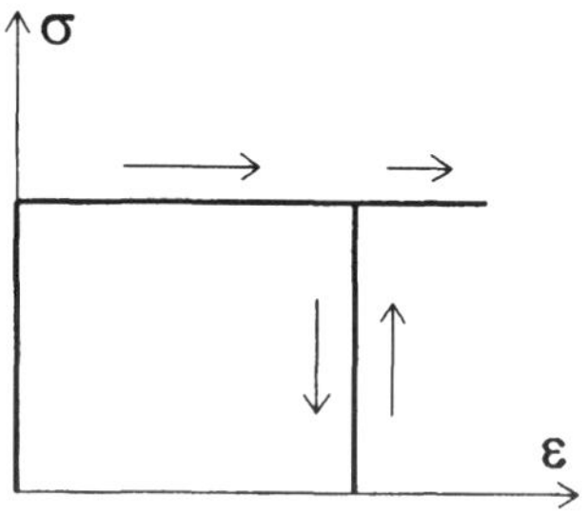


**Figure 3.1**

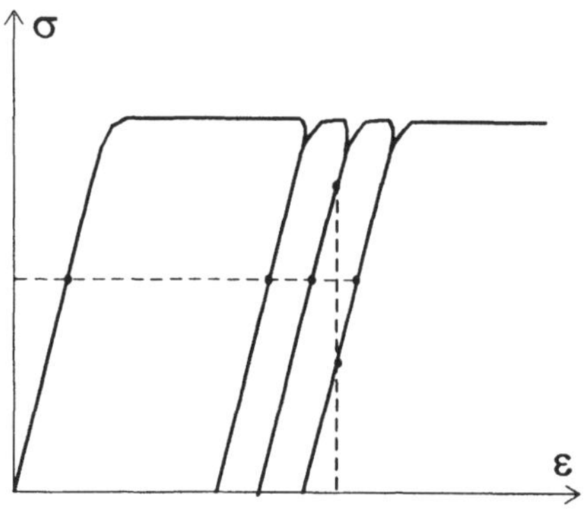


**Figure 3.2**

**3.1. Normality flow rule.** Consider a plastic material with a large yield plateau. Plastic flow along the plateau results in plastic strain much greater than the elastic one. Therefore, in many cases, it is acceptable for the researcher to neglect the elastic strain. By doing this we replace the real stress-strain diagram (like diagram 1 in Figure 1.1) by the idealized diagram shown in Figure 3.1. Models which make use of this assumption are referred to as *rigid-plastic*. We restrict ourselves to these models, thus assuming that the plastic strain is the total strain and the elastic strain is zero. More general elastic-plastic models are considered in Chapter X.

Let the condition $F_x(\boldsymbol{\sigma}) \leq 0$ specify the set $C_x$ of admissible stresses and the equation $F_x(\boldsymbol{\sigma}) = 0$ determine the corresponding yield surface. Let $\mathrm{e}$ denote the strain rate. According to the assumption about the rigid-plastic behavior, no strain occurs if the stress $\boldsymbol{\sigma}$ is inside the yield surface:

$$\mathrm{e} = 0 \quad \text{if} \quad F_x(\boldsymbol{\sigma}) < 0. \tag{3.1}$$

Let us now formulate the plastic constitutive relations in case the stress $\boldsymbol{\sigma}$ is on the yield surface. We start with the following two remarks.

First of all, the strain cannot be a function of the stress in the case of plastic material, and neither can the stress be a function of the strain. Consider, for example, a uniaxial loading like that described in Section 1. Let increasing of the load alternate with unloading as shown in Figure 3.2. Then, various stresses can be observed at the same strain and various strains at the same stress. Therefore, none of these variables is a function of the other.

Secondly, the strain rate cannot be a function of the stress. Indeed, deforming along a yield plateau occurs at the fixed stress $\boldsymbol{\sigma}$ but can be carried out at various strain rates $\mathbf{e}$. Therefore, $\mathbf{e}$ is not a function of $\boldsymbol{\sigma}$. The same conclusion arises from the following formal reasoning. Suppose $\mathbf{e}$ is a function of $\boldsymbol{\sigma}$. This means that the corresponding six constitutive equations for the components $e_{ij}$ are included in the system which describes behavior of the body. The system always includes a momentum balance equation and kinematic formulas which express $\mathbf{e}$ through the velocity components $v_i$. The equation $F_x(\boldsymbol{\sigma}) = 0$ is to be included in the system as well: it indicates that deforming is possible provided the stress is on the yield surface. Thus, we have arrived at an overdetermined system. Hence, the constitutive assumption $e_{ij} = f(\sigma_{kl})$ is not appropriate in the case of a plastic material.

The last remark suggests how to improve the constitutive assumption. Five equations expressing the strain rate *direction* $\mathbf{e}/|\mathbf{e}|$ as a function of $\boldsymbol{\sigma}$ can be used instead of the six equations $e_{ij} = f(\sigma_{kl})$ (this leaves the magnitude $|\mathbf{e}|$ undetermined). To specify $\mathbf{e}/|\mathbf{e}|$, we assume that the strain rate tensor $\mathbf{e}$ is in the direction of the outward normal to the yield surface at the point $\boldsymbol{\sigma}$. This assumption together with relation (3.1) can be written in the following form:

$$e_{ij} = \begin{cases} \dot{\lambda}\,\dfrac{\partial F_x}{\partial \sigma_{ij}}(\boldsymbol{\sigma}), \quad \dot{\lambda} \geq 0 & \text{if } F_x(\boldsymbol{\sigma}) = 0, \\ 0 & \text{if } F_x(\boldsymbol{\sigma}) < 0, \end{cases}$$

or more concisely:

$$\mathbf{e} = \begin{cases} \dot{\lambda}\,\dfrac{\partial F_x}{\partial \boldsymbol{\sigma}}(\boldsymbol{\sigma}), \quad \dot{\lambda} \geq 0 & \text{if } F_x(\boldsymbol{\sigma}) = 0, \\ 0 & \text{if } F_x(\boldsymbol{\sigma}) < 0. \end{cases} \tag{3.2}$$

These constitutive relations are referred to as the *normality flow rule*, the multiplier $\dot{\lambda}$ in (3.2) being an additional unknown. Therefore, the system of equations we arrive at is no longer overdetermined. Notice that $\dot{\lambda}$ should be considered as an indivisible symbol; we just maintain a tradition using this notation. (In all other cases the dot indicates the time differentiation.)

EXAMPLE 1. Consider an explicit form of the normality rule in the case of the Mises yield function

$$F_x(\boldsymbol{\sigma}) = |\boldsymbol{\sigma}^d| - \sqrt{2}k_x = \sqrt{\left(\sigma_{kl} - \frac{1}{3}\sigma_{mm}\delta_{kl}\right)\left(\sigma_{kl} - \frac{1}{3}\sigma_{mm}\delta_{kl}\right)} - \sqrt{2}k_x.$$

Here, $k_x > 0$ does not depend on $\boldsymbol{\sigma}$. Taking into account the formulas

$$\frac{\partial \sigma_{kl}}{\partial \sigma_{ij}} = \delta_{ki}\delta_{lj}, \qquad \frac{\partial \sigma_{mm}}{\partial \sigma_{ij}} = \delta_{ij},$$

we find

$$\frac{\partial F_x}{\partial \sigma_{ij}} = \frac{1}{|\sigma^d|}\left(\sigma_{kl} - \frac{1}{3}\sigma_{mm}\delta_{kl}\right)\left(\delta_{ki}\delta_{lj} - \frac{1}{3}\delta_{kl}\delta_{ij}\right)$$
$$= \frac{1}{|\sigma^d|}\left(\sigma_{ij} - \frac{1}{3}\sigma_{kk}\delta_{ij}\right) = \frac{1}{|\sigma^d|}\sigma^d_{ij}.$$

The expression

$$\dot{\lambda}\frac{\partial F_x}{\partial \sigma_{ij}} = \dot{\lambda}\frac{1}{|\sigma^d|}\sigma_{ij}$$

for the outward normal enters the normality rule. Here, $\dot{\lambda} \geq 0$ is an unknown, which can be conveniently replaced by $\dot{\lambda}^* = \dot{\lambda}/|\boldsymbol{\sigma}|$. Let us redenote the new variable and use again the symbol $\dot{\lambda}$ for $\dot{\lambda}^*$. Then, $\dot{\lambda}\sigma_{ij}$ is written for the outward normal to the yield surface, and the normality flow rule takes the form

$$e_{ij} = \begin{cases} \dot{\lambda}\,\sigma^d_{ij}, \quad \dot{\lambda} \geq 0 & \text{if } |\sigma^d| - \sqrt{2}k_x = 0, \\ 0 & \text{if } |\sigma^d| - \sqrt{2}k_x < 0. \end{cases}$$

REMARK 1. As we see from Example 1, it is possible to use somewhat different multipliers in the normality rule, which corresponds to using different expressions for the outward normal to the given yield surface. It is also possible to represent this yield surface by different yield functions. For example, $F_x(\boldsymbol{\sigma}) = |\boldsymbol{\sigma}^d|^2 - 2k_x^2$ can be used instead of $F_x(\boldsymbol{\sigma}) = |\boldsymbol{\sigma}^d| - \sqrt{2}k_x$ to represent the Mises yield surface. This changes the form of the yield surface equation but affects neither the surface itself nor the set $C_x$ of admissible stresses.

The normality flow rule in (3.2) is only defined for smooth (differentiable) yield surfaces while surfaces with corners and conic vertices are often considered in plasticity. The Tresca yield surface belongs to the class of nonsmooth yield surfaces (see Example 2.2). We now consider a formulation of the normality rule in the case of a piecewise smooth yield surface. Let $F_x^{(1)}, \ldots, F_x^{(m)}$ be convex continuously differentiable functions on the space $Sym$ satisfying the conditions $F_x^{(1)}(0) \leq 0, \ldots, F_x^{(m)}(0) \leq 0$. The functions determine the set of admissible stresses

$$C_x = \left\{\boldsymbol{\sigma} \in Sym : \; F_x^{(1)}(\boldsymbol{\sigma}) \leq 0, \ldots, F_x^{(m)}(\boldsymbol{\sigma}) \leq 0\right\}.$$

Apparently, this set possesses all properties peculiar to a set of admissible stresses: it is convex, closed and contains the origin. We adopt the following form of the normality flow rule in the case of such set $C_x$:

$$\begin{gathered} \mathbf{e} = \dot{\lambda}_1 \frac{\partial F_x^{(1)}}{\partial \boldsymbol{\sigma}}(\boldsymbol{\sigma}) + \cdots + \dot{\lambda}_m \frac{\partial F_x^{(m)}}{\partial \boldsymbol{\sigma}}(\boldsymbol{\sigma}), \\ F_x^{(1)}(\boldsymbol{\sigma}) \leq 0, \ldots, F_x^{(m)}(\boldsymbol{\sigma}) \leq 0, \quad \dot{\lambda}_1\, geq 0, \ldots, \dot{\lambda}_m \geq 0, \\ \dot{\lambda}_1 F_x^{(1)}(\boldsymbol{\sigma}) = 0, \ldots, \dot{\lambda}_m F_x^{(m)}(\boldsymbol{\sigma}) = 0. \end{gathered} \tag{3.3}$$

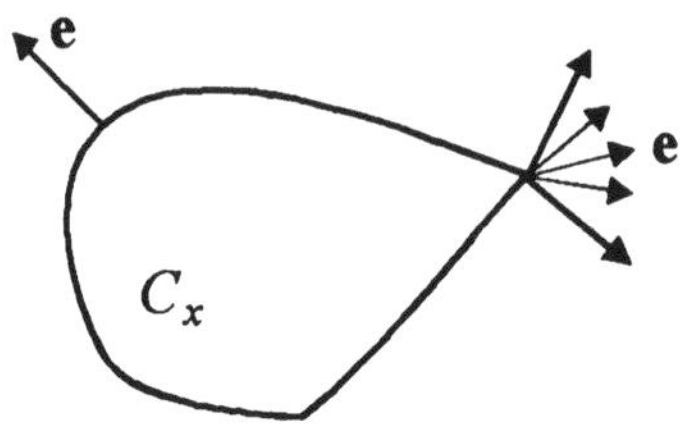


**Figure 3.3**

These relations reduce to formulation (3.2) if $m = 1$, that is, if the yield surface is smooth. In particular, the conditions $F_x(\sigma) \leq 0$, $\dot{\lambda} F_x(\sigma) = 0$ imply that $\dot{\lambda} = 0$ if the stress $\sigma$ is inside the yield surface, that is, if $F_x(\sigma) < 0$. Hence, the strain rate e vanishes in this case. The conditions $F_x^{(i)}(\sigma) \leq 0$, $\dot{\lambda}_i F_x^{(i)}(\sigma) = 0$ at any $i = 1, \ldots, m$ imply that $\dot{\lambda}$ can differ from 0 provided the stress $\sigma$ is on the surface $F_x^{(i)}(\sigma) = 0$.

Hence, the $i$-th summand actually enters the expression (3.3) provided $F_x^{(i)}(\sigma) = 0$. If there is only one $i$ for which this equality is valid, then the yield surface is smooth at the point $\sigma$ and the strain rate e has the direction of the outward normal, the latter being uniquely defined at $\sigma$. If $\sigma$ is a point of intersection of several surfaces $F_x^{(i)}(\sigma) = 0$, then e is a linear combination of the normals to these surfaces. Both cases are shown in Figure 3.3. The strain rate direction $\mathrm{e}/|\mathrm{e}|$ is not uniquely determined by the normality rule in the second case. As far as the magnitude $|\mathrm{e}|$ is concerned, it is not uniquely determined in both cases. For example, relation (3.2) holds if e and $\dot{\lambda}$, satisfying (3.2), are replaced by $a\mathrm{e}$ and $a\dot{\lambda}$, where $a$ is an arbitrary positive number.

**3.2. Constitutive maximum principle.** Formulation (3.3) of constitutive relations is of great convenience since it gives an explicit expression for the strain rate e through the stress $\sigma$. Although this formulation is useful for solving concrete problems, it is better to use another and more general formulation as long as one considers the general properties of plastic bodies. The general formulation does not express e through $\sigma$ or vice versa, but highlights an important extremum property of their relation. Namely, the following constitutive maximum principle is valid.

> The stress $\sigma$ in a rigid perfectly plastic body is admissible and together with the strain rate e satisfy the following condition: the value $\mathrm{e} \cdot \sigma$ is not less than $\mathrm{e} \cdot \sigma_*$ for any admissible stress $\sigma_*$ or, in short,
>
> $$\sigma \in C_x, \qquad \mathrm{e} \cdot (\sigma - \sigma_*) \geq 0 \quad \text{for every} \quad \sigma_* \in C_x, \tag{3.4}$$
>
> where $C_x$ is the set of admissible stresses at the point $x$ of the body.

The principle is equivalent to the normality flow rule in the case of a piecewise smooth yield surface.

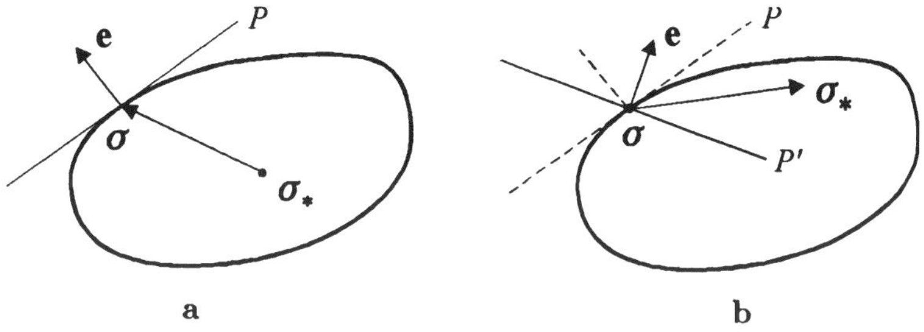


**Figure 3.4**

PROPOSITION 1. Suppose the set of admissible stresses is of the form

$$C_x = \left\{ \sigma \in Sym : \; F_x^{(1)}(\sigma) \leq 0, \ldots, F_x^{(m)}(\sigma) \leq 0 \right\},$$

where $F_x^{(1)}, \ldots, F_x^{(m)}$ are convex continuously differentiable functions, on the space $Sym$. Then (3.4) is equivalent to normality flow rule (3.3) if the interior of the set $C_x$ is not empty.

PROOF. Each of relations (3.3) and (3.4) reduces to the requirement of $\sigma$ admissibility in case $\mathbf{e} = 0$. Therefore, it remains to consider the case of $\mathbf{e} \neq 0$.

Suppose $\mathbf{e} \neq 0$. Consider first a smooth yield surface; then $m = 1$ in (3.3) and the normality rule is of the form (3.2). Let $\sigma$ and $\mathbf{e} \neq 0$ satisfy normality rule (3.2); it will be shown that they possess extremum property (3.4). Note that the point $\sigma$ is on the yield surface since $\mathbf{e} \neq 0$. Consider the plane $P$ tangent to the surface at the point $\sigma$, see Figure 3.4a. Because of convexity, the set $C_x$ lies in one of the two half-spaces separated by $P$. The strain rate $\mathbf{e}$ is directed into the other half-space and $\mathbf{e}$ is orthogonal to the plane $P$ according to the normality rule. Therefore, the scalar product $\mathbf{e} \cdot (\sigma - \sigma_*)$ is nonnegative for any admissible stress $\sigma_*$. In other words, constitutive maximum principle (3.4) is satisfied. Let now the strain rate $\mathbf{e} \neq 0$ and the stress $\sigma$ satisfy principle (3.4). Let us show that they satisfy normality rule (3.2). First, it should be noted that $\sigma$ is on the yield surface. Indeed, otherwise the set $C_x$ would contain a ball centered at $\sigma$. Then it is possible to find such $\sigma_*$ in this ball that the inequality $\mathbf{e} \cdot (\sigma - \sigma_*) < 0$ is valid in contradiction to (3.4). Hence, the stress $\sigma$ is on the yield surface. Now suppose $\mathbf{e}$ violates normality rule (3.2), that is, the direction of $\mathbf{e}$ differs from the direction of the outward normal to the yield surface at $\sigma$. Then the plane $P'$ which passes through $\sigma$ and is orthogonal to $\mathbf{e}$ does not coincide with the tangent plane $P$. Therefore, $P'$ intersects the set $C_x$, see Figure 3.4b. Consider two half-spaces separated by the plane $P'$. It is clear that there exists a stress $\sigma_*$ in $C_x$ such that $\sigma_* - \sigma$ and $\mathbf{e}$ point in the same half-space. Then the inequality $\mathbf{e} \cdot (\sigma - \sigma_*) > 0$ is valid in this case in contradiction to principle (3.4). The contradiction arose from the assumption that $\sigma$ and $\mathbf{e}$ do not satisfy relation (3.2). Hence, (3.2) is valid, that is, $\sigma$ and $\mathbf{e}$ satisfy the normality rule.

Now consider the general case of a piecewise smooth yield surface, that is, $m > 1$ in the definition of $C_x$. Constitutive maximum principle (3.4) can be stated in the following way: the stress $\boldsymbol{\sigma}$, which corresponds to the strain rate $\mathbf{e}$, is a solution to the extremum problem

$$\mathbf{e} \cdot \boldsymbol{\sigma}_* \to \sup; \quad \boldsymbol{\sigma}_* \in C_x$$

or, equivalently, to the problem

$$-\mathbf{e} \cdot \boldsymbol{\sigma}_* \to \inf; \quad \boldsymbol{\sigma}_* \in C_x.$$

(See notations we use in formulations of extremum problems in Subsection C.2.) Necessary and sufficient conditions for $\boldsymbol{\sigma}$ to be a solution to this extremum problem are given by the Kuhn-Tucker theorem (Subsection C.3). We observe that they are relations (3.3), which finishes the proof.

**3.3. Rigid perfectly plastic body.** The constitutive maximum principle is equivalent to normality flow rule (3.3) under the assumption of Proposition 1. The formulation of the normality rule makes sense when the yield surface consists of a finite number of parts of continuously differentiable surfaces $F_x^{(1)}(\boldsymbol{\sigma}) = 0, \ldots, F_x^{(m)}(\boldsymbol{\sigma}) = 0$. At the same time, the formulation of the constitutive maximum principle makes sense in a more general case, for example, in the case of a yield surface with a conic vertex. Therefore, we consider the principle as a generalization of the normality flow rule and use it to define a rigid perfectly plastic body.

A body is referred to as *rigid perfectly plastic* if, for every point $x$ of the body,

> a set $C_x$ of admissible stresses is specified: convex, closed and containing the origin set in the space *Sym* of symmetric second rank tensors,
>
> the relation between the stress $\boldsymbol{\sigma}$ and the strain rate $\mathbf{e}$ is given by the constitutive maximum principle: the stress is admissible and the value $\mathbf{e} \cdot \boldsymbol{\sigma}$ is not less then $\mathbf{e} \cdot \boldsymbol{\sigma}_*$ for any admissible stress $\boldsymbol{\sigma}_*$, that is,

$$\boldsymbol{\sigma} \in C_x, \qquad \mathbf{e} \cdot (\boldsymbol{\sigma} - \boldsymbol{\sigma}_*) \geq 0 \quad \text{for every } \boldsymbol{\sigma}_* \in C_x.$$

We are reminded that, under some additional assumptions, the constitutive maximum principle is reduced to an explicit expression for the strain rate. For example, if the yield surface is smooth, the principle is equivalent to the normality flow rule:

$$\mathbf{e} = \begin{cases} \dot{\lambda} \dfrac{\partial F_x}{\partial \boldsymbol{\sigma}}(\boldsymbol{\sigma}), \quad \dot{\lambda} \geq 0 & \text{if } F_x(\boldsymbol{\sigma}) = 0, \\ 0 & \text{if } F_x(\boldsymbol{\sigma}) < 0. \end{cases}$$

The constitutive maximum principle does not determine the strain rate magnitude. If $\boldsymbol{\sigma}$, $\mathbf{e}$ satisfy the principle, then so do $\boldsymbol{\sigma}$, $a\mathbf{e}$, $a$ being an arbitrary nonnegative number. Such nonuniqueness is not a shortcoming of the theory but reflects the actual behavior of some materials, as deforming along a yield plateau is possible at various strain rates. It is also worth noting that the constitutive principle does not uniquely determine the strain rate direction for a given stress $\boldsymbol{\sigma}$ if the yield surface is not smooth at $\boldsymbol{\sigma}$, see (3.3). Such nonuniqueness is observed for some materials, and models with singular yield surfaces are used to describe their behavior. Different shapes of yield surfaces (or, equivalently, different shapes of admissible stress sets) are what makes the difference between rigid perfectly plastic models. In other words, within the framework of the rigid perfectly plastic modeling the set of admissible stresses represents material properties, likewise, Young's modulus and Poisson's ratio represent the material properties of an isotropic linearly elastic body.

## 4. Dissipation

Dissipation is a function of the strain rate which plays a very important role in limit analysis. In this section, dissipation will be introduced and some of its properties will be studied. It will be shown that a plastic material is completely characterized by its dissipation.

**4.1. Constitutive maximum principle and dissipation.** Constitutive relations of the rigid perfectly plastic model are given by the maximum principle (Subsection 3.3):

$$\boldsymbol{\sigma} \in C_x, \qquad \mathrm{e} \cdot \boldsymbol{\sigma} \geq \mathrm{e} \cdot \boldsymbol{\sigma}_* \quad \text{for every} \quad \boldsymbol{\sigma}_* \in C_x.$$

In particular, we can set $\boldsymbol{\sigma}_* = \boldsymbol{\sigma}$, where $\boldsymbol{\sigma}$ is the actual stress. This results in the equality

$$\boldsymbol{\sigma} \in C_x, \qquad \mathrm{e} \cdot \boldsymbol{\sigma} = \sup\left\{\mathrm{e} \cdot \boldsymbol{\sigma}_* : \boldsymbol{\sigma}_* \in C_x\right\}. \tag{4.1}$$

We observe that (4.1) is equivalent to the constitutive maximum principle. Here, $\boldsymbol{\sigma}$ and $\mathrm{e}$ are the stress and the strain rate at the point $x$ of a plastic body and $C_x$ is a given set of admissible stresses at this point. The expression on the right-hand side in (4.1) only depends on the strain rate $\mathrm{e}$ and, therefore, defines a function on the space $Sym$ of symmetric second rank tensors:

$$d_x(\mathrm{e}) = \sup\left\{\mathrm{e} \cdot \boldsymbol{\sigma}_* : \boldsymbol{\sigma}_* \in C_x\right\}. \tag{4.2}$$

This function is referred to as *dissipation*. Using it we write the constitutive maximum principle in the following form:

$$\boldsymbol{\sigma} \in C_x, \qquad \mathrm{e} \cdot \boldsymbol{\sigma} = d_x(\mathrm{e}). \tag{4.3}$$

Let us consider an example of the explicit formula for dissipation.

EXAMPLE 1. *Dissipation in the case of the Mises yield surface.* Consider the set of admissible stresses specified by the Mises condition $\left|\boldsymbol{\sigma}^d\right| \leq \sqrt{2}k_x$ (see Example 2.1). Then, by the definition of dissipation, we have:

$$d_x(\mathrm{e}) = \sup\left\{\mathrm{e} \cdot \boldsymbol{\sigma}_* : \boldsymbol{\sigma}_* \in Sym, \quad \left|\boldsymbol{\sigma}_*^d\right| \leq \sqrt{2}k_x\right\}.$$

This implies that $d_x(\mathbf{e}) = 0$ if $\mathbf{e} = 0$. Consider next $\mathbf{e} \neq 0$ with non-vanishing spheric part $\mathbf{e}^s \neq 0$. In this case, the value $\mathbf{e} \cdot \boldsymbol{\sigma}_* = \mathbf{e}^d \cdot \boldsymbol{\sigma}_*^d + \mathbf{e}^s \cdot \boldsymbol{\sigma}_*^s$ can be arbitrarily large if the spheric part of admissible stress $\boldsymbol{\sigma}_*$ is chosen appropriately. Therefore, $d_x(\mathbf{e}) = +\infty$ if $\mathbf{e}^s \neq 0$. (This equality shows that the body under consideration is incompressible. Indeed, the constitutive relations cannot be satisfied if $\mathbf{e}^s \neq 0$: equality (4.3) is not valid for such $\mathbf{e}$ since its right-hand side is $+\infty$ while the left-hand side is finite.) Consider finally $\mathbf{e}$ with $\mathbf{e}^s = 0$. In this case, $\mathbf{e} \cdot \boldsymbol{\sigma}_* = \mathbf{e} \cdot \boldsymbol{\sigma}_*^d$ and the inequality $|\mathbf{e} \cdot \boldsymbol{\sigma}_*| \leq \sqrt{2} k_x \, |\mathbf{e}|$ is valid for any admissible stress $\boldsymbol{\sigma}_*$. By the definition of the dissipation, we have: $d_x(\mathbf{e}) \leq \sqrt{2} k_x \, |\mathbf{e}|$. It is easy to see that the equality $d_x(\mathbf{e}) = \sqrt{2} k_x \, |\mathbf{e}|$ is actually valid. The equality is obvious if $\mathbf{e} = 0$. If $\mathbf{e} \neq 0$, consider the admissible stress $\boldsymbol{\sigma}_0 = \sqrt{2} k_x \mathbf{e} / |\mathbf{e}|$. The estimation $d_x(\mathbf{e}) \geq \mathbf{e} \cdot \boldsymbol{\sigma}_0 = \sqrt{2} k_x \, |\mathbf{e}|$ together with the above-mentioned opposite inequality determine the value of the dissipation: $d_x(\mathbf{e}) = \sqrt{2} k_x \, |\mathbf{e}|$. Thus, we arrive to the following formula in the case of the Mises yield surface:

$$d_x(\mathbf{e}) = \begin{cases} \sqrt{2} k_x \, |\mathbf{e}| & \text{if} \quad \mathbf{e}^s = 0, \\ +\infty & \text{if} \quad \mathbf{e}^s \neq 0. \end{cases}$$

The following example establishes a convenient representation for the dissipation in the general case of cylindrical yield surface.

EXAMPLE 2. *Dissipation in the case of cylindrical yield surface.* A cylindrical yield surface is a cylinder in the space *Sym* of symmetric second rank tensors with its axis directed along the unity tensor **I**, see Example 2.3. The corresponding set $C_x$ of admissible stresses contains tensors $\boldsymbol{\sigma}_*$ with arbitrary spheric parts. Therefore, like the previous example, it is obvious that the dissipation $d_x(\mathbf{e})$ is $+\infty$ if $\mathbf{e}^s \neq 0$. On the other hand, $d_x(\mathbf{e})$ is normally finite if $\mathbf{e}^s = 0$ and is bounded for all $\mathbf{e}$ satisfying this condition. It is easily seen that the values $d_x(\mathbf{e})$ are bounded for all $\mathbf{e}$ with $\mathbf{e}^s = 0$ if the set $C_x^d$ is bounded; here, $C_x^d$ is the cross-section of $C_x$ by the deviatoric plane. It is natural to introduce a function $d_x^{(d)}$ such that its value 1) is finite for every $\mathbf{e}$ in *Sym*; 2) does not depend of $\mathbf{e}^s$; 3) equals $d_x(\mathbf{e})$ if $\mathbf{e}^s = 0$. There is an obvious expression for $d_x^{(d)}$ through the dissipation $d_x(\mathbf{e})$: $d_x^{(d)}(\mathbf{e}) = d_x(\mathbf{e}^d)$. For example, $d_x^{(d)}(\mathbf{e}) = \sqrt{2} k_x \left|\mathbf{e}^d\right|$ in the case of the Mises yield surface. On the other hand, the dissipation $d_x$ is expressed through the function $d_x^{(d)}$:

$$d_x(\mathbf{e}) = \begin{cases} d_x^{(d)}(\mathbf{e}) & \text{if} \quad \mathbf{e}^s = 0, \\ +\infty & \text{if} \quad \mathbf{e}^s \neq 0. \end{cases}$$

This representation for the dissipation is convenient since the function $d_x^{(d)}$ is defined and normally finite on the whole *Sym*. In particular, one can make use of its derivatives.

**4.2. Properties of dissipation.** Noteworthy are the properties of dissipation arising immediately from its definition.

1. *Dissipation is nonnegative.* Indeed, the zero stress is admissible, $0 \in C_x$; therefore, the following inequality is valid:

$$d_x(\mathrm{e}) = \sup\{\mathrm{e}\cdot\boldsymbol{\sigma}_* : \boldsymbol{\sigma}_* \in C_x\} \geq 0\cdot\mathrm{e} = 0.$$

2. *Dissipation is a positively homogeneous function of degree 1*, that is, the equality

$$d_x(a\mathrm{e}) = a d_x(\mathrm{e})$$

holds for every $\mathrm{e} \in Sym$ and every number $a$. Indeed, by the definition of dissipation, this equality is equivalent to the relation (with the fixed $a$)

$$\sup\{a\mathrm{e}\cdot\boldsymbol{\sigma}_* : \boldsymbol{\sigma}_* \in C_x\} = a\sup\{\mathrm{e}\cdot\boldsymbol{\sigma}_* : \boldsymbol{\sigma}_* \in C_x\},$$

which is apparently valid.

3. *Dissipation is a convex function.* Indeed, the inequality

$$d_x(\alpha\mathrm{e}_1 + (1-\alpha)\mathrm{e}_2) \leq \alpha d_x(\mathrm{e}_1) + (1-\alpha)d_x(\mathrm{e}_2)$$

for every $\mathrm{e}_1$ and $\mathrm{e}_2$ in $Sym$ and every $\alpha$ satisfying the conditions $0 \leq \alpha \leq 1$ is the criterion for convexity of $d_x$ (see Subsection B.1). To establish this inequality, it is sufficient to observe that the supremum of a sum is not greater than the sum of the supremums:

$$\sup\{(\alpha\mathrm{e}_1 + (1-\alpha)\mathrm{e}_2)\cdot\boldsymbol{\sigma}_* : \boldsymbol{\sigma}_* \in C_x\} \leq$$
$$\leq \sup\{\alpha\mathrm{e}_1\cdot\boldsymbol{\sigma}_* : \boldsymbol{\sigma}_* \in C_x\} + \sup\{(1-\alpha)\mathrm{e}_2\cdot\boldsymbol{\sigma}_* : \boldsymbol{\sigma}_* \in C_x\},$$

and to make use of the homogeneity (the previous property) on the right-hand side.

4. *If the set $C_x$ of admissible stresses contains a ball with its center at the origin:* $|\boldsymbol{\sigma}| \leq r$, $r > 0$, *the dissipation is estimated from below:*

$$d_x(\mathrm{e}) \geq r\,|\mathrm{e}|, \qquad \mathrm{e} \in Sym.$$

Indeed, this inequality is obvious if $\mathrm{e} = 0$. If $\mathrm{e} \neq 0$, we make use of the admissible stress $\boldsymbol{\sigma}_0 = r\mathrm{e}/|\mathrm{e}|$ and, by the definition of dissipation, arrive at the estimation:

$$d_x(\mathrm{e}) = \sup\{\mathrm{e}\cdot\boldsymbol{\sigma}_* : \boldsymbol{\sigma}_* \in C_x\} \geq \mathrm{e}\cdot\boldsymbol{\sigma}_0 = r\,|\mathrm{e}|.$$

COROLLARY. If the set $C_x$ contains a ball with its center at the origin, then vanishing dissipation, $d_x(\mathrm{e}) = 0$, is equivalent to vanishing of the strain rate, $\mathrm{e} = 0$.

Indeed, by the above estimation we have: $0 \geq |\mathrm{e}|$ and, hence, $\mathrm{e} = 0$.

**4.3. Yield surface determined by dissipation.** The set $C_x$ of admissible stresses or, equivalently, its boundary (the yield surface) determines dissipation (4.2). It turns out that, conversely, a given dissipation allows one to restore uniquely the corresponding set $C_x$. First, we consider the following example of the restoration.

EXAMPLE 3. Consider the dissipation

$$d_x(\mathrm{e}) = \begin{cases} \sqrt{2}k_x\,|\mathrm{e}| & \text{if } \ \mathrm{e}^s = 0, \\ +\infty & \text{if } \ \mathrm{e}^s \neq 0 \end{cases}$$

corresponding to the Mises yield surface (see Example 1). Let us evaluate the supremum

$$\psi_x(\boldsymbol{\sigma}) = \sup\{\mathrm{e}\cdot\boldsymbol{\sigma} - d_x(\mathrm{e}) : \mathrm{e} \in Sym\}$$

for any $\boldsymbol{\sigma}$ in $Sym$. Searching for the supremum, we omit e with $\mathrm{e}^s = 0$ since $d_x(\mathrm{e}) = +\infty$ for such e and, therefore, $\mathrm{e}\cdot\boldsymbol{\sigma} - d_x(\mathrm{e}) = +\infty$. For any e with $\mathrm{e}^s = 0$, we have $d_x(\mathrm{e}) = \sqrt{2}k_x\,|\mathrm{e}|$ and, consequently,

$$\psi_x(\boldsymbol{\sigma}) = \sup\left\{\mathrm{e}\cdot\boldsymbol{\sigma} - \sqrt{2}k_x\,|\mathrm{e}| : \mathrm{e} \in Sym^d\right\}.$$

First, consider $\boldsymbol{\sigma}$ satisfying the condition $\left|\boldsymbol{\sigma}^d\right| \leq \sqrt{2}k_x$. In this case, the inequality $\mathrm{e}\cdot\boldsymbol{\sigma} - \sqrt{2}k_x\,|\mathrm{e}| \leq 0$ is valid for every e, and consequently, the supremum $\psi_x(\boldsymbol{\sigma})$ is attained at $\mathrm{e} = 0$ and equals 0. Now consider $\boldsymbol{\sigma}$ with $\left|\boldsymbol{\sigma}^d\right| > \sqrt{2}k_x$. We make use of the strain rate $\mathrm{e}_a$ defined by $\boldsymbol{\sigma}$ and a positive number $a$: $\mathrm{e}_a = a\boldsymbol{\sigma}^d/\left|\boldsymbol{\sigma}^d\right|$. Due to the inequality $\left|\boldsymbol{\sigma}^d\right| > \sqrt{2}k_x$, the value

$$\mathrm{e}_a\cdot\boldsymbol{\sigma} - \sqrt{2}k_x\,|\mathrm{e}_a| = a(\left|\boldsymbol{\sigma}^d\right| - \sqrt{2}k_x)$$

can be made arbitrarily large by choosing an appropriate $a$. This implies $\psi_x(\boldsymbol{\sigma}) = +\infty$, and we arrive at the following formula for the supremum:

$$\psi_x(\boldsymbol{\sigma}) = \begin{cases} 0 & \text{if } \ \left|\boldsymbol{\sigma}^d\right| \leq \sqrt{2}k_x, \\ +\infty & \text{if } \ \left|\boldsymbol{\sigma}^d\right| > \sqrt{2}k_x. \end{cases}$$

The function $\psi_x$ vanishes exactly on the set $C_x$ which determined the Mises dissipation in Example 1.

The coincidence observed in the previous example is not accidental. It will now be shown that, like the example, the set of admissible stresses can be restored in the general case as soon as the dissipation is known. Let $d_x$ be the dissipation corresponding to a set $C_x$ of admissible stresses. Let us evaluate the supremum

$$\psi_x(\boldsymbol{\sigma}) = \sup\{\mathrm{e}\cdot\boldsymbol{\sigma} - d_x(\mathrm{e}) : \mathrm{e} \in Sym\}. \tag{4.4}$$

First, we consider $\boldsymbol{\sigma}$ in $C_x$ and show that $\psi_x(\mathrm{e}) = 0$. In case $\boldsymbol{\sigma} \in C_x$, the inequality $\mathrm{e}\cdot\boldsymbol{\sigma} - d_x(\mathrm{e}) \leq 0$ holds for every e by definition (4.2). Then, supremum (4.4) is attained at $\mathrm{e} = 0$ and equals 0.

Now we consider $\boldsymbol{\sigma}$ not in $C_x$ and show that $\psi_x(\boldsymbol{\sigma}) = +\infty$. By the separation theorem (subsection B.2) for the convex closed set $C_x$ and $\boldsymbol{\sigma} \notin C_x$ there is such $\mathbf{e}$ in $Sym$ that

$$\mathbf{e} \cdot \boldsymbol{\sigma} > \sup\{\mathbf{e} \cdot \boldsymbol{\sigma}_* : \boldsymbol{\sigma}_* \in C_x\} = d_x(\mathbf{e}).$$

We make use of the strain rate $\mathbf{e}_a = a\mathbf{e}$, where $a$ is a positive number. Due to the previous inequality, the value $\mathbf{e}_a \cdot \boldsymbol{\sigma} - d_x(\mathbf{e}_a) = a(\mathbf{e} \cdot \boldsymbol{\sigma} - d_x(\mathbf{e}))$ can be made arbitrarily large by choosing an appropriate $a$. This implies that $\psi_x(\mathbf{e})$ is $+\infty$. Thus, the known dissipation defines function (4.4). The function takes values

$$\psi_x(\boldsymbol{\sigma}) = \begin{cases} 0 & \text{if} \quad \boldsymbol{\sigma} \in C_x, \\ +\infty & \text{if} \quad \boldsymbol{\sigma} \notin C_x, \end{cases}$$

that is, vanishes exactly on the set $C_x$ to which the given dissipation corresponds. This allows us to restore the set $C_x$ of admissible stresses as soon as we know the dissipation $d_x$.

## 5. Rigid-plastic problem

In this section, we formulate a system of equations and boundary conditions describing the response of a rigid perfectly plastic body to a quasistatic loading. A simple example illustrates specific features of this response, in particular, existence of the limit load and failure mechanism, which are the main subjects of the book.

**5.1. Quasistatic problem.** Consider a rigid perfectly plastic body occupying the domain $\Omega$. The body is subjected to body forces with the volume density $\mathbf{f}$ given in $\Omega$ and surface tractions with the density $\mathbf{q}$ given at part $S_q$ of the body boundary, see Figure 5.1. The remaining part $S_v$ of the boundary is fixed, that is, the velocity field $\mathbf{v}$ satisfies the condition $\mathbf{v}|_{S_v} = 0$. It is assumed that the body is in equilibrium at the initial moment $t = 0$ and displacements and strains are small under the action of the load $\mathbf{l} = (\mathbf{f}, \mathbf{q})$. It is also assumed that the load varies slowly enough to result in only negligibly small accelerations.

Under these assumptions the following equilibrium conditions are satisfied (see Subsection A.1):

$$\frac{\partial \sigma_{ij}}{\partial x_j} + f_i = 0, \qquad \sigma_{ij}\nu_j|_{S_q} = q_i.$$

Here, $\boldsymbol{\nu}$ is the unit outward normal. The stress field is assumed to be sufficiently smooth, say, continuously differentiable. Let the velocity field $\mathbf{v}$ be also differentiable and let $\mathbf{e}$ stand for the corresponding strain rate field; then the following kinematic relations are valid:

$$e_{ij} = \frac{1}{2}\left(\frac{\partial v_i}{\partial x_j} + \frac{\partial v_j}{\partial x_i}\right), \qquad \mathbf{v}\,|_{S_v} = 0.$$

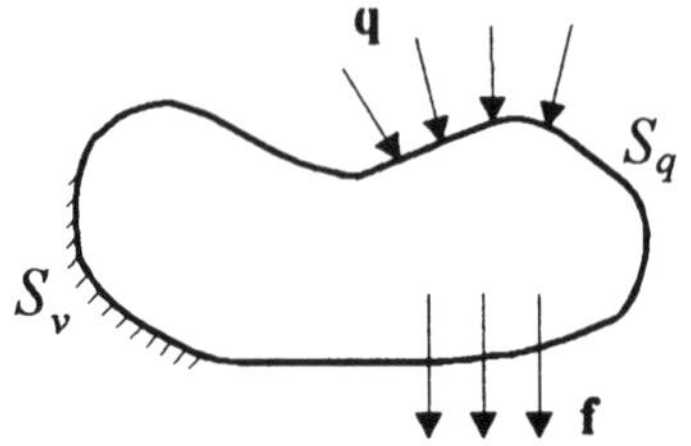


**Figure 5.1**

Apart from these relations, the stress $\boldsymbol{\sigma}$ and the strain rate $\mathbf{e}$ should satisfy the rigid perfectly plastic constitutive relations. In the case of a smooth yield surface, the constitutive relations can be written as

$$F_x(\boldsymbol{\sigma}) \leq 0,$$

$$\mathbf{e} = \begin{cases} \dot{\lambda}\,\dfrac{\partial F_x}{\partial \boldsymbol{\sigma}}(\boldsymbol{\sigma}), \quad \dot{\lambda} \geq 0 & \text{if} \quad F_x(\boldsymbol{\sigma}) = 0, \\ 0 & \text{if} \quad F_x(\boldsymbol{\sigma}) < 0. \end{cases}$$

We will use a more general formulation, that is, the constitutive maximum principle (Subsection 4.1) suitable both for smooth and singular yield surfaces:

$$\boldsymbol{\sigma} \in C_x, \qquad \mathbf{e} \cdot \boldsymbol{\sigma} \geq \mathbf{e} \cdot \boldsymbol{\sigma}_* \quad \text{for every} \quad \boldsymbol{\sigma}_* \in C_x.$$

Here, $C_x$ is a given set of admissible stresses. At last, the displacement $\mathbf{u}$ is related to the velocity: $\dot{\mathbf{u}} = \mathbf{v}$ (the dot indicates the derivative with respect to the time), and the following initial conditions are to be satisfied: $\mathbf{u}|_{t=0} = 0$. The above-mentioned relations constitute a complete system of equations, initial and boundary conditions. Thus, we arrive to the following formulation of the quasistatic problem for a rigid perfectly plastic body:

$$\frac{\partial \sigma_{ij}}{\partial x_j} + f_i = 0, \qquad \sigma_{ij}\nu_j|_{S_q} = q_i, \tag{5.1}$$

$$e_{ij} = \frac{1}{2}\left(\frac{\partial v_i}{\partial x_j} + \frac{\partial v_j}{\partial x_i}\right), \qquad \mathbf{v}\,|_{S_v} = 0, \tag{5.2}$$

$$\boldsymbol{\sigma} \in C_x, \qquad \mathbf{e} \cdot \boldsymbol{\sigma} \geq \mathbf{e} \cdot \boldsymbol{\sigma}_* \quad \text{for every} \quad \boldsymbol{\sigma}_* \in C_x, \tag{5.3}$$

$$\dot{\mathbf{u}} = \mathbf{v}, \qquad \mathbf{u}|_{t=0} = 0, \qquad \mathbf{v}|_{t=0} = 0. \tag{5.4}$$

REMARK 1. We do not include any initial conditions for the stress field $\boldsymbol{\sigma}$ in system (5.1) – (5.4) for the following reason. A rigid perfectly plastic body does not deform under the action of a sufficiently small load (we will see in Chapter 3 that this is one of the main properties of the rigid perfectly plastic body). We always assume that the load $\mathbf{l}(0)$ is small and the body does not deform at the initial moment $t = 0$. Stresses in the (rigid) body

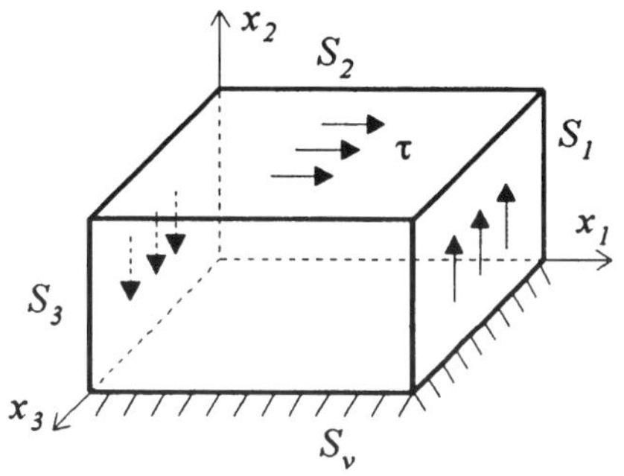


**Figure 5.2**

are not uniquely determined at $t = 0$, and one cannot prescribe initial values of $\boldsymbol{\sigma}$. Generally, one should not expect the uniqueness of the rigid-plastic problem solution at all. (It is worthwhile to bear the nonuniqueness in mind when interpreting a certain solution to such a problem). At the same time, the main aim of the rigid-plastic analysis is to find the limit load rather than the stress and velocity fields, and the limit load is defined uniquely. We discuss the concept of a limit load in the next subsection.

REMARK 2. One can consider a rigid-plastic problem with kinematic constraints different from $\mathbf{v}|_{S_v} = 0$. Limit analysis is developed only under the assumption that kinematic constraints by themselves (that is, when the given load is zero) cannot result in deformation of the body. The kinematic condition $v_\nu|_S = 0$, where $v_\nu$ is the normal component of velocity $\mathbf{v}$ on the part $S$ of the body boundary, satisfies this assumption. For example, the condition $\mathbf{v}|_S = \mathbf{v}_0$, where $\mathbf{v}_0 \neq 0$ is given on $S$ and differs from a rigid motion velocity field, does not satisfy the assumption.

**5.2. First example of the limit load.** We now consider a simple example illustrating the most important characteristic features of a rigid perfectly plastic body.

Consider a rigid perfectly plastic body occupying the domain $\Omega$, which is a rectangular parallelepiped. The base of the parallelepiped is fixed. The body is in equilibrium at the initial moment $t = 0$ and is subjected to quasistatic loading: uniform tangent surface traction with the density $\tau$ on the surfaces $S_1$, $S_2$, $S_3$; see Figure 5.2. The two remaining sides are free. The load $\tau$ increases monotonically starting from $\tau = 0$ at $t = 0$. We adopt the Mises admissibility condition for stresses $\boldsymbol{\sigma}$: $|\boldsymbol{\sigma}^d| = \sqrt{2}k$, $k = \text{const} > 0$. A system of the form (5.1) – (5.4) describes behavior of the body.

To analyze solutions of this system, it is convenient to introduce the stress field $\boldsymbol{\sigma}^0$ with the only nonzero components $\sigma^0_{12} = \sigma^0_{12} = 0$. The field $\boldsymbol{\sigma}^0$ equilibrates the given load and is admissible as long as $\tau(t) \leq k$. Moreover, $\boldsymbol{\sigma}^0$ and the zero displacement field are obviously a solution to the quasistatic rigid-plastic problem under consideration.

Noteworthy are the following peculiarities of the rigid perfectly plastic body behavior.

I. The body does not deform as long as $\tau(t) < k$.

Although $\boldsymbol{\sigma}^0$ as the stress and $\mathbf{u} \equiv 0$ as the displacement are a solution to the problem, we know nothing about the uniqueness of the solution. Actually, it is not unique in displacement. We now consider any solution $\boldsymbol{\sigma}$, $\mathbf{u}$ and show that $\mathbf{u} \equiv 0$ as long as $\tau(t) < k$. To verify this, it is sufficient to show that the velocity field $\mathbf{v}$ is zero or, equivalently, that the strain rate field $\mathbf{e}$ is zero at every moment. Indeed, due to the kinematic constraints, $\mathbf{e} \equiv 0$ implies that $\mathbf{v} \equiv 0$. To prove the equality $\mathbf{e} = 0$, we consider the integral

$$\int_{\Omega} \mathbf{e} \cdot (\boldsymbol{\sigma} - \boldsymbol{\sigma}^0)\, dV = \int_{\Omega} \frac{\partial v_i}{\partial x_j}(\sigma_{ij} - \sigma^0_{ij})\, dV$$

$$= -\int_{\Omega} v_i \frac{\partial}{\partial x_j}(\sigma_{ij} - \sigma^0_{ij})\, dV + \int_{\partial\Omega} v_i \nu_j (\sigma_{ij} - \sigma^0_{ij})\, dS,$$

where $\boldsymbol{\sigma}^0$ is the above-introduced stress field. Both stress fields $\boldsymbol{\sigma}^0$ and $\boldsymbol{\sigma}$ satisfy the equilibrium equations, and the velocity field $\mathbf{v}$ satisfies the kinematic condition $\mathbf{v}|_{S_v} = 0$. Therefore, integrals on the right-hand side of the previous formula vanish and, consequently,

$$\int_{\Omega} \mathbf{e} \cdot (\boldsymbol{\sigma} - \boldsymbol{\sigma}^0)\, dV = 0.$$

We are reminded that $\boldsymbol{\sigma}$ and $\mathbf{e}$ satisfy the constitutive maximum principle; in particular, the inequality $\mathbf{e} \cdot \boldsymbol{\sigma} \geq \mathbf{e} \cdot \boldsymbol{\sigma}^0$ (since the stress $\boldsymbol{\sigma}^0$ is admissible when $\tau(t) < k$). This inequality together with the previous formula results in the equality: $\mathbf{e} \cdot \boldsymbol{\sigma} = \mathbf{e} \cdot \boldsymbol{\sigma}^0$. By this formula we find from (5.3) that the inequality $\mathbf{e} \cdot (\boldsymbol{\sigma}^0 - \boldsymbol{\sigma}_*) \geq 0$ is valid for any admissible stress $\boldsymbol{\sigma}_*$. This implies that $\mathbf{e} = 0$ since the stress $\boldsymbol{\sigma}^0$ is strictly inside the yield surface as long as $\tau < k$ (which is the case under consideration).

II. The body deforms at an arbitrarily large strain rate when the load $\tau$ reaches the value $k$.

Indeed, in the case $\tau = k$, consider 1) the stress field $\boldsymbol{\sigma} = \boldsymbol{\sigma}^0$, $\sigma_{12} = \sigma_{21}$ being its only nonzero components, and 2) the velocity field $\mathbf{v}$ with the components $v_1 = ax_2$, $v_2 = v_3 = 0$, where $a$ is an arbitrary positive number. It is clear that $\boldsymbol{\sigma}$ satisfies the equilibrium equations with $\tau = k$ and $\mathbf{v}$ satisfies the kinematic condition $\mathbf{v}|_{S_v} = 0$. The strain rate field $\mathbf{e}$ corresponding to $\mathbf{v}$ has the components $e_{12} = e_{21} = a/2$, all the rest of the components vanishing. The fields $\boldsymbol{\sigma}$ and $\mathbf{e}$ satisfy the normality flow rule, which has the form $e_{ij} = \dot{\lambda}\sigma^d_{ij}$ for a body with the Mises yield surface (see Example 3.1). Thus, $\mathbf{v}$ and $\boldsymbol{\sigma}$ are a solution to the problem under consideration, an arbitrary positive number $a$ being a parameter in this solution. Thus, we have obtained a collection of solutions, and the strain rate components $e_{12} = e_{21} = a/2$ are arbitrarily large within the collection.

III. If $\tau > k$, the load cannot be equilibrated by any admissible stress field.

Indeed, if the stress field $\boldsymbol{\sigma}$ equilibrates the load, then $\boldsymbol{\sigma}$ satisfies the boundary condition $\sigma_{12}|_{S_2} = \tau$ with $\tau > k$, thus violating the admissibility condition $|\boldsymbol{\sigma}^d| \leq \sqrt{2}k$, and there is no admissible stress field that equilibrates the load.

Thus we see that the body does not deform until the load reaches a certain critical value, that it deforms with an arbitrarily large strain rate under the action of this critical load, and that it cannot bear a greater load. Existence of the *limit load* is shown here by the example. We will see that such a load always exists when a rigid perfectly plastic body is subjected to a proportionally increasing load. It is the most important feature of rigid perfectly plastic bodies.

REMARK 3. As was mentioned, solution of a quasistatic problem for a rigid-plastic body is generally not unique in stress. In particular, the above considered problem has solutions with a stress field different from $\boldsymbol{\sigma}^0$. Indeed, let $\boldsymbol{\sigma}'$ be a smooth bounded stress field satisfying the conditions

$$\frac{\partial \sigma'_{ij}}{\partial x_j} = 0, \qquad \sigma_{ij}\nu_j\big|_{S_q} = 0.$$

Then, for $0 \leq \tau < k$, consider the stress field $\boldsymbol{\sigma} = \boldsymbol{\sigma}^0 + \boldsymbol{\sigma}'/b$. This field equilibrates the given load and it is also clear that $\boldsymbol{\sigma}$ is admissible if the number $b$ is sufficiently large. Therefore, $\boldsymbol{\sigma}$ and the displacement field $\mathbf{u} \equiv 0$ solve the problem under consideration. Thus, there is a wide class of solutions with the various stress fields.

**5.3. Rigid-plastic problem.** We now summarize the features of the body behavior shown in the above-considered example which are peculiar to any rigid perfectly plastic body. Recall that the most important property of perfectly plastic material is the division of stresses into admissible and inadmissible.

At any moment of quasistatic loading, either there is an admissible stress field equilibrating the load or there is none. In the latter case, the body has already lost its bearing capacity. In the former case, there are the following alternatives: the body remains rigid or it deforms (see Figure 5.3). As long as the body remains rigid, it bears the load. If the body deforms, it loses its bearing capacity: the strain rate is arbitrarily large.

It should be noted that it is the current load and not the loading history that determines the type of body response. The current load is an input of the following problem (which is of instantaneous rather than evolutionary type):

$$\frac{\partial \sigma_{ij}}{\partial x_j} + f_i = 0, \qquad \sigma_{ij}\nu_j\big|_{S_q} = q_i, \tag{5.5}$$

$$e_{ij} = \frac{1}{2}\left(\frac{\partial v_i}{\partial x_j} + \frac{\partial v_j}{\partial x_i}\right), \qquad \mathbf{v}\,\big|_{S_v} = 0, \tag{5.6}$$

$$\boldsymbol{\sigma}(x) \in C_x, \quad \mathrm{e}(x)\cdot(\boldsymbol{\sigma}(x) - \boldsymbol{\sigma}_*) \geq 0 \text{ for every } \boldsymbol{\sigma}_* \in C_x. \tag{5.7}$$

Solvability of this problem is the crucial point in estimating the rigid-plastic body response to a given load. Analysis of the solvability in stress answers the first question of the scheme in Figure 5.3; analysis of the solvability in velocity (with nonzero velocity field) answers the second one. We refer to (5.5) - (5.7) as the *rigid-plastic problem*.

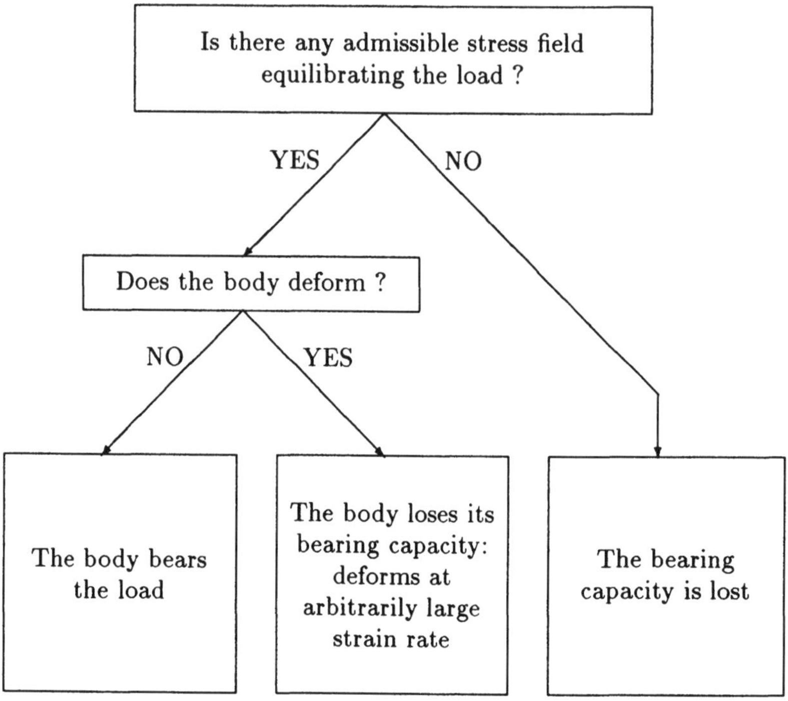


**Figure 5.3**

Limit analysis establishes in the general case the features of the rigid-plastic behavior pointed out here. It also develops methods for finding out which of the response alternatives (see Figure 5.3) corresponds to a given load. These are the main subjects of the subsequent chapters.

**5.4. On the admissibility and equilibrium conditions.** As was shown in the previous subsections, it is of the utmost importance to find out whether a given load can be equilibrated by an admissible stress field. A load is referred to as *admissible* if such a field exists, and as *inadmissible* otherwise. If there exists a differentiable stress field $\boldsymbol{\sigma}$ satisfying (5.5) and the admissibility condition $\boldsymbol{\sigma}(x) \in C_x$ holds for every point $x$, then, apparently, the load $\mathbf{l} = (\mathbf{f}, \mathbf{q})$ is admissible. However, the nonexistence of such $\boldsymbol{\sigma}$ does not imply that $\mathbf{l}$ is inadmissible. Actually, it is possible that $\mathbf{l}$ can be equilibrated by an admissible nondifferentiable stress field. For example, the field may be piecewise differentiable and discontinuous, in which case equilibrium conditions (5.5) are supplemented with the relation $[\sigma_{ij}]\nu_j = 0$ at the discontinuity surface (see Subsection A.1). However, such generalization of the equilibrium conditions is not ample. The existence of nonregular loads that cannot be equilibrated by a piecewise smooth stress field is not the sole factor of importance. What matters most is the nonuniqueness of the stress field $\boldsymbol{\sigma}$ in a body that remains rigid. Any self-equilibrated stress field $\boldsymbol{\sigma}'$ can be added to $\boldsymbol{\sigma}$ without violating equilibrium, and neither is the admissibility condition violated if $\sigma'$ is "sufficiently small"; see Remark 3. It should be emphasized that $\sigma'$ may be nonsmooth, which fundamentally makes no difference, as smooth and nonsmooth stress fields have equal status. Therefore, in connection with the load admissibility problem, the equilibrium conditions are to be generalized in order to make sense for a sufficiently wide class of stress fields. The generalized formulation of equilibrium conditions is well known in continuum mechanics; this is the virtual work principle. We consider this principle in the next chapter.

## 6. Beams and trusses

Three-dimensional bodies are not the only subjects of mechanics in our three-dimensional world. For example, thin-walled structures are often considered as two-dimensional continual systems: fields representing their states depend on two spatial coordinates rather than three. Even discrete models of bodies and structures are efficient enough in many cases. Within the framework of a discrete model, a finite set of numbers represents the state of a system. This set consists, for example, of the values of stress or displacement fields at a number of points, while the fields themselves represent the state of a continual system.

In this section, we consider two examples of rigid perfectly plastic models: a one-dimensional continual system (beam) and a discrete system (frame). The models are constructed like the model of a three-dimensional body, and they possess the same main properties. Therefore, a unified limit analysis theory can and will be developed for them as well as for all other rigid perfectly plastic systems in the following chapters.

**6.1. Beam. Kirchhoff hypothesis.** Structures are often built of rods. A *rod* is a body whose dimensions in two directions are much less than those in the third dimension. We refer to a rod subjected mostly to bending and stretching with negligible torsion as a *beam*.

For simplicity reasons we restrict ourselves to the case of a beam with the rectilinear axis in the undeformed configuration. We only consider small displacements and strains of the beam. We adopt the following Kirchhoff hypothesis: particles that form a cross-section orthogonal to the beam axis in the undeformed configuration, when moved to a deformed configuration, again form the cross-section of the beam by the plane orthogonal to the (deformed) axis. Briefly, a cross-section of the beam remains plane and orthogonal to the beam axis. A cross-section can deform within the plane; however, this deformation will not be considered in what follows. Thus, from the kinematic viewpoint, a deformable curve (the axis) and plane elements (the cross-sections) represent the beam. The axis deforms, and each of the cross-sections turns around its point lying on the axis; the cross-sections remain orthogonal to the axis.

To introduce kinematic variables for a beam, let us place the origin of a Cartesian coordinate frame at the left end of the beam and direct its $x$-axis along the axis of the beam. Let $\boldsymbol{\tau}$ be its basis vector, $\mathbf{n}_y$ and $\mathbf{n}_z$ be the basis vectors of $y$ and $z$ axis, see Figure 6.1. We denote by $S(x)$ the cross-section

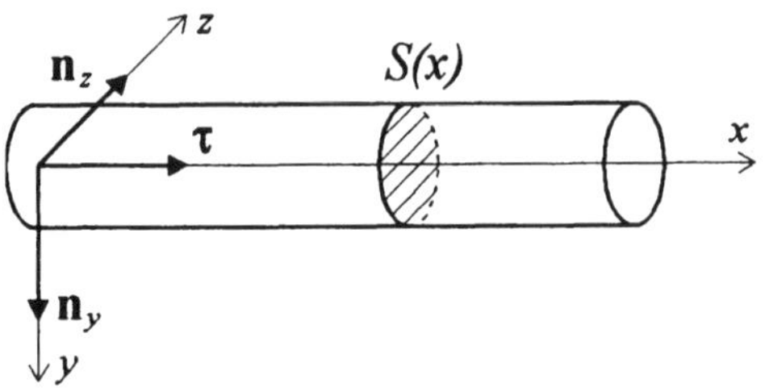


**Figure 6.1**

passing through the point $(x, 0, 0)$ of the beam axis. The cross-section can turn around this point; let $\omega_x(x)$, $\omega_y(x)$, $\omega_z(x)$ denote the components of the angular velocity. By the no-torsion assumption we have $\omega_x \equiv 0$. According to the Kirchhoff hypothesis the element $S(x)$ remains plane. Therefore, to describe a motion of the beam it is sufficient to describe the motion of its axis and the rotations of the cross-sections $S(x)$ around the corresponding points $(x, 0, 0)$. Let $u(x) = u_x(x)$, $u_y(x)$, $u_z(x)$ denote the components of the velocity of the particle $(x, 0, 0)$. Then, the collection of kinematic variables $\mathbf{v} = (\omega_y, \omega_z, u, u_y, u_z)$ represents the motion of the beam. Analogously, the collection

$$\mathbf{e} = (\omega_y', \omega_z', u', u_y' - \omega_z, u_z' + \omega_y) \tag{6.1}$$

plays the role of strain rate (the prime indicates the derivative with respect to $x$). In particular, the axis does not deform and the cross-sections do not rotate if and only if $\mathbf{e} \equiv 0$.

The Kirchhoff hypothesis reduces to certain restrictions on the variables $\mathbf{v} = (\omega_y, \omega_z, u, u_y, u_z)$. Namely, the hypothesis is equivalent to the following relations

$$\omega_y = -u_z', \qquad \omega_z = u_y'. \tag{6.2}$$

Indeed, consider a small segment of the beam axis between the points $(x_1, 0, 0)$ and $(x_2, 0, 0)$. The vectors $\boldsymbol{\rho} = (x_2 - x_1)\boldsymbol{\tau}$ and

$$\boldsymbol{\rho}_* = \boldsymbol{\rho} + \mathbf{u}(x_2)\, dt - \mathbf{u}(x_1)\, dt$$

represent the positions of the segment at the moments $t$ and $t + dt$; here the velocity $\mathbf{u} = u\boldsymbol{\tau} + u_y\mathbf{n}_y + u_z\mathbf{n}_z$ depends on $x$; see Figure 6.2. Retaining the terms of power 1 in $|\boldsymbol{\rho}| = x_2 - x_1$, we obtain

$$\mathbf{u}(x_2) - \mathbf{u}(x_1) = (u'\boldsymbol{\tau} + u_y'\mathbf{n}_y + u_z'\mathbf{n}_z)(x_2 - x_1) = u'\boldsymbol{\rho} + \boldsymbol{\Omega} \times \boldsymbol{\rho},$$

where $\boldsymbol{\Omega} = -u_z'\mathbf{n}_y + u_y'\mathbf{n}_z$. Then the expression for $\boldsymbol{\rho}_*$ can be written as: $\boldsymbol{\rho}_* = (1 + u'dt)\boldsymbol{\rho} + \boldsymbol{\Omega} \times \boldsymbol{\rho}\, dt$. This means that the linear element $\boldsymbol{\rho}$ deforms with the relative elongation rate $u'$ and rotates with the angular velocity $\boldsymbol{\Omega}$.

The Kirchhoff hypothesis states that the rotation of $\boldsymbol{\rho}$ and the rotation of the cross-section $S(x)$ leave the angle formed by $\boldsymbol{\rho}$ and $S(x)$ unchanged.

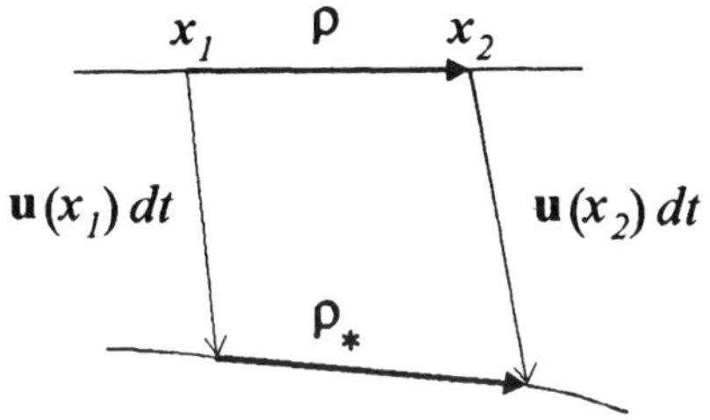


**Figure 6.2**

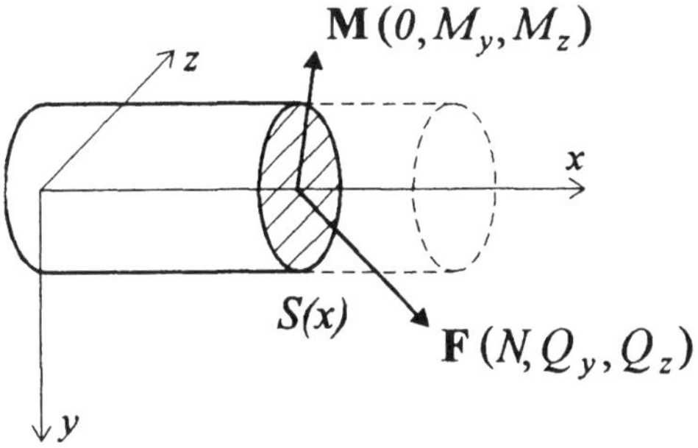


**Figure 6.3**

This is possible if and only if the angular velocities of the rotations are equal: $\boldsymbol{\omega} = \boldsymbol{\Omega}$, that is, relations (6.2) are valid.

**6.2. Internal forces in a beam.** Consider two parts of a beam separated by the cross-section $S(x)$. The bending moment $\mathbf{M}(x) = M_y\mathbf{n}_y + M_z\mathbf{n}_z$ and the force $\mathbf{F}(x) = N\boldsymbol{\tau} + Q_y\mathbf{n}_y + Q_z\mathbf{n}_z$ are assumed to represent the action of the right part of the beam on its left part (see Figure 6.3). We are reminded that, due to the no-torsion assumption, the internal torsional moment is zero. We refer to the components $Q_y$, $Q_z$ of the force $\mathbf{F}$ as *cutting forces*, and to the component $N$ as the *axial force*. The collection of internal forces $\boldsymbol{\sigma} = (M_y, M_z, N, Q_y, Q_z)$ plays the same role as stress plays in the case of a three-dimensional body. Note that $\mathbf{M}(x)$, $\mathbf{F}(x)$ represent the action of the right part of the beam on its left part; the moment $-\mathbf{M}(x)$ and the force $-\mathbf{F}(x)$ represent the backward action.

Let us now derive the equilibrium equations for a beam. Consider a beam subjected to a load distributed along it with the density $\mathbf{q}$. This means that the part of the beam bounded by any two of its cross-sections $S(x_1)$ and $S(x_2)$, $x_2 > x_1$, is subjected to the load

$$\int_{x_1}^{x_2} \mathbf{q}\, dx.$$

In particular, the element of the beam contained between the cross-sections $S(x)$ and $S(x+dx)$ is subjected to the load $\mathbf{q}\,dx$. Apart from this external load, the element is also subjected to the force $-\mathbf{F}(x+dx)$ and the moment $-\mathbf{M}(x)$ on its left boundary $S(x)$ and to the force $\mathbf{F}(x+dx)$ and the moment $\mathbf{M}(x)$ on its right boundary $S(x+dx)$. The element is in equilibrium if and only if the resultant force and moment vanish:

$$-\mathbf{F}(x)+\mathbf{F}(x+dx)+\mathbf{q}(x)\,dx=0, \quad -\mathbf{M}(x)+\mathbf{M}(x+dx)+\boldsymbol{\tau}\times\mathbf{F}(x)\,dx=0.$$

These relations are differential equilibrium equations for a beam:

$$N'=-q, \quad Q_y'=-q_y, \quad Q_z'=-q_z, \quad M_y'=Q_z, \quad M_z'=-Q_y$$

(here, $q$, $q_y$, $q_z$ stand for the components of the load $\mathbf{q}$).

Boundary conditions for a beam are derived in the same way. For example, let the right end of the beam be subjected to the force $\mathbf{f}=f\boldsymbol{\tau}+f_y\mathbf{n}_y+f_z\mathbf{n}_z$ and the moment $\mathbf{m}=m_y\mathbf{n}_y+m_z\mathbf{n}_z$. To find the corresponding boundary conditions, we consider the element of the beam contained between the cross-section $S(a-dx)$ and the right end $S(a)$ of the beam; here, $x=a$ is the coordinate of the beam's right end. The element is in equilibrium if and only if it is subjected to the zero resultant force and moment. This immediately results in the following relations:

$$M_y\big|_{x=a}=m_y, \quad M_z\big|_{x=a}=m_z, \quad N\big|_{x=a}=f, \quad Q_y\big|_{x=a}=f_y, \quad Q_z\big|_{x=a}=f_z.$$

These are the boundary conditions for the above-derived differential equations.

**6.3. Constitutive relations for rigid perfectly plastic beam.** We assign the rigid perfectly plastic properties to a beam following the same scheme as in the case of a three-dimensional body. First we introduce the set $C_x$ of internal forces which are admissible at the cross-section $S(x)$. We assume that $C_x$ is of the form

$$C_x=\{\boldsymbol{\sigma}:\ F_x(N,M_y,M_z)\le 0\},$$

where $F_x$ is a continuous, convex yield function with $F_x(0,0,0)\le 0$. The latter property assures that the zero internal forces are admissible. We adopt the constitutive maximum principle:

$$\boldsymbol{\sigma}\in C_x, \qquad \mathrm{e}(x)\cdot(\boldsymbol{\sigma}(x)-\boldsymbol{\sigma}_*)\ge 0 \quad \text{for every} \quad \boldsymbol{\sigma}_*\in C_x. \tag{6.3}$$

These relations are exactly the same as in the three-dimensional case with the only difference: stress and strain rate are now represented by the collections

$$\boldsymbol{\sigma}=(M_y,\,M_z,\,N,\,Q_y,\,Q_z), \qquad \mathrm{e}=(\omega_y',\,\omega_z',\,u',\,u_y'-\omega_z,\,u_z'+\omega_y).$$

The expression $\mathbf{e}\cdot\boldsymbol{\sigma}$ now denotes the sum of the products of the strain rate and stress components. Recall that all functions now depend on the single coordinate $x$ and that the yield function $F_x$ is assumed independent of cutting forces $Q_y$, $Q_z$. The following simple proposition explains the meaning of this assumption.

*A rigid perfectly plastic beam satisfies the Kirchhoff hypothesis if and only if the yield function $F_x$ is independent of the cutting forces.* Indeed, suppose that $F_x$ does not depend on $Q_y$, $Q_z$; let us show that the Kirchhoff hypothesis is valid in this case. Consider any strain rate e and stress $\boldsymbol{\sigma}$ which satisfy the constitutive maximum principle

$$\mathrm{e}(x) \cdot \boldsymbol{\sigma}(x) \geq \mathrm{e}(x) \cdot \boldsymbol{\sigma}_* \quad \text{for every} \quad \boldsymbol{\sigma}_* \in C_x.$$

Since $F_x$ does not depend on $Q_y$, $Q_z$, any internal forces of the form $\boldsymbol{\sigma}_* = (0, 0, 0, Q_y, Q_z)$ are admissible. This implies that the right-hand side of the constitutive inequality takes arbitrarily large values if one or both of the strain rate components $u_y' - \omega_z$, $u_z' + \omega_y$ do not vanish. In this case, e, $\boldsymbol{\sigma}$ do not satisfy the constitutive maximum principle. Consequently, e can be a strain rate in the beam provided e satisfies the conditions $u_y' - \omega_z = 0$, $u_z' + \omega_y = 0$, that is, the Kirchhoff hypothesis. Now we suppose that the Kirchhoff hypothesis is valid and verify that the yield function does not depend on the cutting forces in this case. Consider, for example, sufficiently smooth yield function $F_x$. If $F_x$ depends on $Q_y$, $Q_z$, there exists a point on the yield surface at which $\frac{\partial F_x}{\partial Q_y}$ and/or $\frac{\partial F_x}{\partial Q_z}$ do not vanish. These are components of the normal to the surface, and the normal determines the direction of the strain rate e. Because of that, at least one of the strain rate components $u_y' - \omega_z$, $u_z' + \omega_y$, which correspond to $\frac{\partial F_x}{\partial Q_y}$, $\frac{\partial F_x}{\partial Q_z}$, does not vanish. This contradicts the Kirchhoff hypothesis. Consequently, if the latter is valid, the assumption about the yield function dependence on $Q_y$, $Q_z$ is wrong. Thus, $F_x$ does not depend on the cutting forces.

**6.4. Rigid-plastic problem for a beam.** Let us formulate the system of equations describing response of a rigid perfectly plastic beam to a quasistatic loading.

Consider a beam subjected to a load distributed along it with the density $\mathbf{q}$ and to a moment $\mathbf{m}$ and force $\mathbf{f}$ applied at the right end of the beam. The left end is fixed, that is, the velocity satisfies the following conditions:

$$\omega_y(0) = \omega_z(0) = 0, \quad u(0) = u_y(0) = u_z(0) = 0.$$

We assume that the beam is in equilibrium at the initial moment $t = 0$ and that the load varies slowly enough to result only in small accelerations. Neglecting the accelerations, we consider the quasistatic description of the beam behavior. Moreover, we restrict ourselves to studying the corresponding instantaneous problem. We will see that the type of solution to this problem is what determines the response of the beam to the current load.

Formulation of the instantaneous problem consists of the above-mentioned equations and boundary conditions:

$$M_y' = Q_z, \quad M_z' = -Q_y, \quad N' = -q, \quad Q_y' = -q_y, \quad Q_z' = -q_z, \tag{6.4}$$

$$M_y|_{x=a} = m_y, \quad M_z|_{x=a} = m_z, \quad N|_{x=a} = f, \quad Q_y|_{x=a} = f_y, \quad Q_z|_{x=a} = f_z, \tag{6.5}$$

$$\mathbf{e} = (\omega_y', \omega_z', u', u_y' - \omega_z', u_z' + \omega_y) \tag{6.6}$$

$$\mathbf{u}|_{x=0} = 0, \quad \omega|_{x=0} = 0, \tag{6.7}$$

$$\boldsymbol{\sigma}(x) \in C_x, \quad \mathbf{e}(x) \cdot (\boldsymbol{\sigma}(x) - \boldsymbol{\sigma}_*) \geq 0 \text{ for every } \boldsymbol{\sigma}_* \in C_x. \tag{6.8}$$

Here, $a$ is the length of the beam and the prime indicates the derivative with respect to $x$. The unknowns in (6.4) – (6.8) are the velocities $\mathbf{v}$ and the internal forces $\boldsymbol{\sigma}$:

$$\mathbf{v} = (\omega_y, \omega_z, u, u_y, u_z), \qquad \boldsymbol{\sigma} = (M_y, M_z, N, Q_y, Q_z).$$

Recall that equations (6.4) and boundary conditions (6.5) are equilibrium conditions, formulas (6.6) express strain rate through the velocities, (6.7) presents kinematic boundary conditions and (6.8) are the constitutive relations. We refer to (6.4) – (6.8) as the *rigid-plastic problem* for the beam. This is the formulation of the problem for typical loading and typical kinematic constraints; other variants can be treated similarly.

The meaning of the relations constituting problem (6.4) – (6.8) is exactly the same in the case of the three-dimensional problem (5.5) – (5.7). On the other hand, equilibrium conditions (6.4), (6.5) and (5.5) are not alike, and there is no resemblance between the expressions for the strain rates (6.6) and (5.6). Therefore, the two formulations seem quite different. However, the above-mentioned particular expressions are not of much importance. What is really important is a certain relation between the equilibrium equations and kinematic formulas, which is the same in cases of both a three-dimensional body and a beam. Moreover, this relation is the same for all other continual and discrete mechanical systems, which will be shown in Chapter 2. As to the constitutive relations, we see that they, being based on the same constitutive maximum principle, are essentially the same for three-dimensional bodies and beams. Thus, there is a substantial similarity between the rigid-plastic problem formulations for bodies and beams, which enables us to study the main features of their responses within the framework of the unified theory (Chapter 4).

In the next subsection, we give an example of a different system that can also be included into the scope of the unified theory.

**6.5. Truss.** Consider a structure consisting of rectilinear rods with each of their two ends either fixed or joined to the end(s) of one or several other rods. In the first case, the rod can rotate around the fixed point. In the second case, each of the joined rods can rotate around the joint. In both cases the moment at the end of the rod is assumed to be zero, so that the only reaction of the joint is force. In particular, rotation of the rod around the joint meets no resistance. We refer to the joint possessing these properties as

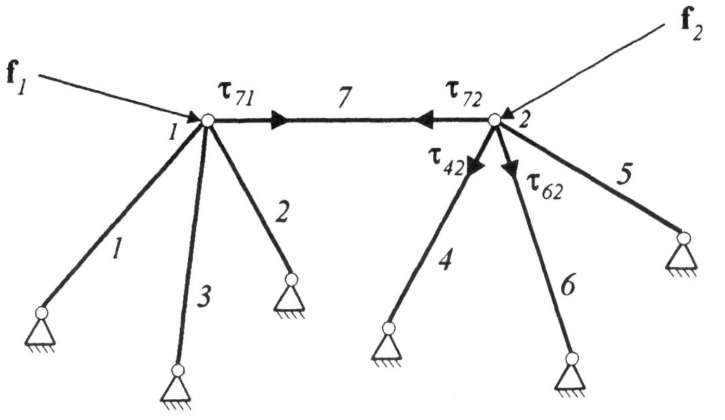


**Figure 6.4**

*perfect hinge*, or *hinge* for short. A hinge itself can be fixed, thus fixing the end of the rod (or the ends of several joined rods). It is assumed that the number of fixed hinges is sufficient to make motion of the structure impossible without deforming at least one of the rods. In other words, velocities of all hinges are zero if the strain rates of all rods vanish. It is also supposed that external loads can only be applied at the ends of the rods, that is, at joints, so that there are no loads distributed along the rods. The structure meeting the above-mentioned requirements is referred to as a *truss*. An example of a truss is shown in Figure 6.4.

We do not consider the rod if both of its ends are fixed, as such a rod would have no effect on motion and stress in the truss. Thus, there is an unfixed hinge at at least one end of each rod in a truss. We number the rods and unfixed hinges of a truss. We also introduce the vectors $\boldsymbol{\tau}_{i\alpha}$ defined as follows: if the $\alpha$-hinge is one of the ends of the $i$-th rod, then $\boldsymbol{\tau}_{i\alpha}$ is the unit vector directed along the $i$-th rod from the $\alpha$-th hinge to the other end of the rod; if the $\alpha$-th hinge is not an end of the $i$-th rod, then $\boldsymbol{\tau}_{i\alpha} = 0$. Thus, if the ends of the rod are the $\alpha$-th and $\beta$-th hinges, then

$$\boldsymbol{\tau}_{i\alpha} = -\boldsymbol{\tau}_{i\beta} \neq 0; \qquad \boldsymbol{\tau}_{i\mu} = 0 \quad \text{if} \quad \mu \neq \alpha,\, \beta. \tag{6.9}$$

If one end of the $i$-th rod is the $\alpha$-th hinge and the other one is fixed, then

$$\boldsymbol{\tau}_{i\alpha} \neq 0; \qquad \boldsymbol{\tau}_{i\mu} = 0 \quad \text{if} \quad \mu \neq \alpha.$$

For example, the vectors $\boldsymbol{\tau}_{42}$, $\boldsymbol{\tau}_{62}$, $\boldsymbol{\tau}_{71} = -\boldsymbol{\tau}_{72}$ are shown in Figure 6.4.

Let $\mathbf{v}_\alpha$ be the velocity of the $\alpha$-th hinge and $k$ be the number of hinges in the truss. Here, the finite collection $\mathbf{v} = (\mathbf{v}_1, \ldots, \mathbf{v}_k)$ plays the same role as the velocity field does in the case of a three-dimensional body. In particular, the velocities $\mathbf{v}_\alpha$ determine elongation rates for all the rods. Indeed, consider the $i$-th rod with ends at the $\alpha$-th and $\beta$-th hinges. Let the vector $\boldsymbol{\rho} = \boldsymbol{\rho}_i(t)$ represent the position of the $i$-th rod at the moment $t$. Elongation of the rod

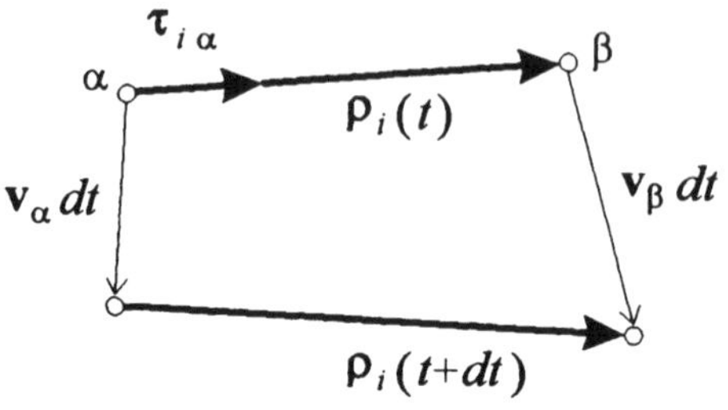


**Figure 6.5**

at $t + dt$ (with respect to the rod's length at the moment $t$) is

$$d\,|\boldsymbol{\rho}_i| = \frac{1}{2}\frac{1}{|\boldsymbol{\rho}_i|}d(\boldsymbol{\rho}_i\boldsymbol{\rho}_i) = \frac{\boldsymbol{\rho}_i}{|\boldsymbol{\rho}_i|}d\boldsymbol{\rho}_i = \boldsymbol{\tau}_{i\alpha}\,d\boldsymbol{\rho}_i.$$

Here, $d\boldsymbol{\rho}_i = (\mathbf{v}_\beta - \mathbf{v}_\alpha)\,dt$ (see Figure 6.5), and, consequently,

$$d\,|\boldsymbol{\rho}_i| = \boldsymbol{\tau}_{i\alpha}(\mathbf{v}_\beta - \mathbf{v}_\alpha)\,dt$$

(no summation over $\alpha$). Taking (6.9) into account, we write the expression $\boldsymbol{\tau}_{i\alpha}(\mathbf{v}_\beta - \mathbf{v}_\alpha)$ in the form $\boldsymbol{\tau}_{i\gamma}\mathbf{v}_\gamma$ with summation over $\gamma = 1, \cdots, k$. Then, the previous formula implies the following expression for the elongation rate of the $i$-th rod:

$$e_i = \frac{d\,|\boldsymbol{\rho}_i|}{dt} = -\boldsymbol{\tau}_{i\gamma}\mathbf{v}_\gamma.$$

This formula also holds for the elongation of a rod if one of its ends is fixed.

REMARK 1. In the case of a truss, we use the elongation $e_i$ as the strain rate of the rod instead of the relative elongation, that is, $e_i$ divided by the length of the rod. In principle, there is no difference between using $e_i$ or the relative elongation. The reason we prefer $e_i$ is that, due to this choice, the lengths of the rods do not enter some important formulas in the sequel.

**6.6. Internal forces in a truss.** Let us find out the type of stress state for the rods in a truss and formulate the equilibrium conditions for a truss. We are reminded that loads can only be applied at the hinges and the rods are not subjected to torsion. Therefore, the internal forces $\boldsymbol{\sigma} = (M_y, M_z, N, Q_y, Q_z)$ in any rod satisfy the equilibrium equations (6.4), which were derived exactly for the rods subjected to bending and stretching only. The load $\mathbf{q}$, whose components enter the right-hand sides in (6.4), vanishes in the case of a truss as there is no load distributed along the rods. Then (6.4) becomes

$$M_y' = Q_z, \quad M_z' = -Q_y, \quad N' = 0, \quad Q_y' = 0, \quad Q_z' = 0,$$

where the $x$-axis has its origin at the end of the rod and is directed along the rod axis. By the definition of a perfect hinge, moments at the ends of the rod are zero:

$$M_y\big|_{x=0} = 0, \quad M_z\big|_{x=0} = 0, \quad M_y\big|_{x=a} = 0, \quad M_z\big|_{x=a} = 0.$$

The equilibrium equations together with the latter conditions result in the equalities

$$M_y = M_z = 0, \quad Q_y = Q_z = 0, \quad N = \text{const}.$$

Consequently, the stress state of a rod in a truss is represented only by the axial force $N$, cutting forces and moments being zero. Thus, all rods in a truss are only subjected to tension (compression). The collection $\boldsymbol{\sigma} = (N_1, \dots, N_m)$ represents internal forces in a truss and plays the same role as the stress field does in the case of a three-dimensional body.

Truss equilibrium conditions apparently consist of vanishing of the resultant force at every hinge. Let $\mathbf{f}_\alpha$ be the load applied at the $\alpha$-th hinge. Apart from $\mathbf{f}_\alpha$, the hinge is subjected to the force exerted on it by each of the rods joined by this hinge. When evaluating the resultant force, the summation in $N_i \boldsymbol{\tau}_{i\alpha}$ over $i$ running through the numbers of these rods can be replaced by summation over all $i = 1, \cdots, m$, where $m$ is the number of rods in the truss. Indeed, $\boldsymbol{\tau}_{i\alpha} = 0$ if the $\alpha$-th hinge is not one of the ends of the $i$-th rod. Finally, the resultant force in the $\alpha$-th hinge is $N_i \boldsymbol{\tau}_{i\alpha} + \mathbf{f}_\alpha$, and the truss equilibrium conditions become:

$$N_i \boldsymbol{\tau}_{i\alpha} + \mathbf{f}_\alpha = 0, \qquad \alpha = 1, \dots, k,$$

where the summation index runs through the values $1, \dots, m$ ($m$ is the number of the rods, $k$ is the number of the unfixed hinges).

**6.7. Rigid-plastic problem for a truss.** Let all rods in a truss be rigid perfectly plastic. This means that a set of admissible stresses is given and the constitutive maximum principle is adopted for every rod. As we saw in the previous subsection, the stress in a truss rod is reduced to an axial force, the latter being represented by one number. Therefore, the convex set of admissible stresses in the $j$-th rod is a segment in $\mathbf{R}$:

$$C_j = \left\{N \in \mathbf{R} : -Y_j^- \leq N \leq Y_j^+\right\}, \qquad j = 1, \dots, m.$$

Here, $Y_j^+$, $Y_j^-$ are the yield stresses in tension and compression. The constitutive maximum principle can be written as

$$N_j \in C_j, \qquad e_j(N_j - N_*) \geq 0 \quad \text{for every} \quad N_* \in C_j$$

(no summation over $j$). Here, $N_j$ and $e_j$ stand for the axial force and the elongation of the $j$-th rod, respectively. Figure 6.6 shows the stress–strain diagram for the rod under tension, compression and unloadings.

The constitutive relations for a rigid-plastic rod subjected only to tension–compression can also be written in the form of the normality flow rule (no summation over $j$):

$$e_j = \dot{\lambda}_j^+ - \dot{\lambda}_j^-, \quad \dot{\lambda}_j^+ \geq 0, \quad \dot{\lambda}_j^- \geq 0,$$
$$N_j - Y_j^+ \leq 0, \quad N_j - Y_j^- \leq 0, \quad \dot{\lambda}_j^+(N_j - Y_j^+) = 0, \quad \dot{\lambda}_j^+(N_j - Y_j^-) = 0.$$

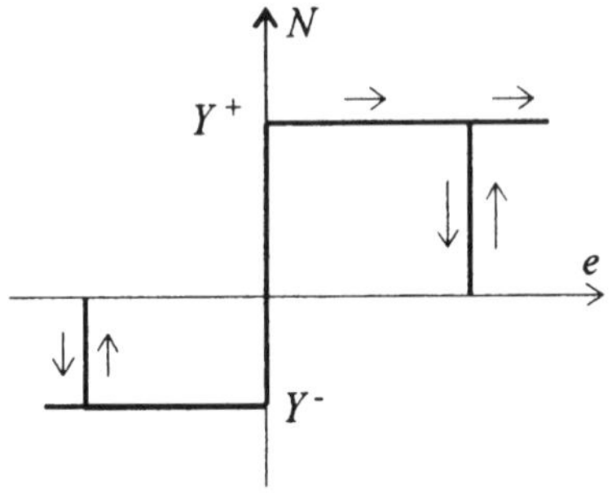


**Figure 6.6**

Let us now formulate the rigid-plastic problem for a truss. Consider a rigid perfectly plastic truss, that is, a truss consisting of rigid perfectly plastic rods. Let $k$ be the number of unfixed hinges and $m$ be the number of rods. The truss is subjected to forces $\mathbf{f}_\alpha$, $\alpha = 1, \ldots, k$; $\mathbf{f}_\alpha$ being the load applied at the $\alpha$-th hinge. We assume that the truss is in equilibrium at the initial moment and that the load varies slowly enough to result only in small accelerations. Neglecting the acceleration, we consider a quasistatic description of the truss behavior. Moreover, we will restrict ourselves to studying the corresponding instantaneous problem as we did in the cases of three-dimensional bodies and beams. Instantaneous *rigid-plastic problem* for the truss consists of searching for the velocities $\mathbf{v} = (\mathbf{v}_1, \ldots, \mathbf{v}_k)$ and internal forces $\boldsymbol{\sigma} = (N_1, \ldots, N_m)$ that satisfy the following equilibrium equations, kinematic and constitutive relations:

$$N_i \boldsymbol{\tau}_{i\alpha} + \mathbf{f}_\alpha = 0, \tag{6.10}$$

$$e_i = -\boldsymbol{\tau}_{i\gamma} \mathbf{v}_\gamma, \tag{6.11}$$

$$N_j \in C_j, \quad e_j(N_j - N_*) \geq 0 \text{ for every } N_* \in C_j. \tag{6.12}$$

The problem (6.10) – (6.12) is *discrete*, that is, the unknowns

$$\mathbf{v} = (\mathbf{v}_1, \ldots, \mathbf{v}_k), \qquad \boldsymbol{\sigma} = (N_1, \ldots, N_m)$$

are finite collections of numbers rather than functions of one or three coordinates, as in the cases of beams and three-dimensional bodies, respectively. However, the rigid-plastic problem formulations are substantially similar in all these cases. Indeed, we adopt the same constitutive maximum principle in all the cases, and we will see in the next chapter that the relations between the equilibrium conditions and kinematic formulas are also the same for all types of mechanical systems. Due to this similarity, a unified limit analysis theory covers all the cases; this theory will be presented in Chapter IV.

## Appendix A. Stress and strain

**A.1. Stress.** Continuum mechanics normally adopts the following assumption about internal forces in a medium. Let $S$ be the surface separating two parts $\Omega^-$ and $\Omega^+$ of a domain occupied by the medium. Then, the force exerted by the medium occupying the domain $\Omega^+$ on the medium occupying $\Omega^-$ is

$$\int_S \mathbf{p}_\nu \, dS.$$

The density $\mathbf{p}_\nu$ of the surface traction obviously depends on the state of the medium and on the surface $S$. It is assumed that the only characteristic of the surface $S$ affecting $\mathbf{p}_\nu$ is the normal $\boldsymbol{\nu}$ to the surface.

Cauchy theorem derives from the basic principles that $\mathbf{p}_\nu$ linearly depends on $\boldsymbol{\nu}$. In other words, at any point of the medium, there exists a linear operator $\boldsymbol{\sigma}$ such that $\mathbf{p}_\nu = \boldsymbol{\sigma}\boldsymbol{\nu}$ or, in components, $(\mathbf{p}_\nu)_i = \sigma_{ij}\nu_j$. The operator only depends on the state of the medium and is referred to as *stress tensor* or *stresses* (the plural reminds us of the components of the stress tensor, which were introduced originally in continuum mechanics). Under some non-restrictive hypotheses the stress tensor is symmetric, that is, $\sigma_{ij} = \sigma_{ji}$, and this is the only type of stress tensors dealt with throughout this book.

Consider a continuous medium occupying domain $\Omega$. The medium is subjected to body forces with the volume density $\mathbf{f}$ given in $\Omega$ and surface traction with the density $\mathbf{q}$ given on the part $S_q$ of $\partial\Omega$. If the medium is in equilibrium and the stress field $\boldsymbol{\sigma}$ is differentiable, then $\boldsymbol{\sigma}$ satisfies the equilibrium conditions

$$\frac{\partial \sigma_{ij}}{\partial x_j} + f_i = 0 \quad \text{in} \quad \Omega, \qquad \sigma_{ij}\nu_j = q_i \quad \text{on} \quad S_q,$$

where $x_1$, $x_2$, $x_3$ are Cartesian coordinates, $\boldsymbol{\nu}$ is the unit outward normal to $\partial\Omega$. If $\boldsymbol{\sigma}$ is piecewise differentiable with a jump on the discontinuity surface $\Gamma$, then equilibrium conditions consist of the same differential equations *off* $\Gamma$, the same boundary conditions, and additional relations at $\Gamma$: $[\sigma_{ij}]\nu_j = 0$. Here, $\boldsymbol{\nu}$ is the unit normal to $\Gamma$ and $[\sigma_{ij}]$ is the jump of $\sigma_{ij}$ on $\Gamma$, that is, the difference between the limit values of $\sigma_{ij}$ at the two sides of $\Gamma$.

**A.2. Deformation.** In order to specify each particle of a continuous medium, we use coordinates $\xi_1$, $\xi_2$, $\xi_3$ of the particle position at the initial moment $t = 0$. Thus $\xi$ is the name of the particle, and "the particle $\xi$" means "the particle that was at the point $\xi = (\xi_1, \dots, \xi_3)$ at the moment $t = 0$". Configuration of the medium at the moment $t$ is described by the mapping $\xi \to (f_1(\xi,t), f_2(\xi,t), f_3(\xi,t))$ or $\xi \to f(\xi,t)$, for short, where $f_i(\xi,t)$ are the coordinates of the position the particle $\xi$ occupies at the moment $t$.

Consider the particles which form the segment with the ends $(\xi_1, \xi_2, \xi_3)$ and $(\xi_1 + d\xi_1, \xi_2 + d\xi_2, \xi_3 + d\xi_3)$ at the initial moment $t = 0$. Such a set is referred to as a *material line element*. To specify the position of the element at $t = 0$, let us introduce the infinitesimal vector $d\boldsymbol{\xi}$ with the components $(d\xi_1, d\xi_2, d\xi_3)$ emanating from the point $\xi = (\xi_1, \dots, \xi_3)$. Similarly, the position $f(\xi, t)$ of the particle $\xi$ and the infinitesimal vector $d\mathbf{x}$ with the components

$$dx_i = \frac{\partial f_i}{\partial \xi_\alpha}(\xi, t) d\xi_\alpha$$

represent the position of the element at the current moment $t$. Variation of the lengths of material line elements and the angles which they form is referred to as *deformation.* At the initial moment, the length of the element corresponding to the vector $d\boldsymbol{\xi}$ is $\sqrt{d\xi_\alpha d\xi_\alpha}$. Consider two elements which correspond to the vectors $d\boldsymbol{\xi}$, $d\boldsymbol{\eta}$ emanating from the same point $\xi$ at $t = 0$. The lengths of the elements and the scalar product $d\boldsymbol{\xi}\, d\boldsymbol{\eta}$ determine the angle $\varphi_0$ which the elements form at $t = 0$: $\cos\varphi_0 = d\boldsymbol{\xi}\, d\boldsymbol{\eta} / |d\boldsymbol{\xi}|\, |d\boldsymbol{\eta}|$. At the current moment $t$, the point $f(\xi, t)$ and the vectors $d\mathbf{x}$, $d\mathbf{y}$ with components

$$dx_i = \frac{\partial f_i}{\partial \xi_\alpha} d\xi_\alpha, \qquad dy_j = \frac{\partial f_j}{\partial \xi_\beta} d\eta_\beta$$

specify the positions of these elements. Their lengths $|d\mathbf{x}|$, $|d\mathbf{y}|$ and the angle which they form at this moment are expressed through

$$|d\mathbf{x}|^2 = \frac{\partial f_i}{\partial \xi_\alpha}\frac{\partial f_i}{\partial \xi_\beta} d\xi_\alpha d\xi_\beta, \qquad |d\mathbf{y}|^2 = \frac{\partial f_i}{\partial \xi_\alpha}\frac{\partial f_i}{\partial \xi_\beta} d\eta_\alpha d\eta_\beta,$$

$$d\mathbf{x}\, d\mathbf{y} = \frac{\partial f_i}{\partial \xi_\alpha}\frac{\partial f_i}{\partial \xi_\beta} d\xi_\alpha d\eta_\beta.$$

Increments of the lengths and of the angle are expressed through

$$\frac{\partial f_i}{\partial \xi_\alpha}\frac{\partial f_i}{\partial \xi_\beta} - \delta_{\alpha\beta}$$

and the components of the vectors $d\boldsymbol{\xi}$, $d\boldsymbol{\eta}$. That is why the tensor $\boldsymbol{\varepsilon}$ with the components

$$\varepsilon_{\alpha\beta} = \frac{1}{2}\left(\frac{\partial f_i}{\partial \xi_\alpha}\frac{\partial f_i}{\partial \xi_\beta} - \delta_{\alpha\beta}\right)$$

is normally used as a measure of deformation. We refer to this tensor as a *strain tensor* or *strain* for short.

There is a nonlinear expression for the components of the strain tensor through the derivatives of the displacement components with respect to the coordinates. However, if the derivatives are small, the nonlinear terms can be neglected, which yields a simple linear formula for $\varepsilon_{ij}$. Namely, consider the particle located at the point $(x_1, x_2, x_3)$ at the current moment $t$. Let $\mathbf{u}(x_1, x_2, x_3, t)$ be the displacement of the particle from the initial configuration to the current one. The above-mentioned linear formula for $\varepsilon_{ij}$ is as follows:

$$\varepsilon_{\alpha\beta} = \frac{1}{2}\left(\frac{\partial f_i}{\partial \xi_\alpha}\frac{\partial f_i}{\partial \xi_\beta} - \delta_{\alpha\beta}\right).$$

A similar formula determines a measure of the deformation rate. Namely, consider the particle located at the point $(x_1, x_2, x_3)$ at the moment $t$. The velocity of the particle at this moment is $\mathbf{v}(x_1, x_2, x_3, t)$. We refer to the tensor $\mathbf{e}$ with the components

$$e_{ij} = \frac{1}{2}\left(\frac{\partial v_i}{\partial x_j} + \frac{\partial v_j}{\partial x_i}\right)$$

as the *strain rate tensor* or *strain rate.* The components $e_{ij}$ determine the rate of the relative elongation of any material line element as well as the rate of the angle formed by any two of the elements emanating from one particle.

**A.3. Elasticity law.** Stresses in a continuous medium are determined by the state of the medium, strain being the most important state variable. Consider a medium under the constant temperature condition. If there is a one-to-one correspondence between stress and strain in this medium, the medium is referred to as *elastic*. In many cases, this correspondence can be regarded as linear: $\boldsymbol{\varepsilon} = \mathbf{A}\boldsymbol{\sigma}$ or, in components, $\varepsilon_{ij} = A_{ijkl}\sigma_{kl}$, which is called the generalized Hooke's law. The fourth rank tensor $\mathbf{A}$ is symmetric: $A_{ijkl} = A_{klij} = A_{jikl} = A_{ijlk}$. It characterizes material properties of the elastic body. If the body is isotopic, the tensor $\mathbf{A}$ is determined by two parameters: Young's modulus E and Poisson's ratio $\nu$, in which case the linear elasticity law is reduced to Hooke's law

$$\varepsilon_{ij} = -\frac{\nu}{\mathrm{E}}\sigma_{kk}\delta_{ij} + \frac{1+\nu}{\mathrm{E}}\sigma_{ij}.$$

For many inelastic bodies, strain can be decomposed into the sum $\boldsymbol{\varepsilon} = \boldsymbol{\varepsilon}^e + \boldsymbol{\varepsilon}^i$ of elastic and inelastic parts $\boldsymbol{\varepsilon}^e$ and $\boldsymbol{\varepsilon}^i$, and the linear elasticity law is valid for the elastic strain $\boldsymbol{\varepsilon}^e$: $\boldsymbol{\varepsilon}^e = \mathbf{A}\boldsymbol{\sigma}$.

## Appendix B. Convex sets and convex functions

**B.1. Definitions and examples.** Let $X$ be a linear space. Consider a set $A$ in $X$ such that it contains any element of the form $\alpha x_1 + (1-\alpha)x_2$ as soon as $x_1$, $x_2$ belong to $A$ and $\alpha$ is a number, satisfying the inequalities $0 \le \alpha \le 1$. Such a set is referred to as *convex*. For example, a ball, a tetrahedron, a plane, a disc, the interior of a triangle are examples of convex sets in three-dimensional space. There is an apparent geometric interpretation of the convex set definition: the set $A$ is convex if and only if the segment joining any two points of $A$ lies entirely in $A$.

Let $f$ be a function defined on a linear space $X$. We consider the linear space $\mathbf{R} \times X$ consisting of couples $(r, x)$, where $r$ belongs to $\mathbf{R}$ and $x$ belongs to $X$, with the following operations:

$$(r_1, x_1) + (r_2, x_2) = (r_1 + r_2, x_1 + x_2), \qquad a(r, x) = (ar, ax)$$

($a$ is a real number). In the space $\mathbf{R} \times X$ we introduce the set consisting of points $(r, x)$ which satisfy the condition $r \ge f(x)$, that is, the point $(r, x)$ lies over the graph of the function $f$. This set is referred to as an *epigraph* of the function $f$ and denoted by the symbol epi $f$:

$$\operatorname{epi} f = \{(r, x) \in \mathbf{R} \times X : r \ge f(x)\}.$$

Figure B.1a illustrates the definitions.

Function $f$ is referred to as *convex* if its epigraph is a convex set. A necessary and sufficient condition for convexity of $f$ is the validity of the inequality

$$f(\alpha x_1 + (1-\alpha)x_2) \le \alpha f(x_1) + (1-\alpha) f(x_2) \qquad \text{(B.1)}$$

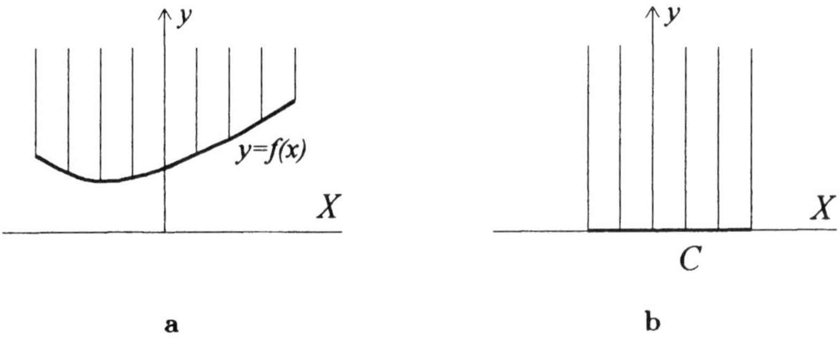


**Figure B.1**

for every $x_1$, $x_2$ in $X$ and every number $\alpha$, satisfying the condition $0 \leq \alpha \leq 1$. For example, the function $y = |x|^p$ is convex if $p \geq 1$ and is not convex if $0 < p < 1$.

Sometimes it is convenient to consider functions that can take the value $+\infty$. The definitions of epigraph and convexity make sense for such functions and the convexity criterion (B.1) stays valid.

EXAMPLE 1. Let $C$ be a set in a linear space $X$ and $\psi_C$ be the function defined as follows:

$$\psi_C(x) = \begin{cases} 0 & \text{if} \quad x \in C, \\ +\infty & \text{if} \quad x \notin C. \end{cases}$$

Figure B.1.b shows the epigraph of $\psi_C$. The function $\psi_C$ is convex if and only if the set $C$ is convex.

If a function $f$ is convex, the set $\{x \in X : f(x) \leq a\}$ is convex for every number $a$. This set may be empty, in which case it is also convex.

**B.2. Separation theorem.** A convex closed set and a point that does not belong to it can be strictly separated by a plane. More precisely, the following proposition is valid.

THEOREM 1. Let $A$ be a convex closed set in the Euclidian space $X$ and $\mathbf{x}_0 \in X$ does not belong to $A$. Then there exists a unit vector $\mathbf{g} \in X$, $|\mathbf{g}| = 1$ and a number $c$ such that

$$\mathbf{gx} < c < \mathbf{gx}_0 \qquad \text{for every} \qquad x \in X. \tag{B.2}$$

PROOF. Since the set $A$ is closed, there exists a point $\mathbf{z}_0$ in $A$ nearest to the point $\mathbf{x}_0$, that is, the following equality is valid:

$$|\mathbf{z}_0 - \mathbf{x}_0| = \min\{|\mathbf{z}_0 - \mathbf{x}_0| : \mathbf{z} \in A\}.$$

Then, we set

$$\mathbf{g} = \frac{\mathbf{x}_0 - \mathbf{z}_0}{|\mathbf{x}_0 - \mathbf{z}_0|}, \qquad c = \mathbf{gx}_0 - \frac{1}{2}|\mathbf{x}_0 - \mathbf{z}_0|$$

and show that inequalities (B.2) are valid. Indeed, let $\mathbf{x}$ be a point in $A$. Due to convexity, the set $A$ also contains the points $\mathbf{y}_\alpha = \alpha\mathbf{x} + (1-\alpha)\mathbf{z}_0$, where $\alpha$ is a parameter, $0 \le \alpha \le 1$.

The distance from $\mathbf{y}_\alpha$ to $\mathbf{x}_0$,

$$\rho(\alpha) = |\alpha\mathbf{x} + (1-\alpha)\mathbf{z}_0 - \mathbf{x}_0|, \qquad 0 \le \alpha \le 1,$$

is minimum when $\alpha = 0$. Therefore, the right derivative of the function $\rho(\alpha)$ is nonnegative at $\alpha = 0$:

$$\frac{\mathbf{z}_0 - \mathbf{x}_0}{|\mathbf{z}_0 - \mathbf{x}_0|}(\mathbf{x} - \mathbf{z}_0) = -\mathbf{g}(\mathbf{x} - \mathbf{z}_0) \ge 0.$$

Consequently, the following inequality is valid:

$$\mathbf{g}\mathbf{x} - \mathbf{g}\mathbf{x}_0 \le \mathbf{g}\mathbf{z}_0 - \mathbf{g}\mathbf{x}_0 = -\mathbf{g}(\mathbf{x}_0 - \mathbf{z}_0) = -|\mathbf{x}_0 - \mathbf{z}_0|.$$

This implies $\mathbf{g}\mathbf{x} < \mathbf{g}\mathbf{x}_0 - |\mathbf{x}_0 - \mathbf{z}_0|/2 < \mathbf{g}\mathbf{x}_0$, which finishes the proof.

COROLLARY. If $A$ is a convex closed set in a Euclidian space and $\mathbf{x}_0$ is a point on its boundary, then there exists a plane passing through $\mathbf{x}_0$ such that the set $A$ lies in one of the two half-spaces separated by this plane. In other words, there exists a unit vector $\mathbf{g} \in X$, $|\mathbf{g}| = 1$, such that the inequality $\mathbf{g}(\mathbf{x} - \mathbf{x}_0) \le 0$ holds for every $\mathbf{x} \in A$.

Indeed, consider a sequence of points $x_n$ not in $A$, $n = 1, 2, \ldots$, converging to $\mathbf{x}_0$. By the separation theorem, for every $n$ there exists $\mathbf{g}_n \in X$ such that $|\mathbf{g}_n| = 1$ and $\mathbf{g}_n(\mathbf{x} - \mathbf{x}_n) < 0$ for every $\mathbf{x} \in A$. The sequence $\{\mathbf{g}_n\}$ is bounded, therefore, it is possible to extract a converging subsequence from it. Let $\mathbf{g}$ be the limit of such subsequence. Then, the previous inequality implies: $\mathbf{g}(\mathbf{x} - \mathbf{x}_0) \le 0$ for every $\mathbf{x} \in A$.

## Appendix C. Extremums

**C.1. Minimum and maximum.** Let $A$ be a set of numbers. If there exists the least of them, this number is referred to as a *minimum* of $A$ and denoted by the symbol $\min A$. For example, 0 is the minimum of the set

$$[0,1] = \{x \in \mathbf{R} : \quad 0 \le x \le 1\};$$

the set

$$(0,1] = \{x \in \mathbf{R} : \quad 0 < x \le 1\}$$

has no minimum as well as the set of all integers. It appears that in the case of the set $(0,1]$ zero plays almost the same role as in the case of $[0,1]$. The following definition appropriately generalizes the concept of minimum.

Even if the set $A$ has no minimum, it can be bounded from below, that is, there exists a number $a_-$ such that $a_- \le a$ for every $a \in A$. We refer to the greatest of the numbers $a_-$ possessing this property as an *infimum* of $A$ and denote it by the symbol $\inf A$. For example, 0 is the infimum of the set $(0,1]$. If the set $A$ contains its minimum, then the minimum is still the infimum of

$A$. The infimum of a set bounded from below always exists in contrast to the minimum.

A number $m$ is the infimum of the set $A$ if and only if

(1) $m \leq a$ for every $a \in A$,

(2) for every $\varepsilon > 0$ there exists such $a_\varepsilon$ in $A$ that $a_\varepsilon < m + \varepsilon$.

We set by definition that 1) $\inf A = -\infty$ if the set $A$ is unbounded from below, 2) the infimum of the empty set is $+\infty$: $\inf \emptyset = +\infty$.

If $A$ is a set of numbers and there exists the greatest of them, this number is referred to as a *maximum* of $A$ and denoted by the symbol $\max A$.

A set $A$ is bounded from above if there exists a number $a_+$ such that $a_+ \geq a$ for every $a \in A$. We refer to the least of the numbers $a_+$ possessing this property as a *supremum* of the set $A$ and denote it by the symbol $\sup A$. The supremum of the set $A$ bounded from above always exists in contrast to the maximum of $A$. If a set $A$ has the maximum it is still the supremum of $A$.

A number $M$ is the supremum of a set $A$ if and only if

(1) $M \geq a$ for every $a \in A$,

(2) for every $\varepsilon > 0$ there exists such $a_\varepsilon$ in $A$ that $a_\varepsilon > M - \varepsilon$.

We set by definition that 1) $\sup A = +\infty$ if the set $A$ is not bounded from above, 2) the supremum of the empty set is $-\infty$: $\sup \emptyset = -\infty$.

REMARK 1. Sometimes it is convenient to consider extended number sets, that is, sets that may contain the members $+\infty$ and $-\infty$ apart from numbers. We set by definition that $\sup A = +\infty$ if $A$ contains the element $+\infty$ and $\inf A = -\infty$ if $A$ contains the element $-\infty$.

**C.2. Extremum problems.** Infimum and supremum of a set are called its *extremums*. The problem of searching for an extremum of a set is referred to as an *extremum problem*. Let $f$ be a function and $X$ be a subset in its domain; then the following extremums are associated with them:

$$\inf \{f(x) : x \in X\}, \qquad \sup \{f(x) : x \in X\}.$$

We use the following symbols to denote the problems of searching these extremums:

$$\begin{aligned} f(x) \quad &\to \quad \inf; \qquad x \in X, \\ f(x) \quad &\to \quad \sup; \qquad x \in X. \end{aligned}$$

If the set $\{f(x) : x \in X\}$ contains its minimum (maximum), we refer to this minimum (maximum) as the *minimum (maximum) of the function $f$ over the set $X$*. If the function takes the value $f(x_0) = \min \{f(x) : x \in X\}$ at the point $x_0$, we say that $f$ *attains* its minimum (or infimum) at $x_0$. Analogously we speak of attaining the maximum (supremum).

EXAMPLE 1. Consider the set of all rectangles and function $S$ defined on it: $S$ is the rectangle's area. Let $X$ be the set of rectangles of the given perimeter $p$. The function $S$ attains its maximum over $X$: $S(x_0) = p^2/16$, where $x_0$ is the square of perimeter $p$. The infimum $\inf\{S(x) : x \in X\} = 0$ is not attained on $X$ (we do not consider a segment as a rectangle).

As we see, it is possible that a function does not attain its infimum. However, due to the criterion for a number $m$ to be the infimum (Subsection C.1), there exists a sequence of the $x_n \in X$ such that the values $f(x_n)$ converge to $m$:

$$\lim_{n\to\infty} f(x_n) = m = \inf\{f(x) : x \in X\} \qquad (x_n \in X).$$

We refer to this sequence as *minimizing*. Analogously, there exists a *maximizing* sequence $\{\widehat{x}_n\}$ of the problem $f(x) \to \sup;\ x \in X$:

$$\lim_{n\to\infty} f(\widehat{x}_n) = M = \sup\{f(x) : x \in X\} \qquad (x_n \in X).$$

EXAMPLE 2. Consider the problem in Example 1. The sequence of rectangles with sides of the length $a_n = p/4 - p/4n$ and $b_n = p/4n$ is a minimizing sequence in this problem.

EXAMPLE 3. Consider the extremum problem $x^2/2 \to \inf;\ x \in [-1,1]$. The sequence of numbers $x_n = 1/n$, $n = 1, 2, \ldots$, is a minimizing sequence in this problem. The sequence of numbers $x_n = 0$, $n = 1, 2, \ldots$, is also a minimizing sequence (members of a minimizing sequence are not supposed to be different). The minimum is attained at $x = 0$.

EXAMPLE 4. Consider the extremum problem $x^2/2 \to \sup;\ x \in [-1,1]$. The sequence of numbers $x'_n = 1 - 1/n$, $n = 1, 2, \ldots$ is a maximizing sequence in this problem. The sequence of numbers

$$x''_n = \begin{cases} 1 & \text{if} \quad n = 1, 3, 5, \ldots, \\ -1 & \text{if} \quad n = 2, 4, 6, \ldots. \end{cases}$$

is also a maximizing sequence (a maximizing or minimizing sequence does not necessarily converge). The maximum is attained at $x = -1$ and at $x = 1$. The problem $x^2/2 \to \sup;\ x \in (-1,1)$ differs from the one in Example 3. The supremum is not attained in the problem under consideration. The sequence $x'_n$ is a maximizing sequence in this problem, the sequence $x''_n$ is not: $x''_n$ does not belong to $(-1,1)$.

**C.3. Conditions for the minimum of a convex function.** Let $f, f_1, \ldots, f_m$ be convex functions on $\mathbf{R}^n$. Consider the extremum problem: find the infimum of the function $f$ over the set specified by the inequalities $f_1(x) \le 0, \ldots, f_m(x) \le 0$. Apparently, the latter set is convex. In connection with this convexity and with the convexity of $f$, this problem is referred to as a *convex extremum problem* or problem of *convex optimization*. The following proposition gives necessary and sufficient conditions for attaining the infimum in a problem of convex optimization.

THEOREM *(Kuhn, Tucker)*. Suppose $f, f_1, \ldots, f_m$ are convex continuously differentiable functions on $\mathbf{R}^n$ and there exists a point $x$ at which the strict inequalities $f_1(x) < 0, \ldots, f_m(x) < 0$ are valid. Then the infimum

$$f(x) \to \inf; \quad f_1(x) \le 0, \ldots, f_m(x) \le 0$$

is attained. The following conditions are necessary and sufficient for the infimum to be attained at $x_0$: there exist numbers $\lambda_1, \ldots, \lambda_m$ such that

$$\frac{\partial f}{\partial x_i}(x_0) + \lambda_1 \frac{\partial f_1}{\partial x_i}(x_0) + \cdots + \lambda_m \frac{\partial f_m}{\partial x_i}(x_0) = 0 \qquad (i = 1, 2, \ldots),$$

$$f_1(x_0) \leq 0, \ldots, f_m(x_0) \leq 0, \qquad \lambda_1 \geq 0, \ldots, \lambda_m \geq 0,$$

$$\lambda_1 f_1(x_0) = 0, \ldots, \lambda_m f_m(x_0) = 0.$$

In the problem under consideration, we look for the infimum over the set $S$ specified by the unilateral constraints $f_1(x) \leq 0, \cdots, f_m(x) \leq 0$. If $x_0$ is in the interior of $S$, the conditions for attainability of the infimum given by the Kuhn-Tucker theorem are reduced to the optimality conditions $\partial f/\partial x_i = 0$ $(i = 1, 2, \ldots)$ well known for unconstrained extremum problems. If $x_0$ is on the boundary of $S$, the multiplier $\lambda_i$ can differ from 0 provided $x_0$ lies on the part of the boundary where $f_i(x) = 0$. Let $f_{i_1}(x) = 0, \cdots, f_{i_k}(x) = 0$ be the complete set of equations specifying the part of the boundary where $x_0$ lies. Then, the Kuhn-Tucker conditions are exactly those given by the Lagrange multipliers rule for the extremum problem with the constraints $f_{i_1}(x) = 0, \cdots, f_{i_k}(x) = 0$. Actually, the Kuhn-Tucker theorem extends the Lagrange multipliers rule to the problem with unilateral constraints.

## Comments

Systematic studies in plasticity originated from a work of Tresca (1864). In this and subsequent articles Tresca presented results of numerous experiments in which he studied the behavior of various materials. He also introduced the yield condition of the maximum tangent stress (Example 2.2). Starting from these experimental results, de Saint Vénant (1870) developed constitutive relations for the rigid perfectly plastic body under the plane strain conditions and formulated the corresponding rigid-plastic problem. Mises (1913) gave a complete formulation of the three-dimensional rigid-plastic problem in the case of the quadratic yield function. Later, Mises (1928) also considered the case of an arbitrary convex smooth yield function. The interesting point is that only rigid-plastic bodies were considered originally, which could be an impact of the Tresca experimental studies: he carried out his experiments mostly at large strain with a negligible elastic part.

The possibility of reformulating the normality flow rule into the form of the constitutive maximum principle was unknown until the works by Drucker (1951) and Bishop and Hill (1951). The maximum principle arose from Drucker's stability postulate in the former work and from the analysis of crystal behavior in the latter. It should be pointed out that using the constitutive maximum principle is more convenient for consideration of general theory rather than for solving concrete problems. In particular, the principle does not presume any smoothness of the yield surface.

Dissipation plays a very important role in studying various problems of plasticity. Gvozdev (1938, 1948, 1949) and Drucker, Prager and Greenberg

(1951) made use of dissipation when developing the limit analysis theory. Ziegler (1963) and Ivlev (1967) showed the possibility of formulating plastic constitutive relations in terms of dissipation. Nayroles (1970) established a one-to-one correspondence between yield surfaces and dissipation functions. In this article, Nayroles also formulated an extremum principle, which is equivalent to the constitutive maximum principle and makes use of dissipation instead of the yield surface. Namely, the following *minimum principle* is valid: in a rigid perfectly plastic body the stress $\boldsymbol{\sigma}$ and the strain rate $\mathbf{e}$ satisfy the condition

$$d_x(\mathbf{e}) - \mathbf{e} \cdot \boldsymbol{\sigma} \leq d_x(\mathbf{e}_*) - \mathbf{e}_* \cdot \boldsymbol{\sigma} \quad \text{for every} \quad \mathbf{e}_* \in Sym.$$

The minimum principle and the constitutive maximum principle are equivalent and dual in a sense. In particular, the maximum principle results in the normality flow rule which expresses the strain rate through the stress. Likewise, the minimum principle results in an expression for the stress through the strain rate. This useful expression is of the following form:

$$\boldsymbol{\sigma} = \frac{\partial d_x}{\partial \mathbf{e}}$$

if the dissipation is differentiable at the point $\mathbf{e}$. However, the dissipation is not differentiable at $\mathbf{e} = 0$ since $d_x$ is a positively homogeneous function of degree 1. Moreover, the dissipation is nondifferentiable at every $\mathbf{e}$ in the case of a body with a cylindrical yield surface: $d_x$ is not even continuous in this case since $d_x(\mathbf{e}) = +\infty$ if $\mathbf{e}^s \neq 0$. That is why it is convenient to use the function $d_x^{(d)}(\mathbf{e}) = d_x(\mathbf{e}^d)$, see Example 4.2. Normally, it is at least piecewise differentiable. If $d_x^{(d)}$ is differentiable, the minimum principle results in the following relations:

$$\mathbf{e}^s = 0,$$
$$\boldsymbol{\sigma} = -p\mathbf{I} + \frac{\partial d_x^{(d)}}{\partial \mathbf{e}} \qquad \text{if } \mathbf{e} \neq 0,$$
$$\boldsymbol{\sigma} \in C_x \qquad \text{if } \mathbf{e} = 0.$$

The pressure $p$ is not determined by the strain rate $\mathbf{e}$. The condition $\boldsymbol{\sigma} \in C_x$ is satisfied by virtue of the formula for $\boldsymbol{\sigma}$ in the case $\mathbf{e} \neq 0$. The admissibility condition $\boldsymbol{\sigma} \in C_x$ is the only requirement of the constitutive relations in case $\mathbf{e} = 0$. It is easy to generalize the above relations to cover the case of piecewise differentiable functions $d_x^{(d)}$.

The reader can find detailed descriptions of plasticity concepts, for example, in the books by Kachanov (1971) and Martin (1975). The main concepts and relations of continuum mechanics (stress, strain, equilibrium equations, elasticity law, etc.) are presented in the books by Sedov (1994) and Germain (1973), concepts of convex analysis and extremum problems theory in books by Rockafellar (1967) and Ioffe and Tikhomirov (1979).

CHAPTER II

# VIRTUAL WORK PRINCIPLE

Equilibrium conditions in continuum mechanics are normally written in the form of differential equations with boundary conditions. Thus, the stress field is presumed differentiable or, at least, piecewise differentiable. When this assumption becomes too restrictive as, for example, in the load admissibility problem (Subsection I.5.4), a need arises for a form of equilibrium conditions that requires less smoothness of stress fields. The virtual work principle suggests the formulation: a distribution of internal forces equilibrates a given load if and only if the total power of the load and the internal forces vanishes for every virtual velocity field.

In this chapter, equilibrium conditions for various mechanical systems are formulated on the basis of the virtual work principle. This makes it possible to consider both regular and irregular stress fields and also results in a unified formulation of equilibrium conditions for mechanical systems of all types, which, in turn, allows development of a unified limit analysis theory for various rigid perfectly plastic systems.

## 1. Bodies under standard loading

In this section, we use the virtual work principle to formulate conditions for equilibrium of a three-dimensional body under a loading of the most frequently encountered type. The body is subjected to body forces and surface traction on a part of the body boundary, while velocity distribution is given on the remaining part of the boundary. It will be shown that the virtual work principle 1) is equivalent to the usual equilibrium conditions if the stress field is regular enough (piecewise smooth), 2) generalizes the usual conditions for the case of stress fields which are only integrable.

**1.1. Virtual velocity fields and power.** Consider a body occupying domain $\Omega$ and subjected to body forces with the volume density $\mathbf{f}$ given in $\Omega$ and surface tractions with the density $\mathbf{q}$ given on a part $S_q$ of the body boundary. Velocity distribution $\mathbf{v}_0$ is given on the remaining part $S_v$ of the body boundary, that is, every kinematically admissible velocity field satisfies the condition $\mathbf{v}|_{S_v} = \mathbf{v}_0$. We specify the set of virtual velocity fields and expressions for the external and internal powers in this case.

According to the general appproach, a virtual velocity field is the difference of two kinematically admissible velocity fields. Under the kinematic constraint $\mathbf{v}|_{S_v} = \mathbf{v}_0$, virtual velocity fields are exactly the ones that vanish on $S_v$. This requirement is not definite enough to determine the set $U$ of virtual velocity fields. Therefore, we have to specify regularity of the fields included in $U$. We restrict ourselves to *smooth* or, more precisely, infinitely differentiable fields, thus assuming that *the set $U$ of virtual velocity fields consists of all smooth fields vanishing on the surface $S_v$*.

The *external power*, that is, the power of the given load $\mathbf{l} = (\mathbf{f}, \mathbf{q})$ on a virtual velocity field $\mathbf{v}$, is

$$\int_{\Omega} \mathbf{f}\mathbf{v}\, dV + \int_{S_q} \mathbf{q}\mathbf{v}\, dS.$$

The *internal power*, that is, the power of a stress field $\boldsymbol{\sigma}$ on $\mathbf{v}$, is

$$-\int_{\Omega} \operatorname{Def} \mathbf{v} \cdot \boldsymbol{\sigma}\, dV,$$

where $\operatorname{Def} \mathbf{v}$ is the strain rate:

$$(\operatorname{Def} \mathbf{v})_{ij} = \frac{1}{2}\left(\frac{\partial v_i}{\partial x_j} + \frac{\partial v_j}{\partial x_i}\right).$$

**1.2. Virtual work principle.** The general formulation of the virtual work principle in the case under consideration takes the following form:

stress field $\boldsymbol{\sigma}$ equilibrates the load $\mathbf{l} = (\mathbf{f}, \mathbf{q})$ if and only if

$$\int_{\Omega} \operatorname{Def} \mathbf{v} \cdot \boldsymbol{\sigma}\, dV = \int_{\Omega} \mathbf{f}\mathbf{v}\, dV + \int_{S_q} \mathbf{q}\mathbf{v}\, dS \quad \text{for every} \quad \mathbf{v} \in U, \tag{1.1}$$

where $U$ is the set of all smooth vector fields vanishing on the surface $S_v$.

The virtual work principle is related to the usual equilibrium conditions, and we will discuss the way they are related. The two formulations can be spoken about provided they both make sense, which means that the stress fields should be at least piecewise regular (as is required by the usual equilibrium conditions). Here, we call $\boldsymbol{\sigma}$ *piecewise regular in domain $\Omega$* if 1) $\Omega$ is subdivided into a finite number of pairwise disjoint domains $\omega_1, \ldots, \omega_N$, $\boldsymbol{\sigma}$ being continuously differentiable in every $\omega_a$; 2) $\boldsymbol{\sigma}$ has limit values on every boundary $\partial\omega_a$. The latter assumption means that the values $\boldsymbol{\sigma}(x)$ tend to a finite limit when $x$ tends from inside of $\partial\omega_a$ to a point $x_0$ belonging to $\partial\omega_a$. Consider the surface $\gamma$ separating two of the domains $\omega_a$. The limit values of $\boldsymbol{\sigma}$ are defined on both sides of $\gamma$, and are, generally speaking, not equal. The difference of these limit values is referred to as the *jump* of $\boldsymbol{\sigma}$ on the surface $\gamma$. The jump is denoted by the symbol $[\boldsymbol{\sigma}]$, and $\gamma$ is referred to as the *discontinuity surface*. The additional requirement that the boundaries $\partial\omega_a$ of the smoothness domains are sufficiently regular is now included in the definition of piecewise smooth field $\boldsymbol{\sigma}$. For example, these boundaries can be piecewise continuously differentiable. Actually, $\partial\omega_a$ should be regular enough to ensure validity of the formula for integration by parts

$$\int_{\omega_a} \tau_{ij} \frac{\partial u_i}{\partial x_j}\, dV = -\int_{\omega_a} \frac{\partial \tau_{ij}}{\partial x_j} u_i \, dV + \int_{\partial\omega_a} \tau_{ij}\nu_j u_i \, dS$$

for any continuously differentiable in $\omega_a$ fields $\tau_{ij}$ and $v_i$. Here $\boldsymbol{\nu}$ stands for the unit outward normal.

A piecewise regular field $\boldsymbol{\sigma}$ can be discontinuous on a certain surface $\Gamma$ (the surface consists of some parts of the surfaces which separate adjacent smoothness domains $\omega_a$). The usual conditions for equilibrium of the stress field $\boldsymbol{\sigma}$ with the given load $\mathbf{l} = (\mathbf{f}, \mathbf{q})$ may be written in the form (Subsection A.1):

$$\frac{\partial \sigma_{ij}}{\partial x_j} + f_i = 0 \qquad \text{in the domain } \omega_a, \quad a = 1, \ldots, N, \tag{1.2}$$

$$[\sigma_{ij}]\nu_j = 0 \qquad \text{on the surface } \Gamma, \tag{1.3}$$

$$\sigma_{ij}\nu_j = q_i \qquad \text{on the surface } S_q. \tag{1.4}$$

The following proposition establishes relation of these conditions to the virtual work principle.

PROPOSITION 1. A piecewise smooth stress field $\boldsymbol{\sigma}$ satisfies equilibrium conditions (1.2) – (1.4) if and only if it satisfies virtual work principle (1.1).

PROOF. To prove the statement we will make use of the following formula for the right-hand side of (1.1):

$$\int_\Omega \operatorname{Def} \mathbf{v} \cdot \boldsymbol{\sigma} \, dV$$

$$= \sum_{a=1}^{N} \left( -\int_{\omega_a} \frac{\partial \sigma_{ij}}{\partial x_j} v_i \, dV \right) - \int_\Gamma [\sigma_{ij}]\nu_j v_i \, dS + \int_{\partial\Omega} \sigma_{ij}\nu_j v_i \, dS. \tag{1.5}$$

The formula will be verified below.

Consider a virtual velocity field $\mathbf{v}$. It vanishes on the surface $S_v$, thus restricting the integration in the last term in (1.5) to $S_q$. This conclusion taken in conjunction with (1.5) allows re-arranging equality (1.1) as

$$\sum_{a=1}^{N} \int_{\omega_a} \left( \frac{\partial \sigma_{ij}}{\partial x_j} + f_i \right) v_i \, dV + \int_\Gamma [\sigma_{ij}]\nu_j v_i \, dS + \int_{S_q} (q_i - \sigma_{ij}\nu_j)\, v_i \, dS = 0. \tag{1.6}$$

Consider now a stress field $\boldsymbol{\sigma}$ satisfying usual equilibrium conditions (1.2) – (1.4). It is clear that equality (1.6) is valid for such $\boldsymbol{\sigma}$ and for every virtual velocity field $\mathbf{v}$. In other words, the stress field $\boldsymbol{\sigma}$ satisfies the virtual work principle.

Suppose now that $\boldsymbol{\sigma}$ satisfies virtual work principle (1.1) or, equivalently, equality (1.6) holds for every virtual velocity field $\mathbf{v} \in U$. Choosing an appropriate $\mathbf{v}$, we show that $\boldsymbol{\sigma}$ satisfies equilibrium conditions (1.2) – (1.4). Consider, for example, fields $\mathbf{v} \in U$ which can take nonzero values only in a certain ball $B \subset \omega_1$. Then, equality (1.6) is reduced to the condition

$$\int_B \left( \frac{\partial \sigma_{ij}}{\partial x_j} + f_i \right) v_i \, dV = 0$$

for every such $\mathbf{v}$. Arbitrariness of $\mathbf{v}$ in $B$ yields

$$\frac{\partial \sigma_{ij}}{\partial x_j} + f_i = 0,$$

which holds everywhere in $B$. This implies that the equality is valid everywhere in $\omega_1$. The equality is similarly proved to hold in every other domain $\omega_a$ as well. Consequently, condition (1.2) is satisfied, and the first term in (1.6) vanishes. Let $x$ be a point on the surface $\gamma$ separating two adjacent smoothness domains. Consider a ball $B$ that only intersects with these two of the domains $\omega_1, \ldots, \omega_N$, and let $\mathbf{v}$ be a virtual velocity field which vanishes in the exterior of $B$. Equality (1.6) is reduced to the condition

$$\int_{\gamma \cap B} [\sigma_{ij}] \nu_j v_i \, dS = 0,$$

for every such $\mathbf{v}$. Due to the arbitrariness of $\mathbf{v}$ in $B$, we arrive to (1.3). In a similar manner it may be shown that equality (1.4) is valid.

To finish the proof, let us verify formula (1.5). The left-hand side of (1.5) can be written as

$$\int_\Omega \operatorname{Def} \mathbf{v} \cdot \boldsymbol{\sigma} \, dV = \int_{\omega_1} \operatorname{Def} \mathbf{v} \cdot \boldsymbol{\sigma} \, dV + \cdots + \int_{\omega_N} \operatorname{Def} \mathbf{v} \cdot \boldsymbol{\sigma} \, dV. \tag{1.7}$$

For further transforming of the expression, we subdivide the boundary of every domain $\omega_a$ into the following parts: 1) $S_a$ which lies on $\partial\Omega$; 2) $\gamma_{(ab)}$ which separates $\omega_a$ from the domain $\omega_b$, $b = 1, \ldots, N$, $b \neq a$. (Some of the surfaces $S_a$, $\gamma_{(ab)}$ may be empty; for example, $S_a = \emptyset$ if $\omega_a$ is on a positive distance from $\partial\Omega$.) Integration by parts results in

$$\int_{\omega_a} \operatorname{Def} \mathbf{v} \cdot \boldsymbol{\sigma} \, dV = \int_{\omega_a} \sigma_{ij} \frac{\partial v_i}{\partial x_j} \, dV =$$

$$= \int_{S_a} \sigma_{ij}^{(a)} \nu_j^{(a)} v_i \, dS + \sum_{\substack{b=1 \\ b \neq a}}^{N} \int_{\gamma_{(ab)}} \sigma_{ij}^{(a)} \nu_j^{(a)} v_i \, dS - \int_{\omega_a} \frac{\partial \sigma_{ij}}{\partial x_j} v_i \, dV. \tag{1.8}$$

The superscript in $\boldsymbol{\sigma}^{(a)}$ indicates the stress value on $\partial\omega_a$ which is the limit value for $\boldsymbol{\sigma}(x)$ when $x$ tends to a point on $\partial\omega_a$ from the interior of $\omega_a$. The

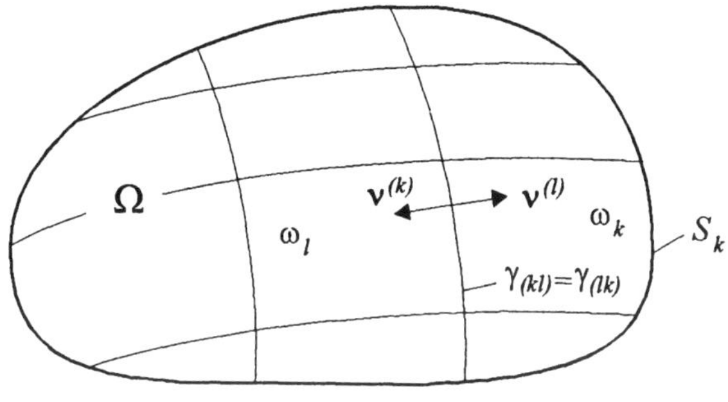


**Figure 1.1**

superscript in $\boldsymbol{\nu}^{(a)}$ indicates the (unit) normal to $\partial\omega_a$ outward with respect to the domain $\omega_a$.

The right-hand side in (1.7) is the sum of all integrals of (1.8) type. It is clear that the sum of the integrals over the surface $S_a$ can be written as

$$\int_{\partial\Omega} \sigma_{ij}\nu_j v_i \, dS.$$

Integrals over the surface $\gamma_{(kl)} = \gamma_{(lk)}$ are met twice while summing integrals (1.8): at $a = k$ and $a = l$. In the first case, it is the normal $\boldsymbol{\nu}^{(k)}$ that enters the integrand; in the second case, it is the normal $\boldsymbol{\nu}^{(l)} = -\boldsymbol{\nu}^{(k)}$ (see Figure 1.1). The sum of the two integrals is

$$\int_{\gamma_{(kl)}} \sigma_{ij}^{(k)}\nu_j^{(k)} v_i \, dS + \int_{\gamma_{(lk)}} \sigma_{ij}^{(l)}\nu_j^{(l)} v_i \, dS = \int_{\gamma_{(kl)}} \left(\sigma_{ij}^{(k)} - \sigma_{ij}^{(l)}\right)\nu_j^{(k)} v_i \, dS. \qquad (1.9)$$

To simplify the form of (1.9), consider the jump $[\boldsymbol{\sigma}]$ *in the direction of the normal* $\boldsymbol{\nu}$ to the surface $\gamma_{(kl)}$, the direction being chosen out of the two possible ones. Thus, we have

$$[\boldsymbol{\sigma}] = \boldsymbol{\sigma}^{(l)} - \boldsymbol{\sigma}^{(k)} \quad \text{if} \quad \boldsymbol{\nu} = \boldsymbol{\nu}^{(k)},$$
$$[\boldsymbol{\sigma}] = \boldsymbol{\sigma}^{(k)} - \boldsymbol{\sigma}^{(l)} \quad \text{if} \quad \boldsymbol{\nu} = \boldsymbol{\nu}^{(l)}.$$

According to this convention the value $[\sigma_{ij}]\nu_j$ does not depend upon whether the normal $\boldsymbol{\nu} = \boldsymbol{\nu}^{(k)}$ or $\boldsymbol{\nu} = \boldsymbol{\nu}^{(l)}$ is chosen, and integral (1.9) can be written as

$$-\int_{\gamma_{(kl)}} [\sigma_{ij}]\nu_j v_i \, dS.$$

The sum of all these integrals entering the right-hand side in (1.7) equals

$$-\int_{\Gamma'} [\sigma_{ij}]\nu_j v_i \, dS,$$

where $\Gamma'$ is the union of all surfaces $\gamma_{(ab)}$. Recall that the discontinuity surface $\Gamma$ is a part of $\Gamma'$ on which $[\sigma] \neq 0$. Therefore, integration in the previous expression is, actually, restricted to the surface $\Gamma$.

Using (1.8) and the above-mentioned remarks, we find that the right-hand side in (1.7) equals

$$\int\limits_{\partial\Omega} \sigma_{ij}\nu_j v_i \, dS - \int\limits_{\Gamma} [\sigma_{ij}]\nu_j v_i \, dS - \sum_{a=1}^{N} \int\limits_{\omega_a} \frac{\partial \sigma_{ij}}{\partial x_j} v_i \, dV.$$

Consequently, formula (1.5) is verified, which finishes the proof.

Thus, in the case of piecewise regular stress fields the virtual work principle is equivalent to the usual equilibrium conditions.

**1.3. Generalized equilibrium conditions.** The set of all piecewise regular stress fields is not wide enough to be used in the analysis of rigid perfectly plastic bodies (see Subsection I.5.4), and has to be extended. At the same time, under the weakened regularity assumptions equilibrium conditions (1.2) – (1.4) do not make sense, thus requiring generalization. Proposition 1 shows that the virtual work principle is an immediate generalization of (1.2) – (1.4), making sense even for stress fields which are only integrable.

Let $\boldsymbol{\sigma}$ be an integrable stress field in a body occupying domain $\Omega$ and subjected to a load $\mathbf{l} = (\mathbf{f}, \mathbf{q})$. We set by definition that *the stress field* $\boldsymbol{\sigma}$ *equilibrates the load* $\mathbf{l}$ *if* (1.1) *holds*. That is, the virtual work principle is a generalized equilibrium condition.

In limit analysis, it is convenient to use the set of all stress fields equilibrating a certain load $\mathbf{l}$, in particular, the load $\mathbf{l} = 0$. A stress field equilibrating the load $\mathbf{l} = 0$ is referred to as *self-equilibrated*. The field $\boldsymbol{\sigma} = 0$ is self-equilibrated, and there are many other self-equilibrated stress fields. For example, a smooth stress field $\boldsymbol{\sigma}$ is self-equilibrated if and only if it satisfies the differential equations $\partial\sigma_{ij}/\partial x_j = 0$ and boundary conditions $\sigma_{ij}\nu_j|_{S_q} = 0$. There are only three equations for six independent components of $\boldsymbol{\sigma}$, and it is evident that the set of self-equilibrated smooth stress fields is rather wide. Normally, we will consider stress fields from a certain collection $\mathcal{S}$ of integrable fields. Then self-equilibrated fields form the set

$$\Sigma = \left\{ \boldsymbol{\sigma} \in \mathcal{S} : \int\limits_{\Omega} \operatorname{Def} \mathbf{v} \cdot \boldsymbol{\sigma} \, dV = 0 \quad \text{for every} \quad \mathbf{v} \in U \right\},$$

where we made use of the virtual work principle to formulate the equilibrium conditions. Recall that $U$ is the set of all smooth vector fields which satisfy the condition $\mathbf{v}|_{S_v} = 0$.

Throughout the book, the collection $\mathcal{S}$ of stress fields is assumed to be a linear space, that is, the sum of any two fields in $\mathcal{S}$ belongs to $\mathcal{S}$ as well as the product of a field in $\mathcal{S}$ and a number. It is obvious that the set $\Sigma$ possesses the same properties, that is, $\Sigma$ is a linear subspace in $\mathcal{S}$ (see Section IV.A below).

Using the subspace $\Sigma$ of self-equilibrated stress fields, it is easy to describe the set of stress fields equilibrating the load $\mathbf{l}$. Indeed, let $\mathbf{s}_l$ be a fixed stress field in $\mathcal{S}$ equilibrating the load. Note that equilibrium conditions (1.1) are linear in stress. Therefore, a stress field $\boldsymbol{\sigma}$ equilibrates the load $\mathbf{l}$ if and only if $\boldsymbol{\sigma} - \mathbf{s}_l$ is self-equilibrated, that is, $\boldsymbol{\sigma} - \mathbf{s}_l \in \Sigma$. The latter relation can be written as $\boldsymbol{\sigma} \in \Sigma + \mathbf{s}_l$, and it is a short form of the equilibrium conditions we will often use.

## 2. Bodies under mixed boundary conditions

In this section, we again use the virtual work principle to formulate the equilibrium conditions for a three-dimensional body. We consider loadings and kinematic constraints of two types that are not standard in terms of the previous section. We change the set of virtual velocity fields and formulas for the external and internal powers, while the principle itself remains unchanged.

**2.1 Tangent load and surface slip.** Consider a body with its boundary divided into three parts: $S_v$, $S_q$ and $S_t$. Velocity distribution and surface tractions are given at $S_v$ and $S_q$, respectively, as in the previous section. The surface $S_t$ is assumed to remain in contact with a perfectly lubricated surface $S_*$, the motion of $S_*$ being given. Due to perfect lubrication, $S_t$ can slip over surface $S_*$ without friction, that is, the tangent force vanishes on $S_t$. We will consider a bit more general situation of the same type: the normal velocity component $u$ and the tangent component $\mathbf{t}$ of the load are given on the surface $S_t$:

$$v_\nu|_{S_t} = u, \qquad \boldsymbol{\sigma}_{\nu\tau}|_{S_t} = \mathbf{t}. \tag{2.1}$$

Here, $v_\nu = \mathbf{v}\boldsymbol{\nu}$, $\boldsymbol{\nu}$ is the unit outward normal to $S_t$, $\boldsymbol{\sigma}_{\nu\tau}$ is the tangent component of the surface traction exerted by the stress $\boldsymbol{\sigma}$ on the surface element with the normal $\boldsymbol{\nu}$ (see Subsection I.A.1), that is, $\boldsymbol{\sigma}_{\nu\tau}$ is the vector with the components $(\boldsymbol{\sigma}_{\nu\tau})_i = \sigma_{ij}\nu_j - \sigma_{kl}\nu_k\nu_l\nu_i$.

The boundary condition $\boldsymbol{\sigma}_{\nu\tau}|_{S_t} = \mathbf{t}$ is to be added to equilibrium conditions (1.2) – (1.4) in the case under consideration. Our aim is to formulate all these conditions into the form of the virtual work principle. This will result in generalization of the equilibrium conditions for the case of nonsmooth stress fields. It is obvious that an appropriate formulation cannot make use of the same set of virtual velocity fields as in Section 1. The general approach suggests the necessary change: a virtual velocity field is the difference of two kinematically admissible fields. Under kinematic constraint (1.2), this means that virtual velocity fields $\mathbf{v} = 0$ satisfy the condition $v_\nu|_{S_t} = 0$ (and the condition $\mathbf{v}|_{S_v} = 0$ as before).

The virtual work principle requires summing powers of all external forces. In particular, we have to add the power of the load applied to the surface $S_t$. Note that boundary condition (2.1) only specifies the tangent component of this load and not the load itself. However, it is possible to evaluate the power of the load using the formula

$$\int_{S_t} \mathbf{t}\mathbf{v}\, dS$$

provided $\mathbf{v}|_{S_t} = 0$. Indeed, the normal component of the load makes no work on a velocity field $\mathbf{v}$ that satisfies the latter condition. Therefore, the previous formula gives exactly the power of the load on the virtual velocity field $\mathbf{v}$ although the normal component of the load is unknown.

Thus, in the case under consideration, the virtual work principle takes the form

$$\int_\Omega \mathrm{Def}\,\mathbf{v}\cdot\boldsymbol{\sigma}\,dV = \int_\Omega \mathbf{f}\mathbf{v}\,dV + \int_{S_q} \mathbf{q}\mathbf{v}\,dS + \int_{S_t} \mathbf{t}\mathbf{v}\,dS \text{ for every } \mathbf{v}\in U. \tag{2.2}$$

Here $U$ is the set of smooth vector fields which satisfy the conditions $\mathbf{v}|_{S_v} = 0$, $v_\nu|_{S_t} = 0$. This formulation of the principle is equivalent to the equilibrium conditions (1.2) – (1.4) and the additional relation $\boldsymbol{\sigma}_{\nu\tau}|_{S_t} = \mathbf{t}$. Of course, we can only speak of the equivalency if all the relations make sense, for example, in the case of a piecewise regular stress field. The equivalence can be established in exactly the same way as in Proposition 1.1.

As before, we use the virtual work principle (2.2) to define generalized equilibrium conditions which make sense even if the stress field is only integrable. A short formulation of the equilibrium conditions is of the form $\boldsymbol{\sigma}\in\Sigma+\mathbf{s}_l$, where $\mathbf{s}_l$ is a fixed stress field equilibrating the given load $\mathbf{l}$ and $\Sigma$ is the set of self-equilibrated stress fields:

$$\Sigma = \left\{\boldsymbol{\sigma}\in\mathcal{S} : \int_\Omega \mathrm{Def}\,\mathbf{v}\cdot\boldsymbol{\sigma}\,dV = 0 \quad \text{for every} \quad \mathbf{v}\in U\right\}.$$

Here, $\mathcal{S}$ is the space of stress fields and $U$ is the same as in formulation (2.2) of the virtual work principle.

**2.2. Rigid punch loading.** Consider body $B$ with its boundary subdivided into three parts: $S_q$, $S_v$, and $S_R$. Surface tractions and velocity distribution are given on $S_q$ and $S_v$ as in Section 1 and the previous subsection. The body is joined to a rigid body $R$ at the surface $S_R$ with no slip or separation at $S_R$. The rigid body $R$ is subjected to a load with the given resultant $\mathbf{F}$ and given resultant moment $\mathbf{M}$, see Figure 2.1. If the body $R$ is in equilibrium, the forces which $R$ exerts on $B$ at the surface $S_R$ are known. Therefore, the boundary conditions

$$\int_{S_R} \boldsymbol{\sigma}_\nu\,dS = \mathbf{F}, \qquad \int_{S_R} \mathbf{r}\times\boldsymbol{\sigma}_\nu\,dS = \mathbf{M} \tag{2.3}$$

are to be added to equilibrium conditions (1.2) – (1.4). Here, $\boldsymbol{\sigma}_\nu$ is the surface traction exerted by the stress $\boldsymbol{\sigma}$ on the surface element with the normal $\boldsymbol{\nu}$ (see Subsection I.A.1); $\mathbf{r}$ stands for the position vector with respect to the point which is used to determine the resultant moment $\mathbf{M}$.

Kinematic constraints imposed on the body $B$ at the surface $S_R$ require that the surface moves with the rigid body $R$. Therefore, velocity distribution at $S_R$ is of the form

$$\mathbf{v}|_{S_R} = \mathbf{V}_0 + \boldsymbol{\Omega}_0\times\mathbf{r}. \tag{2.4}$$

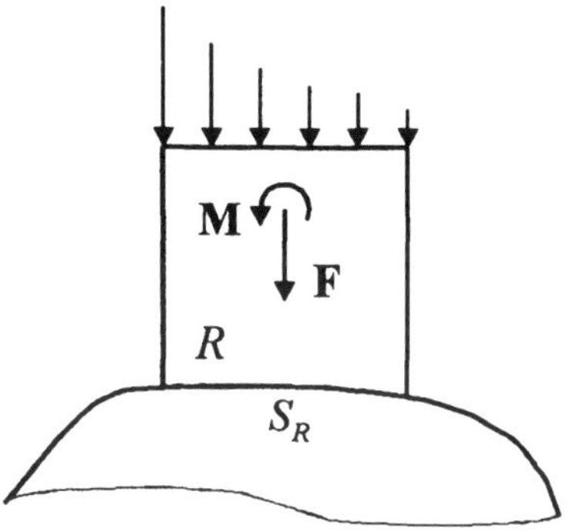


**Figure 2.1**

Here, $\mathbf{V}_0$ and $\mathbf{\Omega}_0$ do not depend on $\mathbf{r}$. At the same time, $\mathbf{V}_0$ and $\mathbf{\Omega}_0$ are, generally, different for different kinematically admissible velocity fields. As to virtual velocity fields, they should have the form of the difference of two kinematically admissible fields. Under kinematic constraint (2.4), this means that a virtual velocity field $\mathbf{v}$ satisfies the condition $\mathbf{v}|_{S_R} = \mathbf{v}_0 + \boldsymbol{\omega}_0 \times \mathbf{r}$ ($\mathbf{v}_0$ and $\boldsymbol{\omega}_0$ do not depend on $\mathbf{r}$, and are, generally, different for different virtual velocity fields). In addition, $\mathbf{v}$ satisfies the condition $\mathbf{v}|_{S_v} = 0$ as before.

The virtual work principle requires summing powers of all external and internal forces. In particular, we have to add the power of the load applied to the surface $S_R$. Note that boundary conditions (2.3) only specify the resultant $\mathbf{F}$ and the resultant moment $\mathbf{M}$ of the load. However, it is possible to evaluate the power of the load at every virtual velocity field $\mathbf{v}$. Indeed, $\mathbf{v}$ satisfies the condition $\mathbf{v}|_{S_R} = \mathbf{v}_0 + \boldsymbol{\omega}_0 \times \mathbf{r}$, where $\mathbf{v}_0$ and $\boldsymbol{\omega}_0$ do not depend on $\mathbf{r}$. This means that $\mathbf{v}$ on $S_R$ is a velocity field of a rigid motion. Only the resultant force and resultant moment on $S_R$ enter the formula for the power of the load on such a velocity field: $\mathbf{F}\mathbf{v}_0 + \mathbf{M}\boldsymbol{\omega}_0$.

Thus, the virtual work principle takes the form

$$\int_{\Omega} \mathrm{Def}\,\mathbf{v} \cdot \boldsymbol{\sigma}\, dV = \int_{\Omega} \mathbf{f}\mathbf{v}\, dV + \int_{S_q} \mathbf{q}\mathbf{v}\, dS + \mathbf{F}\mathbf{v}_0 + \mathbf{M}\boldsymbol{\omega}_0 \quad \text{for every} \quad \mathbf{v} \in U. \tag{2.5}$$

Formulation (2.5) of the virtual work principle is equivalent to the equilibrium conditions consisting of (1.2) – (1.4) and (2.3) as far as piecewise smooth stress fields are concerned. The equivalence can be established in exactly the same way as in Proposition 1.1.

As in Section 1 and the previous subsection, we use the virtual work principle (2.5) to define generalized equilibrium conditions which make sense even if the stress field is only integrable. A short formulation of the equilibrium conditions is the same as in Section 1 and the previous subsection: $\boldsymbol{\sigma} \in \Sigma + \mathbf{s}_l$. Here, $\mathbf{s}_l$ is a fixed stress field equilibrating the given load $\mathbf{l}$, $\Sigma$ is the set of

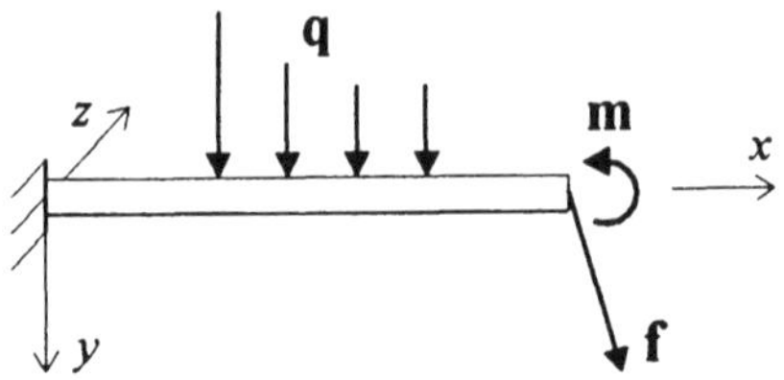


**Figure 3.1**

self-equilibrated stress fields:

$$\Sigma = \left\{ \sigma \in \mathcal{S} : \int_{\Omega} \operatorname{Def} \mathbf{v} \cdot \sigma \, dV = 0 \quad \text{for every} \quad \mathbf{v} \in U \right\};$$

$U$ is the same as in (2.5); and $\mathcal{S}$ stands for the space of stress fields.

## 3. Beams and trusses

In this section, the virtual work principle is used to formulate equilibrium conditions for a beam (a one-dimensional continual system) and a truss (a discrete system). The formulations are essentially the same as in the case of a three-dimensional body.

**3.1. Virtual velocity fields and power for a beam.** Consider a beam occupying the segment $[0, a]$ of the $x$-axis and subjected to a typical loading and typical kinematic constraints. Namely, let the left end of the beam be pinched while the right end is subjected to a given external force $\mathbf{f}$ and bending moment $\mathbf{m}$. There are no kinematic constraints at the right end of the beam. The beam is also subjected to a load distributed along it with a given density $\mathbf{q}$, see Figure 3.1. Thus, the triple $\mathbf{l} = (\mathbf{q}, \mathbf{m}, \mathbf{f})$ represents the load acting on the beam, $\mathbf{q}$ being a vector function, $\mathbf{m}$ and $\mathbf{f}$ vectors.

To describe motion of the beam, we use the kinematic variables $\mathbf{v} = (\omega_y, \omega_z, u, u_y, u_z)$ which are the functions of the coordinate $x$. We refer to $\mathbf{v}$ as the velocity field. Recall that $\omega_y$ and $\omega_z$ are components of the angular velocity of the cross-section rotation and $u = u_x$, $u_y$, $u_z$ are components of the velocity of the beam axis. The following boundary conditions represent the imposed kinematic constraints:

$$\omega_y(0) = \omega_z(0) = 0, \qquad u(0) = u_y(0) = u_z(0) = 0. \tag{3.1}$$

Following the general approach, we consider a virtual velocity field as a difference of two kinematically admissible fields. The kinematic constraints (3.1) imply that every virtual velocity field $\mathbf{v}$ satisfies conditions (3.1). As before, we consider only smooth virtual velocity fields.

The external power, that is, the power of the given load $\mathbf{l}$ on a virtual velocity field $\mathbf{v}$, is

$$\int_0^a (qu + q_y u_y + q_z u_z) \, dx + m_y \omega_y(a) + m_z \omega_z(a) + fu(a) + f_y u_y(a) + f_z u_z(a).$$

Internal forces in a beam are represented by the collection

$$\sigma = (M_y, M_z, N, Q_y, Q_z),$$

where $M_y$ and $M_z$ are components of the bending moment, $N$ is the axial force, $Q_y$ and $Q_z$ are the cutting forces (see Subsection I.6.2); all components are functions of the coordinate $x$. We define the internal power, that is, the power of the stresses $\sigma$ on $\mathbf{v}$, as

$$-\int_0^a \mathbf{e}(\mathbf{v})\sigma\, dx = -\int_0^a \left(M_y\omega_y' + M_z\omega_z' + Nu' + Q_y(u_y' - \omega_z) + Q_z(u_z' + \omega_y)\right) dx.$$

Recall that $\mathbf{e}(\mathbf{v}) = (\omega_y', \omega_z', u', u_y' - \omega_z, u_z' + \omega_y)$ is the strain rate and the prime denotes the derivative with respect to $x$.

**3.2. Virtual work principle for a beam.** The general formulation of the virtual work principle in the case under consideration takes the following form: the stress field $\sigma = (M_y, M_z, N, Q_y, Q_z)$ equilibrates the load $\mathbf{l} = (\mathbf{m}, \mathbf{f}, \mathbf{q})$ if and only if

$$\int_0^a \mathbf{e}(\mathbf{v})\sigma\, dx = \int_0^a (qu + q_y u_y + q_z u_z)\, dx + m_y\omega_y(a) + m_z\omega_z(a)$$
$$+ fu(a) + f_y u_y(a) + f_z u_z(a) \quad \text{for every} \quad \mathbf{v} \in U, \qquad (3.2)$$

where $U$ is the set of all smooth velocity fields $\mathbf{v} = (\omega_y, \omega_z, u, u_y, u_z)$ on $[0, a]$ vanishing at $x = 0$.

The virtual work principle is related to the usual equilibrium conditions which make sense if the stress field $\sigma$ is sufficiently regular, for example, differentiable; the conditions are of the form (I.6.4), (I.6.5):

$$\begin{gathered} M_y' = Q_z, \quad M_z' = -Q_y, \quad N' = -q, \quad Q_y' = -q_y, \quad Q_z' = -q_z, \\ M_y|_{x=a} = m_y, \quad M_z|_{x=a} = m_z, \\ N|_{x=a} = f, \quad Q_y|_{x=a} = f_y, \quad Q_z|_{x=a} = f_z. \end{gathered} \qquad (3.3)$$

More precisely, the following proposition is valid.

PROPOSITION 1. A differentiable stress field $\sigma$ satisfies equilibrium conditions (3.3) if and only if it satisfies the virtual work principle (3.2).

The proof is essentially the same as the proof of Proposition 1.1.

**3.3. Generalized conditions for equilibrium of a beam.** A set of differentiable or piecewise differentiable stress fields is not wide enough to be used in the analysis of rigid perfectly plastic beams. It has to be extended the way the set of stress fields was in the case of three-dimensional bodies. Since equilibrium conditions (3.3) do not make sense for nondifferentiable stress fields, we have to generalize the conditions. Proposition 1 shows that the virtual work principle is an immediate generalization of (3.3), and the principle makes sense even if the stress field is only integrable.

Let $\boldsymbol{\sigma}$ be an integrable stress field in a beam occupying a segment $[0, a]$ and subjected to a load $\mathbf{l} = (\mathbf{q}, \mathbf{m}, \mathbf{f})$. We set by definition that the stress field $\boldsymbol{\sigma}$ equilibrates the load $\mathbf{l}$ if (3.2) holds, that is, we adopt the virtual work principle as the generalized equilibrium condition. In what follows, we only assume that stress fields are integrable and that they form a linear space.

A short formulation of the equilibrium conditions is the same as in Sections 1 and 2: $\boldsymbol{\sigma} \in \Sigma + \mathbf{s}_l$. Here, $\mathbf{s}_l$ is a fixed stress field in $\mathcal{S}$ equilibrating the given load $\mathbf{l}$ and $\Sigma$ is the set of self-equilibrated stress fields:

$$\Sigma = \left\{ \boldsymbol{\sigma} \in \mathcal{S} : \int_0^a \mathbf{e}(\mathbf{v})\boldsymbol{\sigma}\, dx = 0 \quad \text{for every} \quad \mathbf{v} \in U \right\}, \tag{3.4}$$

where $U$ is the same as in formulation (3.2) of the virtual work principle.

Note that strain rate and stress fields have different meanings in the cases of a three-dimensional body and a beam. However, as far as the equilibrium conditions are concerned, they are essentially the same. First of all, they are of the form $\boldsymbol{\sigma} \in \Sigma + \mathbf{s}_l$ in both cases. Secondly, the set $\Sigma$ of self-equilibrated stress fields has the same meaning in both cases: $\Sigma$ is defined by (1.6) or by (3.4) and consists of stress fields with zero power on every virtual velocity field.

**3.4. Virtual work principle for a truss.** Consider a truss consisting of $m$ rods with $k$ unfixed hinges (see Subsection I.6.6). The truss is subjected to the load $\mathbf{l} = (\mathbf{f}_1, \ldots, \mathbf{f}_k)$ where $\mathbf{f}_\alpha$ is an external force applied at the $\alpha$-th hinge.

We use velocities $\mathbf{v} = (\mathbf{v}_1, \ldots, \mathbf{v}_k)$ to describe the motion of the truss, where $\mathbf{v}_\alpha$ is the velocity of the $\alpha$-th (unfixed) hinge. Following the general approach, we consider virtual velocity as a difference of two kinematically admissible velocities. Since no kinematic constraints are imposed on the (unfixed) hinges, any collection $\mathbf{v} = (\mathbf{v}_1, \ldots, \mathbf{v}_k)$ represents virtual velocities for the truss. In other words, the set of the virtual velocities is the space $\mathbf{R}^{3k}$.

The external power of the load $\mathbf{l} = (\mathbf{f}_1, \ldots, \mathbf{f}_k)$ on virtual velocities $\mathbf{v} = (\mathbf{v}_1, \ldots, \mathbf{v}_k)$ is $\mathbf{f}_1\mathbf{v}_1 + \cdots + \mathbf{f}_k\mathbf{v}_k$.

Internal forces in the truss are represented by the collection

$$\boldsymbol{\sigma} = (N_1, \ldots, N_m),$$

where $N_i$ is the axial force in the $i$-th rod. We define the internal power, that is, the power of the internal forces $\boldsymbol{\sigma}$ on velocities $\mathbf{v}$, as

$$-N_1 e_1(\mathbf{v}) - \cdots - N_m e_m(\mathbf{v}).$$

Here, $e_i$ is the elongation rate of the $i$-th rod; $\mathbf{e}_i$ is evaluated by the formula $e_i = -\boldsymbol{\tau}_{i\alpha}\mathbf{v}_\alpha$. (Recall the definition: if the $\alpha$-th hinge is one of the ends of the $i$-th rod, then $\boldsymbol{\tau}_{i\alpha}$ is the unit vector directed along the $i$-th rod from the

$\alpha$-th hinge to the other end of the rod; if the $\alpha$-th hinge is not an end of the $i$-th rod, then $\boldsymbol{\tau}_{i\alpha} = 0$.)

The general formulation of the virtual work principle in the case under consideration takes the following form:

> the stresses $\boldsymbol{\sigma} = (N_1, \ldots, N_m)$ equilibrate the load $\mathbf{l} = (\mathbf{f}_1, \ldots, \mathbf{f}_k)$ if and only if

$$N_1 e_1(\mathbf{v}) + \cdots + N_m e_m(\mathbf{v}) = \mathbf{f}_1\mathbf{v}_1 + \cdots + \mathbf{f}_k\mathbf{v}_k \quad \text{for every} \quad \mathbf{v} \in U,$$

where $U$ is the set of virtual velocities $\mathbf{v} = (\mathbf{v}_1, \ldots, \mathbf{v}_k)$, $U = \mathbf{R}^{3k}$.

The principle is reduced to the usual truss equilibrium conditions (Subsection I.6.6):

$$N_i\boldsymbol{\tau}_{i\alpha} + \mathbf{f}_\alpha = 0, \qquad \alpha = 1, \ldots, k.$$

To make this statement evident, let us write the equality of the virtual work principle as:

$$(N_1\boldsymbol{\tau}_{1\alpha} + \cdots + N_m\boldsymbol{\tau}_{m\alpha} + \mathbf{f}_\alpha)\,\mathbf{v}_\alpha = 0$$

(we used the formula $e_i = -\boldsymbol{\tau}_{i\alpha}\mathbf{v}_\alpha$). In this sum, $\mathbf{v}_1, \ldots, \mathbf{v}_k$ are arbitrary vectors, which yields the equalities $N_i\boldsymbol{\tau}_{i\alpha} + \mathbf{f}_\alpha = 0$.

Internal forces in the truss are represented by the vector $\boldsymbol{\sigma} = (N_1, \ldots, N_m)$ in $\mathbf{R}^m$, and the space of stress fields in a beam is $\mathcal{S} = \mathbf{R}^m$. The virtual work principle specifies the subspace of self-equilibrated stresses in $\mathcal{S}$:

$$\Sigma = \{\boldsymbol{\sigma} = (N_1, \ldots, N_m) \in \mathcal{S} : N_1 e_1(\mathbf{v}) + \cdots + N_m e_m(\mathbf{v}) = 0 \quad \text{for every} \;\; \mathbf{v} \in U\},$$

where $U = \mathbf{R}^{3k}$. This is essentially the same definition as for three-dimensional bodies (Section 1 and 2) and for beams (the previous subsection): a stress field $\boldsymbol{\sigma}$ is self-equilibrated if and only if its internal power on every virtual velocity field is zero. We can also write the equilibrium conditions in exactly the same form as before: $\boldsymbol{\sigma} \in \Sigma + \mathbf{s}_l$ ($\mathbf{s}_l$ is a fixed stress field in $\mathcal{S}$ equilibrating the load $\mathbf{l}$).

## Comments

The virtual work principle was known to the ancient world in a very special case: as the lever rule. The generality of the principle was recognized gradually. Galileo was the first who applied the principle not only to the simplest mechanisms but also to hydrostatics. Lagrange finally assumed the virtual work principle as the basis of mechanics (and also studied its equivalence to other formulations of equilibrium conditions).

The virtual work principle provides continuum mechanics with some important facilities. In particular, it makes it possible to get rid of the smoothness assumption: we can assume stress fields to be only integrable, and not necessarily piecewise differentiable. Thus we avoid, for example, some technical complications in limit analysis.

From the mathematical viewpoint, the virtual work principle is a formulation of the equilibrium conditions in the sense of distributions, see Section VI.A. To obtain this interpretation we narrow the set of virtual velocity fields so that it consists of fields vanishing not only on the fixed part $S_v$ of the body boundary but also near it. This makes the following question inevitable (a thorough reader possibly came across it earlier): do equilibrium conditions depend on the assumptions about the set of virtual velocity fields? Do they depend on one smoothness restriction or another imposed on virtual velocity fields? Do the conditions depend on the above-mentioned assumption about vanishing of virtual fields either on the surface $S_v$ or near it? Fortunately, the equilibrium conditions do not depend on these details at least if the boundary of a body is more or less regular. This statement arises from some propositions about velocity fields' approximation we establish in Chapter VI.

We saw in Section 2 that the virtual work principle is effective, in particular, in the case of mixed boundary conditions. It is understandable that the latter cannot be arbitrary, and we now mention a general requirement which the boundary conditions have to meet. Let $S$ be a part of a body boundary where stress and velocity fields $\boldsymbol{\sigma}$ and $\mathbf{v}$ are to satisfy mixed boundary conditions $L_1\boldsymbol{\sigma}_\nu|_S = \mathbf{p}$, $L_2\mathbf{v}|_S = \mathbf{w}$. Here, $\mathbf{p}$ and $\mathbf{w}$ are given on $S$, $L_1$ and $L_2$ are linear operators, $\boldsymbol{\sigma}_\nu$ is the surface traction exerted by $\boldsymbol{\sigma}$ on $S$. According to the general approach, a virtual velocity field is a difference of two kinematically admissible fields. Hence, it has to satisfy the condition $L_2\mathbf{v}|_S = 0$ in the case under consideration. For every such field, we can evaluate the power of a load $\mathbf{q}$ applied at $S$ if the load is given. However, to use the virtual work principle in the case under consideration, we have to be able to evaluate the power without knowing the complete information about $\mathbf{q} = \boldsymbol{\sigma}_\nu|_S$: it is only $L_1\boldsymbol{\sigma}_\nu|_S = \mathbf{p}$ that we know from the boundary condition $L_1\boldsymbol{\sigma}_\nu|_S = \mathbf{p}$. Thus we arrive at the following requirement: the mixed boundary conditions $L_1\boldsymbol{\sigma}_\nu|_S = \mathbf{p}$, $L_2\mathbf{v}|_S = \mathbf{w}$ should be *consistent*, which means that the equality $L_2\mathbf{v}|_S = 0$ specifies the widest subspace of virtual velocity fields $\mathbf{v}$ such that the power of a load $\mathbf{q}$ applied at $S$ is determined for $\mathbf{v}$ as soon as $L_1\mathbf{q}$ is known. For example, it is possible to evaluate the power for a velocity field $\mathbf{v}$ with vanishing normal component $\mathbf{v}_\nu|_S$ if only the tangent component of the load is known on $S$. All the boundary conditions we considered in this chapter are consistent in the above-mentioned sense.

CHAPTER III

# FUNDAMENTALS OF LIMIT ANALYSIS

Stresses in a perfectly plastic body cannot be arbitrary: they have to meet the admissibility requirement. There are loads that cannot be equilibrated by any admissible stress field. Therefore we divide loads into admissible and inadmissible, and it is a very important question whether a given load is admissible or not.

There are two different types of the rigid perfectly plastic body response to admissible loads. The body either remains rigid, that is, does not deform at all, or it deforms at an arbitrarily large strain rate. The main problem of limit analysis is to find out which of the alternatives corresponds to a given load. Calculation of a so-called safety factor turns out to be an answer to the question. We will see in Section 4 that some other problems of limit analysis are also reduced to the problem of calculating the safety factor.

The propositions and examples in this chapter are very simple and give a good idea of limit analysis and its methods. However, they call for refinement and generalization. Improving these propositions and developing methods for limit analysis are the subjects of the subsequent chapters.

## 1. Rigid-plastic problem

Behavior of a rigid perfectly plastic body under quasistatic loading is determined by a solution of the (instantaneous) rigid-plastic problem, see Subsection I.5.3. To answer some relevant questions it is sufficient to consider a strong formulation of the problem, that is, the formulation assuming velocity fields to be regular. On the other hand, one cannot restrict the formulation to only regular stress fields (see Subsection I.5.4). In this section, we consider an appropriate strong formulation of the rigid-plastic problem: we assume that velocity fields are sufficiently regular and stress fields are only integrable.

**1.1. Stress fields and equilibrium conditions.** We will now summarize some assumptions about stress fields made in the previous sections, in order to use them in the formulation of the rigid-plastic problem as well as throughout the book.

Let a body occupy domain $\Omega$ and $\mathcal{S}$ be a set of stress fields on $\Omega$. We assume that $\mathcal{S}$ is closed under addition of fields and multiplication of a field by a number, that is, $\boldsymbol{\sigma}+\boldsymbol{\tau}$ belongs to $\mathcal{S}$ if $\boldsymbol{\sigma}$ and $\boldsymbol{\tau}$ are in $\mathcal{S}$ and $\alpha\boldsymbol{\sigma}$ belongs to $\mathcal{S}$ for every number $\alpha$ and every $\boldsymbol{\sigma}$ in $\mathcal{S}$. Thus, stress fields form a linear space $\mathcal{S}$. We also assume that all fields in $\mathcal{S}$ are integrable.

We take the virtual work principle as a formulation of equilibrium conditions for such stress fields. More precisely, let the body be subjected to load $\mathbf{l} = (\mathbf{f}, \mathbf{q})$, where $\mathbf{f}$ is the volume density of body forces given in $\Omega$ and $\mathbf{q}$ is the density of surface tractions given on the part $S_q$ of the body boundary. The remaining part $S_v$ of the boundary is fixed. According to the virtual work principle a stress field $\boldsymbol{\sigma}$ in $\mathcal{S}$ equilibrates the load $\mathbf{l}$ if and only if

$$\int_{\Omega} \operatorname{Def} \mathbf{v} \cdot \boldsymbol{\sigma}\, dV = \int_{\Omega} \mathbf{f}\mathbf{v}\, dV + \int_{S_q} \mathbf{q}\mathbf{v}\, dS \quad \text{for every} \quad \mathbf{v} \in U. \tag{1.1}$$

Here, $U$ is the set of virtual velocity fields, that is, smooth vector fields on $\Omega$ vanishing on $S_v$. The symbol Def $\mathbf{v}$ denotes the strain rate field corresponding to a velocity field $\mathbf{v}$:

$$(\operatorname{Def} \mathbf{v})_{ij} = \frac{1}{2}\left(\frac{\partial v_i}{\partial x_j} + \frac{\partial v_j}{\partial x_i}\right).$$

We also write equilibrium condition (1.1) in the form: $\boldsymbol{\sigma} \in \Sigma + \mathbf{s}_l$ (see Subsection II.1.3), where $\mathbf{s}_l$ is a fixed stress field in $\mathcal{S}$ equilibrating the load $\mathbf{l}$ and $\Sigma$ is the subspace of self-equilibrated stress fields

$$\Sigma = \left\{ \boldsymbol{\sigma} \in \mathcal{S} : \int_{\Omega} \operatorname{Def} \mathbf{v} \cdot \boldsymbol{\sigma}\, dV = 0 \quad \text{for every} \quad \mathbf{v} \in U \right\}. \tag{1.2}$$

**1.2 Admissible stress fields.** As long as we deal with continuous fields, we consider a stress field $\boldsymbol{\sigma}$ as admissible if its value $\boldsymbol{\sigma}(x)$ is admissible at every point of the body. As far as noncontinuous fields are concerned, such an idea of admissibility appears inadequate, and we refine it in what follows.

What becomes vague in the case of noncontinuous fields is the requirement that a certain condition should be satisfied at every point. The problem is that a physical field which is not sufficiently regular can be determined only up to redefining its values on any zero-volume set. Actually, one cannot measure directly the value of a field $\boldsymbol{\sigma}$ at a point $x_0$: only the mean value of $\boldsymbol{\sigma}$ over a certain neighborhood of $x_0$ can be determined with an instrument. One can imagine a collection of instruments for measuring mean value of $\boldsymbol{\sigma}$ over a sequence of neighborhoods contracting to the point $x_0$. The sequence of the measured mean values over these neighborhoods is the best information one can obtain regarding the field $\boldsymbol{\sigma}$ near the point $x_0$. This information determines the value $\boldsymbol{\sigma}(x_0)$ if $\boldsymbol{\sigma}$ is continuous and does not determine $\boldsymbol{\sigma}(x_0)$ in case $\boldsymbol{\sigma}$ is not continuous. Indeed, let $\boldsymbol{\sigma}_1$ and $\boldsymbol{\sigma}_2$ be integrable fields with equal values $\boldsymbol{\sigma}_1(x) = \boldsymbol{\sigma}_2(x)$ at every $x$ except points of a zero-volume set; then any measuring results in the same mean values for the two fields, making the fields $\boldsymbol{\sigma}_1$ and $\boldsymbol{\sigma}_2$ indistinguishable. Therefore, when dealing with physical fields whose values are determined using the above-mentioned procedure, we consider any two fields that differ only on a zero-volume set as *equivalent*. No physical statement about a field of the type under consideration can concern values of the field on a zero-volume set. To make sense, the statement has to deal with all equivalent fields.

In particular, consider a body occupying domain $\Omega$. The set $C_x$ of admissible stresses is given for every point $x$ in $\Omega$. A stress field $\boldsymbol{\sigma}$ is *admissible* if its value $\boldsymbol{\sigma}(x)$ is admissible, $\boldsymbol{\sigma}(x) \in C_x$, for every $x$ in $\Omega$ may be with the exception of a zero-volume set. We often use the brief form: "for almost every $x \in \Omega$" or the abbreviation: "for a.e. $x \in \Omega$" instead of the expression "for every $x$ in $\Omega$ may be with the exception of a zero-volume set".

Thus, if a set $C_x$ is given for every $x$ in $\Omega$ (or at least for a.e. $x \in \Omega$), and $\mathcal{S}$ is the space of stress fields, we introduce a set of admissible stress fields in $\mathcal{S}$:

$$C = \{\boldsymbol{\sigma} \in \mathcal{S} : \boldsymbol{\sigma}(x) \in C_x \text{ for a.e. } x \in \Omega\}.$$

Noteworthy are the two following important properties of this set: *$C$ contains the zero stress field* and *$C$ is convex*. Indeed, according to the assumption adopted in Section I.2, $C_x$ contains the zero stress, which implies $0 \in C$. As to the second property, consider two admissible stress fields $\boldsymbol{\sigma}_1$ and $\boldsymbol{\sigma}_2$; the admissibility means that the relations $\boldsymbol{\sigma}_1(x) \in C_x$ and $\boldsymbol{\sigma}_2(x) \in C_x$ are valid for a.e. $x \in \Omega$. Then, at any point $x$ where both relations are valid, we have (by convexity of $C_x$):

$$a\boldsymbol{\sigma}_1(x) + (1-a)\boldsymbol{\sigma}_2(x) \in C_x \qquad (0 \le a \le 1).$$

Therefore, the field $a\boldsymbol{\sigma}_1 + (1-a)\boldsymbol{\sigma}_2$ is admissible, and hence $C$ is convex.

REMARK 1. There are cases in which information is available about values of a physical field on a zero-volume set. Consider, for example, a body composed of two parts separated by a perfectly lubricated surface $\gamma$. Tangent force on $\gamma$ is zero if the body is in equilibrium, which gives information about stresses on zero-volume set $\gamma$. In cases like this, that is, if a body possesses singular properties on some zero-volume sets, the concept of the stress field admissibility calls for revising. We do not consider such cases in this book; however, limit analysis theory can be analogously developed in these cases.

REMARK 2. Equivalence of the stress fields which only differ on a zero-volume set suggests that one cannot determine values of the field, say, on a surface. This is true in case no further information is available about properties of the field. However, analysis of the problem involving a stress field $\boldsymbol{\sigma}$ often results in some regularity properties of $\boldsymbol{\sigma}$, which makes $\boldsymbol{\sigma}$ more definite. For example, if it turns out that $\boldsymbol{\sigma}$ is continuous, there is no other continuous field equivalent to $\boldsymbol{\sigma}$, and $\boldsymbol{\sigma}$ is uniquely defined at every point $x$. Regularity properties that are much weaker than the continuity (like integrability of a stress field $\boldsymbol{\sigma}$ and of the body forces equilibrated by $\boldsymbol{\sigma}$) also narrow the set of equivalent fields. This makes it possible to determine values of $\boldsymbol{\sigma}$ on every regular surface up to redefining them on any zero area part of the surface.

**1.3. Velocity fields.** Consider a body occupying domain $\Omega$. Let part $S_v$ of the body boundary be fixed, no other kinematic constraints being imposed on the body. Then a kinematically admissible velocity field $\mathbf{v}$ should satisfy the condition $\mathbf{v}|_{S_v} = 0$. Under this condition, kinematically admissible velocity fields may form a linear space, and we will always assume they do. We introduce the *linear space* $V_0$ *of sufficiently regular kinematically admissible velocity fields*; we assume that $V_0$ possesses the following properties:

(1) $V_0$ contains all virtual velocity fields: $U \subset V_0$,

(2) for every field $\mathbf{v} \in V_0$ are defined:
its values on the body boundary (with $\mathbf{v}|_{S_v} = 0$),
the strain rate field $\operatorname{Def} \mathbf{v}$,
the external power, that is, the power of any load $\mathbf{l} = (\mathbf{f}, q)$ in a certain class of possible loads:

$$\int_{\Omega} \mathbf{f}\mathbf{v}\, dV + \int_{S_q} \mathbf{q}\mathbf{v}\, dS,$$

the internal power, that is, the power of any stress field $\boldsymbol{\sigma}$ in $\mathcal{S}$:

$$-\int_{\Omega} \operatorname{Def} \mathbf{v} \cdot \boldsymbol{\sigma}\, dV;$$

(3) the equilibrium conditions do not change if the set $U$ of virtual velocity fields is replaced by the wider set $V_0$.

The latter property means that

$$\int_{\Omega} \operatorname{Def} \mathbf{v} \cdot \boldsymbol{\sigma}\, dV = \int_{\Omega} \mathbf{f}\mathbf{v}\, dV + \int_{S_q} \mathbf{q}\mathbf{v}\, dS \quad \text{for every} \quad \mathbf{v} \in V_0, \tag{1.3}$$

if the stress field $\boldsymbol{\sigma}$ satisfies virtual work principle (1.1). Due to this, the set $\Sigma$ of self-equilibrated stress fields has the form

$$\Sigma = \left\{ \boldsymbol{\sigma} \in \mathcal{S} : \int_{\Omega} \operatorname{Def} \mathbf{v} \cdot \boldsymbol{\sigma}\, dV = 0 \quad \text{for every} \quad \mathbf{v} \in V_0 \right\}. \tag{1.4}$$

Note that if the area of $S_v$ is positive, then the only velocity field $\mathbf{v} \in V_0$ with $\operatorname{Def} \mathbf{v} = 0$ is zero, $\mathbf{v} = 0$. Indeed, $\mathbf{v}$ satisfies the equations

$$\frac{1}{2}\left(\frac{\partial v_i}{\partial x_j} + \frac{\partial v_j}{\partial x_i}\right) = 0$$

if $\operatorname{Def} \mathbf{v} = 0$. The general solution to this system is six-parametric. The parameters form two vectors: $\mathbf{v}_0$ and $\boldsymbol{\omega}_0$, and the solution is $\mathbf{v} = \mathbf{v}_0 + \boldsymbol{\omega}_0 \times \mathbf{r}$, where $\mathbf{r}$ is the position vector. Thus, $\mathbf{v}$ is a rigid motion velocity field. The condition $\mathbf{v}|_{S_v} = 0$ implies $\mathbf{v}_0 = 0$, $\boldsymbol{\omega}_0 = 0$, and hence $\mathbf{v} = 0$.

A collection of all continuously differentiable vector fields on $\Omega$ vanishing on $S_v$ is an example of the space $V_0$. We will discuss a more important example dealing with numerical methods in Chapter IX.

**1.4. Formulation of the problem.** Consider a rigid perfectly plastic body occupying domain $\Omega$. The body is subjected to load $\mathbf{l} = (\mathbf{f}, \mathbf{q})$, and part $S_v$ of the body boundary is fixed. We assume displacements and strain to be small and the load to vary slowly enough to result in small accelerations which can be neglected. Then the stress and velocity fields in the body, $\boldsymbol{\sigma}$ and $\mathbf{v}$, are a solution to the quasistatic rigid-plastic problem. This problem was formulated in Section I.5 under the assumption that both fields $\boldsymbol{\sigma}$ and $\mathbf{v}$ are sufficiently regular. The formulation includes the following relations. The stress field $\boldsymbol{\sigma}$ equilibrates the load $\mathbf{l} = (\mathbf{f}, \mathbf{q})$:

$$\frac{\partial \sigma_{ij}}{\partial x_j} + f_i = 0, \qquad \sigma_{ij}\nu_j|_{S_q} = q_i; \tag{1.5}$$

the velocity field $\mathbf{v}$ satisfies the kinematic constraint and determines the strain rate field $\mathbf{e}$:

$$\mathbf{v}|_{S_v} = 0; \qquad e_{ij} = \frac{1}{2}\left(\frac{\partial v_i}{\partial x_j} + \frac{\partial v_j}{\partial x_i}\right); \tag{1.6}$$

the strain rate and stress fields satisfy the constitutive maximum principle:

$$\boldsymbol{\sigma}(x) \in C_x, \qquad \mathbf{e}(x) \cdot (\boldsymbol{\sigma}(x) - \boldsymbol{\sigma}_*) \geq 0 \quad \text{for every} \quad \boldsymbol{\sigma}_* \in C_x. \tag{1.7}$$

To get rid of the stress field regularity assumption, first of all we use virtual work principle (1.3) instead of equilibrium equations (1.5), writing the equilibrium conditions in the form $\boldsymbol{\sigma} \in \Sigma + \mathbf{s}_l$. Secondly, we replace the first of constitutive relations (1.7) with the condition $\boldsymbol{\sigma} \in C$, that is, we assume the relation $\boldsymbol{\sigma}(x) \in C_x$ to be valid for almost every $x$ in $\Omega$ (not necessarily for every $x$ in $\Omega$). Finally, we make the same assumption regarding inequality (1.7). Thus, we arrive to the following *strong* formulation of the rigid-plastic problem with the given load $\mathbf{l} = (\mathbf{f}, \mathbf{q})$: find stress and velocity fields $\boldsymbol{\sigma}$ and $\mathbf{v}$ satisfying the relations

$$\begin{gathered} \boldsymbol{\sigma} \in \Sigma + \mathbf{s}_l, \\ \boldsymbol{\sigma} \in C, \\ \mathbf{e} = \operatorname{Def} \mathbf{v}, \quad \mathbf{v} \in V_0, \\ \mathbf{e}(x) \cdot (\boldsymbol{\sigma}(x) - \boldsymbol{\sigma}_*) \geq 0 \quad \text{for every} \quad \boldsymbol{\sigma}_* \in C_x \end{gathered} \tag{1.8}$$

(the latter relation should be valid for a.e. $x \in \Omega$). Here, $\mathbf{s}_l$ is a fixed stress field equilibrating the load $\mathbf{l}$, $C$ and $\Sigma$ are the sets of admissible and self-equilibrated stress fields in $\mathcal{S}$, respectively. The linear space $\mathcal{S}$ consists of integrable and not necessarily smooth stress fields. On the other hand, the space $V_0$ of kinematically admissible velocity fields only contains sufficiently regular fields and meets requirements (1) – (3). This is why we refer to the formulation of the problem and to its solution as *strong* ones.

## 2. Safety factor

Stresses in a perfectly plastic body cannot be arbitrary as they have to meet the admissibility requirement. Therefore, whenever a certain load is considered, the question arises of whether there is any admissible stress field equilibrating this load. In this section, we show that computing a so-called safety factor answers the question, and the answer is positive if the safety factor is greater than 1 and negative if it is less than 1. The safety factor of the load $\mathbf{l}$ also determines which of all the loads proportional to $\mathbf{l}$ are admissible and which are not.

**2.1. Admissible and inadmissible loads.** A stress field in a rigid perfectly plastic body has to be admissible. As far as quasistatic processes are concerned, the stress field also has to equilibrate the load. These requirements are incompatible for "sufficiently large" loads.

EXAMPLE 1. We return to the example in Subsection I.5.1. We have shown that there is no admissible stress field equilibrating the given load if $\tau > k$ ($\tau$ is the nonnegative parameter specifying the load). However, we only considered differentiable stress fields. Now we consider nonsmooth stress fields only assuming them to be integrable and showing that there is no admissible field equilibrating the load among them. Indeed, let $\boldsymbol{\sigma}$ be an integrable stress field which equilibrates the load. Then $\boldsymbol{\sigma}$ satisfies virtual work principle (1.3), in particular, for the virtual velocity field $\mathbf{v}$ with the components $v_1 = ax_2$, $v_2 = v_3 = 0$ ($a = \text{const} > 0$). The velocity $\mathbf{v}$ is orthogonal to the load on the sides $S_1$, $S_3$ of the parallelepiped (see Figure I.5.2) and is parallel to the load on the side $S_2$. The velocity magnitude equals $aH$ on $S_2$, where $H$ is the height of the parallelepiped. Therefore, the external power on the velocity field $\mathbf{v}$ is

$$\int_{S_2} aH\tau \, dS = \tau a V,$$

where $V$ is the volume of the parallelepiped. It is clear that the only nonzero components of the strain rate $\mathbf{e} = \operatorname{Def} \mathbf{v}$ are $e_{12} = e_{21} = a/2$. Therefore the internal power is

$$-\int_{\Omega} \sigma_{12} a \, dV.$$

Then, the equality of the virtual work principle is of the form

$$a \int_{\Omega} \sigma_{12} \, dV = \tau a V \quad \text{or} \quad \tau = \frac{1}{V} \int_{\Omega} \sigma_{12} \, dV. \tag{2.1}$$

If the stress field $\boldsymbol{\sigma}$ is admissible, it satisfies the inequality $|\sigma_{12}| \leq k$ (we use the Mises admissibility condition for the material of the parallelepiped). Then (2.1) implies $\tau \leq k$. Thus, if the number $\tau$ determining the load is greater than $k$, none of the admissible stress fields can equilibrate the load.

In general, considering a perfectly plastic body, we divide all loads into two types. We refer to a load as *admissible* if there is an admissible stress field which equilibrates the load. Otherwise, we refer to the load as *inadmissible*.

If the load $\mathbf{l}$ is admissible, the rigid-plastic problem has at least one solution. Indeed, let $\boldsymbol{\sigma}$ be an admissible stress field which equilibrates $\mathbf{l}$; then the first two of the relations (1.8) are satisfied. The zero velocity field $\mathbf{v} = 0$ and $\boldsymbol{\sigma}$ satisfy the rest of relations (1.8). In other words, $\boldsymbol{\sigma}$ and $\mathbf{v} = 0$ is a strong solution to the rigid-plastic problem with the load $\mathbf{l}$.

REMARK 1. As we see, a solution with the zero velocity field exists for every admissible load. Note that the solution to the rigid-plastic problem is, generally, not unique. The example in Subsection I.5.2 shows that apart from the solution with the zero velocity field there can exist a solution with a nonzero velocity field. *If load* $\mathbf{l}$ *is inadmissible, the rigid-plastic problem has no solution.* By the definition of load admissibility, there is no admissible stress field equilibrating $\mathbf{l}$, and therefore the first two of relations (1.8) cannot be satisfied. In other words, a perfectly plastic body cannot bear an inadmissible load. From the viewpoint of some applications, this is a merit of the model. For example, a real structure can bear certain loads, however, undergoing a large strain (say, along the yield plateau). If functioning of the structure does not allow such deformation, such loads are *practically* inadmissible. One can associate these loads with those *theoretically* inadmissible within the framework of a perfectly plastic model. Then this model can be applied to analyzing the structure safety, which is the actual idea of limit analysis.

Thus, separating admissible and inadmissible loads, as they are defined in perfect plasticity, is not only an internal problem of the theory but is very important for some engineering purposes. The limit stress state principle forms the ideological basis for linking perfect plasticity and mechanical engineering; we discuss this principle in Section 4 below.

**2.2. Safety factor.** Generally speaking, there are admissible as well as inadmissible loads among those proportional to a given load $\mathbf{l}$. We show that loads of the two types cannot alternate, which results in a very important concept of the safety factor.

Let $\mathbf{l} = (\mathbf{f}, \mathbf{q})$ be a given load, and $m\mathbf{l}$ be the load proportional to $l$: $m\mathbf{l} = (m\mathbf{f}, m\mathbf{q})$, $m \geq 0$. That is, the densities of the body forces and the surface tractions become $m$ times greater, $m$ being the same at every point of the body. There is an admissible load among the loads $m\mathbf{l}$, namely, the zero load, which corresponds to $m = 0$. On the other hand, it is clear that the load $m\mathbf{l}$ is inadmissible for sufficiently large $m$. Actually, there exists a number separating values of the multiplier $m$, for which $m\mathbf{l}$ is admissible, from those for which $m\mathbf{l}$ is inadmissible. We refer to this number as the safety factor of the load $\mathbf{l}$. For example, if the safety factor of the load $\mathbf{l}$ equals 3, the load itself is admissible and remains admissible when increased 2.9 times (generally, increased $m$ times where $m < 3$). On the other hand, 3.1 times increase in load (or, generally, $m > 3$ times increase) results in an inadmissible load. Now, we give a formal definition of the safety factor and establish its above-described property.

Let $\mathbf{s}_l$ be a fixed stress field equilibrating the load $\mathbf{l}$. Then, $\Sigma + m\mathbf{s}_l$ is the set of all stress fields equilibrating the load $m\mathbf{l}$. The load $m\mathbf{l}$ is admissible if and only if there is an admissible stress field in $\Sigma + m\mathbf{s}_l$, that is, if and only if $\Sigma + m\mathbf{s}_l$ intersects the set $C$ of admissible stress fields (see Figure 2.1). We

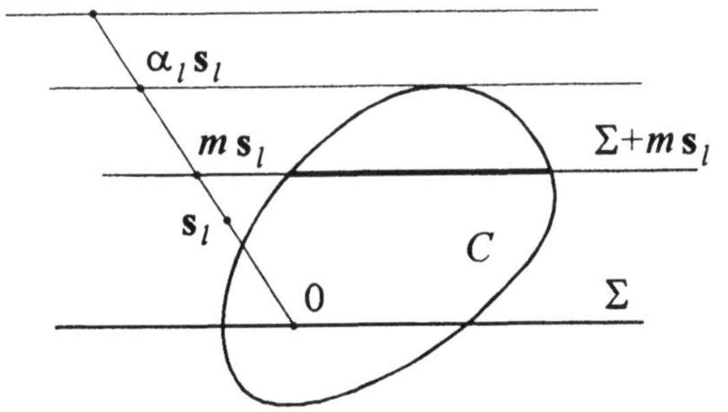


**Figure 2.1**

consider the supremum of the multipliers $m$ for which the intersection

$$(\Sigma + m\mathbf{s}_l) \cap C$$

is not empty, that is, the load $\mathbf{l}$ is admissible:

$$\alpha_l = \sup \{m \geq 0 : (\Sigma + m\mathbf{s}_l) \cap C \neq \emptyset\};$$

we refer to this supremum as the *safety factor* of the load $\mathbf{l}$.

This definition can be written in a slightly different and, sometimes, more convenient form. First, note that the criterion $(\Sigma + m\mathbf{s}_l) \cap C \neq \emptyset$ for admissibility of the load $\mathbf{l}$ is equivalent to

$$m(\Sigma + \mathbf{s}_l) \cap C \neq \emptyset. \tag{2.2}$$

The equivalence is obvious at $m > 0$. At $m = 0$, both relations are still valid since 0 belongs to $C$. Thus, (2.2) is a criterion for admissibility of the load $m\mathbf{l}$. It reads as follows: there exists a self-equilibrated stress field $\boldsymbol{\rho}$ in $\Sigma$ such that the stress field $m(\boldsymbol{\rho} + \mathbf{s}_l)$ is admissible: $m(\boldsymbol{\rho} + \mathbf{s}_l) \in C$. Thus, we can write the definition of the safety factor as

$$\alpha_l = \sup \{m \geq 0 : \text{there exists } \boldsymbol{\rho} \in \Sigma,\ m(\boldsymbol{\rho} + \mathbf{s}_l) \in C\}. \tag{2.3}$$

PROPOSITION 1. The load $m\mathbf{l}$ is inadmissible if $m > \alpha_l$ and is admissible if $0 \leq m < \alpha_l$.

PROOF. The first statement is obviously true due to the definition of the safety factor $\alpha_l$. Now let us show that the second statement is also true. If $0 \leq m < \alpha_l$, then by the definition of $\alpha_l$ as a supremum there exists a number $m_*$ such that the load $m_*\mathbf{l}$ is admissible. In other words, there exists an admissible stress field $\boldsymbol{\sigma}_*$ equilibrating the load $m_*\mathbf{l}$. Then, consider the stress field

$$\boldsymbol{\sigma} = \frac{m}{m_*}\boldsymbol{\sigma}_* + \left(1 - \frac{m}{m_*}\right)0 = \frac{m}{m_*}\boldsymbol{\sigma}_* \qquad \left(0 \leq \frac{m}{m_*} \leq 1\right).$$

The set $C$ is convex and contains both 0 and $\boldsymbol{\sigma}_*$, which implies that $\boldsymbol{\sigma}_*$ is in $C$, that is, $\boldsymbol{\sigma}_*$ is admissible. Recall that $\boldsymbol{\sigma}$ equilibrates the load $m\mathbf{l}$. This means that $m\mathbf{l}$ is an admissible load, which finishes the proof.

We see that among the loads $m\mathbf{l}$ proportional to $\mathbf{l}$ with $m \geq 0$, the admissible and inadmissible ones cannot alternate; the load $\alpha_l\mathbf{l}$ separates them.

By Proposition 1, *load* $\mathbf{l}$ *is admissible if* $\alpha_l > 1$ *and is inadmissible if* $\alpha_l < 1$.

REMARK 2. Proposition 1 says nothing about admissibility of the load $\mathbf{l}$ with $\alpha_l = 1$ (or, which is the same, of the load $m\mathbf{l}$ at $m = \alpha_l$). As we will see in Section VII.1, this load is actually admissible.

## 3. Safe and limit loads

There are two different types of the rigid perfectly plastic body response to an admissible load. The body either remains rigid, that is, does not deform at all, or it deforms at an arbitrarily large strain rate (See Subsection I.5.2 for an example). The load is safe in the former case and is limit in the latter. In this section, we establish a criterion for the safety of a load (judging by the value of its safety factor).

This results in the following picture of behavior of a rigid perfectly plastic body subjected to load $\mathbf{l}$: the body

does not deform if $\alpha_l > 1$,
deforms at an arbitrarily large strain rate if $\alpha_l = 1$,
cannot bear the load if $\alpha_l < 1$.

The load is safe in the first case and limit in the second case, which is plastic failure. In the third case, the load is inadmissible.

Here we establish these properties of the rigid perfectly plastic body within the framework of the strong formulation of the rigid-plastic problem. In Chapter IV we consider these properties within the framework of the rigid-plastic problem general formulation and we also refine some details of the above statements later on. However, the simple formulation, which we use here, correctly reflects the main properties of the rigid perfectly plastic body.

**3.1. Safe loads.** To define a safe load, we use the following concept of a safe stress field. Consider a perfectly plastic body occupying domain $\Omega$; set $C_x$ of admissible stresses is given for every $x$ in $\Omega$. Let $\boldsymbol{\sigma}$ be an admissible stress field, that is, $\boldsymbol{\sigma}(x)$ is either inside or on the yield surface for almost every point $x$ in $\Omega$.

Now, we distinguish those admissible stress fields which do not approach the yield surface. More precisely, consider a stress field $\boldsymbol{\sigma}$ with the distance from $\boldsymbol{\sigma}(x)$ to the yield surface not less than $\delta > 0$ for every $x \in \Omega$ ($\delta$ does not depend on $x$), see Figure 3.1. Recall that the distance in the (six-dimensional) stress space is defined through the scalar multiplication $\mathbf{a} \cdot \mathbf{b} = a_{ij}b_{ij}$, that is, $|\boldsymbol{\sigma} - \boldsymbol{\tau}| = \sqrt{(\boldsymbol{\sigma} - \boldsymbol{\tau}) \cdot (\boldsymbol{\sigma} - \boldsymbol{\tau})}$. We refer to a stress field $\boldsymbol{\sigma}$ as *safe* if for a.e. $x \in \Omega$ the stress $\boldsymbol{\sigma}(x)$ is inside the yield surface and there exists a number $\delta > 0$ independent on $x$ such that the distance from $\boldsymbol{\sigma}(x)$ to the yield surface is not less than $\delta$.

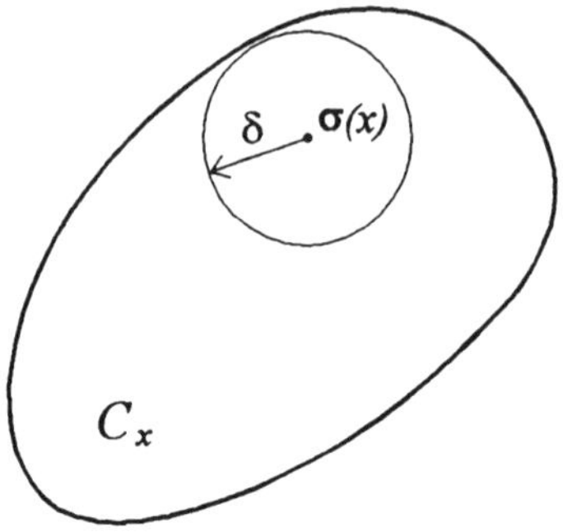


**Figure 3.1**

We refer to a load as *safe* if there is a safe stress field equilibrating the load. The stress field in Example I.5.2 is evidently safe if $\tau < k$; therefore the corresponding load is also safe. Recall that the rigid-plastic problem in that example only has solutions with the zero velocity field if $\tau < k$. This illustrates the main property of a safe load: a rigid perfectly plastic body remains rigid under the action of a safe load (and this is why we call the load safe). The following proposition establishes this property within the framework of the rigid-plastic problem strong formulation (1.8).

PROPOSITION 1. If a rigid perfectly plastic body is subjected to a safe load, then any strong solution to the rigid-plastic problem has the zero velocity field.

PROOF. Let stress field $\boldsymbol{\sigma}$ and velocity field $\mathbf{v}$ be a solution to the rigid-plastic problem with a safe load $\mathbf{l}$. Consider a safe stress field $\boldsymbol{\sigma}_s$ equilibrating $\mathbf{l}$. First, let us show that $\boldsymbol{\sigma}_s$ and the velocity field $\mathbf{v}$ satisfy constitutive maximum principle (I.3.4). Indeed, the solution $\boldsymbol{\sigma}$, $\mathbf{v}$ satisfies the principle, that is,

$$\operatorname{Def}\mathbf{v}\cdot(\boldsymbol{\sigma}(x)-\boldsymbol{\sigma}_*)\geq 0 \quad \text{for every} \quad \boldsymbol{\sigma}_*\in C_x \tag{3.1}$$

for a.e. $x \in \Omega$, where $\Omega$ is the domain occupied by the body. In particular, we set $\boldsymbol{\sigma}_* = \boldsymbol{\sigma}_s(x)$ and obtain from the constitutive maximum principle:

$$\operatorname{Def}\mathbf{v}\cdot(\boldsymbol{\sigma}(x)-\boldsymbol{\sigma}_s(x))\geq 0 \quad \text{for a.e. } x\in\Omega.$$

On the other hand, both fields $\boldsymbol{\sigma}$ and $\boldsymbol{\sigma}_*$ equilibrate the load $\mathbf{l}$. Consequently, the difference $\boldsymbol{\sigma} - \boldsymbol{\sigma}_s$ is self-equilibrated and, by the virtual work principle (1.3), the following equality is valid:

$$\int_\Omega \operatorname{Def}\mathbf{v}\cdot(\boldsymbol{\sigma}-\boldsymbol{\sigma}_s)\,dV = 0.$$

This equality together with (3.1) shows that $\boldsymbol{\sigma}_s$ and $\mathbf{v}$ satisfy the constitutive maximum principle:

$$\operatorname{Def}\mathbf{v}\cdot(\boldsymbol{\sigma}(x)-\boldsymbol{\sigma}_s(x)) = 0 \quad \text{for a.e. } x\in\Omega.$$

Here, the stress field $\boldsymbol{\sigma}_s$ is safe and hence there exists a number $\delta > 0$, such that for a.e. $x \in \Omega$ all stresses $\boldsymbol{\sigma}_*$ which satisfy the inequality $|\boldsymbol{\sigma}_* - \boldsymbol{\sigma}(x)| < \delta$ are admissible. Then the constitutive maximum principle implies $\text{Def}\,\mathbf{v}(x) = 0$ for a.e. $x \in \Omega$, that is, $\mathbf{v}$ is a rigid motion velocity field. Recall that we always assume that kinematic constraints exclude rigid motions of the body; therefore, $\mathbf{v} = 0$ and the proposition is proved.

Note that Proposition 1 is a powerful statement about rigid perfectly plastic body properties: the only condition for the body to remain rigid under the action of a given load is the existence of any safe stress field equilibrating the load.

**3.2. Safety criterion.** We studied admissibility of a load in Section 2, and it turned out that the admissibility is connected with the safety factor value. Now we establish a similar criterion for safety of a load. We make the following additional assumption: the zero stress field is safe for the bodies in question or, in more detail: there exists a number $\delta > 0$ such that the ball of the stress space with its center at the origin and with a radius $\delta$ lies inside the yield surface for a.e. point of the body.

We consider loads $m\mathbf{l}$ proportional to the given load $\mathbf{l}$ at $m \geq 0$. The following proposition connects the safety of the load $m\mathbf{l}$ with the value $\alpha_l$ of the safety factor.

PROPOSITION 2. Suppose the zero stress field is safe for a given rigid perfectly plastic body. Then the load $m\mathbf{l}$ ($m \geq 0$) is safe if and only if $m < \alpha_l$.

PROOF. First consider the load $m\mathbf{l}$ with $0 \leq m < \alpha_l$. For any number $m'$ satisfying the inequalities $m < m' < \alpha_l$, the load $m'\mathbf{l}$ is admissible (Proposition 2.1). Hence there is an admissible stress field $\boldsymbol{\sigma}'$ which equilibrates $m'\mathbf{l}$. Then the stress field

$$\boldsymbol{\sigma} = \frac{m}{m'}\boldsymbol{\sigma}' = \frac{m}{m'}\boldsymbol{\sigma}' + \left(1 - \frac{m}{m'}\right) 0$$

equilibrates the load $m\mathbf{l}$, and $\boldsymbol{\sigma}$ is admissible since $C$ is convex and $0 \in C$. Let us show that $\boldsymbol{\sigma}$ is safe. Indeed, by assumption, there exists a number $\delta > 0$ such that the ball $B_\delta(0)$ of (six-dimensional) stress space with its center at the origin and with a radius $\delta$ lies in $C_x$ for a.e. point $x$ of the body. Then, the set

$$\left(1 - \frac{m}{m'}\right) B_\delta(0) + \frac{m}{m'}\boldsymbol{\sigma}'(x) = \left(1 - \frac{m}{m'}\right) B_\delta(0) + \boldsymbol{\sigma}$$

is also in $C_x$, see Figure 3.2. This is the ball with the center at $\boldsymbol{\sigma}(x)$ and with a radius $\delta' = (1 - m/m')\delta$. Thus the stress field values $\boldsymbol{\sigma}(x)$ are inside the yield surface and the distance from $\boldsymbol{\sigma}(x)$ to the yield surface is not less than $\delta'$ for a.e. point $x$ of the body. Therefore, the load $m\mathbf{l}$ equilibrated by $\boldsymbol{\sigma}$ is safe; note that $0 \leq \mathbf{m} < \alpha_l$ in this case.

Now consider load $m\mathbf{l}$ with $m \geq \alpha_l$. If $m > \alpha_l$ the load is inadmissible (Proposition 2.1) and, consequently, not safe. Let us show that neither is $m\mathbf{l}$ safe in case $m = \alpha_l$. If $\alpha_l\mathbf{l}$ was safe, there would exist a number $\delta > 0$ and a stress field $\boldsymbol{\sigma}$ equilibrating the load $m\mathbf{l}$ such that $\boldsymbol{\sigma}(x)$ lies inside the

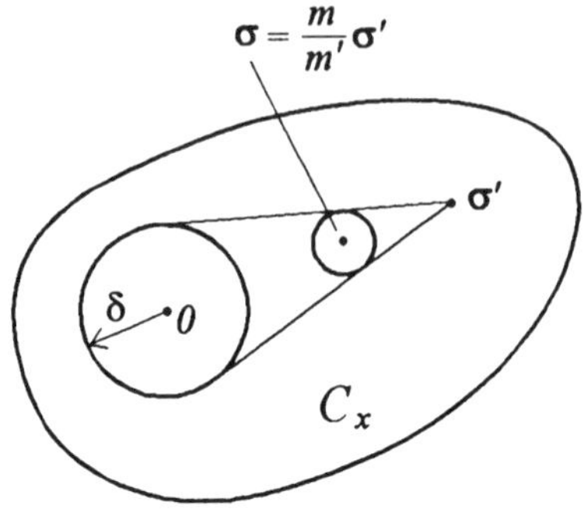


**Figure 3.2**

yield surface and the distance from $\boldsymbol{\sigma}(x)$ to the yield surface is not less than $\delta$ for a.e. point $x$ of the body. Then the stress field $(1+\varepsilon)\boldsymbol{\sigma}$ with $0 < \varepsilon < \delta$ is admissible and equilibrates the load $m'\mathbf{l}$ with $m' = (1+\varepsilon)\alpha_l > \alpha_l$. This contradicts Proposition 2.1, and, therefore, the assumption about safety of the load $m\mathbf{l}$ with $m = \alpha_l$ appears to be wrong, which finishes the proof.

In case $m = 1$, this proposition results in the safety criterion for a given load: if the zero stress field is safe for a given rigid perfectly plastic body, then *load* $\mathbf{l}$ *is safe when* $\alpha_l > 1$ *and is not safe when* $\alpha_l \le 1$.

REMARK 1. Although the zero stress field safety assumption looks natural, there are some models which do not meet this requirement. For example, a conical yield surface with its vertex at the origin is used for modeling soil or granular medium behavior, in which case the zero stress field is not safe.

**3.3. Limit load and failure mechanism.** Again, consider all the loads $m\mathbf{l}$, $m \ge 0$, proportional to the given load $\mathbf{l}$. Those with $m > \alpha_l$ are inadmissible, while those with $m < \alpha_l$ are admissible. The load $\alpha_l\mathbf{l}$ separates the admissible and inadmissible loads; we refer to this load as *limit load*. The load $\mathbf{l}$ is limit if and only if $\alpha_l = 1$.

If the zero stress field is safe for a rigid perfectly plastic body, the loads $m\mathbf{l}$ with $0 \le m < \alpha_l$ are safe (Proposition 2). Then $\alpha_l$ is the only value of the multiplier $m$ at which a solution to a rigid-plastic problem can exist with a nonzero velocity field. In other words, a nonzero velocity field can only exist in a body subjected to a limit load. Moreover, *if the rigid-plastic problem has a solution with a nonzero velocity field for a certain (limit) load, then it also has a solution with an arbitrarily large strain rate.* Indeed, let $\boldsymbol{\sigma}$ and $\mathbf{v} \ne 0$ be a solution to problem (1.8). Then the same stress field $\boldsymbol{\sigma}$ and the velocity field $a\mathbf{v}$ (where $a$ is an arbitrary positive number) also form a solution to (1.8). Note that $\operatorname{Def}\mathbf{v} \ne 0$ if $\mathbf{v} \ne 0$ since we always assume that kinematic constraints exclude rigid motions of the body, that is, $\operatorname{Def}\mathbf{v} = 0$ implies $\mathbf{v} = 0$. Thus, the strain rate $\operatorname{Def} a\mathbf{v} = a \operatorname{Def}\mathbf{v}$ can be arbitrarily large. Because of that we refer to the nonzero velocity field in a solution to the rigid-plastic problem as the *failure mechanism*. The failure mechanism exists for a wide class of rigid perfectly plastic bodies, see Subsection 3.4 and Chapter VII.

The above propositions about safe, limit and inadmissible loads (complemented by the failure mechanism existence theorem in Chapter VII) result in the following statement:

a rigid perfectly plastic body subjected to load $\mathbf{l}$
does not deform if $\alpha_l > 1$ (safe load),
deforms at an arbitrarily large strain rate if $\alpha_l = 1$ (limit load),
cannot bear the load if $\alpha_l < 1$ (inadmissible load).

These are the main properties of the rigid perfectly plastic body.

REMARK 2. There are some assumptions we did not mention about formulating the latter statements. For example, the safety of a load $\mathbf{l}$ with $\alpha_l > 1$ is guaranteed for bodies with the safe zero stress field (Proposition 2). Analogously, existence of the failure mechanism is established in Chapter VII under some (nonrestrictive) hypothesis.

REMARK 3. A simplified model of the plastic behavior we use throughout the book (which we call rigid perfectly plastic) is based on several substantial hypotheses. We assume that there is no hardening, displacements and strains are small, acceleration can be neglected. The failure mechanism violates these assumptions and, therefore, imprecisely describes the body response to a limit load. A more efficient model is required for an accurate description of the response. However, the failure mechanism existence itself definitely indicates that intense deformation starts in the body. Even the beginning of such a process is often inadmissible from a practical viewpoint. Then details of the subsequent behavior are irrelevant, and the safety problem can be solved on the basis of the limit analysis: a body bears load $\mathbf{l}$ as long as $\alpha_l > 1$ and loses its bearing capacity under the action of a limit load, that is, when $\alpha_l = 1$.

**3.4. On the failure mechanism existence.** A failure mechanism in a rigid perfectly plastic body can exist provided the body is subjected to a limit load. We gave an example of a failure mechanism in Subsection I.5.2. At the same time, within the framework of the rigid-plastic problem strong formulation, it is easy to construct an example which shows the nonexistence of a failure mechanism although the body is subjected to a limit load (see Subsection 6.3 ). However, this example does not allow drawing the conclusion that the body remains rigid. In fact, the only velocity fields involved in the strong formulation of the rigid-plastic problem are sufficiently regular fields which form the space $V_0$ (see Subsection 1.3). It is possible that there is no failure mechanism in $V_0$, while it exists in a wider space. Analogously, it is premature to conclude from Proposition 1 the nonexistence of a failure mechanism in case the body is subjected to a safe load. The proposition only states that there is no sufficiently regular failure mechanism, that is, a solution to the rigid-plastic problem with nonzero velocity field in $V_0$. As before, this does not exclude the existence of a failure mechanism in a wider space.

Thus, we arrive at the question of whether or not it is possible to extend the space $V_0$ of velocity fields to ensure that, for every limit load, a failure mechanism exists in the extended space. If the answer is positive and, at the same time, only the zero velocity field in the extended space solves the rigid-plastic problem for a safe load, then the scheme of the rigid-plastic analysis is logically complete. In particular, existence of a failure mechanism explains, within the framework of the theory, why the rigid-plastic problem loses its solution when a load passes from safe to inadmissible.

In Chapter VII, we will consider an extended space $\mathcal{V}$ of velocity fields and show that it meets the above-mentioned requirements. The space is very wide, which makes the corresponding formulation of the rigid-plastic problem rather abstract. Fortunately, neither the definition nor methods for calculating the safety factor are connected with the extended space $\mathcal{V}$, which allows us to move further towards methods for limit analysis. It must be remembered that statements in Chapter VII complete the mathematical basis of the theory, among other things these statements rigorously establish the meaning of the safe and limit loads. We have already found out this meaning: *a rigid perfectly plastic body remains rigid under the action of a safe load and deforms at an arbitrarily large strain rate under the action of a limit load.*

## 4. Problems of limit analysis

In this section, we consider formulations of some problems arising from mechanical engineering. These are the main problems of limit analysis, and we will see that they are reduced to the problem of evaluating the safety factor defined in Section 2.

**4.1. Limit stress state principle.** Engineers often apply the following principle to analyzing the bearing capacity of statically indeterminate systems: *the system bears a certain loading if, at every moment, there exists a safe distribution of internal forces which equilibrates the load.* The principle is called the *limit stress state principle.* Rigid perfectly plastic bodies obey it, the set of safe stress fields being determined by the yield surface (see Subsection 3.1). At the same time, the principle is applicable to systems whose constitutive relations differ from the perfectly plastic ones. For such systems, a set of safe stress fields is not determined automatically as for perfectly plastic bodies. Specifying the set of safe stress fields is of exceptional importance for formulating the problem of bearing capacity analysis and calls for engineering experience and intuition. This book does not concern the important and challenging problem of choosing the set of safe stress fields, but only considers analysis of bearing capacity at a given set.

A particular form of the safe stress fields set does not matter for developing the theory and methods of the limit analysis. However, we always assume that the safe stress fields set possesses the following properties. Let $C_x$ be a set in the (six-dimensional) stress space given for every point $x$ of the body. We assume that $C_x$ is convex and contains the origin. A stress field $\boldsymbol{\sigma}$ is considered as safe if, for almost every point $x$ of the body, $\boldsymbol{\sigma}(x)$ lies in $C_x$ and the distance between $\boldsymbol{\sigma}(x)$ and the frontier of $C_x$ is not less than $\delta > 0$, where $\delta$ does not depend on $x$. To find out whether the body will bear a certain load, the limit stress state principle requires an answer to the following main question: is there a safe stress field equilibrating the given load? It can be easily seen that this is exactly the problem of computing or estimating the safety factor for a certain rigid perfectly plastic body. More precisely, consider the body $\mathcal{B}$ and the problem of its bearing capacity analysis posed on the basis of the limit stress state principle. As mentioned above, this presumes that a set of safe stress fields is defined in terms of the given sets $C_x$. Using these sets, we now introduce an axillary rigid perfectly plastic body $\mathcal{B}'$, which 1) occupies the same domain as $\mathcal{B}$, 2) has the sets $C_x$ of admissible stresses, 3) is subjected to the same loading as $\mathcal{B}$. Then, if the body $\mathcal{B}$ is subjected to load $\mathbf{l}$, the safety of $\mathcal{B}$ in the sense of the limit stress state principle is evidently equivalent to the safety of the load $\mathbf{l}$ for the axillary body $\mathcal{B}'$ in the sense of a rigid-plastic analysis as defined in Subsection 3.1. Therefore, we can replace the former (original) problem by the latter one. Although we only consider analysis of rigid perfectly plastic bodies throughout this book, it should be remembered that those can be fictitious axillary bodies, and the problem of rigid-plastic analysis just represents a certain original problem posed on the basis of the limit stress state principle.

**4.2. Safety of a loading.** Consider a rigid perfectly plastic body occupying domain $\Omega$. The body is subjected to body forces with the volume density $\mathbf{f}$ given in $\Omega$ and surface tractions with the density $\mathbf{q}$ given on the part $S_q$ of the body boundary. The remaining part $S_v$ of the boundary is fixed. We assume that the load $\mathbf{l} = (\mathbf{f}, \mathbf{q})$ varies slowly enough to allow a quasistatic description of the body behavior. The question is whether the body bears a given loading, that is, whether it is safe against the plastic failure.

Recall that the safety factor $\alpha_l$ characterizes the response of the body to the load $\mathbf{l}$. According to Proposition 3.2, the body remains rigid, that is, safe as long as $\alpha_l > 1$, the body loses its bearing capacity at $\alpha_l = 1$, and all the loads $\mathbf{l}$ with $\alpha_l < 1$ are inadmissible. Thus, to find out whether the loading is safe it is sufficient to evaluate the safety factor $\alpha_l$ of the current load $\mathbf{l}$ at every moment during the loading. The safety factor is defined as (see Section 2.2):

$$\alpha_l = \sup\{m \geq 0 : \text{ there exists } \rho \in \Sigma, \ m(\rho + \mathbf{s}_l) \in C\}, \tag{4.1}$$

where $\Sigma$ and $C$ are the sets of self-equilibrated and admissible stress fields respectively; $\mathbf{s}_l$ is a fixed stress field equilibrating the load $\mathbf{l}$. It is often sufficient to find an appropriate estimation for the safety factor. For example, if a lower bound $m_* \leq \alpha_l$ with $m_* > 1$ is established for the safety factor of every load during the loading under consideration, then the loading is safe. Analogously, if an upper bound $m^* \geq \alpha_l$ with $m^* < 1$ is established for the safety factor of at least one of the loads, then the body loses its bearing capacity under the loading.

For this reason *calculating the safety factor $\alpha_l$ is the main problem of limit analysis.* This problem is the subject of almost all of the subsequent chapters.

EXAMPLE 1. *Proportional loading.* Consider a loading during which the load $\mathbf{l}$ is proportional to a certain load $\mathbf{l}_1$: $\mathbf{l} = \kappa\mathbf{l}_1$, where the multiplier $\kappa$ depends only on the time $t$ and not on the point of the body. According to definition (4.1), the safety factor $\alpha_{\kappa l_1}$ is related to $\alpha_{l_1}$:

$$\begin{aligned}
\alpha_{\kappa l_1} &= \sup\{m \geq 0 : \text{ there exists } \rho \in \Sigma,\ m(\rho + \kappa \mathbf{s}_{l_1}) \in C\} \\
&= \sup\{m \geq 0 : \text{ there exists } \rho' \in \Sigma,\ \kappa m(\rho + \mathbf{s}_{l_1}) \in C\} \\
&= \sup\left\{\frac{m'}{\kappa} \geq 0 : \text{ there exists } \rho' \in \Sigma,\ m'(\rho' + \mathbf{s}_{l_1}) \in C\right\} \\
&= \frac{1}{\kappa}\alpha_{l_1}.
\end{aligned}$$

Hence in order to answer the question about safety of the proportional loading, it is sufficient to calculate the safety factor of one load $\mathbf{l}_1$.

**4.3. Limit surface.** A loading is safe if and only if the current load is safe at every moment during the loading. Thus, the safety of a loading does not depend on how the load varies, but only on the set of loads the loading runs through. As far as the loadings running through the same set $\mathcal{L}$ of loads are concerned, either all of the loadings are safe or all of them are not. Therefore, it is often important to find the widest set of safe loads. Every loading within this set is safe, while any loading reaching beyond it is not. In what follows, we consider a slightly more narrow formulation of this problem.

We assume that any load which can be applied to the body is a linear combination of some given loads $\mathbf{l}_1 = (\mathbf{f}_1, \mathbf{q}_1), \ldots, \mathbf{l}_N = (\mathbf{f}_N, \mathbf{q}_N)$. Here, $\mathbf{f}_1, \ldots, \mathbf{f}_N$ and $\mathbf{q}_1, \ldots, \mathbf{q}_N$ are fields of the densities of the body forces and surface tractions, respectively. Note that $\mathbf{q}_1, \ldots, \mathbf{q}_N$ are defined on the same part $S_q$ of the body boundary, and it is due to this that we can consider their linear combinations. The basis loads $\mathbf{l}_1, \ldots, \mathbf{l}_N$ are time-independent, and loadings are defined by coefficients $m_1(t), \ldots, m_N(t)$ depending on the time:

$$\mathbf{l}(t) = m_1(t)\mathbf{l}_1 + \cdots + m_N(t)\mathbf{l}_N \tag{4.2}$$

or, in more detail, $\mathbf{l} = (\mathbf{f}, \mathbf{q})$ and

$$\begin{aligned}
\mathbf{f}(t) &= m_1(t)\mathbf{f}_1(x,t) + \cdots + m_N(t)\mathbf{f}_N(x,t) \quad \text{in} \quad \Omega, \\
\mathbf{q}(t) &= m_1(t)\mathbf{q}_1(x,t) + \cdots + m_N(t)\mathbf{q}_N(x,t) \quad \text{on} \quad S_q.
\end{aligned}$$

We assume that $\mathbf{l}_1, \ldots, \mathbf{l}_N$ are linearly independent. Then these loads form a basis in the $N$-dimensional linear space of loads. We consider the widest set of safe loads in this space and refer to it as the *set of safe loads*. The following statements show that the set of safe loads has a sufficiently simple structure: it is convex and open.

PROPOSITION 1. The set of safe loads is convex.

PROOF. Note that, if the load $\mathbf{l}_s$ is safe and the load $\mathbf{l}_a$ is admissible, all loads

$$\mathbf{l} = (1 - m)\mathbf{l}_s + m\mathbf{l}_a, \quad 0 \leq m < 1 \tag{4.3}$$

are safe. Indeed, let $\boldsymbol{\sigma}_s$ be a safe stress field equilibrating the load $\mathbf{l}$. Then there exists a number $\delta > 0$ such that the ball of the stress space with its center

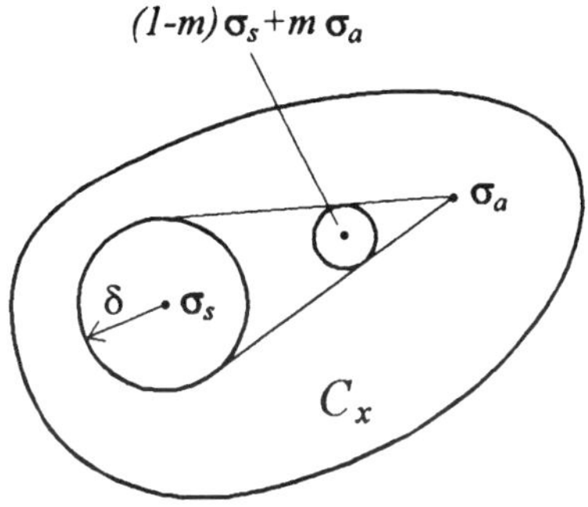


**Figure 4.1**

at $\boldsymbol{\sigma}_s(x)$ and with a radius $\delta$ lies in the set $C_x$ of admissible stresses. There is also an admissible stress field $\boldsymbol{\sigma}_a$ equilibrating the load $\mathbf{l}_a$. Due to convexity, the set $C_x$ contains the ball centered at $\boldsymbol{\sigma}(x) = (1-m)\boldsymbol{\sigma}_s(x) + m\boldsymbol{\sigma}_a(x)$ of the radius $(1-m)\delta$, where $0 \leq m < 1$; see Figure 4.1. Consequently, the field $\boldsymbol{\sigma}$ is safe, and load (4.3) is safe, as $\boldsymbol{\sigma}$ equilibrates it. If the load $\mathbf{l}_a$ is not only admissible but also is safe, then load (4.3) is also safe at $m = 1$. The proposition is proved.

REMARK 1. When proving the latter proposition, we did not make use of the fact that the load space is finite-dimensional. Actually, the proposition holds for any linear space of loads.

PROPOSITION 2. Suppose that for a given rigid-plastic body, 1) the zero stress field is safe, 2) the safety factor does not vanish for every one of the basis loads $\mathbf{l}_1, \cdots, \mathbf{l}_N$ as well as for every one of the loads $-\mathbf{l}_1, \cdots, -\mathbf{l}_N$. Then the set of safe loads contains a neighborhood of the origin in the load space and is open.

PROOF. By the second assumption, there exists a number $\varepsilon > 0$ such that the loads $\varepsilon\mathbf{l}_1, \ldots, \varepsilon\mathbf{l}_N, -\varepsilon\mathbf{l}_1, \ldots, -\varepsilon\mathbf{l}_N$ are admissible. Then, according to Proposition 3.2, for example, the loads

$$\frac{\varepsilon}{2}\mathbf{l}_1, \ldots, \frac{\varepsilon}{2}\mathbf{l}_N, -\frac{\varepsilon}{2}\mathbf{l}_1, \ldots, -\frac{\varepsilon}{2}\mathbf{l}_N \tag{4.4}$$

are safe. Consider the convex polyhedron $P$ with vertices (4.4) in the load space. By Proposition 1, the set of safe loads contains $P$, which is obviously a neighborhood of the origin. The first statement of the proposition is proved. It remains to show that the set of safe loads is open, that is, contains a neighborhood of any safe load. Let $\mathbf{l}$ be a safe load; then there exists a number $\delta > 0$ such that the load $(1+\delta)\mathbf{l}$ is also safe (Proposition 3.2). Consider all segments connecting the point $(1+\delta)\mathbf{l}$ with points of the polyhedron $P$. By Proposition 1, points of these segments correspond to safe loads. Therefore, the loads constituting the set

$$\mathbf{l} + \frac{\delta}{1+\delta}P$$

are safe. We observe that this set is a neighborhood of the load $\mathbf{l}$, which finishes the proof.

Thus, the set of safe loads is convex and open. We refer to its boundary as *the limit surface*. It is clear that the limit surface consists of limit loads. In other words, the following equation with the unknowns $m_1, \ldots, m_N$ specifies the limit surface:

$$\alpha_{m_1 l_1 + \cdots + m_N l_N} = 1.$$

This connects the problem of finding the limit surface with the main problem of limit analysis, evaluating the safety factor. However, to find the limit surface, one does not solve the latter equation since, normally, there is no explicit formula expressing the safety factor in terms of the load. Usually, one evaluates or estimates the safety factor for some particular loads and makes use of these loads to approximate the limit surface from the inside or outside depending on the value of the corresponding safety factor. In particular, Propositions 1 and 2 are very useful for constructing the internal approximation. An example of determining a limit surface will be given in Subsection VIII.5.3.

**4.4. Limit analysis in presence of a permanent load.** Let a rigid-plastic body $\mathcal{B}$ be subjected to a permanent load $\mathbf{l}_0$, say, its weight, and be also subjected to additional loading $\mathbf{l}$. We focus on proportional loadings $\mathbf{l} = \mu \mathbf{l}_1$, which is a case of frequent practical interest. Thus, we consider the total load $\mathbf{l}_0 + \mu \mathbf{l}_1$ where $\mathbf{l}_1$ is fixed as well as $\mathbf{l}_0$ and $\mu \geq 0$ is the loading parameter. The parameter may vary with time, $\mu$ being the same at every point of the body. The aim of limit analysis in this case is to find values of $\mu$ for which the load $\mathbf{l}_0 + \mu \mathbf{l}_1$ is safe. A slightly different problem results in basically the same question. Namely, if a body is subjected to a load consisting of two components: $\mathbf{l}_0 + \mathbf{l}_1$, it is sometimes required to find out how many times the component $\mathbf{l}_1$ can be enlarged while keeping the total load safe. A particular case (at $\mathbf{l}_0 = 0$) is the problem of safety of the load $\mu \mathbf{l}_1$. Calculating the safety factor $\alpha_{\mathbf{l}_1}$ solves the latter problem, see Proposition 3.2. It turns out that in the more general case of the load $\mathbf{l}_0 + \mu \mathbf{l}_1$ there is a number $\alpha(\mathbf{l}_0, \mathbf{l}_1)$ which plays the same role as $\alpha_{\mathbf{l}_1}$ in the particular case. This number can be interpreted and evaluated as the safety factor of the load $\mathbf{l}_1$ for a certain fictitious body $\widehat{\mathcal{B}}$. More precisely, the following statement is valid.

PROPOSITION 3. Suppose the zero stress field is safe for a given rigid-plastic body $\mathcal{B}$. Let $\mathbf{l}_0$ and $\mathbf{l}_1$ be two given loads, the load $\mathbf{l}_0$ being safe. Then there exists a number $\alpha(\mathbf{l}_0, \mathbf{l}_1)$ such that the load $\mathbf{l}_0 + \mu \mathbf{l}_1$ is safe if $0 \leq \mu < \alpha(\mathbf{l}_0, \mathbf{l}_1)$ and is inadmissible if $\mu > \alpha(\mathbf{l}_0, \mathbf{l}_1)$. The number $\alpha(\mathbf{l}_0, \mathbf{l}_1)$ equals the safety factor of the load $\mathbf{l}_1$ for a body which 1) occupies the same domain $\Omega$ as $\mathcal{B}$ and 2) for every $x \in \Omega$ has the set of admissible stresses $\widehat{C}_x = C_x - \mathbf{s}_0(x)$, where $C_x$ is the set of admissible stresses at the point $x$ of the body $\mathcal{B}$ and $\mathbf{s}_0$ is a stress field equilibrating the load $\mathbf{l}_0$.

PROOF. Let us show that the first statement is valid with

$$\alpha(\mathbf{l}_0, \mathbf{l}_1) = \sup \{ m \geq 0 : \text{there exists } \rho \in \Sigma, \ \rho + \mathbf{s}_0 + m \mathbf{s}_1 \in C \}, \qquad (4.5)$$

where $\mathbf{s}_0$ and $\mathbf{s}_1$ are some fixed stress fields which equilibrate the loads $\mathbf{l}_0$ and $\mathbf{l}_1$, respectively. Indeed, if $0 \le \mu < \alpha(\mathbf{l}_0, \mathbf{l}_1)$, by the definition of supremum, there exist a number $\mu_*$, $\mu < \mu_* < \alpha(\mathbf{l}_0, \mathbf{l}_1)$, and a self-equilibrated stress field $\boldsymbol{\rho}_*$ for which $\boldsymbol{\rho}_* + \mathbf{s}_0 + \mu_* \mathbf{s}_1 \in C$. This stress field equilibrates the load $\mathbf{l}_0 + \mu_* \mathbf{l}_1$, and consequently, the latter is admissible. Therefore, the load

$$\mathbf{l}_0 + \mu \mathbf{l}_1 = \left(1 - \frac{\mu}{\mu_*}\right) \mathbf{l}_0 + \frac{\mu}{\mu_*} (\mathbf{l}_0 + \mu_* \mathbf{l}_1)$$

is of the form (4.3) and is safe (as it was shown in the proof of Proposition 1). On the other hand, if $\mu > \alpha(\mathbf{l}_0, \mathbf{l}_1)$ no $\boldsymbol{\rho}$ in $\Sigma$ satisfies the condition $\boldsymbol{\rho} + \mathbf{s}_0 + \mu \mathbf{s}_1 \in C$. Hence there is no admissible stress field equilibrating the load $\mathbf{l}_0 + \mu \mathbf{l}_1$, which finishes the proof of the first statement.

Extremum problem (4.5) can be written as

$$\alpha(\mathbf{l}_0, \mathbf{l}_1) = \sup \{ m \ge 0 : \text{there exists } \boldsymbol{\rho} \in \Sigma, \ \boldsymbol{\rho} + m \mathbf{s}_1 \in C - \mathbf{s}_0 \} .$$

The latter expression is exactly the definition of the safety factor of the load $\mathbf{l}_1$ for the rigid-plastic body $\widehat{\mathcal{B}}$ with the set $\widehat{C} = C - \mathbf{s}_0$ of admissible stress fields of the form

$$\widehat{C} = \{ \mathbf{s} \in \mathcal{S} : \mathbf{s}(x) \in C_x - \mathbf{s}_0(x) \text{ for a.e. } x \in \Omega \} .$$

Consider now the set $\widehat{C}_x = C_x - \mathbf{s}_0(x)$ at every point $x$ of the body $\widehat{\mathcal{B}}$. It is clear that $\widehat{C}_x$ is convex and closed. Without loss of generality $\widehat{C}_x$ may be considered as containing the origin. Indeed, the load $\mathbf{l}_0$ is safe; therefore there is a safe stress field equilibrating $\mathbf{l}_0$ and we consider this field as $\mathbf{s}_0$. Then $\mathbf{s}_0(x) \in C_x$ and, consequently, $0 \in C_x - \mathbf{s}_0(x)$. We observe that the sets $\widehat{C}_x$ possess all regular properties of the admissible stresses sets, and they generate the above introduced set $\widehat{C}$ of admissible stress fields. This finishes the proof of the proposition.

## 5. Basic statements and methods of limit analysis

In this section we consider two basic theorems and three methods for limit analysis. The methods are aimed at solving the main problem of limit analysis: estimating and computing the safety factor of a given load. A static method consists in finding lower bounds for the safety factor and computing its value solely on the basis of the safety factor definition. The first basic theorem of limit analysis suggests another (kinematic) method for estimating the safety factor. The theorem points the way to computing an upper bound for the safety factor starting from any kinematically admissible velocity field. The second basic theorem of limit analysis states that, at least in some cases, the best lower (static) bound for the safety factor equals the best upper (kinematic) bound, and, hence, their common value is $\alpha_l$. Thus the theorem shows that the kinematic method can result in the precise value of the safety factor and not only at its upper bounds. The theorem also results in formulation of the third method of limit analysis, the rigid-plastic solutions method. Unfortunately, assumptions of the theorem are not constructive: they are not expressed explicitly in terms of limit analysis problem inputs. We will improve the theorem in Chapters V and VII replacing its assumptions with simple and constructive ones.

Examples of computing the safety factor by the methods we present here are given in the next section. The reader can find some further examples in Section VIII.5.

**5.1. Lower bound for safety factor: static multiplier.** The main problem of limit analysis is evaluating the safety factor of a given load $\mathbf{l}$ (see Subsection 2.2):

$$\alpha_l = \sup\{m \geq 0 : \text{there exists } \rho \in \Sigma,\ m(\rho + \mathbf{s}_l) \in C\}. \qquad (5.1)$$

Here, $\mathbf{s}_l$ is a fixed stress field equilibrating the load $\mathbf{l}$, and $\Sigma$ and $C$ are the sets of self-equilibrated and admissible stress fields, respectively. Let us also re-write definition (5.1) of the safety factor in another form more traditional in limit analysis. To obtain this form we make use of the concept of the static multiplier. Namely, consider a number $m \geq 0$, such that there exists an admissible stress field $\boldsymbol{\sigma}$ equilibrating the load $m\mathbf{l}$. We refer to such number $m$ as the *static multiplier* of the load $\mathbf{l}$ (corresponding to the stress field $\boldsymbol{\sigma}$); we use the symbol $m_s(\boldsymbol{\sigma})$ to denote the static multiplier. The safety factor of the load is the supremum of its static multipliers. Indeed, if $m_s(\boldsymbol{\sigma})$ is a static multiplier, the number $m = m_s(\boldsymbol{\sigma})$ and the stress field

$$\rho = \frac{1}{m_s(\boldsymbol{\sigma})}\boldsymbol{\sigma} - \mathbf{s}_l$$

satisfy the conditions $\rho \in \Sigma$, $m(\rho + \mathbf{s}_l) \in C$ (in case $m_s(\boldsymbol{\sigma}) = 0$ any self-equilibrated stress field can be taken as $\rho$). Then the inequality $\alpha_l \geq m_s(\boldsymbol{\sigma})$ is valid due to (5.1). On the other hand, consider a sequence of couples $(m_n, \rho_n)$, $n = 1, 2, \ldots$, for which

$$m_n \geq 0, \quad \rho_n \in \Sigma, \quad m_n(\rho_n + \mathbf{s}_l) \in C, \quad \lim_{n\to\infty} m_n = \alpha_l;$$

this is a maximizing sequence of extremum problem (5.1), which always exists (see Subsection I.C.2). The stress field $\boldsymbol{\sigma}_n = m_n(\rho_n + \mathbf{s}_l)$ is admissible and equilibrates the load $m_n\mathbf{l}$; therefore the number $m_n$ is a static multiplier of the load $\mathbf{l}$: $m_n = m_s(\boldsymbol{\sigma}_n)$. Thus,

$$\alpha_l = \lim_{n\to\infty} m_n = \lim_{n\to\infty} m_s(\boldsymbol{\sigma}_n),$$

which results in the required relation $\alpha_l = \sup m_s(\boldsymbol{\sigma})$ since the inequality $\alpha_l \geq m_s(\boldsymbol{\sigma})$ is valid for every static multiplier.

The definition of the safety factor as the supremum of the static multipliers does not provide us with additional means for evaluating the safety factor. At the same time, making use of the static multiplier concept simplifies formulations of limit analysis basic statements.

We refer to the procedure of computing lower bounds or the value of the safety factor with the help of static multipliers,

$$\alpha_l \geq m_s(\boldsymbol{\sigma}), \qquad \alpha_l = \sup m_s(\boldsymbol{\sigma}),$$

as the *static method* of limit analysis.

Note that searching for a static multiplier does not call for selecting a couple $m$, $\boldsymbol{\sigma}$ as the definition of the static multiplier suggests. A static multiplier can be found by starting with any, not necessarily admissible, stress field $\mathbf{s}$ equilibrating the given load. If $\mathbf{s}$ is a such field, the greatest number $m \geq 0$, for which the field $\boldsymbol{\sigma} = m\mathbf{s}$ is admissible, is the static multiplier $m = m_s(\boldsymbol{\sigma})$ for the given load. Consider, for example, the Mises admissibility condition $\left|\boldsymbol{\sigma}^d\right| \leq \sqrt{2}k$, $k = \text{const} > 0$, and a continuous stress field $\mathbf{s}$ in a closed domain $\bar{\Omega}$. The field $\boldsymbol{\sigma} = m\mathbf{s}$ is admissible if and only if $m\left|\mathbf{s}^d(x)\right| \leq \sqrt{2}k$ for every $x$ in $\Omega$. Consequently, the greatest $m$ for which $m\mathbf{s}$ is admissible equals

$$m = \frac{\sqrt{2}k}{\max\left\{\left|\mathbf{s}^d(x)\right| : \ x \in \bar{\Omega}\right\}}, \tag{5.2}$$

and it is a static multiplier of the load equilibrated by $\mathbf{s}$.

This way of computing the static multiplier is simple provided sufficiently regular fields $\mathbf{s}$ are considered. However, the static method calls for searching the supremum of static multipliers over the set of all possible stress fields, which is not an easy task. For example, computing maximum (5.2) is rather complicated if $\mathbf{s}$ is not sufficiently regular. The idea to approximate a nonregular field $\mathbf{s}$ by a sequence of regular fields $\mathbf{s}_n$ leads to nowhere. It would make sense provided the sequence of the maximums $\max\left|\mathbf{s}_n^d(x)\right|$ approximates $\max\left|\mathbf{s}^d(x)\right|$. Actually, there is no appropriate sequence resulting in the latter approximation if $\mathbf{s}$ is nonregular. Therefore, only some lower bounds for the safety factor are normally obtained by the static method using simple and sufficiently regular stress fields. Within the framework of the static method it is also difficult to evaluate the error, that is, the difference between a certain lower bound and the precise value of the safety factor.

Some examples will be presented in the next section and Section VIII.5 to illustrate application of the static method to estimating and computing the safety factor.

**5.2. Upper bound for safety factor: kinematic multiplier.** An opportunity to compute upper bounds for the safety factor is of principle interest for limit analysis. Such estimations together with lower (static) bounds result in the two-sided approximation for the safety factor, which, in particular, allows estimation of the approximation error. Here we present a method for constructing upper bounds for the safety factor.

Consider a rigid-plastic body occupying domain $\Omega$ and subjected to the load $\mathbf{l} = (\mathbf{f}, \mathbf{q})$, while the part $S_v$ of the body boundary is fixed. We consider sufficiently regular kinematically admissible velocity fields, which form the linear space $V_0$ (Subsection 1.3). The strain rate field $\operatorname{Def}\mathbf{v}$ is defined for every $\mathbf{v}$ in $V_0$, and it is easy to compute the total dissipation

$$\int_\Omega d_x(\operatorname{Def}\mathbf{v})\,dV,$$

where the dissipation $d_x$ is defined in Subsection I.4.1,

$$d_x(\mathbf{e}) = \sup\{\boldsymbol{\sigma}_* \cdot \mathbf{e} : \boldsymbol{\sigma}_* \in C_x\}. \tag{5.3}$$

Suppose the external power, that is, the power of $\mathbf{l}$ on $\mathbf{v}$ is positive

$$\int_\Omega \mathbf{f}\mathbf{v}\,dV + \int_{S_q} \mathbf{q}\mathbf{v}\,dS > 0.$$

Then we refer to the number $m_k(\mathbf{v})$ equal to the total dissipation over the external power

$$m_k(\mathbf{v}) = \frac{\int_\Omega d_x(\operatorname{Def}\mathbf{v})\,dV}{\int_\Omega \mathbf{f}\mathbf{v}\,dV + \int_{S_q} \mathbf{q}\mathbf{v}\,dS} \tag{5.4}$$

as the *kinematic multiplier* of the load $\mathbf{l}$ (corresponding to the velocity field $\mathbf{v}$).

The following proposition is the first of two fundamental statements of limit analysis. The theorem establishes the main property of the kinematic multiplier making the latter so important for limit analysis. Note that we defined the kinematic multiplier for sufficiently regular velocity fields and that the following proposition only concerns such fields (we will generalize it later).

THEOREM 1. Every kinematic multiplier of a load is not less than any of its static multipliers: $m_s(\boldsymbol{\sigma}) \le m_k(\mathbf{v})$.

PROOF. Let $\mathbf{v}$ in $V_0$ be a kinematically admissible velocity field, and the power of the load $\mathbf{l}$ on $\mathbf{v}$ be positive. Then the kinematic multiplier $m_k(\mathbf{v})$ is defined and the following equality is valid:

$$\int_\Omega d_x(\operatorname{Def}\mathbf{v})\,dV = m_k(\mathbf{v})\left(\int_\Omega \mathbf{f}\mathbf{v}\,dV + \int_{S_q} \mathbf{q}\mathbf{v}\,dS\right).$$

Let $m_s(\boldsymbol{\sigma}) = m$ be a static multiplier of the load $\mathbf{l}$, that is, the stress field $\boldsymbol{\sigma}$ is admissible and equilibrates the load $m\mathbf{l}$, $m \ge 0$. Using the field $\mathbf{v}$ as a virtual velocity field in the equilibrium conditions of the form (1.3), we obtain:

$$\int_\Omega \operatorname{Def}\mathbf{v} \cdot \boldsymbol{\sigma}\,dV = m\left(\int_\Omega \mathbf{f}\mathbf{v}\,dV + \int_{S_q} \mathbf{q}\mathbf{v}\,dS\right).$$

Subtracting this equality from the previous one results in:

$$\int_{\Omega} (d_x(\mathrm{Def}\,\mathbf{v}) - \mathrm{Def}\,\mathbf{v} \cdot \boldsymbol{\sigma})\, dV = (m_k(\mathbf{v}) - m) \left( \int_{\Omega} \mathbf{f}\mathbf{v}\, dV + \int_{S_q} \mathbf{q}\mathbf{v}\, dS \right). \tag{5.5}$$

The left-hand side of this equality is nonnegative by the definition of the dissipation (5.3) since $\boldsymbol{\sigma}$ is admissible, $\boldsymbol{\sigma} \in C$. The second multiplier on the right-hand side in (5.5) is the external power, which we assumed to be positive. Then the first multiplier is nonnegative, and consequently: $m_k(\mathbf{v}) \geq m = m_s(\boldsymbol{\sigma})$, which finishes the proof.

According to Theorem 1 every kinematic multiplier of load $\mathbf{l}$ is an upper bound for its safety factor: $\alpha_l \leq m_k(\mathbf{v})$. Indeed, let us fix $\mathbf{v}$ in the inequality $m_s(\boldsymbol{\sigma}) \leq m_k(\mathbf{v})$ proved in Theorem 1. The inequality remains valid if we pass to the supremum on its left-hand side: $\sup m_s(\boldsymbol{\sigma}) \leq m_k(\mathbf{v})$ or $\alpha_l \leq m_k(\mathbf{v})$ since the safety factor $\alpha_l$ is defined as the supremum of static multipliers.

It is easier to obtain a kinematic upper bound for the safety factor than a lower bound by the static method. Practically every velocity field $\mathbf{v}$ vanishing on the surface $S_v$ results in an upper bound: if the power of the load $\mathbf{l}$ on $\mathbf{v}$ is negative, $\mathbf{v}$ just has to be replaced by $-\mathbf{v}$. Computing the kinematic multiplier by formula (5.4) creates no problem since an explicit expression for the dissipation $d_x$ is normally known (in particular, Example 4.1 presents such expression in the case of the Mises yield surface).

The static method, at least in principle, results in a value of the safety factor: the latter is the supremum of the static upper bounds. At the same time, Theorem 1 does not allow us to conclude that, analogously, $\alpha_l$ is the infimum of the kinematic upper bounds: $\alpha_l = \inf m_k(\mathbf{v})$. Conditions under which this equality is valid are of principle interest, as they guarantee that $\alpha_l$ can be computed by the (easier) kinematic method and not only by the static method. These conditions will be established in the next subsection; however, they are not constructive. Simple constructive conditions will be given in Chapters V and VI.

The procedure of computing upper bounds for the value of the safety factor by means of kinematic multipliers:

$$\alpha_l \leq m_k(\mathbf{v}), \quad \alpha_l = \inf m_k(\mathbf{v}),$$

is referred to as the *kinematic method* of limit analysis. A two-sided estimation for the safety factor arises from combined application of both methods:

$$m_s(\boldsymbol{\sigma}) \leq \alpha_l \leq m_k(\mathbf{v}). \tag{5.6}$$

This results in an approximation of the safety factor with the error estimation, and the approximation can be improved by narrowing the gap between the upper and lower bounds in (5.6).

We present some examples illustrating application of the kinematic method to estimating and computing the safety factor in the next section and in Section VIII.5.

**5.3. Criterion for static and kinematic multipliers equality.** If a certain static multiplier of the load **l** equals its kinematic multiplier, $m_s(\boldsymbol{\sigma}) = m_k(\mathbf{v})$, then due to (5.6) their common value is the safety factor $\alpha_l$. That is why conditions for this equality are of substantial interest for limit analysis. They allow formulation of the problem of searching for $\boldsymbol{\sigma}$ and $\mathbf{v}$ with $m_s(\boldsymbol{\sigma})$ equal to $m_k(\mathbf{v})$, whose solution provides evaluation of the safety factor.

Conditions which we establish here for the equality $m_s(\boldsymbol{\sigma}) = m_k(\mathbf{v})$ make use of rigid-plastic problem solutions. In that connection, the strong formulation of this problem should be recalled (Subsection 1.4). The problem consists of searching for a stress field $\boldsymbol{\sigma}$ (in a certain space $\mathcal{S}$ of integrable fields) and a velocity field $\mathbf{v}$ (in the space $V_0$ of sufficiently regular kinematically admissible fields) which satisfy the following relations:

$$\begin{gathered}\boldsymbol{\sigma} \in \Sigma + \mathrm{s}_l,\\ \boldsymbol{\sigma} \in C,\\ \mathrm{e} = \operatorname{Def} \mathbf{v}, \quad \mathbf{v} \in V_0,\\ \mathrm{e}(x) \cdot (\boldsymbol{\sigma}(x) - \boldsymbol{\sigma}_*) \geq 0 \quad \text{for every} \quad \boldsymbol{\sigma}_* \in C_x.\end{gathered}$$

Here, $\mathrm{s}_l$ is a fixed stress field which equilibrates the given load **l**. The first of these relations is the equilibrium condition; the second one is the stress admissibility condition. The next two relations are of a kinematic nature and represent the expression for strain rate **e** through the velocity field **v** and the kinematic admissibility condition (recall that, normally, **v** is subjected to some kinematic constraints). The last one in the above system of relations is assumed to be valid for a.e. $x$ in $\Omega$, $\Omega$ being the domain occupied by the body. This inequality together with the stress admissibility condition form the rigid perfectly plastic constitutive relations.

Using dissipation (5.3), we can also write the constitutive relations as

$$\operatorname{Def} \mathbf{v}(x) \cdot \boldsymbol{\sigma}(x) = d_x(\mathrm{e}(x)).$$

The following proposition is the second of two fundamental theorems of limit analysis. The theorem establishes a criterion for equality of static and kinematic multipliers.

THEOREM 2. Let $\boldsymbol{\sigma}$ be a stress field and $\mathbf{v}$ be a sufficiently regular kinematically admissible velocity field. The static and kinematic multipliers of the load **l** are equal, $m_s(\boldsymbol{\sigma}) = m_k(\mathbf{v}) = m > 0$, if and only if $\boldsymbol{\sigma}$ and $\mathbf{v}$ are a strong solution to the rigid-plastic problem for the load $m\mathbf{l}$, $m > 0$, and the power of **l** on **v** is nonzero (in case the zero stress field is safe for the body: if and only if $\boldsymbol{\sigma}$ and $\mathbf{v}$ are a strong solution to this problem and $\mathbf{v} \neq 0$).

PROOF. Suppose $m_s(\boldsymbol{\sigma}) = m_k(\mathbf{v}) = m > 0$. Let us show that $\boldsymbol{\sigma}$ and $\mathbf{v}$ are a strong solution to the rigid-plastic problem for the load $m\mathbf{l}$ (note that we do not need to verify positiveness of the power: existence of the kinematic multiplier $m_k(\mathbf{v})$ presumes that the power is positive). Since $m = m_s(\boldsymbol{\sigma})$ is a static multiplier corresponding to $\boldsymbol{\sigma}$, we have

$$\boldsymbol{\sigma} \in \Sigma + m\mathrm{s}_l, \qquad \boldsymbol{\sigma} \in C. \tag{5.7}$$

The equality $m = m_k(\mathbf{v})$ and formula (5.5) derived in the proof of Theorem 1 imply

$$\int_{\Omega} (d_x(\operatorname{Def}\mathbf{v}) - \operatorname{Def}\mathbf{v}\cdot\boldsymbol{\sigma})\, dV = 0.$$

On the other hand, the inequality $d_x(\operatorname{Def}\mathbf{v}) \geq \operatorname{Def}\mathbf{v}\cdot\boldsymbol{\sigma}$ is valid due to the definition of dissipation (5.3) and the admissibility of the stress field $\boldsymbol{\sigma}$. Therefore, the integrand in the previous formula vanishes, that is,

$$d_x\left(\operatorname{Def}\mathbf{v}(x)\right) = \operatorname{Def}\mathbf{v}(x)\cdot\boldsymbol{\sigma}(x). \tag{5.8}$$

Thus, $\boldsymbol{\sigma}$ and $\mathbf{v} \in V_0$ satisfy relations (5.7), (5.8); in other words, $\boldsymbol{\sigma}$ and $\mathbf{v}$ are a strong solution to the rigid-plastic problem for the load $m\mathbf{l}$.

Suppose now that $\boldsymbol{\sigma}$ and $\mathbf{v}$ are strong solutions to the rigid-plastic problem for the load $m\mathbf{l}$, $m > 0$, that is, $\boldsymbol{\sigma}$, $\mathbf{v}$ satisfy relations (5.7), (5.8). Suppose also that

$$\int_{\Omega} \mathbf{f}\mathbf{v}\, dV + \int_{S_q} \mathbf{q}\mathbf{v}\, dS \neq 0.$$

Let us show that this power is positive (consequently, the kinematic multiplier $m_k(\mathbf{v})$ is defined) and verify the equality $m_k(\mathbf{v}) = m_s(\boldsymbol{\sigma})$. Note that $\boldsymbol{\sigma}$ equilibrates the load $m\mathbf{l}$ (see the first of relations (5.7)). Using the field $\mathbf{v}$ as a virtual velocity field in the equilibrium conditions of the form (1.3), we obtain

$$\int_{\Omega} \operatorname{Def}\mathbf{v}\cdot\boldsymbol{\sigma}\, dV = m\left(\int_{\Omega} \mathbf{f}\mathbf{v}\, dV + \int_{S_q} \mathbf{q}\mathbf{v}\, dS\right). \tag{5.10}$$

Here, $m$ is positive by assumption, power (5.9) is nonzero and the left-hand side is nonnegative due to (5.8) and nonnegativeness of the dissipation. Therefore, power (5.9) in (5.10) can be only positive. Then, the kinematic multiplier is defined for the velocity field $\mathbf{v}$, and (5.8), (5.10) results in equality $m_k(\mathbf{v}) = m$. On the other hand, (5.7) also implies that $m_s(\boldsymbol{\sigma}) = m$. Thus, we arrive at the required equality $m_s(\boldsymbol{\sigma}) = m_k(\mathbf{v})$.

It remains to show that, in case the zero stress field is safe, assumption (5.9) can be replaced by the relation $\mathbf{v} \neq 0$ in the above reasoning. That is, more precisely, (5.9) is valid as soon as the velocity field $\mathbf{v}$ of the solution $\boldsymbol{\sigma}$, $\mathbf{v}$ to rigid-plastic problem (5.7), (5.8) is nonzero. Indeed, let $\boldsymbol{\sigma}$, $\mathbf{v} \neq 0$ be such a solution. Then $\operatorname{Def}\mathbf{v} \neq 0$ (see Subsection 1.3). Since the zero stress field is safe, the dissipation $d_x(\mathbf{e})$ does not vanish for $\mathbf{e} \neq 0$, see property 4 in Subsection I.4.2. Thus, $d_x\left(\operatorname{Def}\mathbf{v}(x)\right) > 0$ for the field $\mathbf{v}$ under consideration, and, consequently,

$$\int_{\Omega} d_x\left(\operatorname{Def}\mathbf{v}(x)\right) dV > 0.$$

Recall that (5.10) is valid for a solution of problem (5.7), (5.8). The left-hand side in (5.10) is positive according to the previous inequality. Consequently, the power on the right-hand side in (5.10) is positive, that is, (5.9) is valid, and this finishes the proof.

**5.4. Rigid-plastic solution method.** Theorem 2 results in formulation of the rigid-plastic solutions method for evaluating the safety factor $\alpha_l$ of the load $\mathbf{l}$. The method suggests solving the rigid-plastic problem for loads $m\mathbf{l}$ proportional to the given one. If for a certain $m = m_0$ there exists a solution to this problem with a nonzero velocity field, then $\alpha_l = m_0$ according to Theorem 2. The method is not very effective since it is not easy to solve the rigid-plastic problem. However, the method is useful, for example, in the case of two-dimentional problems, for which it is easier to solve the rigid-plastic problem. Apart from that, the method results in accumulating a store of limit load examples, as the solution to any rigid-plastic problem results in such an example. This collection is helpful in limit analysis as a source of approximate solutions: looking through it allows sometimes finding a load sufficiently close to the one to be analyzed.

An example of calculating the safety factor by the rigid-plastic solutions method will be presented in Subsection 6.2.

REMARK 1. Applications of the rigid-plastic solutions method are restricted not only practically, that is, by complications in solving the rigid-plastic problem, but also in principle. The point is that the method makes use of strong solutions, and it is quite possible that such a solution does not exist even if the problem has smooth inputs (see Subsection 6.3 for an example). Considering discontinuous solutions of the rigid-plastic problem extends capabilities of the method, see Chapter VIII.

**5.5. On the kinematic method.** As far as sufficiently regular velocity fields are only considered, the best estimation for the safety factor that can be obtained by the kinematic method is $\alpha_l \leq \inf m_k(\mathbf{v})$, where the infimum is taken over the space $V_0$. An important question is whether the best upper bound equals the safety factor or is there a gap between them. In the first case, the safety factor can be computed by the kinematic method, which is simpler than the static one. Theorem 2 provides us with new and very important information concerning this question. It shows that the equality $m_s(\boldsymbol{\sigma}) = m_k(\mathbf{v})$ is realizable for a wide range of loads, and consequently, $\alpha_l = m_k(\mathbf{v})$ in that case.

Conditions that Theorem 2 establishes for the equality $m_s(\boldsymbol{\sigma}) = m_k(\mathbf{v})$ are not constructive. The theorem only connects validity of the equality with existence of a strong solution to the rigid-plastic problem and does not specify conditions under which such a solution exists. Actually, the problem does not necessarily have a strong solution even if its inputs (the body properties and the load) are smooth; see example in Subsection 6.2. However, this is not a reason for a pessimistic attitude towards the kinematic method. The method results in a value of the safety factor if there is no gap between $\alpha_l$ and $\inf m_k(\mathbf{v})$ regardless of whether the infimum is attained or not. To find $\alpha_l$ in case it equals $\inf m_k(\mathbf{v})$, it is sufficient to construct a sequence of fields $\mathbf{v}_1, \mathbf{v}_2, \dots$ for which

$$\lim_{n\to\infty} m_k(\mathbf{v}_n) = \inf m_k(\mathbf{v}).$$

Thus, we can avoid answering the question about attainability of the infimum or, which is more or less the same, about existence of a strong solution to the rigid-plastic problem. Constructive conditions under which the infimum equals the safety factor are given in Chapters V, VI.

The kinematic method involving only sufficiently regular velocity fields is focused on in this chapter. Making use of discontinuous fields extends capabilities of the method. However, the formula (5.4) determining the kinematic multiplier is to be generalized first, as it contains the derivatives $\partial v_i/\partial x_j$ and, therefore, makes no sense for discontinuous velocity fields. We derive an appropriate generalized formula and prove the corresponding extended version of Theorems 1 and 2 in Chapter VIII.

## 6. Examples of limit analysis

This section presents two examples of limit analysis illustrating the application of the static, kinematic and rigid-plastic solutions methods to evaluaton of the safety factor. One of the examples also outlines the peculiarity of limit analysis: the infimum of kinematic multipliers can be unattainable at a sufficiently regular velocity field even if inputs of the problem are smooth. In case the infimum is unattainable, the rigid-plastic problem for the limit load has no solution with a sufficiently regular nonzero velocity field (failure mechanism).

Both examples concern a pipe under axially symmetric plane strain conditions.

**6.1. Formulation of the axially symmetric plane strain problem.** Consider a rigid perfectly plastic, infinitely long, thick-walled pipe. Let the material of the pipe be homogeneous and isotropic. The pipe is subjected to a load and some kinematic constraints. We assume that the load and the constraints are axially symmetric; then a solution to the rigid-plastic problem for the pipe is also axially symmetric.

In this connection, it is convenient to use the cylindrical coordinate system $\rho$, $\theta$, $z$ with its $z$-axis directed along the pipe axis. We also use the local Cartesian basis $\mathbf{e}_\rho$, $\mathbf{e}_\theta$, $\mathbf{e}_z$ associated with the cylindrical coordinate system, see Figure 6.1. The local basis is defined at every point, and components of vector and tensor fields are considered with respect to this basis. We use letters rather than numbers to indicate these components. For example, $v_\rho$, $v_\theta$, $v_z$ are velocity components, $e_\rho$, $e_\theta$, $e_z$ are the diagonal components of strain rate, and $e_{\rho\theta} = e_{\theta\rho}$, $e_{\rho z} = e_{z\rho}$, $e_{\theta z} = e_{z\theta}$ are its off-diagonal components.

In what follows, we also assume that the body forces are zero while the given surface load and kinematic constraints are axially symmetrical and do not depend on $z$. Then we can restrict ourselves to the velocity fields whose components $v_\rho$, $v_\theta$ only depend on $\rho$, $v_z$ being zero. The corresponding strain rate components are as follows:

$$e_\rho = v'_\rho, \quad e_{\rho\theta} = \frac{1}{2}\left(v'_\theta - \frac{1}{\rho}v_\theta\right), \quad e_z = e_{\theta z} = e_{\rho z} = 0, \quad e_\theta = \frac{1}{\rho}v_\rho \qquad (6.1)$$

(prime indicates the derivative with respect to $\rho$).

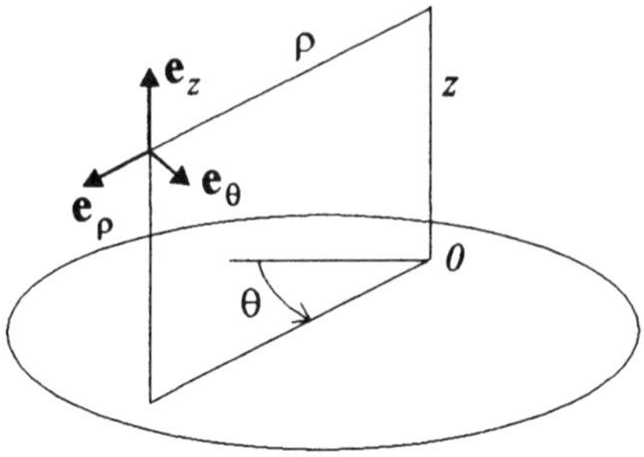


**Figure 6.1**

We adopt the Mises yield criterion, $|\sigma^d| = \sqrt{2}k$, $k = \text{const} > 0$, to describe material properties of the pipe. Then the normality flow rule is of the form $e_{ij} = \dot{\lambda}\sigma^d_{ij}$ (see Example I.3.1) or, in more detail,

$$e_\rho = \frac{1}{3}\dot{\lambda}\left(2\sigma_\rho - \sigma_\theta - \sigma_z\right), \quad e_{\rho\theta} = \dot{\lambda}\sigma_{\rho\theta}, \quad e_\theta = \frac{1}{3}\dot{\lambda}\left(2\sigma_\theta - \sigma_\rho - \sigma_z\right),$$
$$e_{\rho z} = \dot{\lambda}\sigma_{\rho z}, \quad e_{\theta z} = \dot{\lambda}\sigma_{\theta z}, \quad e_z = \dot{\lambda}\frac{1}{3}\left(2\sigma_z - \sigma_\rho - \sigma_\theta\right).$$

Taking into account (6.1), we write these relations as:

$$\sigma_{\rho z} = \sigma_\theta z = 0, \quad \sigma_z = \frac{1}{2}\left(\sigma_\rho + \sigma_\theta\right),$$
$$e_\rho = \frac{1}{2}\dot{\lambda}\left(\sigma_\rho - \sigma_\theta\right) \quad e_{\rho\theta} = \dot{\lambda}\sigma_{\rho\theta}, \quad e_\theta = -\frac{1}{2}\dot{\lambda}\left(\sigma_\rho - \sigma_\theta\right), \tag{6.2}$$

and the yield criterion as

$$\left(\sigma_\rho - \sigma_\theta\right)^2 + 4\sigma_{\rho\theta}^2 = 4k^2.$$

The corresponding dissipation is

$$d(\mathrm{e}) = \sqrt{2}k\,|\mathrm{e}| = k\sqrt{\left(e_\rho - e_\theta\right)^2 + 4e_{\rho\theta}^2} \quad \text{if} \quad e_\rho + e_\theta = 0$$

and $d(\mathrm{e}) = +\infty$ if $e_\rho + e_\theta \neq 0$ (see Example I.4.1).

Finally, note that due to (6.2) there are only three independent components of the stress tensor: $\sigma_\rho$, $\sigma_\theta$, $\sigma_{\rho\theta} = \sigma_{\theta\rho}$. They do not depend on $\theta$ and $z$; therefore, the equilibrium conditions are reduced to

$$\sigma_\rho' + \frac{1}{\rho}\left(\sigma_\rho - \sigma_\theta\right) = 0, \qquad \sigma_{\rho\theta}' + \frac{2}{\rho}\sigma_{\rho\theta} = 0.$$

It is easy to find their general solution:

$$\sigma_\rho = \frac{1}{\rho}\Phi(\rho), \qquad \sigma_{\rho\theta} = \frac{A}{\rho^2}, \qquad \sigma_\theta = \Phi'(\rho), \tag{6.3}$$

where $\Phi$ is an arbitrary function and $A$ is an arbitrary constant.

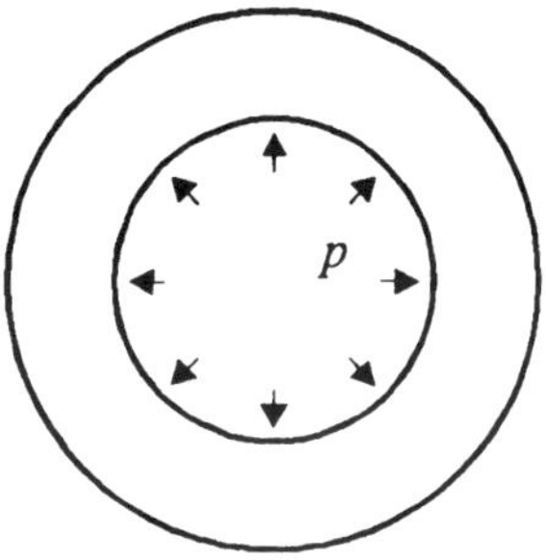


**Figure 6.2**

**6.2. A pipe under internal pressure.** Consider a rigid perfectly plastic pipe subjected to the pressure $p > 0$ uniformly distributed on its inner surface. The outer boundary of the pipe is free, see Figure 6.2. In other words, the stress should satisfy the following boundary conditions:

$$\sigma_\rho\big|_{\rho=r_0} = -p, \quad \sigma_{\rho\theta}\big|_{\rho=r_0} = 0, \quad \sigma_{\rho 2}\big|_{\rho=r_0} = 0,$$
$$\sigma_\rho\big|_{\rho=r_1} = 0, \quad \sigma_{\rho\theta}\big|_{\rho=r_1} = 0, \sigma_{\rho z}\big|_{\rho=r_1} = 0$$

where $r_0$ and $r_1$ are the inner and the outer radii of the pipe, $p = \text{const} > 0$. We adopt the Mises yield criterion for the material of the pipe and evaluate the safety factor of the given load $p$.

We apply the rigid-plastic solutions method to computing the safety factor. That is, we have to find a number $m > 0$ such that the rigid-plastic problem for the pipe under the internal pressure $mp$ has a solution with a nonzero velocity field; according to Theorem 5.2 the safety factor equals this number.

Let us look for a smooth solution of the rigid-plastic problem assuming that the stress satisfies the yield criterion at every point of the body. The problem under consideration is an axially symmetric plane strain problem, and we use the rigid-plastic problem formulation given in the previous subsection:

$$\sigma_\rho = \frac{1}{\rho}\Phi(\rho), \qquad \sigma_{\rho\theta} = \frac{A}{\rho^2}, \qquad \sigma_\theta = \Phi'(\rho),$$
$$\sigma_\rho\big|_{\rho=r_0} = -mp, \quad \sigma_{\rho\theta}\big|_{\rho=r_0} = 0, \quad \sigma_\rho\big|_{\rho=r_1} = 0, \quad \sigma_{\rho\theta}\big|_{\rho=r_1} = 0,$$
$$(\sigma_\rho - \sigma_\theta)^2 + 4\sigma_{\rho\theta}^2 = 4k^2,$$
$$v_\rho' = \dot{\lambda}\left(\sigma_\rho - \sigma_\theta\right), \quad \frac{1}{2}\left(v_\theta' - \frac{1}{\rho}v_\theta\right) = \dot{\lambda}\sigma_{\rho\theta}, \quad \frac{1}{\rho}v_\rho = -\frac{1}{2}\dot{\lambda}\left(\sigma_\rho - \sigma_\theta\right), \quad \dot{\lambda} \geq 0.$$

Note that the boundary conditions for $\sigma_{\rho\theta}$ determine the constant $A$: $A = 0$. Then the yield criterion is reduced to the equality

$$\left|\frac{1}{\rho}\Phi - \Phi'\right| = 2k.$$

Making use of the boundary conditions for $\sigma_\rho$, we obtain from the latter equation that

$$\Phi(\rho) = 2k\rho \ln \frac{\rho}{r_1}, \qquad m = \frac{2k}{p} \ln \frac{r_1}{r_0}.$$

This determines all components of the stress field:

$$\sigma_\rho = 2k \ln \frac{\rho}{r_1}, \quad \sigma_{\rho\theta} = 0, \quad \sigma_\theta = 2k \left( \ln \frac{\rho}{r_1} + 1 \right), \qquad \sigma_{\rho z} = \sigma_{\theta z} = \sigma_{zz} = 0. \tag{6.4}$$

The normality rule for such stresses is reduced to the following relations:

$$v'_\rho = -2k\dot{\lambda}, \quad v'_\theta - \frac{1}{\rho} v_\theta = 0, \quad \frac{1}{\rho} v_\rho = -k\dot{\lambda}, \quad \dot{\lambda} \geq 0.$$

They are evidently satisfied if we set

$$v_\rho = \frac{B}{\rho^2}, \quad \dot{\lambda} = \frac{B}{k\rho^2}, \quad B = \text{const} > 0, \quad v_\theta = 0. \tag{6.5}$$

Thus, we have found solution (6.4), (6.5) to the rigid-plastic problem for the pipe under internal pressure $mp$ with

$$m = \frac{2k}{p} \ln \frac{r_1}{r_0}.$$

Note that velocity field (6.5) is nonzero and that $m$ is the safety factor of the load $p$ by Theorem 5.2.

**6.3. A pipe under internal torsion.** Consider now a rigid perfectly plastic pipe subjected to a torsional load uniformly distributed on its inner surface with the density $q$. The outer boundary of the pipe is fixed, see Figure 6.3. In other words, the stress and velocity fields should satisfy the following boundary conditions:

$$\sigma_\rho\big|_{\rho=r_0} = 0, \quad \sigma_{\rho\theta}\big|_{\rho=r_0} = -q, \quad \sigma_{\rho 2}\big|_{\rho=r_0} = 0,$$
$$v_\rho\big|_{\rho=r_1} = 0, \quad v_\theta\big|_{\rho=r_1} = 0, \; v_z\big|_{\rho=r_1} = 0,$$

where $r_0$ and $r_1$ are the inner and outer radii of the pipe and $q = \text{const}$. We adopt the Mises yield criterion for the material of the pipe and evaluate the safety factor of the load $q$. The problem under consideration is an axially symmetric plane strain problem, and we use the rigid-plastic problem formulation given in Subsection 6.1.

First, apply the static method to estimating the safety factor $\alpha_q$. We restrict ourselves to sufficiently regular stress fields with components of the form (6.3) and with $\sigma_{\rho z} = \sigma_{\theta z} = \sigma_{zz} = 0$. We are looking for multipliers $m > 0$ such that the load $mq$ is equilibrated by an admissible stress field. Each of such numbers is a static multiplier of the load $q$ and gives a lower

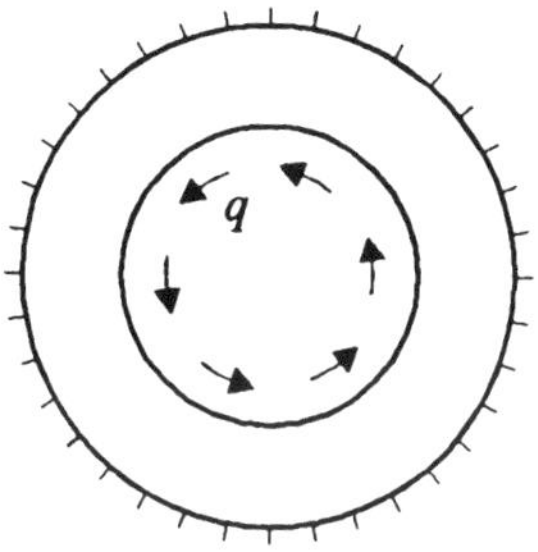


**Figure 6.3**

bound for the safety factor: $m \le \alpha_q$. Due to (6.3), the above introduced stress field also has to satisfy the boundary conditions and the admissibility condition, which are reduced to the following relations:

$$\Phi(r_0) = 0, \ \frac{A}{r_0^2} = -mq; \ \left(\frac{1}{\rho}\Phi(\rho) - \Phi'(\rho)\right)^2 + \frac{4A^2}{\rho^4} \le 4k^2 \text{ for } r_0 \le \rho \le r_1.$$

A function $\Phi = 0$ and a constant $A = -mqr_0^2$ satisfy these relations provided $m \le k/q$ (the corresponding stress field has components $\sigma_{\rho\theta} = \sigma_{\theta\rho} = -mqr_0^2/\rho^2$, the rest of the components being zero). This results in a lower bound for the safety factor: $k/q \le \alpha_q$.

Let us now apply the kinematic method to estimating $\alpha_q$ from the above. Only sufficiently regular velocity fields $\mathbf{v}$ are used, so formula (5.4) for the kinematic multiplier is of the form

$$m_k(\mathbf{v}) = \frac{\int_{r_0}^{r_1} d\,(\mathrm{Def}\,\mathbf{v})\, 2\pi\rho\, d\rho}{\int_0^{2\pi} q v_\theta r_0\, d\theta}.$$

Let us now compute the kinematic multiplier of the velocity field with the components $v_\rho = 0$, $v_z = 0$, $v_\theta$ being a nonincreasing function of the coordinate $\rho$. In this case formulas (6.1) yield that the only nonvanishing components of the strain rate are

$$e_{\rho\theta} = e_{\theta\rho} = \frac{\rho}{2}\left(\frac{v_\theta}{\rho}\right)' \le 0.$$

Using the expression for the dissipation (Subsection 6.1), we obtain

$$\int_{r_0}^{r_1} d\,(\mathrm{Def}\,\mathbf{v})\, 2\pi\rho\, d\rho = 2\pi k \int_{r_0}^{r_1} \left|\left(\frac{v_\theta}{\rho}\right)'\right| \rho^2\, d\rho$$

$$= 2\pi k \left( r_0 v_\theta(r_0) + 2\int_{r_0}^{r_1} v_\theta(\rho)\, d\rho \right),$$

where the boundary condition $v_\theta(r_1) = 0$ is taken into account. Then

$$m_k(\mathbf{v}) = \frac{k}{q}\left(1 + \frac{2}{r_0 v_\theta(r_0)} \int\limits_{r_0}^{r_1} v_\theta(\rho)\, d\rho\right).$$

We fix a certain value $v_\theta(r_0)$ and consider a function $v_\theta$ which does not vanish only near the inner surface of the pipe, that is, at $r_0 \le \rho < r_0 + \delta$. Then the second term in the previous formula for $m_k(\mathbf{v})$ can be arbitrarily small if $\delta$ is small enough. Thus, the kinematic multiplier gives the following upper bound for the safety factor $\alpha_q$:

$$\alpha_q \le m_k(\mathbf{v}) \le \frac{k}{q}(1+\varepsilon) \quad \text{for every} \quad \varepsilon > 0. \tag{6.6}$$

We have already found the lower bound $k/q \le \alpha_q$, which provides the two-sided estimation:

$$\frac{k}{q} \le \alpha_q \le \frac{k}{q(1+\varepsilon)},$$

that is, results in the precise value $\alpha_q = k/q$.

Note that we did not find the velocity field $\mathbf{v}_0$, at which the kinematic multiplier attains its minimum: $m_k(\mathbf{v}_0) = \inf m_k(\mathbf{v}) = \alpha_q$. This is not because we were unlucky with the velocity fields which we used. In general, even if the inputs of the problem are smooth,

> the infimum of kinematic multipliers can be unattainable at a sufficiently regular velocity field;
>
> the rigid-plastic problem for a limit load may have no solution with a sufficiently regular nonzero velocity field (failure mechanism).

An example of such a situation can be easily found in the problem under consideration. Indeed, consider the load $q = k$ in the internal torsion problem for the pipe. This is the limit load since $\alpha_q = q/k = 1$ if $q = k$. The stress field we used applying the static method has the following components:

$$\sigma_\rho = 0, \qquad \sigma_{\rho\theta} = -k\frac{r_0^2}{\rho^2}, \qquad \sigma_\theta = \sigma_{\rho z} = \sigma_{\theta z} = \sigma_{zz} = 0 \tag{6.7}$$

when $q = k$. It equilibrates the limit load, and its static multiplier is $m_s(\boldsymbol{\sigma}) = 1$. Due to inequality (6.6), the infimum of kinematic multipliers also equals 1. Let us now verify that the infimum is unattainable if only sufficiently regular velocity fields are considered. Suppose this statement is wrong and $\mathbf{m}_k(\mathbf{v}_0) = 1$ for a sufficiently regular velocity field $\mathbf{v}_0$. Then this velocity field and stress field (6.7) form a solution to the rigid-plastic problem for the limit load $q = k$ (Theorem 5.2). In particular, they satisfy the constitutive maximum principle, that is,

$$\operatorname{Def} \mathbf{v}_0(\rho) \cdot (\boldsymbol{\sigma}(\rho) - \boldsymbol{\sigma}_*) \ge 0 \ \text{ for every } \ \boldsymbol{\sigma}_* \ \text{ with } \ \left|\boldsymbol{\sigma}_*^d\right| \le \sqrt{2}k$$

for almost every $\rho \in (r_0, r_1)$. The inequality $|\sigma^d| < \sqrt{2}k$ is valid for $\rho$ within the interval $r_0 \le \rho \le r_1$, that is, the stress $\boldsymbol{\sigma}(\rho)$ is strictly inside the yield surface. Then the constitutive maximum principle implies $\mathrm{Def}\, \mathbf{v}_0(\rho) = 0$ for a.e. $\rho \in (r_0, r_1)$. Thus, $\mathbf{v}_0$ has the zero strain rate field; in other words, $\mathbf{v}_0$ is a rigid motion velocity field. Together with the boundary condition $\mathbf{v}|_{\rho=r_1} = 0$ this yields that the velocity field is zero, which contradicts the assumption that $m_k(\mathbf{v}_0) = 1$. Therefore, the infimum of the kinematic multiplier is unattainable in the set of sufficiently regular velocity fields. In particular, this means that the rigid-plastic problem has no solution with the nonzero sufficiently regular velocity field (failure mechanism): the kinematic multiplier would attain its infimum at such a field.

## Comments

We find anticipation of the limit analysis idea back to the XVIII century. The early history of limit analysis is described by Prager (1974). Kazinczy (1914) and Kist (1917) came close enough to the modern understanding of the problem and methods to solve it, even though only for particular mechanical systems. However, all those achievements remained practically unnoticed. A systematic approach was developed by Gvozdev (1938) and independently established by Drucker, Prager and Greenberg (1951). A monograph by Gvozdev (1948) and an article by Feinberg (1948) also made valuable contributions to limit analysis.

Starting from the above-mentioned works, limit analysis was widely recognized as a powerful method for designing and optimizing plastic bodies and structures; see books and reviews by Rzhanitsin (1954, 1983), Ovechkin (1961), Neal (1963), Reitman and Shapiro (1964), Chiras (1971), Bolotin, Goldenblat and Smirnov (1972), Prager (1974), Salençon (1974), Hodge (1981), Protsenko (1982), Save (1985), and Gudramovich, Gerasimov and Demenkov (1990). Another important field of the rigid plastic analysis application is metal processing; books by Hill (1950), Tarnovsky and Pozdeev (1963), Thomsen, Yang and Kobayashi (1965), Avitzur (1980), and many other works deal with this problem.

Limit analysis is normally implemented by numerical methods, Charnes and Greenberg (1951), Koopman and Lance (1965), Chiras (1969), Hodge and Belytschko (1970), van Rij and Hodge (1978), van Rij (1979), Kukudzhanov, Lyubimov and Myshev (1984), and Frietas (1985), see also the above-mentioned books). Engineers and researchers most commonly make use of the kinematic method, although Rzhanitsin (1954), Koopman and Lance (1965), and Kukudzhanov, Lyubimov and Myshev (1984) also used the static method.

A number of solutions to rigid-plastic problems are described in books by Nadai (1949), Hill (1950), Freidental and Geiringer (1958), Ivlev (1966), Prager and Hodge (1968), Sokolovsky (1969), Kachanov (1971), Martin (1975), and Klyushnikov (1979). Some new analytical solutions appear from time to time; see, for example, Annin, Bytev and Senashov (1985), and Zadoian (1992). Several problems are solved by the asymptotic method; see Ivlev and Ershov (1978).

Limit analysis theory is normally discussed in terms of the rigid perfectly plastic model: yield surface, dissipation, rigid zone, etc. (and we do not break with the tradition in this book). However, the domain of limit analysis applications is not restricted to materials accurately obeying the rigid perfectly plastic constitutive relations. First of all, engineering practice shows that limit analysis is applicable to estimating bearing capacity of structures made of materials whose constitutive relations substantially differ from the perfectly plastic ones. An appropriate choice of an axillary rigid-plastic body, which replaces the real one, is what makes limit analysis effective in this case. The limit stress state principle forms the basis of this approach. Secondly, the safety factor determined by limit analysis plays an important role within the framework of some plasticity models, which however are not rigid perfectly plastic. Consider, for example, a rigid *visco*-plastic body $\mathcal{B}_{vp}$. There is a rigid perfectly plastic body $\mathcal{B}$ which is naturally associated with $\mathcal{B}_{vp}$: $\mathcal{B}$ occupies the same domain as $\mathcal{B}_{vp}$ and has at every point the same set of safe stresses as $\mathcal{B}_{vp}$. If the visco-plastic body is subjected to the load $\mathbf{l}$, the safety factor $\alpha_l$ of the load is defined in the sense of the limit analysis of the associated body $\mathcal{B}$. It turns out that $\alpha_l$ characterizes the response of the visco-plastic body to $\mathbf{l}$. Namely, if the velocity field is zero at the initial moment and boundary values for the velocity field also are zero, then $\mathcal{B}_{vp}$ deforms under a quasistatic loading if and only if $\alpha_l < 1$; see Mosolov and Myasnikov (1971). Similarly, consider an elastic perfectly plastic body $\mathcal{B}_{ep}$ subjected to a quasistatic loading and the rigid perfectly plastic body $\mathcal{B}$ naturally associated with $\mathcal{B}_{ep}$ (occupying the same domain and having the same sets of admissible stresses). The limit analysis of $\mathcal{B}$ determines the safety factor $\alpha_l$ of the load $\mathbf{l}$ applied to $\mathcal{B}_{ep}$. Johnson (1976) showed that the inequality $\alpha_l > 1+\delta$ satisfied with constant $\delta > 0$ for every load $\mathbf{l}$ during loading is the condition for solvability of the elastic-plastic problem. If $\mathbf{l}$ is a limit load for the rigid-plastic body $\mathcal{B}$ (that is, $\alpha_l = 1$), the elastic-plastic body $\mathcal{B}_{ep}$ behaves like $\mathcal{B}$ under the action of $\mathbf{l}$: $\mathcal{B}_{ep}$ deforms at an arbitrarily large strain rate.

Thus, limit analysis makes sense far beyond the framework of the rigid perfectly plastic model it originated from. What is more, limit analysis allows us to glance at the very model from a new viewpoint. Consider the limit analysis problem as arising from the limit stress state principle rather than from the rigid perfectly plastic model. The problem consists of evaluating the safety factor of a given load and its formulation involves a certain given set of safe stress fields, the equilibrium conditions and no constitutive relations. This is an extremum problem: the safety factor is the supremum of the static multipliers. The concept of the static multiplier can lead to the associated concept of the kinematic multiplier without even mentioning the rigid-plastic model. The outstanding intuition, which resulted in invention of the kinematic multiplier, can now be replaced by familiarity with a state-of-the-art mathematical procedure (described in Chapter V). Applying this procedure to the static multiplier immediately results in what we call a kinematic multiplier, which appear to be an upper bound for the safety factor. Since static multipliers are lower bounds for the safety factor, a question arises about conditions for the equality of a static $m_s(\boldsymbol{\sigma})$ and kinematic $m_k(\mathbf{v})$ multipliers. While answering this question, we establish, among necessary and sufficient conditions for the equality $m_s(\boldsymbol{\sigma}) = m_k(\mathbf{v})$, the following one: the stress field $\boldsymbol{\sigma}$ and the velocity field $\mathbf{v}$ satisfy what we call the normality flow rule. Note that the way we have just obtained the normality rule has nothing to do with postulating a constitutive relation. Actually, we have answered the question by Koiter (1960) of whether the adoption of the normality flow rule is necessary for proving the fundamental statements of limit analysis. Yes, it is. If the perfectly plastic model with the normality rule had not existed, it should have been invented.

CHAPTER IV

# LIMIT ANALYSIS: GENERAL THEORY

We have already encountered rigid-plastic problems for a three-dimensional body, a one-dimensional continual system (a beam) and a discrete system (a truss). Although at first glance the three formulations appear rather different, the analogy between them is so close that it enables us to establish a unified formulation of the problems. The general formulation covers not only the above-mentioned cases but also all cases of rigid perfectly plastic systems. The general, unified formulation of the rigid-plastic problem will be given in Subsection 1.4.

Within the framework of the unified formulation, we introduce concepts and establish basic statements of limit analysis previously considered in the case of a three-dimensional body (Chapter III). These statements appear to hold for rigid perfectly plastic systems of all types. Ideas of the proofs remain the same and the proofs become more distinct after getting rid of some particular three-dimensional details.

We keep using the three-dimensional terminology in the unified theory to avoid introducing new terms. We take the term body to mean not only three-dimensional bodies but also beams, shells and other systems, and the term stress to mean all kinds of internal forces such as cutting force or moments. We speak about fields although we consider not only continuous but also discrete systems, whose state is described by functions of a discrete variable.

## 1. Rigid-plastic problem: general formulation

A unified formulation of the rigid-plastic problem is presented in this section. It covers rigid perfectly plastic systems of all types: three-dimensional bodies, plates, rods, etc. To obtain a unified formulation of the equilibrium conditions, kinematic and constitutive relations, we start with discussing their particular forms. Finding out basic structures of these relations, we arrive at their general formulations. After establishing the unified formulation, we consider formulations of the problem for three-dimensional bodies and trusses as examples of a particular realization of the general formulation. The reader will find two more examples in the next section.

The major reason for finding out whether a certain formulation of the rigid-plastic problem is a particular case of the general formulation is that this would allow applying all statements of the limit analysis unified theory to the concrete type of the problem.

**1.1. Equilibrium conditions.** The virtual work principle is the unified formulation of the equilibrium conditions. According to the principle, *a distribution of internal forces equilibrates a given load if and only if the total external and internal power vanishes at every virtual velocity field.*

We start with considering realizations of the virtual work principle for three-dimensional bodies under standard loading and trusses (Subsections II.1.4 and II.3.4).

The load $\mathbf{l}$ acting on a three-dimensional body is represented by the density $\mathbf{f}$ of body forces and the density $\mathbf{q}$ of the surface tractions. The densities $\mathbf{f}$ and $\mathbf{q}$ are given in the domain $\Omega$ occupied by the body and on the part $S_q$ of its boundary, respectively. The load $\mathbf{l}$ acting on a truss is represented by the forces $\mathbf{f}_1, \ldots, \mathbf{f}_k$ applied to the unfixed hinges of the truss. In both cases we assume that the sum $\mathbf{l}_1 + \mathbf{l}_2$ of any two possible loads is also a possible load. For example, if $\mathbf{l}_1 = (\mathbf{f}_1, \mathbf{q}_1)$ and $\mathbf{l}_2 = (\mathbf{f}_2, \mathbf{q}_2)$ are loads applicable to a three-dimensional body, then $\mathbf{l}_1 + \mathbf{l}_2 = (\mathbf{f}_1 + \mathbf{f}_2, \mathbf{q}_1 + q_2)$ can also be considered as a load. It should be remembered that the densities $\mathbf{q}_1$ and $\mathbf{q}_2$ of the surface tractions are given on the same part $S_q$ of the body boundary, and this is what allows us to add the loads. Analogously, we assume that the product $\alpha\mathbf{l}$ of any possible load $\mathbf{l}$ and any number $\alpha$ is also a possible load. Thus, *loads form a linear space* (see the definition and examples of linear spaces in Subsection A.1).

A virtual velocity field is the difference of two kinematically admissible velocity fields. Thus, the set of virtual velocity fields for a three-dimensional body under standard boundary conditions consists of smooth vector fields which vanish on the fixed part $S_v$ of the body boundary. The set of virtual velocities for a truss consists of collections $\mathbf{v} = (\mathbf{v}_1, \ldots, \mathbf{v}_k)$, where $\mathbf{v}_\alpha$ is the velocity of the $\alpha$-th (unfixed) hinge, $\alpha = 1, \ldots, k$; $k$ is the number of the unfixed hinges. In both cases *virtual velocity fields form a linear space* $U$.

The external power, that is, the power of the load $\mathbf{l}$ on the velocity field $\mathbf{v}$, is calculated for a body and for a truss by the formulas

$$\int_{\Omega} \mathbf{f}\mathbf{v}\, dV + \int_{S_q} \mathbf{q}\mathbf{v}\, dS \quad \text{and} \quad \mathbf{f}_1\mathbf{v}_1 + \cdots + \mathbf{f}_k\mathbf{v}_k,$$

respectively. In both cases the power linearly depends on $\mathbf{l}$ and on $\mathbf{v}$. In other words, the *external power of a load on a velocity field is the value of a certain bilinear form* (see Subsection A.4).

The strain rate field $\operatorname{Def}\mathbf{v}$ is defined for any virtual velocity field $\mathbf{v}$. In the case of a three-dimensional body, $\operatorname{Def}\mathbf{v}$ is a symmetric second rank tensor with the components

$$(\operatorname{Def}\mathbf{v})_{ij} = \frac{1}{2}\left(\frac{\partial v_i}{\partial x_j} + \frac{\partial v_j}{\partial x_i}\right). \tag{1.1}$$

In the case of a truss, the strain rate is the collection $\operatorname{Def}\mathbf{v} = (e_1, \ldots, e_m)$, where $e_i$ is the elongation rate of the $i$-th rod:

$$e_i = -\boldsymbol{\tau}_{i\alpha}\mathbf{v}_\alpha. \tag{1.2}$$

Here, $m$ is the number of the rods in the truss; summation index $\alpha$ runs through the values $1, \dots, k$. (Recall the definition of $\boldsymbol{\tau}_{i\alpha}$: if the $\alpha$-th hinge is one of the ends of the $i$-th rod, then $\boldsymbol{\tau}_{i\alpha}$ is the unit vector directed along the $i$-th rod from the $\alpha$-th hinge to the other end of the rod; if the $\alpha$-th hinge is not an end of the $i$-th rod, then $\boldsymbol{\tau}_{i\alpha} = 0$.) The *strain rate field is a value of a linear operator* (see Subsection A.3) in the case of a truss as well as in the case of a three-dimensional body.

We consider stress fields forming a certain linear space $\mathcal{S}$. For example, $\mathcal{S}$ is a finite dimensional space of collections $\boldsymbol{\sigma} = (N_1, \dots, N_m)$ in the case of a frame; here, $N_i$ is the axial force in the $i$-th rod, $i = 1, \dots, m$.

The internal power, that is, the power of the stress field $\boldsymbol{\sigma}$, is defined for every virtual velocity field $\mathbf{v}$. The internal power for a body and a truss is given by the formulas

$$-\int_{\Omega} \mathrm{Def}\,\mathbf{v} \cdot \boldsymbol{\sigma}\, dV \qquad \text{and} \qquad -N_1 e_1 - \cdots - N_m e_m,$$

respectively. In both cases the internal power linearly depends on $\boldsymbol{\sigma}$ and $\mathrm{Def}\,\mathbf{v}$, that is, the *internal power of a stress field* $\boldsymbol{\sigma}$ *on a strain rate field* $\mathrm{Def}\,\mathbf{v}$ *is the value of a certain bilinear form.*

According to the virtual work principle, a stress field in a body equilibrates the given load $\mathbf{l} = (\mathbf{f}, \mathbf{q})$ if and only if

$$\int_{\Omega} \mathrm{Def}\,\mathbf{v} \cdot \boldsymbol{\sigma}\, dV = \int_{\Omega} \mathbf{f}\mathbf{v}\, dV + \int_{S_q} \mathbf{q}\mathbf{v}\, dS,$$

for every virtual velocity field $\mathbf{v}$.

Analogously a stress field in a truss equilibrates the given load $\mathbf{l} = (\mathbf{f}_1, \dots \mathbf{f}_k)$ if and only if

$$N_1 e_1 + \cdots + N_m e_m = \mathbf{f}_1 \mathbf{v}_1 + \cdots + \mathbf{f}_k \mathbf{v}_k$$

for every virtual velocity field $\mathbf{v}$.

Based on the above-mentioned examples we now formulate the following idea of the unified equilibrium conditions. (Note that we keep using the three-dimensional terminology in the unified theory.) A body and loads are characterized by:

- the linear space $\mathcal{L}$ of possible loads;
- the linear space $U$ of virtual velocity field;
- the external power bilinear form $\langle\langle \cdot, \cdot \rangle\rangle$ defined on $U \times \mathcal{L}$; its value $\langle\langle \mathbf{v}, \mathbf{l} \rangle\rangle$ is the power of the load $\mathbf{l}$ on the virtual velocity field $\mathbf{v}$;
- the linear space $\mathcal{S}$ of stress fields;
- the linear space $\mathcal{E}$ of strain rate fields;
- the internal power bilinear form $\langle \cdot, \cdot \rangle$ defined on $\mathcal{E} \times \mathcal{S}$; the value $-\langle \mathbf{e}, \mathbf{s} \rangle$ is the power of the stress field $\mathbf{s}$ at the strain rate field $\mathbf{e}$;

- the linear operator Def, defined on $U$ and taking values in $\mathcal{E}$; its value $\operatorname{Def}\mathbf{v}$ is the strain rate field corresponding to the velocity field $\mathbf{v}$;
- the virtual work principle: a stress field $\boldsymbol{\sigma}$ equilibrates the load $\mathbf{l}$ if and only if

$$\langle \operatorname{Def}\mathbf{v}, \boldsymbol{\sigma} \rangle = \langle\langle \mathbf{v}, \mathbf{l} \rangle\rangle \quad \text{for every} \quad \mathbf{v} \in U.$$

The short form of these equilibrium conditions can be written using the set $\Sigma$ of self-equilibrated stress fields, that is, those equilibrating the zero load. Let $\mathbf{s}_l$ be a fixed stress field in $\mathcal{S}$ equilibrating the load $\mathbf{l}$. It is clear that $\boldsymbol{\sigma}$ equilibrates the load $\mathbf{l}$ if and only if the field $\boldsymbol{\sigma} - \mathbf{s}_l$ is self-equilibrated. Thus, we can write the equilibrium conditions in the form: $\boldsymbol{\sigma} \in \Sigma + \mathbf{s}_l$. Here, the set $\Sigma$ of self-equilibrated stress fields is defined through the virtual work principle:

$$\Sigma = \{\boldsymbol{\sigma} \in \mathcal{S} : \langle \operatorname{Def}\mathbf{v}, \boldsymbol{\sigma} \rangle = 0 \ \text{for every} \ \mathbf{v} \in U\}.$$

This definition will normally be written in a short form. First of all, let us denote the set of all values which the operator Def takes on velocity fields in $U$ by the symbol $\operatorname{Def} U$. Secondly, if $\Pi$ is a subspace in $\mathcal{E}$, we denote by $\Pi^0$ the set

$$\{\mathbf{s} \in \mathcal{S} : \langle \mathbf{e}, \mathbf{s} \rangle = 0 \ \text{for every} \ \mathbf{e} \in \Pi\};$$

this set is called the *polar set* of $\Pi$ (see Subsection A.4). Thus, the definition of $\Sigma$ can be written in the form $\Sigma = (\operatorname{Def} U)^0$.

**1.2. Kinematic relations.** Let us start with a standard example of kinematic constraints considering a three-dimensional body with a fixed part $S_v$ of its boundary. In this case, the kinematic boundary conditions are of the form $\mathbf{v}|_{S_v} = 0$ and we assume that kinematically admissible velocity fields form a linear space $\mathcal{V}$ containing the subspace $U$ of virtual velocity fields. Within the framework of the strong formulation of the rigid-plastic problem, one can set $\mathcal{V} = V_0$, where $V_0$ is the set of sufficiently regular velocity fields introduced in Subsection III.1.3. The operator Def, which maps a velocity field on its strain rate, is defined not only on $U$ but also on $V_0$. An important fact is that *the only velocity field with the zero strain rate field is zero* (provided $S_v$ is of positive area).

Many other types of kinematic constraints for three-dimensional bodies and other mechanical systems possess the latter property. Consider, for example, the boundary conditions of the form $v_\nu|_S = 0$, where $S$ is a part of the body boundary and $v_\nu$ is the normal component of velocity $\mathbf{v}$. This constraint admits a slip over the surface $S$. However, the slip is only possible if the body deforms (except in cases of some particular forms of the surface $S$). In other words, under the condition $v_\nu|_S = 0$ velocity field $\mathbf{v}$ is zero if the strain rate field $\operatorname{Def}\mathbf{v}$ is zero. Analogously, consider a frame with a sufficient number of fixed hinges. Under this kinematic constraint, velocities of all hinges are zero if the elongation rate of each rod vanishes.

An important fact about extension of the set $U$ is worth noting here. For example, in the case of the rigid-plastic problem strong formulation (Section

III.1), extending the set $U$ of virtual velocity fields to the space $V_0$ of sufficiently regular kinematically admissible velocity fields does not change the equilibrium conditions. In other words, if the stress field $\boldsymbol{\sigma}$ satisfies the equality of the virtual work principle for every $\mathbf{v} \in U$, then $\boldsymbol{\sigma}$ also satisfies this equality for every $\mathbf{v} \in V_0$.

The above-mentioned examples allow formulating the following idea of the unified kinematic description. From the kinematic viewpoint, the body is characterized by:

- the linear space of kinematically admissible velocity fields containing the subspace $U$ of virtual velocity fields;
- the linear operator Def which maps the space $\mathcal{V}$ into the space $\mathcal{E}$; the value Def $\mathbf{v}$ is the strain rate field corresponding to $\mathbf{v}$; the only velocity field with the zero strain rate field is zero.

We note that the static and kinematic concepts are connected: the space $U$ of virtual velocity fields and the operator Def enter the equilibrium conditions. We adopt the following assumption about retaining this connection in case the wider space $\mathcal{V}$ replaces $U$:

if the stress field $\boldsymbol{\sigma}$ satisfies the virtual work principle, that is,

$$\langle \operatorname{Def} \mathbf{v}, \boldsymbol{\sigma} \rangle = \langle\langle \mathbf{v}, \mathbf{l} \rangle\rangle \quad \text{for every } \mathbf{v} \text{ in } U,$$

then the latter equality also holds for every $\mathbf{v}$ in $\mathcal{V}$.

Within the framework of this general approach, the kinematic relations can be written in the form

$$\mathbf{v} \in \mathcal{V}, \quad \mathbf{e} = \operatorname{Def} \mathbf{v}.$$

The first relation means that the velocity field $\mathbf{v}$ is kinematically admissible; the second one determines the strain rate field which corresponds to $\mathbf{v}$.

**1.3. Constitutive relations.** We will now transform the rigid perfectly plastic constitutive relations to a form admitting a natural generalization. Let us start with a simple example of a discrete system.

EXAMPLE 1. Consider a truss consisting of $m$ rods. We adopt the constitutive maximum principle (Subsection I.6.6) for every rod:

$$N_j \in C_j, \quad e_j(N_j - N_*) \geq 0 \text{ for every } N_* \in C_j \tag{1.3}$$

(no summation over $j$). Here, $C_j$ is the set of admissible stresses in the $j$-th rod:

$$C_j = \{N \in \mathbf{R} : -Y_j^- \leq N \leq Y_j^+\},$$

$Y_j^+$, $Y_j^-$ being yield stresses in tension and compression, respectively.

In the $m$-dimensional space $\mathcal{S}$, we introduce the set

$$C = \{\boldsymbol{\sigma} = (N_1, \ldots, N_m) \in \mathcal{S} : N_1 \in C_1, \ldots, N_m \in C_m\}.$$

Then the collection of the admissibility conditions $N_j \in C_j$, $j = 1, \ldots, m$, can be written as $(N_1, \ldots, N_m) \in C$. To obtain a consequence of inequalities

(1.3), let us set $N_* = N*\mu$ for the $\mu$-th rod ($N_{*1}, \ldots, N_{*m}$ are independent variables) and sum all inequalities (1.3); this yields

$$\begin{gathered}(N_1, \ldots, N_m) \in C, \\ e_\mu(N_\mu - N_{*\mu}) \geq 0 \text{ for every } (N_{*1}, \ldots, N_{*m}) \in C,\end{gathered} \tag{1.4}$$

with summation over $\mu$ running through the values $1, \ldots, m$.

It can be easily seen that (1.4) in turn implies (1.3). Indeed, inequality (1.4) means precisely that

$$e_\mu N_\mu \geq \sup \{e_\mu N_{*\mu} : (N_{*1}, \ldots, N_{*m}) \in C\}.$$

Here we can set $N_{*1} = N_1, \ldots, N_{*m} = N_m$ with real stresses $N_1, \ldots, N_m$. This results in the equality

$$e_\mu N_\mu = \sup \{e_\mu N_{*\mu} : (N_{*1}, \ldots, N_{*m}) \in C\},$$

and we see that it is equivalent to the previous inequality. The variables $N_{*1}, \ldots, N_{*m}$ are independent, and we find that the supremum is

$$\begin{aligned}&\sup \{e_\mu N_{*\mu} : (N_{*1}, \ldots, N_{*m}) \in C\} \\ &\qquad = \sup \left\{e_1 \widetilde{N}_1 : \widetilde{N}_1 \in C_1\right\} + \cdots + \sup \left\{e_m \widetilde{N}_m : \widetilde{N}_m \in C_m\right\}.\end{aligned} \tag{1.5}$$

Therefore, the previous equality takes the form

$$\begin{aligned}&\left(e_1 N_1 - \sup \left\{e_1 \widetilde{N}_1 : \widetilde{N}_1 \in C_1\right\}\right) + \cdots \\ &\qquad + \left(e_m N_m - \sup \left\{e_m \widetilde{N}_m : \widetilde{N}_m \in C_m\right\}\right) = 0.\end{aligned}$$

Each of the expressions in parentheses is nonnegative since $N_1, \ldots, N_m$ are admissible, so that, if the sum vanishes, each of these expressions also vanishes, that is, (1.3) holds at $j = 1, \ldots, m$. Thus, the total (summed) maximum principle (1.4) is equivalent to the collection of the constitutive maximum principles written for every rod of the truss.

Note that the left-hand side in (1.4) is expressed through values of the bilinear form

$$\langle \mathbf{e}, \boldsymbol{\sigma} \rangle = e_\mu N_\mu, \quad \mathbf{e} = (e_1, \ldots, e_m), \quad \boldsymbol{\sigma} = (N_1, \ldots, N_m).$$

This is the very form we use in the formulation of the equilibrium conditions (Subsection 1.1).

Finally, let us write the constitutive relations for the truss in the following equivalent form:

$$\boldsymbol{\sigma} \in C, \quad \langle \mathbf{e}, \boldsymbol{\sigma} - \boldsymbol{\sigma}_* \rangle \geq 0 \quad \text{for every} \quad \boldsymbol{\sigma}_* \in C.$$

Let us now consider the constitutive maximum principle for a three-dimensional body (Subsection I.3.2) occupying domain $\Omega$:

$$\boldsymbol{\sigma}(x) \in C_x, \quad \mathrm{e}(x) \cdot (\boldsymbol{\sigma}(x) - \boldsymbol{\sigma}_*) \geq 0 \quad \text{for every} \quad \boldsymbol{\sigma}_* \in C_x. \tag{1.6}$$

These relations should hold for almost every $x$ in $\Omega$. We now transform this local (point-wise) formulation to an integral one. The first of relations (1.6) can be written in the form $\boldsymbol{\sigma} \in C$, where

$$C = \{\boldsymbol{\sigma} \in \mathcal{S} : \boldsymbol{\sigma}(x) \in C_x \ \text{for a.e.}\ x \in \Omega\}.$$

By integrating inequality (1.6) over the domain $\Omega$ and setting $\boldsymbol{\sigma}_*$ at every point $x$ equal to the value $\boldsymbol{\sigma}(x)$ of a certain admissible stress field $\boldsymbol{\sigma}$, we arrive at the following consequence of the constitutive maximum principle (1.6):

$$\boldsymbol{\sigma} \in C, \quad \int_\Omega \mathrm{e}(x) \cdot (\boldsymbol{\sigma}(x) - \boldsymbol{\sigma}_*(x))\, dV \geq 0 \ \text{for every} \ \boldsymbol{\sigma}_* \in C. \tag{1.7}$$

The question is whether the integral and point-wise formulations (1.7) and (1.6) are equivalent.

A similar question has just been discussed for a discrete system in Example 1. It has been established that the summed constitutive maximum principle (1.4) is equivalent to the collection of the constitutive relations for all elements of the system. We now use similar reasoning for the three-dimensional case. Namely, we write inequality (1.7) in the form

$$\int_\Omega \mathrm{e}(x) \cdot \boldsymbol{\sigma}(x)\, dV \geq \sup \left\{ \int_\Omega \mathrm{e}(x) \cdot \boldsymbol{\sigma}_*(x)\, dV : \boldsymbol{\sigma}_* \in C_x \right\}.$$

Here we can set $\boldsymbol{\sigma}_* = \boldsymbol{\sigma}$ with the actual stress field $\boldsymbol{\sigma}$, which results in the equality

$$\int_\Omega \mathrm{e}(x) \cdot \boldsymbol{\sigma}(x)\, dV = \sup \left\{ \int_\Omega \mathrm{e}(x) \cdot \boldsymbol{\sigma}_*(x)\, dV : \boldsymbol{\sigma}_* \in C_x \right\}, \tag{1.8}$$

apparently equivalent to the previous inequality.

We now suppose that the following relation is valid

$$\sup \left\{ \int_\Omega \mathrm{e}(x) \cdot \boldsymbol{\sigma}_*(x)\, dV : \boldsymbol{\sigma}_* \in C \right\} = \int_\Omega \sup \{\mathrm{e}(x) \cdot \tilde{\boldsymbol{\sigma}} : \tilde{\boldsymbol{\sigma}} \in C_x\}\, dV. \tag{1.9}$$

Note that 1) the supremum on the left-hand side is taken over admissible stress fields while the supremum on the right-hand side over admissible stresses at each point $x$ separately, 2) on the left-hand side the first operation is integration, the second is maximization, and on the right-hand side it is vice versa. Thus, (1.9) is analogous to (1.5) and states that summation is permutable with maximization.

If (1.9) is valid, equality (1.8) takes the form

$$\int_{\Omega} (\mathrm{e}(x) \cdot \boldsymbol{\sigma}(x) - \sup \{\mathrm{e}(x) \cdot \widetilde{\boldsymbol{\sigma}} : \widetilde{\boldsymbol{\sigma}} \in C_x\}) \, dV = 0.$$

Here, the integrand is nonpositive for a.e. $x \in \Omega$ if $\boldsymbol{\sigma}$ is an admissible stress field. Consequently, the integrand is zero since the integral vanishes. Vanishing of the integrand

$$\mathrm{e}(x) \cdot \boldsymbol{\sigma}(x) - \sup \{\mathrm{e}(x) \cdot \widetilde{\boldsymbol{\sigma}} : \widetilde{\boldsymbol{\sigma}} \in C_x\} = 0 \text{ for a.e. } x \in \Omega$$

together with admissibility of $\boldsymbol{\sigma}$ are equivalent to relations (1.6). Thus, integral relation (1.7) implies point-wise relation (1.6) for a.e. $x \in \Omega$. Together with the above-mentioned converse statement, this means that the formulations (1.6) and (1.7) of the constitutive relations are equivalent.

REMARK 1. A simple reasoning like that in Example 1 is not sufficient for verifying equality (1.9). Indeed, the possibility of choosing the variables $N_{*1}, \ldots, N_{*m}$ independently was most substantial for that reasoning. Now, the values $\boldsymbol{\sigma}_*(x)$ at various points are analogous to the variables $N_{*1}, \ldots, N_{*m}$, and the values $\boldsymbol{\sigma}_*(x)$ are not absolutely independent: the field $\boldsymbol{\sigma}_*$ has to belong to the space $\mathcal{S}$, in particular, to be integrable. Apart from that, $\boldsymbol{\sigma}_*$ should be admissible: $\boldsymbol{\sigma}_* \in C$. However, these requirements are not too restrictive, and it turns out that equality (1.9) is valid if dependence $C_x$ on $x$ is sufficiently regular and the space $\mathcal{S}$ is wide enough. Although we delay a more accurate formulation of these conditions until Section 5, we would still like to point out that the conditions are satisfied for all limit analysis problems which are of practical interest.

Note that the left-hand side in (1.7) is expressed through values of the bilinear form

$$\langle \mathrm{e}, \boldsymbol{\sigma} \rangle = \int_{\Omega} \mathrm{e}(x) \cdot \boldsymbol{\sigma}(x) \, dV.$$

This is the very form we use in the formulation of equilibrium conditions (Subsection 1.1). Thus, if equality (1.9) is valid, the constitutive maximum principle can be equivalently written as

$$\boldsymbol{\sigma} \in C, \quad \langle \mathrm{e}, \boldsymbol{\sigma} - \boldsymbol{\sigma}_* \rangle \geq 0 \text{ for every } \boldsymbol{\sigma}_* \in C.$$

We see that the above-mentioned formulations of the constitutive relations are the realizations of the following general scheme:

- the material properties of a rigid perfectly plastic body are specified by the set $C$ of admissible stress fields in the space $\mathcal{S}$; the set is convex and contains 0;
- the constitutive maximum principle relates the strain rate and stress fields e and $\boldsymbol{\sigma}$ in the body:

$$\boldsymbol{\sigma} \in C, \quad \langle \mathrm{e}, \boldsymbol{\sigma} - \boldsymbol{\sigma}_* \rangle \geq 0 \text{ for every } \boldsymbol{\sigma}_* \in C.$$

**1.4. General formulation of the problem.** We now summarize the above assumptions about the rigid-plastic problem and give its general, unified formulation.

A rigid perfectly plastic body and loads are characterized by:

(I) the linear space $\mathcal{V}$ of kinematically admissible velocity fields with the subspace $U$ of virtual velocity fields;
the linear space $\mathcal{L}$ of loads;
the external power bilinear form $\langle\langle\cdot,\cdot\rangle\rangle$ defined on $\mathcal{V}\times\mathcal{L}$; its value $\langle\langle \mathbf{v},\mathbf{l}\rangle\rangle$ is the power of the load $\mathbf{l}$ on the velocity field $\mathbf{v}$;

(II) the linear space $\mathcal{S}$ of stress fields;
the linear space $\mathcal{E}$ of strain rate fields;
the internal power bilinear form $\langle\cdot,\cdot\rangle$ defined on $\mathcal{E}\times\mathcal{S}$; the value $-\langle \mathrm{e},\boldsymbol{\sigma}\rangle$ is the power of the stress field $\boldsymbol{\sigma}$ on the strain rate field e;
if $\langle \mathrm{e},\boldsymbol{\sigma}_0\rangle = 0$ for every $\mathrm{e}\in\mathcal{E}$, then $\boldsymbol{\sigma}_0$;
if $\langle \mathrm{e}_0,\boldsymbol{\sigma}\rangle = 0$ for every $\boldsymbol{\sigma}\in\mathcal{S}$, then $\mathrm{e}_0 = 0$;

(III) the linear operator Def on $\mathcal{V}$ with values in $\mathcal{E}$; its value $\mathrm{Def}\,\mathbf{v}$ is the strain rate field corresponding to the velocity field $\mathbf{v}$;
the only velocity field with the zero strain rate field is zero;

(IV) the virtual work principle: a stress field $\boldsymbol{\sigma}$ equilibrates the load $\mathbf{l}$ if and only if

$$\langle \mathrm{Def}\,\mathbf{v},\boldsymbol{\sigma}\rangle = \langle\langle \mathbf{v},\mathbf{l}\rangle\rangle \quad \text{for every} \quad \mathbf{v}\in U;$$

if $\boldsymbol{\sigma}$ satisfies the virtual work principle, the equality $\langle \mathrm{Def}\,\mathbf{v},\boldsymbol{\sigma}\rangle = \langle\langle \mathbf{v},\mathbf{l}\rangle\rangle$ also holds for every $\mathbf{v}$ in $\mathcal{V}$;

(V) the set $C$ of admissible stress fields in $\mathcal{S}$; $C$ is convex and contains the zero stress field;

(VI) the constitutive maximum principle: the stress and strain rate fields $\boldsymbol{\sigma}$ and e satisfy constitutive relations if and only if

$$\boldsymbol{\sigma}\in C, \quad \langle \mathrm{e},\boldsymbol{\sigma}-\boldsymbol{\sigma}_*\rangle \geq 0 \quad \text{for every} \quad \boldsymbol{\sigma}_*\in C.$$

*Solution to the rigid-plastic problem* for a given load $\mathbf{l}$ is the stress and velocity fields $\boldsymbol{\sigma}$ and $\mathbf{v}$ satisfying the following relations:

$$\begin{gathered}\boldsymbol{\sigma}\in\Sigma+\mathbf{s}_l,\\ \mathrm{e}=\mathrm{Def}\,\mathbf{v}, \quad \mathbf{v}\in\mathcal{V},\\ \boldsymbol{\sigma}\in C, \quad \langle \mathrm{e},\boldsymbol{\sigma}-\boldsymbol{\sigma}_*\rangle \geq 0 \quad \text{for every} \quad \boldsymbol{\sigma}_*\in C.\end{gathered} \tag{1.10}$$

Here, $\mathbf{s}_l$ is a fixed stress field equilibrating the given load $\mathbf{l}$ and $\Sigma$ is the subspace of self-equilibrated stress fields.

In (1.10) we used the formulation of the equilibrium conditions which is just another form of the virtual work principle, see Subsection 1.1.

REMARK 2. Due to (IV), the definition of the self-equilibrated stress fields set $\Sigma$ can also be written in the form $\Sigma = (\mathrm{Def}\,\mathcal{V})^0$. In particular, the equality $\langle \mathrm{e},\boldsymbol{\sigma}\rangle = 0$ holds for every kinematically admissible strain rate field $\mathrm{e} = \mathrm{Def}\,\mathbf{v}$, $\mathbf{v}\in\mathcal{V}$ and every self-equilibrated stress field $\boldsymbol{\sigma}\in\Sigma$.

Formulation of the rigid-plastic problem for a certain type of mechanical system is a realization of the general unified formulation. The following examples present two previously encountered formulations as particular cases of the general formulation.

EXAMPLE 2. *Three-dimensional body (strong formulation of the rigid-plastic problem).* The strong formulation of the problem was given in Subsection III.1.4. Let us verify that this is a particular case of the unified formulation. We set $\mathcal{V} = V_0$, where $V_0$ is the space of sufficiently regular velocity fields (Subsection III.1.3). Consider a sufficiently regular field of body forces $\mathbf{f}$ and surface tractions $\mathbf{q}$ and let $\mathcal{L}$ be a linear space of the loads $\mathbf{l} = (\mathbf{f}, \mathbf{q})$. We choose the spaces $V_0$ and $\mathcal{L}$ in such a way that the external power $\langle\langle\, \mathbf{v}, \mathbf{l} \,\rangle\rangle$ is determined by the regular formula

$$\langle\langle\, \mathbf{l}, \mathbf{v} \,\rangle\rangle = \int_{\Omega} \mathbf{f}\mathbf{v}\, dV + \int_{S_q} \mathbf{q}\mathbf{v}\, dS.$$

We choose certain spaces of integrable stress and strain rate fields $\mathcal{S}$ and $\mathcal{E}$ in such a way that the scalar field $\mathbf{e} \cdot \boldsymbol{\sigma}$ is integrable for every $\mathbf{e}$ in $\mathcal{E}$ and every $\boldsymbol{\sigma}$ in $\mathcal{S}$, and the internal power $-\langle\, \mathbf{e}, \boldsymbol{\sigma} \,\rangle$ is determined by the regular formula

$$\langle\, \mathbf{e}, \boldsymbol{\sigma} \,\rangle = \int_{\Omega} \mathbf{e} \cdot \boldsymbol{\sigma}\, dV.$$

The spaces $\mathcal{S}$, $\mathcal{E}$ possess properties (II) if $\mathcal{S}$ and $\mathcal{E}$ contain at least all smooth fields. Anyway, we can take the space of all integrable fields as $\mathcal{E}$ and the space of all integrable essentially bounded fields as $\mathcal{S}$. The latter choice is substantial since it makes $\mathcal{S}$ sufficiently wide, which is important in connection with the load admissibility problem (Subsection I.5.4) and with the integral formulation of the constitutive relations (Remark 1).

The operator Def is defined on the space $\mathcal{V} = V_0$ by (1.1) and possesses property (III) as it was verified in Subsection III.1.3.

Every velocity field $\mathbf{v}$ in $V_0$ can be approximated by smooth fields in $U$ (see Corollary 1 in Subsection VI.B.1 below) and that is why replacing the set $U$ of virtual velocity fields by the space $\mathcal{V} = V_0$ does not change the equilibrium conditions as (IV) requires.

The set $C$ of admissible stress fields in $\mathcal{S}$ is determined as usuall by the sets $C_x$ of admissible stresses given for all points $x$ of the body, and $C$ always possesses properties (V). The point-wise constitutive maximum principle which we use in the strong formulation of the rigid-plastic problem is equivalent to (4.7); see Subsection 1.3, in particular, Remark 1. The latter formulation involves an integral which is a value of the bilinear form $\langle\, \cdot, \cdot \,\rangle$ we adopt here. Therefore, (1.7) is exactly the constitutive maximum principle (VI).

Thus, strong formulation (III.1.8) of the rigid-plastic problem meets requirements (I) – (VI) and has the form (1.10), that is, the strong formulation is a particular case of the unified one.

EXAMPLE 3. *Truss.* Let us show that the formulation of the rigid-plastic problem for a truss in Subsection I.6.7 is a particular case of the unified formulation. Indeed, velocity fields and loads are represented by collections $\mathbf{v} = (\mathbf{v}_1, \ldots, \mathbf{v}_k)$ and $\mathbf{l} = (\mathbf{f}_1, \ldots, \mathbf{f}_k)$, where $\mathbf{v}_\alpha$ is the velocity of the $\alpha$-th (unfixed) hinge and $\mathbf{f}_\alpha$ is the external force applied to this hinge; $k$ is the number of unfixed hinges. This suggests setting

$$\mathcal{V} = \mathbf{R}^n, \quad \mathcal{L} = \mathbf{R}^n \quad (n = 3k), \qquad \langle\langle \mathbf{v}, \mathbf{l} \rangle\rangle = \mathbf{f}_1 \mathbf{v}_1 + \cdots + \mathbf{f}_k \mathbf{v}_k$$

within the framework of the unified formulation. Similarly, stress and strain rate fields are represented by collections

$$\boldsymbol{\sigma} = (N_1, \ldots, N_m) \quad \text{and} \quad \mathbf{e} = (e_1, \ldots, e_m),$$

where $N_i$ is the axial force in the $i$-th rod, $e_i$ is the elongation of this rod; $m$ is the number of rods; and we set

$$\mathcal{S} = \mathbf{R}^m, \quad \mathcal{E} = \mathbf{R}^m, \quad \langle \mathbf{e}, \boldsymbol{\sigma} \rangle = N_1 e_1 + \cdots + N_m e_m.$$

The finite dimensional spaces $\mathcal{S}$ and $\mathcal{E}$ obviously possess properties (II). The operator Def defined by (1.2) possesses property (III) since the considered trusses cannot move if none of the truss rods deforms. Conditions for equilibrium of the truss are equivalent to the virtual work principle (IV), see Subsection I.3.4. The question about the possibility of replacing $U$ by $\mathcal{V}$ in (IV) does not arise in the case of a truss (and, generally, in the case of any discrete system): if the space $\mathcal{V}$ of kinematically admissible velocity fields is finite dimensional, then $\mathcal{V} = U$. As to the constitutive relations, we have already shown in Example 1 that the collection of the constitutive maximum principles written for every rod of the truss is equivalent to the total constitutive maximum principle (VI). Thus, the regular formulation of the rigid-plastic problem for a truss meets requirements (I) – (VI), has the form (1.10) and, therefore, is a particular case of the unified formulation.

The next section contains further examples showing some realizations of the unified formulation of the rigid-plastic problem.

**1.5. Local description of material properties.** One can develop a theory of limit analysis based on properties (I) – (VI) of the rigid-plastic problem unified formulation. Within the framework of this approach we establish some general results. To obtain their final and practicall useful form, we specify these statements taking into account an additional assumption. This assumption is satisfied in all cases of practical interest and concerns the possibility of local description of material properties.

Consider a body occupying domain $\Omega$. We are reminded that, within the framework of the unified theory, "body" means a three-dimensional body as well as any other mechanical system, for example, a rod or a shell, and $\Omega$ has the corresponding dimension. We say that the description of material properties of a rigid-plastic body is *local* if (1) stress fields are functions on $\Omega$ taking values in a certain Euclidean space $\mathbf{S}$; (2) the set $C_x$ of admissible stresses is specified for every point $x$ of the body; $C_x$ is a convex, closed set in **S** and contains the origin; (3) stress field $\boldsymbol{\sigma}$ is admissible if and only if its values $\boldsymbol{\sigma}(x)$ are admissible for almost every point $x$ of the body.

For example, **S** is the six-dimensional space *Sym* of second rank symmetric tensors in the case of a three-dimensional body. In the case of a beam, **S** is the five-dimensional space of internal forces $(M_y, M_z, N, Q_y, Q_z)$ defined in Subsection I.6.2.

REMARK 3. In the case of a discrete rigid plastic system, a description of material properties is normally also local. The set of admissible stresses depends on a discrete variable in this case, for example, the number of rods is the case of a frame. If this dependence meets requirements (1) – (3), the description of material properties is local.

## 2. Examples

Formulations of the rigid-plastic problem for mechanical systems of two types are considered in this section to illustrate the realization of the general formulation of the rigid-plastic problem. We are reminded that there is a strong reason to find out that a certain formulation of the rigid-plastic problem is a particular case of the unified formulation. As soon as this is established, all statements of limit analysis general theory can be applied to that concrete type of problem.

**2.1. Rigid-plastic problem for a beam.** Consider a beam subjected to a load distributed along it with the density $\mathbf{q}$ and to the moment $\mathbf{m}$ and the force $\mathbf{f}$ applied to the right end of the beam. The left end of the beam is fixed. We restrict ourselves to in-plane bending of the beam. Let $x$, $y$, $z$ be the Cartesian coordinate system with its origin at the left end of the beam and the $x$-axis directed along the axis of the beam. Consider loads $\mathbf{l} = (\mathbf{q}, \mathbf{m}, \mathbf{f})$ with zero components $q_y = 0$, $m_z = 0$, $f_y = 0$ and $m_x = 0$ (see Figure 2.1). The latter equality is the usual no-torsion condition; the former ones normally ensure that bending the beam takes place only in the $(x, z)$-plane. Then, among the kinematic variables and internal forces used to describe behavior of the beam (see Sections I.6 and II.3), only the following ones can be nonzero: the component $\omega_y$ of the cross-section angular velocity, the components $u_x = u$ and $u_z$ of the axis velocity, the component $M_y$ of the bending moment, the axial force $N$ and the cutting force $Q_z$.

We adopt the rigid perfectly plastic constitutive relations assuming that the beam is homogeneous and its yield function does not depend on cutting forces (see Subsection I.6.3). For example, the following admissibility condition is normally used for stresses $\boldsymbol{\sigma} = (M_y, N, Q_z)$:

$$\frac{|M_y|}{M_0} + \frac{N^2}{N_0^2} \leq 1 \qquad (M_0 = \text{const} > 0,\ N_0 = \text{const} > 0) \tag{2.1}$$

for beams of rectangular cross-section.

The rigid-plastic problem for a beam has already been set up in Subsection I.6.4. Let us now show that its formulation is a particular case of the unified formulation by specifying the spaces $\mathcal{V}$ and $\mathcal{S}$, the operator Def, etc. and verifying that they meet requirements (I) – (VI) in Subsection 1.4.

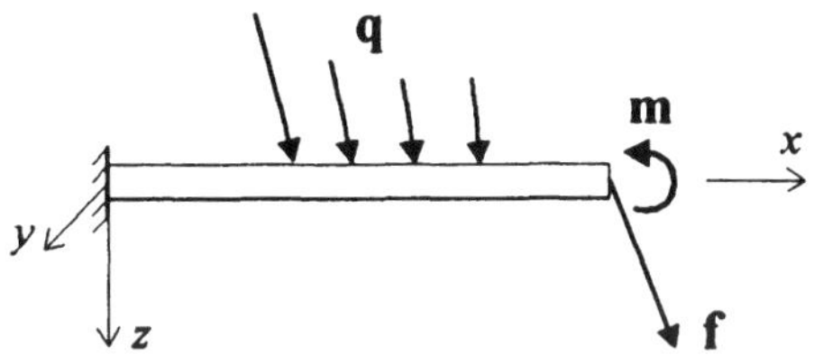


**Figure 2.1**

Note that the kinematic variables and stresses $\mathbf{v} = (\omega_y, u, u_z)$ and $\boldsymbol{\sigma} = (M_y, N, Q_z)$ are functions of $x$ on the interval $[0, a]$, where $a$ is the length of the beam. We choose the set of triples of continuously differentiable functions $(\omega_y, u, u_z)$ on $[0, a]$ that vanish at $x = 0$ as the space $\mathcal{V}$ of kinematically admissible velocity fields. Thus, every $\mathbf{v} = (\omega_y, u, u_z)$ in $\mathcal{V}$ satisfies the conditions $\omega_y|_{x=0} = 0$, $u|_{x=0} = 0$, $u_z|_{x=0} = 0$, which represent the kinematic constraint imposed on the beam. The subspace $U$ of virtual velocity fields in $\mathcal{V}$ consists of smooth (infinitely differentiable) fields $\mathbf{v} = (\omega_y, u, u_z)$. The load applied to the beam is represented by the collection $\mathbf{l} = (q, q_y, m_y, f, f_z)$, where $q$, $q_y$ are functions on $[0, a]$ and $m_y$, $f$, $f_z$ are numbers. We consider the set of such collections with continuously differentiable $q$ and $q_y$ as the space $\mathcal{L}$ of loads. The power of the load $\mathbf{l}$ on the velocity field $\mathbf{v} = (\omega_y, u, u_z)$ is

$$\int_0^a (qu + q_z u_z)\, dx + m_y \omega_y(a) + fu(a) + f_z u_z(a)$$

(see Subsection II.3.1). This expression linearly depends on both $\mathbf{l}$ and $\mathbf{v}$ and this is the value $\langle\langle \mathbf{l}, \mathbf{v} \rangle\rangle$ of the bilinear form $\langle\langle \cdot, \cdot \rangle\rangle$. Condition (2.1) implies that the components $M_y$, $N$ of any admissible stress field $\boldsymbol{\sigma} = (M_y, N, Q_z)$ are bounded; at the same time, (2.1) imposes no restrictions on $Q_z$. This allows consideration of stress fields $\boldsymbol{\sigma} = (M_y, N, Q_z)$ whose components $M_y$, $N$ are integrable bounded functions and $Q_z$ is a function integrable, say, in power 2. We take the set of such fields $\boldsymbol{\sigma} = (M_y, N, Q_z)$ as the space $\mathcal{S}$ on stress fields. This is an appropriate choice in the case of condition (2.1) as well as in the case of any other admissibility condition which requires boundness of $M_y$, $N$ and does not restrict $Q_z$. The internal power of the stress field $\boldsymbol{\sigma}$, by definition in Subsection II.3.1, is

$$-\int_0^a \left(M_y \omega_y' + N u' + Q_z(u_z' + \omega_y)\right)\, dx,$$

where prime indicates the derivatives with respect to $x$.

Recall that $\omega_y'$, $u'$, $u_z' + \omega_y$ are components of the strain rate field $\mathbf{e}(\mathbf{v})$, which corresponds to the velocity field $\mathbf{v} = (\omega_y, u, u_z)$, see Subsection I.6.1. Then the internal power can be written as $-\langle \mathbf{e}, \boldsymbol{\sigma} \rangle$, where we introduced bilinear form $\langle \cdot, \cdot \rangle$:

$$\langle \mathrm{e}, \boldsymbol{\sigma} \rangle = \int_0^a (e_1\sigma_1 + e_2\sigma_2 + e_3\sigma_3)\, dx.$$

We have to choose the space $\mathcal{E}$ in such a way that this form is defined for every $\mathbf{e}$ in $\mathcal{E}$ and every $\boldsymbol{\sigma}$ in $\mathcal{S}$. We consider the triples $(e_1, e_2, e_3)$ where $e_1$, $e_2$ are integrable functions and $e_3$ is a function integrable in power 2 (if the similar assumption is adopted for $Q_z$), so that $\mathbf{e} = (e_1, e_2, e_3)$ is a field on $[0, 1]$. By taking the set of such fields as $\mathcal{E}$ we assure that $\langle \mathrm{e}, \boldsymbol{\sigma} \rangle$ is defined for every $\mathbf{e} \in \mathcal{E}$ and every $\boldsymbol{\sigma} \in \mathcal{S}$.

We now introduce a linear operator Def: it maps a velocity field $\mathbf{v} = (\omega_y, u, u_z)$ on the corresponding strain rate field

$$\mathbf{e} = \operatorname{Def} \mathbf{v} = (\omega_y', u', u_z' + \omega_y).$$

The operator is defined on the space $\mathcal{V}$, maps $\mathcal{V}$ into $\mathcal{E}$ and, obviously, possesses property (III) in Subsection 1.4: if $\operatorname{Def} \mathbf{v} = 0$, $\mathbf{v} \in \mathcal{V}$, then, due to the condition $\mathbf{v}(0) = 0$, we have $\mathbf{v} = 0$.

The virtual work principle formulates the equilibrium condition for a beam (see Subsection II.3.3). The principle has exactly the form (IV) of Subsection 1.4 and makes use of the above specified subspace $U$ of virtual velocity fields. Replacing $U$ by $\mathcal{V}$, obviously, does not change the equilibrium conditions as (IV) requires.

The set $C$ of admissible stress field in $\mathcal{S}$ is as usuall determined by the sets $C_x$ of admissible stresses given for all points $x$ of the beam. It is clear that $C$ possesses properties (V) in Subsection 1.4. The point-wise formulation of constitutive maximum principle (I.6.3) is equivalent to the integral formulation

$$\int_0^a \mathrm{e}(x)(\boldsymbol{\sigma}(x) - \boldsymbol{\sigma}_*)\, dx \geq 0 \quad \text{for every} \quad \boldsymbol{\sigma}_* \in C$$

(the equivalence arises from an equality similar to (1.9) that will be established in Section 5). The latter formulation involves the integral which is the value of the above specified bilinear form $\langle \cdot, \cdot \rangle$. Therefore, this formulation is exactly of the form (VI) in Subsection 1.4.

Thus, the formulation of the rigid-plastic problem for beams in Subsection I.6.4 meets requirements (I) – (VI) and is a particular case of the unified formulation in Subsection 1.4.

**2.2. Rigid-plastic problem for a discrete system.** Let us consider the formulation of the rigid-plastic problem for an arbitrary discrete system as a particular case of the unified formulation in Subsection 1.4. A *discrete system* is the one admitting description with finite-dimensional spaces of velocity, stress and strain rate fields. For example, rod structures are often considered as discrete systems under the assumption that a finite collection of numbers represents the state of every rod. In particular, under conditions of Section I.6, a truss is a discrete system.

Here, we only consider discrete systems, and because of that we temporarily leave the above terminology convention simply speaking of stress or velocity (and not about fields). Each of the variables is a collection of numbers. These numbers, the components, may have different meanings. For example, angular velocities or generalized velocities of various kinds may be components of the velocity.

Consider a discrete system with $n$-dimensional linear space $\mathcal{V}$ of kinematically admissible velocities and the subspace $U$ of virtual velocities is the whole of $\mathcal{V}$: $U = \mathcal{V}$. We represent a collection of kinematic variables, that is, the velocity $\mathbf{v}$ in $\mathcal{V}$ by a column matrix $(v_1, \ldots, v_n)^T$, where the superscript $T$ indicates transposition. We assume as usuall that possible loads form a linear space $\mathcal{L}$. There are various ways for specifying the loads; however, the external power admits a standard description. Namely, the power of the fixed load $\mathbf{l}$ on the velocity $\mathbf{v}$ linearly depends on $\mathbf{v}$, and like any linear function on a finite dimensional space $\mathcal{V}$, has the form

$$\mathbf{v} \to \langle\langle \mathbf{v}, \mathbf{l} \rangle\rangle = (l_1, \ldots, l_n)(v_1, \ldots, v_n)^T,$$

that is, its value is the product of the raw matrix representing the function by the column matrix representing the argument. Thus, in the case of a discrete system, we identify the load with a column matrix $\mathbf{l} = (l_1, \ldots, l_n)^T$, its power on the virtual velocity $\mathbf{v}$ being $\langle\langle \mathbf{v}, \mathbf{l} \rangle\rangle = \mathbf{l}^T\mathbf{v}$.

Let $\mathcal{S}$ and $\mathcal{E}$ be $m$-dimensional spaces of stresses and strain rates in the discrete system under consideration. Their dimension $m$, generally, differs from the dimension $n$ of the spaces $\mathcal{V}$ and $\mathcal{L}$. For example, in the case of a frame, $m$ is the number of rods and $n$ is determined by the number of unfixed hinges. The stress $\boldsymbol{\sigma}$ is represented by the column matrix $\boldsymbol{\sigma} = (\sigma_1, \ldots, \sigma_m)^T$ and, similarly, the strain rate by $\mathbf{e} = (e_1, \ldots, e_m)^T$. The internal power of the stress $\boldsymbol{\sigma}$ on the strain rates $\mathbf{e}$ is $-\langle \mathbf{e}, \boldsymbol{\sigma} \rangle$ where $\langle \cdot, \cdot \rangle$ is a certain bilinear form. We always write this form as $\langle \mathbf{e}, \boldsymbol{\sigma} \rangle = \mathbf{e}^T\boldsymbol{\sigma}$ (a linear change of the original variables may be needed to arrive at this expression).

A linear operator Def is defined on the space $\mathcal{V}$ and Def maps a velocity $\mathbf{v}$ on the corresponding strain rate $\mathbf{e}$ in $\mathcal{E}$. Any linear mapping from $\mathcal{V}$ into $\mathcal{E}$ can be represented by $m \times n$-matrix. In particular, let $D$ be such a matrix representing Def: $\mathbf{e} = \operatorname{Def}\mathbf{v} = D\mathbf{v}$, where $D\mathbf{v}$ is the product of the $m \times n$-matrix by the column matrix. Thus, the kinematic relations for the discrete systems under consideration are as follows: $\mathbf{e} = D\mathbf{v}$, $\mathbf{v} \in \mathcal{V}$. The former relation determines the strain rate for the velocity $\mathbf{v}$; the latter just means that $\mathbf{v}$ is kinematically admissible.

The equilibrium conditions for the discrete system are determined by the virtual work principle (IV) in Subsection 1.4:

$$(D\mathbf{v})^T \boldsymbol{\sigma} = \mathbf{l}^T\mathbf{v} \quad \text{for every} \quad \mathbf{v} \in \mathcal{V},$$

where we took into account the expressions $-\mathbf{e}^T\boldsymbol{\sigma}$ and $\mathbf{l}^T\mathbf{v}$ for the internal and external powers. The identity $(D\mathbf{v})^T\boldsymbol{\sigma} = (D^T\boldsymbol{\sigma})^T\mathbf{v}$ allows re-writing the previous equality in the form $(D^T\boldsymbol{\sigma} - \mathbf{l})^T\mathbf{v} = 0$. Since the equality should be satisfied for every $\mathbf{v}$ in $\mathcal{V}$, the virtual work principle is equivalent to the equation $D^T\boldsymbol{\sigma} = \mathbf{l}$.

REMARK 1. The equilibrium equation $D^T\boldsymbol{\sigma} = \mathbf{l}$ can, of course, be written in the form $\boldsymbol{\sigma} \in \Sigma + \mathbf{s}_l$ (where $\mathbf{s}_l$ is a fixed stress equilibrating the load $\mathbf{l}$ and $\Sigma$ is the set of self-equilibrated stresses in $\mathcal{S}$). Indeed, $\mathbf{s}_l$ is a particular solution to the linear system of equations $D^T\boldsymbol{\sigma} = \mathbf{l}$ and

$$\Sigma = \left\{\boldsymbol{\sigma} \in \mathcal{S} : (D\mathbf{v})^T\boldsymbol{\sigma} = 0 \ \text{ for every } \ \mathbf{v} \in \mathcal{V}\right\}$$

is the set of all solutions to the homogeneous system $D^T\boldsymbol{\sigma} = 0$. Therefore, the sum $\Sigma + \mathbf{s}_l$ is the set of all solutions to the equation $D^T\boldsymbol{\sigma} = \mathbf{l}$.

Let $C$ be the given set of admissible stresses in the space $\mathcal{S}$. We assume as always that $C$ is convex and contains the origin. Constitutive maximum principle (VI) (Subsection 1.4) together with the above-mentioned kinematic relations and equilibrium conditions form the following formulation of the rigid-plastic problem for the discrete system:

$$\begin{gathered} D^T\boldsymbol{\sigma} = \mathbf{l}, \\ \mathbf{e} = D\mathbf{v}, \quad \mathbf{v} \in \mathcal{V}, \\ \boldsymbol{\sigma} \in C, \quad \mathbf{e}^T(\boldsymbol{\sigma} - \boldsymbol{\sigma}_*) \geq 0 \quad \text{for every} \quad \boldsymbol{\sigma}_* \in C. \end{gathered} \tag{2.2}$$

It is clear that this is a realization of the unified formulation of the rigid-plastic problem. Recall that here velocities and loads are $n \times 1$ column matrices, strain rates are $m \times 1$ column matrices and $D$ is a $m \times n$ matrix.

## 3. Safe, limit and inadmissible loads

We have already established certain properties of rigid perfectly plastic bodies within the framework of the strong formulation of the three-dimensional rigid-plastic problem (see Sections III.2, III.3). This resulted in the following idea of the body response to the load $\mathbf{l}$. The safety factor $\alpha_l$ characterizes the load $\mathbf{l}$ and

- the body does not deform if $\alpha_l > 1$,
- the body deforms at arbitrarily large strain rate if $\alpha_l = 1$,
- the body cannot bear the load if $\alpha_l < 1$.

In this section, we establish these statements within the framework of the general, unified formulation of the rigid-plastic problem

$$\boldsymbol{\sigma} \in \Sigma + \mathbf{s}_l, \tag{3.1}$$

$$\mathbf{e} = \operatorname{Def}\mathbf{v}, \quad \mathbf{v} \in \mathcal{V}, \tag{3.2}$$

$$\boldsymbol{\sigma} \in C, \quad \langle \mathbf{e}, \boldsymbol{\sigma} - \boldsymbol{\sigma}_* \rangle \geq 0 \quad \text{for every} \quad \boldsymbol{\sigma}_* \in C. \tag{3.3}$$

(The formulation was discussed in detail in Section 1.) Note that unified formulation (3.1) – (3.3) covers particular formulations of the problem for rigid perfectly plastic systems of all types. Therefore, any of these systems possesses each of the properties established within the framework of formulation (3.1) - (3.3). This is the most important reason to reconsider the rigid-plastic body properties within the framework of the general formulation.

Recall that we keep using three-dimensional terminology in the unified theory to avoid introducing new terms.

**3.1. Safety factor.** Consider a rigid-plastic body subjected to a load. We refer to the load as *admissible* if there is an admissible stress field equilibrating it; otherwise, we refer to the load as *inadmissible*. The rigid-plastic problem (3.1) - (3.3) has no solution if the load $\mathbf{l}$ is inadmissible. Recall that $\mathbf{s}_l$ in (3.1) stands for a fixed stress field equilibrating the given load $\mathbf{l}$; $\Sigma$ stands for the subspace of self-equilibrated stress fields. Denoting the set of admissible stress fields by $C$, let us write the definition of the load admissibility in the following form: $\mathbf{l}$ is admissible if and only if there is a stress field $\boldsymbol{\sigma}$ such that $\boldsymbol{\sigma} \in \Sigma + \mathbf{s}_l$ and $\boldsymbol{\sigma} \in C$, that is, the intersection $(\Sigma + \mathbf{s}_l) \cap C$ is not empty. (The relation $\boldsymbol{\sigma} \in \Sigma + \mathbf{s}_l$ means as usuall that $\boldsymbol{\sigma}$ equilibrates the load $\mathbf{l}$, and $\boldsymbol{\sigma} \in C$ means that $\boldsymbol{\sigma}$ is admissible.) Similarly, we write the admissibility criterion for the load $m\mathbf{l}$, $m \geq 0$, as

$$(\Sigma + m\mathbf{s}_l) \cap C \neq \emptyset,$$

where $m\mathbf{s}_l$ is obviously a stress field equilibrating the load $m\mathbf{l}$ proportional to $\mathbf{l}$.

This suggests introducing a concept of the safety factor; the *safety factor* of the load $\mathbf{l}$ is the supremum of multipliers $m \geq 0$ for which $m\mathbf{l}$ is an admissible load:

$$\alpha_l = \sup \{m \geq 0 : (\Sigma + m\mathbf{s}_l) \cap C \neq \emptyset\}.$$

This definition can also be written in the form

$$\alpha_l = \sup \{m \geq 0 : \text{there exists } \boldsymbol{\rho} \in \Sigma, \;\; m(\boldsymbol{\rho} + \mathbf{s}_l) \in C\}.$$

(see Subsection III.2.2 for more details). The following proposition clarifies the meaning of the safety factor.

PROPOSITION 1. The load $m\mathbf{l}$ is inadmissible if $m > \alpha_l$ and is admissible if $0 \leq m < \alpha_l$.

Proof of this proposition given in Subsection III.2.2 holds in the case under consideration, that is, within the framework of the unified formulation of the rigid-plastic problem.

A load with the safety factor equal to 1 is referred to as *limit load*.

According to Proposition 1:

- among the loads $m\boldsymbol{\sigma}$, $m \geq 0$ proportional to the given load $\mathbf{l}$, admissible and inadmissible ones do not alternate; loads of the two types are separated by the limit load;
- the safety factor indicates how many times one can increase a given load still obtaining an admissible load;
- the load $\mathbf{l}$ is admissible if $\alpha_l > 1$ and inadmissible if $\alpha_l < 1$.

Under a certain nonrestrictive assumption, the inequality $\alpha_l > 1$ not only guarantees that the load $\mathbf{l}$ is admissible but also implies that $\mathbf{l}$ possesses a much more important property: the load is safe. This will be shown in the next two subsections.

**3.2. Safe stress fields.** The concept of the safe stress field previously introduced in Subsection III.3.1 will now be generalized. To extend that definition to the general case, let us introduce the concept of absorbing set: set $Q$ in a linear space $\mathcal{S}$ is *absorbing* if for every $\mathbf{s}$ in $\mathcal{S}$ there exists a number $r > 0$ such that $r\mathbf{s}$ belongs to $Q$. Setting $\mathbf{s} = 0$ in the above definition immediately yields that an absorbing set contains the origin.

EXAMPLE 1. A ball with its center at the origin and of radius $R > 0$ is an absorbing set in the Euclidean space.

Consider a stress field $\boldsymbol{\sigma}$ which is admissible, that is, belongs to $C$. Suppose also that there is an absorbing set $Q$ in $\mathcal{S}$ such that $\boldsymbol{\sigma} + Q$ is a subset in $C$; in this case $\boldsymbol{\sigma}$ is said to be *safe*. In particular, the zero stress field is safe if and only if $C$ contains an absorbing set.

EXAMPLE 2. Consider a body occupying domain $\Omega$. Let stress fields in the body be functions on $\Omega$ taking their values in a certain Euclidean space $\mathbf{S}$, for example, $\mathbf{S} = Sym$ in the case of a three-dimensional body. The set $C_x$ of admissible stresses is given for every point $x$ of the body. The sets $C_x$ are assumed to be bounded, and so is the collection of these sets. In other words, there is a ball in $\mathbf{S}$ containing all the sets $C_x$. Then there is a natural choice for the space $\mathcal{S}$ of stress fields: the space of essentially bounded integrable fields. The latter will be defined in Subsection V.A.2; at the moment, the reader may skip the word "essentially" and think about $\mathcal{S}$ just as a collection of bounded integrable functions. Let $\delta$ be any positive number and $Q_\delta$ be a set in $\mathcal{S}$ consisting of fields which satisfy the condition $|\boldsymbol{\sigma}(x)| \leq \delta$ for a.e. $x \in \Omega$. It is clear that $Q_\delta$ is an absorbing set in $\mathcal{S}$.

We can now discuss a connection between the definition of the stress field safety given in Subsection III.3.1 and the one we use here. Let a stress field $\boldsymbol{\sigma}$ be safe in the former sense, that is, $\boldsymbol{\sigma}(x) \in C_x$ and the distance from $\boldsymbol{\sigma}(x)$ to the yield surface is not less than $\delta > 0$ (with $\delta$ not depending on $x$). Consider for a.e. $x \in \Omega$ the field $\boldsymbol{\sigma} + \boldsymbol{\tau}$ with $\boldsymbol{\tau} \in Q_\delta$; it is obvious that $\boldsymbol{\sigma}(x) + \boldsymbol{\tau}(x)$ lies inside the yield surface for a.e. $x \in \Omega$. In other words, $\boldsymbol{\sigma} + \boldsymbol{\tau}$ is in $C$ for every $\boldsymbol{\tau} \in Q_\delta$, that is, $\boldsymbol{\sigma} + Q_\delta$ is a subset in $C$. Thus, a stress field which is safe in the sense of Subsection III.3.1 is also safe in the sense we adopt here.

The following simple proposition points out an important property of safe stress fields connected with the rigid perfectly plastic constitutive relations.

PROPOSITION 2. Suppose a strain rate field $\mathbf{e}$ and stress field $\boldsymbol{\sigma}$ satisfy the constitutive maximum principle

$$\langle \mathbf{e}, \boldsymbol{\sigma} - \boldsymbol{\sigma}_* \rangle \geq 0 \quad \text{for every} \quad \boldsymbol{\sigma}_* \in C$$

and $\boldsymbol{\sigma}$ is a safe stress field, then $\mathbf{e} = 0$.

PROOF. Since $\boldsymbol{\sigma}$ is safe, there is an absorbing set $Q$ such that $\boldsymbol{\sigma} + Q \subset C$ or, equivalently, $\boldsymbol{\sigma} + \boldsymbol{\tau}$ is an admissible stress field for every $\boldsymbol{\tau}$ in $Q$. Setting

$\boldsymbol{\sigma}_* = \boldsymbol{\sigma} + \boldsymbol{\tau}$ in the constitutive maximum principle we obtain $\langle \mathbf{e}, \boldsymbol{\tau} \rangle \leq 0$ for every $\boldsymbol{\tau} \in Q$. By the absorbing set definition, for every $\mathbf{s}$ in $\mathcal{S}$ there are numbers $r_+, r_- > 0$ such that $\boldsymbol{\tau}_+ = r_+\mathbf{s}$ and $\boldsymbol{\tau}_- = r_-(-\mathbf{s})$ belong to $Q$ and, consequently, $\langle \mathbf{e}, r_+\mathbf{s} \rangle \leq 0$ and $\langle \mathbf{e}, r_-(-\mathbf{s}) \rangle \leq 0$. This yields that $\langle \mathbf{e}, \mathbf{s} \rangle = 0$ for every $\mathbf{s} \in \mathcal{S}$; then property (II) in Subsection 1.4 implies $\mathbf{e} = 0$, which finishes the proof.

**3.3. Safe loads.** We refer to a load as *safe* if there is a safe stress field equilibrating this load. Note that the rigid-plastic problem has a solution if the load is safe. Indeed, let $\boldsymbol{\sigma}_s$ be a safe stress field equilibrating the given load $\mathbf{l}$. Then $\boldsymbol{\sigma} = \boldsymbol{\sigma}_s$ and $\mathbf{v} = 0$ obviously satisfy relations (3.1) – (3.3), that is, form a solution to the rigid-plastic problem with the load $\mathbf{l}$. Although this problem may have other solutions, the following proposition shows that the velocity field of any solution is zero, that is, the *rigid-plastic body remains rigid under the action of any safe load.*

PROPOSITION 3. If a rigid-plastic body is subjected to the safe load $\mathbf{l}$, then any solution to rigid-plastic problem (3.1) – (3.3) has the zero velocity field.

PROOF. Let the stress field $\boldsymbol{\sigma}$ and the velocity field $\mathbf{v}$ be a solution to problem (3.1) – (3.3) for the safe load $\mathbf{l}$. Consider the safe stress field $\boldsymbol{\sigma}_s$ equilibrating the load. First let us show that $\boldsymbol{\sigma}_s$ and the strain rate field $\operatorname{Def}\mathbf{v}$ satisfy the constitutive maximum principle. Indeed, the solution $\boldsymbol{\sigma}$, $\mathbf{v}$ satisfies the principle, that is, $\langle \operatorname{Def}\mathbf{v}, \boldsymbol{\sigma} - \boldsymbol{\sigma}_* \rangle \geq 0$ for every $\boldsymbol{\sigma}_* \in C$. On the other hand, each of the fields $\boldsymbol{\sigma}$ and $\boldsymbol{\sigma}_s$ equilibrates the load $\mathbf{l}$ and, consequently, the field $\boldsymbol{\sigma} - \boldsymbol{\sigma}_s$ is self-equilibrated. Then, due to Remark 1.2, the equality $\langle \operatorname{Def}\mathbf{v}, \boldsymbol{\sigma} - \boldsymbol{\sigma}_s \rangle = 0$ holds. Together with the previous inequality this implies that $\boldsymbol{\sigma}_s$ and $\operatorname{Def}\mathbf{v}$ satisfy the constitutive maximum principle:

$$\langle \operatorname{Def}\mathbf{v}, \boldsymbol{\sigma}_s - \boldsymbol{\sigma}_* \rangle \geq 0 \quad \text{for every} \quad \boldsymbol{\sigma}_* \in C.$$

Then, by Proposition 2, $\operatorname{Def}\mathbf{v} = 0$ and hence $\mathbf{v} = 0$ due to property (III) in Subsection 1.4. The proposition is proved.

**3.4. Safety criterion.** Consider loads $m\mathbf{l}$, $m \geq 0$, proportional to the given load $\mathbf{l}$. The following proposition connects the safety of the load $m\mathbf{l}$ with the value $\alpha_l$ of the safety factor.

PROPOSITION 4. Suppose the zero stress field is safe for a given rigid-plastic body. Then the load $m\mathbf{l}$, $m \geq 0$ is safe if and only if $0 \leq m < \alpha_l$.

PROOF. First consider the load $m\mathbf{l}$ with $0 \leq m < \alpha_l$. For number $m'$ satisfying the inequalities $m < m' < \alpha_l$, the load $m'\mathbf{l}$ is admissible (Proposition 1). Hence there exists an admissible stress field $\boldsymbol{\sigma}'$ which equilibrates the load $m'\mathbf{l}$. Then the stress field

$$\boldsymbol{\sigma} = \frac{m}{m'}\boldsymbol{\sigma}' + \left(1 - \frac{m}{m'}\right)0 = \frac{m}{m'}\boldsymbol{\sigma}'$$

equilibrates the load $m\mathbf{l}$, and $\boldsymbol{\sigma}$ is admissible since $C$ is convex and contains $\boldsymbol{\sigma}'$ and 0. Let us show that $\boldsymbol{\sigma}$ is safe. Indeed, by assumption, the zero stress

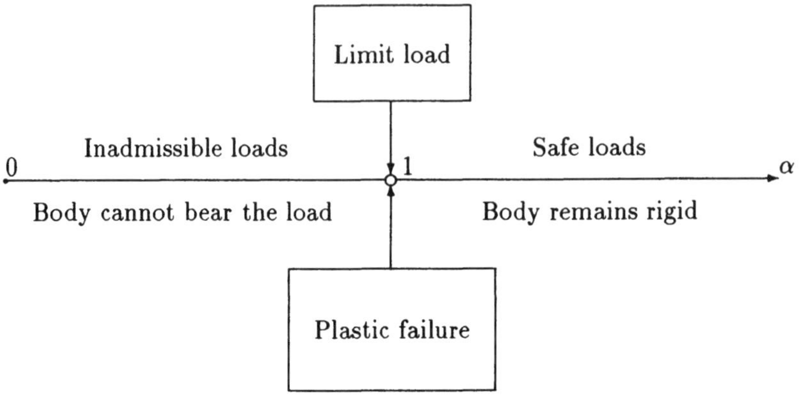


**Figure 3.1**

field is admissible, that is, an absorbing set $Q$ lies in $C$. By convexity of $C$, the set

$$\frac{m}{m'}\boldsymbol{\sigma} + \left(1 - \frac{m}{m'}\right) Q = \boldsymbol{\sigma} + \left(1 - \frac{m}{m'}\right) Q$$

is also in $C$. Here, the set $(1 - m/m')Q$ is evidently absorbing and, therefore, $\boldsymbol{\sigma}$ is a safe stress field.

Now consider the load $m\mathbf{l}$ with $m \geq \alpha_l$. If $m > \alpha_l$, the load is inadmissible (Proposition 1) and, consequently, not safe. Let us show that $m\mathbf{l}$ is not safe in case $m = \alpha_l$ either. Suppose the load $\alpha_l\mathbf{l}$ is safe; then there exists a safe stress field $\boldsymbol{\sigma}$ equilibrating this load. The safety means that there is an absorbing set $Q$ such that $\boldsymbol{\sigma} + Q \subset C$. In particular, by the absorbing set definition, there is a number $r > 0$ such that $r\boldsymbol{\sigma} \in Q$. Then the stress field $\boldsymbol{\sigma} + r\boldsymbol{\sigma}$ belongs to $\boldsymbol{\sigma} + Q \subset C$ and, therefore, is admissible. On the other hand, $\boldsymbol{\sigma} + r\boldsymbol{\sigma}$ equilibrates the load $\alpha_l\mathbf{l} + r(\alpha_l\mathbf{l}) = (1+r)\alpha_l\mathbf{l}$, and for this reason $(1+r)\alpha_l\mathbf{l}$ is admissible. This contradicts Proposition 1. Therefore, the assumption about the safety of $\alpha_l\mathbf{l}$ appears to be wrong, which finishes the proof.

One may apply the safety criterion in case $m = 1$. It reads as follows: if the zero stress field is safe for a given rigid-plastic body, then

the load $\mathbf{l}$ is safe when $\alpha_l > 1$ and not safe when $\alpha_l \leq 1$.

**3.5. Limit analysis.** Among the loads $m\mathbf{l}$, $m \geq 0$, the ones with $m > \alpha_l$ are inadmissible and those with $0 \leq m < \alpha_l$ are admissible (Proposition 1). We refer to the load $\alpha_l\mathbf{l}$ separating admissible and inadmissible loads as the *limit load*. In particular, the load $\mathbf{l}$ is limit if and only if $\alpha_l = 1$.

In what follows, we restrict ourselves to rigid-plastic bodies for which the zero stress field is safe. A body of this type deforms only under the action of a limit load (Proposition 3). If rigid-plastic problem (3.1) – (3.3) has a solution $\boldsymbol{\sigma}$, $\mathbf{v}$ with nonzero velocity field for a certain (limit) load, then it also has a solution with an arbitrarily large strain rate. Indeed, it is obvious that the same stress field $\boldsymbol{\sigma}$ and the velocity field $a\mathbf{v}$ with an arbitrary positive number $a$ are again a solution to (3.1) – (3.3). According to (III) in Subsection 1.4, the nonzero strain rate field $\operatorname{Def}\mathbf{v} \neq 0$ corresponds to the nonzero velocity field $\mathbf{v}$. Therefore, the strain rate $a\,\mathrm{Def}\,\mathbf{v}$ can be arbitrarily large. Because of that we refer to the nonzero velocity field in a solution of the rigid-plastic problem as the *failure mechanism.*

We say that *plastic failure* occurs if there exists a solution to the rigid-plastic problem with a nonzero velocity field. The failure mechanism does exist in a sufficiently wide space $\mathcal{V}$ of velocity fields (see Chapter VII and also discussion in Subsection III.3.4). Then the main properties of the rigid perfectly plastic response to a load are described by the scheme in Figure 3.1. The type of response is connected to the value of the safety factor $\alpha_l$ of the load $\mathbf{l}$ applied to the body. The body remains rigid while $\alpha_l > 1$, loses its bearing capacity and undergoes the plastic failure when $\alpha_l = 1$ and cannot bear the load with $\alpha_l < 1$. Therefore, the main problem of *limit analysis* is evaluating the safety factor for a given load.

## 4. Static and kinematic multipliers

We introduced the static and kinematic multipliers and studied their properties within the framework of the strong formulation of the rigid-plastic problem in Section III.5. Two basic theorems established these properties and also resulted in methods for limit analysis. We now extend these propositions to the unified formulation of the rigid-plastic problem. The first basic theorem states that every kinematic multiplier of a load is not less than any of its static multipliers. The second theorem establishes a criterion for equality of a static and a kinematic multiplier. This implies that three methods for limit analysis introduced in Section III.5 are also applicable to any problem within the framework of the unified formulation of the rigid-plastic problem.

**4.1. Static multiplier.** Definition of the static multiplier within the framework of the rigid-plastic problem unified formulation replicates that given in Section III.3 for the case of three-dimensional bodies within the framework of the strong formulation. Namely, let $\boldsymbol{\sigma}$ be an admissible stress field equilibrating the load $m\mathbf{l}$, where $\mathbf{l}$ is a given load and $m$ is a nonnegative number. In this case, the number $m$ is referred to as a *static multiplier* of the load $\mathbf{l}$ (corresponding to the stress field $\boldsymbol{\sigma}$) and is denoted by the symbol $m_s(\boldsymbol{\sigma})$. Thus, $\boldsymbol{\sigma}$ satisfies the admissibility condition $\boldsymbol{\sigma} \in C$ and equilibrates the load $m_s(\boldsymbol{\sigma})\mathbf{l}$: $\boldsymbol{\sigma} \in \Sigma + m_s(\boldsymbol{\sigma})\mathbf{s}_l$, where $\mathbf{s}_l$ is a fixed stress field equilibrating the load $\mathbf{l}$. Then, according to the safety factor definition (Subsection 3.1), we have

$$m_s(\boldsymbol{\sigma}) \leq \alpha_l, \quad \alpha_l = \sup m_s(\boldsymbol{\sigma}).$$

*Every static multiplier of a load is a lower bound for its safety factor and the safety factor equals the supremum of static multipliers.*

**4.2. Kinematic multiplier.** The kinematic multiplier for a three-dimensional body (Subsection III.5.2) was defined as the total dissipation corresponding to a velocity field over external power. To extend this definition to the rigid-plastic problem unified formulation, we only have to generalize the concept of the dissipation. Namely, let us define (total) dissipation corresponding to $\mathbf{e} \in \mathcal{E}$ as

$$D(\mathbf{e}) = \sup \{ \langle \mathbf{e}, \boldsymbol{\sigma}_* \rangle : \boldsymbol{\sigma}_* \in C \}. \tag{4.1}$$

REMARK 1. This definition resembles that of the local dissipation (Subsection I.4.9):

$$d_x(\mathbf{e}) = \sup \{ \mathbf{e} \cdot \boldsymbol{\sigma}_* : \boldsymbol{\sigma}_* \in C \}.$$

However, $\mathbf{e}$ denotes the value of a strain rate field at the point $x$ in this formula while $\mathbf{e}$ stands for the field itself in (4.1). Actually, (4.1) determines the summarized dissipation and, in the case of a three-dimensional body, results in the following formula for evaluating $D(\mathbf{e})$ if the strain rate field $\mathbf{e}$ is integrable:

$$D(\mathbf{e}) = \int_\Omega d_x(\mathbf{e}(x))\, dV.$$

Here, $\Omega$ is a domain occupied by the body and $d_x$ is the (local) dissipation. It is not very easy to prove this formula or, in more detail, the equality

$$\sup \left\{ \int_\Omega \mathbf{e} \cdot \boldsymbol{\sigma}_* \, dV : \boldsymbol{\sigma}_* \in C \right\} = \int_\Omega \sup \{ \mathbf{e}(x) \cdot \tilde{\boldsymbol{\sigma}} : \tilde{\boldsymbol{\sigma}} \in C \} \, dV$$

("summation is permutable with maximization"). We have already met this relation in Subsection 1.3. Simple and nonrestrictive conditions for its validity will be formulated in Section 5 and proved in Section 6.

Consider, within the framework of the unified formulation of the rigid-plastic problem, a body subjected to the load $\mathbf{l}$. Let $\mathbf{v}$ be a kinematically admissible velocity field and the external power $\langle\langle \mathbf{v}, \mathbf{l} \rangle\rangle$ be positive. Then we refer to the number $m_k(\mathbf{v})$ equal to the total dissipation over the external power,

$$m_k(\mathbf{v}) = \frac{D(\operatorname{Def} \mathbf{v})}{\langle\langle \mathbf{v}, \mathbf{l} \rangle\rangle},$$

as the *kinematic multiplier* of the load $\mathbf{l}$ (corresponding to the velocity field $\mathbf{v}$). We recall that $\operatorname{Def} \mathbf{v}$ stands for the strain rate field corresponding to $\mathbf{v}$. Note that we can use any fixed stress field $\mathbf{s}_l$ which equilibrates the load $\mathbf{l}$ instead of $\mathbf{l}$ while evaluating the external power: $\langle\langle \mathbf{v}, \mathbf{l} \rangle\rangle = \langle \mathbf{e}, \mathbf{s}_l \rangle$ due to (IV) in Subsection 1.4. Therefore, the definition of the kinematic multiplier can be re-written as

$$m_k(\mathbf{v}) = \frac{D(\operatorname{Def} \mathbf{v})}{\langle \operatorname{Def} \mathbf{v}, \mathbf{s}_l \rangle}.$$

EXAMPLE 1. Consider a three-dimensional body occupying domain $\Omega$. Set $C_x$ of admissible stresses is specified for a.e. $x \in \Omega$; let $d_x$ be the corresponding (local) dissipation. The total dissipation $D(\mathbf{e})$ for any integrable strain rate field $\mathbf{e}$ can be evaluated by the formula

$$D(\mathbf{e}) = \int_\Omega d_x(\mathbf{e}(x))\, dV$$

(see Remark 1). Therefore the general definition of the kinematic multiplier for a sufficiently regular velocity field $\mathbf{v}$ takes the form

$$m_k(\mathbf{v}) = \frac{D(\operatorname{Def}\mathbf{v})}{\langle\langle \mathbf{v}, \mathbf{l} \rangle\rangle} = \frac{\int\limits_{\Omega} d_x(\mathbf{e}(x))\, dV}{\int\limits_{\Omega} \mathbf{f}\mathbf{v}\, dV \int\limits_{S_q} q\mathbf{v}\, dS},$$

that is, the definition is reduced to the particular definition in Section III.

The following proposition is the first basic theorem of limit analysis. The proposition practically reproduces the formulation and proof of Theorem III.5.1 within the framework of the unified approach to rigid-plastic problems.

THEOREM 1. Every kinematic multiplier of the load $\mathbf{l}$ is not less than any of its static multipliers: $m_s(\boldsymbol{\sigma}) \leq m_k(\mathbf{v})$.

PROOF. Let $\mathbf{v} \in \mathcal{V}$ be a kinematically admissible velocity field with the positive external power $\langle\langle \mathbf{v}, \mathbf{l} \rangle\rangle$. Then the kinematic multiplier $m_k(\mathbf{v})$ corresponds to $\mathbf{v}$ and the equality

$$D(\operatorname{Def}\mathbf{v}) = m_k(\mathbf{v})\langle\langle \mathbf{v}, \mathbf{l} \rangle\rangle \tag{4.2}$$

is valid. Let $m = m_s(\boldsymbol{\sigma})$ be a static multiplier of the load $\mathbf{l}$, that is, the stress field $\boldsymbol{\sigma}$ is admissible and equilibrates the load $m\mathbf{l}$. Using $\mathbf{v}$ as a virtual velocity field in equilibrium conditions (IV) in Subsection 1.4 yields $\langle \operatorname{Def}\mathbf{v}, \boldsymbol{\sigma} \rangle = \langle\langle \mathbf{v}, m\mathbf{l} \rangle\rangle$. Subtracting this equality from (4.2), we arrive at

$$D(\operatorname{Def}\mathbf{v}) - \langle \operatorname{Def}\mathbf{v}, \boldsymbol{\sigma} \rangle = (m_k(\mathbf{v}) - m)\,\langle\langle \mathbf{v}, \mathbf{l} \rangle\rangle. \tag{4.3}$$

The left-hand side of this equality is nonnegative by the definition of the dissipation (since $\boldsymbol{\sigma} \in C$). The second multiplier on the right-hand side in (4.3) is positive by the above assumption. Then the first multiplier is nonnegative, and consequently $m_k(\mathbf{v}) \geq m = m_s(\boldsymbol{\sigma})$, which proves the theorem.

According to Theorem 1 *every kinematic multiplier of the load* $\mathbf{l}$ *is an upper bound for the safety factor*: $\alpha_l \leq m_k(\mathbf{v})$ (since $\alpha_l$ is the supremum of static multipliers).

**4.3. Criterion for static and kinematic multipliers equality.** Under certain circumstances, a static multiplier of the load $\mathbf{l}$ equals one of its kinematic multipliers, in which case Theorem 1 implies that their common value is the supremum of static multipliers, that is, the safety factor $\alpha_l$. In particular, this means that the kinematic method can result in the precise value of the safety factor and not only in its upper bounds. On the other hand, this suggests looking for a criterion for equality of the static and kinematic multipliers. This criterion allows formulation of the problem of searching for appropriate velocity and stress fields $\mathbf{v}$ and $\boldsymbol{\sigma}$ satisfying the criterion and, consequently, the equality $m_s(\boldsymbol{\sigma}) = m_k(\mathbf{v}) = \alpha_l$. Thus, solving this problem is a way to compute the safety factor $\alpha_l$, which is the main reason why we are looking for a criterion for the equality $m_s(\boldsymbol{\sigma}) = m_k(\mathbf{v})$. Such criterion has already been established for the case of the strong formulation of the rigid-plastic problem;

this time it will be obtained within the framework of the unified formulation. Recall that this problem (Subsection 1.4) consists of searching for the stress field $\boldsymbol{\sigma} \in \mathcal{S}$ and the velocity field $\mathbf{v} \in \mathcal{V}$ satisfying the relations:

$$\boldsymbol{\sigma} \in \Sigma + \mathbf{s}_l,$$
$$\mathbf{e} = \operatorname{Def} \mathbf{v}, \quad \mathbf{v} \in \mathcal{V},$$
$$\boldsymbol{\sigma} \in C, \quad \langle \mathbf{e}, \boldsymbol{\sigma} - \boldsymbol{\sigma}_* \rangle \geq 0 \quad \text{for every} \quad \boldsymbol{\sigma}_* \in C.$$

The following proposition is the second of the two basic theorems of limit analysis. The proposition is a close analog to Theorem III.5.2 and establishes a criterion for equality of the static and kinematic multipliers.

THEOREM 2. Let $\boldsymbol{\sigma} \in \mathcal{S}$ be a stress field and $\mathbf{v} \in \mathcal{V}$ be a kinematically admissible velocity field in a rigid perfectly plastic body. The static and kinematic multipliers $m_s(\boldsymbol{\sigma})$ and $m_k(\mathbf{v})$ of the load $\mathbf{l}$ are equal, $m_s(\boldsymbol{\sigma}) = m_k(\mathbf{v}) = m > 0$, if and only if $\boldsymbol{\sigma}$ and $\mathbf{v}$ are a solution to the rigid-plastic problem for the load $m\mathbf{l}$, $m > 0$, and the external power of $\mathbf{l}$ on $\mathbf{v}$ is nonzero (in case the zero stress field is safe for the body: if and only if $\boldsymbol{\sigma}$ and $\mathbf{v}$ are a solution to this problem and $\mathbf{v} \neq 0$).

PROOF. Suppose $m_s(\boldsymbol{\sigma}) = m_k(\mathbf{v}) = m > 0$. Let us show that $\boldsymbol{\sigma}$ and $\mathbf{v}$ are a solution to the rigid-plastic problem for the load $m\mathbf{l}$ (note that we do not need to verify positiveness of the external power: existence of the kinematic multiplier $m_k(\mathbf{v})$ presumes that $\langle\langle \mathbf{v}, \mathbf{l} \rangle\rangle$ is positive). Since $m = m_s(\boldsymbol{\sigma})$ is the static multiplier corresponding to the stress field $\boldsymbol{\sigma}$, we have

$$\boldsymbol{\sigma} \in \Sigma + m\mathbf{s}_l, \quad \boldsymbol{\sigma} \in C. \tag{4.4}$$

Apart from that, the equality $m = m_k(\mathbf{v})$ and formula (4.3) derived in the proof of Theorem 1 imply

$$D(\operatorname{Def} \mathbf{v}) = \langle \operatorname{Def} \mathbf{v}, \boldsymbol{\sigma} \rangle. \tag{4.5}$$

According to the definition of the dissipation (4.1) and admissibility of the stress field $\boldsymbol{\sigma}$, the previous equality is equivalent to the relation

$$\langle \operatorname{Def} \mathbf{v}, \boldsymbol{\sigma}_* \rangle \geq \langle \operatorname{Def} \mathbf{v}, \boldsymbol{\sigma} \rangle \quad \text{for every} \quad \boldsymbol{\sigma}_* \in C,$$

which means that $\boldsymbol{\sigma}$ and $\mathbf{v}$ satisfy the constitutive maximum principle. Together with (4.4) this implies that $\boldsymbol{\sigma}$, $\mathbf{v}$ are a solution to the rigid-plastic problem for the load $m\mathbf{l}$.

Suppose now that the stress field $\boldsymbol{\sigma}$ and the velocity field $\mathbf{v}$ are a solution to the rigid-plastic problem for the load $m\mathbf{l}$, $m > 0$, that is, $\boldsymbol{\sigma}$ and $\mathbf{v}$ satisfy relations (4.4), (4.5). The external power $\langle\langle \mathbf{v}, \mathbf{l} \rangle\rangle$ is assumed to be nonzero. Let us show that the power is positive (consequently, the kinematic multiplier $m_k(\mathbf{v})$ is defined) and verify the equality $m_k(\mathbf{v}) = m_s(\boldsymbol{\sigma})$. Note that $\boldsymbol{\sigma}$ equilibrates the load $m\mathbf{l}$ (see the first of relations (4.4)). Using $\mathbf{v}$ as a virtual velocity field in the equilibrium condition (IV) in Subsection 1.4 results in

$$\langle \operatorname{Def} \mathbf{v}, \boldsymbol{\sigma} \rangle = m\langle\langle \mathbf{v}, \mathbf{l} \rangle\rangle. \tag{4.6}$$

Here, $m$ is positive by assumption, while the left-hand side is nonnegative due to (4.5) and nonnegativeness of the dissipation. Moreover, we have $\langle\langle \mathbf{v}, \mathbf{l} \rangle\rangle \neq 0$ by assumption and hence both sides in (4.6) are positive. Then $\langle\langle \mathbf{v}, \mathbf{l} \rangle\rangle$ is positive, the kinematic multiplier is defined for the velocity field $\mathbf{v}$, and (4.5) together with (4.6) result in the equality $m_k(\mathbf{v}) = m$. On the other hand, (4.4) implies that $m_s(\boldsymbol{\sigma})$ also equals $m$. Thus, the equality $m_k(\mathbf{v}) = m_s(\boldsymbol{\sigma})$ is valid.

It remains to show that, in case the zero stress field is safe, the assumption $\langle\langle \mathbf{v}, \mathbf{l} \rangle\rangle \neq 0$ can be replaced by $\mathbf{v} \neq 0$ in the above reasoning. More precisely, let us verify for the velocity field $\mathbf{v}$ of the solution $\boldsymbol{\sigma}$, $\mathbf{v}$ to the rigid-plastic problem that the condition $\mathbf{v} \neq 0$ implies $\langle\langle \mathbf{v}, \mathbf{l} \rangle\rangle \neq 0$. Note that as $\boldsymbol{\sigma}$, $\mathbf{v}$ satisfy relations (4.5), (4.6), $\langle\langle \mathbf{v}, \mathbf{l} \rangle\rangle \neq 0$ is equivalent to $D(\operatorname{Def} \mathbf{v}) \neq 0$. To verify the latter relation, note that $\operatorname{Def} \mathbf{v} \neq 0$ since $\mathbf{v} \neq 0$ (see (III) in Subsection 1.4) and use the following general property of the dissipation: $\mathbf{e} \neq 0$ implies $D(\mathbf{e}) \neq 0$ in case the zero stress field is safe.

To finish the proof let us verify the latter property or, in other words, show that the equality $D(\mathbf{e}) = 0$ implies $\mathbf{e} = 0$. Indeed, the set $C$ of admissible stress fields contains an absorbing set $Q$ since the zero stress field is safe. Because of that the equality $D(\mathbf{e}) = 0$ results in

$$0 = D(\mathbf{e}) = \sup\{\mathbf{e} \cdot \boldsymbol{\sigma}_* : \boldsymbol{\sigma}_* \in C\} \geq \sup\{\mathbf{e} \cdot \boldsymbol{\sigma}_* : \boldsymbol{\sigma}_* \in Q\}. \tag{4.7}$$

Let us show that $\langle \mathbf{e}, \mathbf{s} \rangle = 0$ for every $s \in \mathcal{S}$ in this case. Suppose $\langle \mathbf{e}, \mathbf{s} \rangle \neq 0$ for a certain $\mathbf{s} \in \mathcal{S}$. Without loss of generality we assume $\langle \mathbf{e}, \mathbf{s} \rangle > 0$ ($\mathbf{s}$ can be replaced by $-\mathbf{s}$ in case $\langle \mathbf{e}, \mathbf{s} \rangle$ is negative). Since $Q$ is absorbing, there exists $\boldsymbol{\sigma}_* \in Q$ such that $\langle \mathbf{e}, \boldsymbol{\sigma}_* \rangle > 0$, which contradicts inequality (4.7). Thus, the above assumption is wrong and $\langle \mathbf{e}, \mathbf{s} \rangle = 0$ for every $\mathbf{s} \in \mathcal{S}$. Then $\mathbf{e} = 0$ according to (III) in Subsection 1.4, which proves the theorem.

**4.4. On methods for limit analysis.** The definitions of the safety factor and the static multiplier together with Theorems 1 and 2 imply that the safety factor can be estimated and computed by the following methods.

- *Static method*: any static multiplier gives a lower bound for the safety factor; the safety factor equals the supremum of static multipliers.
- *Kinematic method*: any kinematic multiplier gives an upper bound for the safety factor.
- *Rigid-plastic solutions method*: if a solution with a nonzero velocity field is found to the rigid-plastic problem for a load $m_0\mathbf{l}$, $m_0 > 0$, then the safety factor of the load $\mathbf{l}$ equals $m_0$.

The methods are applicable to all types of the rigid perfectly plastic systems: three-dimensional bodies, two- and one-dimensional continual systems (plates, shells, beams), discrete systems (trusses, frames), etc.

The following important remark concerning the kinematic method should be made. Theorem 2 shows that the kinematic method results not only in estimations for the safety factor but, sometimes, also in its precise value. However, the theorem specifies no conditions for the latter possibility. Such conditions will be established in Chapters V and VI.

## 5. Integral formulation of constitutive maximum principle

We have already introduced the integral version of the constitutive maximum principle in Section 1. In this section, we present conditions under which this formulation is equivalent to the original, point-wise one.

These conditions are nonrestrictive and are satisfied for all rigid-plastic problems of practical interest.

**5.1. Integral formulation.** The rigid perfectly plastic constitutive relations have the form of the constitutive maximum principle (see Subsection I.3.2):

$$\begin{gathered}\boldsymbol{\sigma}(x) \in C_x, \quad \mathrm{e}(x)\cdot(\boldsymbol{\sigma}(x)-\boldsymbol{\tau}) \geq 0 \text{ for every } \boldsymbol{\tau} \in C_x \\ \text{for a.e. point } x \text{ of the body.}\end{gathered} \tag{5.1}$$

This point-wise formulation implies the integral inequality

$$\boldsymbol{\sigma} \in C, \quad \int_\Omega \mathrm{e}(x)\cdot(\boldsymbol{\sigma}(x)-\boldsymbol{\sigma}_*(x))\,dV \geq 0 \text{ for every } \boldsymbol{\sigma}_* \in C, \tag{5.2}$$

where $\Omega$ is the domain occupied by the body. To derive (5.2), consider an admissible stress field $\boldsymbol{\sigma}_*$, set $\boldsymbol{\tau} = \boldsymbol{\sigma}(x) \in C_x$ in (5.1) and integrate the inequality

$$\mathrm{e}(x)\cdot(\boldsymbol{\sigma}(x)-\boldsymbol{\sigma}_*(x)) \geq 0$$

over $\Omega$. It is remarkable that not only does the point-wise formulation imply the integral one but also vice versa, that is, these are the equivalent formulations of the constitutive maximum principle. We present conditions for the equivalence in this section and prove the equivalence in the next section. We refer to (5.2) as the *integral formulation of the constitutive maximum principle.*

Other, sometimes more convenient, forms of maximum principles (5.1) and (5.2) are worth mentioning. Let us re-write (5.1) and (5.2) as

$$\boldsymbol{\sigma}(x) \in C_x, \quad \mathrm{e}(x)\cdot\boldsymbol{\sigma}(x) \geq \sup\{\mathrm{e}(x)\cdot\boldsymbol{\tau} : \boldsymbol{\tau} \in C_x\} \quad \text{for a.e. } x \in \Omega$$

and

$$\boldsymbol{\sigma} \in C, \quad \int_\Omega \mathrm{e}(x)\cdot\boldsymbol{\sigma}(x)\,dV \geq \sup\left\{\int_\Omega \mathrm{e}(x)\cdot\boldsymbol{\sigma}_*(x)\,dV : \boldsymbol{\sigma}_* \in C\right\}.$$

Since the real stress field $\boldsymbol{\sigma}$ is admissible, it can be used as $\boldsymbol{\sigma}_*$ and its value $\boldsymbol{\sigma}(x)$ as $\boldsymbol{\tau}$ in the previous relations. This allows us to conclude that equality and not strict inequalities are always realized in these relations:

$$\boldsymbol{\sigma}(x) \in C_x, \quad \mathrm{e}(x)\cdot\boldsymbol{\sigma}(x) = \sup\{\mathrm{e}(x)\cdot\boldsymbol{\tau} : \boldsymbol{\tau} \in C_x\} \quad \text{for a.e. } x \in \Omega \tag{5.3}$$

and

$$\boldsymbol{\sigma} \in C, \quad \int_\Omega \mathrm{e}(x)\cdot\boldsymbol{\sigma}(x)\,dV = \sup\left\{\int_\Omega \mathrm{e}(x)\cdot\boldsymbol{\sigma}_*(x)\,dV : \boldsymbol{\sigma}_* \in C\right\}. \tag{5.4}$$

Taking into account that the expression on the right-hand side in (5.3) is the dissipation

$$d_x(\varepsilon) = \sup\{\varepsilon \cdot \tau : \tau \in C\}, \tag{5.5}$$

and that on the right-hand side in (5.4) is the total dissipation

$$D(\mathrm{e}) = \sup\left\{\int_\Omega \mathrm{e}(x)\cdot\sigma_*(x)\,dV : \sigma_* \in C_x\right\}. \tag{5.6}$$

Therefore constitutive relations (5.3), (5.4) can be written as

$$\sigma(x) \in C_x, \quad \mathrm{e}(x)\cdot\sigma(x) = d_x(\mathrm{e}) \quad \text{for a.e. } x \in \Omega$$

and

$$\sigma \in C, \quad \int_\Omega \mathrm{e}(x)\cdot\sigma(x)\,dV = D(\mathrm{e}).$$

We state that these two relations are equivalent or, which is the same, (5.1) is equivalent to (5.2), and (5.3) is equivalent to (5.4). The first of the two relations in each of these pairs is a *point-wise* formulation of the constitutive maximum principle; the second is the *integral* formulation of the principle. As we have just seen, the point-wise formulation implies the integral one, while the converse statement is by far not obvious. We reduced its proof to establishing a certain extremum property of the dissipation, see Subsection 1.4. This property and a way to verify it are discussed in the next subsection.

**5.2. Extremum property and computation of dissipation.** The total dissipation $D$ in Subsection 4.2 was defined without any reference to the local dissipation $d_x$. However, the following formula relates them:

$$\int_\Omega \sup\{\mathrm{e}(x)\cdot\tilde{\sigma} : \tilde{\sigma} \in C_x\}\,dV = \sup\left\{\int_\Omega \mathrm{e}(x)\cdot\sigma_*(x)\,dV : \sigma_* \in C\right\}$$

or, which is the same,

$$\int_\Omega d_x(\mathrm{e})\,dV = \sup\left\{\int_\Omega \mathrm{e}(x)\cdot\sigma_*(x)\,dV : \sigma_* \in C\right\} \tag{5.7}$$

(recall that the right-hand side is $D(\mathrm{e})$).

Equality (5.7) represents an extremum property of the dissipation: the summarized local dissipation (on the left-hand side) is the supremum of an integral; the supremum is taken over admissible stress fields $\sigma_*$ and the integral depends linearly on $\sigma_*$ and on the parameter e. There are at least two reasons why (5.7) is of fundamental interest: if (5.7) is valid,

- the point-wise and integral formulations of the constitutive maximum principle are equivalent (see Subsection 1.3);
- the total dissipation $D(\mathrm{e})$ can be calculated by simply integrating the local dissipation $d_x(\mathrm{e}(x))$,

$$D(\mathrm{e}) = \int\limits_{\Omega} d_x(\mathrm{e}(x))\, dV.$$

This equality is nothing but (5.7): the right-hand side in (5.7) is exactly $D(\mathrm{e})$ as it is defined in Subsection 4.2. Using the above formula for computing the dissipation $D$ reduces the general definition of the kinematic multiplier, in case $\mathbf{e}$ is an integrable field, to the constructive definition previously given in Subsection III.5.2.

Let us now discuss a way to establish conditions for validity of (5.7). The inequality

$$\int\limits_{\Omega} d_x(\mathrm{e}(x))\, dV \geq \sup \left\{ \int\limits_{\Omega} \mathrm{e}(x) \cdot \boldsymbol{\sigma}_*(x)\, dV \,:\, \boldsymbol{\sigma}_* \in C \right\} \tag{5.8}$$

is obviously valid due to the definition of $d_x$. The following reasoning looks suitable for establishing the inequality

$$\int\limits_{\Omega} d_x(\mathrm{e}(x))\, dV \leq \sup \left\{ \int\limits_{\Omega} \mathrm{e}(x) \cdot \boldsymbol{\sigma}_*(x)\, dV \,:\, \boldsymbol{\sigma}_* \in C \right\}, \tag{5.9}$$

and this inequality is what we are looking for, since together with (5.8) it results in (5.7). Let $\varepsilon$ be an arbitrary positive number. By definition (5.5) of the dissipation $d_x$, there exists stress $\boldsymbol{\sigma}_\varepsilon(x)$ such that

$$\boldsymbol{\sigma}_\varepsilon(x) \in C_x, \quad \mathrm{e}(x) \cdot \boldsymbol{\sigma}_\varepsilon(x) \geq d_x(\mathrm{e}(x)) - \varepsilon. \tag{5.10}$$

Consider the stress field $\boldsymbol{\sigma}_\varepsilon$ which takes the above chosen values $\boldsymbol{\sigma}_\varepsilon(x)$ at a.e. $x \in \Omega$. Suppose the field $\boldsymbol{\sigma}_\varepsilon$ belongs to the space $\mathcal{S}$. Then it also belongs to $C$ since its values are admissible: $\boldsymbol{\sigma}_\varepsilon(x) \in C_x$, which together with (5.10) results in

$$\sup \left\{ \int\limits_{\Omega} \mathrm{e}(x) \cdot \boldsymbol{\sigma}_*(x)\, dV \,:\, \boldsymbol{\sigma}_* \in C \right\} \geq \int\limits_{\Omega} \mathrm{e}(x) \cdot \boldsymbol{\sigma}_\varepsilon(x)\, dV$$
$$\geq \int\limits_{\Omega} d_x(\mathrm{e}(x))\, dV - O(\varepsilon).$$

Since $\varepsilon > 0$ is arbitrary here, we arrive to (5.9) and hence to (5.7). In this way, we obtained (5.7) by assuming that the values $\boldsymbol{\sigma}_\varepsilon(x)$ can be chosen so that the field $\boldsymbol{\sigma}_\varepsilon$ belongs to the space $\mathcal{S}$. Although the values $\boldsymbol{\sigma}_\varepsilon(x)$ are admissible, the admissibility itself does not ensure that $\boldsymbol{\sigma}_\varepsilon$ is in $\mathcal{S}$; for example, $\boldsymbol{\sigma}_\varepsilon$ may be nonmeasurable while all fields in $\mathcal{S}$ possess this property. Thus, it is reasonable to look for conditions under which the appropriate choice of $\boldsymbol{\sigma}_\varepsilon$ is possible. It is clear that the wider the space $\mathcal{S}$ the easier it is to find an appropriate $\boldsymbol{\sigma}_\varepsilon$ in it. Another thing which makes the choice of $\boldsymbol{\sigma}_\varepsilon$ easier is a certain regularity of the mapping $x \to C_x$. We start now by describing a property of this mapping which implies the required regularity. After that, in Subsection 5.4, we discuss the sense in which $\mathcal{S}$ appears sufficiently wide for the appropriate choice of $\boldsymbol{\sigma}_\varepsilon$.

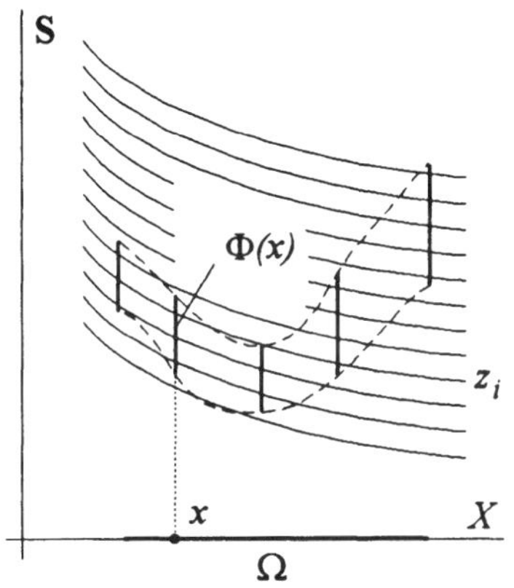


**Figure 5.1**

**5.3. Admissible stresses: a set-valued mapping.** A set of admissible stresses in six-dimensional space $Sym$ corresponds to every point of a three-dimensional rigid-plastic body. Similarly, a set of admissible stresses (bending moments, axial and cutting forces) in a five-dimensional space corresponds to every point of a rigid perfectly plastic beam. We consider these and all other cases within the framework of the unified theory: the body occupies domain $\Omega$ in a one-, two- or three-dimensional space, stresses belong to the corresponding finite-dimensional space $\mathbf{S}$, and the set of admissible stresses $C_x$ in $\mathbf{S}$ is given for every point of the body.

If the subset $\Phi(x)$ in $\mathbf{S}$ is given for every point $x \in \Omega$, we say that $\Phi$ (or $x \to \Phi(x)$) is a *set-valued mapping* from $\Omega$ into $\mathbf{S}$. In particular, the set $C_x$ of admissible stresses corresponds to every point $x$ of a rigid perfectly plastic body, and this determines a set-valued mapping. The concept of measurability of the sets and functions, explained in Appendix B, will now be employed to describe a class of set-valued mappings $x \to C_x$ such that the dissipation $d_x$ possesses extremum property (5.7). We refer to a set-valued mapping $\Phi$ from $\Omega$ into $\mathbf{S}$ as *measurable* if there is a countable collection of measurable functions $\mathbf{z}_1, \mathbf{z}_2, \ldots$ which are defined on $\Omega$ and take their values in $\mathbf{S}$, such that

(M1) the set $\{x \in \Omega : \mathbf{z}_i(x) \in \Phi(x)\}$ is measurable for every $i = 1, 2, \ldots$,
(M2) for a.e. $x$ in $\Omega$ and every $\mathbf{s}$ in $\Phi(x)$, there exists a subsequence $\{\mathbf{z}_{i_n}(x)\}$ which converges to $\mathbf{s}$ when $n \to \infty$.

We say that the collection $\mathbf{z}_1, \mathbf{z}_2, \ldots$ *approximates* the set-valued mapping $\Phi$. Figure 5.1 illustrates this definition. The other way to describe the property (M2) is as follows: for a.e. $x$ in $\Omega$ the set $\Psi(x)$ lies in the closure of the intersection $\Phi(x) \cap \{\mathbf{z}_1(x), \mathbf{z}_2(x), \ldots\}$.

EXAMPLE 1. *Homogeneous body.* Let sets $C_x$ of admissible stresses be the same for every point $x$ of a body, $C_x = C_o$, that is, the body is homogeneous. Then a simple set-valued mapping $x \to C_o$ is defined on the domain $\Omega$ occupied by the body. It is clear that the mapping is measurable. Indeed, let us choose a countable subset $\{\mathbf{s}_1, \mathbf{s}_2, \dots\} \subset C_o$ dense in $C_o$ and define a constant function on $\Omega$ which will be denoted by the same symbol $\mathbf{s}_i$. The collection of the functions $\mathbf{s}_1, \mathbf{s}_2, \dots,$ obviously approximates the set-valued mapping $x \to C_o$. Therefore, the latter is measurable.

EXAMPLE 2. Consider a three-dimensional body occupying domain $\Omega$. Let the set of admissible stresses at every point $x \in \Omega$ be of the form

$$C_x = \{\boldsymbol{\sigma} \in Sym : \left|\boldsymbol{\sigma}^d - \boldsymbol{\sigma}_0^d(x)\right| \le \sqrt{2}k(x)\},$$

where $\boldsymbol{\sigma}_0^d$ and $k$ are measurable functions on $\Omega$, and $k(x) > 0$ for every $x \in \Omega$. The superscript $d$ as usuall denotes the deviatoric part of a tensor. In particular, if $\boldsymbol{\sigma}_0^d \equiv 0$, the set $C_x$ of admissible stresses corresponds to the Mises yield criterion; and if $k$ is not constant, the body is inhomogeneous. Let us show that the mapping $x \to C_x$ is measurable. Consider a countable set $\mathbf{s}_1, \mathbf{s}_2, \dots$ dense in the stress space $Sym$. We define a function which is constant on $\Omega$ and takes the value $\mathbf{s}_i$; and we use the same symbol to denote this function.

The collection of the functions $\mathbf{s}_1, \mathbf{s}_2, \dots$ possesses property (M1) as far as the mapping $x \to C_x$ is concerned. Indeed, consider the set

$$\{x \in \Omega : \mathbf{s}_i(x) \in C_x\} = \{x \in \Omega : \left|\mathbf{s}_i^d(x) - \boldsymbol{\sigma}_0^d(x)\right|^2 - 2k^2(x) \le 0\}$$

that is, a point $x \in \Omega$ belongs to this set if and only if the inequality $F(x) \le 0$ holds, where

$$F(x) = \left|\mathbf{s}_i^d(x) - \boldsymbol{\sigma}_0^d(x)\right|^2 - 2k^2(x),$$

so that $F(x)$ is the result of multiplying and adding measurable functions. Note that $F$ is a measurable function (see Appendix B). Then the set $\{x \in \Omega : F(x) \le 0\}$ is measurable, which means that the collection of the functions $\mathbf{s}_1, \mathbf{s}_2, \dots$ possesses property (M1). Let us now verify that this collection approximates the mapping $x \to C_x$. Consider first $\boldsymbol{\sigma}$ in the interior of $C_x$. Then there is a ball $B(\boldsymbol{\sigma})$ of a positive radius and with its center at $\boldsymbol{\sigma}$ such that $B(\boldsymbol{\sigma})$ is a subset in $C_x$. Since the set of points $\mathbf{s}_1, \mathbf{s}_2, \dots$ is dense in $Sym$, there exists a subsequence $\{\mathbf{s}_{i_n}\}$ converging to $\boldsymbol{\sigma}$ when $n \to \infty$. Members of this subsequence belong to $B(\boldsymbol{\sigma})$ (and, therefore, to $C_x$) if $n$ is sufficiently large. In other words, the values $\mathbf{s}_{i_n}(x)$ of the functions $\mathbf{s}_{i_n}$ converge to $\boldsymbol{\sigma}$ when $n \to \infty$, and $s_{i_n}(x) \in C_x$ for sufficiently large $n$. Thus we arrive at property (M2) if only the interior of $C_x$ is considered instead of $C_x$. We now similarly verify (M2) when $\boldsymbol{\sigma}$ belongs to the boundary of $C_x$. Let $\varepsilon > 0$ be an arbitrarily small number and consider $\boldsymbol{\sigma}_\varepsilon$ in the interior of $C_x$ with the distance between $\boldsymbol{\sigma}$ and $\boldsymbol{\sigma}_\varepsilon$ less than $\varepsilon/2$. As it was established above, there is a member of the set $\{\mathbf{s}_1, \mathbf{s}_2, \dots\} \cap C_x$ with the distance from $\boldsymbol{\sigma}_\varepsilon$ less then $\varepsilon/2$ and, therefore, with the distance from $\boldsymbol{\sigma}$ less than $\varepsilon$. Thus, there exists a subsequence $\{\mathbf{s}_{i_n}\}$ with $\mathbf{s}_{i_n} \in C_x$, which converges to $\boldsymbol{\sigma}$ also in case $\boldsymbol{\sigma}$ belongs to the boundary of $C_x$. This means that the collection $\{\mathbf{s}_1, \mathbf{s}_2. \dots\}$ approximates the mapping $x \to C_x$. Consequently, the mapping is measurable.

EXAMPLE 3. *Composite body.* Consider body $\mathcal{B}$ occupying domain $\Omega$. Let $\Omega$ consist of pair-wise disjointed subdomains $\omega_\mu$, $\mu = 1, \dots, M$. The material properties in each of the domains $\omega_\mu$ are rigid perfectly plastic and are given by the set-valued mapping $\Phi_\mu$ from $\omega_\mu$ into $\mathbf{S}$: $C_x = \Phi_\mu(x)$. We assume that the parts $\omega_\mu$ of the body are "glued" at the surfaces that separate them and that the "glue" is strong enough to allow considering a stress field $\boldsymbol{\sigma}$ as admissible if and only if $\boldsymbol{\sigma}$ is admissible in each of the subdomains $\omega_\mu$. Thus, $\mathcal{B}$ is a composite rigid perfectly plastic body with the set

$$C_x = \begin{cases} \Phi_1(x) & \text{if } x \in \omega_1, \\ \dots \\ \Phi_M(x) & \text{if } x \in \omega_M \end{cases}$$

specified for $x \in \Omega$. In other words, the set-valued mapping $x \to C_x$ is defined on $\Omega$ and determines the material properties of $\mathcal{B}$. It is easily seen that this mapping is measurable if all the mappings $x \to \Phi_\mu(x)$ are measurable. Indeed, let $\{\mathbf{s}_{\mu 1}, \mathbf{s}_{\mu 2}, \dots\}$ be a collection of measurable functions which possesses the properties (M1) and (M2) with respect to the mapping $\Phi_\mu$. Then we introduce the following functions on $\Omega$:

$$\mathbf{s}_i(x) = \begin{cases} \mathbf{s}_{1i}(x) & \text{if } x \in \omega_1, \\ \dots \\ \mathbf{s}_{Mi}(x) & \text{if } x \in \omega_M, \end{cases}$$

$(i = 1, 2, \dots)$. It is clear that every $\mathbf{s}_i$ is a measurable function and that the collection $\{s_1, \mathbf{s}_2, \dots\}$ approximates the set-valued mapping $x \to C_x$. Thus, this mapping is measurable.

Normally, dependence $C_x$ upon $x$ is rather simple, and, like the above examples, it is easy to verify that the mapping $x \to C_x$ is measurable. Actually, this mapping is measurable in all limit analysis problems of practical interest.

**5.4. Conditions for equivalence of constitutive principle formulations.** The above-introduced concept of measurable set-valued mapping allows formulation of conditions under which the dissipation possesses extremum property (5.7). As we know, this property implies that the point-wise and integral formulations of the constitutive maximum principle are equivalent.

To present the final formulation of the equivalence conditions, one more assumption concerning the space of stress fields is needed. It should be remembered that stress fields and strain rate fields are considered in the linear spaces $\mathcal{S}$ and $\mathcal{E}$, respectively, both spaces consisting of functions defined on the domain $\Omega$ with their values in a finite-dimensional space $\mathbf{S}$. For example, $\mathbf{S} = Sym$ in the case of a three-dimensional body. We assume that all functions in $\mathcal{S}$ are integrable and also that for every $\boldsymbol{\sigma} \in \mathcal{S}$ and every integrable $\mathbf{e} \in \mathcal{E}$ the function $\mathbf{e} \cdot \boldsymbol{\sigma}$ is integrable.

The space $\mathcal{S}$ is assumed to be sufficiently wide in the following sense: 1) $\mathcal{S}$ contains all integrable bounded functions, 2) for every subdomain $\omega$ in $\Omega$ and every function $\mathbf{s} \in \mathcal{S}$ the space $\mathcal{S}$ contains the function

$$\mathbf{s}_\omega(x) = \begin{cases} \mathbf{s}(x) & \text{if } x \in \omega, \\ 0 & \text{if } x \notin \omega. \end{cases}$$

The following proposition presents conditions under which the dissipation possesses extremum property (5.7).

THEOREM 1. Let the set-valued mapping $x \to C_x$ be measurable and the space $\mathcal{S}$ be sufficiently wide. Then the equality

$$\int_\Omega d_x(\mathrm{e})\,dV = \sup\left\{\int_\Omega \mathrm{e}(x)\cdot\boldsymbol{\sigma}_*(x)\,dV \,:\, \boldsymbol{\sigma}_* \in C\right\}$$

holds for every integrable field $\mathrm{e} \in \mathcal{E}$.

The proposition will be proved in the next section.

Theorem 1 implies the following statement about conditions for equivalency of the point-wise and integral formulations of the constitutive maximum principle.

THEOREM 2. Let the set-valued mapping $x \to C_x$ be measurable and the space $\mathcal{S}$ be sufficiently wide. Then for every $\boldsymbol{\sigma} \in \mathcal{S}$ and any integrable e in $\mathcal{E}$, the point-wise relation

$$\boldsymbol{\sigma}(x) \in C_x, \quad \mathrm{e}(x)\cdot\boldsymbol{\sigma}(x) = d_x(\mathrm{e}) \quad \text{for a.e. } x \in \Omega$$

is equivalent to the integral relation

$$\boldsymbol{\sigma} \in C, \quad \int_\Omega \mathrm{e}(x)\cdot\boldsymbol{\sigma}(x)\,dV = D(\mathrm{e}),$$

that is, the point-wise and integral formulations of the constitutive maximum principle are equivalent (as far as integrable strain rate fields are concerned).

The proof of the proposition employs equality (5.7) valid by Theorem 1 and has already been given in Subsection 5.2.

## 6. Extremum of integral functional

The extremum property of the dissipation,

$$\int_\Omega d_x(\mathrm{e}(x))\,dV = \sup\left\{\int_\Omega \mathrm{e}(x)\cdot\boldsymbol{\sigma}_*(x)\,dV \,:\, \boldsymbol{\sigma}_* \in C\right\},$$

plays an important role in 1) deriving a formula for computing the kinematic multiplier, 2) establishing conditions for equivalence of the point-wise and integral formulations of the constitutive maximum principle. Theorem 5.1, presenting conditions for validity of equality (6.1), will now be proved as a corollary of a somewhat more general (and very useful in applications) proposition which provides a formula for evaluating the extremum of an integral functional. This formula generalizes the obvious equality

$$\begin{aligned}&\inf\{\phi_1(\mathrm{s}_1)+\phi_2(\mathrm{s}_2) \,:\, \mathrm{s}_1,\mathrm{s}_2 \in \mathbf{S}\}\\ &\qquad = \inf\{\phi_1(\mathrm{s}_1) \,:\, \mathrm{s}_1 \in \mathbf{S}\} + \inf\{\phi_2(\mathrm{s}_2) \,:\, \mathrm{s}_2 \in \mathbf{S}\}\end{aligned}$$

to the case of summing "infinitely many" functions $\phi_x$ (with the index $x$ ranging over a domain $\Omega$) instead of summing two functions $\phi_1$ and $\phi_2$; it is clear that sums should be replaced by integrals in the general case.

**6.1. Integral functional.** Consider the integral on the left-hand side in (6.1). The strain rate field e in its integrand is a function on $\Omega$ taking values in a finite-dimensional linear space $\mathbf{S}$; for example, $\mathbf{S} = Sym$ in the case of a three-dimensional body. Computing the integral maps a strain rate field into a number, that is, defines a function

$$\mathrm{e} \rightarrow \int_{\Omega} d_x(\mathrm{e}(x))\, dV$$

on the space $\mathcal{E}$ of strain rate fields.

In general, consider point $x \in \Omega$ and function $\phi_x$ which depends on the parameter $x$ and is defined on a finite-dimension linear space $\mathbf{S}$. (For example, if a three-dimensional rigid-plastic body occupies domain $\Omega$, the dissipation $d_x$ depends on the parameter $x$ and is a function on $\mathbf{S} = Sym$.) Consider now a linear space $\mathcal{S}$ of fields on $\Omega$ with their values in $\mathbf{S}$. Let $\boldsymbol{\sigma}$ be a field in $\mathcal{S}$, then expression $\phi_x(\boldsymbol{\sigma}(x))$ makes sense and can be integrated over $\Omega$. This generates the mapping

$$\boldsymbol{\sigma} \rightarrow \int_{\Omega} \phi_x(\boldsymbol{\sigma}(x))\, dV, \tag{6.2}$$

which is a function defined on the space $\mathcal{S}$. We emphasize that the argument of this function is a field $\boldsymbol{\sigma} \in \mathcal{S}$. A function of the form (6.2) is often referred to as an *integral functional* and $\phi_x$ as its *integrand.*

REMARK 1. In the above definition, $\mathcal{S}$ is not necessarily the space of stress fields. For example, we can consider space $\mathcal{E}$ of strain rate fields as the domain of an integral functional. The expression on the left-hand side in (6.1) is a value of such a functional with the integrand $d_x$.

REMARK 2. Integral (6.2) may not exist in the regular sense. For example, if the dissipation $d_x$ plays the role of integrand, then it is possible that $d_x(\mathrm{e}(x)) = +\infty$ for every $x$ in a certain subdomain of non-zero volume in $\Omega$ (see Subsection I.4.1). To cover cases like this, let us use the following rule to ascribe a value to integral functional (6.2). Consider all integrable functions $a$ on $\Omega$ which satisfy the condition $a(x) \geq \phi(\boldsymbol{\sigma}(x))$ for a.e. $x \in \Omega$. We define the value of (6.2) as the infimum of the integrals

$$\int_{\Omega} a(x)\, dV$$

over the set of all the above described functions $a$. If, for a certain $\boldsymbol{\sigma} \in \mathcal{S}$, there is no integrable function $a$ which satisfies the above-mentioned conditions, the infimum is $\inf \emptyset = +\infty$. We ascribe the value $+\infty$ to the integrable functional in this case. In case the function $\phi_x(\boldsymbol{\sigma}(x))$ is integrable, we return to the original definition. Indeed, we can use $\phi_x(\boldsymbol{\sigma}(x))$ as $a(x)$ while minimizing the integral

$$\int_{\Omega} a(x)\, dV.$$

This implies that the infimum cannot be greater than

$$\int_\Omega \phi_x(\sigma(x))\, dV.$$

At the same time, the infimum cannot be less than this number since the infimum is sought over functions $a$ satisfying the inequality $a(x) \geq \phi_x(\sigma(x))$ for a.e. $x \in \Omega$. Thus, the infimum equals

$$\int_\Omega \phi_x(\sigma(x))\, dV,$$

and the definition in this remark is equivalent to the original definition of integral functional values in case $\phi_x(\sigma(x))$ is integrable. We keep denoting values of the integral functional by integral (6.2) regardless of whether they are finite or not.

We restrict ourselves to integrands $\phi_x$ satisfying the following additional condition. Consider the epigraph of the function $\phi_x$ defined on the space $\mathbf{S}$, that is, the set

$$\operatorname{epi} \phi_x = \{(r, \mathrm{s}) \in \mathbf{R} \times \mathbf{S} : r \geq \phi_x(\mathrm{s})\}$$

in $\mathbf{R} \times \mathbf{S}$ (see Appendix I.B). We associate with the integrand $\phi_x$ the set-valued mapping $x \to \operatorname{epi} \phi_x$. In what follows, we only consider the integrands $\phi_x$, for which the mapping $x \to \operatorname{epi} \phi_x$ from $\Omega$ into $\mathbf{R} \times \mathbf{S}$ is measurable in the sense defined in Section 5.

EXAMPLE 1. Consider the integral functional with its values defined by the expression on the right-hand side in (6.1). It is more convenient to write this expression as

$$-\inf\left\{\int_\Omega -\mathrm{e}(x) \cdot \boldsymbol{\sigma}_*(x)\, dV : \boldsymbol{\sigma}_* \in C\right\}, \tag{6.3}$$

which is possible due to the obvious relation $\sup A = -\inf(-A)$. Let $C_x$ be the set of admissible stresses at the point $x$ of a rigid-plastic body. Consider the indicator function of $C$, that is, the function $\psi_{C_x}$ on $\mathbf{S}$ which takes the values

$$\psi_{C_x}(\mathrm{s}) = \begin{cases} 0 & \text{if } \mathrm{s} \in C_x, \\ +\infty & \text{if } \mathrm{s} \notin C_x. \end{cases}$$

If $\boldsymbol{\sigma}$ is an admissible stress field, that is, $\psi_{C_x}(\boldsymbol{\sigma}(x)) = 0$ for a.e. $x \in \Omega$, the integral in (6.3) can be written as

$$\int_\Omega \{\psi_{C_x}(\boldsymbol{\sigma}(x)) - \mathrm{e}(x) \cdot \boldsymbol{\sigma}_*(x)\}\, dV.$$

If the stress field $\boldsymbol{\sigma}$ is not admissible, then $\psi_{C_x}(\boldsymbol{\sigma}(x)) = +\infty$ on a set of nonzero volume and the previous expression is $+\infty$ according to Remark 2.

Because of that, the infimums of the expression under consideration over $C$ and over $\mathcal{S}$ are equal, which allows us to re-write (6.3) as

$$-\inf\left\{\int_\Omega \mathrm{e}(x)\cdot\boldsymbol{\sigma}_*(x)\,dV : \boldsymbol{\sigma}_*\in C\right\}$$

$$=\inf\left\{\int_\Omega (\psi_{C_x}(\boldsymbol{\sigma}(x)) - \mathrm{e}(x)\cdot\boldsymbol{\sigma}_*(x))\,dV : \boldsymbol{\sigma}_*\in \mathcal{S}\right\}.$$

Here, the integral functional with integrand $\phi_x$,

$$\phi_x(\mathrm{s}) = \psi_{C_x}(\mathrm{s}) - \mathrm{e}(x)\cdot\mathrm{s}, \qquad \mathrm{s}\in\mathbf{S}, \tag{6.4}$$

is to be minimized over the space $\mathcal{S}$.

Let us show that this integrand satisfies the above condition, that is, the set-valued mapping $x \to \operatorname{epi}\phi_x$ from $\Omega$ into $\mathbf{R}\times\mathbf{S}$ is measurable (if the mapping $x\to C_x$ is measurable). Note that the epigraph $\operatorname{epi}\phi_x$ is of the form

$$\operatorname{epi}\phi_x = \{(r,\mathrm{s})\in\mathbf{R}\times\mathbf{S} : r\geq -\mathrm{e}(x)\cdot\mathrm{s},\ \mathrm{s}\in C_x\}.$$

To verify measurability of the mapping $x\to\operatorname{epi}\phi_x$, let us construct a collection of functions approximating this mapping. Let $\{\mathbf{z}_1,\mathbf{z}_2,\dots\}$ be a collection of measurable functions on $\Omega$ which approximates the set-valued mapping $x\to C_x$. Let $\{r_1,r_2,\dots\}$ be a set of numbers dense in the set of all nonnegative numbers. We introduce functions $\zeta_{ij}$ on $\Omega$ with their values in $\mathbf{R}\times\mathbf{S}$:

$$\zeta_{ij}(x) = (r_i - \mathrm{e}(x)\cdot\mathbf{z}_j(x),\ \mathbf{z}_j(x)).$$

These functions are measurable, since $\mathbf{z}_j$ are measurable by assumption and the functions $r_i - \mathrm{e}\cdot\mathbf{z}_j$ are obtained by adding and multiplying measurable functions (see Appendix B). The following relation is obviously valid according to the definition of $\zeta_{ij}$ and $\phi_x$:

$$\{x\in\Omega : \zeta_{ij}(x)\in\operatorname{epi}\phi_x\} = \{x\in\Omega : \mathbf{z}_j(x)\in C_x\}.$$

This set is measurable since the collection $\{\mathbf{z}_1,\mathbf{z}_2,\dots\}$ possesses the property (M1) with respect to the mapping $x\to C_x$. The measurability of this set means that the collection of functions $\zeta_{ij}$ possesses property (M1) with respect to the mapping $x\to\operatorname{epi}\phi_x$ (see the left-hand side of the previous formula). Let us show that this collection approximates the mapping $x\to\operatorname{epi}\phi_x$. Indeed, let $(r,\mathrm{s})$ be in $\operatorname{epi}\phi_x$, which means that the number $r$ and $\mathrm{s}\in\mathbf{S}$ satisfy the conditions $r-\mathrm{e}(x)\cdot\mathrm{s}\geq 0$, $\mathrm{s}\in C_x$. Due to the latter relation and measurability of the mapping $x\to C_x$, there exists a subsequence $\mathbf{z}_{j_n}(x)$, $n=1,2,\dots$, such that

$$\mathbf{z}_{j_n}(x)\in C_x, \qquad \mathbf{z}_{j_n}(x)\to\mathrm{s} \quad\text{when}\quad n\to\infty. \tag{6.5}$$

We also choose a subsequence of numbers $r_{i_n}$ which converges to $r$ when $n\to\infty$, and consider values of the functions $\zeta_{i_nj_n}$:

$$\zeta_{i_nj_n}(x) = (r_{i_n} - \mathrm{e}(x)\cdot\mathbf{z}_{j_n}(x),\ \mathbf{z}_{j_n}(x))\in\mathbf{R}\times\mathbf{S}.$$

It is clear that $\zeta_{i_nj_n}$, $n=1,2,\dots$, belongs to $\operatorname{epi}\phi_x$ and, due to relations (6.5), converges to $(r,\mathrm{s})$ when $n\to\infty$. Thus, the collection of the functions $\zeta_{ij}$, $i,j=1,2,\dots$, approximates the mapping $x\to\operatorname{epi}\phi_x$, and the mapping is measurable.

**6.2. Evaluating the extremum.** Measurability of the mapping $x \to \operatorname{epi}\phi_x$ results in a simple formula for evaluating the extremum of an integral functional. It turns out that, if the space $\mathcal{S}$ is sufficiently wide in the sense of Subsection 5.4, then "the infimum of the integral equals the integral of the integrand infimum" (likewise the infimum of a sum equals the sum of the summands' infimums in case one minimizes each of the summands over independent variables). More precisely, the following proposition is valid.

THEOREM 1. Suppose 1) the function $\psi_x$ is given for every point $x$ in the domain $\Omega$; the function is defined on a finite-dimensional linear space $\mathbf{S}$, and the set-valued mapping $x \to \operatorname{epi}\phi_x$ from $\Omega$ into $\mathbf{R} \times \mathbf{S}$ is measurable, 2) the linear space $\mathcal{S}$ of functions, which are defined on $\Omega$ and take their values in $\mathbf{S}$, is sufficiently wide, and 3) there is a certain measurable function $\mathrm{s}_0$ in $\mathcal{S}$ such that the integral functional with the integrand $\phi_x$ takes a finite value at $\mathrm{s}_0$:

$$-\infty < \int\limits_{\Omega} \phi_x(\mathrm{s}_0(x))\, dV < +\infty.$$

Then the following equality is valid:

$$\inf\left\{\int\limits_{\Omega} \phi_x(\boldsymbol{\sigma}(x))\, dV \,:\, \boldsymbol{\sigma} \in \mathcal{S}\right\} = \int\limits_{\Omega} \inf\left\{\phi_x(\mathrm{s}) \,:\, \mathrm{s} \in \mathbf{S}\right\} dV. \tag{6.6}$$

The theorem will be proved somewhat later, as is its corollary, formula (6.1) in Theorem 5.1, which is of most interest for limit analysis.

COROLLARY. *Proof of Theorem 5.1.* According to the reasoning in Example 1, the expression on the right-hand side in (6.1) can be written as

$$\sup\left\{\int\limits_{\Omega} \mathrm{e}(x)\cdot\boldsymbol{\sigma}_*(x)\, dV \,:\, \boldsymbol{\sigma}_* \in C\right\} = -\inf\left\{\int\limits_{\Omega} \phi_x(\boldsymbol{\sigma}(x))\, dV \,:\, \boldsymbol{\sigma} \in \mathcal{S}\right\},$$

where $\phi_x(\mathrm{s}) = \psi_{C_x}(\mathrm{s}) - \mathrm{e}(x)\cdot\mathrm{s}$. It was also established in Example 1 that the set-valued mapping $x \to \operatorname{epi}\phi_x$ is measurable if the mapping $x \to C_x$ is. Then, under the assumptions of Theorem 5.1, those of Theorem 1 are satisfied. In particular, $\phi_x(0) = 0$, and the third assumption of Theorem 1 is satisfied at $\mathrm{s}_0 = 0$.

Then, by Theorem 1, we have

$$-\inf\left\{\int\limits_{\Omega} \phi_x(\boldsymbol{\sigma}(x))\, dV \,:\, \boldsymbol{\sigma} \in \mathcal{S}\right\} = -\int\limits_{\Omega} \inf\left\{\phi_x(\mathrm{s}) \,:\, \mathrm{s} \in \mathbf{S}\right\} dV.$$

Using (6.4), the obvious equality $-\inf A = \sup(-A)$ and the definition of the dissipation, we find out that the expression on the right-hand side equals to

$$-\int\limits_{\Omega} \inf\left\{\psi_{C_x}(\mathrm{s}) - \mathrm{e}(x)\cdot\mathrm{s} \,:\, \mathrm{s} \in \mathbf{S}\right\} dV$$

$$= \int\limits_{\Omega} \sup\left\{\mathrm{e}(x)\cdot\mathrm{s} \,:\, \mathrm{s} \in C_x\right\} dV = \int\limits_{\Omega} d_x(\mathrm{e}(x))\, dV,$$

which finishes the proof of Theorem 5.1.

Let us now prove Theorem 1 starting with establishing two auxiliary propositions.

PROPOSITION 1. Let $\Phi$ be a set-valued measurable mapping from $\Omega$ to S. Then one can choose a collection of functions $\mathbf{z}_1, \mathbf{z}_2, \ldots$ involved in the definition of a set-valued measurable mapping so that the values $\mathbf{z}_1(x), \mathbf{z}_2(x), \ldots$ are in $\Phi(x)$ for a.e. $x \in \Omega$ for which $\Phi(x)$ is not empty.

PROOF. Let $\zeta_1, \zeta_2, \ldots$ be a collection of functions which approximates the set-valued mapping $\Phi$ (see Subsection 5.3). By property (M1), the sets

$$\Omega_i = \{x \in \Omega : \zeta_i(x) \in \Phi(x)\}, \quad i = 1, 2, \ldots$$

are measurable. Therefore, the function $\mathbf{z}_1$ defined by the following formula

$$\mathbf{z}_1(x) = \begin{cases} \zeta_1(x) & \text{if } x \in \Omega_1 \text{ or } \Phi(x) = \emptyset, \\ \zeta_2(x) & \text{if } x \in \Omega_2 \setminus \Omega_1, \\ \zeta_3(x) & \text{if } x \in \Omega_3 \setminus (\Omega_1 \cup \Omega_2), \\ \zeta_4(x) & \text{if } x \in \Omega_4 \setminus (\Omega_1 \cup \Omega_2 \cup \Omega_3), \\ \ldots \end{cases}$$

is measurable. By property (M2) almost every $x$ in $\Omega$ belongs to at least one of the sets $\Omega_i$, which yields that $\mathbf{z}_1(x) \in \Phi(x)$ for a.e. $x \in \Omega$.

Let us now introduce the functions $\mathbf{z}_k$, $k = 2, 3, \ldots$:

$$\mathbf{z}_k(x) = \begin{cases} \zeta_k(x) & \text{if } x \in \Omega_k, \\ \mathbf{z}_1(x) & \text{if } x \notin \Omega_k. \end{cases}$$

It is obvious that these functions are measurable. If $\Phi(x) \neq \emptyset$ and $x \in \Omega_k$, it is clear that $\mathbf{z}_k(x) = \zeta_k(x)$ belongs to $\Phi(x)$. If $\Phi(x) \neq \emptyset$ and $x \notin \Omega_k$, then $\mathbf{z}_k(x) = \mathbf{z}_1(x)$. As was shown above, $\mathbf{z}_1(x)$ belongs to $\Phi(x)$ for a.e. $x \in \Omega$. Thus, $\mathbf{z}_k(x)$ with $k = 1, 2, \ldots$ belongs to $\Phi(x)$ for $x \in \Omega$ such that $\Phi(x) \neq \emptyset$. Therefore, in particular, the collection $\mathbf{z}_1, \mathbf{z}_2, \ldots$ possesses property (M1).

It remains to verify that this collection also possesses property (M2). Since the collection $\zeta_1, \zeta_2, \ldots$ approximates the set-valued mapping $\Phi$, for a.e. $x \in \Omega$ and for every $\mathbf{s} \in \Phi(x)$, there exists a subsequence of values $\zeta_{i_n}(x)$, $n = 1, 2, \ldots$, such that

$$\zeta_{i_n}(x) \in \Phi(x), \qquad \zeta_{i_n}(x) \to \mathbf{s} \text{ when } n \to \infty.$$

The first of these relations implies that $\mathbf{z}_{i_n}(x) = \zeta_{i_n}(x)$, so that $\zeta_{i_n}(x)$ can be replaced with $\mathbf{z}_{i_n}(x)$ in the second relation. This means that the collection $\{\mathbf{z}_1, \mathbf{z}_2, \ldots\}$ approximates the mapping $\Phi$. The proposition is proved.

The following proposition is the last step toward the proof of the Theorem 1. It shows that it is possible to find a measurable function $\sigma$ on $\Omega$, for which the values $\phi_x(\sigma(x))$ do not differ too much from their lower bounds.

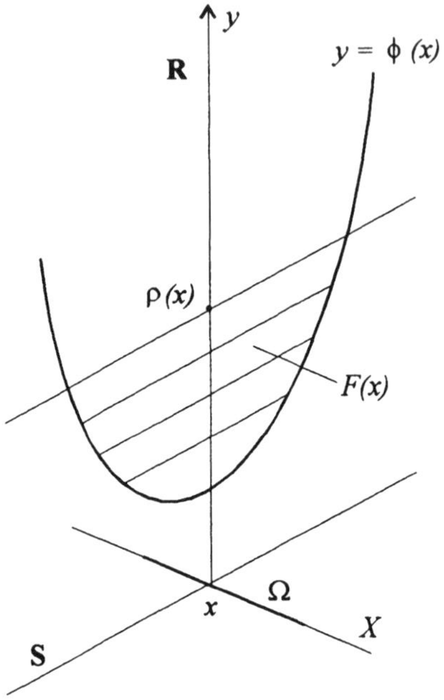


**Figure 6.1**

PROPOSITION 2. Suppose 1) the function $\phi_x$ is given for every point $x$ in the domain $\Omega$; $\phi_x$ is defined on a finite-dimensional linear space $\mathbf{S}$, and the set-valued mapping $x \to \operatorname{epi}\phi_x$ from $\Omega$ into $\mathbf{R} \times \mathbf{S}$ is measurable, and 2) $\rho$ is a measurable function on $\Omega$ and the inequality

$$\rho(x) > \inf\{\phi_x(\mathrm{s}) : \mathrm{s} \in \mathbf{S}\} \tag{6.7}$$

holds for a.e. $x \in \Omega$. Then there exists a measurable function $\boldsymbol{\sigma}$ on $\Omega$ with values in $\mathbf{S}$ such that $\rho(x) > \phi_x(\boldsymbol{\sigma}(x))$ for a.e. $x \in \Omega$.

PROOF. Let $x$ be a point in $\Omega$ at which (6.7) holds. Consider the set $F(x)$ of pairs $(r, \mathrm{s})$ in $\mathbf{R} \times \mathbf{S}$ which satisfy the inequalities $\phi_x(\mathrm{s}) \le r < \rho(x)$:

$$\operatorname{epi}\phi_x \cap \{(r, \mathrm{s}) \in \mathbf{R} \times \mathbf{S} : r < \rho(x)\},$$

see Figure 6.1.

Let us now show that the set-valued mapping $F$ is measurable. More precisely, consider a collection of functions $\{\zeta_1, \zeta_2, \dots\}$, which approximates the set-valued mapping $x \to \operatorname{epi}\Phi(x)$, and show that this collection also approximates $F$. Indeed, the set

$$\{x \in \Omega : \zeta_i(x) \in F(x)\} = \{x \in \Omega : \zeta_i(x) \in \operatorname{epi}\phi_x\} \cap \{x \in \Omega : r_i(x) < \rho(x)\}$$

is measurable: the first set on the right-hand side is measurable since the collection $\{\zeta_1, \zeta_2, \dots\}$ possesses the property (M1) with respect to the mapping $x \to \operatorname{epi}\phi_x$. The second set is also measurable since the functions $r_i$

and $\mathbf{s}$ are measurable. Thus, the collection $\{\zeta_1, \zeta_2, \dots\}$ possesses property (M1) with respect to the mapping $F$. Let us now verify that this collection approximates $F$. Consider $(r, \mathbf{s})$ in $F(x)$, then $(r, \mathbf{s}) \in \operatorname{epi} \phi_x$ and hence there is a subsequence $\{\zeta_{i_n}\}$, $n = 1, 2, \dots$, such that

$$\zeta_{i_n}(x) \in \operatorname{epi} \phi_x, \quad n = 1, 2, \dots, \qquad \zeta_{i_n}(x) \to (r, \mathbf{s}) \text{ when } n \to \infty.$$

In particular, $r_{i_n}(x) \to r < \rho(x)$. Therefore, the inequality $r_{i_n}(x) < \rho(x)$ is valid for sufficiently large $n$ and, consequently, $\zeta_{i_n}(x) \in F(x)$. Together with the relation $\zeta_{i_n} \to (r, \mathbf{s})$, this means that $\{\zeta_{i_n}\}$, $n = 1, 2, \dots$, approximates the mapping $F$, and $F$ is measurable.

Then, by Proposition 1, there exists a measurable function $\zeta$ with values in $\mathbf{R} \times \mathbf{S}$ such that $\zeta(x) = (\alpha(x), \boldsymbol{\sigma}(x)) \in F(x)$, for a.e. $x \in \Omega$. This implies, in particular, that $\phi_x(\boldsymbol{\sigma}(x)) < \rho(x)$ for a.e. $x \in \Omega$. The proposition is proved.

Using the latter proposition it is easy to prove Theorem 1.

PROOF OF THEOREM 1. Let us prove equality (6.6). It is clear that, for every function $\boldsymbol{\sigma} \in \mathcal{S}$ and every $x \in \Omega$, the inequality

$$\phi_x(\boldsymbol{\sigma}(x)) \geq \inf \{\phi_x(\mathbf{s}) : \mathbf{s} \in \mathbf{S}\}$$

is valid, and hence the left-hand side in (6.6) is not less than the right-hand side. Therefore, it remains to establish the inequality

$$\inf \left\{ \int_\Omega \phi_x(\boldsymbol{\sigma}(x))\, dV : \boldsymbol{\sigma} \in \mathcal{S} \right\} \leq \int_\Omega \inf \{\phi_x(\mathbf{s}) : \mathbf{s} \in \mathbf{S}\}\, dV.$$

To prove this inequality it is sufficient to verify that for every $\varepsilon > 0$ there exists $\boldsymbol{\sigma}_\varepsilon$ in $\mathcal{S}$ such that

$$\int_\Omega \phi_x(\boldsymbol{\sigma}_\varepsilon(x))\, dV \leq \int_\Omega \bar{\phi}(x)\, dV + O(\varepsilon), \tag{6.8}$$

where the following notation is used:

$$\bar{\phi}(x) = \inf \{\phi_x(\mathbf{s}) : \mathbf{s} \in \mathbf{S}\}.$$

To prove that the function $\bar{\phi}(x)$ is measurable, consider a collection of functions $\{\mathbf{z}_1, \mathbf{z}_2, \dots\}$ which approximates the measurable set-valued mapping $x \to \operatorname{epi} \phi_x$. According to Proposition 1, this collection is chosen in such a way that $\mathbf{z}_i = (r_i, \mathbf{s}_i)$ is in $\operatorname{epi} \phi_x$ for a.e. $x \in \Omega$ for which $\operatorname{epi} \phi_x \neq \emptyset$. Then the value

$$\bar{\phi}(x) = \inf \{\phi_x(\mathbf{s}) : \mathbf{s} \in \mathbf{S}\} = \inf \{r \in \mathbf{R} : (r, \mathbf{s}) \in \operatorname{epi} \phi_x\}$$

equals $\inf \{r_i(x) : i = 1, 2, \dots\}$, and the function $\bar{\phi}$ is measurable (see Subsection B.2).

To construct the function $\boldsymbol{\sigma}_\varepsilon$ for which inequality (6.8) is valid, we first make use of Proposition 2, with $\rho(x) = \bar{\phi}_N(x) + \varepsilon$, where $N > 0$ and

$$\bar{\phi}_N(x) = \begin{cases} \bar{\phi}(x) & \text{if } \bar{\phi}(x) > -N, \\ -N, & \text{if } \bar{\phi}(x) \le -N. \end{cases}$$

It is clear that $\bar{\phi}_N(x) \ge \bar{\phi}(x)$ and that the function $\rho$ satisfies the assumption of Proposition 2 since $\varepsilon > 0$ and $\bar{\phi}_N$ is measurable (together with $\bar{\phi}$). By Proposition 2, there exists a measurable function $\boldsymbol{\sigma}$ such that

$$\bar{\phi}_N(x) + \varepsilon > \phi_x(\boldsymbol{\sigma}(x)) \quad \text{for a.e. } x \in \Omega.$$

This function does not necessarily belong to the space $\mathcal{S}$. To improve $\boldsymbol{\sigma}$, we consider, for an arbitrary $\delta > 0$, a number $M_\delta$ such that

$$\text{mes}\,\{x \in \Omega : |\boldsymbol{\sigma}(x)| > M_\delta\} < \delta. \tag{6.9}$$

(It is always possible to find such $M_\delta$. Assuming it is not, we conclude from boundness of $\Omega$ that $|\boldsymbol{\sigma}(x)| = +\infty$ on a set of positive measure. However, $\boldsymbol{\sigma}$ takes values in $\mathbf{S}$ and hence $|\boldsymbol{\sigma}(x)|$ is defined and finite for a.e. $x \in \Omega$. Thus, the above assumption is wrong.)

We now replace $\boldsymbol{\sigma}$ by the following function $\boldsymbol{\tau}$:

$$\boldsymbol{\tau}(x) = \begin{cases} \boldsymbol{\sigma}(x) & \text{if } |\boldsymbol{\sigma}(x)| \le M_\delta, \\ \mathbf{s}_0(x), & \text{if } |\boldsymbol{\sigma}(x)| > M_\delta. \end{cases}$$

The second assumption of Theorem 1 implies that $\boldsymbol{\tau}$ is in $\mathcal{S}$. Let us estimate from above the value of the integral functional at $\boldsymbol{\tau}$:

$$\int_\Omega \phi_x(\boldsymbol{\tau}(x))\, dV.$$

Note that

$$\phi_x(\boldsymbol{\tau}(x)) = \begin{cases} \phi_x(\boldsymbol{\sigma}(x)) & \text{if } x \in \Omega_\delta, \\ \phi_x(\mathbf{s}_0(x)) & \text{if } x \in \Omega \setminus \Omega_\delta, \end{cases}$$

where

$$\Omega_\delta = \{x \in \Omega : |\boldsymbol{\sigma}(x)| > M_\delta\}.$$

Then

$$\int_\Omega \phi_x(\boldsymbol{\tau}(x))\, dV \le \int_{\Omega_\delta} (\bar{\phi}_N(x) + \varepsilon)\, dV + \int_{\Omega\setminus\Omega_\delta} \phi_x(\mathbf{s}_0(x))\, dV. \tag{6.10}$$

Here, the left-hand side is understood in the sense of Remark 2, while the integrals on the right-hand side have the regular sense. In particular, the function $\bar{\phi}_N$ is integrable since 1) $\bar{\phi}_N(x) = -N$ at points $x$ where $\bar{\phi}(x) \le -N$, 2) the integrable function $\phi_x(\mathrm{s}_0(x))$ estimates $\bar{\phi}_N(x)$ from above:

$$\bar{\phi}_N(x) = \bar{\phi}(x) \le \phi_x(\mathrm{s}_0(x)) \tag{6.11}$$

at points $x$ where $\bar{\phi}(x) > -N$.

The function $\boldsymbol{\tau}$ in (6.10), as we constructed it, depends upon three parameters: $\varepsilon$, $N$ and $\delta$. Let us choose $N$ and $\delta$ to obtain estimation (6.8), needed to complete the proof.

Consider first the case when the integral

$$\int_\Omega \bar{\phi}(x)\, dV$$

has a finite value. This value is the limit of the integral

$$\int_\Omega \bar{\phi}_N(x)\, dV$$

when $N \to \infty$, and for any $\varepsilon > 0$ we choose $N = N_\varepsilon$ such that

$$\int_\Omega \bar{\phi}_{N_\varepsilon}(x)\, dV - \int_\Omega \bar{\phi}(x)\, dV < \varepsilon. \tag{6.12}$$

Note that, due to (6.9), $\operatorname{mes}\Omega_\delta \to \operatorname{mes}\Omega$ when $\delta \to 0$, which implies that for the integrable functions $\bar{\phi}_{N_\varepsilon}(x)$ and $\phi_x(\mathrm{s}_0(x))$ there is a number $\delta = \delta_\varepsilon$ such that

$$\begin{gathered} \int_{\Omega_{\delta_\varepsilon}} \bar{\phi}_{N_\varepsilon}(x)\, dV - \int_\Omega \bar{\phi}_{N_\varepsilon}(x)\, dV < \varepsilon, \\ \int_{\Omega\setminus\Omega_{\delta_\varepsilon}} \phi(\mathrm{s}_0(x))\, dV < \varepsilon. \end{gathered} \tag{6.13}$$

Now consider the function $\boldsymbol{\tau}$ which corresponds to the values $\varepsilon$, $N = N_\varepsilon$, $\delta = \delta_\varepsilon$; we denote this function by $\boldsymbol{\sigma}_\varepsilon$. Then inequality (6.10) takes the form

$$\int_\Omega \phi_x(\boldsymbol{\sigma}_\varepsilon(x))\, dV \le \int_{\Omega_{\delta_\varepsilon}} (\bar{\phi}_{N_\varepsilon}(x) + \varepsilon)\, dV + \int_{\Omega\setminus\Omega_{\delta_\varepsilon}} \phi(\mathrm{s}_0(x))\, dV$$

and, being summed up with inequalities (6.12), (6.13), results in the required estimation (6.8).

It remains to consider the case of the infinite value of the integral

$$\int_\Omega \bar{\phi}(x)\, dV.$$

Since this integral cannot be $+\infty$ due to (6.11), the only case to discuss is

$$\int_\Omega \bar{\phi}(x)\, dV = -\infty. \tag{6.14}$$

In this case,

$$\int_\Omega \bar{\phi}_N(x)\, dV \to -\infty \quad \text{when} \quad N \to \infty$$

and for any $\varepsilon > 0$ there is a number $N = N_\varepsilon$ such that

$$\int_\Omega \bar{\phi}_N(x)\, dV < -\frac{1}{\varepsilon}.$$

Then, in the above reasoning, we replace (6.12) by the latter inequality and arrive to the inequality

$$\int_\Omega \phi_x(\boldsymbol{\sigma}_\varepsilon(x))\, dV < -\frac{1}{\varepsilon} + O(\varepsilon),$$

which yields

$$\inf\left\{\int_\Omega \phi_x(\boldsymbol{\sigma}(x))\, dV : \boldsymbol{\sigma} \in \mathcal{S}\right\} = -\infty.$$

In the case under consideration, that is, when (6.14) holds, this means that equality (6.6) is valid, which finishes the proof.

## Appendix A. Linear spaces

**A.1. Definitions and examples.** We consider the set $X$ with two operations: addition of its members ($\mathbf{x}_1 + \mathbf{x}_2$ is the sum of $\mathbf{x}_1, \mathbf{x}_2 \in X$) and multiplication of its member by a number ($\alpha\mathbf{x}$ is the product of a number $\alpha$ and $\mathbf{x} \in X$). We assume that the operations possess the following regular properties:

$$\mathbf{x}_1 + \mathbf{x}_2 = \mathbf{x}_2 + \mathbf{x}_1, \quad (\mathbf{x}_1 + \mathbf{x}_2) + \mathbf{x}_3 = \mathbf{x}_1 + (\mathbf{x}_2 + \mathbf{x}_3),$$
$$\alpha(\mathbf{x}_1 + \mathbf{x}_2) = \alpha\mathbf{x}_1 + \alpha\mathbf{x}_2, \quad (\alpha + \beta)\mathbf{x} = \alpha\mathbf{x} + \beta\mathbf{x}, \quad \alpha(\beta\mathbf{x}) = (\alpha\beta)\mathbf{x}, \quad 1\mathbf{x} = \mathbf{x},$$

and there is a zero member 0 in $X$, that is, for every $\mathbf{x} \in X$, the equality $\mathbf{x} + 0 = \mathbf{x}$ is valid. We also assume that for every $\mathbf{x} \in X$ there exists a member of $X$, which we, as usuall, denote by the symbol $-\mathbf{x}$, such that $\mathbf{x} + (-\mathbf{x}) = 0$. Under these assumptions $X$ is referred to as a *linear space*.

EXAMPLE 1. The space *Sym* of symmetric second rank tensors is a linear space with the regular operations of adding tensors and multiplying a tensor by a number. Recall that the sum of $\mathbf{a}$ and $\mathbf{b}$ in *Sym* is the tensor with the components $a_{ij} + b_{ij}$ and the product of $\mathbf{a}$ by a number $r$ is the tensor with the components $ra_{ij}$.

If there exists a set of members $\mathbf{e}_1, \ldots, \mathbf{e}_m$ of the linear space $X$ such that any $\mathbf{x} \in X$ can be written as their linear combination: $\mathbf{x} = x_1\mathbf{e}_1 + \cdots + x_m\mathbf{e}_m$, we say that $X$ is *finite-dimensional*. Here, the numbers $x_1, \ldots, x_m$ depend on both $\mathbf{x}$ and the collection $\mathbf{e}_1, \ldots, \mathbf{e}_m$ (the latter is not unique). If $n$ is the minimum number of $\mathbf{e}_1, \ldots, \mathbf{e}_n$ allowing the representation $\mathbf{x} = x_1\mathbf{e}_1 + \cdots + x_n\mathbf{e}_n$ for every $\mathbf{x} \in X$, the space $X$ is referred to as $n$-dimensional and to the collection $\mathbf{e}_1, \ldots, \mathbf{e}_n$ as its basis. For example, $Sym$ is a 6-dimensional space. Any $n+1$ members $\mathbf{x}_1, \ldots, \mathbf{x}_{n+1}$ of an $n$-dimensional space are linearly dependent, that is, the equality $\alpha_1\mathbf{x}_1 + \ldots \alpha_{n+1}\mathbf{x}_{n+1} = 0$ is valid, where $\alpha_1, \ldots, \alpha_{n+1}$ are certain numbers, at least one of them being nonzero.

The definition of linear space does not assume that the space is finite-dimensional.

EXAMPLE 2. Consider the set of all continuously differentiable functions on $[0,1]$. This set is a linear space with the regular operations of adding functions and multiplying a function by a number. For example, the functions $f_k$, $f_k(x) = x^k$, $k = 1, 2, \ldots$, are members of this space. It is easily seen that $f_1, f_2, \ldots, f_N$ are not linearly dependent no matter how large $N$ is. Thus, the space under consideration is not finite-dimensional.

EXAMPLE 3. Consider a symmetric second rank tensor field on the domain $\Omega$, that is, the function $\mathbf{t}$ defined on $\Omega$ with values $\mathbf{t}(x)$ in the space $Sym$. In other words, we consider six number-valued functions $t_{ij}$ being the components of $\mathbf{t}$ with respect to a certain coordinate system. Consider now the set $\mathcal{E}$ of all symmetric second rank tensor fields on $\Omega$ which are integrable (a field $\mathbf{t}$ is integrable if all its components $t_{ij}$ are). The sum $\mathbf{t}_1 + \mathbf{t}_2$ of $\mathbf{t}_1, \mathbf{t}_2 \in \mathcal{E}$ and the product $\alpha\mathbf{t}$ of $\mathbf{t} \in \mathcal{E}$ by a number $\alpha$ are the tensor fields

$$(\mathbf{t}_1 + \mathbf{t}_2)(x) = \mathbf{t}_1(x) + \mathbf{t}_2(x), \qquad (\alpha\mathbf{t})(x) = \alpha\mathbf{t}(x),$$

which are obviously integrable; hence, $(\mathbf{t}_1+\mathbf{t}_2)$ and $\alpha\mathbf{t}$ are in $\mathcal{E}$. The operations possess all the regular properties, and $\mathcal{E}$ is a linear space with respect to these operations.

**A.2. Subspace.** Let $L$ be a subset in a linear space $X$. If adding any two numbers of $L$ or multiplying any member of $L$ by any number results in a member of $L$, then $L$ itself is a linear space with respect to these operations, and we call $L$ a (linear) *subspace* in $X$. For example, the subset $Sym^d$ of all deviatoric second rank symmetric tensors is a subspace in $Sym$.

**A.3. Linear operator.** Let $X$ and $Y$ be linear spaces. The mapping $A$ from $X$ into $Y$ is referred to as a *linear operator* if $A(\mathbf{x}_1 + \mathbf{x}_2) = A(\mathbf{x}_1) + A(\mathbf{x}_2)$ for every $\mathbf{x}_1, \mathbf{x}_2 \in X$ and $A(\alpha\mathbf{x}) = \alpha A(\mathbf{x})$ for every $\mathbf{x} \in X$ and every number $\alpha$. Normally we denote the value $A(x)$ of the linear operator $A$ by the symbol $A\mathbf{x}$. We use the symbol $AU$ to denote the set of all values which $A$ takes on a subset $U$ in its domain. In particular, for a linear operator $A$ defined on a linear space $X$ the symbol $AX$ denotes the range of $A$.

EXAMPLE 4. Let $V$ be a space of velocity fields $\mathbf{v}$ which are defined on the domain $\Omega$ in the three-dimensional space and possess integrable derivatives $\partial v_i/\partial x_j$. The strain rate field $\operatorname{Def}\mathbf{v}$ with the components

$$(\operatorname{Def}\mathbf{v})_{ij} = \frac{1}{2}\left(\frac{\partial v_i}{\partial x_j} + \frac{\partial v_j}{\partial x_i}\right)$$

corresponds to $\mathbf{v}$ and belongs to the space $\mathcal{E}$ of integrable tensor fields (see Example 3). It is clear that the mapping $\mathbf{v} \to \operatorname{Def}\mathbf{v}$ from $V$ into $\mathcal{E}$ is a linear operator; its range $\operatorname{Def} V$ is a subspace in $V$.

**A.4. Pairing between linear spaces.** Consider two linear spaces $X$, $X^*$ and a function of two arguments: $\mathbf{x} \in X$ and $\mathbf{x}^* \in X^*$, the function being linear in each one of them. Values of this function are denoted by the symbol $\langle \mathbf{x}^*, \mathbf{x} \rangle$ and the function is referred to as a *bilinear form*. Its linearity in each of the arguments means that the equalities

$$\begin{aligned} \langle \mathbf{x}^*, \mathbf{x}_1 + \mathbf{x}_2 \rangle &= \langle \mathbf{x}^*, \mathbf{x}_1 \rangle + \langle \mathbf{x}^*, \mathbf{x}_2 \rangle, & \langle \mathbf{x}^*, \alpha\mathbf{x} \rangle &= \alpha\langle \mathbf{x}^*, \mathbf{x} \rangle, \\ \langle \mathbf{x}_1^* + \mathbf{x}_2^*, \mathbf{x} \rangle &= \langle \mathbf{x}_1^*, \mathbf{x} \rangle + \langle \mathbf{x}_2^*, \mathbf{x} \rangle, & \langle \alpha\mathbf{x}^*, \mathbf{x} \rangle &= \alpha\langle \mathbf{x}^*, \mathbf{x} \rangle \end{aligned}$$

hold for every $\mathbf{x}$, $\mathbf{x}_1$, $\mathbf{x}_2$ in $X$, every $\mathbf{x}^*$, $\mathbf{x}_1^*$, $\mathbf{x}_2^*$ in $X^*$ and every number $\alpha$. We say that the bilinear form $\langle \cdot, \cdot \rangle$ is a *pairing* between the spaces $X$, $X^*$.

EXAMPLE 5. Let $\mathcal{E}$ and $\mathcal{S}$ be linear spaces of symmetric second rank tensor fields on the domain $\Omega$. All fields in $\mathcal{S}$ are assumed to be integrable and bounded, so that the integral

$$\int_{\Omega} \mathbf{e}(x) \cdot \mathbf{s}(x)\, dV$$

makes sense for every $\mathbf{e} \in \mathcal{E}$ and every $\mathbf{s} \in \mathcal{S}$. It is clear that the integral is a bilinear form of the arguments $\mathbf{e} \in \mathcal{E}$, $\mathbf{s} \in \mathcal{S}$, and this form is a pairing between the spaces $\mathcal{E}$, $\mathcal{S}$.

EXAMPLE 6. Let $X$ be a Euclidean space, that is, a finite-dimensional space with a scalar multiplication of its members. The symbol $\mathbf{x}_1 \cdot \mathbf{x}_2$ is used to denote the scalar product of $\mathbf{x}_1$ and $\mathbf{x}_2$ in $X$. Consider a function of two arguments $\mathbf{x}_1$ and $\mathbf{x}_2$ in $X$ which assigns to $\mathbf{x}_1$, $\mathbf{x}_2$ their scalar product $\mathbf{x}_1 \cdot \mathbf{x}_2$. The function is obviously linear in each of its arguments, that is, the bilinear form $\langle \mathbf{x}_1, \mathbf{x}_2 \rangle$ is a pairing between the space $X$ and itself.

Let $X$ and $X^*$ be linear spaces, $L$ be a linear subspace in $X$ and $\langle \cdot, \cdot \rangle$ be a pairing between $X$ and $X^*$.

Consider the $L^0$ set in $X^*$ such that:

$$L^0 = \{\mathbf{x}^* \in X^* : \langle \mathbf{x}^*, \mathbf{x} \rangle = 0 \text{ for every } \mathbf{x} \in L\};$$

$L^0$ is referred to as a *polar* set of the subspace $L$ (with respect to the pairing $\langle \cdot, \cdot \rangle$). If $\mathbf{x}_1^*$ and $\mathbf{x}_2^*$ are in $L^0$, it is obvious that $\mathbf{x}_1^* + \mathbf{x}_2^*$ is also in $L^0$.

Analogously, if $\mathbf{x}^*$ is in $L^0$ and $\alpha$ is an arbitrary number, $\alpha\mathbf{x}^*$ is in $L^*$. Thus, a polar set is a linear subspace in $X^*$.

EXAMPLE 7. Let $X$ be a Euclidean space and $\langle \cdot, \cdot \rangle$ be the pairing between $X$ and itself determined by the scalar multiplication (see Example 6). Let $L$ be a subspace in $X$. Then the polar set $L^0$ is its orthogonal complement.

When defining a polar set, one can start with a subspace $\Pi$ in $X^*$ instead of $L$ in $X$. Then the polar set of $\Pi$ is

$$\Pi^0 = \{\mathbf{x} \in X : \langle \mathbf{x}^*, \mathbf{x} \rangle = 0 \text{ for every } \mathbf{x}^* \in \Pi\}.$$

## Appendix B. Measurable sets and measurable functions

**B.1. Measure.** In this section we describe a structure which allows us to define volume (measure) for a wide class of sets in Euclidean space. For brevity, we restrict ourselves to a plane, that is, the two-dimensional space. In the general case of $n$-dimensional space, it is sufficient to replace rectangles with parallelepipeds in what follows: the definition and all properties of measure remain unchanged.

A set specified by the inequalities $a \le x \le b$, $c \le y \le d$ in $(x, y)$-plane is referred to as a rectangle; here, $a$, $b$, $c$, $d$ are arbitrary nonnegative numbers. (It may occur that a rectangle is empty, for example, if $a > b$). We introduce a set of rectangles which consists of the above defined rectangles and those without one, two, three, or four sides. In what follows, we refer to any member of this class as a *rectangle*. We define measure $\text{mes}P$ for every rectangle $P$ as its area if $P \neq \emptyset$ and 0 if $P = \emptyset$. Consider now a union of finite number pairwise disjoint rectangles; we call it an *elementary set*. We define the measure of an elementary set as the sum of the measures of the rectangles which compose this set. Of course, there are various partitions of an elementary set into rectangles; however, it is easily seen that its measure does not depend on the partition.

Let us now extend the definition of the measure to a class of sets much wider than that of elementary sets. Consider the set $A$ covered by a finite or countable collection $\mathcal{P}$ of rectangles $P_k$. A reasonable definition of measure should meet the following requirement: measure of $A$ is less than the sum $m_{\mathcal{P}}$ of the measures $\text{mes}P_k$ and, consequently, is also less than the infimum of $m_{\mathcal{P}}$ over all various coverings $\mathcal{P}$. This infimum is referred to as an *outer measure* of $A$ and denoted by the symbol $\text{mes}^*A$. The outer measure is defined for any set $A$, but it is not reasonable to take $\text{mes}^*A$ as a measure of $A$ for any $A$. However, this will make sense if $A$ can be approximated by elementary sets. More precisely, let the set $A$ be such that, for every $\varepsilon > 0$, there exists an elementary set $B_\varepsilon$ for which $\text{mes}^*(A \Delta B_\varepsilon) < \varepsilon$. Here, $A\Delta B_\varepsilon$ stands for the union of the two sets: the set of points in $A$ which does not belong to $B_\varepsilon$ and the set of points in $B_\varepsilon$ which does not belong to $A$. This difference between $A$ and $B_\varepsilon$, under the above assumption, becomes infinitely small when $\varepsilon \to 0$, and we say that *$A$ is approximated by elementary sets $B_\varepsilon$*. We say that the set $A$ is *measurable* (in Lebesque's sense) if $A$ can be approximated by elementary sets. In this case, we define the (Lebesque's) measure of $A$ as its outer measure $\text{mes}^* A$. We denote the measure of $A$ by the symbol $\text{mes}A$.

The class of measurable sets is characterized by the following statements:

1. Union and intersection of a finite or countable collection of measurable sets are measurable.

2. Measure of a finite or countable union of pair-wise disjoint measurable sets equals the sum of their measures.

The complement of a measurable set is measurable. Every open set is measurable as well as every closed one.

**B.2. Measurable functions.** We refer to the function $f$ defined on the domain $\Omega$ as *measurable* if for every number $a$ the set of points $x$ in $\Omega$, where the inequality $f(x) < a$ is valid, is measurable.

The following statements point out some cases when the measurability of a function is guaranteed.

1. A function continuous on $\Omega$ is measurable.
2. The sum and the product of two measurable functions are measurable.
3. If functions $f_i$, $i = 1, 2, \ldots$, are measurable, then the function $f$,

$$f(x) = \inf \{f_i(x) : i = 1, 2, \ldots\},$$

is measurable.

Consider the function $\mathbf{f}$ on $\Omega$ with values in a finite-dimensional linear space $\mathbf{S}$. Let $\mathbf{e}_1, \ldots, \mathbf{e}_n$ be a basis in $\mathbf{S}$; then $\mathbf{f}$ can be written in the form $\mathbf{f} = f_1\mathbf{e}_1 + \cdots + f_n\mathbf{e}_n$, where $f_1, \ldots, f_n$ are (number-valued) functions on $\Omega$, the components of $\mathbf{f}$. We refer to $\mathbf{f}$ as a *measurable function* if $f_1, \ldots, f_n$ are measurable (it is easily seen that measurability of $\mathbf{f}$ does not depend on the basis we choose in $\mathbf{S}$).

## Comments

There is nothing surprising in the fact that rigid-plastic problems admit a unified formulation for rigid perfectly plastic systems of all types. Indeed, introducing generalized velocities and forces, considering the external power as their bilinear form, and using the virtual work principle unify equilibrium conditions for all mechanical systems. Therefore, these systems only differ in the mechanical meaning of their generalized velocities and forces (in particular, in dimension of the spaces they belong to) and, of course, in their constitutive relations. Thus, the unified formulation of the rigid-plastic problems is possible for a certain class of systems if their constitutive relations can be written in a unified form. Gvozdev (1938, 1948, 1949) established the unified formulation of the rigid-plastic problems for discrete systems. However, it was not easy to find a unified formulation of constitutive relations in the case of continual systems. To overcome the difficulty, Nayroles (1970) made use of a theorem proved by Rockafellar (1968); the theorem concerns the extremum of an integral functional (Theorem 6.2 here). Nayroles found out that the point-wise formulation of the rigid-plastic constitutive relations is equivalent to the integral maximum principle (Theorem 5.2) in case the strain rate field is integrable. The integral principle induces a unified formulation of the constitutive relations for all rigid perfectly plastic systems; only the expression for the internal and external power bilinear forms differs for different concrete systems.

It should be emphasized that the integral formulation of the constitutive relations is established only for systems whose local constitutive law is the normality flow rule. We study only these systems throughout the book, leaving undiscussed the following important question. Consider, for example, a shell made of a material that obeys the normality rule. Does an analogous normality law relate (generalized) internal forces and velocities within the framework of a two-dimensional model of the shell? Strictly speaking, to answer the question, one has to analyze the asymptotic behavior of the system when $\varepsilon \longrightarrow 0$, $\varepsilon$ being a small parameter, the ratio of the shell thickness to its other two dimensions. Mosolov and Myasnikov (1971, 1977, 1981) considered the question in a number of cases, and the answer was positive. It is worth mentioning that a similar problem arises in the case of a composite body. If the ratio $\varepsilon$ of the inhomogeneity characteristic dimension to the body dimension is small, there arises the well-known homogenization problem. The question is whether it is possible to replace effectively the inhomogeneous body by a fictitious homogeneous one when $\varepsilon \longrightarrow 0$. It is only natural to anticipate that if the original inhomogeneous body is rigid perfectly plastic, then the corresponding fictitious body is too. Suquet (1983), Bouchitté (1985), Jikov (1986), Demengel and Tang (1986), Barabanov (1989) and others studied the homogenization problem; see also the book by Jikov, Kozlov and, Oleinik (1994).

CHAPTER V

# EXTREMUM PROBLEMS OF LIMIT ANALYSIS

The main problem of limit analysis is evaluating the safety factor of a given load. The safety factor was earlier defined as the supremum of static multipliers, so that any of the static multipliers estimates it from below, while any of the kinematic multipliers does so from above. The infimum of kinematic multipliers is the best upper bound for the safety factor obtainable by the kinematic method. Of fundamental interest are the conditions under which there is no gap between this best upper bound and the value of the safety factor: in this case, the latter can be calculated by solving the kinematic extremum problem, which is simpler than the static one. This chapter deals with establishing these conditions. We formulate the extremum problems and describe the main results in Section 1; the reader can proceed to the following chapters upon reading Section 1.

We establish the main results concerning limit analysis problems using the theory of convex extremum problems. Methods of this theory allow for construction of an extremum problem dual to a given one. The functions involved in formulations of both problems always provide a two-sided estimation for the extremum of the original problem. Moreover, under certain conditions, the extremums of both problems are equal and it is possible to evaluate the extremum of the original problem solving the dual problem, which is in many cases simpler than the original one. We present a general scheme for constructing the dual problem and the conditions for equality of the extremums in Section 3. The scheme enjoys wide application, including the limit analysis in the present chapter and the elastic-plastic shakedown theory in Chapter X.

Some basic concepts of the extremum problems theory are presented in Appendix I.C.

## 1. Static and kinematic extremum problems

This section covers formulations of limit analysis extremum problems determining the limit static multiplier (the safety factor) and the limit kinematic multiplier, the latter being the best kinematic upper bound for the safety factor. We also review the main results of this chapter, which deals mostly with conditions for equality of the limit static and kinematic multipliers. Under these conditions, the safety factor can be computed not only by the static method but also by the more effective kinematic method.

**1.1. Limit static and kinematic multipliers.** The main problem of limit analysis in evaluation of the safety factor $\alpha_l$ for a given load $\mathbf{l}$: the load is safe if $\alpha_l > 1$ and inadmissible if $\alpha_l < 1$. The safety factor of the load is the supremum of its static multipliers and, therefore, is also called the *limit static multiplier*,

$$\alpha_l = \sup m_s(\sigma) = \sup\{m \geq 0 : \text{there exists } \rho \in \Sigma,\ m(\rho + \mathbf{s}_l) \in C\} \quad (1.1)$$

(see also Subsection III.2.2 and IV.4.1). Here, $\mathbf{s}_l$ stands for a fixed stress field equilibrating the load $\mathbf{l}$. We refer to (1.1) as the *static extremum problem of limit analysis.* The formulation of the static problem involves the admissibility condition $\sigma \in C$, which means that the stress field $\sigma$ should satisfy the relation $\sigma(x) \in C_x$ for a.e. $x \in \Omega$, where $\Omega$ is the domain occupied by the body and $C_x$ is the set of admissible stresses at the point $x$. This is a point-wise condition, and it is rather difficult to find out whether a certain stress field meets the requirement. This makes the static method not very convenient for practical computations, and the alternative kinematic method for computing the safety factor gains much importance.

The kinematic method makes use of the kinematic multipliers $m_k$ of the given load, defined for every kinematically admissible velocity field $\mathbf{v}$ provided the power of the load is positive on $\mathbf{v}$. Recall that, in the case of a three-dimensional body occupying domain $\Omega$ and subjected to load $\mathbf{l} = (\mathbf{f}, \mathbf{q})$, the kinematic multiplier for a sufficiently regular velocity field $\mathbf{v}$ is defined as

$$m_k(\mathbf{v}) = \frac{\int\limits_{\Omega} d_x(\mathrm{Def}\,\mathbf{v})\, dV}{\int\limits_{\Omega} \mathbf{f}\mathbf{v}\, dV + \int\limits_{S_q} \mathbf{q}\mathbf{v}\, dS}\,.$$

Here, $\mathbf{f}$ is the density of the body forces in $\Omega$ and $\mathbf{q}$ is the density of the surface tractions given on the part $S_q$ of the body boundary; as usual, $d_x$ stands for the dissipation, and $\mathrm{Def}\,\mathbf{v}$ for the strain rate field:

$$(\mathrm{Def}\,\mathbf{v})_{ij} = \frac{1}{2}\left(\frac{\partial v_i}{\partial x_j} + \frac{\partial v_j}{\partial x_i}\right).$$

Within the framework of the general formulation of the rigid-plastic problem the kinematic multiplier of the load $\mathbf{l}$ is determined for every velocity field $\mathbf{v}$ as

$$m_k(\mathbf{v}) = \frac{D(\mathrm{Def}\,\mathbf{v})}{\langle\langle \mathbf{v}, \mathbf{l} \rangle\rangle} = \frac{D(\mathrm{Def}\,\mathbf{v})}{\langle \mathrm{Def}\,\mathbf{v}, \mathbf{s}_l \rangle}$$

provided the external power $\langle\langle \mathbf{v}, \mathbf{l} \rangle\rangle$ is positive. Here, $\mathbf{s}_l$ is a fixed stress field equilibrating the load $\mathbf{l}$.

Every kinematic multiplier is an upper bound for the safety factor: $\alpha_l \leq m_k(\mathbf{v})$. The best upper bound obtainable by kinematic multipliers is

$$\beta_l = \inf m_k = \inf\left\{\frac{D(\mathbf{e})}{\langle \mathbf{e}, \mathbf{s}_l \rangle} : \mathbf{e} \in E,\ \langle \mathbf{e}, \mathbf{s}_l \rangle > 0\right\}, \quad (1.2)$$

where $E = \mathrm{Def}\,\mathbf{v}$ is the set of kinematically admissible strain rate fields. We refer to (1.2) as the *limit kinematic multiplier* and the extremum problem in (1.2) is called the *kinematic extremum problem of limit analysis.*

As we see from Theorems III.5.2 and IV.4.2, there are some cases when the limit kinematic multiplier equals the limit static multiplier: $\beta_l = \alpha_l$. Two important questions arise in this connection:

- under what conditions is the equality $\beta_l = \alpha_l$ valid?
- under what conditions is it possible to evaluate the extremum in the kinematic extremum problem using only sufficiently regular velocity fields?

Under the conditions of the first type, the safety factor can be calculated by both static and kinematic methods, while the second type makes calculating the safety factor by the kinematic method rational. Therefore, answering the above questions is the main task of this and the next chapters. The answers we find are constructive, which means that only the inputs of the limit analysis problem are involved in the conditions and also that it is easy to verify whether or not the inputs of a certain problem satisfy the conditions.

**1.2. Main results.** Here, we review the main results of this chapter concerning limit analysis. To be more definite, we consider three-dimensional bodies. However, all statements are established within the framework of the rigid-plastic problem unified formulation, and these results are also applicable to continual and discrete systems of all types (plates, beams, trusses, etc.). We formulate the final results for bodies with a local description of material properties, assuming that the set of admissible stresses is given for every point $x$ of the body and the set-valued mapping $x \longrightarrow C_x$ is measurable (see Subsection IV.4.3). These assumptions are satisfied for all practical interesting problems of limit analysis.

We consider plastic bodies of two types: 1) *bodies with bounded yield surfaces*, that is, with bounded sets $C_x$ of admissible stresses, and 2) *three-dimensional bodies with cylindrical yield surfaces*, that is, with cylindrical sets $C_x$, the axis of $C_x$ being directed along the unity tensor $\mathbf{I}$ in the space $Sym$ (see Subsection I.2.3). Bodies of the latter type are of much interest since cylindrical yield surfaces are normally used for describing behavior of metals and alloys.

The main result of this chapter is the conditions for the equality $\beta_l = \alpha_l$ under which the safety factor can be computed not only by the static method but also by the kinematic one.

The safety factor can be computed by both static and kinematic methods, that is,

$$\alpha_l = \sup\{m \geq 0 : \text{there exists } \rho \in \Sigma,\ m(\rho + \mathbf{s}_l) \in C\}$$
$$= \inf\left\{\frac{D(\mathbf{e})}{\langle \mathbf{e}, \mathbf{s}_l \rangle} : \mathbf{e} \in E,\ \langle \mathbf{e}, \mathbf{s}_l \rangle > 0\right\} = \beta_l$$

in the following cases:

(I) for a body with bounded yield surfaces, if for every point of the body the yield surface lies in the same ball of the stress space; to find the infimum in the kinematic extremum problem, in this case, one may use only sufficiently regular velocity fields;

(II) for a body with cylindrical yield surfaces, if there is a cylinder, $|\sigma^d| \leq R$, $R > 0$, which lies inside the yield surface for every point of the body; the infimum in the kinematic extremum problem is attainable in this case if one looks for the infimum over a certain extended space of velocity fields.

In the case of bodies with bounded yield surfaces, this statement answers both questions asked in Subsection 1.1: it establishes conditions under which 1) the safety factor may be evaluated by the kinematic method, and 2) it is sufficient to use only sufficiently regular velocity fields when applying the kinematic method for this calculation. In the case of bodies with cylindrical yield surfaces, the statement establishes conditions for the equality $\beta_l = \alpha_l$ and does not specify the conditions for the possibility of using only sufficiently regular velocity fields when computing the safety factor by the kinematic method.

Special attention will be paid to the latter question in Chapter VI. As to the extended space of velocity fields mentioned in (II), it plays an important role in answering a question about the existence of failure mechanism for a perfectly plastic body under the action of a limit load (see Subsection III.3.4 and Chapter VII).

The importance of conditions (I), (II) in the above statement is illustrated with counterexamples in Section 8. The first example shows that the equality $\beta_l = \alpha_l$ may be invalid if conditions (I), (II) are not satisfied (although they are not necessary conditions for the equality). The second example concerns bodies with cylindrical yield surfaces and shows that, even in case $\beta_l = \alpha_l$, it may occur that the set of sufficiently regular velocity fields is too narrow for computing the safety factor $\alpha_l$ by the kinematic method, as there remains a gap between the kinematic multipliers of such fields and $\alpha_l$.

We obtain the main results in this chapter within the framework of a general scheme for constructing the dual extremum problem, the scheme being presented in Section 3. Together with the theorem on conditions for equality of the dual problems extremums, it has numerous applications for various domains. The scheme is applicable when the original extremum problem is of a certain standard form, in which connection we transform the extremum problems of limit analysis to this form in the next section. Then we apply the scheme of constructing the dual problem to these extremum problems twice: first, we consider the kinematic problem and then the static problem as the original one for constructing the dual problem. This results in the two parts (I) and (II) of the above statement in Sections 4 and 5 and Sections 6 and 7, respectively.

## 2. Static and kinematic extremum problems: standard formulation

In this section, we transform the extremum problems of limit analysis to a standard form. We refer to the extremum problem

$$f(x) \to \inf; \quad x \in L \qquad (-g(x) \to \sup; \quad x \in L)$$

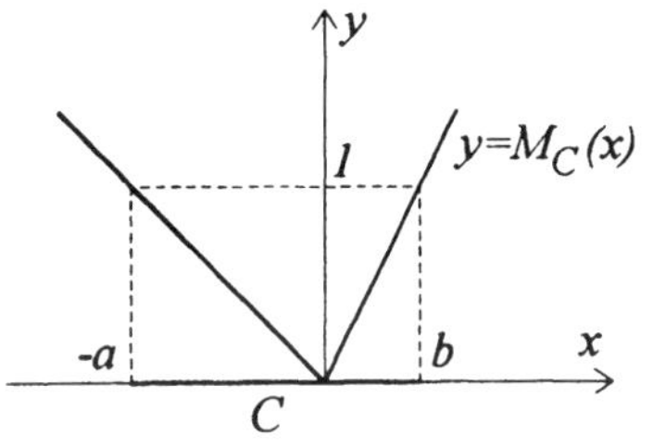


**Figure 2.1**

as standard if $f$ or, respectively, $g$ is a convex function and $L$ is a subspace in a linear space.

**2.1. Minkowski function.** To write the extremum problems of limit analysis in the standard form we use the Minkowski function defined for a subset $C$ in a linear space $\mathcal{S}$ assuming that 0 is in $C$. Consider member $\mathbf{s}$ in $\mathcal{S}$ (it either belongs to $C$ or not). Suppose first that there exists a number $\mu_0 > 0$ such that $\mathbf{s}/\mu_0$ belongs to $C$. Then let us consider the infimum of all numbers satisfying this condition and define the value of the Minkowski function $M_C$ as

$$M_C(\mathbf{s}) = \inf\left\{\mu > 0 : \frac{1}{\mu}\mathbf{s} \in C\right\}. \tag{2.1}$$

Note, in particular, that $M_C(0) = 0$. Consider now the case of $\mathbf{s} \in \mathcal{S}$ for which no number $\mu > 0$ satisfying the condition $\mathbf{s}/\mu \in C$ exists. Then the value of the Minkowski function at $\mathbf{s}$ is defined as $M_C(\mathbf{s}) = +\infty$. Note that the expression on the right-hand side in (2.1) is $\inf \emptyset = +\infty$, so that (2.1) can be used to define the value of $M_C(\mathbf{s})$ in this case as well. The function defined by (2.1) is referred to as the *Minkowski function* corresponding to the set $C$.

EXAMPLE 1. Consider a segment $[-a, b]$ in $\mathbf{R}$. Figure 2.1 shows the graph of the Minkowski function corresponding to this set.

EXAMPLE 2. Let $\mathcal{S}$ be a space with the norm $\|\cdot\|$ and $C$ be the ball specified by the condition $\|\mathbf{s}\| \leq R$. Then the Minkowski function values are

$$M_C(\mathbf{s}) = \inf\left\{\mu > 0 : \left\|\frac{1}{\mu}\mathbf{s}\right\| \leq R\right\} = \frac{1}{R}\|\mathbf{s}\|.$$

EXAMPLE 3. Consider a rigid-plastic body occupying a bounded domain $\Omega$ in a three-dimensional space. Let the set $C$ of admissible stress fields be of the form

$$C = \{\sigma \in \mathcal{S} : \left|\sigma^d(x)\right| \leq \sqrt{2}k \text{ for a.e. } x \in \Omega\}, \quad k = \text{const} > 0,$$

that is, $C$ is defined by the Mises admissibility condition. Here, $\mathcal{S}$ is the space of the stress fields; we assume for example that $\mathcal{S}$ contains all integrable fields.

For the stress field s, which is continuous on $\bar{\Omega} = \Omega \cup \partial\Omega$, the value of the Minkowski function is

$$M_C(\mathbf{s}) = \inf\left\{\mu > 0 : \frac{1}{\mu}\mathbf{s} \in C\right\}$$
$$= \inf\{\mu > 0 : \left|\frac{1}{\mu}\mathbf{s}^d(x)\right| \le \sqrt{2}k \text{ for a.e. } x \in \Omega\} = \frac{1}{\sqrt{2}k}\max\{\left|\mathbf{s}^d(x)\right| : x \in \bar{\Omega}\}.$$

Let us list some simple and important properties of the Minkowski function, which we will often use.

(I) The Minkowski function corresponding to the convex set $C$ is convex.

Indeed, consider any $\boldsymbol{\sigma}$ and $\boldsymbol{\tau}$ in $\mathcal{S}$ and numbers $\alpha$, $\beta \ge 0$, $\alpha + \beta = 1$. Let us verify the inequality $M_C(\alpha\boldsymbol{\sigma} + \beta\boldsymbol{\tau}) \le \alpha M_C(\boldsymbol{\sigma}) + \beta M_C(\boldsymbol{\tau})$, which is the criterion for convexity of $M_C$. For arbitrary positive number $\varepsilon$, we set $p = M_C(\boldsymbol{\sigma}) + \varepsilon$, $q = M_C(\boldsymbol{\tau}) + \varepsilon$. By the definition of the Minkowski function, $\boldsymbol{\sigma}/p$ and $\boldsymbol{\tau}/q$ are in $C$, and due to convexity of $C$,

$$\frac{\alpha p}{\alpha p + \beta q}\frac{\boldsymbol{\sigma}}{p} + \frac{\beta p}{\alpha p + \beta q}\frac{\boldsymbol{\tau}}{q} = \frac{1}{\alpha p + \beta q}(\alpha\boldsymbol{\sigma} + \beta\boldsymbol{\tau})$$

is in $C$ as well. Then, by the definition of the Minkowski function, the inequality $M_C(\alpha\boldsymbol{\sigma} + \beta\boldsymbol{\tau}) \le \alpha p + \beta q$ is valid, where the right-hand side obviously equals $\alpha M_C(\boldsymbol{\sigma}) + \beta M_C(\boldsymbol{\tau}) + \varepsilon$. Since $\varepsilon$ is an arbitrary positive number, this yields

$$M_C(\alpha\boldsymbol{\sigma} + \beta\boldsymbol{\tau}) \le \alpha M_C(\boldsymbol{\sigma}) + \beta M_C(\boldsymbol{\tau}),$$

that is, $M_C$ is convex.

(II) The Minkowski function is a positive homogeneous function of degree one.

This means that for every number $a > 0$ and every $\mathbf{s} \in \mathcal{S}$ the equality $M_C(a\mathbf{s}) = aM_C(\mathbf{s})$ is valid. The equality arises immediately from definition (2.1).

By (II), the graph of the Minkowski function $M_C$ is a cone. The next proposition specifies the cone relating it to the set $C$.

(III) If a convex set $C$ is closed along rays and contains the origin, then the following relation is valid for every $a > 0$:

$$\{\mathbf{s} \in \mathcal{S} : M_C(\mathbf{s}) \le a\} = aC,$$

the set of $\mathbf{s} \in \mathcal{S}$ for which $M_C(\mathbf{s}) \le a$ is homothetic to $C$ with the coefficient $a$. Here, we refer to the (convex and containing the origin) set $C$ as *closed along rays* if the point $\mathbf{s} = T\mathbf{s}_0$ of any ray $\mathbf{s} = t\mathbf{s}_0$, $t \ge 0$, belongs to $C$ as soon as $C$ contains the segment $\mathbf{s} = t\mathbf{s}_0$, $0 \le t < T$.

Let us verify the relation $\{\mathbf{s} \in \mathcal{S} : M_C(\mathbf{s}) \le a\} = aC$. If $\mathbf{s}$ belongs to $aC$, then $\mathbf{s} \in C/a$ and $M_C(\mathbf{s}) \le a$ by the definition of the Minkowski function. Conversely, if $M_C(\mathbf{s}_0) \le a$, then $\mathbf{s}_0/(a+\varepsilon) \in C$ for every $\varepsilon > 0$ and the set $C$ contains the segment $\mathbf{s} = t\mathbf{s}_0$, $0 \le t < 1/a$. Since $C$ is closed along rays, we have $\mathbf{s}/a \in C$, that is $\mathbf{s}_0 \in C$, which finishes the proof.

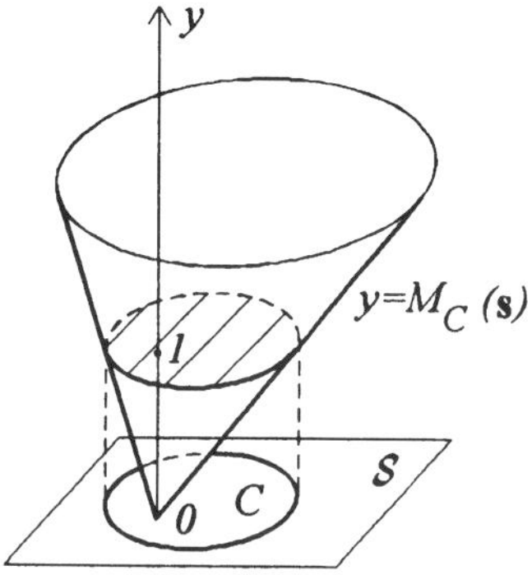


**Figure 2.2**

Property (III) implies that for every convex set $C$ closed along rays and containing the origin

the relation $\mathbf{s} \in C$ is equivalent to $M_C(\mathbf{s}) \leq 1$.

By (III), the graph of the Minkowski function $M_C$ has the form shown in Figure 2.2.

(IV) For every $\mathbf{s}_1$ and $\mathbf{s}_2$ in $C$, the following inequality is valid for the Minkowski function of the set $C$: $M_C(\mathbf{s}_1 + \mathbf{s}_2) \leq M_C(\mathbf{s}_1) + M_C(\mathbf{s}_2)$. Indeed, using first (II) and then (I), we obtain

$$\frac{1}{2} M_C(\mathbf{s}_1 + \mathbf{s}_2) = M_C(\frac{1}{2}\mathbf{s}_1 + \frac{1}{2}\mathbf{s}_2) \leq \frac{1}{2} M_C(\mathbf{s}_1) + \frac{1}{2} M_C(\mathbf{s}_2).$$

(V) If $\mathcal{S}$ is a normed space and $C$ contains the ball of radius $r > 0$ with its center at 0, then, for every $\mathbf{s} \in \mathcal{S}$, the value of the Minkowski function is estimated from above: $M_C(\mathbf{s}) \leq |\mathbf{s}|/r$.

(VI) If $\mathcal{S}$ is a normed space and $C$ is bounded and lies in the ball of radius $R > 0$ with its center at 0, then, for every $\mathbf{s} \in \mathcal{S}$, the value of the Minkowski function is estimated from below: $M_C(\mathbf{s}) \geq |\mathbf{s}|/R$.

Properties (V) and (VI) arise immediately from definition (2.1) of the Minkowski function.

**2.2. Static extremum problem standard form.** Consider the static extremum problem introduced in Subsection 1.1:

$$m \to \inf; \quad m \geq 0, \text{ there exists } \rho \in \Sigma, \ m(\rho + \mathbf{s}_l) \in C. \qquad (2.2)$$

Here, $\Sigma$ is the subspace of self-equilibrated stress fields and $\mathbf{s}_l$ is a fixed stress field equilibrating the given load $\mathbf{l}$. Problem (2.2) determines the safety factor of the load:

$$\alpha_l = \sup\{m \geq 0 : \text{there exists } \rho \in \Sigma, \ m(\rho + \mathbf{s}_l) \in C\}. \qquad (2.3)$$

Let us now transform the static extremum problem to the standard form by formulating a problem which is equivalent to (2.3) and determines the value

$1/\alpha_l$. We will show that the function $f(\rho) = M_C(\rho + \mathbf{s}_l)$ is to be minimized in this problem over the subspace $\Sigma$. Since the Minkowski function is convex, the function $f$ is convex as well, and the above-mentioned extremum problem is standard. The following proposition establishes equivalence of this problem and (2.3).

PROPOSITION 1. Suppose a convex set $C$ is closed along rays and contains 0. If the limit static multiplier of a load $\mathbf{l}$ is nonzero, $\alpha_l > 0$, then static extremum problem (2.2) is equivalent to the standard problem

$$M_C(\rho + \mathbf{s}_l) \to \inf; \quad \rho \in \Sigma, \tag{2.4}$$

and the equality

$$\inf \{M_C(\rho + \mathbf{s}_l) : \rho \in \Sigma\} = \frac{1}{\alpha_l} \tag{2.5}$$

is valid. This equality holds in case $\alpha_l = 0$ as well, that is, the infimum of (2.4) is $+\infty$ when $\alpha_l = 0$.

PROOF. Consider first the case $\alpha_l > 0$. Let $\{m_n\}$, $n = 1, 2, \ldots$, be a maximizing sequence in problem (2.2). This means that $m_n \to \alpha_l$ when $n \to \infty$ and, for every $n$, there is a stress field $\rho_n$ which, together with $m_n$, satisfies the restrictions of problem (2.2). Due to the condition $\alpha_l > 0$, we have $m_n > 0$ for sufficiently large $n$. Since $m_n > 0$ and $m_n(\rho_n + \mathbf{s}_l) \in C$, the inequality $M_C(\rho_n + \mathbf{s}_l) \le 1/m_n$ holds and the following estimation is valid for the infimum in (2.4):

$$\inf \{M_C(\rho_n + \mathbf{s}_l) : n = 1, 2, \ldots\} \le \lim_{n\to\infty} \frac{1}{m_n} = \frac{1}{\alpha_l}. \tag{2.6}$$

Let us show that the inequality

$$\inf \{M_C(\rho_n + \mathbf{s}_l) : n = 1, 2, \ldots\} \ge \lim_{n\to\infty} \frac{1}{m_n} = \frac{1}{\alpha_l}$$

is also valid and, hence, equality (2.5) holds. Let $\{\rho'_n\}$, $n = 1, 2, \ldots$, be a minimizing sequence of the problem (2.4), in particular, $\rho'_n \in \Sigma$. By estimation (2.6) the infimum (2.4) is not greater than $1/\alpha_l < +\infty$ (recall that $\alpha_l \ne 0$ in the case under consideration). Therefore, $M_C(\rho'_n + \mathbf{s}_l) < +\infty$ for sufficiently large $n$. Let us now introduce a sequence of positive numbers

$$m'_n = \begin{cases} \dfrac{1}{M_C(\rho'_n + \mathbf{s}_l)} & \text{if } M_C(\rho'_n + \mathbf{s}_l) > 0, \\ n & \text{if } M_C(\rho'_n + \mathbf{s}_l) = 0, \end{cases} \tag{2.7}$$

$(n = 1, 2, \ldots)$. In case $M_C(\rho'_n + \mathbf{s}_l) > 0$, property (III) of the Minkowski function yields $m'_n(\rho'_n + \mathbf{s}_l) \in C$. This relation is also valid in case $M_C(\rho'_n + \mathbf{s}_l) = 0$ simply by definition of the Minkowski function. Thus in both cases $m'_n$ and $\rho'_n$ satisfy the restrictions of problem (2.2) and, therefore, $\alpha_l \ge m + n'$. This implies

$$\frac{1}{\alpha_l} \le \lim_{n\to\infty} \frac{1}{m'_n} = \lim_{n\to\infty} M_C(\rho'_n + \mathbf{s}_l) = \inf \{M_C(\rho + \mathbf{s}_l) : \rho \in \Sigma\}, \tag{2.8}$$

where the latter equality holds since $\{\rho'_n\}$ is a minimizing sequence of problem (2.4). Inequalities (2.6) and (2.8) together imply (2.5), and the equality is realized in both formulas (2.6), (2.8). This means that 1) if $\{m_n\}$ is a minimizing sequence in problem (2.2) and $\rho_n$ is a stress field satisfying, together with $m_n$, the restrictions in (2.2), $n = 1, 2, \ldots$, then $\{\rho_n\}$ is a minimizing sequence in problem (2.4), and 2) if $\{\rho'_n\}$ is a minimizing sequence in problem (2.4), then $\{m'_n\}$ defined by (2.7) is a minimizing sequence in problem (2.2). Briefly, we say that problems (2.2) and (2.4) are equivalent.

It remains to consider the case $\alpha_l = 0$. The latter equality implies that, if $\rho$ is in $\Sigma$, then there is no number $\mu$ for which $(\rho + \mathbf{s}_l)/\mu \in C$ (otherwise we would have $\alpha_l \geq 1/\mu > 0$). This means that $M_C(\rho + \mathbf{s}_l) = +\infty$ for every $\rho \in \Sigma$. Then the infimum in (2.4) is $+\infty$, which finishes the proof.

**2.3. Kinematic extremum problem standard form.** Consider the kinematic extremum problem introduced in Subsection 1.1:

$$\frac{D(\mathbf{e})}{\langle \mathbf{e}, \mathbf{s}_l \rangle} \to \inf; \quad \mathbf{e} \in E, \ \langle \mathbf{e}, \mathbf{s}_l \rangle > 0. \tag{2.9}$$

Here, $E$ is the subspace of kinematically admissible strain rate fields. Problem (2.9) determines the limit kinematic multiplier of the load $\mathbf{l}$:

$$\beta_l = \inf \left\{ \frac{D(\mathbf{e})}{\langle \mathbf{e}, \mathbf{s}_l \rangle} : \mathbf{e} \in E, \ \langle \mathbf{e}, \mathbf{s}_l \rangle > 0 \right\}. \tag{2.10}$$

Let us now transform the kinematic extremum problem to the standard form by formulating a problem which is equivalent to (2.9) and determines the value $1/\beta_l$. It will now be shown that the function $g(\mathbf{e}) = -\langle \mathbf{e}, \mathbf{s}_l \rangle + M_C^*(\mathbf{e})$ with

$$M_C^*(\mathbf{e}) = \begin{cases} 0 & \text{if } D(\mathbf{e}) \leq 1, \\ +\infty & \text{if } D(\mathbf{e}) > 1 \end{cases}$$

is to be maximized in this problem over the subspace $E$. By convexity of the dissipation $D$, the set on which $D(\mathbf{e}) \leq 1$ is convex. Then the function $M_C^*$ is convex, and, consequently, $g$, which is its sum with a linear function, is also convex. Thus, the above-mentioned extremum problem is standard. The following proposition establishes equivalence of this problem and (2.9).

PROPOSITION 2. If the limit kinematic multiplier is finite, $\beta_l < +\infty$, then kinematic extremum problem (2.9) is equivalent to the standard problem

$$\langle \mathbf{e}, \mathbf{s}_l \rangle - M_C^*(\mathbf{e}) \to \sup; \quad \mathbf{e} \in E, \tag{2.11}$$

and the equality

$$\sup \{\langle \mathbf{e}, \mathbf{s}_l \rangle - M_C^* : \mathbf{e} \in E\} = \frac{1}{\beta_l} \tag{2.12}$$

is valid. This equality also holds in case $\beta_l = +\infty$, that is, the supremum in (2.11) is zero.

PROOF. Consider first the case $\beta_l = +\infty$. Let $\{\mathbf{e}_n\}$, $n = 1, 2, \ldots$, be a minimizing sequence in problem (2.9), that is,

$$\mathbf{e}_n \in E, \quad \langle \mathbf{e}_n, \mathbf{s}_l \rangle > 0, \quad \frac{\langle \mathbf{e}_n, \mathbf{s}_l \rangle}{D(\mathbf{e}_n)} \to \frac{1}{\beta_l} \quad \text{when } n \to \infty.$$

We modify $\mathbf{e}_n$ introducing the strain rate fields

$$\mathbf{e}'_n = \begin{cases} \dfrac{\mathbf{e}_n}{D(\mathbf{e}_n)} & \text{if } D(\mathbf{e}_n) > 0, \\ n\dfrac{\mathbf{e}_n}{\langle \mathbf{e}_n, \mathbf{s}_l \rangle} & \text{if } D(\mathbf{e}_n) = 0 \end{cases} \tag{2.13}$$

(here, $D(\mathbf{e}_n) = 0$ provided $\beta_l = 0$). Due to the first degree homogeneity of the dissipation, we have $D(\mathbf{e}'_n) = 1$ in the first case and $D(\mathbf{e}'_n) = 0$ in the second case. In both cases $M^*_C(\mathbf{e}'_n) = 0$ and $\mathbf{e}'_n$ satisfies the restriction of problem (2.11): $\mathbf{e}'_n \in E$. Thus, we obtain a lower bound for the supremum in (2.11):

$$\sup\{\langle \mathbf{e}, \mathbf{s}_l \rangle - M^*_C(\mathbf{e}) : \mathbf{e} \in E\} \\ \geq \sup\{\langle \mathbf{e}'_n, \mathbf{s}_l \rangle : n = 1, 2, \ldots\} \geq \lim_{n\to\infty} \langle \mathbf{e}'_n, \mathbf{s}_l \rangle. \tag{2.14}$$

Let us evaluate the latter limit. First, consider the case $\beta_l = 0$. It may occur that $D(\mathbf{e}_n) > 0$ or $D(\mathbf{e}_n) = 0$ in this case, and according to (2.13), we have

$$\langle \mathbf{e}'_n, \mathbf{s}_l \rangle = \begin{cases} \dfrac{\langle \mathbf{e}_n, \mathbf{s}_l \rangle}{D(\mathbf{e}_n)} & \text{if } D(\mathbf{e}_n) > 0, \\ n & \text{if } D(\mathbf{e}_n) = 0. \end{cases}$$

Therefore $\langle \mathbf{e}'_n, \mathbf{s}_l \rangle$ tends to $+\infty$ when $n \to \infty$ (recall that

$$\lim_{n\to\infty} \frac{\langle \mathbf{e}_n, \mathbf{s}_l \rangle}{D(\mathbf{e}_n)} = \frac{1}{\beta_l} = +\infty$$

in case $\beta_l = 0$). Then, by (2.14), we find out that the supremum in (2.11) is $+\infty$, which proves (2.12) in case $\beta_l = 0$.

Consider now the case $\beta_l \neq 0$, that is, $\beta_l > 0$. The latter inequality implies $D(\mathbf{e}_n) > 0$, $n = 1, 2, \ldots$. Then according to (2.13) we have

$$\lim_{n\to\infty} \langle \mathbf{e}'_n, \mathbf{s}_l \rangle = \lim_{n\to\infty} \frac{\langle \mathbf{e}_n, \mathbf{s}_l \rangle}{D(\mathbf{e}_n)} = \frac{1}{\beta_l},$$

and (2.14) results in the lower bound for the supremum in (2.11):

$$\sup\{\langle \mathbf{e}, \mathbf{s}_l \rangle - M^*_C(\mathbf{e}) : \mathbf{e} \in E\} \geq \frac{1}{\beta_l}.$$

Let us now verify that $1/\beta_l$ is also an upper bound for the supremum in (2.11) and hence equality (2.12) is valid. Let $\{\mathbf{e}'_n\}$, $n = 1, 2, \ldots$, be a maximizing sequence in problem (2.11). Since $M_C^*$ only takes the values 0 and $+\infty$, we have $M_C^*(\mathbf{e}'_n) = 0$ for sufficiently large $n$ and, therefore,

$$\sup\{\langle \mathbf{e}, \mathbf{s}_l\rangle - M_C^*(\mathbf{e}) : \mathbf{e} \in E\} = \lim_{n\to\infty}(\langle \mathbf{e}'_n, \mathbf{s}_l\rangle - M_C^*(\mathbf{e}'_n)) = \lim_{n\to\infty}\langle \mathbf{e}'_n, \mathbf{s}_l\rangle. \quad (2.15)$$

We have already established that this supremum is not less than $1/\beta_l$; consequently, it is positive (we consider the case $0 < \beta_l < +\infty$), which implies $\langle \mathbf{e}'_n, \mathbf{s}_l\rangle > 0$ for sufficiently large $n$. Thus, $\mathbf{e}'_n$ satisfies the restriction of problem (2.9), and consequently $D(\mathbf{e}'_n)/\langle \mathbf{e}'_n, \mathbf{s}_l\rangle \geq \beta_l$ (for sufficiently large $n$). Note that $D(\mathbf{e}'_n) \leq 1$ since $M_C^*(\mathbf{e}'_n) = 0$. Then the previous inequality can be written as $\langle \mathbf{e}'_n, \mathbf{s}_l\rangle \leq 1/\beta_l$, and (2.15) results in the upper bound $1/\beta_l$ for the supremum in problem (2.11). As we mentioned before, this proves (2.12) in case $0 < \beta_l < +\infty$. We have also seen that (2.13) determines a maximizing sequence $\{\mathbf{e}'_n\}$ in problem (2.11) if $\{\mathbf{e}_n\}$ is a minimizing sequence in (2.9), and vice versa. Thus, problems (2.9) and (2.11) are equivalent if $0 < \beta_l < +\infty$.

It remains to consider the case $\beta_l = +\infty$. Definition (2.10) implies that, for every $\mathbf{e} \in E$ satisfying the condition $\langle \mathbf{e}, \mathbf{s}_l\rangle > 0$, the dissipation $D(\mathbf{e})$ is $+\infty$. For such $\mathbf{e}$, the function which we maximize in (2.11) takes the value $-\infty$. Since this function only takes nonpositive values for $\mathbf{e}$ satisfying the condition $\langle \mathbf{e}, \mathbf{s}_l\rangle \leq 0$, the supremum in (2.11) is nonpositive, and hence equal to zero: this value is attained at $\mathbf{e} = 0$. This is what we had to verify in case $\beta_l = +\infty$, and the proposition is proved.

## 3. Dual extremum problem

The main problem of limit analysis is evaluating the safety factor of a given load $\mathbf{l}$. The safety factor can be determined as extremum

$$a = \inf\{f(\mathbf{x}) : \mathbf{x} \in L\},$$

where $f$ is a certain convex function and $L$ is a subspace in a linear space (see Section 2). Extremum problems of this form often arise in different fields of science and technology; we call these problems standard. In this section, we present a general scheme allowing us to obtain lower bounds for the infimum $a$ in a standard extremum problem. This is of much importance since using the function $f$ involved in the original formulation only results in upper bounds for $a$: $a \leq f(\mathbf{x}_0)$ for any $\mathbf{x}_0$ in $L$. A regular method for obtaining lower bounds arises from constructing a dual extremum problem to the original one. It is remarkable that, under some conditions, the extremums in the original and dual problems are equal. In this case, it is possible to evaluate the extremum $a$ using the dual problem, which, in many cases, appears to be simpler than the original one.

We apply the general scheme to limit analysis in the following sections and to the elastic-plastic shakedown theory in Chapter X.

**3.1. Fenhel transformation.** We start with a definition and the simplest examples of a transformation which plays a key role in constructing the dual

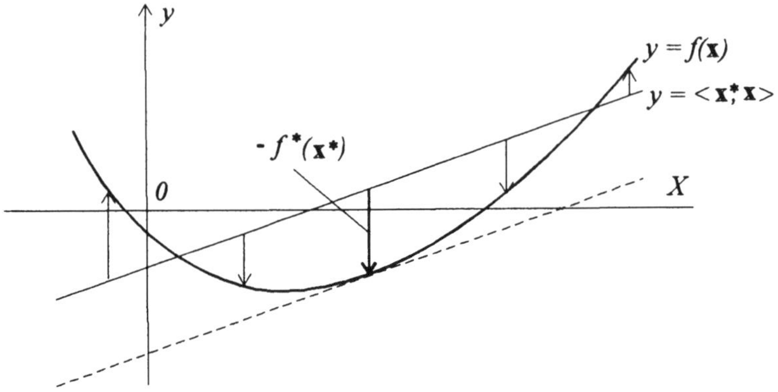


**Figure 3.1**

extremum problem. Consider two linear spaces $X$, $X^*$ and a function of two arguments: $\mathbf{x} \in X$ and $\mathbf{x}^* \in X^*$, the function being linear in each of them. Such function is referred to as a *bilinear form* which *pairs* the spaces $X$ and $X^*$ or as a *pairing* between $X$ and $X^*$; its values are denoted by the symbol $\langle \mathbf{x}^*, \mathbf{x} \rangle$. For more details and examples of pairings see Subsection IV.A.4. Within the framework of the unified formulation of the rigid-plastic problem the spaces of stress and strain rate fields $\mathcal{S}$ and $\mathcal{E}$ are paired by the internal power bilinear form; see Subsection IV.1.4.

Given the pairing between $X$ and $X^*$ and a function $f$ on $X$, it is always possible to construct the function $f^*$ on $X^*$ with its values defined as

$$f(\mathbf{x}^*) = \sup \{ \langle \mathbf{x}^*, \mathbf{x} \rangle - f(\mathbf{x}) : \mathbf{x} \in X \}. \tag{3.1}$$

This function is called the *Fenhel transformation* of the function $f$. A simple geometrical interpretation of this definition can be obtained by re-writing (3.1) in the form

$$-f(\mathbf{x}^*) = \inf \{ f(\mathbf{x}) - \langle \mathbf{x}^*, \mathbf{x} \rangle : \mathbf{x} \in X \},$$

where we made use of the obvious equality $\sup A = -\inf(-A)$. The meaning of the value $f(\mathbf{x}) - \langle \mathbf{x}^*, \mathbf{x} \rangle$ is the altitude of the point $f(\mathbf{x})$ in the graph of $f$ above the point $(\mathbf{x}, \langle \mathbf{x}^*, \mathbf{x} \rangle)$ in the graph of the linear function $\mathbf{x} \to \langle \mathbf{x}^*, \mathbf{x} \rangle$, see Figure 3.1. In the figure, the altitude is depicted by a segment with an arrow, which assigns the segment a direction: the upward direction corresponds to positive values of $f(\mathbf{x}) - \langle \mathbf{x}^*, \mathbf{x} \rangle$, the downward direction to negative values. Speaking about altitude, we always take into account its sign or, equivalently, the direction of the above segment. The minimum (or, more precisely, the infimum) of the altitude is exactly $-f^*(\mathbf{x}^*)$, which interprets geometrically the definition of the Fenhel transformation.

EXAMPLE 1. Let $X = X^*$ be $\mathbf{R}$ and $\langle x^*, x \rangle$ be the product $x^* x$ of real numbers $x^*$ and $x$. This bilinear form pairs $\mathbf{R}$ with itself. Let us find the Fenhel transformation of the function $f$, $f(x) = x^2$. The supremum $f^*(x^*) =$

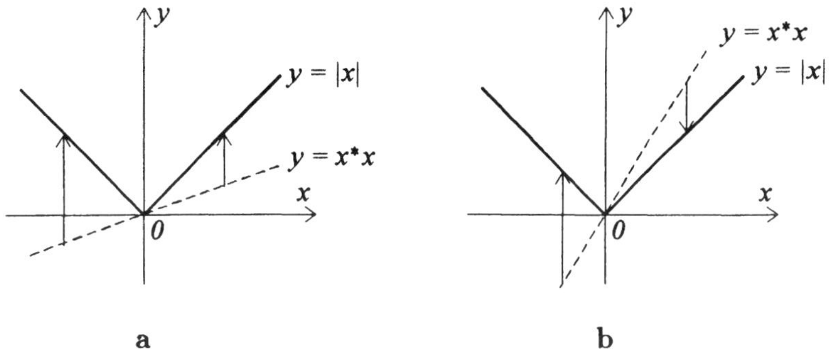


**Figure 3.2**

sup $\{x^*x - x^2 : x \in \mathbf{R}\}$ is the maximum of the quadratic polynomial attained at $x = x^*/2$ and equals $f^*(x^*) = (x^*)^2/2$.

EXAMPLE 2. Consider again $X = X^* = \mathbf{R}$ and $\langle x^*, x \rangle = x^*x$ for $x$, $x^*$ in $\mathbf{R}$. Let us find the Fenhel transformation of the function $f$, $f(x) = |x|$. The altitude $|x| - x^*x$ is nonnegative if $|x^*| \leq 1$; then the minimum of the altitude is 0 and is attained at $x = 0$ (see Figure 3.2a). Thus, $f^*(x^*) = 0$ if $|x^*| \leq 1$. If $|x^*| > 1$, the altitude also takes negative values of an arbitrary large absolute value (see Figure 3.2b); then the infimum of the altitude is $-\infty$, which results in

$$f^*(x^*) = \begin{cases} 0 & \text{if } |x^*| \leq 1, \\ +\infty & \text{if } |x^*| > 1. \end{cases}$$

EXAMPLE 3. *Fenhel transformation of Minkowski function.* Consider the set $C$ in a linear space $\mathcal{S}$ and its Minkowski function $M_C$, see (2.1). If a bilinear form $\langle \cdot, \cdot \rangle$ pairs $\mathcal{S}$ and another space $\mathcal{E}$, then the Fenhel transformation of the Minkowski function

$$M_C^*(\mathrm{e}) = \sup\{\langle \mathrm{e}, \mathrm{s} \rangle - M_C(\mathrm{s}) : \mathrm{s} \in \mathcal{S}\}, \quad \mathrm{e} \in \mathcal{E},$$

is defined as a function on $\mathcal{E}$. Let us show that if $C$ is convex, closed along rays and contains 0, then

$$M_C^*(\mathrm{e}) = \begin{cases} 0 & \text{if } D(\mathrm{e}) \leq 1, \\ +\infty & \text{if } D(\mathrm{e}) > 1, \end{cases} \tag{3.2}$$

where $D(\mathrm{e}) = \sup\{\langle \mathrm{e}, \boldsymbol{\tau} \rangle : \boldsymbol{\tau} \in C\}$. In particular, within the framework of the unified formulation of the rigid-plastic problem, the internal power bilinear form pairs the spaces $\mathcal{S}$ and $\mathcal{E}$ of the stress and strain rate fields. If $C$ is the set of admissible stress fields, then $D$ is the dissipation.

Let us show first that $M_C^*(\mathrm{e}) = +\infty$ if $D(\mathrm{e}) > 1$. Indeed, the inequality $D(\mathrm{e}) > 1$ implies existence of $\boldsymbol{\tau} \in C$ such that $\langle \mathrm{e}, \boldsymbol{\tau} \rangle > 1$. Since $\boldsymbol{\tau}$ is in $C$, we have $M_C(\boldsymbol{\tau}) \leq 1$ due to (2.1). The two latter inequalities result in $\langle \mathrm{e}, \boldsymbol{\tau} \rangle - M_C(\boldsymbol{\tau}) > 0$. Then, we consider $\mathrm{s} = a\boldsymbol{\tau}$, where $a$ is a positive number.

The value $\langle \mathbf{e}, \mathbf{s} \rangle - M_C(\mathbf{s}) = a(\langle \mathbf{e}, \boldsymbol{\tau} \rangle - M_C(\boldsymbol{\tau}))$ is arbitrarily large when $a \to +\infty$ and, consequently, $M_C^*(\mathbf{e}) = +\infty$.

Now let us show now that $M_C^*(\mathbf{e}) = 0$ if $D(\mathbf{e}) \leq 1$. It is sufficient to verify that

$$\langle \mathbf{e}, \mathbf{s} \rangle - M_C(\mathbf{s}) \leq 0 \quad \text{for every} \quad \mathbf{s} \in \mathcal{S}. \tag{3.3}$$

Indeed, this implies

$$M_C^*(\mathbf{e}) = \sup \{ \langle \mathbf{e}, \mathbf{s} \rangle - M_C(\mathbf{s}) : \mathbf{s} \in \mathcal{S} \} \leq 0,$$

where $\langle \mathbf{e}, \mathbf{s} \rangle - M_C(\mathbf{s})$ vanishes at $\mathbf{s} = 0$ and, therefore, the supremum equals zero and $M_C^*(\mathbf{s}) = 0$. It remains to verify (3.3).

To prove (3.3) we consider the following cases: 1) $\mathbf{s} = 0$, 2) $M_C(\mathbf{s}) = +\infty$, 3) $+\infty > M_C(\mathbf{s}) = m > 0$, and 4) $M_C(\mathbf{s}) = 0$ with $\mathbf{s} \neq 0$. Validity of (3.3) is obvious in the first two cases. In the third case, we consider $\mathbf{s}_1 = \mathbf{s}/m$. By property (III) of the Minkowski function (Subsection 2.1), $\mathbf{s}_1$ is in $C$. Then, $\langle \mathbf{e}, \mathbf{s}_1 \rangle \leq D(\mathbf{e})$ due to the definition of the dissipation, from which it follows that $\langle \mathbf{e}, \mathbf{s}_1 \rangle \leq 1$ in case $D(\mathbf{e}) \leq 1$. Thus,

$$\langle \mathbf{e}, \mathbf{s} \rangle - M_C(\mathbf{s}) = \langle \mathbf{e}, \mathbf{s} \rangle - m = m(\langle \mathbf{e}, \mathbf{s}_1 \rangle - 1) \leq 0,$$

and (3.3) holds. In the fourth of the above-mentioned cases, (3.3) takes the form $\langle \mathbf{e}, \mathbf{s} \rangle \leq 0$. Let us show that this inequality is valid. Indeed, the equality $M_C(\mathbf{s}) = 0$ implies that $\mathbf{s}/\mu$ is in $C$ for every $\mu > 0$. Then $D(\mathbf{e}) \geq \langle \mathbf{e}, \mathbf{s} \rangle / \mu$ due to the definition of the dissipation. In case $\langle \mathbf{e}, \mathbf{s} \rangle > 0$, we set $\mu \to 0$ in the latter inequality and obtain $D(\mathbf{e}) = +\infty$, which contradicts the condition $D(\mathbf{e}) \leq 1$. Thus, $\langle \mathbf{e}, \mathbf{s} \rangle > 0$ is impossible, and $\langle \mathbf{e}, \mathbf{s} \rangle$ is nonpositive, which proves (3.3). Recall that this was the last step in verifying formula (3.2) for evaluating the Fenhel transformation of the Minkowski function.

REMARK 1. The denotation by the symbol $M_C^*$ for the function involved in the kinematic extremum problem (Subsection 2.3) now proves to have been reasonable: the function appears to be the Fenhel transformation of the Minkowski function.

Noteworthy is the following important property of the Fenhel transformation.

PROPOSITION 1. Fenhel transformation of any function is a convex function.

PROOF. Consider $\mathbf{x}_1^*$ and $\mathbf{x}_2^*$ in the space $X^*$ and numbers $\alpha_1$, $\alpha_2$, $0 \leq \alpha_1, \alpha_2 \leq 1$, $\alpha_1 + \alpha_2 = 1$. Due to the latter equality, we find by the definition of Fenhel transformation:

$$\begin{aligned} f^*(\alpha_1 \mathbf{x}_1^* + \alpha_2 \mathbf{x}_2^*) &= \sup \{ \langle \alpha_1 \mathbf{x}_1^* + \alpha_2 \mathbf{x}_2^*, \mathbf{x} \rangle : \mathbf{x} \in X \} \\ &= \sup \{ \alpha_1 (\langle \mathbf{x}_1^*, \mathbf{x} \rangle - f(\mathbf{x})) + \alpha_2 (\langle \mathbf{x}_2^*, \mathbf{x} \rangle - f(\mathbf{x})) : \mathbf{x} \in X \} \\ &\leq \alpha_1 f^*(\mathbf{x}_1^*) + \alpha_2 f^*(\mathbf{x}_2^*). \end{aligned}$$

We make use of the obvious inequality $\sup(A + B) \leq \sup A + \sup B$ and estimate the latter expression from above by

$$\begin{aligned} \alpha_1 \sup \{ \langle \mathbf{x}_1^*, \mathbf{x} \rangle - f(\mathbf{x}) : \mathbf{x} \in X \} + \alpha_2 \sup \{ \langle \mathbf{x}_2^*, \mathbf{x} \rangle - f(\mathbf{x}) : \mathbf{x} \in X \} \\ = \alpha_1 f(\mathbf{x}_1^*) + \alpha_2 f(\mathbf{x}_2^*). \end{aligned}$$

Thus,

$$f^*(\alpha_1 \mathbf{x}_1^* + \alpha_1 \mathbf{x}_1^*) \le \alpha_1 f(\mathbf{x}_1^*) + \alpha_2 f(\mathbf{x}_2^*),$$

which shows that $f^*$ is a convex function and finishes the proof.

**3.2. Constructing the dual problem.** Let $X$ be a linear space, $L$ be a subspace in $X$ and $f$ be a function on $X$. We consider the extremum problem

$$f(\mathbf{x}) \to \inf; \quad \mathbf{x} \in L. \tag{3.4}$$

We now present a general scheme for constructing another extremum problem

$$\varphi(\mathbf{x}^*) \to \sup; \quad \mathbf{x}^* \in L^0 \tag{3.5}$$

such that the following relation

$$\sup\left\{\varphi(\mathbf{x}^*) : \mathbf{x}^* \in L^0\right\} \le \inf\left\{f(\mathbf{x}) : \mathbf{x} \in L\right\} \tag{3.6}$$

is valid. Problem (3.5) which possesses this property is very useful: it allows obtainment of the two-sided estimation for the extremum

$$a = \inf\left\{f(\mathbf{x}) : \mathbf{x} \in L\right\}$$

in the original problem:

$$\varphi(\mathbf{x}^*) \le a \le f(\mathbf{x}) \text{ for every } \mathbf{x} \in L \text{ and every } \mathbf{x}^* \in L^0.$$

To construct the required extremum problem another linear space $X^*$ and bilinear form $\langle \cdot, \cdot \rangle$ which pairs $X$ and $X^*$ are needed. Provided there are such space $X^*$ and pairing $\langle \cdot, \cdot \rangle$, consider the Fenhel transformation of the function $f$. By the definition of the Fenhel transformation:

$$f^*(\mathbf{x}^*) \ge \langle \mathbf{x}^*, \mathbf{x} \rangle \text{ for every } \mathbf{x} \in X \text{ and every } \mathbf{x}^* \in X^*. \tag{3.7}$$

Consider now the polar set $L^0$ for the subspace $L$:

$$L^0 = \{\mathbf{x}^* \in X^* : \langle \mathbf{x}^*, \mathbf{x} \rangle = 0 \text{ for every } \mathbf{x} \in L\}.$$

Restricting (3.7) to $\mathbf{x}$ in $L$ and $\mathbf{x}^*$ in $L^0$ ($\langle \mathbf{x}^*, \mathbf{x} \rangle = 0$ in this case) we obtain:

$$f(\mathbf{x}) \ge -f^*(\mathbf{x}^*) \text{ for every } \mathbf{x} \in L \text{ and every } \mathbf{x}^* \in L^0.$$

This relation is of the form (3.6) with $\varphi = -f^*$:

$$\sup\left\{-f^*(\mathbf{x}^*) : \mathbf{x}^* \in L^0\right\} \le \inf\left\{f(\mathbf{x}) : \mathbf{x} \in L\right\}. \tag{3.8}$$

We refer to the extremum problem

$$-f^*(\mathbf{x}^*) \to \sup; \quad \mathbf{x}^* \in L^0 \tag{3.9}$$

as the *dual extremum problem* of (3.4).

Sometimes, the original extremum problem is considered in the form

$$-g(x) \to \sup; \quad \mathbf{x} \in L \tag{3.10}$$

instead of (3.4). For example, the kinematic extremum problem of limit analysis is traditionally written in this form. In the case of original extremum problem (3.10), we refer to the problem

$$g^*(\mathbf{x}^*) \to \inf; \quad \mathbf{x}^* \in L^0 \tag{3.11}$$

as dual of (3.10).

Applications of the dual problem can be effective provided the Fenhel transformation of $f$ and the polar set for $L$ have constructive descriptions. Those can be obtained in many cases; in particular, note that extremum problem (3.1), which is to be solved for determining $f^*$, is simpler than the original problem (3.4). Indeed, (3.1) is a problem of nonrestricted optimization: minimization of $f$ proceeds over the whole space $X$ while the restriction $\mathbf{x} \in L$ is involved in (3.4).

**3.3. Applying the dual problem.** Due to inequality (3.8), the dual problem is always applicable to estimating the extremum $a$ of the original problem from below:

$$-f^*(\mathbf{x}^*) \leq a \text{ for every } \mathbf{x}^* \in L^0.$$

The best of these lower bounds is

$$\sup\left\{-f^*(\mathbf{x}^*) : \mathbf{x}^* \in L^0\right\} \leq a.$$

Of the most interest is application of the dual problem in case the equality

$$\sup\left\{-f^*(\mathbf{x}^*) : \mathbf{x}^* \in L^0\right\} = \inf\left\{f(\mathbf{x}) : \mathbf{x} \in L\right\}$$

is guaranteed. Then, the gap between lower and upper bounds

$$-f^*(\mathbf{x}^*) \leq a \leq f(\mathbf{x}) \quad (\text{for every } \mathbf{x} \in L \text{ and every } \mathbf{x}^* \in L^0)$$

can, in principle, be made arbitrarily small. In this case, the extremum of the original problem can be calculated not only by solving this problem but also by solving the dual one.

We will see that, for example, the kinematic extremum problem of limit analysis is dual for the static extremum problem and we will obtain conditions for equality of their extremums. Under such conditions, the safety factor can be evaluated by the kinematic method, which is simpler than the static one.

**3.4. Conditions for extremums equality.** We point out conditions under which the extremum of the original and dual extremum problems are equal. We assume that the linear space $X$ in which the original extremum problem (3.4) is set up is a normed space (see Appendix A). Using the space $X'$ of continuous linear functions on $X$ (see Subsection A.4) as $X^*$, we define the following pairing between $X$ and $X^*$: $\langle \mathbf{x}', \mathbf{x} \rangle$ is the value of the functional $\mathbf{x}' \in X'$ at $\mathbf{x} \in X$. The pairing allows construction of the Fenhel transformation transformation and the polar set using this pairing.

THEOREM 1. Suppose $X$ is a complete normed linear space, $L$ is its subspace and $f$ is a convex function on $X$ continuous at point $\mathbf{x}_0 \in L$. Then the infimum in the problem

$$f(\mathbf{x}) \to \inf; \quad \mathbf{x} \in L$$

and the supremum in the dual problem

$$-f^*(\mathbf{x}^*) \to \sup; \quad \mathbf{x}^* \in L^0$$

are equal:

$$\inf \{f(\mathbf{x}) : \mathbf{x} \in L\} = \sup \{-f^*(\mathbf{x}^*) : \mathbf{x}^* \in L^0\}.$$

The extremum in the dual problem is attainable if $f$ is bounded from below on $L$.

Proof is given in Appendix B.

Theorem 1 will be applied to limit analysis in the following sections and to the elastic-plastic shakedown theory in Chapter X.

REMARK 2. If the original extremum problem is written in the form

$$-g(\mathbf{x}) \to \sup; \quad \mathbf{x} \in L,$$

the proposition analogous to Theorem 1 obviously holds. To obtain this proposition, it is sufficient to replace $f$ by $-g$ in Theorem 1. Then, the original problem is (3.10) and dual of (3.10) is (3.11).

## 4. Conditions for equality of limit multipliers – I

Conditions for equality of limit static and limit kinematic multipliers specify the cases in which the safety factor can be evaluated by both static and kinematic methods.

In this section, we establish one type of such conditions within the framework of the unified formulation of the rigid-plastic problem. The next section contains a more detailed version of these conditions and outlines a wide class of plastic bodies, which satisfy the conditions.

To obtain the conditions for equality of limit multipliers, we construct the extremum problem dual of the kinematic extremum problem of limit analysis and show that this problem is exactly the static extremum problem which determines the safety factor. Then, Theorem 3.1 results immediately in conditions for equality of the extremums in the original and dual problems.

**4.1. Static extremum problem dual of the kinematic problem.** Let us construct an extremum problem dual to the kinematic extremum problem of limit analysis (2.11):

$$\langle \mathbf{e}, \mathbf{s}_l \rangle - M_C^*(\mathbf{e}) \to \sup; \quad \mathbf{e} \in E. \tag{4.1}$$

Here, $M_C$ is the Minkowski function of the set $C$ of admissible stress fields; $M_C^*$ is the Fenhel transformation of $M_C$. The supremum in (4.1) equals $1/\beta_l$, $\beta_l$ standing for the limit kinematic multiplier of the load $\mathbf{l}$. We consider problem (4.1) within the framework of the rigid-plastic problem unified formulation. We adopt the assumption made in Subsection IV.1.4 and use the notation introduced in that subsection. In particular, Def $\mathcal{V}$ is the subspace of kinematically admissible strain rate fields and $\mathbf{s}_l$ is a fixed stress field equilibrating the load $\mathbf{l}$.

Formulation (4.1) of the kinematic extremum problem is standard in the sense of Section 2: the problem consists of maximizing the function $-g$ over the subspace $E$, where $g(\mathbf{e}) = M_C^*(\mathbf{e}) - \langle \mathbf{e}, \mathbf{s}_l \rangle$ is convex. According to the general scheme in Subsection 3.2, the dual of this problem is

$$g^*(\rho) \to \inf; \quad \rho \in E^0. \tag{4.2}$$

Here, $g^*$ is the Fenhel transformation of $g$ and $E^0$ is the polar set of $E$. Both are determined in terms of the integral power bilinear form $\langle \cdot, \cdot \rangle$ which pairs the spaces $\mathcal{S}$ and $\mathcal{E}$ within the framework of the unified formulation of the rigid-plastic problem.

To explicitly formulate dual problem (4.2), we recall that $E^0 = \Sigma$, where $\Sigma$ is the subspace of self-equilibrated stress fields in $\mathcal{S}$ (see Remark IV.1.2) and consider the Fenhel transformation:

$$\begin{aligned} g^*(\rho) &= \sup \{\langle \mathbf{e}, \rho \rangle - M_C^*(\mathbf{e}) + \langle \mathbf{e}, \mathbf{s}_l \rangle : \mathbf{e} \in \mathcal{E}\} \\ &= \sup \{\langle \mathbf{e}, \rho + \mathbf{s}_l \rangle - M_C^*(\mathbf{e}) : \mathbf{e} \in \mathcal{E}\} = M_C^{**}(\rho + \mathbf{s}_l). \end{aligned}$$

Here, $M_C^{**}$ is the Fenhel transformation of $M_C^*$, that is, the repeated Fenhel transformation of $M_C$. Validity of the equality $M_C^{**} = M_C$ in cases of practical interest is established below. Consequently, $g^*(\rho) = M_C(\rho + \mathbf{s}_l)$ and problem (4.2) assumes the form

$$M_C(\rho + \mathbf{s}_l) \to \inf; \quad \rho \in \Sigma.$$

This is exactly the static extremum problem (2.4). Thus, in case $M_C^{**} = M_C$, the static extremum problem which determines the limit static multiplier

$$\frac{1}{\alpha_l} = \inf \{M_C(\rho + \mathbf{s}_l) : \rho \in \Sigma\}$$

is a dual of kinematic extremum problem (4.1), which determines the limit kinematic multiplier

$$\frac{1}{\beta_l} = \sup \{\langle \mathbf{e}, \mathbf{s}_l \rangle - M_C^*(\mathbf{e}) : \mathbf{e} \in E\}.$$

Now, our nearest objective is to find out what properties of the set $C$ imply the equality $M_C^{**} = M_C$, which is a substantial condition for the duality of the static and kinetic problems. Then we will turn to the discussion of conditions for equality of the extremums in these problems, that is, limit static and kinematic multipliers.

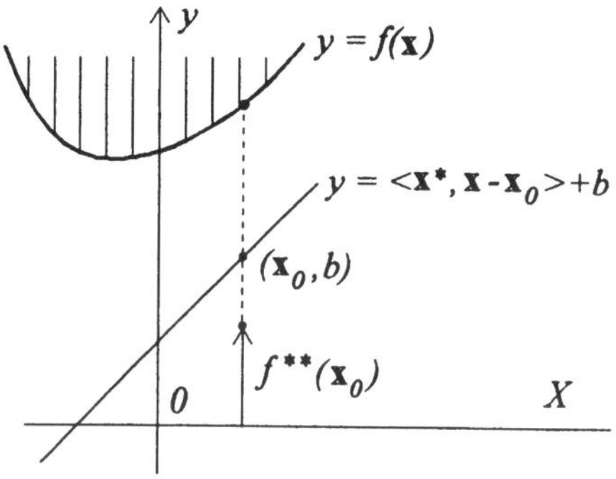


**Figure 4.1**

**4.2. Repeated Fenhel transformation.** We now describe the conditions under which the repeated Fenhel transformation of function $f$ results in the original function: $f^{**} = f$.

Let $\langle \cdot, \cdot \rangle$ be a bilinear form which pairs the linear spaces $X$ and $X^*$ (see Subsections IV.A and 3.1). The Fenhel transformation of the function $f$ defined on $X$ is the function $f^*$ defined on $X^*$ as:

$$f^*(\mathbf{x}^*) = \sup \{\langle \mathbf{x}^*, \mathbf{x} \rangle - f(\mathbf{x}) : \mathbf{x} \in X\}.$$

Analogously, the Fenhel transformation of the function $\varphi$ defined on $X^*$ is the function $\varphi^*$ defined on $X$ as:

$$\varphi(\mathbf{x}^*) = \sup \{\langle \mathbf{x}^*, \mathbf{x} \rangle - \varphi(\mathbf{x}^*) : \mathbf{x}^* \in X^*\}.$$

In particular, consider the transformation of the function $\varphi = f^*$, that is, the repeated Fenhel transformation of $f$:

$$f^{**}(\mathbf{x}) = \sup \{\langle \mathbf{x}^*, \mathbf{x} \rangle - f^*(\mathbf{x}^*) : \mathbf{x}^* \in X^*\}. \tag{4.3}$$

It can be easily seen that the inequality $f \geq f^{**}$ is valid for every function $f$. Indeed, by the definition of the Fenhel transformation, we have

$$f(\mathbf{x}) \geq \langle \mathbf{x}^*, \mathbf{x} \rangle - f^*(\mathbf{x}^*)$$

and, therefore:

$$f(\mathbf{x}) \geq \sup \{\langle \mathbf{x}^*, \mathbf{x} \rangle - f^*(\mathbf{x}^*) : \mathbf{x}^* \in X^*\}.$$

Actually, if $f$ is convex and satisfies an additional condition, not only the above-mentioned inequality $f \geq f^{**}$ is valid, but also the equality $f = f^{**}$. Figure 4.1 illustrates this statement. The figure shows the epigraph of $f$ and the value $f^{**}(\mathbf{x}_0)$ of the function $f^{**}$ at the point $\mathbf{x}_0$ while the strict inequality $f(\mathbf{x}_0) > f^{**}(\mathbf{x}_0)$ is assumed to be valid. It looks almost obvious that a convex set, the epigraph of $f$, and the point $(\mathbf{x}_0, f^{**}(\mathbf{x}_0))$ not in the epigraph can be separated by the graph of a certain linear function $y = \langle \mathbf{x}^*, \mathbf{x} - \mathbf{x}_0 \rangle + b$. In other words, there exists $\mathbf{x}^* \in X^*$ such that the inequalities

$$f^{**}(\mathbf{x}_0) < b; \quad \langle \mathbf{x}^*, \mathbf{x} - \mathbf{x}_0 \rangle + b < f(\mathbf{x}) \quad \text{for every } \mathbf{x} \in X \tag{4.4}$$

are valid.

We write the latter inequality in the form

$$\langle \mathbf{x}^*, \mathbf{x} \rangle - f(\mathbf{x}) < \langle \mathbf{x}^*, \mathbf{x}_0 \rangle - b$$

and consider the supremum with respect to $\mathbf{x}$ on its left-hand side. The supremum is $f^*(\mathbf{x}^*)$, and we arrive at the inequality

$$f^*(\mathbf{x}^*) \leq \langle \mathbf{x}^*, \mathbf{x}_0 \rangle - b < \langle \mathbf{x}^*, \mathbf{x}_0 \rangle - f^{**}(\mathbf{x}_0).$$

Since $f^{**} < b$, the latter inequality implies $f^*(\mathbf{x}^*) < \langle \mathbf{x}^*, \mathbf{x}_0 \rangle - f^{**}(\mathbf{x}_0)$, which contradicts (4.3). Thus, the assumption $f(\mathbf{x}_0) > f^{**}(\mathbf{x}_0)$ we started with is wrong, which narrows the above established relation $f(\mathbf{x}_0) \leq f^{**}(\mathbf{x}_0)$ to the equality $f(\mathbf{x}_0) = f^{**}(\mathbf{x}_0)$.

This equality is what we need, and the above reasoning contains the main idea of its proof. However, the reasoning calls for significant improvement. Namely, we assumed that a convex set and a point not belonging to it can be strictly separated by a plane as is shown in Figure 4.1. This is true if the set is closed and is not necessarily true otherwise. That is why to arrive at $f^{**} = f$ the epigraph of $f$ is assumed to be closed. There are various concepts of closure in case the space $X$ is not finite-dimensional. We now introduce a variant of this concept, weak closure, which allows formulating effective conditions for the equality $f^{**} = f$.

Let $A$ be a subset in $X$ and $\langle \cdot, \cdot \rangle$ be a pairing between $X$ and $X^*$. The point $\mathbf{x}_0 \in X$ is referred to as the *weak limit point* for $A$ if for every $\varepsilon > 0$ and every finite collection $\{\mathbf{x}_1^*, \ldots, \mathbf{x}_K^*\} \subset X^*$ there is $\mathbf{x}$ in $A$ such that

$$|\langle \mathbf{x}_1^*, \mathbf{x} - \mathbf{x}_0 \rangle| < \varepsilon, \; \ldots, \; |\langle \mathbf{x}_K^*, \mathbf{x} - \mathbf{x}_0 \rangle| < \varepsilon. \tag{4.5}$$

We emphasize that the number $K$ is not fixed in this definition. The set $A$ is referred to as *weak closed* if $A$ contains all its weak limit points. The function $f$ on $X$ is referred to as *weak closed* if, for every number $a$, the set of points $\mathbf{x} \in X$, at which $f(\mathbf{x}) \leq a$, is weak closed.

EXAMPLE 1. Let us show that, in a finite-dimensional space, a set is weak closed if and only if it is closed in a standard sense. Indeed, consider the spaces $X = \mathbf{R}^n$ and $X^* = \mathbf{R}^n$ with the pairing

$$\langle \mathbf{x}^*, \mathbf{x} \rangle = x_1^* x_1 + \cdots + x_n^* x_n \quad (\mathbf{x} = (x_1, \ldots, x_n), \; \mathbf{x}^* = (x_1^*, \ldots, x_n^*)).$$

We denote by $\mathbf{x}_1^*$ the member $(1, 0, \ldots, 0)$ of $X$; then the equality $\langle \mathbf{x}_1^*, \mathbf{x} \rangle = x_1$ holds for every $\mathbf{x} = (x_1, \ldots, x_n)$ in $X$. We introduce analogously $\mathbf{x}_2^*, \ldots, \mathbf{x}_n^*$ in $X^*$ such that $\langle \mathbf{x}_2^*, \mathbf{x} \rangle = x_2, \; \ldots, \; \langle \mathbf{x}_n^*, \mathbf{x} \rangle = x_n$ for every $\mathbf{x} \in X$. If $\mathbf{x}_0 =$

$(\mathbf{x}_{01}, \ldots, \mathbf{x}_{0n})$ is a weak limit point for a set $A$, then for every $\varepsilon > 0$ and for the collection $\{\mathbf{x}_1^*, \ldots, \mathbf{x}_n^*\}$ there is $\mathbf{x}$ in $A$ satisfying (4.5). The latter conditions are of the form

$$|x_1 - x_{01}| < \varepsilon, \ldots, |x_n - x_{0n}| < \varepsilon$$

in the case under consideration. Consequently, $\mathbf{x}_0$ is a limit point for $A$ in the standard sense. It is obvious that, conversely, a limit point for $A$ in the standard sense is also its weak limit point.

The following well known statement of convex analysis specifies conditions for the equality $f^{**} = f$. In this proposition, $\langle \cdot, \cdot \rangle$ is a pairing between the linear spaces $X$ and $X^*$. We also assume that the pairing meets the following quite natural requirement: if $\mathbf{x}_0 \in X$ satisfies the condition $\langle \mathbf{x}^*, \mathbf{x}_0 \rangle = 0$ for every $\mathbf{x}^* \in X^*$, then $\mathbf{x}_0 = 0$.

THEOREM 1 (*Fenhel, Moreau*). Suppose function $f$ on $X$ does not take the value $-\infty$. Then $f^{**} = f$ if and only if $f$ is convex and weak closed.

This theorem allows us to find out what properties of the set $C$ guarantee the equality $M_C^{**} = M_C$. This equality is a condition for the static extremum problem to be a dual of the kinematic problem (see the previous subsection). We are reminded that, within the framework of limit analysis, $C$ is the subset of admissible stress fields. The internal power bilinear form $\langle \cdot, \cdot \rangle$ is a pairing between $\mathcal{S}$ and the space $\mathcal{E}$ of strain rate fields. The Fenhel transformation and the weak closure in $\mathcal{S}$ are defined in terms of $\langle \cdot, \cdot \rangle$. The following proposition establishes conditions for the equality $M_C^{**} = M_C$.

PROPOSITION 1. Suppose the set $C$ of admissible stress fields is weak closed in $\mathcal{S}$. If the set of those $\mathbf{s} \in \mathcal{S}$, for which $M_C(\mathbf{s}) = 0$, is also weak closed in $\mathcal{S}$, then $M_C^{**} = M_C$.

PROOF. The Minkowski function $M_C$ is nonnegative and convex (Subsection 2.1). By Theorem 1, to prove the equality $M_C^{**} = M_C$, it remains to verify that $C$ is weak closed. Recall that $C$ is weak closed if the set

$$\{\mathbf{s} \in \mathcal{S} : M_C(\mathbf{s}) \leq a\}$$

is weak closed for every number $a$. If $a > 0$, this set is $aC$ according to (III) in Subsection 2.1, and $aC$ is weak closed together with $C$. If $a = 0$, we have

$$\{\mathbf{s} \in \mathcal{S} : M_C(\mathbf{s}) \leq 0\} = \{\mathbf{s} \in \mathcal{S} : M_C(\mathbf{s}) = 0\}$$

since $M_C$ is nonnegative. The latter set is assumed to be weak closed. If $a < 0$, the set $\{\mathbf{s} \in \mathcal{S} : M_C(\mathbf{s}) \leq a\}$ is empty, and, consequently, weak closed, which finishes the proof.

COROLLARY. If the set $C$ of admissible stress fields is weak closed and bounded along every ray emanating from 0, then $M_C^{**} = M_C$.

Indeed, the boundness of $C$ along every ray implies that $M_C(\mathbf{s}) = 0$ only if $\mathbf{s} = 0$. Thus, the set $\{\mathbf{s} \in \mathcal{S} : M_C(\mathbf{s}) = 0\}$ contains a single member and, consequently, is weak closed. Then $M_C^{**} = M_C$ by Proposition 1.

REMARK 1. We will see (in Subsection 5.1) that $C$ is weak closed in all limit analysis problems of practical interest.

**4.3. Limit multipliers equality.** Suppose the set $C$ of admissible stress fields is weak closed in $\mathcal{S}$ and bounded along every ray emanating from 0 (and we always mean $C$ is convex and contains 0). Then, by Subsection 4.1 and the Corollary of Proposition 1, the static extremum problem, which determines the limit static multiplier (the safety factor),

$$\frac{1}{\alpha_l} = \inf\{M_C(\rho + \mathbf{s}_l) : \rho \in \Sigma\}, \tag{4.6}$$

is a dual of the kinematic extremum problem, which determines the limit kinematic multiplier

$$\frac{1}{\beta_l} = \sup\{\langle \mathbf{e}, \mathbf{s}_l \rangle - M_C^*(\mathbf{e}) : \mathbf{e} \in E\}. \tag{4.6}$$

We will now apply Theorem 3.1 to these problems and establish conditions for the limit multipliers equality. We consider the static and kinematic extremum problems within the framework of the unified formulation of the rigid-plastic problem and adopt the following additional assumptions: the space $\mathcal{E}$ of strain rate fields is a complete normed space, the space $\mathcal{S}$ of stress fields is conjugate of $\mathcal{E}$, that is, $\mathcal{S}$ is the space of continuous linear functionals on $\mathcal{E}$, $\mathcal{S} = \mathcal{E}'$, $\langle \mathbf{e}, \boldsymbol{\sigma} \rangle$ is the value of a functional $\boldsymbol{\sigma} \in \mathcal{S}$ at $\mathbf{e} \in \mathcal{E}$: $\langle \mathbf{e}, \boldsymbol{\sigma} \rangle = \boldsymbol{\sigma}(\mathbf{e})$ (see definitions of basic concepts related to normed spaces in Appendix A). The above assumptions are obviously satisfied for the discrete system, the spaces $\mathcal{S}$, $\mathcal{E}$ being finite-dimensional. We will see that it is possible to choose appropriate spaces $\mathcal{S}$, $\mathcal{E}$ for continual systems too (Section 5).

In the following proposition, $U$ as usual denotes the set of virtual velocity fields.

THEOREM 2. Suppose $\mathcal{S}$ and $\mathcal{E}$ are complete normed spaces, $\mathcal{S}$ being conjugate of $\mathcal{E}$: $\mathcal{S} = \mathcal{E}'$ (thus, $\langle \mathbf{e}, \boldsymbol{\sigma} \rangle = \boldsymbol{\sigma}(\mathbf{e})$ for every $\mathbf{e} \in \mathcal{E}$ and every $\boldsymbol{\sigma} \in \mathcal{S}$). Let $C$ be a convex set in $\mathcal{S}$ containing 0. Then

1) the limit static and limit kinematic multipliers of the load $\mathbf{l}$ are equal, $\alpha_l = \beta_l$,

2) the extremum $1/\alpha_l$ in static problem (4.6) is attainable,

3) the extremum in kinematic problem (4.7) remains the same if the set $E$ in its formulation is replaced by any subspace $\widehat{E}$ in $\mathcal{E}$ satisfying the condition $\widehat{E}^0 = \Sigma$; in particular, $E$ can be replaced by $\operatorname{Def} U$ or by any linear subspace in $\mathcal{E}$ whose closure is the same as the closure of $\operatorname{Def} U$.

PROOF. Since $C$ is bounded in the norm of $\mathcal{S}$, it is also bounded along every ray. Then the static problem (4.6) is a dual of the kinematic problem (4.7) as shown above. The latter problem is of the standard form

$$-g(\mathbf{e}) \to \sup; \quad \mathbf{e} \in E$$
$$(g(\mathbf{e}) = M_C^*(\mathbf{e}) - \langle \mathbf{e}, \mathbf{s}_l \rangle).$$

To prove the equality $\alpha_l = \beta_l$, we use Theorem 3.1 taking into account Remark 3.2. This only calls for verifying of the assumption of Theorem 3.1 about continuity of $g$ at a point $\mathbf{e}_0 \in E$. Since the function $\mathbf{e} \to \langle \mathbf{e}, \mathbf{s}_l \rangle$ is linear and continuous on $\mathcal{E}$, the continuity of $g$ is equivalent to the continuity of $M_C^*$. We considered the latter function in Example 3.3; in particular, we established that $M_C^*(\mathbf{e}) = 0$ if $D(\mathbf{e}) \leq 1$. This implies that $M_C^*(\mathbf{e}) = 0$ on a certain ball with its center at 0. Indeed, $C$ is bounded, that is, there exists a number $R > 0$ such that $\|\boldsymbol{\sigma}\| \leq R$ for every $\boldsymbol{\sigma} \in C$. Then, by the definition of the dissipation, $D(\mathbf{e}) \leq R|\mathbf{e}|$ for every $\mathbf{e} \in E$ and, consequently, $D(\mathbf{e}) \leq 1$ if $\|\mathbf{e}\| < 1/R$. Therefore, $M_C^*(\mathbf{e}) = 0$ on the ball $\|\mathbf{e}\| < 1/R$. In particular, $M_C^*$ is continuous at $\mathbf{e}_0 = 0 \in E$. Thus, the assumptions of Theorem 3.1 are satisfied and hence $\alpha_l = \beta_l$.

Due to the latter statement, the safety factor $\alpha_l$ can be evaluated using the kinematic extremum problem; in particular, we have

$$\inf\{g(\mathbf{e}) : \mathbf{e} \in E\} = -\sup\{-g(\mathbf{e}) : \mathbf{e} \in E\} = -\frac{1}{\alpha_l} > -\infty$$

in case $\alpha_l \neq 0$. This shows that $g$ is bounded from below on $E$, and Theorem 3.1 states that the extremum in the kinematic problem (dual of the original static one) is attained.

It remains to prove the statement about the possibility of replacing $E$ by another subspace in $\mathcal{E}$. Note that the only properties of $E$ involved in the above proof were: 1) $E^0 = \Sigma$ and 2) $0 \in E$, the latter being the property of any subspace $\widehat{E}$. Consequently, $\alpha_l = \beta_l$ if the subspace $\widehat{E}$ with $\widehat{E}^0 = \Sigma$ is used instead of $E$ in the kinematic extremum problem which determines $\beta_l$.

Let us now verify that $\widehat{E}^0 = \Sigma$ for any subspace $\widehat{E}$ with the closure equal to that of $\operatorname{Def} U$. We note that the polar set $\Pi^0$ for any subspace $\Pi$ is the same as the polar set $[\Pi]^0$ for the closure $[\Pi]$. Indeed, $\Pi \subset [\Pi]$ and, therefore, $[\Pi]^0 \subset \Pi^0$. On the other hand, for an arbitrary $\mathbf{e}$ in $[\Pi]$, there is a sequence of $\mathbf{e}_n \in \Pi$, $n = 1, 2, \ldots$, which converges to $\mathbf{e}$. Then, if $\boldsymbol{\sigma}$ is in $\Pi^0$, we have $\langle \mathbf{e}_n, \boldsymbol{\sigma} \rangle = 0$ and passage to the limit in this equality results in $\langle \mathbf{e}, \boldsymbol{\sigma} \rangle = 0$. Since $\mathbf{e}$ is an arbitrary member of $[\Pi]$, this means that $\boldsymbol{\sigma}$ is in $[\Pi]^0$. We established the relation $\Pi^0 \subset [\Pi]^0$, which together with the above-mentioned relation $\Pi^0 \supset [\Pi]^0$ results in $\Pi^0 = [\Pi]^0$. In particular, we have $\widehat{E}^0 = [\widehat{E}]^0$ and $(\operatorname{Def} U)^0 = [\operatorname{Def} U]^0$. Since $[\widehat{E}] = [\operatorname{Def} U]$, these relations imply $\widehat{E}^0 = (\operatorname{Def} U)^0$ or, which is the same, $\widehat{E}^0 = \Sigma$. This finishes the proof.

According to Theorem 2 the safety factor can be evaluated by the kinematic method as well as by the static. The possibility of various choices for the set $E$, over which the extremum is searched, makes the kinematic method very convenient from the computational viewpoint, which will be discussed in detail in Chapter IX.

REMARK 2. With the expression for $M_C^*$ (Example 3.3) the kinematic extremum problem can be written in the form

$$\frac{1}{\beta_l} = \sup\{\langle \mathbf{e}, \mathbf{s}_l \rangle : \mathbf{e} \in E,\ D(\mathbf{e}) \leq 1\}$$

or, equivalently, in the form

$$\frac{1}{\beta_l} = \sup\{\langle \mathbf{e}, \mathbf{s}_l \rangle : \mathbf{e} \in E,\ D(\mathbf{e}) \leq 1\}$$

(Proposition 2.2).

Conditions for the limit multipliers equality given by Theorem 2 can be further specified. This means specifying the spaces $\mathcal{S}$, $\mathcal{E}$ and formulating assumptions about the set $C$ in the form of easily verifiable conditions. We will do this in the next section, now restricting ourselves to a simple example of a Theorem 2 application.

EXAMPLE 2. *Truss.* Consider a truss consisting of $m$ rods. Let $N_j$ be the axial force and $C_j$ be the set of admissible stresses in the $j$-th rod. As always, $C_j$ is convex, closed and contains the origin; therefore, it is of one of the following forms:

$$-Y_j^- \leq N_j \leq Y_j^+, \quad -\infty < N_j \leq Y_j^+, \quad -Y_j^- \leq N_j^+ < +\infty, \quad -\infty < N_j < +\infty$$
$$\left(0 \leq -Y_j^-, Y_j^+ < +\infty\right).$$

Here, the yield stresses in tension and compression are either finite and equal to $Y_j^-$, $Y_j^+$, or one of the yield stresses is infinite. Both yield stresses can also be infinite, and this corresponds to a rigid body. Let us find conditions under which the set of admissible stresses

$$C = \{\sigma = (\sigma_1, \ldots, \sigma_m) \in \mathcal{S} : \sigma_1 \in C_1, \ldots, \sigma_m \in C_m\} \qquad (\mathcal{S} = \mathbf{R}^m)$$

satisfies the assumptions of Theorem 3. The space $\mathcal{S}$ of stress fields in the truss is finite-dimensional. The set $C$ is obviously closed no matter whether the yield stresses are finite or infinite. Then, $C$ is also weak closed, as in the finite-dimensional case, weak closure is the same as the closure in the regular sense (see Example 1). Apart from that, Theorem 2 requires the boundness of $C$, which makes yield stresses finite. Thus, by Theorem 2, the limit static and kinematic multipliers for a truss are equal if yield stresses in tension and compression are finite for all rods of the truss. It should be remarked that this admits vanishing of the yield stress.

## 5. Bodies with bounded yield surfaces

In the previous section, we established conditions for the limit multipliers equality within the framework of the unified formulation of the rigid-plastic problem. Here, we specify a class of plastic bodies which satisfy these conditions. The safety factor for such bodies can be evaluated not only by the static method but also by the kinematic method, the latter being more convenient from the computational viewpoint.

In this section we restrict ourselves to bodies with bounded yield surfaces. Another important class of bodies with cylindrical yield surfaces is considered in Section 7.

**5.1. Set of admissible stress fields.** Theorem 4.2 establishes the conditions for limit multipliers equality. We now make this condition more constructive by replacing the assumption of Theorem 4.2 about the weak closure of $C$ with some effective conditions which guarantee that $C$ meets this requirement. We restrict ourselves to the case of material properties which admit local description.

Consider a body occupying domain $\Omega$. Recall that, within the framework of the unified formulation of the rigid-plastic problem, "body" means not only a three-dimensional body but also a plate, a shell, a beam, etc. Stress fields in a body are functions on $\Omega$ with values in a certain finite-dimensional space $\mathbf{S}$, for example, $\mathbf{S} = Sym$ in the case of a three-dimensional body. We consider a linear space $\mathcal{S}$ of stress fields assuming that it only contains integrable fields. We are reminded that the internal power bilinear form $\langle \cdot, \cdot \rangle$ is a pairing between $\mathcal{S}$ and another linear space $\mathcal{E}$ of strain rate fields (see Subsections IV.1.4 and IV.A.4). The strain rate fields are defined on $\Omega$ and take their values in $\mathbf{S}$. We assume that, for every $\boldsymbol{\sigma}$ in $\mathcal{S}$ and every integrable field $\mathbf{e}$ in $\mathcal{E}$, the value of the bilinear form is defined as

$$\langle \mathbf{e}, \boldsymbol{\sigma} \rangle = \int_{\Omega} \mathbf{e}(x) \cdot \boldsymbol{\sigma}(x) \, dV$$

(the dot indicates the scalar multiplication in $\mathbf{S}$). The pairing between $\mathcal{S}$ and $\mathcal{E}$ allows us to consider the Fenhel transformation and weak closed sets in $\mathcal{S}$ (see Subsections 3.1 and 4.2).

The set $C$ of admissible stress fields is

$$C = \{\boldsymbol{\sigma} \in \mathcal{S} : \boldsymbol{\sigma}(x) \in C_x \text{ for every } x \in \Omega\},$$

where $C_x$ is the set of admissible stresses given for every point $x \in \Omega$; $C_x$ is a convex, closed set in $\mathbf{S}$, and $C_x$ contains the origin.

The next proposition establishes conditions for the set $C$ to be weak closed. The proposition uses the concept of measurable set-valued mapping $x \to C_x$ (see Subsection IV.5.4 and Appendix IV.B). The mapping is measurable in all limit analysis problems of practical interest. Apart from measurability of the mapping $x \to C_x$, we will assume that the spaces $\mathcal{S}$ and $\mathcal{E}$ are *sufficiently wide*, which means that 1) they contain all bounded measurable functions, and 2) if a function $\mathbf{s}$ belongs to the space, then for any subdomain $\omega \subset \Omega$, the space also contains the function

$$\mathbf{s}_\omega(x) = \begin{cases} \mathbf{s}(x) & \text{if } x \in \omega, \\ 0 & \text{if } x \notin \omega. \end{cases}$$

It is obvious that the space of integrable functions in Examples A.1.3, A.1.5 and the space of essentially bounded functions in Subsection A.2 are sufficiently wide.

PROPOSITION 1. If a set-valued mapping $x \to C_x$ is measurable and the spaces $\mathcal{S}$, $\mathcal{E}$ are sufficiently wide, the set $C$ is weak closed.

PROOF. Consider the function

$$\Psi_C(\boldsymbol{\sigma}) = \begin{cases} 0 & \text{if } \boldsymbol{\sigma} \in C, \\ +\infty & \text{if } \boldsymbol{\sigma} \notin C. \end{cases}$$

It is clear that the set $C$ is weak closed if and only if $\Psi_C$ is weak closed. The function $\Psi_C$ is the Fenhel transformation of the dissipation $D$:

$$\Psi_C(\sigma) = \sup\{\langle \mathrm{e}, \sigma \rangle - D(\mathrm{e}) : \mathrm{e} \in \mathcal{E}\}$$

(see Proposition 3 below). Thus, $\Psi_C$ is the supremum of a collection of the functions $f_{\mathrm{e}}$, $\mathrm{e} \in \mathcal{E}$, each of the functions $f_{\mathrm{e}}(\sigma) = \langle \mathrm{e}, \sigma \rangle - D(\mathrm{e})$ being linear in its argument $\sigma$ and, therefore, weak closed. Then the supremum of the collection is also a weak closed function (see Proposition 4 below), which finishes the proof.

The remaining part of this subsection deals with the auxiliary propositions we used when proving Proposition 1. Let us first prove that $\Psi_C = D^*$, where $D$ is the dissipation

$$D(\mathrm{e}) = \sup\{\langle \mathrm{e}, \sigma_* \rangle : \sigma_* \in C\}.$$

We verify the equality $\Psi_C = D^*$ using the following well known proposition of convex analysis. The proposition concerns the Fenhel transformation of functions on a finite-dimensional space $\mathbf{S}$ with scalar multiplication; the latter is considered as a pairing between $\mathbf{S}$ and itself (see Example IV.A.6).

PROPOSITION 2. Suppose the function $f_x$ is given for every point $x$ in a domain $\Omega$; $f_x$ is defined on a finite-dimensional space $\mathbf{S}$. If, for every $x$ in $\Omega$, the function $f_x$ is convex, closed, does not take the value $-\infty$, takes a finite value at least at one point, and the set-valued mapping $x \to \operatorname{epi} f_x$ is measurable, then the Fenhel transformation

$$f_x^*(\mathrm{e}) = \sup\{\mathrm{e} \cdot \mathrm{s} - f_x(\mathrm{s}) : \mathrm{s} \in \mathbf{S}\}$$

has the same properties, that is, it is convex, closed, does not take the value $-\infty$, takes a finite value at least at one point $\mathrm{e} \in \mathbf{S}$, and the set-valued mapping $x \to \operatorname{epi} f_x^*$ is measurable.

This proposition results in conditions for the equality $\Psi_C = D^*$.

PROPOSITION 3. If the set-valued mapping $x \to C_x$ is measurable and the spaces $\mathcal{S}$, $\mathcal{E}$ are sufficiently wide, then $D^* = \Psi_C$.

PROOF. We first verify the equality $D^*(\sigma) = \Psi_C(\sigma)$ in case $\sigma$ belongs to $C$, that is,

$$\sup\{\langle \mathrm{e}, \sigma \rangle - D(\mathrm{e}) : \mathrm{e} \in \mathcal{E}\} = 0 \quad \text{if} \quad \sigma \in C.$$

Indeed, if $\sigma \in C$, then $D(\mathrm{e}) \geq \langle \mathrm{e}, \sigma \rangle$ for every e in $\mathcal{E}$. Therefore, the supremum in the previous formula is nonpositive and, hence, equals 0: the zero value is attained at $\mathrm{e} = 0$.

It remains to show that $D^*(\sigma) = \Psi_C(\sigma)$ if $\sigma$ does not belong to $C$, that is,

$$\sup\{\langle \mathrm{e}, \sigma \rangle - D(\mathrm{e}) : \mathrm{e} \in \mathcal{E}\} = +\infty \quad \text{if} \quad \sigma \notin C.$$

It is sufficient to verify that

$$\sup\{\langle \mathrm{e}, \sigma \rangle - D(\mathrm{e}) : \mathrm{e} \in \mathcal{E}_1\} = +\infty,$$

where the subspace $\mathcal{E}_1$ in $\mathcal{E}$ consists only of integrable functions. We evaluate the bilinear form which pairs $\mathcal{S}$ and $\mathcal{E}$ and the dissipation for integrable e using the formulas

$$\langle \mathrm{e}, \boldsymbol{\sigma} \rangle = \int_\Omega \mathrm{e}(x) \cdot \boldsymbol{\sigma}(x)\, dV, \qquad D(\mathrm{e}) = \int_\Omega d_x(\mathrm{e}(x))\, dV$$

(the second formula is valid by Theorem IV.5.1). Therefore,

$$\sup \{\langle \mathrm{e}, \boldsymbol{\sigma} \rangle - D(\mathrm{e}) : \mathrm{e} \in \mathcal{E}_1\}$$
$$= \sup \left\{ \int_\Omega (\mathrm{e}(x) \cdot \boldsymbol{\sigma}(x) - d_x(\mathrm{e}(x)))\, dV : \mathrm{e} \in \mathcal{E}_1 \right\}. \quad (5.1)$$

We express the local dissipation $d_x$ through the function

$$\psi_{C_x}(\mathrm{s}) = \begin{cases} 0 & \text{if } \mathrm{s} \in C_x, \\ +\infty & \text{if } \mathrm{s} \notin C_x. \end{cases}$$

Namely:

$$d_x(\boldsymbol{\varepsilon}) = \sup \{\mathrm{e} \cdot \mathrm{s} : \mathrm{s} \in C_x\} = \sup \{\mathrm{e} \cdot \mathrm{s} - \psi_{C_x} : \mathrm{s} \in \mathbf{S}\} = \psi^*_{C_x}(\mathrm{e}).$$

The function $f_x = \psi_{C_x}$ satisfies the assumptions of Proposition 2 since the mapping $x \to C_x$ is measurable. By Proposition 2, the mapping $x \to \operatorname{epi} f^*_x \equiv \operatorname{epi} d_x$ is measurable. Applying Theorem IV.6.1 we find out that (5.1) equals

$$\int_\Omega \sup \{\boldsymbol{\varepsilon} \cdot \boldsymbol{\sigma}(x) - \psi^*_{C_x}(\boldsymbol{\varepsilon}) : \boldsymbol{\varepsilon} \in \mathbf{S}\}\, dV = \int_\Omega \psi^{**}_{C_x}(\boldsymbol{\sigma}(x))\, dV.$$

Here, $\psi^{**}_{C_x} = \psi_{C_x}$ by Theorem 4.1, and $\psi_{C_x}(\boldsymbol{\sigma}(x)) = +\infty$ since $\boldsymbol{\sigma}(x) \notin C_x$ for $x$ in a set of positive measure because of $\boldsymbol{\sigma} \notin C$. Therefore, the integral in the previous formula is $+\infty$, which finishes the proof.

We now verify another statement used when proving Proposition 1.

PROPOSITION 4. Suppose functions $f_i$, $i \in I$, where $I$ is an arbitrary set of indices, are defined on the space $\mathcal{S}$ and weak closed in $\mathcal{S}$. Then the supremum of this collection $f(\mathrm{s}) = \sup \{f_i(\mathrm{s}) : i \in I\}$ is weak closed.

PROOF. We have to verify that, for every number $a$, the set

$$\{\mathrm{s} \in \mathcal{S} : f(\mathrm{s}) \le a\} = \bigcap_{i \in I} \{\mathrm{s} \in \mathcal{S} : f_i(\mathrm{s}) \le a\}$$

is weak closed. Each of the sets $\{\mathrm{s} \in \mathcal{S} : f_i(\mathrm{s}) \le a\}$ is weak closed by the assumption. We prove the proposition by showing that the intersection $A$ of weak closed sets $A_i$, $i \in I$, is weak closed. Indeed, if $\mathbf{x}_0$ is a weak limit point for $A$, then for every $\varepsilon > 0$ and any finite collection of $\{\mathbf{x}^*_1, \dots, \mathbf{x}^*_K\}$ in $\mathcal{E}$, there exists $\mathbf{x} \in A$ satisfying conditions (4.5). Since $\mathbf{x}$ belongs to $A_i$ and $A_i$ is weak closed, then $\mathbf{x}_0$ also belongs to $A_i$, $i \in I$. Then $\mathbf{x}_0$ is in $A$, and the proposition is proved.

Thus, we accomplished justifying Proposition 1, which provides the conditions for the set $C$ to be weak closed. This is the first assumption of Theorem 4.2. Let us now describe a class of bodies and an appropriate choice of the spaces $\mathcal{S}$ and $\mathcal{E}$ satisfying the other assumption of Theorem 4.2. This will result in establishing simple and constructive conditions for the equality of the limit static and kinematic multipliers.

**5.2. Spaces of stress and strain rate fields.** Consider a rigid-plastic body occupying domain $\Omega$ and suppose the yield surface for every $x$ in $\Omega$ is bounded. In other words, the sets $C_x$ of admissible stresses are bounded in the stress space $\mathbf{S}$. Moreover, the collection of all $C_x$ is assumed to be bounded, that is, the set $C_x$ for every $x$ in $\Omega$ lies in the same fixed ball in $\mathbf{S}$. Let us now specify the space of stress fields for such bodies. First of all, note that, due to the boundness assumption, the value $|\boldsymbol{\sigma}(x)|$ is bounded by the same number for every admissible stress field $\boldsymbol{\sigma}$ and for a.e. $x \in \Omega$. On the other hand, there is no reason to require any smoothness of the stress field. For example, the stress field is not uniquely determined while the body remains rigid, that is, while the load is safe: the stress field only satisfies equilibrium and admissibility conditions. It is clear that among these fields there are nonsmooth ones. Therefore, it is natural to take the space of all essentially bounded measurable fields with values in $\mathbf{S}$ as the stress fields space: $\mathcal{S} = L_\infty(\Omega; \mathbf{S})$. This is a normed space with the norm $\|\boldsymbol{\sigma}\|_{\mathcal{S}} = \|\,|\boldsymbol{\sigma}|\,\|_{L_\infty(\Omega;\mathbf{S})}$, see Subsection A.2.

Let us take the space of integrable fields on $\Omega$ with values in $\mathbf{S}$ as the space $\mathcal{E}$ of strain rate fields: $\mathcal{E} = L_1(\Omega; \mathbf{S})$. This is a normed space with the norm $\|\mathrm{e}\|_{\mathcal{E}} = \|\,|\mathrm{e}|\,\|_{L_1(\Omega)}$, see Example A.5. Note that $\mathcal{E}$ is the dual space of $\mathcal{S}$, that is, the space of linear continuous functionals on $\mathcal{S}$: $\mathcal{E} = \mathcal{S}'$. There is a standard pairing between $\mathcal{E}$ and $\mathcal{S}$: $\langle \mathrm{e}, \boldsymbol{\sigma} \rangle$ is the value $\boldsymbol{\sigma}(\mathrm{e})$ of the functional $\boldsymbol{\sigma} \in \mathcal{S}$ at $\mathrm{e} \in \mathcal{E}$:

$$\langle \mathrm{e}, \boldsymbol{\sigma} \rangle = \boldsymbol{\sigma}(\mathrm{e}) = \int_\Omega \mathrm{e}(`x) \cdot \boldsymbol{\sigma}(x)\, dV.$$

It is clear that the above chosen $\mathcal{S}$ and $\mathcal{E}$ meet the requirements of the rigid-plastic problem unified formulation (Subsection IV.1.4) that concern the spaces of stress and strain rate fields. We assume that there is a space $\mathcal{V}$ of velocity fields which meets the rest of the requirements. For example, in the case of three-dimensional bodies, it is sufficient to set $\mathcal{V} = U$ or $\mathcal{V} = V_0$, where $U$ is the space of smooth vector fields and $V_0$ is a space of sufficiently regular vector fields on $\Omega$ vanishing on the fixed part $S_v$ of the body boundary, see Subsection III.1.3.

**5.3. Equality of limit multipliers.** The equality of the limit static and kinematic multipliers, $\alpha_l = \beta_l$, means that the safety factor $\alpha_l$ can be evaluated not only as the extremum in the static problem (which is always possible just according to the definition of $\alpha_l$) but also as the extremum in the kinematic problem. We are now ready to establish one of the concrete forms of constructive conditions for the equality $\alpha_l = \beta_l$. In the following proposition,

for bodies with bounded yield surfaces, we use the spaces

$$\mathcal{S} = L_\infty(\Omega; \mathbf{S}), \quad \mathcal{E} = L_1(\Omega; \mathbf{S})$$

and the pairing

$$\langle \mathrm{e}, \boldsymbol{\sigma} \rangle = \boldsymbol{\sigma}(\mathrm{e}) = \int_\Omega \mathrm{e}(x) \cdot \boldsymbol{\sigma}(x)\, dV$$

between them;

the mapping $x \to C_x$ is measurable, as usual.

The reader can find definitions of these spaces in Appendix A; measurability of a set-valued mapping is defined in Subsection IV.5.3. Actually, the mapping $x \to C_x$ is measurable in all limit analysis problems of practical interest.

THEOREM 1. Suppose the foregoing assumptions are satisfied and for every point of the body the yield surface lies in the same ball of the stress space. Then

1) the safety factor can be evaluated by the kinematic as well as by the static method:

$$\frac{1}{\alpha_l} = \inf \{ M_C(\boldsymbol{\rho} + \mathrm{s}_l) : \boldsymbol{\rho} \in \Sigma \} = \sup \{ \langle \mathrm{e}, \mathrm{s}_l \rangle : \mathrm{e} \in E,\ D(\mathrm{e}) \leq 1 \},$$

2) the extremum in the kinematic problem is attained if the safety factor is nonzero, $\alpha_l \neq 0$,

3) the extremum in the kinematic problem remains the same if the set $E$ in its formulation is replaced by any subspace $\widehat{E}$ in $\mathcal{E}$ satisfying the condition $\widehat{E}^0 = \Sigma$; in particular, $E$ can be replaced by $\operatorname{Def} U$ or by any linear subspace in $\mathcal{E}$ whose closure is the same as the closure of $\operatorname{Def} U$.

PROOF. Let us write down the static and kinematic extremum problems in their standard forms (2.4) and (2.11), and then apply the formula for evaluating $M_C^*$ to the kinematic problem formulation (see Example 3.3). To prove the present proposition, we apply Theorem 4.2. It is clear that the above chosen spaces $\mathcal{E}$, $\mathcal{S}$ satisfy the first assumption of Theorem 4.2 and also that the spaces are sufficiently wide and the mapping $x \to C_x$ is measurable; hence, the set $C$ is weak closed (Proposition 1). At last, by assumption, there is a ball $B_R(0)$ in the stress space $\mathbf{S}$ such that $C_x \subset B_R(0)$ for a.e. $x \in \Omega$. Then $C$ is bounded in the norm of $\mathcal{S}$. Thus, the second assumption of Theorem 4.2 is also satisfied. All three statements of this proposition follow from Theorem 4.2, which finishes the proof.

To evaluate the safety factor, one can use any equivalent formulation of the static or kinematic extremum problems, see Propositions 2.1 and 2.2.

Consider, in particular, evaluating the safety factor as the problem of the kinematic multiplier minimization. We can restrict ourselves to minimizing only over integrable strain rate fields (Theorem 1). Dissipation $D(\mathbf{e})$ for such field $\mathbf{e}$ is computed by integrating the function $d_x(\mathbf{e}(x))$, see Theorem IV.5.1. This means that, say, in the case of a three-dimensional body, the safety factor is the infimum of the kinematic multipliers, which are computed using the explicit formula:

$$m_k(\mathbf{v}) = \frac{\int\limits_{\Omega} d_x(\operatorname{Def}\mathbf{v})\,dV}{\int\limits_{\Omega}\mathbf{fv}\,dV + \int\limits_{S_q}\mathbf{qv}\,dS},$$

and only sufficiently regular velocity fields are involved in the minimization. According to Theorem 1, it is even sufficient to minimize the kinematic multipliers over only smooth velocity fields, which is rather convenient for theoretical studying of limit analysis problems. However, another choice is preferable from the computational viewpoint. It will be considered while introducing the finite element method, see Chapter IX, particularly Proposition IX.1.1.

REMARK 2. It is clear that a statement analogous to Theorem 1 is valid for discrete systems, that is, in case the spaces $\mathcal{S}$ and $\mathcal{E}$ are finite-dimensional. (Recall that the weak closure and the closure in the regular sense are the same in this case, see Example 4.1.) The limit static and limit kinematic multipliers are equal, if the yield surface for every member of a discrete system lies in the same fixed ball of the stress space.

## 6. Conditions for equality of limit multipliers – II

Conditions for equality of the limit static and limit kinematic multipliers specify the cases in which the safety factor can be evaluated by both static and kinematic methods. In this section, we establish one type of such conditions within the framework of the unified formulation of the rigid-plastic problem. These conditions complement those considered in Section 4, as they cover different classes of plastic bodies. In particular, the conditions we establish in this section are applicable to bodies with cylindrical yield surfaces.

As in Section 4, the general scheme of constructing the dual extremum problem is used to obtain the new conditions for the equality of the limit multipliers. However, this time we start with the static extremum problem and obtain the kinematic one as dual. Conditions for the limit multipliers equality arise immediately from Theorem 3.1.

Consideration in this section allows a fresh look at the kinematic method of limit analysis. In contrast to the static extremum problem, which is "naturally" generated by simply the division of stresses into admissible and inadmissible, the kinematic extremum problem was "conjectured". In particular, the nature of the formula which determines the kinematic multiplier (dissipation over external power) is not clear. This makes the whole picture dim, as we do not understand the origins of the kinematic method and do not know whether the analogous approach is applicable to similar problems. In this section, we will see how the kinematic extremum problem naturally arises from the static one.

**6.1. Kinematic extremum problem dual of the static problem.** Let us construct the extremum problem dual of the static extremum problem of limit analysis (2.4):

$$M_C(\boldsymbol{\rho} + \mathbf{s}_l) \to \inf; \qquad \boldsymbol{\rho} \in \Sigma. \tag{6.1}$$

Here, $\Sigma$ is the subspace of self-equilibrated stress fields, $\mathbf{s}_l$ is a fixed stress field equilibrating the given load $\mathbf{l}$, $C$ is the set of admissible stress fields and $M_C$ is its Minkowski function. The infimum in problem (6.1) equals $1/\alpha_l$ (Proposition 2.1), where $\alpha_l$ is the safety factor of the load $\mathbf{l}$. We consider problem (6.1) within the framework of the unified formulation of the rigid-plastic problem, adopting the assumptions and using the notation introduced in Subsection IV.1.4.

Formulation (6.1) of the static extremum problem is standard in the sense of Section 2: the problem consists in minimizing the convex function $f(\boldsymbol{\sigma}) = M_C(\boldsymbol{\sigma} + \mathbf{s}_l)$ over the subspace $\Sigma$ in the stress fields space. According to the general scheme in Subsection 3.2, the dual of this problem is

$$-f^*(\mathbf{e}) \to \sup; \qquad \mathbf{e} \in \Sigma^0. \tag{6.2}$$

Here $f^*$ is the Fenhel transformation of $f$ and $\Sigma^0$ is the polar set for $\Sigma$; both are defined in terms of the internal power bilinear form $\langle \cdot, \cdot \rangle$, which pairs the spaces $\mathcal{S}$ and $\mathcal{E}$ within the framework of the unified formulation of the rigid-plastic problem.

To formulate the dual problem (6.2) explicitly, let us express the Fenhel transformation of $f$ in terms of the dissipation

$$D(\mathbf{e}) = \sup \{\langle \mathbf{e}, \boldsymbol{\tau} \rangle : \boldsymbol{\tau} \in C\}$$

and the fixed stress field $\mathbf{s}_l$ (representing the load). Below is the formula for evaluating $M_C^*$ from Example 3.3:

$$M_C^*(\mathbf{e}) = \begin{cases} 0 & \text{if } D(\mathbf{e}) \le 1, \\ +\infty & \text{if } D(\mathbf{e}) > 1, \end{cases} \tag{6.3}$$

whereby the evaluation for $f^*$ can be easily obtained:

$$\begin{aligned} f^*(\mathbf{e}) &= \sup \{\langle \mathbf{e}, \mathbf{s} \rangle - M_C(\mathbf{s} + \mathbf{s}_l) : \mathbf{s} \in \mathcal{S}\} \\ &= \sup \{\langle \mathbf{e}, \boldsymbol{\tau} - \mathbf{s}_l \rangle - M_C(\boldsymbol{\tau}) : \boldsymbol{\tau} \in \mathcal{S}\} = -\langle \mathbf{e}, \mathbf{s}_l \rangle + M_C^*(\mathbf{e}) \\ &= \begin{cases} -\langle \mathbf{e}, \mathbf{s}_l \rangle & \text{if } D(\mathbf{e}) \le 1, \\ +\infty & \text{if } D(\mathbf{e}) > 1. \end{cases} \end{aligned}$$

Thus, problem (6.2) dual of the static extremum problem (6.1) is of the form

$$\langle \mathbf{e}, \mathbf{s}_l \rangle - M_C^*(\mathbf{e}) \to \sup; \qquad \mathbf{e} \in \Sigma^0. \tag{6.4}$$

We compare it with kinematic extremum problem (2.12) and find out that, in case $\Sigma^0 = E$, the kinematic extremum problem, which determines the limit kinematic multiplier

$$\frac{1}{\beta_l} = \sup\left\{\langle \mathbf{e}, \mathbf{s}_l \rangle - M_C^*(\mathbf{e}) : \mathbf{e} \in E\right\},$$

is dual of static extremum problem (6.1), which determines the limit static multiplier (the safety factor)

$$\frac{1}{\alpha_l} = \inf\left\{M_C(\boldsymbol{\rho} + \mathbf{s}_l) : \boldsymbol{\rho} \in \Sigma\right\}.$$

REMARK 1. Any discrete system satisfies the condition $\Sigma^0 = E$, see Remark 7.2 below. The relation $\Sigma^0 = E$ also holds for continual systems if the space of kinematically admissible velocity fields is sufficiently extended, see Section VII.2 below.

REMARK 2. The general scheme for constructing the dual problem appears to result in the "invention" of the kinematic multiplier without any additional reasoning. Indeed, the dual problem always allows estimation of the extremum of the original problem (Subsection 3.2). In the case under consideration, the original problem determines the value $1/\alpha_l$ and the corresponding estimation is

$$\langle \mathbf{e}, \mathbf{s}_l \rangle - M_C^*(\mathbf{e}) \leq \frac{1}{\alpha_l} \quad \text{for every} \quad \mathbf{e} \in \Sigma^0.$$

In particular,

$$\langle \mathbf{e}, \mathbf{s}_l \rangle - M_C^*(\mathbf{e}) \leq \frac{1}{\alpha_l} \quad \text{for every} \quad \mathbf{e} \in E$$

since every kinematically admissible strain rate field $\mathbf{e} = \operatorname{Def} \mathbf{v}$ belongs to $\Sigma^0$ (Remark IV.1.2). Using formula (6.3), we write this estimation in the form

$$\alpha_l \leq \frac{D(\mathbf{e})}{\langle \mathbf{e}, \mathbf{s}_l \rangle} \quad \text{for every} \quad \mathbf{e} \in E, \ \langle \mathbf{e}, \mathbf{s}_l \rangle > 0.$$

The expression on the right-hand side is nothing but the kinematic multiplier $m_k(\mathbf{v})$ for the velocity field whose strain rate field is $\mathbf{e}$.

**6.2. Continuity of convex functions.** We are now going to establish conditions for the equality of limit multipliers by applying Theorem 3.2 to problems (6.1) and (6.4). The following general property of convex functions is useful in this respect.

PROPOSITION 1. Suppose $f$ is a convex function on the normed linear space $X$ and $f$ does not take the value $-\infty$. Then the boundness of $f$ on a neighborhood of a point $\mathbf{x}_0$ is sufficient for continuity of $f$ at $\mathbf{x}_0$.

PROOF. Since $f(\mathbf{x}_0)$ is a finite number, we will verify the continuity of the function $\mathbf{x} \to f(\mathbf{x}) - f(\mathbf{x}_0)$, which vanishes at $\mathbf{x}_0$. Without loss of generality we can also assume that $\mathbf{x}_0 = 0$. Thus, we will prove the continuity at $\mathbf{x}_0 = 0$

of a convex function $f$ which vanishes at this point and is bounded on its neighborhood $\mathcal{U}$. Let a positive constant $c$ be an upper bound for $f$ on $\mathcal{U}$: $f(\mathbf{x}) \leq c < +\infty$ for every $\mathbf{x}$ in $\mathcal{U}$. Consider an arbitrary $\varepsilon$, $0 < \varepsilon \leq 1$, and the neighborhood $V_\varepsilon = (\varepsilon\mathcal{U}) \cap (-\varepsilon\mathcal{U})$ of 0. We now show that, for every $\mathbf{x} \in V_\varepsilon$, the inequality $|f(\mathbf{x})| \leq c\varepsilon$ holds, which means exactly that $f$ is continuous at $\mathbf{x} = 0$. Indeed, let $\mathbf{x}$ belong to $V_\varepsilon$. Then $\mathbf{x} \in \varepsilon\mathcal{U}$, that is, $\mathbf{x} = \varepsilon\mathbf{x}_+$ with $\mathbf{x}_+$ in $\mathcal{U}$. Due to the convexity of $f$:

$$f(\mathbf{x}) \leq \varepsilon f(x_+) + (1-\varepsilon)f(0) = \varepsilon f(\mathbf{x}_+) \leq \varepsilon c.$$

On the other hand, $\mathbf{x} \in V_\varepsilon$ also implies $\mathbf{x} \in -\varepsilon\mathcal{U}$ and, consequently, $\mathbf{x} = \varepsilon\mathbf{x}_-$ with $\mathbf{x}_-$ in $\mathcal{U}$. Note that

$$0 = \frac{1}{1+\varepsilon}\mathbf{x} + \frac{\varepsilon}{1+\varepsilon}\mathbf{x}_-$$

and, therefore, due to the convexity of $f$, we obtain:

$$0 = f(0) \leq \frac{1}{1+\varepsilon} f(\mathbf{x}) + \frac{\varepsilon}{1+\varepsilon} f(\mathbf{x}_-).$$

This implies the inequality $f(\mathbf{x}) \geq -\varepsilon f(\mathbf{x}_-) \geq -c\varepsilon$, which together with the above inequality $f(\mathbf{x}) \leq \varepsilon c$, results in $|f(\mathbf{x})| \leq \varepsilon c$ and finishes the proof.

**6.3. Equality of limit multipliers.** Suppose the subspaces of kinematically admissible strain rate fields $E$ and self-equilibrated stress fields $\Sigma$ satisfy the condition $E = \Sigma^0$. Then, by Subsection 6.1, the kinematic extremum problem, which determines the limit kinematic multiplier $m_k(\mathbf{v})$, is dual of the static extremum problem, which determines the limit static multiplier (the safety factor) $\alpha_l$. Let us apply Theorem 3.1 to these problems and establish conditions for the limit multipliers equality. We consider the static and kinematic extremum problems within the framework of the unified formulation of the rigid-plastic problem and adopt the following additional assumptions: the space $\mathcal{S}$ of stress fields is a complete normed space, the space $\mathcal{E}$ of strain rate fields is a conjugate of $\mathcal{S}$, that is, $\mathcal{E}$ is the space of continuous linear functionals on $\mathcal{S}$, $\mathcal{E} = \mathcal{S}'$, $\langle \mathbf{e}, \boldsymbol{\sigma} \rangle$ is the value of a functional $\mathbf{e} \in \mathcal{E}$ at $\boldsymbol{\sigma} \in \mathcal{S}$: $\langle \mathbf{e}, \boldsymbol{\sigma} \rangle = \mathbf{e}(\boldsymbol{\sigma})$. See definitions of basic concepts related to normal spaces in Appendix A.

THEOREM 1. Suppose $\mathcal{S}$ and $\mathcal{E}$ are complete normed spaces, $\mathcal{E}$ being a conjugate of $\mathcal{S}$: $\mathcal{E} = \mathcal{S}'$, and $\langle \mathbf{e}, \boldsymbol{\sigma} \rangle = \mathbf{e}(\boldsymbol{\sigma})$ for every $\mathbf{e} \in \mathcal{E}$ and every $\boldsymbol{\sigma} \in \mathcal{S}$. Let $C$ be a convex set in $\mathcal{S}$ closed along rays and containing the ball $B_r(0)$ of radius $r > 0$ centered at the origin. If the subspaces of kinematically admissible strain rate fields and self-equilibrated stress fields $E$ and $\Sigma$ satisfy the condition $E = \Sigma^0$, then

1) the limit static and kinematic multipliers of the load $\mathbf{l}$ are equal, $\alpha_l = \beta_l$,

2) the extremum in the kinematic problem

$$\frac{1}{\beta_l} = \sup\left\{\langle \mathbf{e}, \mathbf{s}_l \rangle - M_C^*(\mathbf{e}) : \mathbf{e} \in E\right\}$$

is attainable.

PROOF. Static extremum problem (6.1) is of the standard form

$$f(\rho) \to \inf; \quad \mathrm{e} \in \Sigma^0$$

with $f(\rho) = M_C(\rho + \mathrm{s}_l)$, and its extremum is $1/\alpha_l$. If $E = \Sigma^0$, the dual of the static problem is the kinematic problem

$$-f^*(\mathrm{e}) \to \sup; \quad \mathrm{e} \in E$$

(Subsection 6.1); its extremum is $1/\beta_l$. To prove the equality $\alpha_l = \beta_l$, we appeal to Theorem 3.1. This calls for verifying that $f$ satisfies the continuity assumption of the theorem, which requires that $f$ be continuous at some point $\rho_0 \in \Sigma$, and we will show this for $\rho = 0$. By Proposition 1, it is sufficient to show that $f$ is bounded on a neighborhood of $\rho_0 = 0 \in \Sigma$. To establish the boundness, we use property (v) of the Minkowski function (Subsection 2.1), which implies

$$|f(\rho)| = |M_C(\rho + \mathrm{s}_l)| \leq \frac{1}{r} \|\rho + \mathrm{s}_l\|,$$

that is, $f$ is bounded on every bounded neighborhood of the origin. Thus, the equality $\alpha_l = \beta_l$ is proved.

Note that the function $f$ we minimize in the original problem (6.1) is non-negative, $M_C \geq 0$, and, therefore, is bounded from below. Then, by Theorem 3.1, the extremum in the kinematic problem, dual problem of (6.1), is attainable, which finishes the proof.

REMARK 3. In view of the expression for $M_C^*$ from Example 3.3, the kinematic extremum problem may be re-written in the form

$$\frac{1}{\beta_l} = \sup \{ \langle \mathrm{e}, \mathrm{s}_l \rangle : \mathrm{e} \in E, \; D(\mathrm{e}) \leq 1 \}$$

or, equivalently, in the form

$$\beta_l = \inf \left\{ \frac{D(\mathrm{e})}{\langle \mathrm{e}, \mathrm{s}_l \rangle} : \mathrm{e} \in E, \; \langle \mathrm{e}, \mathrm{s}_l \rangle > 0 \right\}$$

(Proposition 2.2).

It should be emphasized that Theorem 1 complements Theorem 4.1 by specifying a different class of plastic bodies for which the equality $\alpha_l = \beta_l$ holds.

EXAMPLE 1. (*Truss*). Consider the same truss as in Example 4.2; recall that spaces $\mathcal{S}$ and $\mathcal{E}$ are $m$-dimensional for it. Subspaces of self-equilibrated stress fields and kinematically admissible strain rate fields $\Sigma$ and $E$ as always satisfy the relation $\Sigma = E^0$. In the finite dimensional case this implies $E = \Sigma^0$ (see Remark 7.2 below), that is, one of the assumptions of Theorem 1 is satisfied. The other assumption ($B_r(0) \subset C$, $r > 0$) is satisfied if and only if none of the yield stresses $Y_j^+$, $Y_j^-$ vanishes. If it is so, the limit static and limit kinematic multipliers are equal by Theorem 1.

It should be noted that this condition for the limit multipliers equality differs from that arising in Example 4.2 from application of Theorem 4.1 (boundness of the sets $C_j$).

## 7. Bodies with cylindrical yield surfaces

In the previous section, we established conditions for the limit multipliers equality within the framework of the unified formulation of the rigid-plastic problem. Here, we specify a class of plastic bodies which satisfy these conditions. The safety factor for such bodies can be evaluated not only by the static method but also by the kinematic method, which is more convenient from the computational viewpoint.

In this section, we consider bodies with cylindrical yield surfaces, this class being of most interest since cylindrical yield surfaces are normally used for modelling metal and alloy behavior.

**7.1. Cylindrical yield surfaces.** Consider a rigid-plastic body occupying domain $\Omega$. We are reminded that, within the framework of the rigid-plastic problem unified formulation, "body" means not only a three-dimensional body but also a plate, a shell, a beam, etc. Stress fields in the body are functions on $\Omega$ with values in a finite-dimensional space $\mathbf{S}$; for example, $\mathbf{S} = Sym$ in the case of three-dimensional bodies. The set $C_x$ of admissible stresses in $\mathbf{S}$ is given for every $x \in \Omega$, $C_x$ being convex, closed and containing the origin.

Cylindrical sets of admissible stresses are often used in applications. In rigid-plastic models of three-dimensional bodies made of metals or alloys, $C_x$ is normally a cylinder in $Sym$ with the axis directed along unity tensor $\mathbf{I}$. That is, $C_x$ is of the form

$$C_x = \{\boldsymbol{\sigma} \in Sym : \boldsymbol{\sigma}^d \in C_x^d\}, \tag{7.1}$$

where the superscript $d$ indicates the deviatoric part of the tensor and $C_x^d$ is a given set in the deviatoric plane $Sym^d$; see Examples I.2.1 – I.2.3. The following example shows a different type of a cylindrical set of admissible stresses.

EXAMPLE 1. *Beams.* Stresses in a beam are represented by the collection $(M_y, M_z, N, Q_y, Q_z)$, where $M_y$, $M_z$ are the bending moments, $Q_y$, $Q_z$ are the cutting forces, and $N$ is the axial force (Subsection I.6.2). The stresses depend on coordinate $x$, $x$-axis being directed along the axis of the beam. The set $C_x$ of admissible stresses also depends on $x$ in case the beam is inhomogeneous. The Kirchhoff hypothesis is adopted in beam models, and, in this connection, $C_x$ has the form (Subsection I.6.3):

$$C_x = \{\boldsymbol{\sigma} \in \mathbf{S} : F_x(M_y, M_z, N) \leq 0\}.$$

Here, $F_x$ is a convex yield function, $\boldsymbol{\sigma} = (M_y, M_z, N, Q_y, Q_z)$, and $\mathbf{S} = \mathbf{R}^5$. Consider the subspace

$$\mathbf{S}^d = \{\boldsymbol{\sigma} \in \mathbf{S} : Q_y = Q_z = 0\}$$

in $\mathbf{S}$ and, for every $\boldsymbol{\sigma} \in \mathbf{S}$, the projection $\boldsymbol{\sigma}^d$ of $\boldsymbol{\sigma}$ on $\mathbf{S}^d$. Then the set of admissible stresses can be written in the form

$$C_x = \{\boldsymbol{\sigma} \in \mathbf{S} : \boldsymbol{\sigma}^d \in C_x^d\},$$

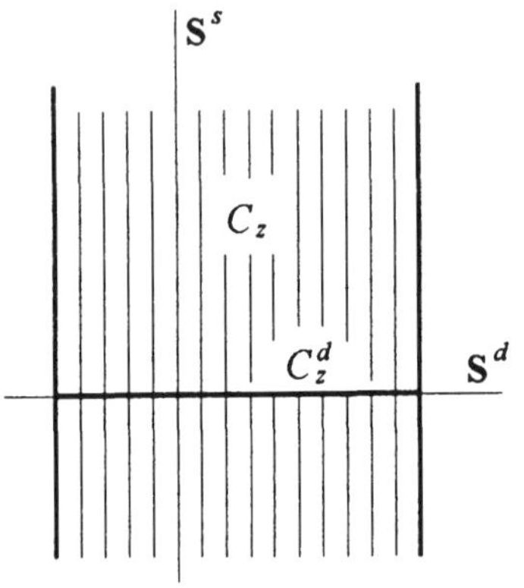


**Figure 7.1**

where $C_x^d$ is a given set in the subspace $\mathbf{S}^d$; $C_x^d$ is the intersection of $C_x$ and $\mathbf{S}^d$. In other words, $C_x$ is a cylinder in $\mathbf{S}$ with the three-dimensional cross-section $C_x^d \subset \mathbf{S}^d$ and the two-dimensional generatrix $\mathbf{S}^s$: the plane passing through the $Q_y$- and $Q_z$-axis; see Figure 7.1.

In what follows, we consider a body with a finite-dimensional stress space $\mathbf{S}$. The dimensions of the stress spaces are different for various types of bodies (three-dimensional bodies, plates, beams, etc.). We assume that two subspaces $\mathbf{S}^d$ and $\mathbf{S}^s$ are specified in $\mathbf{S}$; they are orthogonal complements of each other, neither $\mathbf{S}^d$ nor $\mathbf{S}^s$ being the whole $\mathbf{S}$. We denote the orthogonal projections of $\mathbf{s} \in \mathbf{S}$ on these subspaces by the symbols $\mathbf{s}^d$ and $\mathbf{s}^s$, respectively. It is clear that

$$\mathbf{s} = \mathbf{s}^d + \mathbf{s}^s, \quad |\mathbf{s}|^2 = \left|\mathbf{s}^d\right|^2 + |\mathbf{s}^s|^2, \quad \boldsymbol{\sigma}^d \cdot \boldsymbol{\tau}^s = 0$$

for every $\boldsymbol{\sigma}, \boldsymbol{\tau} \in \mathbf{S}$ (the dot stands for the scalar multiplication in $\mathbf{S}$). Suppose the set $C_x^d$ in $\mathbf{S}$ is given for every point $x$ of the body, and the set $C_x$ of admissible stresses at this point is $C_x = \{\mathbf{s} \in \mathbf{S} : \mathbf{s}^d \in C_x^d\}$. In this case, we say that we deal with a body with *cylindrical yield surfaces.*

**7.2. Spaces of stress and strain rate fields.** First, we choose the space $\mathcal{S}$ of stress fields for a body with cylindrical yield surfaces. Let the body occupy domain $\Omega$ and stress fields take their values in a finite-dimensional space $\mathbf{S}$. The set $C_x$ of admissible stresses is specified by its cross-section $C_x^d$ with a certain subspace $\mathbf{S}^d$ in $\mathbf{S}$: $C_x$ is a cylinder with the generatix $\mathbf{S}^s$, the latter being an orthogonal complement of $\mathbf{S}^d$. We assume that the collection of all $C_x^d$ is bounded, that is, for every $x \in \Omega$ the set $C_x^d$ lies in the same ball of the subspace $S^d$. Then $\left|\boldsymbol{\sigma}^d(x)\right|$ is bounded by the same number for every admissible stress field $\boldsymbol{\sigma}$ and for a.e. $x \in \Omega$. We do not require any smoothness of stress fields, cf. Subsection I.5.4 and 5.2. Therefore, it is natural to take the space of all essentially bounded measurable functions on $\Omega$ with values in $\mathbf{S}^d$ as the space of stress fields components $\mathbf{s}^d$: $\mathcal{S}^d = L_\infty(\Omega; \mathbf{S}^d)$. This is a normed space with the norm $\|\tau\|_{\mathcal{S}^d} = \|\,|\tau|\,\|_{L_\infty(\Omega)}$, see Subsection A.2. On the other hand, the admissibility condition imposes no restrictions on the stress field component $\mathbf{s}^s$. Therefore, we only assume that $\mathbf{s}^s$ is integrable, say, in power 2, and take the space of all square integrable fields as the space of stress fields' components $\mathbf{s}^s$: $\mathcal{S}^s = L_2(\Omega; \mathbf{S}^s)$. This is a normed space with the norm

$$\|\tau\|_{\mathcal{S}^s} = \left(\int_\Omega |\tau(x)|^2 \, dV\right)^{\frac{1}{2}}.$$

Thus, we assume that the stress fields space $\mathcal{S}$ is the space of functions on $\Omega$ with values in $\mathbf{S}$, the components $\sigma^d$ and $\sigma^s$ of $\sigma$ belonging to $L_\infty(\Omega; \mathbf{S}^d)$ and $L_2(\Omega; \mathbf{S}^s)$, respectively. This is a complete normed space with the norm

$$\|\sigma\|_{\mathcal{S}} = \sqrt{\|\sigma^d\|^2_{\mathcal{S}^d} + \|\sigma^s\|^2_{\mathcal{S}^s}}.$$

We denote the space by the symbol $\mathcal{S} = \mathcal{S}^d \times \mathcal{S}^s$.

In the case of three-dimensional bodies we only use the decomposition of stresses into the deviatoric and spheric parts. Therefore, the symbols $Sym^d$ and $Sym^s$ are only used to denote the deviatoric plane and the subspace of spheric tensors. According to the above assumption, we consider deviatoric parts of stress fields in $\mathcal{S}^d = L_\infty(\Omega; Sym^d)$ and spheric parts in $\mathcal{S}^s = L_2(\Omega; Sym^s)$, which makes

$$\mathcal{S} = L_\infty(\Omega; Sym^d) \times L_2(\Omega; Sym^s)$$

the space of stress fields in a three-dimensional body occupying the domain $\Omega$.

We take the space of linear continuous functionals on $\mathcal{S}$ as the space $\mathcal{E}$ of strain rate fields: $\mathcal{E} = \mathcal{S}'$ (see Subsection A.4). Every functional $\mathbf{e}$ in $\mathcal{E}$ is represented by the pair $(\mathbf{e}^d, \mathbf{e}^s)$, where $\mathbf{e}^d$ and $\mathbf{e}^s$ are linear continuous functionals of $\mathcal{S}^d$ and $\mathcal{S}^s$, respectively: $\mathbf{e}^d \in \mathcal{E}^d = (\mathcal{S}^d)'$, $\mathbf{e}^s \in \mathcal{E}^s = (\mathcal{S}^s)'$. The value of $\mathbf{e} = (\mathbf{e}^d, \mathbf{e}^s) \in \mathcal{E} = \mathcal{E}^d \times \mathcal{E}^s$ at $\sigma = (\sigma^d, \sigma^s) \in \mathcal{S} = \mathcal{S}^d \times \mathcal{S}^s$ is $\mathbf{e}(\sigma) = \mathbf{e}^d(\sigma^d) + \mathbf{e}^s(\sigma^s)$. $\mathcal{E}$ is a normed space with the norm

$$\|\mathbf{e}\|_{\mathcal{E}} = \sqrt{\|\mathbf{e}^d\|^2_{\mathcal{E}^d} + \|\mathbf{e}^s\|^2_{\mathcal{E}^s}}.$$

There is a standard pairing $\langle\cdot,\cdot\rangle$ between $\mathcal{E}$ and $\mathcal{S}$: $\langle \mathbf{e}, \sigma \rangle$ is the value of the functionals $\mathbf{e} \in \mathcal{E}$ at $\sigma \in \mathcal{S}$.

It is clear that the above chosen $\mathcal{S}$ and $\mathcal{E}$ meet those of the requirements of the rigid-plastic problem unified formulation (Subsection IV.1.4) that concern spaces of stress and strain rate fields. We also assume that there is space $\mathcal{V}$ of velocity fields which satisfies the rest of the requirements. For example, in the case of three-dimensional bodies, it is sufficient to set $\mathcal{V} = U$ or $\mathcal{V} = V_0$, where $U$ is the space of smooth vector fields vanishing on the fixed part $S_v$ of the body boundary and $V_0$ is the space of sufficiently regular vector fields vanishing on $S_v$ (see Subsection III.1.3).

**7.3. Equality of limit multipliers.** The equality of the limit static and kinematic multipliers, $\alpha_l = \beta_l$, means that the safety factor $\alpha_l$ can be evaluated not only as the extremum in the static problem (which is always possible due to the definition of $\alpha_l$ alone) but also as the extremum in the kinematic problem. We now establish one of the concrete forms of constructive conditions for the equality $\alpha_l = \beta_l$. In the following proposition,

for bodies with cylindrical yield surfaces, we use the spaces

$$\mathcal{S} = \mathcal{S}^d \times \mathcal{S}^s, \ \ \mathcal{S}^d = L_\infty(\Omega; \mathbf{S}^d), \ \ \mathcal{S}^s = L_2(\Omega; \mathbf{S}^s), \ \ \mathcal{E} = \mathcal{S}'$$

and the pairing

$$\langle \mathrm{e}, \boldsymbol{\sigma} \rangle = \mathrm{e}(\boldsymbol{\sigma}) = \mathrm{e}^d(\boldsymbol{\sigma}^d) + \mathrm{e}^s(\boldsymbol{\sigma}^s)$$

between them;

the mapping $x \to C_x$ is measurable, as usual.

The above-mentioned spaces are defined in Appendix A; measurability of a set-valued mapping is defined in Subsection IV.5.3. The mapping $x \to C_x$ is measurable in all limit analysis problems of practical interest.

THEOREM 1. Suppose the foregoing assumptions are satisfied and there is a cylinder $|\mathrm{s}^d| \leq r$, $r > 0$, which lies inside the yield surface for every point of the body. If the subspaces of kinematically admissible strain rate fields and of self-equilibrated stress fields $E$ and $\Sigma$ satisfy the relation $E = \Sigma^0$, then

1) the safety factor can be evaluated by the kinematic as well as by the static method:

$$\frac{1}{\alpha_l} = \inf \{ M_C(\boldsymbol{\rho} + \mathrm{s}_l) : \boldsymbol{\rho} \in \Sigma \} = \sup \{ \langle \mathrm{e}, \mathrm{s}_l \rangle : \mathrm{e} \in E, \ D(\mathrm{e}) \leq 1 \},$$

2) the extremum in the kinematic problem is attainable.

PROOF. To prove the statement, we appeal to Theorem 6.1. It is clear that the above chosen spaces $\mathcal{S}$, $\mathcal{E}$ satisfy the assumptions of Theorem 6.1. It remains to verify that the set $C$ is closed along rays and contains a ball of the space $\mathcal{S}$. First, we observe that $C$ is weak closed by Proposition 5.1 and, consequently, is closed along rays. Secondly, every stress field $\boldsymbol{\sigma} \in \mathcal{S}$ which satisfies the condition $|\boldsymbol{\sigma}^d| \leq r$ for a.e. point $x$ of the body is admissible by the assumption. According to the definition of the norm in $\mathcal{S}^d = L_\infty(\Omega; \mathbf{S}^d)$ (Subsection A.2), this means that $\boldsymbol{\sigma} \in \mathcal{S}$ is admissible if $\|\boldsymbol{\sigma}^d\|_{\mathcal{S}^d} \leq r$. Then the ball $B_r(0)$ of the radius $r$ centered at the origin lies in $C$. Thus, all assumptions of Theorem 6.1 are satisfied; this theorem implies both statements of the present propositions, which finishes the proof.

Any equivalent formulation of the static or kinematic extremum problems can be used to evaluate the safety factor, see Propositions 2.1 and 2.2. In Theorem 1, we wrote the problems in standard forms (2.4) and (2.11). The formulation of the kinematic problem involves the formula for $M_C^*$ derived in Example 3.3. Traditionally, the kinematic problem is written in the form

$$\beta_l = \inf \left\{ \frac{D(\mathrm{e})}{\langle \mathrm{e}, \mathrm{s}_l \rangle} : \mathrm{e} \in E, \ \langle \mathrm{e}, \mathrm{s}_l \rangle > 0 \right\}$$

(Proposition 2.2).

REMARK 1. The statement analogous to Theorem 1 is obviously valid for discrete systems, that is, in case spaces $\mathcal{S}$ and $\mathcal{E}$ are finite-dimensional.

REMARK 2. The assumption $E = \Sigma^0$ is always satisfied for discrete systems. Indeed, let $m$ be the dimension of the stress fields space $\mathcal{S}$ and of the strain rate fields space $\mathcal{E}$ (see Subsection IV.2.2), and $k$ be the dimension of the subspace $E$ of kinematically admissible strain rate fields. We choose a basis $\mathbf{e}_1, \ldots, \mathbf{e}_k$ in $E$. Then the subspace $\Sigma$ of self-equilibrated stress fields is specified by the system of linear equations $\langle \mathbf{e}_1, \boldsymbol{\sigma} \rangle = 0 \ldots, \langle \mathbf{e}_k, \boldsymbol{\sigma} \rangle = 0$. The rank of the system is $k$ and, hence, the dimension of $\Sigma$ is $m - k$. We now choose a basis $\boldsymbol{\sigma}_1, \ldots, \boldsymbol{\sigma}_{m-k}$ in $\Sigma$ and consider the polar set of $\Sigma$:

$$\Sigma^0 = \{\mathbf{e} \in \mathcal{E} : \langle \mathbf{e}, \boldsymbol{\sigma} \rangle = 0 \text{ for every } \boldsymbol{\sigma} \in \Sigma\}.$$

Thus, the system of linear equations $\langle \mathbf{e}, \boldsymbol{\sigma}_1 \rangle = 0, \ldots, \langle \mathbf{e}, \boldsymbol{\sigma}_{m-k} \rangle = 0$ describes the subspace $\Sigma^0$, its rank being $m - k$. Consequently, the dimension of $\Sigma^0$ is $m - (m - k) = k$. Since $\mathbf{e}_1, \ldots, \mathbf{e}_k$ satisfy this system and are linearly independent, the subspace $E$ with the basis $\mathbf{e}_1, \ldots, \mathbf{e}_k$ is the set of all solutions to the system, so that $\Sigma^0 = E$.

Continual systems satisfy the condition $\Sigma^0 = E$ too if one sufficiently extends the space of kinematically admissible velocity fields (see Section VII.2 below).

Let us now compare the two types of conditions for the limit multipliers equality established in Theorems 1 and 5.1. First of all, these conditions specify different classes of plastic bodies and, in this respect, complement each other. Secondly, the conclusions of Theorems 1 and 5.1 are different. By Theorem 5.1, the safety factor can be evaluated as the extremum in the kinematic extremum problem taken over kinematically admissible strain rate fields in the space $\mathcal{E} = L_1(\Omega; \mathbf{S})$. These fields are sufficiently regular and the formulation of the kinematic extremum problem is quite appropriate for practical calculations. In contrast to that, Theorem 1 suggests, in the case of bodies with cylindrical yield surfaces, taking the extremum over strain rate fields with their deviatoric parts in the space $\mathcal{E}^d = \big(L_\infty(\Omega; \mathbf{S}^d)\big)'$. This space contains generalized functions of a rather complicated nature, and the formulation of the kinematic extremum problem is hardly suitable for practical limit analysis. In this connection, it is desirable to narrow the space $\mathcal{E}^d$ keeping the equality of the static multipliers valid, so that the safety factor can be effectively calculated by the kinematic method. This plan will be carried out in the next section, after which the first statement of Theorem 1 will lose its importance since the revised version of the limit multipliers equality deals with the more convenient space of velocity fields. At the same time, the second statement of Theorem 1 is of principle interest for limit analysis theory, as the attainability of the extremum in the kinematic problem means that a plastic failure mechanism exists in a perfectly plastic body subjected to a limit load. It should be recalled that Theorem 5.1 does not guarantee existence of the failure mechanism or, which is the same, attainability of the extremum in the

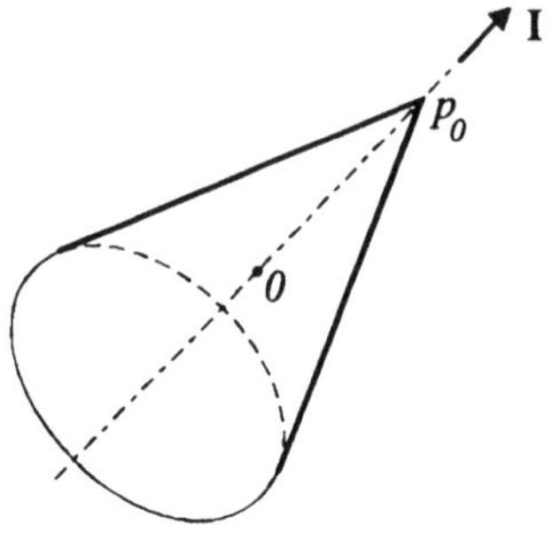


**Figure 7.2**

kinematic extremum problem. It is quite possible that the kinematic multipliers actually do not attain the infimum if the space of velocity fields is not extended widely enough; see Example 8.2 below.

**7.4. Another case of limit multipliers equality.** Theorem 6.1 results in conditions for the limit multipliers equality not only in the case of the bodies with cylindrical yield surfaces and will now be applied to bodies for which the zero stress field is safe. More precisely, we consider bodies for which the origin of the stress space **S** is inside the yield surface for every point $x$ of the body and the distance from 0 to the yield surface is not less than $\delta > 0$, where $\delta$ does not depend on $x$. The yield surfaces of such bodies may be unbounded and noncylindrical. For example, the stress admissibility condition

$$\sigma_{ij}^d \sigma_{ij}^d \leq k^2(\sigma_{kk} - p_0)^2, \qquad \sigma_{kk} \leq p_0 \tag{7.2}$$

and some others analogous to it are often used for modelling granular media, soil and rock materials. As usual, the superscript $d$ in (7.2) indicates the deviatoric part of a tensor; $k^2$ and $p_0$ are constants specifying the material properties. According to (7.2), the all-side tensile stress $\sigma_{kk} \geq 0$ is admissible provided $\sigma_{kk}$ is not greater than $p_0$. On the other hand, the all-sides compression of any magnitude is admissible, and the larger $|\sigma_{kk}|$ is the larger is the admissible magnitude of the shear stress. The yield surface corresponding to the admissibility condition (7.2) is a cone; see Figure 7.2.

Neither Theorem 1 nor Theorem 5.1 is applicable to the above described body. However, Theorem 6.1 results in conditions for the equality $\alpha_l = \beta_l$ which cover this case. To obtain these conditions, we consider a body for which the zero stress field is safe, restricting ourselves to the case when the load **l** the body is subjected to can be equilibrated by a bounded stress field.

In the following proposition,

we use the spaces

$$\mathcal{S} = L_\infty(\Omega; \mathbf{S}) \quad \mathcal{E} = \mathcal{S}'$$

and the pairing $\langle \mathrm{e}, \boldsymbol{\sigma} \rangle = \mathrm{e}(\boldsymbol{\sigma})$ between them;
the mapping $x \to C_x$ is measurable, as usual.

THEOREM 2. Suppose the foregoing assumptions are satisfied, the zero stress field is safe for the body under consideration and the subspaces of kinematically admissible strain rate fields and of self-equilibrated stress fields $E$ and $\Sigma$ satisfy the relation $E = \Sigma^0$. Then

1) the safety factor can be evaluated by the kinematic method as well as by the static method:

$$\frac{1}{\alpha_l} = \inf\{M_C(\rho + \mathrm{s}_l) : \rho \in \Sigma\} = \sup\{\langle \mathrm{e}, \mathrm{s}_l \rangle : \mathrm{e} \in E,\ D(\mathrm{e}) \leq 1\},$$

2) the extremum in the kinematic problem is attainable.

Proof consists in appealing to Theorem 6.1.

The above Remarks 1 and 2 hold for Theorem 2.

REMARK 3. Both Theorems 2 and 5.2 are applicable to bodies with bounded yield surfaces. Note that the former guarantees attainability of the extremum in the kinematic problem, while the latter does not.

## 8. Counterexamples

In the previous sections, the conditions for the equality of the limit static and limit kinematic multipliers were established by Theorems 4.2, 6.1 and, in more constructive and convenient forms, by Theorems 5.1, 7.1, and 7.2. If the assumptions of these statements are satisfied, the safety factor can be evaluated not only by the static but also by the kinematic method. We have also established additional conditions under which it is possible to determine the safety factor as the extremum in the kinematic problem using only regular velocity fields (Theorem 5.1). In this section, we present two examples which show that the above-mentioned conditions are essential. First, we will see that the limit static and limit kinematic multipliers may be unequal if these conditions are not satisfied. Secondly, even if the conditions for the limit multipliers' equality are satisfied (but the additional conditions are not), it may occur that making use of only regular velocity fields does not allow evaluating the safety factor as the extremum in the kinematic problem.

**8.1. Unequality of limit multipliers.** Let us consider a simple example of a discrete system for which the limit static and limit kinematic multipliers are not equal, $\alpha_l \neq \beta_l$. Such an example can be easily found provided the assumptions of Theorem 4.2 and 6.1 are not satisfied. The examples show that the conditions for the equality $\alpha_l = \beta_l$ established in these theorems are essential (although not necessary).

Consider a beam occupying the segment $0 \leq x \leq 1$. The left end of the beam is pinched, that is, the components $u$, $u_y$, $u_z$ of the velocity and the components $\omega_y$, $\omega_z$ of the angular velocity satisfy the conditions

$$u(0) = u_y(0) = u_z(0) = 0, \quad \omega_y(0) = \omega_z(0) = 0$$

(we use the notation introduced in Subsection I.6.1). Let the right end of the beam be subjected to the kinematic constraint

$$\omega_y(0) = 0 \tag{8.1}$$

and loaded by the external tensile force $f$ and the external bending moment $m_z$. This results in the boundary conditions

$$M_z(1) = m_z, \quad N(1) = f, \quad Q_y(1) = Q_z(1) = 0, \tag{8.2}$$

where $Q_y$, $Q_z$, $M_z$ are the cutting forces and the bending moment, and $N$ is the axial force.

Since no load is applied to the beam at $0 < x < 1$, the equilibrium conditions for the beam (Subsection I.6.2) are of the form

$$M_y' = Q_z, \quad M_z' = -Q_y, \quad N' = 0, \quad Q_y' = 0, \quad Q_z' = 0,$$

where the prime indicates the derivative with respect to $x$. Using boundary conditions (8.2), we obtain from the equilibrium equations that, at $0 \geq x \geq 1$,

$$M_y(x) = \text{const}, \quad M_z(x) = m_z, \quad N(x) = f, \quad Q_y(x) = 0, \quad Q_z(x) = 0. \tag{8.3}$$

This means that a stress field in the beam is characterized by three numbers $M_y$, $M_z$, $N$. The given load determines two of them, $M_z$ and $N$, while $M_y$, the reaction of the constraint, is not statically determined. Thus, we consider the beam as a discrete system with the three-dimensional space $\mathcal{S} = \mathbf{R}^3$ of stress fields.

We adopt the following stress admissibility condition:

$$N^2 \leq M_y M_z, \quad M_y \geq 0. \tag{8.4}$$

The set of admissible stress fields is a cone shown in Figure 8.1. Let us find the limit static and limit kinematic multipliers of the load $\mathbf{l}$ given by $f = 1$, $m_z = 1$.

It is easy to evaluate the limit static multiplier using its definition:

$$\alpha_l = \sup \{ m \geq 0 : (\Sigma + m \mathbf{s}_l) \cap C \neq \emptyset \} .$$

Here,

$$\mathbf{s}_l = (N_l, M_{ly}, M_{lz}), \quad M_{ly} = 0, \quad M_{lz} = 0, \quad N_l = 1, \tag{8.5}$$

is a fixed stress field equilibrating the load $\mathbf{l}$ and $\Sigma$ is the subspace of self-equilibrated stress fields, which according to (8.3) is of the form

$$\Sigma = \{(N, M_y, M_z) \in \mathcal{S} : N = 0, M_z = 0\}, \qquad \mathcal{S} = \mathbf{R}^3. \tag{8.6}$$

Figure 8.1. shows $\Sigma$ as well as the set $\Sigma_l = \Sigma + \mathbf{s}_l$ of all stress fields equilibrating the load $\mathbf{l}$, $\Sigma$ and $\Sigma_l$ being straight lines.

The line $\Sigma + m\mathbf{s}_l$ has common points with the set $C$ of admissible stress fields only if $m = 0$. Indeed, every stress field in $\Sigma + m\mathbf{s}_l$ satisfies the conditions $M_z = 0$, $N = m$, then inequality (8.4) is valid for the stress field provided $m = 0$. Thus, the safety factor of the load $\mathbf{l}$ is $\alpha_l = 0$.

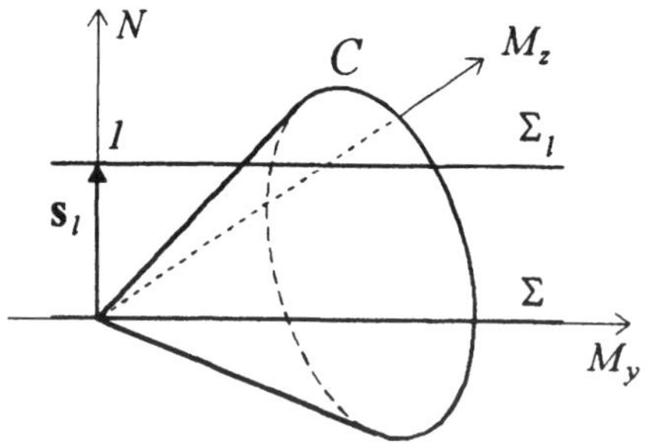


**Figure 8.1**

Let us now evaluate the limit kinematic multiplier. Consider the space $\mathcal{E}$ conjugate of $\mathcal{S} = \mathbf{R}^3$: $\mathcal{E} = \mathbf{R}^3$. The space consists of the triples $\mathbf{e} = (e, e_y, e_z)$, and the internal power bilinear form

$$\langle \mathbf{e}, \boldsymbol{\sigma} \rangle = Ne + M_y e_y + M_z e_z$$

is a pairing between $\mathcal{S}$ and $\mathcal{E}$. The limit kinematic multiplier $\beta_l$ is determined by the kinematic extremum problem (Subsection IV.4.2)

$$\beta_l = \sup \left\{ \frac{D(\mathbf{e})}{\langle \mathbf{e}, \mathbf{s}_l \rangle} : \mathbf{e} \in E, \ \langle \mathbf{e}, \mathbf{s}_l \rangle > 0 \right\}.$$

Here, $D(\mathbf{e})$ is the dissipation and $E$ is the subspace of the kinematically admissible strain rate fields. In the case of discrete systems, both relations $\Sigma = E^0$ and $E = \Sigma^0$ are valid. In particular, the subspace $E = \Sigma^0$ is described by the equation $e_y = 0$ in the problem under consideration. We also take into account that $\langle \mathbf{e}, \mathbf{s}_l \rangle = e$ according to (8.5); then the kinematic extremum problem is written as

$$\beta_l = \sup \left\{ \frac{1}{e} D(\mathbf{e}) : \mathbf{e} = (e, 0, e_z) \in \mathbf{R}^3, \ e > 0 \right\}.$$

Let us show that $\beta_l = +\infty$ by verifying that, for every $\mathbf{e} = (e, 0, e_z)$, $e > 0$, the dissipation $D(\mathbf{e})$ is $+\infty$. Consider first the case $e > 0$, $e_z \neq 0$. Let $\boldsymbol{\sigma}$ be the stress field with the components

$$N = a\sqrt{2}\frac{|e_z|}{e}, \quad M_y = 2a\left(\frac{e_z}{e}\right)^2, \quad M_z = a,$$

where $a$ is a positive number. It is obvious that $\boldsymbol{\sigma}$ satisfies admissibility condition (8.4). Then $D(\mathbf{e}) \geq \langle \mathbf{e}, \boldsymbol{\sigma} \rangle = a(e_z + \sqrt{2}\,|e_z|)$, where $a > 0$ is arbitrary; this implies $D(\mathbf{e}) = +\infty$ if $e_z \neq 0$. Consider now the remaining case $e > 0$, $e_z = 0$. This time we use the stress field with the components $N = a$, $M_y = a$, $M_z = a$. Analogously to the above consideration, we obtain $D(\mathbf{e}) > ae$ and, consequently, $D(\mathbf{e}) = +\infty$. Thus, for every $\mathbf{e}$ satisfying the constraints $\mathbf{e} = (e, 0, e_z)$, $e > 0$ of the kinematic extremum problem, we have $D(\mathbf{e}) = +\infty$, and, consequently, $\beta_l = +\infty$.

We now see that the load under consideration has the limit kinematic multiplier $\beta_l = +\infty$ while the limit static multiplier is $\alpha_l = 0$. The limit multipliers are not equal, which is definitely linked to the fact that the assumptions of Theorems 4.2 and 6.2 are not satisfied: the set $C$ is unbounded and contains no ball centered at the origin.

**8.2. Unattainability of extremums over smooth fields.** We now present an example showing that using only regular velocity fields to evaluate the safety factor as the extremum in the kinematic extremum problem may be insufficient. (At the same time, searching for the extremum over an extended space of velocity fields results in the safety factor value.)

Consider a beam occupying the segment $-1 \leq x \leq 1$ and subjected to external bending moments at the ends, see Figure 8.2; the magnitude of the moments is $m_y$. The beam is also subjected to kinematic constraints

$$u_z(-1) = u_z(1) = 0, \quad u_z(0) = 0, \tag{8.7}$$

where $u_z$ is the velocity component (we use the notation introduced in Subsection I.6.1).

Let the set of admissible stresses be of the form

$$C_x = \{(M_y, M_z, N, Q_y, Q_z) \in \mathbf{R}^5 : M_y \leq M_*\}$$

for every point $x$, $M_*$ being a positive constant. Let us find the limit static multiplier $\alpha_l$ of the load $\mathbf{l}$ given by $m_y = 1$. We will also find the best upper bound for $\alpha_l$, which can be obtained by the kinematic method using only regular velocity fields.

Consider the space $\mathcal{S} = L_\infty(\Omega; \mathbf{R}^5)$ of essentially bounded measurable functions on $\Omega = (-1, 1)$ with values in $\mathbf{R}^5$ as the space of stress fields (see Subsection A.2). The stress field $\mathbf{s}_l$ with the components

$$M_{ly} = 1, \quad M_{lz} = 0, \quad N_l = 0, \quad Q_{ly} = 0, \quad Q_{lz} = 0$$

equilibrates the given load. The set $C$ of admissible stress fields is of the form

$$C = \{\boldsymbol{\sigma} \in \mathcal{S} : \boldsymbol{\sigma}(x) \in C_x \text{ for a.e. } x \in [-1, 1]\}.$$

Apart from $\mathbf{s}_l$ and $C$, the subspace $\Sigma$ of self-equilibrated stress fields enters the definition of the safety factor:

$$\alpha_l = \sup\{m \geq 0 : (\Sigma + m\mathbf{s}_l) \cap C \neq \emptyset\}.$$

The subspace $\Sigma$ in $\mathcal{S}$ is determined by the virtual work principle (Subsection II.3.2): the stress field $\boldsymbol{\sigma} = (M_y, M_z, N, Q_y, Q_z)$ is self-equilibrated if and only if

$$\int_{-1}^{1} [M_y\omega_y' + M_z\omega_z' + Nu' + Q_y(u_y' - \omega_z) + Q_z(u_z' + \omega_y)]\, dx = 0 \quad \text{for every} \quad (\omega_y, \omega_z, u, u_y, u_z) \in U. \tag{8.8}$$

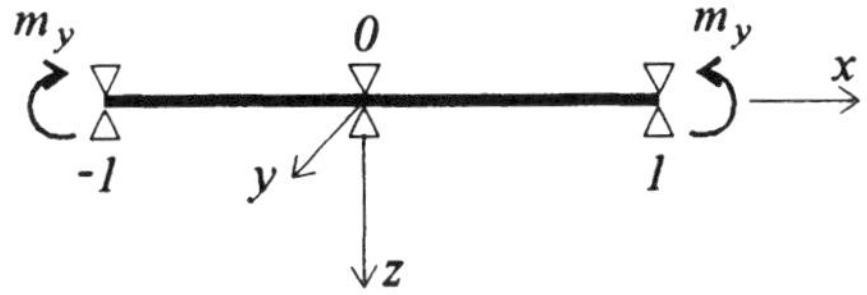


**Figure 8.2**

Here, $U$ is the set of smooth functions $\mathbf{u} = (\omega_y, \omega_z, u, u_y, u_z)$ on $[-1,1]$ with values in $\mathbf{R}^5$, the component $u_z$ satisfying kinematic conditions (8.7).

We will now give an explicit description of $\Sigma$. Let us consider a self-equilibrated stress field $\boldsymbol{\sigma} = (M_y, M_z, N, Q_y, Q_z) \in \Sigma$ and write (8.8) for virtual velocity fields with $\omega_y = \omega_z = 0$, $u_y = u = 0$:

$$N(1)u(1) - N(-1)u(-1) - \int_{-1}^{1} N'u\,dx = 0.$$

Here, $u$ is an arbitrary smooth function, and, therefore, the previous equality results in $N(x) = 0$ for every $x \in [-1,1]$. Analogously, (8.8) written for virtual velocity fields with $\omega_y = \omega_z = 0$, $u_z = 0$ and arbitrary $\omega_z$ results in $Q_y(x) = 0$ for every $x \in [-1,1]$. The same reasoning with $\omega_y = 0$, $u_z = 0$ and arbitrary $\omega_z$ implies $M_z(x) = 0$ for every $x \in [-1,1]$. Then (8.8) with $N = 0$, $Q_y = 0$, $M_z = 0$ is reduced to

$$\int_{-1}^{1} [M_y\omega_y' + Q_z(u_z' + \omega_y)]\,dx = 0 \text{ for every } (\omega_y, u_z) \in U, \tag{8.9}$$

where $U$ stands for the set consisting of pairs of smooth functions on $\omega_y$, $u_z$ on $[-1,1]$, $u_z$ satisfying kinematic conditions (8.7). Let us further simplify (8.9) by writing it with $\omega_y = 0$:

$$\int_{-1}^{1} Q_z u_z'\,dx = 0,$$

which should hold for every smooth function $u_z$ on $[-1,1]$ satisfying (8.7). In particular, we consider functions $u_z$ vanishing on $(-1,0)$, and then the previous equality implies $Q_z(x) = Q_+ = \text{const}$ for every $x \in (0,1)$. Analogously, $Q_z(x) = Q_- = \text{const}$ for $x \in (-1,0)$.

Let us now consider (8.9) with

$$Q_z(x) = \begin{cases} Q_- & \text{if } -1 < x < 0, \\ Q_+ & \text{if } \quad 0 < x < 1, \end{cases}$$

$u_z = 0$ and arbitrary $\omega_y$. This results in the following relations:

$$-M_y'(x) + Q_z(x) = 0 \text{ for every } x \in (-1,1), \qquad M_y(-1) = M_y(1) = 0.$$

Taking into account the above formula for $Q_z$ we arrive at

$$M_y(x) = M_0(1-|x|), \qquad Q_z(x) = \begin{cases} M_0 & \text{if } -1 < x < 0, \\ -M_0 & \text{if } \ 0 < x < 1, \end{cases} \tag{8.10}$$

where $M_0 = \text{const}$.

Thus, the stress field $\boldsymbol{\sigma} = (M_y, M_z, N, Q_y, Q_z)$ is self-equilibrated if and only if $M_z = 0$, $N = 0$, $Q_y = 0$ and $M_y$, $Q_z$ are determined by (8.10) with an arbitrary constant $M_0$.

We observe that $\Sigma$ is in the subspace specified by the relations $M_z = 0$, $N = 0$, $Q_y = 0$ in $\mathcal{S} = L_\infty(\Omega; \mathbf{R}^5)$. Therefore, in order to evaluate $\alpha_l$, we can narrow down the stress fields space to this subspace, that is, introduce a new stress fields space $\mathcal{S} = L_\infty(\Omega; \mathbf{R}^2)$ consisting of pairs $(M_y, Q_z)$ of essentially bounded functions on $\Omega = (-1,1)$. In particular, the stress field $\mathbf{s}_l$ which represents the load $\mathbf{l}$ can be considered as a member of this space. According to the admissibility condition $M_y \le M_*$, we define the set of admissible stress fields in the new $\mathcal{S}$ and again denote it by $C$:

$$C = \{\boldsymbol{\sigma} \in \mathcal{S} : \boldsymbol{\sigma}(x) \in C_x \text{ for a.e. } x \in [-1,1]\},$$
$$C_x = \{(M_y, Q_z) \in \mathbf{R}^2 : M_y \le M_*\}.$$

Let us now write the extremum problem which determines the safety factor in the form

$$\begin{aligned} \alpha_l &= \sup\{m \ge 0 : \ (\Sigma + m\mathbf{s}_l) \cap C \ne \emptyset\} \\ &= \sup\{m \ge 0 : \text{ there exists } \boldsymbol{\sigma} \in \Sigma, \ \Sigma + m\mathbf{s}_l \in C\} \\ &= \sup\{m \ge 0 : \text{ there exists } M_0 \in \mathbf{R}, \\ &\qquad\qquad M_0(1-|x|) + m \le M_* \text{ for every } x \in [-1,1]\}. \end{aligned}$$

The latter extremum can be easily evaluated as $\alpha_l = M_*$.

Let us now find the best upper bound for $\alpha_l$ obtainable by the kinematic method when only regular velocity fields are used. Recall that the method consists of estimating the safety factor from above by the kinematic multipliers. The definition of the kinematic multiplier (Subsections III.5.2, IV.4.2) is reduced for an integrable strain rate field $\mathbf{e}$ to a simple formula

$$m_k(\mathbf{v}) = \frac{\int_{-1}^{1} d_x(e_y, e_z)\,dx}{\int_{-1}^{1} (M_{ly}e_y + M_{lz}e_z)\,dx}.$$

Here, $e_y = \omega_y'$, $e_z = u_z' + \omega_y$ are components of the strain rate corresponding to the velocity field $\mathbf{u} = (\omega_y, u_z)$, and the external power is supposed to be positive:

$$\int_{-1}^{1} (M_{ly} e_y + Q_{lz} e_z)\, dx = \int_{-1}^{1} e_y\, dx > 0. \tag{8.11}$$

We consider velocity fields $\mathbf{u} = (\omega_y, u_z)$ with integrable components such that the strain rate components $e_y = \omega_y'$, $e_z = u_z' + \omega_y$ are integrable too. In this section we call such velocity fields *regular*.

Let us show that the kinematic multiplier for any regular velocity field is $+\infty$. Indeed, the dissipation

$$\begin{aligned} d_x(e_y, e_z) &= \sup \{M_y e_y + Q_z e_z : (M_y, Q_z) \in C_x\} = \\ &= \sup \{M_y e_y + Q_z e_z : (M_y, Q_z) \in \mathbf{R}^2, M_y \le M_*\} \end{aligned}$$

corresponds to the given set $C_x$ of admissible stresses. It is clear that

$$d_x(e_y, 0) = \begin{cases} M_* e_y & \text{if } e_y \ge 0, \\ +\infty & \text{if } e_y < 0 \end{cases}$$

when $e_z = 0$ and $d_x(e_y, e_z) = +\infty$ when $e_z \neq 0$. Thus, the kinematic multiplier may be finite only for the velocity field $(\omega_y, u_z)$ which satisfies the conditions $e_z(x) = 0$, $e_y(x) \ge 0$ for a.e. $x \in [-1, 1]$. With the expressions $e_z(x) = u_z'(x) + \omega_y(x)$ and $e_y(x) = \omega_y'(x)$ these conditions can be written in the form

$$\omega_y(x) = -u_z'(x) \quad \text{and} \quad -u_z''(x) \ge 0 \quad \text{for a.e. } x \in [-1, 1].$$

Now let us show that the only velocity field meeting these requirements is zero. Indeed, the inequality $-u_z''(x) \ge 0$ means that the function $-u_z$ is convex on $[-1, 1]$. Since $-u_z$ vanishes at $x = -1$ and $x = 1$, we have $-u_z(x) \le 0$ due to the convexity. Note that $-u_z$ does not take negative values: if $-u_z(x_0) < 0$ for $x_0 \in (-1, 1)$, then, at the point $x = 0$, which lies between $x_0$ and either the left or the right end of $[-1, 1]$, the inequality $-u_z < 0$ is valid due to the convexity. This contradicts condition (8.7) and, therefore, $u_z(x) = 0$ for a.e. $x \in [-1, 1]$. Here, $u_z$ is the component of a velocity field $\mathbf{u} = (\omega_y, u_z)$ satisfying (8.11) and, hence, $\mathbf{u}(x) = 0$ for a.e. $x \in [-1, 1]$. Recall that this $\mathbf{u}$ is the only velocity field for which the kinematic multiplier could be finite. However, it turned out that $\mathbf{u} = 0$, and the kinematic multiplier is not defined for this velocity field since $\mathbf{u}$ does not satisfy the condition (8.11). Consequently, the kinematic multiplier is $+\infty$ for every kinematically admissible velocity field which satisfies the condition (8.11).

We can see that kinematic multipliers of regular velocity fields do not approximate the safety factor $\alpha_l = M_*$ and only result in the trivial estimation: $\alpha_l \le +\infty$. The gap between the limit static multiplier $\alpha_l$ and the best kinematic upper bound for $\alpha_l$ is possible since the assumption of Theorem 5.1

is not satisfied in our example: the yield surface is not bounded. In case it is, Theorem 5.1 guarantees that $\alpha_l$ can be approximated by kinematic multipliers of regular velocity fields and our example shows that this boundness assumption is essential. At the same time, boundness of $C$ is not a necessary condition for such an approximation. In the next chapter we will establish another condition which guarantees that $\alpha_l$ can be evaluated by the kinematic method using only regular velocity fields, the condition being applicable to bodies with cylindrical yield surfaces.

## Appendix A. Normed spaces

**A.1. Definitions and examples.** Let a nonnegative function $\mathbf{x} \to ||\mathbf{x}||$ be defined on a linear space $X$ possessing the following properties:

$||\mathbf{x}|| = 0$ if and only if $\mathbf{x} = 0$,

$||a\mathbf{x}|| = |a|\,||\mathbf{x}||$ for every number $a$ and every $\mathbf{x}$ in $X$,

$||\mathbf{x}_1 + \mathbf{x}_2|| \leq ||\mathbf{x}_1|| + ||\mathbf{x}_2||$ for every $\mathbf{x}_1$ and $\mathbf{x}_2$ in $X$.

This function is referred to as a *norm* and $X$ is referred to as a *normed space*. We also use the symbol $|\cdot|$ to denote norms in some spaces.

EXAMPLE 1. Consider a linear space $\mathbf{R}^n$ whose members are collections of $n$ numbers $\mathbf{x} = (x_1, \ldots, x_n)$. The function

$$||\mathbf{x}|| = \sqrt{x_1^2 + \cdots + x_n^2} \tag{A.1}$$

is a norm in $\mathbf{R}^n$. This norm is not unique; for example,

$$||\mathbf{x}|| = |x_1| + \cdots + |x_n| \quad \text{or} \quad ||\mathbf{x}|| = \max\{|x_1|, \cdots, |x_n|\} \tag{A.2}$$

are also norms in $\mathbf{R}^n$. All these norms are equivalent in the following sense: two norms $||\cdot||_1$ and $||\cdot||_2$ are *equivalent* if there is a constant $c$ such that the inequalities $||\mathbf{x}||_1 \leq c\,||\mathbf{x}||_2$, $||\mathbf{x}||_2 \leq c\,||\mathbf{x}||_1$ are valid for every member $\mathbf{x}$ of the space.

Let $X$ be a linear space and let a function mapping every two members $\mathbf{u}$, $\mathbf{v}$ of $X$ onto a number $(\mathbf{u}, \mathbf{v})$ possess the following properties:

$(\mathbf{u}, \mathbf{v}) = (\mathbf{v}, \mathbf{u})$ $\quad (\alpha_1\mathbf{u}_1 + \alpha_2\mathbf{u}, \mathbf{v}) = \alpha_1(\mathbf{u}_1, \mathbf{v}) + \alpha_2(\mathbf{u}_2, \mathbf{v})$ for every $\mathbf{u}_1$, $\mathbf{u}_2$, $\mathbf{v}$ in $X$ and every numbers $\alpha_1$, $\alpha_2$,

$(\mathbf{u}, \mathbf{u}) > 0$ if $\mathbf{u} \neq 0$.

Then the function $\mathbf{u}, \mathbf{v} \to (\mathbf{u}, \mathbf{v})$ is referred to as $(\mathbf{u}, \mathbf{v})$ as *scalar multiplication*, $(\mathbf{u}, \mathbf{v})$ as the *scalar product* of $\mathbf{u}$ and $\mathbf{v}$, and to $X$ as the space with scalar multiplication. We also use the symbols $\mathbf{u}\cdot\mathbf{v}$ and $\mathbf{uv}$ to denote the scalar products in some spaces. A space with scalar multiplication is always a normed space, namely, $(\mathbf{u}, \mathbf{u})^{\frac{1}{2}}$ is the norm $||\mathbf{u}||$ of $\mathbf{u}$. For example, $\mathbf{R}^n$ is a space with scalar multiplication $\mathbf{uv} = u_1v_1 + \cdots + u_nv_n$, where $\mathbf{u} = (u_1, \ldots, u_n)$, $\mathbf{v} = (v_1, \ldots, v_n)$, and (A.1) is the norm corresponding to this scalar multiplication.

EXAMPLE 2. Consider the space *Sym* of the symmetric second rank tensor. A scalar multiplication $\mathbf{r} \cdot \mathbf{s} = r_{ij}s_{ij}$ is defined in *Sym* and $|\mathbf{s}| = (\mathbf{s} \cdot \mathbf{s})^{\frac{1}{2}}$ is the corresponding norm.

EXAMPLE 3. Consider a set of all those measurable functions on a domain $\Omega$ in $\mathbf{R}^n$ for which the integral

$$\int_{\Omega} f^2(x)\, dV$$

is finite. It is clear that the set is a linear space with the usual operations of adding functions and multiplying a function by a number. A scalar multiplication and the corresponding norm are defined on this space as follows:

$$(f,g) = \int_{\Omega} f(x)g(x)\, dV, \qquad \|f\| = \left(\int_{\Omega} f^2(x)\, dV\right)^{\frac{1}{2}}.$$

They are analogous to the scalar multiplication and the norm (A.1) in $\mathbf{R}^n$. We denote the space by the symbol $L_2(\Omega)$. This space is not finite-dimensional.

EXAMPLE 4. Consider fields of symmetric second rank tensors on domain $\Omega$ in $\mathbf{R}^3$, that is, functions on $\Omega$ with values in *Sym*. We restrict ourselves to those of the fields for which the integral

$$\int_{\Omega} |\,\mathbf{s}\,|^2\, dV \tag{A.3}$$

is finite. It is clear that the set of such fields is a linear space. A scalar multiplication and the corresponding norm are defined on this space as follows:

$$(\mathbf{r},\mathbf{s}) = \int_{\Omega} \mathbf{r}(x)\cdot\mathbf{s}(x)\, dV, \qquad \|\mathbf{s}\| = \left(\int_{\Omega} |\,\mathbf{s}\,|^2(x)\, dV\right)^{\frac{1}{2}}. \tag{A.4}$$

The space is denoted by the symbol $L_2(\Omega; Sym)$.

Let us similarily define the space $L_2(\Omega;\mathbf{S})$. Consider measurable functions on $\Omega$ with values in the space $\mathbf{S}$, the latter being a space with a scalar multiplication $\mathbf{r}\cdot\mathbf{s}$ ($\mathbf{r},\mathbf{s}\in\mathbf{S}$). The space $L_2(\Omega;\mathbf{S})$ consists of those functions which have a finite integral (A.3). A scalar multiplication and corresponding norm are defined in $L_2(\Omega;\mathbf{S})$ by (A.4).

EXAMPLE 5. Consider a set of all those measurable functions on the domain $\Omega$ for which the integral

$$\int_{\Omega} |f(x)|\, dV$$

is finite. This set is a linear space with the usual operations of adding functions and multiplying a function by a number. The norm

$$\|f\| = \int_{\Omega} |f(x)|\, dV$$

is defined in this space and is analogous to the first norm in (A.2). The space is denoted by the symbol $L_1(\Omega)$.

Similarly, measurable functions with finite integral

$$\int_\Omega |f(x)|^p \, dV \qquad (p \geq 1)$$

form a linear space with the norm

$$\|f\| = \left( \int_\Omega |f(x)|^p \, dV \right)^{\frac{1}{p}}. \tag{A.5}$$

Replacing, in the latter example, the number-valued functions by functions with values in the normed space $\mathbf{S}$, we define the normed space $\mathbf{L}_p(\Omega; \mathbf{S})$. It consists of measurable functions from $\Omega$ into $\mathbf{S}$ with finite integrals (A.5), where $|\cdot|$ now stands for the norm in $\mathbf{S}$; (A.5) is the norm in $L_p(\Omega; \mathbf{S})$.

For example, $L_1(\Omega; Sym)$ is the space of tensor-valued functions on $\Omega$ with the norm

$$\|\mathbf{s}\| = \int_\Omega |\mathbf{s}(x)| \, dx.$$

If $\mathbf{S}$ is a space with scalar multiplication, the scalar multiplication is also defined in the space $\mathbf{L}_2(\Omega; \mathbf{S})$. No scalar multiplication is defined in the spaces $\mathbf{L}_p(\Omega)$, $\mathbf{L}_p(\Omega; \mathbf{S})$ if $p \neq 2$.

It should be noted that the norm $\|\mathbf{u}\|_{L_1(\Omega)}$ (and, similarly, the norms in the other spaces of the type $L_p(\Omega; \mathbf{S})$) is insensitive to changing the values of $\mathbf{u}$ on a zero measure set. Therefore, considering the function $\mathbf{u}$ as a member of $L_p(\Omega; \mathbf{S})$ leads to identifying $\mathbf{u}$ with any function whose values differ from values of $\mathbf{u}$ only on a zero measure set.

**A.2. Space of essentially bounded functions.** There are some unbounded functions for which changing their values only on a zero measure set makes them bounded. For example, the function $f$ on $(0, 1)$

$$f(x) = \begin{cases} 1 & \text{if } \ x \in (0,1), \ x \neq \dfrac{1}{n}, \ \ n = 1, 2, \ldots, \\ \dfrac{1}{n} & \text{if } \ x = \dfrac{1}{n}, \ \ n = 1, 2, \ldots, \end{cases}$$

becomes bounded after its values at the points $x = 1/n$, $n = 1, 2, \ldots$, are replaced, for example, by 1. At the same time, it is impossible to change values of the function $g(x) = 1/x$ only on a zero-measure set to obtain a bounded function.

Normally, one cannot distinguish between two stress fields whose values are the same almost everywhere, that is, at all points except the points of a zero volume (See Subsection III.1.2). A lot of fields considered in physics and mechanics are of this kind. It is clear that, in case the above functions $g$ and

$f$ represent such fields, there is no way to consider $g$ as a bounded function while $f$ is actually equivalent to a bounded function. We refer to a function as *essentially bounded* if changing its values on a zero-measure set results in a bounded function.

Consider a set of measurable essentially bounded functions on the domain $\Omega$. This is a linear space with respect to the regular operations of adding functions and multiplying a function by a number. Let us now define a norm in this space. For the function $f$ belonging to the space there is a zero-measure set $N$ such that the values $f(x)$ are bounded on $\Omega \setminus N$, that is,

$$\sup\{|f(x)| : x \in \Omega \setminus N\}$$

is finite. This number depends on $N$ but it is easy to get rid of this dependence considering the infimum of such numbers over the class of all zero-measure subsets in $\Omega$:

$$\inf_N \sup\{|f(x)| : x \in \Omega \setminus N\}.$$

This number is referred to as the *essential supremum* of the function $f$ on the domain $\Omega$. We set a norm of $f$ in the space of essentially bounded functions on $\Omega$ equal to the essential supremum of $f$ over $\Omega$. The norm is analogous to the second of the norms of the finite-dimensional space in (A.2). We denote by the symbol $L_\infty(\Omega)$ the space of essentially bounded functions and by $\|\cdot\|_{L_\infty(\Omega)}$ the above defined norm in this space.

Analogously, we define the space $L_\infty(\Omega; \mathbf{S})$ consisting of all measurable essentially bounded functions on $\Omega$ with values in a normed space $\mathbf{S}$, and a norm in the space is defined by simply replacing $|f(x)|$ by $\|f(x)\|_\mathbf{S}$ in the previous definition.

REMARK 1. Norms (A.1) and (A.2) in $\mathbf{R}^n$ are equivalent, see Example 1. In contrast to that, the norms $\|\cdot\|_{L_2(\Omega)}$, $\|\cdot\|_{L_1(\Omega)}$ and $\|\cdot\|_{L_\infty(\Omega)}$ are defined in different spaces and are not equivalent on the intersection of the spaces. For example, $L_\infty(\Omega)$ is a subset in $L_2(\Omega)$, which in turn is a subset in $L_1(\Omega)$, and the inequalities

$$\|\mathbf{u}\|_{L_1(\Omega)} \le c_1 \|\mathbf{u}\|_{L_2(\Omega)} \le c_2 \|\mathbf{u}\|_{L_\infty(\Omega)}$$

are valid for every $\mathbf{u} \in L_\infty(\Omega)$ with $c_1$ and $c_2$ not depending on $\mathbf{u}$. However, for example, there is no constant $c$ such that the inequality $\|\mathbf{u}\|_{L_2(\Omega)} \le c\|\mathbf{u}\|_{L_1(\Omega)}$ holds for every $\mathbf{u}$ in $L_2(\Omega)$.

**A.3. Convergency. Closure. Continuity.** A norm allows determination of the *distance* between any two members $\mathbf{u}$ and $\mathbf{v}$ of a normed linear space $X$ as $\|\mathbf{u} - \mathbf{v}\|$. Convergency in $X$ can now be defined as follows: a sequence $\mathbf{x}_1, \mathbf{x}_2, \ldots$ in $X$ *converges* to $\mathbf{x}_0 \in X$ if $\|\mathbf{x}_n - \mathbf{x}_0\| \to 0$ when $n \to \infty$; $\mathbf{x}_0$ is called the *limit of the sequence*. A sequence either does not have a limit or has a unique limit.

Let $A$ be a subset in a normed linear space $X$ and, for $\mathbf{x}_0 \in X$, there exists a sequence $a_n$, $n = 1, 2, \ldots$, which converges to $\mathbf{x}_0$; then $\mathbf{x}_0$ is referred to as the *limit point for A*. A limit point of a set either belongs to it or not. The set $A$ is referred to as *closed* if it contains all its limit points. For example, a segment $[a,b]$ is closed in $\mathbf{R}$; the sets $(a,b]$, $[a,b)$, $(a,b)$ are not. The whole space $X$ is closed. Consequently, for every set $A$ there are closed sets in $X$, which contain $A$. The "smallest" closed set containing $A$, that is, the intersection of closed sets each of which contains $A$, is referred to as the *closure* of $A$. In other words, the closure of $A$ is $A$ complemented with all its limit points.

Let $A$ be a set in a normed linear space $X$ and $\widehat{A}$ be a subset in $A$. We say that $\widehat{A}$ is *dense in A* if every $a \in A$ can be approximated by members of $\widehat{A}$, that is, there exists a sequence $\widehat{a}_1, \widehat{a}_2, \ldots$ in $\widehat{A}$, which converges to $a$.

Consider a sequence $\mathbf{x}_1, \mathbf{x}_2, \ldots$ in a normed linear space $X$ with the distance between $\mathbf{x}_n$ and $\mathbf{x}_m$ tending to 0: $\|\mathbf{x}_n - \mathbf{x}_m\| \to 0$ when $n, m \to \infty$. If every sequence which possesses this property has a limit, $X$ is referred to as a *complete* or *Banach* space. All normed spaces throughout this book are complete.

For functions on a space $X$, the concept of continuity can be introduced once the norm is defined in $X$. The function $f$ on $X$ is called *continuous at the point* $\mathbf{x}_0 \in X$ if, for every $\varepsilon > 0$, there exists a number $\delta_\varepsilon > 0$ such that

$$|f(\mathbf{x}) - f(\mathbf{x}_0)| \leq \varepsilon \quad \text{if} \quad \|\mathbf{x} - \mathbf{x}_0\| < \delta_\varepsilon .$$

**A.4. Conjugate space.** There is a very important class of continuous linear functions on a normed linear space $X$, which are also called *continuous linear functionals*. (The function $f$ on $X$ is referred to as linear if $f(\mathbf{x}_1 + \mathbf{x}_2) = f(\mathbf{x}_1) + f(\mathbf{x}_2)$ for every $\mathbf{x}_1, \mathbf{x}_2 \in X$ and $f(\alpha\mathbf{x}) = \alpha f(\mathbf{x})$ for every $\mathbf{x} \in X$ and every number $\alpha$.) It is clear that continuous linear functionals on $X$ form a linear space; we denote this space by $X'$. The norm $\|\cdot\|_X$ in $X$ naturally generates a norm $\|\cdot\|_{X'}$ in $X'$. Namely, for $\mathbf{x}' \in X'$ we set

$$\|\mathbf{x}'\|_{X'} = \sup\left\{\frac{|\mathbf{x}'(\mathbf{x})|}{\|\mathbf{x}\|_X} : \mathbf{x} \in X\right\}. \tag{A.6}$$

Thus, the inequality $|\mathbf{x}'(\mathbf{x})| \leq \|\mathbf{x}'\|_{X'} \|\mathbf{x}\|_X$ holds for every $\mathbf{x} \in X$ and every $\mathbf{x}' \in X'$. The space $X'$ is referred to as the *conjugate* of $X$. If $X$ is complete, then $X'$ is complete as well.

A general form of a continuous linear functional is known for many normed spaces. Consider, for example, the space $X = L_1(\Omega)$ and the description of its conjugate $(L_1(\Omega))'$. For every functional in $(L_1(\Omega))'$ there exists a function $g \in L_\infty(\Omega)$ such that the value of the functional at $f \in L_1(\Omega)$ is

$$\int_\Omega g(x) f(x)\, dV. \tag{A.7}$$

This correspondence of functionals on $L_1(\Omega)$ and functions in $L_\infty(\Omega)$ turns out to be a one-to-one correspondence. Moreover, the norm of a functional on $L_1(\Omega)$ defined according to (A.6) equals the norm of the corresponding function $g$ in $L_\infty(\Omega)$, which allows identification of $(L_1(\Omega))'$ with $L_\infty(\Omega)$. Similarly, the conjugate $(L_2(\Omega))'$ of $L_2(\Omega)$ is identified with $L_2(\Omega)$ itself. Generally, the conjugate of $L_p(\Omega)$, $1 < p < +\infty$ is the space $L_{p'}(\Omega)$, where $p' = p/(p-1)$; the value of the functional $g \in (L_p(\Omega))' = L_{p'}(\Omega)$ at $f \in L_p(\Omega)$ can be computed by formula (A.7).

The situation with functionals on the space $L_\infty(\Omega)$ is different. Every function $g \in L_1(\Omega)$ induces a functional on $L_\infty(\Omega)$, its value at any $f \in L_\infty(\Omega)$ being defined by (A.7). The norm of this functional defined according to (A.6) equals $||g||_{L_\infty(\Omega)}$. This resembles the above examples of conjugate spaces. However, in contrast to those examples, there are functionals on $L_\infty(\Omega)$ which cannot be identified with functions in $L_1(\Omega)$; $L_1(\Omega)$ is embedded in $(L_\infty(\Omega))'$ and the latter is wider than the former.

## Appendix B. Duality theorem

Here we prove the theorem which establishes conditions for equality of the extremums in the original and dual extremum problems (Theorem V.1).

Recall that the original problem

$$f(\mathbf{x}) \to \inf; \quad \mathbf{x} \in L \tag{B.1}$$

is considered in the linear space $X$, $L$ being a subspace in $X$. The bilinear form $\langle \cdot, \cdot \rangle$ is a pairing between $X$ and another linear space $X^*$. The pairing allows us to introduce the Fenhel transformation $f^*$ of any function $f$ on $X$:

$$f^*(\mathbf{x}^*) = \sup\{\langle \mathbf{x}^*, \mathbf{x} \rangle - f(\mathbf{x}) : \mathbf{x} \in X\}$$

and the polar set $L^0$ of any subspace in $X$:

$$L^0 = \{\mathbf{x}^* \in X^* : \langle \mathbf{x}^*, \mathbf{x} \rangle = 0 \text{ for every } \mathbf{x} \in L\}.$$

The dual problem of (B.1) is

$$-f^*(\mathbf{x}^*) \to \sup; \quad \mathbf{x}^* \in L^0. \tag{B.2}$$

The following theorem makes use of the space $X'$ of continuous linear functionals on the normed space $X$: $X'$ is the space which is paired with $X$ ($X^* = X'$); the pairing between them is standard: $\langle \mathbf{x}^*, \mathbf{x} \rangle$ is the value of the functional $\mathbf{x}^* \in X'$ at $\mathbf{x} \in X$.

THEOREM (*duality theorem*). Suppose $X$ is a complete normed linear space, $L$ is its subspace and $f$ is a convex function on $X$ continuous at the point $\mathbf{x}_0 \in L$. Then the infimum in problem (B.1) and the supremum in problem (B.2) are equal:

$$\inf\{f(\mathbf{x}) : \mathbf{x} \in L\} = \sup\left\{-f^*(\mathbf{x}^*) : \mathbf{x}^* \in L^0\right\}.$$

The extremum in the dual problem is attainable if $f$ is bounded from below on $L$.

PROOF. First af all, recall that the inequality

$$\sup\left\{-f^*(\mathbf{x}^*) : \mathbf{x}^* \in L^0\right\} \le \inf\{f(\mathbf{x}) : \mathbf{x} \in L\} \tag{B.3}$$

is always valid, see (V.3.8). Because of that, in case

$$\inf\{f(\mathbf{x}) : \mathbf{x} \in L\} = -\infty,$$

we also have

$$\sup\left\{-f^*(\mathbf{x}^*) : \mathbf{x}^* \in L^0\right\} = -\infty,$$

and the statement of the theorem is true. Thus, it remains to consider the case

$$\inf\{f(\mathbf{x}) : \mathbf{x} \in L\} > -\infty,$$

and verify the inequality

$$\sup\left\{-f^*(\mathbf{x}^*) : \mathbf{x}^* \in L^0\right\} \ge \inf\{f(\mathbf{x}) : \mathbf{x} \in L\} \tag{B.4}$$

for this case. In order to do so, let us introduce the set

$$\mathcal{H} = \{(r, \mathbf{x}) \in \mathbf{R} \times X : \text{there exists } \mathbf{l} \in L,\ \ r \ge f(\mathbf{x} + \mathbf{l})\}.$$

It possesses the following properties (we will verify them later):

(i) $\mathcal{H}$ is convex,

(ii) the interior of $\mathcal{H}$ is not empty, in particular, $(r_0, 0) \in \mathbf{R} \times X$ is in the interior of $\mathcal{H}$ if $r_0$ satisfies the condition $r_0 > f(\mathbf{x}_0)$,

(iii) the point

$$(m, 0) \in \mathbf{R} \times X, \ \text{ where } m = \inf\{f(\mathbf{x}) : \mathbf{x} \in L\} > -\infty,$$

is not in the interior of $\mathcal{H}$,

(iv) any point $(\alpha, 0) \in \mathbf{R} \times X$, where $\alpha > m$ and $\mathbf{x} \in L$, is in $\mathcal{H}$.

According to these properties, the interior of $\mathcal{H}$ is a nonempty convex set and the point $(m, 0)$ is not in it. We apply the following separation theorem (a corollary of the Hahn-Banach theorem) to them: let $A$ be a convex set in Banach space $E$, the interior of $A$ being not empty, and $\mathbf{b} \in E$ is not in $A$; then there exists a continuous linear functional $\varphi \in E'$, $\varphi \ne 0$, such that

$$\langle \varphi, \mathbf{a} \rangle \ge \langle \varphi, \mathbf{b} \rangle \quad \text{for every} \quad \mathbf{a} \in A.$$

In the case under consideration, this means that there is a number $\alpha^* \in \mathbf{R}$ and $\mathbf{x}^* \in X^*$ such that $(\alpha^*, \mathbf{x}^*) \ne (0, 0)$ and

$$\alpha^*\alpha + \langle \mathbf{x}^*, \mathbf{x} \rangle \ge \alpha^* m + \langle \mathbf{x}^*, 0 \rangle \quad \text{for every} \quad (\alpha, \mathbf{x}) \in \mathcal{H}. \tag{B.5}$$

In particular, due to (iv), this implies that

$$\alpha^*\alpha + \langle \mathbf{x}^*, \mathbf{x} \rangle \ge \alpha^* m \quad \text{for every } \alpha > m \text{ and every } \mathbf{x} \in L. \tag{B.6}$$

Setting $\mathbf{x} = 0$, we obtain $\alpha^*(\alpha - m) \geq 0$ for every $\alpha > m$, that is, $\alpha^* \geq 0$. Then, it can be easily seen that $\mathbf{x}^* \in L^0$. Indeed, we choose a concrete $\alpha > m$, any $\mathbf{x}_1$ in $L$ and an arbitrary number $c$. Note that $c\mathbf{x}_1$ is in $L$ and (B.6) yields

$$\alpha^*(\alpha - m) + c\langle \mathbf{x}^*, \mathbf{x}_1 \rangle \geq 0 \quad \text{for every} \quad c \in \mathbf{R}.$$

This relation is valid provided $\langle \mathbf{x}^*, \mathbf{x}_1 \rangle = 0$ and, consequently, $\mathbf{x}^* \in L^0$ since the previous equality holds for every $\mathbf{x}_1$ in $L$.

Finally, we observe that $\alpha^*$ is positive. Indeed, under the assumption that $\alpha^* = 0$, (B.5) takes the form

$$\langle \mathbf{x}^*, \mathbf{x} \rangle \geq 0 \quad \text{for every} \quad (\alpha, \mathbf{x}) \in \mathcal{H}.$$

We choose a number $r_0 > f(\mathbf{x}_0)$; then, due to (ii), there exists $\varepsilon > 0$ such that $(r_0, \mathbf{x})$ is in $\mathcal{H}$ as soon as $\|\mathbf{x}\| < \varepsilon$. Then, the previous relation implies

$$\langle \mathbf{x}^*, \mathbf{x} \rangle \geq 0 \quad \text{for every} \quad \mathbf{x} \in X \quad \text{with} \quad \|\mathbf{x}\| < \varepsilon,$$

and, hence, $\mathbf{x}^* = 0$. Together with $\alpha^* = 0$, this contradicts the vadility of the relation $(\alpha^*, \mathbf{x}^*) \neq (0, 0)$ by virtue of the separation theorem. Therefore, $\alpha^* \neq 0$ and, consequently, $\alpha^* > 0$ according to the above inequality $\alpha^* \geq 0$.

Dividing (B.5) by $\alpha^* > 0$ and denoting $\mathbf{x}^*/\alpha^*$ by $\bar{\mathbf{x}}^*$ results in the relation

$$\alpha + \langle \bar{\mathbf{x}}^*, \mathbf{x} \rangle \geq m \quad \text{for every} \quad (\alpha, \mathbf{x}) \in \mathcal{H}.$$

The inequality is valid, in particular, for $\alpha = f(\mathbf{x})$, $\mathbf{x} \in X$, since the point $(f(\mathbf{x}), \mathbf{x})$ is obviously in $\mathcal{H}$. Thus, we have

$$f(\mathbf{x}) + \langle \bar{\mathbf{x}}^*, \mathbf{x} \rangle \geq m \quad \text{for every} \quad \mathbf{x} \in X$$

or, which is the same,

$$\inf \{\langle \bar{\mathbf{x}}^*, \mathbf{x} \rangle + f(\mathbf{x}) : \mathbf{x} \in X\} \geq m.$$

Recall that by the definition of the Fenhel transformation:

$$f^*(-\bar{\mathbf{x}}^*) = \sup \{\langle -\bar{\mathbf{x}}^*, \mathbf{x} \rangle - f(\mathbf{x}) : \mathbf{x} \in X\},$$

which obviously equals $-\inf \{\langle \mathbf{x}^*, \mathbf{x} \rangle + f(\mathbf{x}) : \mathbf{x} \in X\}$. This allows re-writing the previous inequality as $-f^*(-\bar{\mathbf{x}}^*) \geq m$. Recall that $m$ stands for

$$\inf \{f(\mathbf{x}) : \mathbf{x} \in L\}$$

and hence the relation $-f^*(-\bar{\mathbf{x}}^*) \geq m$, where $-\bar{\mathbf{x}}^*$ belongs to $L^0$, proves inequality (B.4) and the main statement of the theorem that the extremums in the original and dual problems are equal.

Moreover, in case $f$ is bounded on $L$ from below, that is, $m > -\infty$, the above reasoning determines the point $-\bar{\mathbf{x}}^* \in L^0$ at which $-f^*(-\bar{\mathbf{x}}^*) \geq m$.

Due to (B.3), the supremum of $-f^*$ over $L^0$ is not greater than $m$, which means that

$$-f^*(-\bar{\mathbf{x}}^*) = \sup\left\{-f^*(\mathbf{x}^*) : \mathbf{x}^* \in L^0\right\}.$$

This proves the second statement of the theorem: the extremum in the dual problem is attainable.

It remains to verify the above listed properties of the set $\mathcal{H}$.

Let us, first, show that $\mathcal{H}$ is convex. Indeed, let $(r_1, \mathbf{x}_1)$ and $(r_2, \mathbf{x}_2)$ be in $\mathcal{H}$, that is, for certain $\mathbf{l}_1, \mathbf{l}_2$ in $L$ the inequalities $r_1 \geq f(\mathbf{x}_1+\mathbf{l}_1)$, $r_2 \geq f(\mathbf{x}_2+\mathbf{l}_2)$ are valid. Let us consider $r = \alpha r_1 + \beta r_2$ and $\mathbf{x} = \alpha\mathbf{x}_1 + \beta\mathbf{x}_2$, where $\alpha, \beta \geq 0$ and $\alpha + \beta = 1$, and show that $(r, \mathbf{x})$ is in $\mathcal{H}$. We introduce $\mathbf{l} = \alpha\mathbf{l}_1 + \beta\mathbf{l}_2$ belonging to $L$ and obtain that

$$f(\mathbf{x}+\mathbf{l}) = f(\alpha(\mathbf{x}_1+\mathbf{l}_1)+\beta(\mathbf{x}_2+\mathbf{l}_2)) \leq \alpha f(\mathbf{x}_1+\mathbf{l}_1)+\beta f(\mathbf{x}_2+\mathbf{l}_2) \leq \alpha r_1+\beta r_2 = r.$$

Thus, $(r, \mathbf{x})$ is in $\mathcal{H}$ and, hence, $\mathcal{H}$ is convex.

To verify (ii), let us consider the point $\mathbf{x}_0$ in the formulation of the theorem and any number $r_0$ such that $r_0 > f(\mathbf{x}_0)$. In order to show that the point $(r_0, 0)$ is in the interior of $\mathcal{H}$, it has to be proved that there exist numbers $\varepsilon_1 > 0$, $\varepsilon_2 > 0$ such that the inequalities

$$|\, r - r_0 \,| < \varepsilon_1, \quad \|\mathbf{x}\| < \varepsilon_2 \tag{B.7}$$

imply $(r, \mathbf{x}) \in \mathcal{H}$. In other words, we have to verify that, for any $(r, \mathbf{x})$ satisfying (B.7), there is such $\mathbf{l}$ in $L$ that $r \geq f(\mathbf{x} + \mathbf{l})$. To show this, we choose $\varepsilon_1 > 0$ which satisfies the inequality

$$r_0 - 2\varepsilon_1 > f(\mathbf{x}_0). \tag{B.8}$$

Then, according to the continuity of $f$ at $\mathbf{x}_0$, we choose $\varepsilon_2 > 0$ such that

$$|\, f(\mathbf{x}_0 + \mathbf{x}) - f(\mathbf{x}_0) \,| < \varepsilon_1 \quad \text{for every } \mathbf{x} \in X \text{ with } \|\mathbf{x}\| < \varepsilon_2. \tag{B.9}$$

Finally, we verify that, for every $(r, \mathbf{x})$ satisfying conditions (B.7), the required inequality $r \geq f(\mathbf{x} + \mathbf{l})$ is valid if $\mathbf{l}$ takes the value of $\mathbf{x}_0$. Indeed, the second of relations (B.7) and (B.9) imply

$$f(\mathbf{x} + \mathbf{x}_0) \leq f(\mathbf{x}_0) + \varepsilon_1.$$

The estimation of the right-hand side according to (B.8) is as follows:

$$f(\mathbf{x}_0) + \varepsilon_1 \leq r_0 - 2\varepsilon_1 + \varepsilon_1 = r_0 - \varepsilon_1,$$

and the estimation of $r_0 - \varepsilon_1$ according to the first of relations (B.7) is $r_0 - \varepsilon_1 < r$. Thus, we arrive at the required inequality $f(\mathbf{x} + \mathbf{x}_0) < r$.

To verify (iii), we assume that the infimum $m$ of $f$ over $L$ is finite. Then, if $(m, 0)$ is in the interior of $\mathcal{H}$, the point $(m - \delta, 0)$ is also in this interior when $\delta > 0$ is sufficiently small. This means that, for some $\mathbf{l}$ in $L$, $f(0 + \mathbf{l}) \leq m - \delta$ and, consequently, $f(\mathbf{l}) < m$. This contradicts the fact that $m$ is the infimum. Therefore, the assumption that $(m, 0)$ is in the interior of $\mathcal{H}$ is wrong, which proves (iii).

Finally, we note that (iv) is valid. Indeed, for any number $\alpha > m$, by the definition of the infimum, there is $\mathbf{x}_\alpha \in L$ such that $\alpha > f(\mathbf{x}_\alpha)$. This means that $(\alpha, \mathbf{x}_\alpha)$ belongs to $\mathcal{H}$, which finishes the proof of the theorem.

## Comments

The recognition of the convex extremum problems theory as the mathematical basis of limit analysis was explicitly expressed by Nayroles (1970), and this recognition played a very important role in further development of limit analysis. In particular, the way to formulate conditions for the limit multipliers equality seems to have been first proposed in the article by Nayroles. Some refining of Nayroles' reasoning in Kamenjarzh (1979) resulted in the two kinds of such condition; see Theorems 4.2 and 6.1. These conditions proved essential by the examples of Frémond and Friâa (1982), and Kamenjarzh and Merzljakov (1985) reproduced in Section 8.

The convex functions theory originated from studies by Fenchel (1949, 1951), Moreau (1962, 1964, 1966) and Rockafellar (1967). The duality theory for convex extremum problems was developed by Fenchel (1951), Goldstein (1967, 1968, 1971) and Rockafellar (1967, 1969). Theorem 3.1 represents a comparatively simple proposition about conditions for the equality of the extremum in dual extremum problems. The reader can find more general statements in the books by Goldstein (1971) and Ekeland and Temam (1976). Limit analysis is one of the numerous fields of application of this theorem. Its various applications to continuum mechanics are considered in the articles and books by Nayroles (1971), Halphen and Nguen (1975), Berdichevsky (1983), and Panagiatopoulos (1985).

Nayroles (1970) proved that the set of admissible stress fields is weak closed (Proposition 5.1). His reasoning is based on the theorem proved by Rockafellar (1968), see Proposition 5.2.

The reader can find detailed descriptions of the basic concepts and propositions of convex analysis in the books by Rockafellar (1968), Ekeland and Temam (1976), and Ioffe and Tikhomirov (1979). For information about normed spaces see the books by Kolmogorov and Fomin (1970) and Edwards (1965).

CHAPTER VI

# REDUCTION OF LIMIT ANALYSIS EXTREMUM PROBLEMS

This chapter presents further study of the extremum problems of limit analysis in the case of plastic bodies with cylindrical yield surfaces. Both static and kinematic problems are reduced. Five stress deviator components are involved in the reduced formulation of the static problem instead of the six components in the original formulation. The reduced kinematic problem allows using only sufficiently regular velocity fields for evaluating the safety factor (note that Theorem V.7.1 suggests using generalized fields for this purpose). Reduction of the kinematic problem is very important since the reduced formulation leads to rational numerical methods for calculating of the safety factor.

The reader can proceed to the following chapters upon reading Section 1 which contains the main results of this chapter. The rest of the chapter deals with proving the results, in which connection restoring the pressure field and approximating the solenoidal vector fields are considered. Solutions to these problems play an important role not only in reducing the limit analysis extremum problems but also in studying some other questions. In particular, we apply the above-mentioned results to the finite element formulation of the limit analysis problem in Chapter IX and to the elastic-plastic shakedown theory in Chapter X.

We use some simple concepts and results from the theory of distributions and Sobolev spaces when studying the pressure field restoration problem. A survey of the necessary information is presented in Appendices A and B.

## 1. Reduction of static and kinematic extremum problems

In this section we reduce the extremum problems of limit analysis and also formulate conditions under which the safety factor can be evaluated as the extremum in the reduced static or kinematic problems.

**1.1. Static problem reduction.** The safety factor of the load $\mathbf{l}$ is determined by static extremum problem (V.1.1)

$$\alpha_l = \sup\{m \geq 0 : \text{there exists } \rho \in \Sigma,\ m(\rho + \mathbf{s}_l) \in C\}.$$

Here, $\mathbf{s}_l$ is a fixed stress field which equilibrates the load $\mathbf{l}$; $C$ and $\Sigma$ are the sets of admissible and self-equilibrated stress fields.

In the case of three-dimensional bodies with cylindrical yield surfaces, we decompose a stress field $\boldsymbol{\sigma}$ into the deviatoric and spheric parts $\boldsymbol{\sigma}^d$ and $\boldsymbol{\sigma}^s$. According to this, we represent the stress fields space $\mathcal{S}$ in the form $\mathcal{S} = \mathcal{S}^d \times \mathcal{S}^s$, where $\mathcal{S}^d$ and $\mathcal{S}^s$ are certain spaces of functions on the domain $\Omega$ occupied by the body (Subsection V.7.1). Functions in $\mathcal{S}^d$ and $\mathcal{S}^s$ take the values in $Sym^d$ and $Sym^s$, respectively. The set of admissible stress fields is of the form

$$C = \{\boldsymbol{\sigma} \in \mathcal{S} : \boldsymbol{\sigma}^d(x) \in C_x^d \text{ for a.e. } x \in \Omega\}$$

in the case of bodies with cylindrical yield surfaces. Here, $C_x^d$ is a convex, closed set in the deviatoric plane $Sym^d$ and contains the origin; $C_x^d$ is given for every $x \in \Omega$.

We now introduce the set of admissible deviatoric stress fields:

$$C^d = \{\boldsymbol{\tau} \in \mathcal{S}^d : \boldsymbol{\tau}(x) \in C_x^d \text{ for a.e. } x \in \Omega\}$$

and write the definition of the safety factor in the following form:

$$\alpha_l = \sup\left\{m \geq 0 : \text{there exists } \boldsymbol{\rho} \in \Sigma,\ m(\boldsymbol{\rho}^d + \mathbf{s}_l^d) \in C^d\right\}. \tag{1.1}$$

Here, the spheric part of the stress $\boldsymbol{\rho}$, that is, the hydrostatic pressure field, only enters the relation $\boldsymbol{\rho} \in \Sigma$, which is the equilibrium condition. The idea of our approach consists of 1) excluding the pressure field from the equilibrium condition, which results in the deviatoric reduced problem; 2) studying the reduced problem; and 3) establishing the possibility of restoring the pressure field for the solution of the reduced problem. Such an approach guarantees equivalence of the deviatoric original problems and it also make sense since the reduced deviatoric problem is easier to study than the original one.

We are reminded that the set $\Sigma$ of self-equilibrated stress fields is defined in terms of the virtual work principle as

$$\Sigma = (\operatorname{Def} U)^0 = \{\boldsymbol{\sigma} \in \mathcal{S} : \int_\Omega \operatorname{Def} \mathbf{v} \cdot \boldsymbol{\sigma}\, dV = 0 \text{ for every } \mathbf{v} \in U\}.$$

Here, $U$ is the set of (smooth) virtual velocity fields. To exclude the spheric part $\boldsymbol{\rho}^s$ from the equilibrium condition $\boldsymbol{\rho} \in \Sigma$, we replace $U$ by its subspace consisting of solenoidal fields:

$$U_d = \{\mathbf{v} \in U : \operatorname{div} \mathbf{v} = 0\}.$$

This results in the following definition of the set of *weak self-equilibrated deviatoric stress fields*:

$$\Sigma_d = (\operatorname{Def} U_d)^0 = \{\boldsymbol{\tau} \in \mathcal{S}^d : \int_\Omega \operatorname{Def} \mathbf{v} \cdot \boldsymbol{\tau}\, dV = 0 \text{ for every } \mathbf{v} \in U_d\}.$$

We use the condition $\boldsymbol{\sigma}^d \in \Sigma_d$ instead of $\boldsymbol{\sigma} \in \Sigma$ in the formulation of static extremum problem (1.1) and arrive to the following problem:

$$\alpha_{ld} = \sup\left\{m \geq 0 : \text{there exists } \boldsymbol{\tau} \in \Sigma_d,\ m(\boldsymbol{\tau} + \mathbf{s}_l^d) \in C^d\right\}. \tag{1.2}$$

It is referred to as a *reduced* or *deviatoric* static extremum problem.

The supremums in the original and reduced problems satisfy the inequality $\alpha_{ld} \geq \alpha_l$. Indeed, consider a maximizing sequence $m_1, m_2, \ldots$ in problem (1.1): $m_n \to \alpha_l$ when $n \to \infty$, and for every $n = 1.2. \ldots$, there exists a self-equilibrated stress field $\boldsymbol{\rho}_n$ which, together with $m_n$, satisfies the constraints in (1.1). Then $\boldsymbol{\tau}_n = \boldsymbol{\rho}_n^d$ and $m_n$ also satisfy the constraints in (1.2) and, consequently, $\alpha_{ld} \geq \alpha_l$. Actually, the equality $\alpha_{ld} = \alpha_l$ is valid in many cases, and the safety factor can be evaluated as the extremum in the reduced problem. Formulation of this problem involves five stress deviator components instead of the six stress components involved in the original formulation. We describe conditions under which the original and reduced problems are equivalent and $\alpha_{ld} = \alpha_l$ in Subsection 1.3 and justify them in Section 4.

**1.2. Kinematic problem reduction.** Under certain conditions, the safety factor can be evaluated as the limit kinematic multiplier:

$$\alpha_l = \beta_l = \inf \left\{ \frac{D(\mathbf{e})}{\langle \mathbf{e}, \mathbf{s}_l \rangle} : \mathbf{e} \in E, \ \langle \mathbf{e}, \mathbf{s}_l \rangle > 0 \right\} \tag{1.3}$$

(Theorem V.7.1 for bodies with cylindrical yield surfaces). Here, $E = \operatorname{Def} \mathcal{V}$ is the subspace of kinematically admissible strain rate fields in a certain space $\mathcal{E}$, $\mathcal{V}$ is the space of kinematically admissible velocity fields, $D$ is the dissipation

$$D(\mathbf{e}) = \sup \{ \langle \mathbf{e}, \mathbf{s} \rangle : \mathbf{s} \in C \} \tag{1.4}$$

and $\langle \cdot, \cdot \rangle$ is the internal power bilinear form. The latter pairs the space $\mathcal{E}$ and the stress fields space $\mathcal{S}$.

In the case of three-dimensional bodies with cylindrical yield surfaces, we decompose a strain rate field $\mathbf{e}$ into the deviatoric and spheric parts $\mathbf{e}^d$ and $\mathbf{e}^s$. This allows presentation of the space $\mathcal{E}$ in the form $\mathcal{E} = \mathcal{E}^d \times \mathcal{E}^s$, where $\mathcal{E}^d$ and $\mathcal{E}^s$ are certain spaces of deviatoric and spheric symmetric tensor fields, respectively. It should be recalled that in the case of bodies with cylindrical yield surfaces $D(\mathbf{e}) = +\infty$ if $\mathbf{e}^s \neq 0$. Therefore, the fields with $\mathbf{e}^s \neq 0$ may be ignored when solving (1.3), that is, $\mathcal{E}$ and $\mathcal{V}$ may be replaced by $\mathcal{E}^d$ and

$$\mathcal{V}_d = \{ \mathbf{v} \in \mathcal{V} : (\operatorname{div} \mathbf{v})^s = 0 \}$$

in the kinematic extremum problem (we use the symbol $\operatorname{div} \mathbf{v}$ to denote $(\operatorname{Def} \mathbf{v})_{kk}$). In particular, $E = \operatorname{Def} \mathcal{V}$ is replaced by $E_d = \operatorname{Def} \mathcal{V}_d$ and the kinematic problem can be written as

$$\alpha_l = \beta_l = \inf \left\{ \frac{D(\mathbf{e})}{\langle \mathbf{e}, \mathbf{s}_l^d \rangle} : \mathbf{e} \in E_d, \ \langle \mathbf{e}, \mathbf{s}_l^d \rangle > 0 \right\}, \tag{1.5}$$

where the equality $\alpha_l = \beta_l$ is valid, for example, under conditions of Theorem V.7.1. We refer to this formulation of the kinematic problem as *deviatoric*. In contrast to the deviatoric static problem, (1.5) is not a reduction of the original problem (1.3) but its equivalent formulation.

Reducing the kinematic problem consists of narrowing the spaces $\mathcal{E}$ and $\mathcal{V}$. Recall that Theorem V.7.1 suggests using a rather abstract space $\mathcal{E}$ and a considerably wide set $E$ in it. This guarantees the equality $\alpha_l = \beta_l$, but results in a formulation of the kinematic extremum problem which is inconvenient for practical calculations. We refer to (1.5) as a *reduced kinematic problem* if the formulation involves the space of integrable strain rate fields or, equivalently, the space of sufficiently regular velocity fields. In the case of integrable strain rate fields there is a simple explicit formula for computing the values of the function to be minimized in (1.5) (the kinematic multiplier: dissipation over external power). In particular the dissipation $D(\mathbf{e})$ can be evaluated by integrating the local dissipation $d_x(\mathbf{e}(x))$ over the domain $\Omega$ occupied by the body, see Theorem IV.5.1. Then the reduced kinematic problem has the form

$$\beta_{ld} = \inf\left\{ \frac{\int\limits_{\Omega} d_x(\mathbf{e}(x))\,dV}{\int\limits_{\Omega} \mathbf{e}(x)\cdot\mathbf{s}_l^d(x)\,dV} : \mathbf{e}\in E_d,\ \int\limits_{\Omega} \mathbf{e}(x)\cdot\mathbf{s}_l^d(x)\,dV > 0 \right\}, \tag{1.6}$$

where $E_d$ is the set of kinematically admissible strain rate fields corresponding to sufficiently regular velocity fields.

Any value of the function to be minimized in the reduced kinematic problem is a kinematic multiplier of some velocity field. Then the inequality $\beta_{ld} \geq \alpha_l$ is valid due to Theorem III.5.1. Actually, the equality $\beta_{ld} = \alpha_l$ is valid in many cases, and the safety factor can be evaluated as the extremum in a rather simple reduced kinematic problem. Establishing constructive conditions for this equality is the primary goal of this chapter. We describe these conditions in the next subsection and justify them in Subsection 1.4.

**1.3. Reduced extremum problems: main results.** The reduced static and kinematic problems are most effective for limit analysis if their extremums equal the safety factor: $\alpha_{ld} = \beta_{ld} = \alpha_l$. Below are the main results concerning conditions for these equalities.

These conditions have to do with the solvability of the following problem of pressure field restoration. Let $\boldsymbol{\tau}$ be a deviatoric stress field in $\mathcal{S}^d$ weak self-equilibrated, that is, satisfying the condition

$$\int\limits_{\Omega} \operatorname{Def}\mathbf{v}\cdot\boldsymbol{\tau}\,dV = 0 \ \text{ for every } \ \mathbf{v}\in U_d, \quad \text{where} \quad U_d = \{\mathbf{v}\in U : \operatorname{div}\mathbf{v} = 0\}$$

($\Omega$ as usual stands for the domain occupied by the body). The question is whether there exists a pressure field $p$ such that the stress field $\boldsymbol{\sigma} = \boldsymbol{\tau} - p\mathbf{I}$ is in $\mathcal{S}$ and is self-equilibrated, that is, satisfies the condition

$$\int\limits_{\Omega} \operatorname{Def}\mathbf{v}\cdot\boldsymbol{\sigma}\,dV = 0 \quad \text{for every} \quad \mathbf{v}\in U.$$

We refer to this problem as a *pressure field restoration problem.* Solvability of the problem implies the equality $\alpha_{ld} = \alpha_l$. Indeed, consider a maximizing sequence $m_1, m_2, \ldots$ in the reduced static problem (1.2). This means that $m_n \geq 0$, $n = 1, 2, \ldots$, $m_n \to \alpha_{ld}$ when $n \to \infty$, and for every $n$, there exists a weak self-equilibrated deviatoric stress field $\boldsymbol{\tau}_n$ which, together with $m_n$, satisfies the constraints in (1.2). Then, for every $\boldsymbol{\tau}_n$ we consider the solution $p_n$ to the pressure field restoration problem. The stress field $\boldsymbol{\rho}_n = \boldsymbol{\tau}_n - p_n\mathrm{I}$ and the number $m_n$ obviously satisfy the constraints in original static problem (1.1), and, consequently, $\alpha_l \leq \alpha_{ld}$. Since $\alpha_{ld}$ is always not greater than $\alpha_l$ (see Subsection 1.1), we have $\alpha_l = \alpha_{ld}$. Thus,

> the safety factor can be evaluated as the extremum in the reduced static problem, $\alpha_l = \alpha_{ld}$, if the pressure field restoration problem is solvable.

We will show that the pressure field restoration problem is solvable if the body boundary is sufficiently regular and the free part of the boundary is its sufficiently regular part. Here, as usual, we refer to the part $S_q$ of the boundary, where surface tractions are given, as free. We will comment on the regularity requirements later.

Under certain conditions, which we establish in Subsection 1.4, the extremums in the reduced static and kinematic problems are equal. Together with the above statements, this implies that

> the safety factor $\alpha_l$ can be evaluated by both the static and kinematic methods using the reduced extremum problems: $\alpha_l = \alpha_{ld} = \beta_{ld}$ if
>
> 1) the body boundary is sufficiently regular and the free surface $S_q$ is its sufficiently regular part and
>
> 2) for every point of the body the cylindrical yield surface lies in the same cylinder $|\mathbf{s}^d| \leq R$, $R > 0$.

The boundary and its free part $S_q$ meet the regularity requirements, for example, if the boundary consists of a finite number of continuously differentiable surfaces which intersect at nonzero angles and $S_q$ is bounded by a finite number of continuously differentiable curves which intersect at nonzero angles. The reduced static problem makes use only of deviatoric stress fields, the reduced kinematic problem only of regular velocity fields.

We prove this main result in the next subsection. The proof is based on solvability of the pressure field restoration problem. The remaining part of this chapter deals with this problem and a closely related question about solenoidal vector field approximation.

**1.4. Safety factor as extremum in the reduced problems.** We now prove the conditions which allow to evaluate the safety factor for bodies with cylindrical yield surfaces as the extremum in the reduced static and kinematic problems. In the following proposition,

> for bodies with cylindrical yield surface, we use the spaces

$$\mathcal{S} = \mathcal{S}^d \times \mathcal{S}^s, \ \ \mathcal{S}^d = L_\infty(\Omega; Sym^d), \ \ \mathcal{S}^s = L_2(\Omega; Sym^s),$$
$$\mathcal{E} = \mathcal{E}^d \times \mathcal{E}^s, \ \ \mathcal{E}^d = L_1(\Omega; Sym^d), \ \ \mathcal{E}^s = L_2(\Omega; Sym^s)$$

and the pairing

$$\langle \mathrm{e}, \boldsymbol{\sigma} \rangle = \int_{\Omega} \mathrm{e}(x) \cdot \boldsymbol{\sigma}(x)\, dV$$

between them;

the mapping $x \longrightarrow C_x$ is measurable, as usual.

The above-mentioned spaces are defined in Appendix V.A (see also Subsections V.7.2 and 1.2); measurability of a set-valued mapping is defined in Subsection IV.5.3. Actually, the mapping $x \longrightarrow C_x$ is measurable in all limit analysis problems of practical interest.

THEOREM 1. Suppose the foregoing assumptions are satisfied and the body occupies domain $\Omega$ with a sufficiently regular boundary, $S_q$ being its sufficiently regular part. Suppose also that for every point of the body the cylindrical yield surface lies in the same cylinder $|\mathrm{s}^d| \leq R$, $R > 0$. Then

1) the safety factor can be evaluated by both static and kinematic methods using the reduced extremum problems:

$$\alpha_l = \sup\left\{m \geq 0 : \text{there exists } \boldsymbol{\tau} \in \Sigma_d,\ m(\boldsymbol{\tau} + \mathrm{s}_l^d) \in C^d\right\}$$
$$= \inf\left\{ \frac{\int_{\Omega} d_x(\mathrm{e}(x))\, dV}{\int_{\Omega} \mathrm{e}(x) \cdot \mathrm{s}_l^d(x)\, dV} : \mathrm{e} \in E_d,\ \int_{\Omega} \mathrm{e}(x) \cdot \mathrm{s}_l^d(x)\, dV > 0 \right\}; \qquad (1.7)$$

2) the extremum in the reduced kinematic problem remains the same if the set $E_d$ in its formulation is replaced by any subspace $\widehat{E}_d$ in $\mathcal{E}^d$ satisfying the condition $(\widehat{E}_d)^0 = \Sigma_d$; in particular, $E_d$ can be replaced by $\operatorname{Def} U_d$ or by any linear subspace in $\mathcal{E}^d$ whose closure is the same as the closure of $\operatorname{Def} U_d$.

REMARK 1. The assumption about regularity of $\partial\Omega$ and $S_q$ is to ensure that the pressure field restoration problem is solvable. We establish the solvability in the next section, see Theorems 2.3 and 2.4. The surfaces $\partial\Omega$, $S_q$ meet the regularity requirements, for example, if $\partial\Omega$ consists of a finite number of continuously differentiable surfaces which intersect at nonzero angles and $S_q$ is bounded by a finite number of continuously differentiable curves which intersect at nonzero angles.

PROOF. Due to the solvability of the pressure field restoration problem the first equality in (1.7) is valid, as it was shown in the previous subsection. We now show that the second equality in (1.7) is also valid, that is, the extremums in the reduced static and kinematic problems are equal. Note that these problems have the exact form of the original extremum problems (1.1) and (1.3) with the spaces $\mathcal{S}$, $\mathcal{E}$, $\mathcal{V}$ replaced by $\mathcal{S}^d$, $\mathcal{E}^d$, $\mathcal{V}_d$ and the sets $C$, $U$, $\Sigma$, $E$ replaced by $C^d$, $U_d$, $\Sigma_d$, $E_d$. We observe that the reduced problems meet all requirements of the rigid-plastic problem unified formulation given in Subsection IV.1.4. In particular, it is clear that the relations $E_d = \operatorname{Def} \mathcal{V}_d$, $\Sigma_d = (\operatorname{Def} U_d)^0$ similar to $E = \operatorname{Def} \mathcal{V}$, $\Sigma = (\operatorname{Def} U)^0$ are valid. The relation $\Sigma_d = (\operatorname{Def} \mathcal{V}_d)^0$ replacing $\Sigma = (\operatorname{Def} \mathcal{V})^0$ is also valid if $\mathcal{V}$ is chosen appropriately; it is sufficient, for example, to set $\mathcal{V} = U$. The mapping $x \longrightarrow C_x^d$ is measurable since $x \to C_x$ is a measurable mapping. The set $C_x^d$, for a.e. point $x$ of the body, lies inside the same ball in $Sym^d$. Thus, the reduced problems (1.7) satisfy the assumptions of Theorem V.5.1 with $\mathbf{S} = Sym^d$. Application of this theorem proves both statements of the present proposition.

REMARK 2. It is clear that the statement analogous to Theorem 1 is valid for discrete systems, that is, in case the spaces $\mathcal{S}$ and $\mathcal{E}$ are finite dimensional.

REMARK 3. The second statement of Theorem 1 allows evaluating the safety factor as the extremum of the kinematic multipliers over only smooth velocity fields. However, another choice is more convenient from the computational viewpoint. Namely, the safety factor is the extremum

$$\alpha_l = \inf \left\{ \int_\Omega d_x(\operatorname{Def} \mathbf{v})\, dV \,:\, \mathbf{v} \in V,\ \int_\Omega \operatorname{Def} \mathbf{v} \cdot \mathbf{s}_l \, dV = 1 \right\},$$

where $V$ is the closure of $U$ in $\mathbf{W}_2^1(\Omega)$. The computational aspect of the limit analysis will be discussed in Chapter IX.

## 2. Pressure field restoration

Every weak self-equilibrated deviatoric stress field can be supplemented with a pressure field in order to obtain a self-equilibrated stress field. This proposition played an important role in the previous section and will now be proved first for the case when the whole body boundary is fixed and then for the case when a part of the boundary is free. In the first case, the proposition arises immediately from the theorems about restoring a distribution from its derivatives (we present formulations of these theorems and related concepts of the distribution theory in Subsection 2.2 and Appendix B). In the second case, the proposition is based on the possibility of appropriate approximating of solenoidal vector fields. We consider the approximations in Section 4; the condition for approximability turns out to be nonrestrictive and is normally satisfied in continuum mechanics problems.

**2.1. Regularity of body boundary.** We now introduce certain concepts required to describe the body boundary. We will use these concepts to accurately formulate conditions for solvability of the pressure field restoration problem.

First, it should be recalled that function $f$ on domain $\mathcal{O}$ in $\mathbf{R}^n$ satisfies the the *Lipschitz condition* in $\mathcal{O}$ if there is the number $L$ such that

$$|f(\mathbf{x}_1) - f(\mathbf{x}_2)| \le L\,|\mathbf{x}_1 - \mathbf{x}_2|$$

for every $x_1$, $x_2$ in $\mathcal{O}$; in this case $f$ is referred to as the *Lipschitz function* (on $\mathcal{O}$). In particular, $f$ is a Lipschitz function if the closure $\bar{\mathcal{O}} = \mathcal{O} \cup \partial\mathcal{O}$ lies in the domain $\mathcal{O}_1$ on which $f$ is continuously differentiable.

The continuous differentiability of $f$ on the open domain $\mathcal{O}$ is not sufficient for $f$ to be a Lipschitz function on $\mathcal{O}$. At the same time, the differentiability is not a necessary condition for $f$ to be a Lipschitz function. Figure 2.1 shows

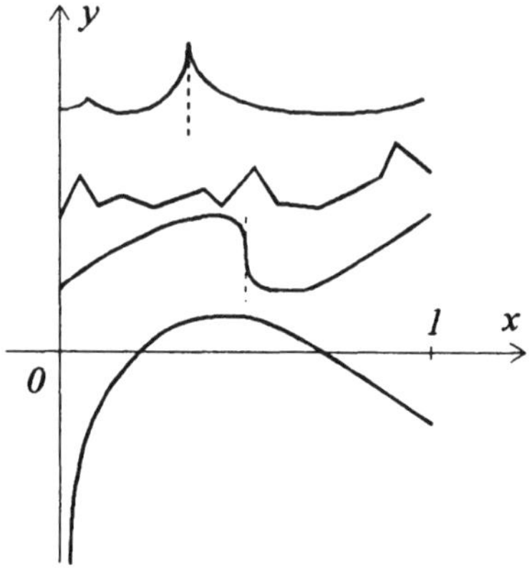


**Figure 2.1**

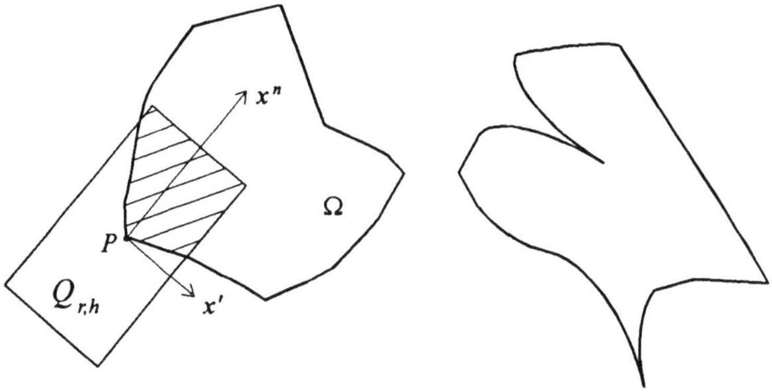


**Figure 2.2**

three graphs of functions which do not satisfy the Lipschitz condition and one graph of a Lipschitz function.

We consider domains in $\mathbf{R}^n$ whose boundary is locally represented by the equation $x_n = f(x_1, \ldots, x_{n-1})$, where $f$ is a Lipschitz function. We now give a more accurate formulation of this property using local coordinate systems. Namely, let $\boldsymbol{\xi}^0 = (\xi_1^0, \ldots, \xi_n^0)$ be a point in $\mathbf{R}^n$ and $R_{ij}$ be the components of an orthogonal matrix. We assign to every point $(\xi_1^0, \ldots, \xi_n^0)$ in $\mathbf{R}^n$ the collection of numbers $x_1, \ldots x_n$, where $x_i = R_{ij}(\xi_j - \xi_j^0)$, and we say that $x_1, \ldots x_n$ is a Cartesian coordinate system with its origin at $\boldsymbol{\xi}^0$. We use the notation $x' = (x_1, \ldots, x_{n-1})$, and consider a cylinder $Q_{r,h}$ in $\mathbf{R}^n$ specified by the conditions $|x'| < r$, $|x_n| \leq h$. Of course, the cylinder as well as the coordinate system $x_1, \ldots, x_n$ depends on the point $\boldsymbol{\xi}^0$ and on the matrix $R$.

Let $\Omega$ be a bounded domain in $\mathbf{R}^n$. We assume that there exist numbers $r > 0$ and $h > 0$ such that for every point $P$ in $\partial\Omega$ there is a Cartesian coordinate system $x_1, \ldots, x_n$ with its origin at $P$ such that the set $\Omega \cap Q_{r,h}$ can be described by the conditions $|x'| < r$, $f_P(x') < x^n < h$, where $f_P$ is a Lipschitz function in the domain $|x'| < r$ in $\mathbf{R}^{n-1}$, the constant $L$ in the Lipschitz condition for $f_P$ being less than a certain number $L_0 < h/r$ ($L_0$ does not depend on $P$). We refer to $\Omega$ possessing this property as the *Lipschitz domain*; Figure 2.2a illustrates this definition. If the domain $\Omega$ is bounded by a finite number of continuously differentiable surfaces and the surfaces intersect at nonzero angles, then $\Omega$ is a Lipschitz domain. In case some parts of the domain boundary are tangent to each other, the domain is not Lipschitzian (see Figure 2.2b), and neither is a circle with a removed segment. Normally, subjects of limit analysis are bodies which occupy Lipschitz domains; an important exclusion to the rule are bodies with cracks.

**2.2. Distribution restoration.** The pressure field restoration problem is closely related to the following problem. Let $X_1, \ldots, X_n$ be given distributions on domain $\Omega$ (see Appendix A); what are the conditions under which there exists a distribution such that $X_1 = \partial p/\partial x_1, \ldots, X_n = \partial p/\partial x_n$? What can be said about regularity of $p$, if $X_1, \ldots, X_n$ are regular in a certain sense? We now answer the first question and, partially, the second one.

The system of equations $\partial p/\partial x_i = X_i$, $i = 1, \ldots, n$, with the unknown $p$ has a solution provided $X_i$ satisfy certain compatibility conditions. In case $X_i$ are sufficiently smooth functions, these are well known conditions: the system is solvable in a simply connected domain $\Omega$ if and only if

$$\frac{\partial X_i}{\partial x_j} = \frac{\partial X_j}{\partial x_i}, \qquad i, j = 1, \ldots, n. \tag{2.1}$$

In case $\Omega$ is not simply connected, (2.1) should be supplemented with some conditions which are not local in contrast to (2.1). Conditions of both types admit a unified formulation which holds not only when $X_1, \ldots, X_n$ are regular functions but also when they are distributions. Namely, the following well known proposition is valid.

THEOREM 1. Let $X_1, \ldots, X_n$ be distributions on domain $\Omega$ in $\mathbf{R}^n$. The system of equations $\partial p/\partial x_1 = X_1, \ldots, \partial p/\partial x_n = X_n$ has the solution $p$ if and only if for every collection $\varphi_1, \ldots, \varphi_n$ of test functions satisfying the condition

$$\frac{\partial \varphi_1}{\partial x_1} + \cdots + \frac{\partial \varphi_n}{\partial x_n} = 0$$

the equality $\langle X_1, \varphi_1 \rangle + \cdots + \langle X_n, \varphi_n \rangle = 0$ is valid. The solution $p$ is unique up to the addition of a constant.

It is easily seen that the condition given by this theorem can be simply reduced to (2.1) in case the domain $\Omega$ is simply connected. In case $\Omega$ is not simply connected, the condition is not as constructive as (2.1); however, it is quite suitable for our purposes.

If the derivatives $\partial p/\partial x_1 = X_1, \ldots, \partial p/\partial x_n = X_n$ have some regularity properties, it is possible to conclude that $p$ is a regular function and even to estimate its norm. The following proposition formulates one of the results of this type in terms of Sobolev space $W_2^1(\Omega)$ (see Subsection B.1).

THEOREM 2. Let $X_1, \ldots, X_n, p$ be distributions on a bounded Lipschitz domain in $\mathbf{R}^n$ and

$$\frac{\partial p}{\partial x_1} = X_1, \ldots, \frac{\partial p}{\partial x_n} = X_n.$$

Let the estimations

$$|\langle X_1, \varphi \rangle| \leq c_1 \|\varphi\|_{W_2^1(\Omega)}, \ldots, |\langle X_n, \varphi \rangle| \leq c_n \|\varphi\|_{W_2^1(\Omega)},$$

where $c_1, \ldots, c_n$ does not depend on $\varphi$, be valid for every test function $\varphi$. Then the distribution $p$ is a regular function in $L_2(\Omega)$, and there exists a constant $p_0$ such that $\|p - p_0\|_{L_2(\Omega)} \leq c(c_1 + \cdots + c_n)$, where $c$ only depends on $\Omega$.

**2.3. Pressure field in a body with fixed boundary.** Consider the pressure field restoration problem in case the whole boundary of a body is fixed. In this case, the set $U$ involved in the virtual work principle consists of all smooth vector fields on $\Omega$ that vanish on $\partial\Omega$ ($\Omega$ is the domain occupied by the body). Recall that the weak equilibrium conditions (Subsection 1.1) make use of the subspace

$$U_d = \{\mathbf{v} \in U : \operatorname{div} \mathbf{v} = 0\}$$

instead of $U$.

THEOREM 3. Let $\Omega$ be a bounded Lipschitz domain. If a deviatoric stress field $\boldsymbol{\tau} \in L_2(\Omega; Sym^d)$ is weak self-equilibrated, that is,

$$\int_\Omega \operatorname{Def} \mathbf{v} \cdot \boldsymbol{\tau} \, dV = 0 \quad \text{for every} \quad \mathbf{v} \in U_d, \tag{2.2}$$

then there exists a function (pressure field) $p \in L_2(\Omega)$ such that the stress field $\boldsymbol{\sigma} = \boldsymbol{\tau} - p\mathbf{I}$ is self-equilibrated, that is,

$$\int_\Omega \operatorname{Def} \mathbf{v} \cdot \boldsymbol{\sigma} \, dV = 0 \quad \text{for every} \quad \mathbf{v} \in U.$$

The function $p$ is defined up to the addition of a constant, and there exists a number $p_0$ such that

$$\|p - p_0\|_{L_2(\Omega)} \leq c \, \|\boldsymbol{\tau}\|_{L_2(\Omega; Sym^d)}, \tag{2.3}$$

where $c$ only depends on $\Omega$.

PROOF. Consider a collection of test functions $v_1, \ldots, v_n$ on $\Omega$ which satisfy the condition

$$\frac{\partial v_1}{\partial x_1} + \cdots + \frac{\partial v_n}{\partial x_n} = 0.$$

They can be taken as the components of a virtual velocity field $\mathbf{v}$ in $U_d$ (which vanishes not only on $\partial\Omega$ but also in its neighborhood). Then (2.2) can be written in the form $\langle \partial\tau_{ij}/\partial x_j, v_i \rangle = 0$, where $\partial\tau_{ij}/\partial x_k$ are the derivatives in the sense of the distributions theory, and Theorem 1 implies that there exists the distribution $p$ that solves the system

$$\frac{\partial p}{\partial x_i} = \frac{\partial \tau_{ij}}{\partial x_j}, \tag{2.4}$$

the solution being defined up to the addition of a constant. The following estimation is obviously valid for the right-hand side in (2.4), that is, for the distributions $X_i = \partial\tau_{ij}/\partial x_j$:

$$\begin{aligned} |\langle X_i, \varphi \rangle| &= \left| \langle \tau_{ij}, \frac{\partial \varphi}{\partial x_j} \rangle \right| = \left| \int\limits_{\Omega} \tau_{ij} \frac{\partial \varphi}{\partial x_j}\, dV \right| \\ &\leq \left( \int\limits_{\Omega} |\tau|^2\, dV \right)^{\frac{1}{2}} \left( \int\limits_{\Omega} \left[ \left( \frac{\partial \varphi}{\partial x_1} \right)^2 + \cdots + \left( \frac{\partial \varphi}{\partial x_n} \right)^2 \right] dV \right)^{\frac{1}{2}} \\ &\leq \|\tau\|_{L_2(\Omega;Sym)} \|\varphi\|_{W_2^1(\Omega)}, \end{aligned}$$

where $\varphi$ is any test function. Then, by Theorem 2, the distribution $p$ is actually a regular function in $L_2(\Omega)$ and the estimation (2.3) is valid with a certain constant $p_0$.

Consider now the stress field $\boldsymbol{\sigma} = \boldsymbol{\tau} - p\mathbf{I}$. According to (2.4), the components $\sigma_{ij}$ satisfy the equations $\partial\sigma_{ij}/\partial x_j = 0$ in the sense of the distributions theory. This means that, for every collection $v_i$, $i = 1, \ldots, n$, of test functions (which vanish in a certain neighborhood of $\partial\Omega$), we have $\langle \sigma_{ij}, \partial v_i/\partial x_j \rangle = 0$ or, in more detail,

$$\int\limits_{\Omega} \sigma_{ij} \frac{\partial v_i}{\partial x_j}\, dV = 0.$$

Since the set of smooth vector fields vanishing near $\partial\Omega$ is dense in $\overset{\circ}{\mathbf{W}}{}_2^1(\Omega)$ (see Corollary B.1), the previous equality also holds for every $\mathbf{v} \in \overset{\circ}{\mathbf{W}}{}_2^1(\Omega)$, in particular, for every $\mathbf{v} \in U$, which finishes the proof.

REMARK 1. The theorem also holds if (2.2) is replaced by the condition

$$\int\limits_{\Omega} \operatorname{Def} \mathbf{v} \cdot \boldsymbol{\tau}\, dV = 0 \quad \text{for every} \quad \mathbf{v} \in \mathcal{U}_d,$$

where $\mathcal{U}_d$ is a subspace in $U_d$ and $\mathbf{u}$ belongs to $\mathcal{U}_d$ provided $\mathbf{u}$ vanishes not only on $\partial\Omega$ but also in a neighborhood of $\partial\Omega$. Actually, we only made use of this property of virtual velocity fields while proving the above proposition.

**2.4. Pressure field in a body with fixed part of boundary.** Consider the pressure field restoration problem in case only the part $S_v$ of the body boundary is fixed. In this case, the set $U$ involved in the virtual work principle consists of all smooth vector fields on $\Omega$, which vanish on $S_v$ ($\Omega$ is the domain occupied by the body). Recall that the weak equilibrium conditions (Subsection 1.1) make use of the subspace

$$U_d = \{\mathbf{v} \in U : \operatorname{div} \mathbf{v} = 0\}$$

instead of $U$.

We now start working with vector fields in Sobolev space $\mathbf{W}_2^1(\Omega)$ (see Subsection B.1), in which we will approximate solenoidal vector fields vanishing near $S_v$. The latter condition means that, for a vector field $\mathbf{v}$, there is a number $\delta > 0$ such that $\mathbf{v}$ vanishes at every point $x$ in $\Omega$ with the distance between $x$ and $\partial\Omega$ less than $\delta$. We will establish sufficient conditions for the approximation of solenoidal vector fields (Theorems 4.1 and 4.3). These are the very conditions which determine constructive formulations of the assumptions under which the pressure field restoration problem has a solution. We now only mention that these assumptions are satisfied if 1) $\partial\Omega$ consists of a finite number of continuously differentiable surfaces which intersect at nonzero angles, and 2) the free part $S_q = \partial\Omega \setminus S_v$ of the boundary is bounded with a finite number of continuously differentiable curves which intersect at nonzero angles.

We have already considered the case of the fixed body boundary in Theorem 3. We assume in the following proposition therefore that the free part $S_q = \partial\Omega \setminus S_v$ of the boundary has a positive measure in $\partial\Omega$ (in three-dimensional case, this means $S_q$ has a positive area).

THEOREM 4. Suppose 1) a body occupies a bounded Lipschitz domain $\Omega$ in $\mathbf{R}^n$ and the free part $S_q$ of its boundary has a positive measure in $\partial\Omega$, and 2) every solenoidal vector field $\mathbf{v} \in \mathbf{W}_2^1(\Omega)$ with $\mathbf{v}|_{S_v} = 0$ can be approximated in $\mathbf{W}_2^1(\Omega)$ by smooth solenoidal fields each of which vanishes near $S_v$. Then, if a deviatoric stress field $\boldsymbol{\tau} \in L_2(\Omega; Sym^d)$ is weak self-equilibrated, that is,

$$\int_\Omega \operatorname{Def} \mathbf{v} \cdot \boldsymbol{\tau} \, dV = 0 \quad \text{for every} \quad \mathbf{v} \in U_d,$$

then there exists the function (pressure field) $p \in L_2(\Omega)$ such that the stress field $\boldsymbol{\sigma} = \boldsymbol{\tau} - p\mathbf{I}$ is self-equilibrated, that is,

$$\int_\Omega \operatorname{Def} \mathbf{v} \cdot \boldsymbol{\sigma} \, dV = 0 \quad \text{for every} \quad \mathbf{v} \in U.$$

This function $p$ is unique and the following estimation is valid:

$$\|p\|_{L_2(\Omega)} \leq c \, \|\boldsymbol{\tau}\|_{L_2(\Omega; Sym^d)},$$

where the constant $c$ does not depend on $\boldsymbol{\tau}$.

PROOF. For the given $\boldsymbol{\tau}$, the equality

$$\int_\Omega \operatorname{Def} \mathbf{v} \cdot \boldsymbol{\tau} \, dV = 0$$

is valid if $\mathbf{v}$ is in $U_d$, in particular, if $\mathbf{v} \in U_d$ and vanishes not only on $S_v$, but also near the whole boundary $\partial\Omega$. Then, by Theorem 3 and Remark 1, there is a function $p_* \in L_2(\Omega)$ such that the stress field $\boldsymbol{\sigma} = \boldsymbol{\tau} - p_*\mathbf{I}$ satisfies the condition

$$\int_\Omega \operatorname{Def} \mathbf{v} \cdot \boldsymbol{\sigma}_* \, dV = 0 \tag{2.5}$$

for every smooth vector field $\mathbf{v}$ with $\mathbf{v}|_{\partial\Omega} = 0$. Now let us verify that (2.5) also holds for every virtual velocity field $\mathbf{v}$ (which means that $\mathbf{v}$ vanishes on $S_v$ and not necessarily on the whole of $\partial\Omega$) if $\mathbf{v}$ meets the additional requirement

$$\int_\Omega \operatorname{div} \mathbf{v} \, dV = 0. \tag{2.6}$$

Indeed, in this case there exists the vector field $\mathbf{v}_0$ which possesses the following properties:

$$\mathbf{v}_0 \in \mathbf{W}_2^1(\Omega), \quad \mathbf{v}_0|_{\partial\Omega} = 0, \quad \operatorname{div} \mathbf{v}_0 = \operatorname{div} \mathbf{v} \tag{2.7}$$

(see Theorem 4.2 below). Let us consider the integral

$$\int_\Omega \operatorname{Def} \mathbf{v} \cdot \boldsymbol{\sigma}_* \, dV = \int_\Omega \operatorname{Def} \mathbf{v}_0 \cdot \boldsymbol{\sigma}_* \, dV + \int_\Omega \operatorname{Def}(\mathbf{v} - \mathbf{v}_0) \cdot \boldsymbol{\sigma}_* \, dV. \tag{2.8}$$

Here, $\mathbf{v}_0$ can be approximated in $\mathbf{W}_2^1(\Omega)$ by smooth vector fields which vanish near $\partial\Omega$ (see Corollary B.1 below). Therefore, due to (2.5), the first integral on the right-hand side in (2.8) is zero. Analogously, the field $\mathbf{v} - \mathbf{v}_0$ possesses the following properties:

$$\mathbf{v} - \mathbf{v}_0 \in \mathbf{W}_2^1(\Omega), \quad (\mathbf{v} - \mathbf{v}_0)|_{S_v} = 0, \quad \operatorname{div}(\mathbf{v} - \mathbf{v}_0) = 0,$$

and, therefore, $\mathbf{v} - \mathbf{v}_0$ can be approximated in $\mathbf{W}_2^1(\Omega)$ by smooth solenoidal vector fields $\mathbf{w}_n \in U_d$, $n = 1, 2, \ldots$, which vanish near $S_v$: $\mathbf{w}_n \to \mathbf{v} - \mathbf{v}_0$ when $n \to \infty$. Since $\operatorname{div} \mathbf{w}_n = 0$, we have

$$\int_\Omega \operatorname{Def} \mathbf{w}_n \cdot \boldsymbol{\sigma}_* \, dV = \int_\Omega \operatorname{Def} \mathbf{w}_n \cdot \boldsymbol{\tau} \, dV,$$

and the right-hand side integral is zero (since $\boldsymbol{\tau}$ is weak self-equilibrated by assumption). Passing to the limit in the equality

$$\int_\Omega \operatorname{Def} \mathbf{w}_n \cdot \boldsymbol{\sigma}_* \, dV = 0,$$

we find out that the second integral on the right-hand side in (2.8) is zero, and we have already seen that the first one is zero. Therefore, (2.5) is valid for every $\mathbf{v} \in U$ which satisfies condition (2.6).

Let us show that we can get rid of condition (2.6) in the previous statement: adding a certain constant $c_*$ to $p_*$ we arrive to the stress field $\boldsymbol{\sigma} = \boldsymbol{\tau} - (p_* + c_*)\mathbf{I}$ which is self-equilibrated, that is,

$$\int_\Omega \operatorname{Def} \mathbf{v} \cdot \boldsymbol{\sigma}\, dV = 0 \quad \text{for every} \quad \mathbf{v} \in U.$$

Indeed, consider the field $\mathbf{v}_1 \in U$ such that

$$\int_\Omega \operatorname{div} \mathbf{v}_1\, dV = 1.$$

The existence of such a field follows immediately from the Stokes formula,

$$\int_\Omega \operatorname{div} \mathbf{v}_1\, dV = \int_{\partial\Omega} \mathbf{v}_1 \boldsymbol{\nu}\, dS,$$

and the fact that $S_q$ has a positive measure in $\partial\Omega$ allows us to make the latter integral equal to 1 by choosing the suitable $\mathbf{v}_1|_{S_q}$. Then consider the number

$$c_* = \int_\Omega \operatorname{Def} \mathbf{v}_1 \cdot \boldsymbol{\sigma}_*\, dV \tag{2.9}$$

and the stress field $\boldsymbol{\sigma} = \boldsymbol{\sigma}_* - c_*\mathbf{I} = \boldsymbol{\tau} - (p_* + c_*)\mathbf{I}$. By formula (2.9), we have

$$\int_\Omega \operatorname{Def} \mathbf{u} \cdot c_*\mathbf{I}\, dV = \int_\Omega \operatorname{Def} \mathbf{v}_1 \cdot \boldsymbol{\sigma}_*\, dV \int_\Omega \operatorname{div} \mathbf{u}\, dV = \int_\Omega \operatorname{Def} \left( \mathbf{v}_1 \int_\Omega \operatorname{div} \mathbf{u}\, dV \right) \cdot \boldsymbol{\sigma}_*\, dV$$

for every sufficiently regular $\mathbf{u}$. Using this equality, we obtain

$$\int_\Omega \operatorname{Def} \mathbf{u} \cdot \boldsymbol{\sigma}\, dV = \int_\Omega \operatorname{Def} \mathbf{u} \cdot \boldsymbol{\sigma}_*\, dV - \int_\Omega \operatorname{Def} \mathbf{u} \cdot c_*\mathbf{I}\, dV =$$

$$= \int_\Omega \operatorname{Def} \left( \mathbf{u} - \mathbf{v}_1 \int_\Omega \operatorname{div} \mathbf{u}\, dV \right) \cdot \boldsymbol{\sigma}_*\, dV. \tag{2.10}$$

It is clear that the field

$$\mathbf{v} = \mathbf{u} - \mathbf{v}_1 \int_\Omega \operatorname{div} \mathbf{u}\, dV \in U$$

satisfies condition (2.6). Therefore (2.5) is valid and integral (2.10) is zero. Thus, $\boldsymbol{\sigma}$ is a self-equilibrated stress field, and we obtained it by adding the spheric part $-(p_* + c_*)\mathbf{I}$ to the given deviatoric stress field $\boldsymbol{\tau}$, that is, $p_* + c_*$ is a solution to the pressure field restoration problem.

It remains to verify the uniqueness of the restored pressure field and estimate its norm. Note that the function $p$, which results in a self-equilibrated stress field $\boldsymbol{\tau} - p\mathbf{I}$, is defined up to the addition of a constant (Theorem 3). Therefore, if $\boldsymbol{\sigma}_1 = \boldsymbol{\tau} - p_1\mathbf{I}$ and $\boldsymbol{\sigma}_2 = \boldsymbol{\tau} - p_2\mathbf{I}$ are self-equilibrated stress fields, then $\boldsymbol{\sigma}_1 - \boldsymbol{\sigma}_2$ equals $P\mathbf{I}$ with a certain constant $P$. Since $\boldsymbol{\sigma}_1 - \boldsymbol{\sigma}_2$ is a self-equilibrated stress field, we have for every $\mathbf{v} \in U$:

$$0 = \int_\Omega \operatorname{Def}\mathbf{v} \cdot (\boldsymbol{\sigma}_1 - \boldsymbol{\sigma}_2)\, dV = P \int_\Omega \operatorname{div}\mathbf{v}\, dV = P \int_{S_q} \mathbf{v}\boldsymbol{\nu}\, dS,$$

which implies $P = 0$. Consequently, for a given $\boldsymbol{\tau}$, there exists a unique self-equilibrated stress field of the form $\boldsymbol{\tau} - p\mathbf{I}$. To estimate the norm of $p$, note that the inequality $\|p_*\|_{L_2(\Omega)} \leq \|\boldsymbol{\tau}\|_{L_2(\Omega;Sym^d)}$ is valid due to Theorem 3. Therefore, analogous estimations hold for $|\boldsymbol{\sigma}_*| = |\boldsymbol{\tau} - p_*\mathbf{I}|$, then, for the constant $c_*$ due to its definition (2.9), and, finally, for $p = p_* + c_*$, which finishes the proof.

REMARK 2. The theorem remains valid if we replace $U_d$ in the condition

$$\int_\Omega \operatorname{Def}\mathbf{v} \cdot \boldsymbol{\tau}\, dV = 0 \quad \text{for every} \quad \mathbf{v} \in U_d$$

by the subspace $\mathcal{U}_d$ of those fields in $U_d$ which vanish not only on the surface $S_v$ but also near it. Actually we only made use of the condition

$$\int_\Omega \operatorname{Def}\mathbf{v} \cdot \boldsymbol{\tau}\, dV = 0 \quad \text{for every} \quad \mathbf{v} \in \mathcal{U}_d$$

while proving the above proposition.

## 3. Approximations to vector fields

A kinematically admissible velocity field $\mathbf{v}$ vanishes on the fixed part $S_v$ of the body boundary. In this section, we give conditions under which $\mathbf{v}$ can be approximated by smooth vector fields vanishing *near* $S_v$. Based on this result, we show that a solenoidal vector field can be analogously approximated by smooth solenoidal fields vanishing near $S_v$. We have already used this approximation in Section 2, and it also plays an important role in formulating numerical methods, in particular, for limit analysis (Chapter IX).

We approximate vector fields in Sobolev space $\mathbf{W}_2^1(\Omega)$; see Appendix B for definition and some properties of this space.

**3.1. Regularity of the free part of the body boundary.** Consider the domain $\Omega$ occupied by a body and let its boundary be divided into non-intersecting parts $S_v$ and $S_q$. We now formulate an assumption under which we will study approximations of a vector field on $\Omega$ vanishing on $S_v$.

Let $\Omega$ be a bounded Lipschitz domain in $\mathbf{R}^n$ (see Subsection 2.1), $P$ be a point in $\partial\Omega$, $x_1, \ldots, x_n$ be a Cartesian coordinate system with its origin at $P$, and $x'$ stands for $(x_1, \ldots, x_{n-1})$. Consider a cylinder

$$Q_{r,h} = \{x \in \mathbf{R}^n : |x'| < r,\ |x_n| < h\}$$

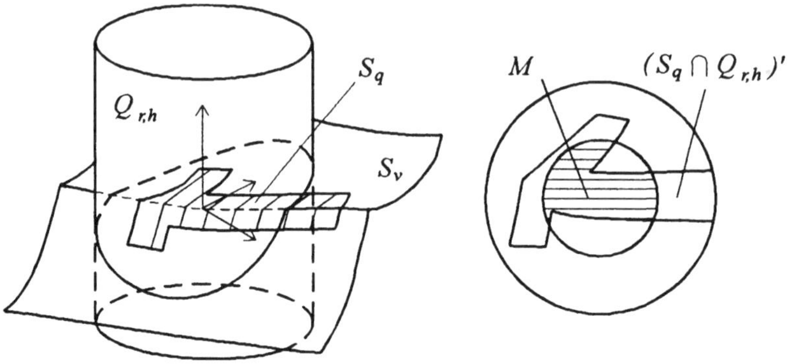


**Figure 3.1**

centered at $P$. Due to the definition of Lipschitz domain, it is possible to describe the set $\Omega \cap Q_{r,h}$ by the relations $|x'| < r$, $f_P(x') < x_n < h$, where $f$ is a Lipschitz function. Collection of cylinders $Q_{r,h}$ constructed for all points of the boundary covers $\partial\Omega$; we refer to this collection as $Q_{r,h}$-covering for $\partial\Omega$. Let us also consider the cylinder

$$Q_{r/2,h} = \{x \in \mathbf{R}^n : |x'| < r/2, \ |x_n| < h\}$$

for every $P \in \partial\Omega$. Collection of these cylinders is a $Q_{r/2,h}$-covering for $\partial\Omega$. We extract a finite number of cylinders from this covering so that the selected cylinders also form a covering for $\partial\Omega$; we refer to it as finite $Q_{r/2,h}$-covering for $\partial\Omega$. A finite $Q_{r/2,h}$-covering always exists since $\partial\Omega$ is a bounded closed set in $\mathbf{R}^n$. We emphasize that $\partial\Omega$ is represented by the equation $x_n = f_P(x')$ not only within the cylinder centered at $P$ of the radius $r/2$, but also within a larger cylinder of the radius $r$ centered at the same point.

Consider a set $S_q \cap Q_{r,h}$ in each of the cylinders $Q_{r,h}$. If $S_q \cap Q_{r,h}$ is not empty, consider its projection $(S_q \cap Q_{r,h})'$ on the subspace $x_n = 0$ and the intersection $M$ of this projection with the ball $|x'| < r/2$ in the subspace $\mathbf{R}^{n-1}$

$$M = \{x' \in \mathbf{R}^{n-1} : x' \in (S_q \cap Q_{r,h})', \ |x'| < r/2\}. \tag{3.1}$$

(see Figure 3.1). If there exists a finite $Q_{r,h/2}$-covering for $\partial\Omega$ such that in each of the larger cylinders $Q_{r,h}$ set (3.1) is a Lipschitz domain or is empty, we say that $S_q$ is a *regular part* of the boundary $\partial\Omega$.

In applications, one often considers domains bounded by a finite number of continuously differentiable surfaces which intersect at nonzero angles; these domains are Lipschitzian. If $\Omega$ is such a domain and $S_q$ is a part of $\partial\Omega$ bounded by a finite number of continuously differentiable curves which intersect at nonzero angles, then $S_q$ is a regular part of $\partial\Omega$.

**3.2. Conditions for approximation.** We now establish conditions under which every sufficiently regular kinematically admissible velocity field $\mathbf{v}$ on $\Omega$ can be approximated by smooth fields vanishing near $S_v$. We refer to the function $f$ as *smooth* on $\bar{\Omega}$ if $f$ is infinitely differentiable on $\Omega$ and can be extended to the whole space as such a function. We denote the set of smooth functions on $\bar{\Omega}$ by the symbol $C^\infty(\bar{\Omega})$. We say that the function $f$ vanishes *near* $S_v$ if there is a number $\delta > 0$ such that $f(x) = 0$ for every $x$ in $\Omega$ whose distance from $S_v$ is less than $\delta$.

We now formulate and prove an approximation theorem for functions on $\partial\Omega$. The theorem also holds for vector-functions, in which case, $W_2^1(\Omega)$ is to be replaced by the space $\mathbf{W}_2^1(\Omega) = (W_2^1(\Omega))^3$ of vector functions on $\Omega$, see Example B.1.

THEOREM 1. Suppose $\Omega$ is a bounded Lipschitz domain and $S_q$ is a regular part of $\partial\Omega$. Then for every function $v \in W_2^1(\Omega)$ with zero trace on $S_v$, $v|_{S_v} = 0$, there is a sequence of smooth functions $v_1, v_2, \ldots$ on $\bar{\Omega}$ which converges in $v$ to $W_2^1(\Omega)$, every $v_n$ vanishing near $S_v$.

PROOF. Consider a finite $Q_{r/2,h}$-covering for $\partial\Omega$ introduced in Subsection 3.1. We enumerate cylinders in this covering denoting them by $Q_{*i}$, $i = 1, \ldots, N$. It is easily seen that there is a number $\delta > 0$ such that $Q_{*1}, \ldots, Q_{*N}$ cover not only $\partial\Omega$ but also a set of all points in $\Omega$ whose distance from $\partial\Omega$ is less than $\delta$. We cover the remaining part of $\Omega$ with an additional finite number of domains $Q_{*i}$, $i = N+1, \ldots, M$. According to the previous remark, we choose these sets so that the distance between $Q_{*i}$, $i = N+1, \ldots, M$, and $\partial\Omega$ is positive. Let us now choose for every $i = 1, \ldots, M$ a smooth function $\alpha_i \geq 0$ such that $\alpha_i$ vanishes in the exterior of the closed set $\operatorname{supp}\alpha_i \subset Q_{*i}$, whose distance from $\partial Q_{*i}$ is positive and the sum $\alpha_1 + \cdots + \alpha_M$ takes the value 1 at every point in $\Omega$. We introduce the functions $w_i = \alpha_i v$ and write $v$ as $v = w_1 + \cdots + w_M$ on $\Omega$. To prove the theorem it is sufficient to show that, for every $\varepsilon > 0$, $w_i$ can be approximated by the function $w_{i,\varepsilon} \in W_2^1(\Omega)$ such that 1) $\|w_i - w_{i,\varepsilon}\|_{W_2^1(\Omega)} < c_i\varepsilon$, where the constant $c_i$ does not depend on $w_i$, $w_{i,\varepsilon}$ and $\varepsilon$, and 2) the trace $w_{i,\varepsilon}|_{\partial\Omega}$ vanishes in a certain neighborhood $U_{i,\varepsilon}$ of $\bar{S}_v$ in $\partial\Omega$ (the symbol $\bar{A}S$ stands for the closure of $A$). Indeed, it is clear that the sum $w_{1,\varepsilon} + \cdots + w_{M,\varepsilon}$ approximates the function $w_1 + \cdots + w_M = v$ on $\Omega$:

$$\|v - w_{1,\varepsilon} - \cdots - w_{M,\varepsilon}\|_{W_2^1(\Omega)} < (c_1 + \cdots + c_M)\varepsilon \tag{3.2}$$

and vanishes in the neighborhood $U_{1,\varepsilon} \cap \cdots \cap U_{M,\varepsilon}$ of $\bar{S}_v$ in $\partial\Omega$. The latter property allows approximation of the function $w_{1,\varepsilon} + \cdots + w_{M,\varepsilon}$ by smooth function $v_\varepsilon$ such that

$$\|w_{1,\varepsilon} + \cdots + w_{M,\varepsilon} - v_\varepsilon\|_{W_2^1(\Omega)} < \varepsilon$$

and $v_\varepsilon$ vanishes in a neighborhood of $\bar{S}_v$ in $\mathbf{R}^n$ (see Corollary B.1). Together with inequality (3.2), this means that $\mathbf{v}_\varepsilon$ is a suitable approximation to $v$, and the theorem is valid.

To finish the proof, it remains to construct functions $w_{i,\varepsilon}$ which possess the above-mentioned properties. If $\operatorname{supp}\alpha_i$ does not intersect $\bar{S}_v$ (which is the case at least for $i = N+1, \ldots, M$), we take the very function $w_i = \alpha_i v$ as $w_{i,\varepsilon}$, which is possible since $w_i$ vanishes in the exterior of $\operatorname{supp}\alpha_i$ and, hence, in a certain neighborhood of $\bar{S}_v$ in $\partial\Omega$. If $\operatorname{supp}\alpha_i$ intersects $\bar{S}_v$ and does not intersect $\bar{S}_q$, then $\alpha_i v|_{\partial\Omega} = 0$ since $\alpha_i v|_{S_v} = 0$ together with $v|_{S_v} = 0$ and $\alpha_i v|_{S_q} = 0$ together with $\alpha_i|_{S_q} = 0$. Therefore, we again take $w_i = \alpha_i v$ as $w_{i,\varepsilon}$.

Let us now construct a suitable function $w_{i,\varepsilon}$, in case $\operatorname{supp}\alpha_i$ intersects both $\bar{S}_v$ and $\bar{S}_q$. Consider the cylinder $Q_{*i}$ in the $Q_{r/2,h}$-covering for $\partial\Omega$ chosen at the beginning of the proof. Let $x_1, \dots, x_n$ be the corresponding local coordinate system and

$$Q_i = \{x \in \mathbf{R}^n : |x'| < r, \ |x_n| < h\}$$

be the larger cylinder within which $\partial\Omega$ is represented by the equation $x_n = f_i(x')$, where $f$ is a Lipschitz function (see Subsection 2.1). The set $\Omega_i = \Omega \cap Q_i$ is described by the relations $|x'| < r$, $f_i(x') < x_n < h$. Of course, the local coordinate systems are different in different cylinders $Q_i$, although this fact is not indicated in the notation $x_1, \dots, x_n$. For brevity, we will also omit the index $i$ in the symbols $Q_i$, $\alpha_i$, $f_i$, $w_i$ and others.

Thus, it is required to approximate $w = \alpha v$ by the function $w_\varepsilon \in W_2^1(\Omega)$ such that 1) $\|w - w_\varepsilon\|_{W_2^1(\Omega)} < c\varepsilon$, where the constant $c$ does not depend on $w$, $w_\varepsilon$ and $\varepsilon$, and 2) the trace $w_\varepsilon|_{\partial\Omega}$ vanishes in a neighborhood of $\bar{S}_v$ in $\partial\Omega$. To construct $w_\varepsilon$, we use the mapping $F$ which maps the point $x \in \omega = \Omega \cap Q$ on the point

$$y = (y_1, \dots, y_n), \quad \text{where} \quad y' = x', \quad y_n = x_n - f(x')$$

(we use the notation $y' = (y_1, \dots, y_{n-1})$). Note that $F$ maps $\omega = \Omega \cap Q$ on a certain domain $\widetilde{\omega}$ and the part $\Gamma = \partial\Omega \cap Q$ of $\partial\Omega$ on the set

$$\widetilde{\Gamma} = \{y \in \mathbf{R}^n : |y'| < r, \ y_n = 0\},$$

see Figure 3.2.

It is obvious that $F$ is a one-to-one mapping and the inverse mapping $F^{-1}$ from $\widetilde{\omega}$ on $\omega$ maps a point $y \in \widetilde{\omega}$ on the point $x$: $x' = y'$, $x_n = y_n + f(y')$. This induces a correspondence between functions on $\omega$ and functions on $\widetilde{\omega}$. Namely, if $u$ is a function on $\omega$, the corresponding function $\widetilde{u}$ on $\widetilde{\omega}$ is determined as follows:

$$\widetilde{u}(y) = u(F^{-1}(y)) = u(y', y_n + f(y')).$$

Conversely, if $\widetilde{u}$ is a function on $\widetilde{\omega}$, the function

$$u(x) = \widetilde{u}(F(x)) = \widetilde{u}(x', x_n - f(x')) \tag{3.3}$$

on $\omega$ corresponds to $\widetilde{u}$. We use this correspondence to construct a suitable approximation $w_\varepsilon$ to $w$. The approximation has to improve $w$, whose trace $w|_{\partial\Omega}$ vanishes in $S_v$: the trace of the improved function is to vanish in a neighborhood of $\bar{S}_v$ in $\partial\Omega$. Actually, we will first construct an approximation to the function $\widetilde{w}$ on $\widetilde{\omega}$ and then consider the function on $\omega$ which corresponds

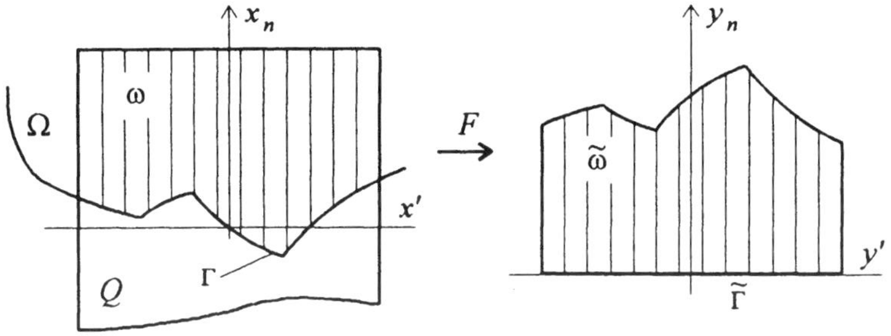


**Figure 3.2**

to the approximation. This function will be a suitable approximation to $w$, in particular, it will vanish near $S_v$.

Thus, we consider the function $\tilde{w} = \tilde{\alpha}\tilde{v}$. Since $w \in W_2^1(\omega)$ and the mappings $F$ and $F^{-1}$ are Lipschitz functions, $\tilde{w}$ is in $W_2^1(\tilde{\omega})$ (Theorem B.6). The trace $\tilde{w}|_{\tilde{\Gamma}}$ vanishes at $|y'| > r/2$ together with $\tilde{\alpha}$, which allows extension of $\tilde{w}|_{\tilde{\Gamma}}$ to all of $\mathbf{R}^{n-1}$, setting the extension equal zero in the exterior of $\tilde{\Gamma}$. We denote the extension by $t$; it belongs to $W_2^{\frac{1}{2}}(\mathbf{R}^{n-1})$ (see Subsection B.2) and takes the values

$$t(y',0) = \begin{cases} \tilde{\alpha}(y',0)\tilde{v}(y',0) & \text{if } y' \in \tilde{\Gamma}, \\ 0 & \text{if } y' \in \mathbf{R}_{n-1} \setminus \tilde{\Gamma}. \end{cases} \tag{3.4}$$

Recall that $\mathbf{v}|_{S_v} = 0$ and, hence, $t$ vanishes in the exterior of the set $F(S_q \cap Q)$ and also in the exterior of

$$G = \{y' \in \mathbf{R}_{n-1} : y' \in F(S_q \cap Q), |y'| < r/2\},$$

as $\tilde{\alpha}(y') = 0$ at $|y'| > r/2$. Since $S_q$ is a regular part of $\partial\Omega$, $G$ is a Lipschitz domain in $\mathbf{R}^{n-1}$. Then, for the function

$$t \in W_2^{\frac{1}{2}}(\mathbf{R}^{n-1}),$$

which vanishes in the exterior of $G$, there exists a function

$$t_\varepsilon \in C^\infty(\mathbf{R}^{n-1})$$

such that $\|t - t_\varepsilon\|_{W_2^{\frac{1}{2}}(\mathbf{R}^{n-1})} < \varepsilon$ and $t_\varepsilon$ vanishes in the exterior of a certain closed set $\operatorname{supp} t_\varepsilon$ (Theorem B.8), see Figure 3.3. This means that the function $t_\varepsilon$ improves the trace $\tilde{w}|_{\tilde{\Gamma}}$ as desired: $t_\varepsilon$ approximates $\tilde{w}|_{\tilde{\Gamma}}$ and vanishes in the exterior of a closed set which is in $F(S_q \cap Q)$.

We now return to the functions on $\omega$. We start with the function $t - t_\varepsilon \in W_2^{\frac{1}{2}}(\mathbf{R}^{n-1})$, which vanishes at $|x'| > r/2$:

$$t(x') - t_\varepsilon(x') = 0 \quad \text{at} \quad |x'| > r/2 \tag{3.5}$$

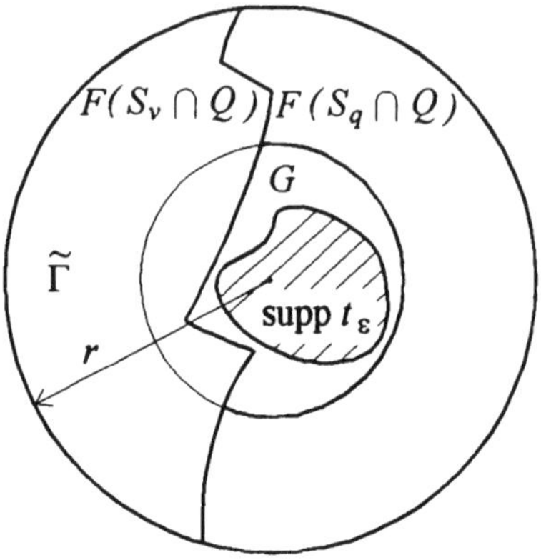


**Figure 3.3**

due to the relations $\operatorname{supp}\alpha \subset Q_*$, $\operatorname{supp} t_\varepsilon \subset G$ and (3.4). It follows from Theorem B.7 that there exists a function $\widetilde{u}_\varepsilon \in W_2^1(\widetilde{\omega})$ for which

$$\widetilde{u}_\varepsilon|_{\widetilde{\Gamma}} = (t - t_\varepsilon)|_{\widetilde{\Gamma}}, \quad \widetilde{u}_\varepsilon|_{\partial\widetilde{\omega}\setminus\widetilde{\Gamma}} = 0, \tag{3.6}$$

$$\|\widetilde{u}_\varepsilon\|_{W_2^1(\widetilde{\omega})} \leq C_1 \|t - t_\varepsilon\|_{W_2^{\frac{1}{2}}(\mathbf{R}^{n-1})} < C_1\varepsilon, \tag{3.7}$$

where the constant $C_1$ does not depend on $t$, $t_\varepsilon$ and $\varepsilon$. Consider then the function $u_\varepsilon$ on $\omega$ defined by (3.3). Due to (3.5) and the first of relations (3.6), the trace $u_\varepsilon|_{\partial\omega}$ may take nonzero values only on a part of $\partial\omega$, namely, on

$$\{x \in \Gamma : |x'| < r/2\}.$$

We now extend $u_\varepsilon$ from $\omega$ to all of $\Omega$, setting the extension equal to zero in the exterior of $\omega$. Due to the previous property of $u_\varepsilon|_{\partial\omega}$, this extension belongs to $W_2^1(\Omega)$, see Theorem B.4. We keep denoting the extension by the symbol $u_\varepsilon$.

Let us show that the function $w_\varepsilon = w - u_\varepsilon$ is the desired approximation to $w$. Indeed,

$$\|w - w_\varepsilon\|_{W_2^1(\Omega)} = \|u_\varepsilon\|_{W_2^1(\Omega)} = \|u_\varepsilon\|_{W_2^1(\omega)}$$

since $u_\varepsilon$ vanishes on $\Omega\setminus\omega$. Moreover, the inequality $\|u_\varepsilon\|_{W_2^1(\omega)} \leq C_2 \|\widetilde{u}_\varepsilon\|_{W_2^1(\widetilde{\omega})}$ is valid by Theorem B.6, where the constant $C_2$ does not depend on $\widetilde{u}_\varepsilon$. Together with (3.7) this results in the estimation $\|w - w_\varepsilon\|_{W_2^1(\Omega)} < C_1C_2\varepsilon$, that is, $w_\varepsilon$ approximates the function $w$ in $W_2^1(\Omega)$.

To finish the proof, it remains to show that the trace $w_\varepsilon|_{\partial\Omega}$ vanishes in a neighborhood of $\bar{S}_v$ in $\partial\Omega$. To verify this property, we consider separately the parts $\partial\Omega \setminus \Gamma$ and $\Gamma$ of the boundary $\partial\Omega$ (recall that $\Gamma = \partial\Omega \cap Q$). The equality

$$w_\varepsilon|_{\partial\Omega\setminus\Gamma} = 0 \tag{3.8}$$

immediately follows from the fact that the first term in the expression $w_\varepsilon = \alpha v - u_\varepsilon$ vanishes in the exterior of $\operatorname{supp}\alpha \subset Q$, and the second in the exterior of $\omega = \Omega \cap Q$. On the other hand, we have on $\Gamma$:

$$\begin{aligned} w_\varepsilon|_\Gamma(x', f(x')) &= \tilde{w}_\varepsilon|_{\tilde{\Gamma}}(x') = (\tilde{w} - \tilde{u}_\varepsilon)|_{\tilde{\Gamma}}(x') \\ &= \tilde{\alpha}(x', 0)\tilde{v}(x', 0) - t + t_\varepsilon|_{\tilde{\Gamma}}(x') = t_\varepsilon|_{\tilde{\Gamma}}(x'), \end{aligned} \tag{3.9}$$

where the last inequality follows from (3.4). Recall that $t_\varepsilon|_{\tilde{\Gamma}}$ vanishes in the exterior of the closed set

$$\operatorname{supp} t_\varepsilon \subset G \subset F(S_q \cap Q),$$

which is at a positive distance from the boundary of $G$ and, hence, at a positive distance from the boundary of $F(S_q \cap Q)$. Therefore, (3.9) implies that $w_\varepsilon|_\Gamma$ vanishes in the exterior of a certain set which lies in $S_q \cap Q$ at a positive distance from its boundary and, hence, at a positive distance from $S_v$. Together with (3.8) this means that $w_\varepsilon|_\Gamma$ vanishes in a neighborhood of $\bar{S}_v$ in $\partial\Omega$, which finishes the proof.

REMARK 1. If $\mathbf{v}$ is a vector field in $\mathbf{W}_2^1(\Omega)$ vanishing on the part $S_v$ of $\partial\Omega$, Theorem 1 is applicable to each of its components. This results in the following statement: the set $\mathcal{U}$ is dense in the subspace of sufficiently regular kinematically admissible velocity fields $\{\mathbf{u} \in \mathbf{W}_2^1(\Omega) : \mathbf{u}|_{S_v} = 0\}$ with respect to the $\mathbf{W}_2^1(\Omega)$-norm. Here, $\mathcal{U}$ is the set of all smooth vector fields $\mathbf{v}$ vanishing near $S_v$. The latter condition means that, for every $\mathbf{v} \in \mathcal{U}$, there exists a number $\delta > 0$ such that $\mathbf{v}(x) = 0$ at every $x \in \Omega$, whose distance from $S_v$ is less than $\delta$.

## 4. Approximations to solenoidal vector fields

Velocity fields in incompressible bodies are solenoidal and, in case a part $S_v$ of the body boundary is fixed, velocity vanishes on $S_v$. In this section, we establish conditions under which such fields can be approximated by smooth solenoidal fields vanishing *near* $S_v$. The possibility of this approximation played an important role in solving the pressure field restoration problem in Section 2. We will also make use of the approximation in the finite element formulation of the limit analysis problem in Chapter IX.

Here, we show first that the approximation is possible in case the whole of the body boundary is fixed. Then, we establish two types of conditions for approximation in case only a part of the body boundary is fixed. The conditions are quite simple, and it is easy to verify whether a concrete problem satisfies them. The conditions of both types together seem to cover all limit analysis problems of practical interest.

We consider approximations to a vector field in Sobolev space $\mathbf{W}_2^1(\Omega)$ (see Appendix B).

**4.1. Approximation in case of fixed boundary.** Consider an incompressible body with a fixed boundary. Velocity fields in the body are solenoidal and have zero traces on the boundary. Let us establish approximation of such vector field by fields which vanish near the boundary. Recall that the vector field $\mathbf{v}$ in the domain $\Omega$ satisfies the latter condition if there is a number $\delta > 0$ such that $\mathbf{v}(x) = 0$ for every $x \in \Omega$ whose distance from $\partial\Omega$ is less than $\delta$.

THEOREM 1. Suppose $\Omega$ is a bounded Lipschitz domain. Then for every solenoidal vector field $\mathbf{v} \in \mathbf{W}_2^1(\Omega)$ with the zero trace $\mathbf{v}|_{\partial\Omega} = 0$, there exists a sequence of smooth solenoidal vector fields vanishing near $S_v$ and converging to $\mathbf{v}$ in $\mathbf{W}_2^1(\Omega)$.

PROOF. The field $\mathbf{v}$ belongs to the subspace $\overset{\circ}{\mathbf{W}}{}_2^1(\Omega)$ of the fields in $\mathbf{W}_2^1(\Omega)$ with the zero trace on the boundary $\partial\Omega$. This subspace is a normed space with the norm equivalent to the $\mathbf{W}_2^1(\Omega)$-norm on $\overset{\circ}{\mathbf{W}}{}_2^1(\Omega)$; the norm is generated by the scalar multiplication:

$$(\mathbf{u}, \mathbf{v}) = \int_\Omega \left( \frac{\partial \mathbf{u}}{\partial x_1} \frac{\partial \mathbf{v}}{\partial x_1} + \cdots + \frac{\partial \mathbf{u}}{\partial x_n} \frac{\partial \mathbf{v}}{\partial x_n} \right) dV$$

(Theorem B.5). This is the norm we use in the sequel. The statement to be proved can now be formulated in the following way: the field $\mathbf{v}$ belongs to the closure $\bar{\mathcal{U}}_d$ of the set $\mathcal{U}_d$ with respect to the above norm; here, $\mathcal{U}_d$ is the set of smooth solenoidal fields vanishing near $\partial\Omega$.

Let us represent $\mathbf{v}$ as the sum $\mathbf{v} = \mathbf{v}_d + \mathbf{w}$ of $\mathbf{v}_d$ which is in $\bar{\mathcal{U}}_d$ and $\mathbf{w}$ which is in the orthogonal complement of $\bar{\mathcal{U}}_d$ in $\overset{\circ}{\mathbf{W}}{}_2^1(\Omega)$. It is clear that the component $\mathbf{w}$ satisfies the following conditions:

$$\operatorname{div} \mathbf{w} = 0, \quad \mathbf{w}|_{\partial\Omega} = 0, \quad \int_\Omega \frac{\partial w_i}{\partial x_j} \frac{\partial u_i}{\partial x_j}\, dV = 0 \text{ for every } \mathbf{u} \in \mathcal{U}_d \tag{4.1}$$

(the first equality is valid since $\mathbf{v}$ and $\mathbf{v}_d$ are solenoidal). To prove the theorem, it is sufficient to verify that $\mathbf{w} = 0$. To establish this equality, we introduce the tensor field $\boldsymbol{\tau}$ with the components

$$\tau_{ij} = \frac{\partial w_i}{\partial x_j} + \frac{\partial w_j}{\partial x_i} \in L_2(\Omega).$$

It is a deviatoric field: $\boldsymbol{\tau} = \boldsymbol{\tau}^d$ since $\operatorname{div} \mathbf{w} = 0$. Consider now the following integral

$$\int_\Omega \operatorname{Def} \mathbf{u} \cdot \boldsymbol{\tau}\, dV = \int_\Omega \frac{\partial w_i}{\partial x_j} \frac{\partial u_i}{\partial x_j}\, dV + \int_\Omega \frac{\partial w_j}{\partial x_i} \frac{\partial u_i}{\partial x_j}\, dV,$$

where $\mathbf{u} \in \mathcal{U}_d$. Both terms on the right-hand side are zero: the first due to (4.1), the second due to the formula

$$\int_\Omega \frac{\partial w_j}{\partial x_i} \frac{\partial u_i}{\partial x_j}\, dV = \int_{\partial\Omega} w_j \nu_i \frac{\partial u_i}{\partial x_j}\, dS - \int_\Omega w_j \frac{\partial}{\partial x_i} \frac{\partial u_i}{\partial x_j}\, dS$$

and the equalities $\mathbf{w}|_{\partial\Omega} = 0$, $\operatorname{div}\mathbf{u} = 0$. Thus, we have

$$\int_\Omega \operatorname{Def}\mathbf{u}\cdot\boldsymbol{\tau}\,dV = 0 \quad \text{for every} \quad \mathbf{u}\in\mathcal{U}_d,$$

and we can appeal to Theorem 2.3 (see also Remark 2.1). This implies existence of the function $p \in L_2(\Omega)$ such that the equality

$$\int_\Omega \operatorname{Def}\mathbf{u}\cdot(\boldsymbol{\tau} - p\mathbf{I})\,dV = 0$$

holds for every smooth vector field $\mathbf{u}$ which vanishes near $\partial\Omega$. Let us write the latter equality in a more detailed form:

$$\int_\Omega \frac{\partial w_i}{\partial x_j}\frac{\partial u_i}{\partial x_j}\,dV + \int_\Omega \frac{\partial w_j}{\partial x_i}\frac{\partial u_i}{\partial x_j}\,dV - \int_\Omega p\operatorname{div}\mathbf{u}\,dV = 0.$$

Here, the second integral can be interpreted as the value of the distribution $\partial w_j/\partial x_i$ at the test function $\partial u_i/\partial x_j$:

$$\langle\frac{\partial w_j}{\partial x_i}, \frac{\partial u_i}{\partial x_j}\rangle = -\langle\frac{\partial}{\partial x_j}\left(\frac{\partial w_j}{\partial x_i}\right), u_i\rangle = -\langle\frac{\partial}{\partial x_i}\left(\frac{\partial w_j}{\partial x_j}\right), u_i\rangle.$$

This value is zero due to $\operatorname{div}\mathbf{w} = 0$. Thus, the equality

$$\int_\Omega \frac{\partial w_i}{\partial x_j}\frac{\partial u_i}{\partial x_j}\,dV - \int_\Omega p\operatorname{div}\mathbf{u}\,dV = 0$$

is valid for every smooth vector field $\mathbf{u}$ which vanishes near $\partial\Omega$. By Theorem 3.1 the equality also holds for every $\mathbf{u} \in \mathbf{W}_2^1(\Omega)$ with the zero trace on $\partial\Omega$. In particular, setting $\mathbf{w} = \mathbf{u}$ and making use of $\operatorname{div}\mathbf{w} = 0$, we arrive at the relation

$$\int_\Omega \frac{\partial w_i}{\partial x_j}\frac{\partial w_i}{\partial x_j}\,dV = 0,$$

that is, $\|\mathbf{w}\|^2 = (\mathbf{w}, \mathbf{w}) = 0$, which finishes the proof.

Theorem 1 is immediately applicable to the case of a fixed body boundary. We will also make use of the proposition while proving the analogous approximation theorem in case the fixed surface $S_v$ is only a part of $\partial\Omega$. The next subsection presents another important statement, which we will need in case $S_v \neq \partial\Omega$.

**4.2. Vector fields with a given divergence.** Suppose the function $\varphi$ is given on the domain $\Omega$; is there a vector field $\mathbf{u}$ on $\Omega$ with $\operatorname{div}\mathbf{u} = \varphi$?

It is clear that if $\mathbf{u}$ has to meet the additional requirement $\mathbf{u}|_{\partial\Omega} = 0$, the divergence $\varphi$ should satisfy a certain condition for existence of $\mathbf{u}$ with $\operatorname{div}\mathbf{u} = \varphi$. Namely, we have by the Stokes formula:

$$\int\limits_{\Omega} \operatorname{div}\mathbf{u}\, dV = \int\limits_{\partial\Omega} \mathbf{u}\boldsymbol{\nu}\, dS$$

($\boldsymbol{\nu}$ is the unit outward normal to $\partial\Omega$), which implies

$$\int\limits_{\Omega} \varphi\, dV = 0.$$

The following well known proposition shows that, under certain regularity assumptions, the latter equality is also a sufficient condition for existence of $\mathbf{u}$ with $\operatorname{div}\mathbf{u} = \varphi$.

THEOREM 2. Suppose the function $\varphi_0 \in L_2(\Omega)$ is given on a bounded Lipschitz domain $\Omega$ and

$$\int\limits_{\Omega} \varphi_0\, dV = 0.$$

Then there exists a vector field $\mathbf{u} \in \mathbf{W}_2^1(\Omega)$ satisfying the conditions

$$\operatorname{div}\mathbf{u} = \varphi_0, \quad \mathbf{u}|_{\partial\Omega} = 0, \quad \|\mathbf{u}\|_{\mathbf{W}_2^1(\Omega)} \leq C\,\|\varphi_0\|_{L_2(\Omega)}, \tag{4.2}$$

where constant $C$ does not depend on $\varphi_0$.

In case vanishing of $\mathbf{u}$ is required only on a part of $\partial\Omega$, no additional condition should be imposed on $\varphi_0$. More precisely, the following proposition is valid.

COROLLARY. Suppose $S$ is a part of the boundary of a bounded Lipschitz domain $\Omega$, $S$ is of positive measure in $\partial\Omega$, and $\varphi \in L_2(\Omega)$ is a given function. Then there exists the vector field $\mathbf{v} \in \mathbf{W}_2^1(\Omega)$ satisfying the conditions

$$\operatorname{div}\mathbf{v} = \varphi, \quad \mathbf{u}|_{\partial\Omega\setminus S} = 0, \quad \|\mathbf{v}\|_{\mathbf{W}_2^1(\Omega)} \leq c\,\|\varphi\|_{L_2(\Omega)} \tag{4.3}$$

where constant $c$ does not depend on $\varphi$.

Indeed, we choose a smooth vector field $\mathbf{v}_1$ on $\Omega$ with

$$\mathbf{v}_1|_{\partial\Omega\setminus S} = 0, \qquad \int\limits_{\Omega} \operatorname{div}\mathbf{v}_1\, dV = 1$$

(it is easily seen from the Stokes formula that the second equality is guanteed by assigning $\mathbf{v}_1$ suitable values on $S$). By Theorem 1, for the function

$$\varphi_0 = \varphi - \operatorname{div}\mathbf{v}_1 \int\limits_{\Omega} \varphi\, dV,$$

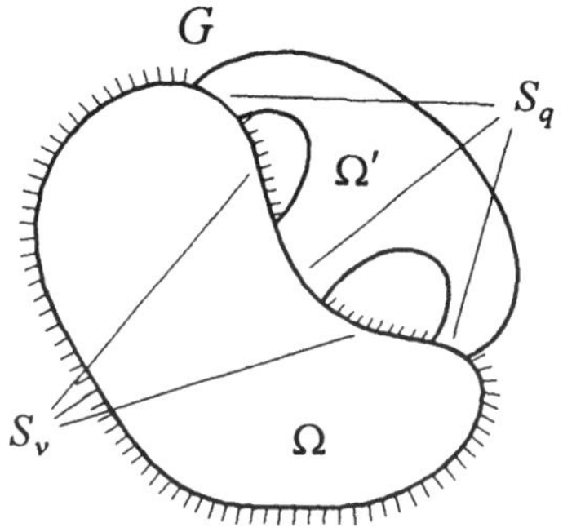


**Figure 4.1**

there exists the vector field $\mathbf{u}$ for which (4.2) is valid. Then it is clear that the field

$$\mathbf{v} = \mathbf{u} + \mathbf{v}_1 \int_\Omega \varphi \, dV$$

satisfies the relations $\operatorname{div} \mathbf{v} = \varphi$, $\mathbf{v}|_{\partial\Omega \setminus S} = 0$. Moreover, the inequalities

$$\|\mathbf{v}\|_{\mathbf{W}_2^1(\Omega)} \leq \|\mathbf{u}\|_{\mathbf{W}_2^1(\Omega)} + \left| \int_\Omega \varphi \, dV \right| \|\mathbf{v}_1\|_{\mathbf{W}_2^1(\Omega)}$$

$$\leq C \left\| \varphi - \operatorname{div} \mathbf{v}_1 \int_\Omega \varphi \, dV \right\|_{L_2(\Omega)} + \left| \int_\Omega \varphi \, dV \right| \|\mathbf{v}_1\|_{\mathbf{W}_2^1(\Omega)}$$

imply the required estimation (4.3).

**4.3. Approximation conditions – I.** We now establish one type of condition under which any field $\mathbf{v} \in \mathbf{W}_2^1(\Omega)$ with $\operatorname{div} \mathbf{v} = 0$ and $\mathbf{v}|_{S_v} = 0$ can be approximated by smooth solenoidal fields vanishing near $S_v$. The smoothness and vanishing *near* $S_v$ are understood in the sense of Subsection 3.2. We assume that the boundary $\partial\Omega$ consists of two non-intersecting parts: $S_v$ and $S_q$. Figure 4.1 illustrates assumptions of the following proposition.

THEOREM 3. Suppose 1) $\Omega$ is a bounded Lipschitz domain, $S_q$ is a part of its boundary, $S_v$ is of positive measure in $\partial\Omega$, 2) $S_q$ is the common part of the boundaries of $\Omega$ and another bounded Lipschitz domain $\Omega'$ with $\Omega \cap \Omega' = \emptyset$, 3) the domain $G$ composed of $\Omega$ and $\Omega'$ is a Lipschitz domain, and 4) every vector field in $\mathbf{W}_2^1(\Omega)$ with the zero trace on $S_v$ can be approximated in $\mathbf{W}_2^1(\Omega)$ by smooth fields vanishing near $S_v$ (for example, assumptions of Theorem 3.1 are satisfied). Then, for every solenoidal vector field $\mathbf{u} \in \mathbf{W}_2^1(\Omega)$ with $\mathbf{u}|_{S_v} = 0$, there exists a sequence of smooth solenoidal vector fields vanishing near $S_v$ and converging to $\mathbf{u}$ in $\mathbf{W}_2^1(\Omega)$.

PROOF. By assumption, for every $\varepsilon > 0$ there exists $\mathbf{w}_\varepsilon \in \mathbf{W}_2^1(\Omega)$ such that

$$\|\mathbf{u} - \mathbf{w}_\varepsilon\|_{\mathbf{W}_2^1(\Omega)} < \varepsilon \tag{4.4}$$

and $\mathbf{w}_\varepsilon$ vanishes near $S_v$. The latter means, in particular, that the trace of $\mathbf{w}_\varepsilon$ in a certain neighborhood $\gamma$ of $\bar{S}_v$ in $\partial\Omega$ is zero: $\mathbf{w}_\varepsilon|_\gamma = 0$. Without loss of generality we assume that $\partial\Omega \setminus \gamma$ has a positive measure in $\partial\Omega$. Note that our assumption only ensures existence of such $\mathbf{w}_\varepsilon$, and does not guarantee that $\mathbf{w}_\varepsilon$ is solenoidal. However, the condition $\operatorname{div}\mathbf{u} = 0$ implies that $\operatorname{div}\mathbf{w}_\varepsilon$ is small in a sense and, therefore, we can improve $\mathbf{w}_\varepsilon$ and make it solenoidal. Indeed, in case $\operatorname{div}\mathbf{u} = 0$, inequality (4.4) implies

$$\|\operatorname{div}\mathbf{w}_\varepsilon\|_{L_2(\Omega)} < c_1 \|\mathbf{w}_\varepsilon - \mathbf{u}\|_{\mathbf{W}_2^1(\Omega)} < c_1\varepsilon$$

with a constant $c_1$, which does not depend on $\mathbf{w}_\varepsilon$, $\mathbf{u}$. Then, by the Corollary in the previous subsection, there exists a field $\mathbf{w}_{*\varepsilon}$ in $\mathbf{W}_2^1(\Omega)$ for which

$$\operatorname{div}\mathbf{w}_{*\varepsilon} = \operatorname{div}\mathbf{w}_\varepsilon, \quad \mathbf{w}_{*\varepsilon}|_\gamma = 0, \quad \|\mathbf{w}_{*\varepsilon}\|_{\mathbf{W}_2^1(\Omega)} \leq \|\operatorname{div}\mathbf{w}_\varepsilon\|_{L_2(\Omega)}\,.$$

It is a clear that the field $\mathbf{v}_\varepsilon = \mathbf{w}_\varepsilon - \mathbf{w}_{*\varepsilon}$ possesses the following properties:

$$\mathbf{v}_\varepsilon \in \mathbf{W}_2^1(\Omega), \quad \operatorname{div}\mathbf{v}_\varepsilon = 0, \quad \mathbf{v}_\varepsilon|_\gamma = 0, \quad \|\mathbf{v}_\varepsilon - \mathbf{u}\|_{\mathbf{W}_2^1(\Omega)} < c_2\varepsilon, \tag{4.5}$$

where constant $c_2$ does not depend on $\mathbf{u}$, $\mathbf{v}_\varepsilon$.

To finish the proof, it remains to approximate $\mathbf{v}_\varepsilon$ in $\mathbf{W}_2^1(\Omega)$ by the field $\mathbf{u}_\varepsilon \in C^\infty(\bar{\Omega})$ which vanishes near $S_v$, $\mathbf{u}_\varepsilon$ being a solenoidal field. Note that it is sufficient to construct such extension $\mathbf{v}_\varepsilon^c$ of $\mathbf{v}_\varepsilon$ to the domain $\Omega$ that the following conditions are satisfied:

$$\mathbf{v}_\varepsilon^c \in \mathbf{W}_2^1(G), \quad \operatorname{div}\mathbf{v}_\varepsilon^c = 0, \quad \mathbf{v}_\varepsilon^c|_{\partial G} = 0. \tag{4.6}$$

Indeed, according to Theorem 1, such a field can be approximated by the field $\mathbf{u}_\varepsilon^c$ which vanishes near $\partial G$ and possesses the following properties:

$$\mathbf{u}_\varepsilon^c \in C^\infty(\bar{G}), \quad \operatorname{div}\mathbf{u}_\varepsilon^c = 0, \quad \|\mathbf{v}_\varepsilon^c - \mathbf{u}_\varepsilon^c\|_{\mathbf{W}_2^1(G)} < c_3\varepsilon,$$

where constant $c_3$ does not depend on $\mathbf{u}_\varepsilon^c$, $\mathbf{v}_\varepsilon^c$. Then, we introduce the restriction $\mathbf{u}_\varepsilon = \mathbf{u}_\varepsilon^c|_\Omega$. The field $\mathbf{u}_\varepsilon^c$ is solenoidal and approximates $\mathbf{u}$: due to the equalities $\mathbf{u}_\varepsilon = \mathbf{u}_\varepsilon^c|_\Omega$, $\mathbf{v}_\varepsilon = \mathbf{v}_\varepsilon^c|_\Omega$, the estimation

$$\|\mathbf{u} - \mathbf{u}_\varepsilon\|_{W_2^1(\Omega)} \leq \|\mathbf{u} - \mathbf{v}_\varepsilon\|_{W_2^1(\Omega)} + \|\mathbf{v}_\varepsilon^c - \mathbf{u}_\varepsilon^c\|_{W_2^1(\Omega)} < (c_2 + c_3)\varepsilon$$

is valid. Also note that $\mathbf{u}_\varepsilon$ vanishes near $S_v$ since $\bar{S}_v \subset \partial G$.

Thus, it remains to construct extension (4.6) of the field $\mathbf{v}_\varepsilon$. We start by considering an arbitrary extension $\mathbf{v}_{\varepsilon,1} \in \mathbf{W}_2^1(\mathbf{R}^n)$ to all of $\mathbf{R}^n$ (the extension exists according to Theorem B.1). We observe that the bounded closed sets $\partial\Omega \setminus \gamma$ and $\partial\Omega' \setminus S_q$ do not intersect; therefore, there are neighborhoods $\mathcal{O}(\partial\Omega \setminus \gamma)$ and $\mathcal{O}(\partial\Omega' \setminus S_q)$ of these sets in $\mathbf{R}^n$ which do not intersect either. Consider then a smooth function $\alpha$ which equals 1 on $\partial\Omega \setminus \gamma$, equals 0 on $\mathcal{O}(\partial\Omega' \setminus S_q)$ and vanishes in the exterior of a certain ball. We note that $\alpha$ equals

1 on $\partial\Omega \setminus \gamma$ and $\mathbf{v}_{\varepsilon 1}$ (together with $\mathbf{v}_\varepsilon$) has a zero trace on $\gamma$; consequently, $\alpha \mathbf{v}_{\varepsilon,1}|_{\partial\Omega} = \mathbf{v}_\varepsilon|_{\partial\Omega}$. We now introduce a new extension $\mathbf{v}_{\varepsilon,2}$ of $\mathbf{v}_\varepsilon$:

$$\mathbf{v}_{\varepsilon,2}(x) = \begin{cases} \mathbf{v}_\varepsilon(x) & \text{if } x \in \Omega, \\ \alpha(x)\mathbf{v}_{\varepsilon,1}(x) & \text{if } x \in \mathbf{R}^n \setminus \Omega; \end{cases}$$

this extension is in $\mathbf{W}_2^1(\mathbf{R}^n)$ according to Theorem B.4. The relations

$$\alpha \mathbf{v}_{\varepsilon,1}|_{\mathcal{O}(\partial\Omega' \setminus S_q)} = 0, \quad \mathbf{v}_\varepsilon|_\gamma = 0$$

imply $\mathbf{v}_{\varepsilon,2}|_{\partial G} = 0$. We also have $\operatorname{div} \mathbf{v}_{\varepsilon,2} = 0$ on $\Omega$ due to (4.5). Thus, $\mathbf{v}_{\varepsilon,2}$ meets all requirements (4.6) to the field $\mathbf{v}_\varepsilon^c$ with the exception of the condition $\operatorname{div} \mathbf{v}_\varepsilon^c = 0$ in $\Omega'$. Let us now improve $\mathbf{v}_{\varepsilon 2}$ to satisfy this condition. We observe that, by the Stokes formula and the relations $\mathbf{v}_{\varepsilon,2}|_{\partial\Omega' \setminus S_q} = 0$ and $\mathbf{v}_{\varepsilon,2}|_{S_v} = 0$, the following equalities are valid:

$$\int_{\Omega'} \operatorname{div} \mathbf{v}_{\varepsilon,2}\, dV = \int_{\partial\Omega'} \mathbf{v}_{\varepsilon,2}\boldsymbol{\nu}'\, dS = \int_{S_q} \mathbf{v}_{\varepsilon,2}\boldsymbol{\nu}'\, dS = -\int_{S_q} \mathbf{v}_{\varepsilon,2}\boldsymbol{\nu}\, dS = -\int_{\partial\Omega} \mathbf{v}_{\varepsilon,2}\boldsymbol{\nu}\, dS$$
$$= \int_{\Omega} \operatorname{div} \mathbf{v}_{\varepsilon,2}\, dV$$

($\boldsymbol{\nu}$, $\boldsymbol{\nu}'$ are the unit outward normals to $\partial\Omega$, $\partial\Omega'$). Since $\mathbf{v}_{\varepsilon,2}$ is solenoidal in $\Omega$, this implies

$$\int_{\Omega'} \operatorname{div} \mathbf{v}_{\varepsilon,2}\, dV = 0,$$

which allows us to appeal to Theorem 2 and conclude that there is the field $\mathbf{v}'_{\varepsilon,2} \in \mathbf{W}_2^1(\Omega')$ for which $\operatorname{div} \mathbf{v}'_{\varepsilon,2} = \operatorname{div} \mathbf{v}_{\varepsilon,2}$ and $\mathbf{v}'_{\varepsilon,2}|_{\partial\Omega'} = 0$. Then we introduce the field $\mathbf{v}'_\varepsilon = \mathbf{v}_{\varepsilon,2} - \mathbf{v}'_{\varepsilon,2}$, which possesses the following obvious properties:

$$\mathbf{v}'_\varepsilon \in \mathbf{W}_2^1(\Omega'), \quad \operatorname{div} \mathbf{v}'_\varepsilon = 0, \quad \mathbf{v}'_\varepsilon|_{\partial\Omega'} = \mathbf{v}_{\varepsilon,2}|_{\partial\Omega'}\,. \tag{4.7}$$

The latter equality means, in particular, that the traces $\mathbf{v}'_\varepsilon|_{S_q}$ and $\mathbf{v}_\varepsilon|_{S_q} = \mathbf{v}_{\varepsilon,2}|_{S_q}$ on the common part $S_q$ of the boundaries $\partial\Omega$ and $\partial\Omega'$ are equal. Then, by Theorem B.4, the field

$$\mathbf{v}_\varepsilon^c(x) = \begin{cases} \mathbf{v}_\varepsilon(x) & \text{if } x \in \Omega, \\ \mathbf{v}'_\varepsilon(x) & \text{if } x \in \Omega' \end{cases}$$

is in $\mathbf{W}_2^1(G)$. The field is solenoidal due to (4.5), (4.7), and it has the zero traces

$$\mathbf{v}_\varepsilon^c|_\gamma = \mathbf{v}_\varepsilon|_\gamma = 0, \quad \mathbf{v}_\varepsilon^c|_{\partial\Omega' \setminus S_q} = \mathbf{v}'_\varepsilon|_{\partial\Omega' \setminus S_q} = \mathbf{v}_{\varepsilon,2}|_{\partial\Omega' \setminus S_q} = 0.$$

This implies $\mathbf{v}_\varepsilon^c|_{\partial G} = 0$. Thus, $\mathbf{v}_\varepsilon^c$ meets all requirements (4.6), which finishes the proof.

REMARK 1. Theorem 3, in particular, covers the case of the free boundary, that is, $\partial\Omega = S_q$, $S_v = \emptyset$.

**4.4 Approximation conditions – II.** We now establish conditions which differ from those in Theorem 3 and, at the same time, guarantee that every vector field $\mathbf{v} \in \mathbf{W}_2^1(\Omega)$ with $\mathbf{v}|_{S_v} = 0$ and $\operatorname{div} \mathbf{v} = 0$ can be approximated by smooth solenoidal fields vanishing near $S_v$.

Suppose $\Omega'$ and $\Omega''$ are bounded domains whose intersection consists of domains $\Omega_1, \dots, \Omega_N$, the distance between any two of them being positive; we also assume that the distance between $\Omega \setminus \Omega'$ and $\Omega \setminus \Omega''$ is positive. In this case, we say that $\Omega = \Omega' \cap \Omega''$ is *regularly composed* of $\Omega$ and $\Omega'$. Figure 4.2 shows an example of a regularly composed domain; the domains $\Omega'$ and $\Omega''$ are hatched with different slopes; their intersection consists of two domains: $\Omega_1$ and $\Omega_2$.

We denote the part of the boundary $\partial\Omega_i$ which lies in $\Omega''$ by the symbol $S_i'$: $S_i' = \partial\Omega_i \cap \Omega''$ (see Figure 4.2) and denote the union of these surfaces by $S'$: $S' = S_1' \cup \dots S_N'$. The surface $S'$ is the common part of the frontiers of the sets $\Omega'$ and $\Omega \setminus \Omega'$; $S'$ separates the domain $\Omega'$ from the remaining part of the regularly composed domain $\Omega$. We analogously introduce $S_i'' = \partial\Omega_i \cap \Omega'$ and $S'' = S_1'' \cup \dots S_N''$.

Let the boundary $\partial\Omega$, as usual, consist of the two nonintersecting parts $S_v$ and $S_q$. The surface $S_v$ generally has both of the two parts $S_v \cap \partial\Omega'$ and $S_v \cap \partial\Omega''$, which lie in $\partial\Omega'$ and $\partial\Omega''$, respectively, and possibly have a nonempty intersection. We consider the surfaces

$$S_v' = (S_v \cap \partial\Omega') \cup S', \qquad S_v'' = (S_v \cap \partial\Omega'') \cup S''. \tag{4.8}$$

It turns out that solenoidal vector fields in $\Omega$ with the zero trace on $S_v$ can be approximated by smooth solenoidal fields vanishing near $S_v$ if solenoidal fields in $\Omega'$ and $\Omega''$ with the zero traces on $S_v'$ and $S_v''$, respectively, can be approximated in the analogous sense. The following proposition gives a more accurate formulation of this property; the proposition uses the above introduced notation, in particular, (4.8).

THEOREM 4. Suppose 1) the domain $\Omega$ is regularly composed of domains $\Omega'$ $\Omega''$; 2) $\Omega$, $\Omega'$, $\Omega''$, $\Omega_i$, $i = 1, \dots, N$, are Lipschitz domains; 3) the boundary $\partial\Omega_i$, $i = 1, \dots, N$ contains a set of a positive measure in $\partial\Omega_i$ and this set is a part of $S_q$; and 4) for the domain $\Omega'$ and the part $S_v'$ of its boundary, every solenoidal vector field $\mathbf{v} \in \mathbf{W}_2^1(\Omega')$ with the zero trace on $S_v'$ can be approximated in $\mathbf{W}_2^1(\Omega')$ by a smooth solenoidal vector field vanishing near $S_v'$, and, analogously, for the domain $\Omega''$ and the part $S_v''$ of its boundary, every solenoidal vector field $\mathbf{v} \in \mathbf{W}_2^1(\Omega'')$ with the zero trace on $S_v''$ can be approximated in $\mathbf{W}_2^1(\Omega'')$ by the smooth solenoidal vector field vanishing near $S_v''$. Then, for every solenoidal vector field $\mathbf{w} \in \mathbf{W}_2^1(\Omega)$ with zero trace on $S_v$, there exists a sequence of smooth solenoidal fields vanishing near $S_v$ and converging to $\mathbf{w}$ in $\mathbf{W}_2^1(\Omega)$.

PROOF. Let us first show that the proposition is valid in case $\mathbf{w}$ can be written in the form $\mathbf{w} = \mathbf{w}' + \mathbf{w}''$ with

$$\begin{aligned} &\mathbf{w}' \in \mathbf{W}_2^1(\Omega'), \quad \mathbf{w}'|_{\Omega\setminus\Omega'} = 0, \quad \mathbf{w}'|_{S_v'} = 0, \quad \operatorname{div} \mathbf{w}' = 0, \\ &\mathbf{w}'' \in \mathbf{W}_2^1(\Omega''), \quad \mathbf{w}''|_{\Omega\setminus\Omega''} = 0, \quad \mathbf{w}''|_{S_v''} = 0, \quad \operatorname{div} \mathbf{w}'' = 0. \end{aligned} \tag{4.9}$$

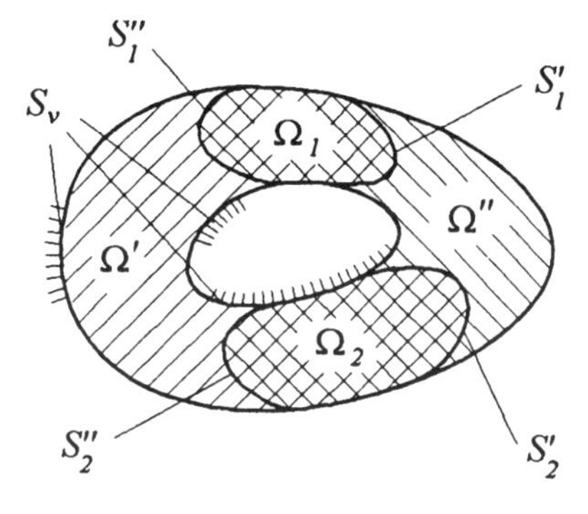


**Figure 4.2**

In this case the theorem can be proved by constructing suitable approximations to $\mathbf{w}'$ in $\Omega'$ and $\mathbf{w}''$ in $\Omega''$. We start with the approximation to $\mathbf{w}'$ which exists due to assumption (4). Namely, for every $\varepsilon > 0$ there is a field $\mathbf{u}' \in C^\infty(\bar{\Omega}')$ such that

$$\operatorname{div} \mathbf{u}' = 0, \quad \|\mathbf{w}' - \mathbf{u}'\|_{\mathbf{W}_2^1(\Omega')} < \varepsilon \tag{4.10}$$

and $\mathbf{u}'$ vanishes near $S_v'$. Let us consider the following extension of $\mathbf{u}'$ to all of $\Omega$:

$$\mathbf{u}_c'(x) = \begin{cases} \mathbf{u}'(x) & \text{if } x \in \Omega', \\ 0 & \text{if } x \in \Omega'', \end{cases}$$

and show that $\mathbf{u}_c'$ is a suitable approximation to $\mathbf{w}'$ in $\Omega$. First, we observe that $\mathbf{u}_c'$ is smooth: $\mathbf{u}_c' \in C^\infty(\bar{\Omega})$. Indeed, consider the common part $S'$ of the boundaries of $\Omega$ and $\Omega \setminus \Omega'$; we have $S' \subset S_v'$. Then $\mathbf{u}_c'$ vanishes in a certain neighborhood of $\bar{S}_v'$ in $\Omega'$ since $\mathbf{u}'$ vanishes near $S_v'$, from which it follows that $\mathbf{u}_c'$ is in $C^\infty(\bar{\Omega})$. It is clear that $\mathbf{u}_c'$ is a solenoidal field: $\operatorname{div} \mathbf{u}_c' = 0$. It is also an approximation to $\mathbf{w}'$ in $\Omega$, as both $\mathbf{w}'$ and $\mathbf{u}_c'$ vanish on $\Omega \setminus \Omega'$ and, therefore, the estimation

$$\|\mathbf{w}' - \mathbf{u}_c'\|_{\mathbf{W}_2^1(\Omega)} < \varepsilon \tag{4.11}$$

arises from (4.10).

For our purposes, we also have to show that $\mathbf{u}_c'$ vanishes near $S_v$ in $\Omega$, that is, on the intersection of $\Omega$ with a certain neighborhood of $\bar{S}_v$. To find such a neighborhood, consider the frontier $\gamma$ of the set $\Omega'' \setminus \Omega'$ and the union of $\gamma$ with $\bar{S}_v'$. Let us show that there is a neighborhood $N$ of the set $\bar{S}_v' \cup \gamma$ such that $\mathbf{u}_c'$ vanishes on $N \cap \Omega$. This implies that $\mathbf{u}_c'$ vanishes near $S_v$ in $\Omega$ since $\bar{S}_v' \cup \gamma \supset S_v$. We now verify that there is a number $\delta > 0$ such that

$$N = \{x \in \Omega : \rho(x, \bar{S}_v' \cup \gamma) < \delta\}$$

is the suitable neighborhood (as usual, $\rho(x, A)$ denotes the distance between the point $x$ and the set $A$). Recall that $\mathbf{u}_c'$ vanishes on $\Omega''$ and $\mathbf{u}_c' = \mathbf{u}$ on $\Omega'$; then $\mathbf{u}_c'$ vanishes on $N \cap \Omega$ if

$$\mathbf{u}'(x) = 0 \quad \text{for every} \quad x' \in \Omega' \quad \text{with} \quad \rho(x, \bar{S}_v' \cup \gamma) < \delta. \tag{4.12}$$

To find suitable $\delta$, recall that $\mathbf{u}'$ vanishes near $S'_v$, that is, there is a number $\delta_0$ such that $\mathbf{u}'(x) = 0$ for every $x' \in \Omega'$ with $\rho(x, S'_v) < \delta_0$. We choose such $\delta$, $\delta_0 > \delta > 0$, that the following condition is satisfied:

$$\rho(x, S'_v) < \delta, \quad x \in \Omega' \quad \text{implies} \quad \rho(x, S') < \delta_0.$$

(Such $\delta$ does exist. Otherwise, there would exist a sequence of points $x_1$, $x_2, \ldots$ in $\Omega'$ for which $\rho(x_n, \gamma) \to 0$ and $\rho(x_n, S') > \delta_0$. The first of these relations implies that a limit point of the sequence belongs to $\partial\Omega' \cap \gamma = \bar{S}_v$, which contradicts the inequality $\rho(x_n, S') > \delta_0$.) For the above $\delta$, the set

$$\begin{aligned} \Omega'_\delta &= \{x \in \Omega' : \rho(x, S'_v \cup \gamma) < \delta\} \\ &= \{x \in \Omega' : \rho(x, S'_v) < \delta\} \cup \{x \in \Omega' : \rho(x, \gamma) < \delta\} \end{aligned}$$

is contained in

$$\begin{aligned} &\{x \in \Omega' : \rho(x, S'_v) < \delta_0\} \cup \{x \in \Omega' : \rho(x, S') < \delta_0\} \\ &\qquad\qquad = \{x \in \Omega' : \rho(x, S'_v \cup S') < \delta_0\}. \end{aligned}$$

Here, $S'_v \cup S' = S'_v$ due to (4.8), and, consequently, $\mathbf{u}'(x) = 0$ for every $x$ in the latter set. This implies, in particular, that $\mathbf{u}'(x) = 0$ for every $x$ in $\Omega'_\delta$. In other words, (4.12) is valid, and $\mathbf{u}'_c$ vanishes near $S_v$ in $\Omega$.

Thus, $\mathbf{u}'_c$ is a smooth vector field which approximates $\mathbf{w}'$ in $\Omega$ with estimation (4.11), and $\mathbf{u}'_c$ vanishes near $S_v$ in $\Omega$. We similarly construct a smooth approximation $\mathbf{u}''_c$ to $\mathbf{w}''$ in $\Omega$ with the estimation

$$\|\mathbf{w}'' - \mathbf{u}''_c\|_{\mathbf{W}^1_2(\Omega)} < \varepsilon,$$

$\mathbf{u}''_c$ vanishing near $S_v$ in $\Omega$. Then, the sum $\mathbf{u} = \mathbf{u}'_c + \mathbf{u}''_c$ is obviously the required approximation to $\mathbf{w} = \mathbf{w}' + \mathbf{w}''$.

To finish the proof, it remains to verify that $\mathbf{w}$ can be written in the form $\mathbf{w} = \mathbf{w}' + \mathbf{w}''$ with $\mathbf{w}'$, $\mathbf{w}''$ satisfying (4.9). It should be noted that, since the distance between $\Omega \setminus \Omega'$ and $\Omega \setminus \Omega''$ is positive, there is a smooth function $\alpha$ which equals 1 on $\Omega \setminus \Omega'$, equals 0 on $\Omega \setminus \Omega''$ and vanishes in the exterior of some ball. Then, $\alpha\mathbf{w}$ is in $\mathbf{W}^1_2(\Omega)$, vanishes on $\Omega \setminus \Omega'$ and $\alpha\mathbf{w}|_{\Omega'}$ is in $\mathbf{W}^1_2(\Omega')$. It can be easily seen that $\alpha\mathbf{w}|_{S'_v} = 0$. Indeed, we have $S'_v = (S_v \cap \partial\Omega') \cup S'$, and $\alpha\mathbf{w}$ vanishes on $S_v$ together with $\mathbf{w}$ and vanishes on $S'$ together with $\alpha$ (since $S'$ is a part of the frontier of $\omega \setminus \Omega'$). As to the condition $\operatorname{div} \alpha\mathbf{w} = 0$, it is valid on $\Omega \setminus \Omega'$, where $\alpha\mathbf{w} = 0$, and on $\Omega \setminus \Omega''$, where $\alpha\mathbf{w} = \mathbf{w}$. Thus, $\alpha\mathbf{w}$ meets all requirements (4.9) which $\mathbf{w}'$ has to satisfy, with the exception of the condition $\operatorname{div} \alpha\mathbf{w} = 0$ on $\Omega_i$, $i = 1, \ldots, N$. Let us now improve the field $\alpha\mathbf{w}$ in the domains $\Omega_i$ to make it solenoidal. Consider the set $\Gamma_i = (S_v \cap \partial\Omega_i) \cup S'_i \cup S''_i$ in $\partial\Omega_i$. By assumption (3), $\partial\Omega_i \setminus \Gamma_i$ has a positive measure in $\partial\Omega_i$. Then the Corollary in Subsection 4.3 implies that there is the vector field $\mathbf{v}_i$ which possesses the following properties:

$$\mathbf{v}_i \in \mathbf{W}^1_2(\Omega_i), \quad \operatorname{div} \mathbf{v}_i = -\operatorname{div} \alpha\mathbf{w}, \quad \mathbf{v}_i|_{\Gamma_i} = 0.$$

The last relation means, in particular, that $\mathbf{v}_i$ has zero traces on $S_i'$ and $S_i''$. Therefore, the extension $\mathbf{w}_i$ of $\mathbf{v}_i$ to all of $\Omega$,

$$\mathbf{w}_i(x) = \begin{cases} \mathbf{v}_i(x) & \text{if } x \in \Omega_i, \\ 0 & \text{if } x \in \Omega \setminus \Omega_i, \end{cases}$$

belongs to $\mathbf{W}_2^1(\Omega)$, see Theorem B.4. We now improve $\alpha\mathbf{w}$ by adding $\mathbf{w}_1, \ldots, \mathbf{w}_N$; it is clear that $\mathbf{w}' = \alpha\mathbf{w} + \mathbf{w}_1 + \cdots + \mathbf{w}_N$, which obviously meets requirements (4.9).

We also define $\mathbf{w}''$ as $\mathbf{w}'' = (1-\alpha)\mathbf{w} + \mathbf{w}_1 + \cdots + \mathbf{w}_N$, that is, $\mathbf{w} = \mathbf{w}' + \mathbf{w}''$. It is clear that $\mathbf{w}''$ satisfies (4.9); in particular, $\operatorname{div} \mathbf{w}'' = 0$ since $\operatorname{div} \mathbf{w} = 0$ and $\operatorname{div}(\alpha\mathbf{w} + \mathbf{w}_1 + \cdots + \mathbf{w}_N) = 0$. This finishes the proof.

REMARK 2. The statement of Theorems 1 and 3 can also be formulated in the following way: the set $\mathcal{U}_d$ is dense in the subspace of sufficiently regular kinematically admissible solenoidal velocity fields

$$\{\mathbf{u} \in \mathbf{W}_2^1(\Omega) : \operatorname{div} \mathbf{u} = 0, \ \mathbf{u}|_{S_v} = 0\}$$

with respect to the $\mathbf{W}_2^1(\Omega)$-norm. Here, $\mathcal{U}_d$ is the set of all smooth solenoidal velocity fields $\mathbf{v}$ vanishing near $S_v$. The latter condition means that, for every $\mathbf{v} \in \mathcal{U}_d$, there exists a number $\delta > 0$ such that $\mathbf{v}(x) = 0$ at every $x \in \Omega$ whose distance from $S_v$ is less than $\delta$.

## Appendix A. Distributions

Normally, one cannot measure directly the value $f(x_0)$ of a physical field $f$ at a point $x_0$. Only a certain mean value

$$\int_\Omega f(x)\varphi(x)\, dV$$

can be determined with an instrument. Here, $\varphi$ is a function which characterizes the instrument. In case the measurement is taken in a neighborhood of $x_0$, the graph of such a test function looks like the one shown in Figure A.1. When $f$ is sufficiently regular and $\varepsilon$ is a small number, the value

$$\frac{\int_\Omega f(x)\varphi(x)\, dV}{\int_\Omega \varphi(x)\, dV}$$

is a good approximation to $f(x_0)$. If $f$ is a continuous function and a collection of "instruments", that is, the test functions $\varphi_\varepsilon$ is available, where the parameter $\varepsilon > 0$ takes arbitrarily small values, then $f(x_0)$ can be determined as

$$f(x_0) = \lim_{\varepsilon \to 0} \frac{\int_\Omega f(x)\varphi_\varepsilon(x)\, dV}{\int_\Omega \varphi_\varepsilon(x)\, dV}.$$

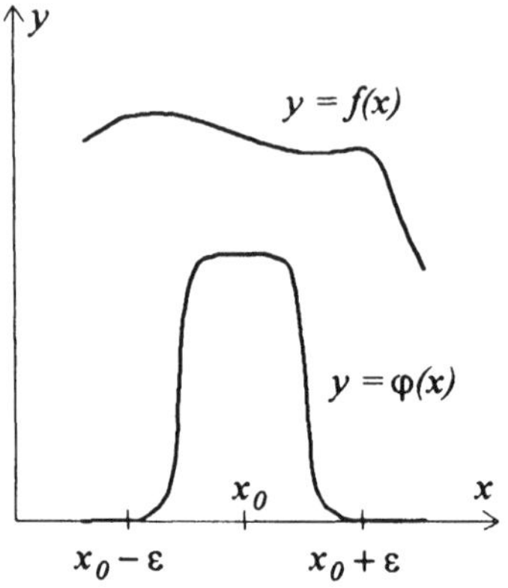


**Figure A.1**

Suppose one can "move the instrument", that is, the collection of test functions analogous to $\varphi_\varepsilon$ is available in a neighborhood of any point in $\Omega$. Then the previous formula can be considered for any $x_0 \in \Omega$ and it determines the values of the continuous field $f$ everywhere in $\Omega$. In case $f$ is only integrable (and noncontinuous), the above procedure determines its values up to redefinition on an arbitrary set of zero volume.

To determine the values of $f$, the integrals

$$\int_\Omega f(x)\varphi(x)\, dV$$

need to be known for a sufficiently wide class of functions $\varphi$. Normally, one makes use of the class $D(\Omega)$ which consists of smooth finite functions on $\Omega$. A function $f$ is called *finite on* $\Omega$ if it vanishes in the exterior of a bounded closed set $C_\varphi \subset \Omega$. We refer to $\varphi$ in $D(\Omega)$ as a *test function.*

Any integrable function $f$ generates a mapping

$$\varphi \rightarrow \int_\Omega f(x)\varphi(x)\, dV, \tag{A.1}$$

which maps a test-function on a number. This mapping is linear and continuous (in a sense we will discuss below). Of course, there are mappings from $D(\Omega)$ into numbers, which are not of the type (A.1). The best known example of such mapping, not generated by a regular function, is Dirac's "delta-function" $\delta_{x_0}$. Here, $x_0$ is a point in $\Omega$ and $\delta_{x_0}$ maps a test-function $\varphi$ on a number $\varphi(x_0)$, $\delta_{x_0} : \varphi \rightarrow \varphi(x_0)$. It looks attractive to complement the collection of regular functions with "generalized functions", that is, with linear continuous mappings from $D(\Omega)$ into $\mathbf{R}$, like the Dirac's "delta-function". This idea turned out to be very fruitful and we will now give an accurate definition of the generalized function.

Consider the set $D(\Omega)$ of test functions, that is, the set of smooth (infinitely differentiable) finite functions on the domain $\Omega$ in $\mathbf{R}^n$. A sequence of test functions $\varphi_1, \varphi_2, \dots$ *converges in* $D(\Omega)$ *to* $\varphi \in D(\Omega)$ if

1) all of $\varphi_1, \varphi_2, \dots$ vanish in the exterior of the same bounded closed subset in $\omega$,
2) the sequence $\varphi_1, \varphi_2, \dots$ uniformly converges to $\varphi$ on $\Omega$, that is

$$\max\{|\varphi_k(x) - \varphi(x)| : x \in \Omega\} \to 0 \quad \text{when} \quad k \to \infty,$$

3) analogously, the sequence of the derivatives

$$\frac{\partial^{\alpha_1+\cdots+\alpha_n}\varphi_k}{\partial x_1^{\alpha_1}\dots\partial x_n^{\alpha_n}}, \qquad k = 1, 2, \dots$$

uniformly converges to the corresponding derivative of $\varphi$ whatever the integers $\alpha_1, \dots, \alpha_n$ are.

In case the sequence $\varphi_1, \varphi_2, \dots$ converges to $\varphi$ in the above sense, we write: $\varphi_k \xrightarrow{D} \varphi$ when $k \to \infty$. The set $D(\Omega)$ equipped with this structure is called the *space of test functions.*

Let $F$ be a linear mapping from $D(\Omega)$ into $\mathbf{R}$; we denote the value $F(\varphi)$ by the symbol $\langle F, \varphi \rangle$. We say that $F$ is *continuous* if, for every sequence of test functions $\varphi_1, \varphi_2, \dots$ which converges in $D(\Omega)$, $\varphi_k \xrightarrow{D} \varphi$ when $k \to \infty$, the equality

$$\lim_{k\to\infty} \langle F, \varphi_k \rangle = \langle F, \varphi \rangle$$

is valid. We refer to a continuous linear mapping $F$ from $D(\Omega)$ into $\mathbf{R}$ as a *distribution* or *generalized function* on $\Omega$.

A regular function $f$ is a distribution:

$$\langle f, \varphi \rangle = \int_\Omega f(x)\varphi(x)\, dV. \tag{A.2}$$

The Dirac's "delta-function" is also a distribution: $\langle \delta_{x_0}, \varphi \rangle = \varphi(x_0)$. Another example of a distribution close to the previous one is the "delta-function" $\delta_S$ associated with the surface $S$ in $\Omega$:

$$\langle \delta_S, \varphi \rangle = \int_S \varphi(x)\, dS.$$

There is a natural definition of a distribution derivative. Indeed, consider distribution (A.2) generated by a differentiable function $f$ on $\Omega$. The derivative $\partial f/\partial x_i$ analogously generates a distribution

$$\langle \frac{\partial f}{\partial x_i}, \varphi \rangle = \int_\Omega \frac{\partial f}{\partial x_i}(x)\varphi(x)\, dV.$$

Since every test-function $\varphi$ vanishes on $\partial\Omega$, we have

$$\langle \frac{\partial f}{\partial x_i}, \varphi(x) \rangle = \int_\Omega \frac{\partial f}{\partial x_i} \varphi(x)\, dV = -\int_\Omega f(x) \frac{\partial \varphi}{\partial x_i}(x)\, dV = -\langle f, \frac{\partial \varphi}{\partial x_i} \rangle.$$

This equality induces the following definition: *derivative of the distribution $F$ with respect to $x_i$* is the distribution (we denote it by the symbol $\partial F/\partial x_i$) whose value at $\varphi \in D(\Omega)$ is

$$\langle \frac{\partial F}{\partial x_i}, \varphi \rangle = -\langle F, \frac{\partial \varphi}{\partial x_i} \rangle.$$

Thus every distribution has derivatives of the first order and, consequently, derivatives of any other order too.

EXAMPLE 1. Consider the domain $(-1,1)$ in $\mathbf{R}$. Let $h$ be the Heaviside function:

$$h(x) = \begin{cases} 0 & \text{if } x < 0, \\ 1 & \text{if } x > 0. \end{cases}$$

This function can also be considered as a distribution

$$\langle h, \varphi \rangle = \int_{-1}^{1} h(x)\varphi(x)\, dx = \int_0^1 \varphi(x)\, dx.$$

By the definition, the derivative $dh/dx$ is the distribution:

$$\langle \frac{dh}{dx}, \varphi \rangle = -\langle h, \frac{d\varphi}{dx} \rangle = \int_0^1 \frac{d\varphi}{dx}\, dx = \varphi(0)$$

(where we took into account that $\varphi(1) = 0$). Thus $\langle dh/dx, \varphi \rangle = \varphi(0)$, and $dh/dx$ is the Dirac's "delta-function" $\delta_0$. It is easy to find the second and subsequent derivatives; for example,

$$\langle \frac{d^2 h}{dx^2}, \varphi \rangle = \langle \frac{d\delta_0}{dx}, \varphi \rangle = -\langle \delta_0, \frac{d\varphi}{dx} \rangle = -\frac{d\varphi}{dx}(0).$$

EXAMPLE 2. One can consider collections of distributions numerated with indices, for example, $\sigma_{ij}$, $i, j = 1, \ldots, n$, and various linear combinations of distributions and their derivatives, for example, $\partial \sigma_{ij}/\partial x_j$ (with the usual summation over repeated indices). One can also consider linear differential equations with unknown distributions, for example, the system of equations

$$\frac{\partial \sigma_{1j}}{\partial x_j} + f_1 = 0, \ \ldots, \ \frac{\partial \sigma_{nj}}{\partial x_j} + f_n = 0,$$

where $\sigma_{ij}$ are unknown and $f_i$ are given distributions. According to the definition of derivative, the system is equivalent to the following condition:

$$\langle -\sigma_{1j}, \frac{\partial \varphi_1}{\partial x_j} \rangle + \langle f_1, \varphi \rangle_1 = 0, \ \ldots, \langle -\sigma_{nj}, \frac{\partial \varphi_n}{\partial x_j} \rangle + \langle f_n, \varphi \rangle_n = 0$$
$$\text{for every} \quad \varphi_1 \in D(\Omega), \ldots \varphi_n \in D(\Omega).$$

Here, the test functions $\varphi_1, \ldots, \varphi_n$ are independent from each other. Therefore, the previous relations can be written as

$$\langle -\sigma_{ij}, \frac{\partial \varphi_i}{\partial x_j} \rangle + \langle f_i, \varphi_i \rangle = 0$$
$$\text{for every collection of} \quad \varphi_1, \ldots, \varphi_n \quad \text{in} \quad D(\Omega),$$

where summation is taken over $i$ and over $j$.

## Appendix B. Sobolev spaces

**B.1. Definition and main properties.** Let $\Omega$ be a domain in $\mathbf{R}^n$ and $L_p = L_p(\Omega)$ be the space of functions on $\Omega$ integrable in $p$-th power, $p \geq 1$. The space is equipped with the norm

$$\|v\|_{L_p(\Omega)} = \left( \int_\Omega |v|^p \, dV \right)^{\frac{1}{p}}$$

(see Appendix V.A). The function $v \in L_p(\Omega)$ is not necessarily differentiable in the regular sense; however, one can consider this function as a distribution on $\Omega$ (Appendix A), and then $v$ possesses derivatives of any order in the sense of distributions. For some functions $v$ these generalized derivatives turn out to be regular functions. We consider now all functions $v$ in $L_p(\Omega)$ whose derivatives $\partial v/\partial x_1, \ldots, \partial v/\partial x_n$ are in $L_p(\Omega)$ too. The set is a linear space and the following norm is defined in it:

$$\|v\|_{W_p^1(\Omega)} = \left( \int_\Omega \left( |v|^p + \left| \frac{\partial v}{\partial x_1} \right|^p + \cdots + \left| \frac{\partial v}{\partial x_n} \right|^p \right) dV \right)^{\frac{1}{p}}. \tag{B.1}$$

We denote this space by the symbol $W_p^1(\Omega)$. The space is complete with respect to norm (B.1) and is referred to as a *Sobolev space*. We also consider Sobolev spaces of vector fields on $\Omega$: $\mathbf{W}_p^1(\Omega) = W_p^1(\Omega) \times \cdots \times W_p^1(\Omega)$.

EXAMPLE 1. The norm in the Sobolev space $\mathbf{W}_2^1(\Omega)$,

$$\|\mathbf{v}\|_{\mathbf{W}_2^1(\Omega)} = \left( \int_\Omega \left( |\mathbf{v}|^2 + \left| \frac{\partial \mathbf{v}}{\partial x_1} \right|^2 + \cdots + \left| \frac{\partial \mathbf{v}}{\partial x_n} \right|^2 \right) dV \right)^{\frac{1}{2}}, \quad \Omega \subset \mathbf{R}^n,$$

corresponds to scalar multiplication defined in $\mathbf{W}_2^1(\Omega)$:

$$(\mathbf{u}, \mathbf{v}) = \int\limits_{\Omega} \left( \mathbf{u}\mathbf{v} + \frac{\partial \mathbf{u}}{\partial x_1}\frac{\partial \mathbf{v}}{\partial x_1} + \cdots + \frac{\partial \mathbf{u}}{\partial x_n}\frac{\partial \mathbf{v}}{\partial x_n} \right) dV.$$

Under some conditions concerning regularity of the boundary $\partial\Omega$, Sobolev spaces possess a number of remarkable properties. We give a brief review of the most important properties assuming that $\Omega$ is a Lipschitz domain (see Subsection 2.1).

THEOREM 1. Suppose $\Omega$ is a Lipschitz domain in $\mathbf{R}^n$. Then every function $v \in W_p^1(\Omega)$, $p \geq 1$, can be extended to all of $\mathbf{R}^n$ as the function $v^c$ in $W_p^1(\mathbf{R}^n)$ satisfying the condition $\|v^c\|_{W_p^1(\mathbf{R}^n)} \leq C\|v\|_{W_p^1(\Omega)}$ with constant $C$ which only depends on the domain $\Omega$.

Using this theorem it is easy to obtain the following proposition.

THEOREM 2. If $\Omega$ is a Lipschitz domain, then the set $C^\infty(\bar{\Omega})$ of infinitely differentiable functions is dense in $W_p^1(\Omega)$, $p \geq 1$.

A function $v$ as a member of $L_p(\Omega)$ is defined up to changing its values on any set of zero measure (see Subsection III.1.2 and Appendix A). Neither does changing values of $v$ affect its derivatives $\partial v/\partial x_i$ in the sense of distributions. Thus, we identify two functions whose values differ only on a zero measure set: they correspond to the same member of $L_p(\Omega)$ and the same member of $W_p^1(\Omega)$. Nevertheless, the following proposition allows us to speak about values of a function $v \in W_p^1(\Omega)$ on a Lipschitz surface (we refer to an $(n-1)$-dimensional manifold in $\mathbf{R}^n$ as a Lipschitz surface if the manifold possesses the same properties as the boundary $\partial\Omega$ in the definition of the Lipschitz domain).

THEOREM 3. Suppose a sequence of smooth functions $v_k \in C^\infty(\bar{\Omega})$, $k = 1, 2, \ldots$, converges to $v$ in $W_p^1(\Omega)$, $p \geq 1$. Let $S$ be a Lipschitz surface in $\Omega$ or in $\partial\Omega$. Then the sequence of the restrictions $v_k|_S$ converges in $L_p(S)$ to a function $v_S \in L_p(S)$.

It is clear that $v_S$ does not depend on the sequence converging to $v$ in the formulation of Theorem 3. We refer to $v_S$ as a *trace* of $v$ on the surface $S$ and denote this function by the symbol $v|_S$.

Theorem 3 is very important for the Sobolev spaces theory. In particular, it results in the following proposition.

THEOREM 4. Suppose a Lipschitz domain $\Omega$ is composed of two Lipschitz domains $\Omega^+$, $\Omega^-$ with the common part $S$ of their boundaries. If $v^+$, $v^-$ are functions in $W_2^1(\Omega^+)$, $W_2^1(\Omega^-)$, respectively, and their traces on $S$ are equal, $v^+|_S = v^-|_S$, then the function

$$v(x) = \begin{cases} v^+(x) & \text{if } x \in \Omega^+, \\ v^-(x) & \text{if } x \in \Omega^-, \end{cases}$$

belongs to $W_p^1(\Omega)$.

In turn, this theorem implies two corollaries which improve the result of Theorem 2 in case $v$ has the zero trace on the part $S$ of the domain boundary. In the formulations of the following propositions, the symbol $\bar{S}$, as always, denotes the closure of the set $S$.

COROLLARY 1. Suppose $\Omega$ is a Lipschitz domain in $\mathbf{R}^n$, $S$ is a part of $\partial\Omega$, and $\Gamma_S$ is a neighborhood of $\bar{S}$ in $\partial\Omega$. If $v \in W_p^1(\Omega)$ and $v|_{\Gamma_S} = 0$, then there is a sequence of smooth functions $v_k \in C^\infty(\bar{\Omega})$, $k = 1, 2, \ldots$, that converges to $v$ in $W_p^1(\Omega)$, each of the functions vanishing on a certain neighborhood of $\bar{S}$ in $\mathbf{R}^n$.

This proposition means, in particular, that the function $v \in W_p^1(\Omega)$ with zero trace $v|_{\partial\Omega} = 0$ can be approximated by smooth functions vanishing near $\partial\Omega$.

COROLLARY 2. Suppose $S$ is a Lipschitz domain in $\mathbf{R}^{n-1}$ and $\Omega$ is a cylinder in $\mathbf{R}^n$:

$$\Omega = \{x \in \mathbf{R}^n : \ (x_1, \ldots, x_{n-1}) \in S, \ 0 < x_n < 1\}.$$

If $v \in W_p^1(\Omega)$ and $v|_S = 0$, then there is a sequence of smooth functions $v_k \in C^\infty(\bar{\Omega})$, $k = 1, 2, \ldots$, that converges to $v$ in $W_p^1(\Omega)$, each of the functions vanishing on a certain neighborhood of $\bar{S}$ in $\mathbf{R}^n$.

Note that, in contrast to the previous proposition, the trace $v|_{\partial\Omega}$ is assumed to vanish on $S$ and not necessarily in its neighborhood in $\partial\Omega$.

By Theorem 3, the subspace of functions with zero trace on the boundary,

$$\overset{\circ}{W}{}_p^1(\Omega) = \{v \in W_p^1(\Omega) : \ v|_{\partial\Omega} = 0\},$$

is closed in $W_p^1(\Omega)$. Thus, $\overset{\circ}{W}{}_p^1(\Omega)$ is a complete space with norm (B.1). The following proposition shows that there is a norm equivalent to (B.1). Recall that norms $||\cdot||_1$ and $||\cdot||_2$ in the space $X$ are equivalent if there are positive constants $c_1$ and $c_2$ such that $c_1||\mathbf{x}||_2 \le ||\mathbf{x}||_1 \le c_2||\mathbf{x}||_2$ for every $\mathbf{x} \in X$. It is clear that a space, which is complete with respect to a certain norm, is also complete with respect to any equivalent norm.

THEOREM 5. Suppose $\Omega$ is a bounded Lipschitz domain in $\mathbf{R}^n$. Then, the norm

$$||v|| = \left( \left\| \frac{\partial u}{\partial x_1} \right\|_{L_p(\Omega)}^p + \cdots + \left\| \frac{\partial u}{\partial x_n} \right\|_{L_p(\Omega)}^p \right)^{\frac{1}{p}}$$

in the space $\overset{\circ}{W}{}_p^1(\Omega)$ is equivalent to norm (B.1).

We note that, in case $p = 2$, this norm is generated by the scalar multiplication

$$(u, v) = \int\limits_\Omega \left( \frac{\partial u}{\partial x_1} \frac{\partial v}{\partial x_1} + \cdots + \frac{\partial u}{\partial x_n} \frac{\partial v}{\partial x_n} \right) dV.$$

The following proposition establishes conditions under which changing of variables maps function $v \in W_p^1(\Omega)$ on a function in an analogous Sobolev space.

THEOREM 6. Suppose $\Omega$ and $\widehat{\Omega}$ are domains in $\mathbf{R}^n$, $f$ is one-to-one mapping from $\widehat{\Omega}$ on $\Omega$, and both $f$ and the inverse mapping $f^{-1}$ satisfy the Lipschitz condition. Then for every $v \in W_p^1(\Omega)$ the function $\widehat{v} : x \to v(f(x))$ is in $W_2^1(\widehat{\Omega})$ and the estimation $\|\widehat{v}\|_{W_2^1(\widehat{\Omega})} \le c\,\|v\|_{W_p^1(\Omega)}$ is valid with constant $c$ which does not depend on $v$.

REMARK 1. Analogously to $W_p^1(\Omega)$ one defines the space of functions whose derivatives up to a certain order are regular functions. For example, let $v$ be a regular function and its derivatives in the sense of distributions $\partial v/\partial x_i$, $\partial^2 v/\partial x_i \partial x_j$, $i, j = 1, \dots, n$, be regular functions integrable in the $p$-th power, $p \ge 1$. Such functions form a complete normed linear space with the norm

$$\|v\|_{W_p^2(\Omega)} = \left( \int_\Omega \left( |v|^p + \sum_{i=1}^n \left| \frac{\partial v}{\partial x_i} \right|^p + \sum_{i,j=1}^n \left| \frac{\partial^2 v}{\partial x_i \partial x_j} \right|^p \right) dV \right)^{\frac{1}{p}} .$$

It is also called the Sobolev space, and we denote it by the symbol $W_p^2(\Omega)$. Analogously one defines the spaces $W_p^l(\Omega)$ with any integer $l$ replacing 2 in the above definition, that is, $l$ being the maximum order of the derivatives which are assumed to be regular functions. Immediate generalizations of Theorems 1 – 6 are valid for the spaces $W_p^l(\Omega)$.

**B.2. Spaces of traces.** Theorem 3 states that every function $v \in W_2^1(\Omega)$ has a trace on the boundary: $v|_{\partial\Omega} \in L_2(\Omega)$. However, the theorem does not answer the question about properties of the function $u$ defined on the domain boundary $\partial\Omega$ which would guarantee that $u$ is a trace of a certain function $v \in W_2^1(\Omega)$: $u = v|_{\partial\Omega}$. We now describe a structure which answers this question.

Consider a set of functions $u$ in $L_2(\mathbf{R}^m)$ for which the integral

$$J(u) = \iint \frac{|u(x) - u(y)|^2}{|x - y|^{m+1}}\, dV_x\, dV_y$$

is finite. These functions form a linear space, and it turns out that the space can be equipped with the norm

$$\|u\|_{W_2^{\frac{1}{2}}(\mathbf{R}^m)} = \left( \int_\Omega u^2\, dV + J(u) \right)^{\frac{1}{2}} .$$

The space is complete with respect to this norm and denoted by the symbol $W_2^{\frac{1}{2}}(\mathbf{R}^m)$.

Using local coordinate systems, one introduces an analogous space of functions defined on a manifold, in particular, on the boundary of a domain in

$\mathbf{R}^n$. Namely, let $\Omega$ be a Lipschitz domain in $\mathbf{R}^m$ (see Subsection 2.1). Consider the cylinders $Q_{r,h}$ mentioned in the definition of the Lipschitz domain. We assume that in everyone of these cylinders the function $f$, involved in the equation $x_n = f(x_1, \ldots, x_{m-1})$ representing $\partial\Omega$ within the cylinder, is continuously differentiable. In this case, we refer to $\partial\Omega$ as the domain with the *continuously differentiable boundary*. Let us choose a finite number of the cylinders $Q_1, \ldots, Q_K$ which forms a covering of $\partial\Omega$. Let $\alpha_1, \ldots, \alpha_K$ be smooth functions such that 1) $\alpha_k$ vanishes in the exterior of a bounded closed set $\operatorname{supp} \alpha_k \subset Q_k$, and 2) the sum $\alpha_1 + \cdots + \alpha_K$ equals 1 on $\partial\Omega$. Consider a square integrable function $u$ on $\partial\Omega$. The function $u_k = \alpha_k u$ is also in $L_2(\partial\Omega)$ and vanishes in the exterior of $\operatorname{supp} \alpha_k \cap \partial\Omega$ in $\partial\Omega$. We now define function $u_{*k}$ which corresponds to $u_k$ and is in $W_2^{\frac{1}{2}}(\mathbf{R}^{m-1})$. Namely, let $x_1, \ldots, x_m$ be a local coordinate system in $Q_k$ and let $x_m = f(x')$, where $x' = (x_1, \ldots, x_{m-1})$ and $|x'| \leq r$, be the equation of $\partial\Omega$ within the cylinder $Q_k$ ($r$ is the radius of the cylinder). Let $P(x')$ denote the point in $\partial\Omega$ with local coordinates $x_1, \ldots, x_{m-1}$, $x_m = f(x_1, \ldots, x_{m-1})$. Then the function $u_{*k}$ is defined on the set $|x'| < r$ in $\mathbf{R}^{m-1}$: $u_{*k}(x') = u_k(P(x'))$. The function vanishes near the boundary $|x'| = r$ together with $\alpha_k$. We extend $u_{*k}$ to all of $\mathbf{R}^{m-1}$ setting $u_{*k}(x') = 0$ if $|x'| > r$; we keep denoting the extension by the same symbol $u_{*k}$. Suppose that the function $u$ is such that $u_{*k} \in W_2^{\frac{1}{2}}(\mathbf{R}^{m-1})$ for every $k = 1, \ldots, K$. It easy to verify that this property of $u$ does not depend on the choice of cylinders $Q_k$ and functions $\alpha_k$. Functions $u$ which possess this property form a complete normed linear space with the norm

$$\|u\|_{W_2^{\frac{1}{2}}(\partial\Omega)} = \left( \|u_{*1}\|^2_{W_2^{\frac{1}{2}}(\mathbf{R}^{m-1})} + \cdots + \|u_{*K}\|^2_{W_2^{\frac{1}{2}}(\mathbf{R}^{m-1})} \right)^{\frac{1}{2}}.$$

We denote this space by the symbol $W_2^{\frac{1}{2}}(\partial\Omega)$. The following proposition shows that $W_2^{\frac{1}{2}}(\partial\Omega)$ is the space of traces on $\partial\Omega$ of functions in $W_2^1(\Omega)$. Namely, the following proposition is valid.

THEOREM 7. Suppose $\Omega$ is a bounded domain in $\mathbf{R}^n$ with a continuously differentiable boundary. Then, for every $v \in W_2^1(\Omega)$ the trace $v|_{\partial\Omega}$ belongs to $W_2^{\frac{1}{2}}(\partial\Omega)$ and the following estimation is valid:

$$\|v|_{\partial\Omega}\|_{W_2^{\frac{1}{2}}(\partial\Omega)} \leq c_1 \|v\|_{W_2^1(\Omega)}$$

with constant $c_1$ which does not depend on $u$. On the contrary, there is a linear operator $R$ on $W_2^{\frac{1}{2}}(\partial\Omega)$ which maps $u \in W_2^{\frac{1}{2}}(\partial\Omega)$ on the function $Ru \in W_2^1(\Omega)$ with the trace $(Ru)|_{\partial\Omega} = u$, the operator being continuous, that is,

$$\|Ru\|_{W_p^1(\Omega)} \leq c_2 \|u\|_{W_2^{\frac{1}{2}}(\partial\Omega)}$$

with constant $c_2$ which does not depend on $u$.

The following useful proposition about functions in $W_2^{\frac{1}{2}}(\mathbf{R}^m)$ should also be mentioned.

THEOREM 8. Suppose $G$ is a Lipschitz domain in $\mathbf{R}^m$ and the function $u \in W_2^{\frac{1}{2}}(\mathbf{R}^m)$ vanishes in the exterior of $G$. Then there exists a sequence of smooth functions $u_k \in C^\infty(\mathbf{R}^m)$, $k = 1, 2, \ldots$, which converges to $u$ in $W_2^{\frac{1}{2}}(\mathbf{R}^m)$, each of the functions vanishing in the exterior of a certain closed subset in $G$.

Propositions analogous to Theorems 7 and 8 hold in a more general case. For example, traces of functions in the Sobolev space $W_p^1(\Omega)$, $p \geq 1$, $\Omega \subset \mathbf{R}^m$, form the space $W_p^{\frac{1}{p'}}(\partial\Omega)$, where $p' = p/(p-1)$. The space is defined exactly as $W_2^{\frac{1}{2}}(\partial\Omega)$, the expression for $J(u)$ and the norm being replaced by

$$J(u) = \iint \frac{|u(x) - u(y)|^p}{|x - y|^{m+p/p'}} \, dV_x \, dV_y,$$

$$\|u\|_{W_p^{\frac{1}{p'}}(\mathbf{R}^m)} = \left( \int_\Omega |u|^p \, dV + J(u) \right)^{\frac{1}{p}},$$

respectively.

## Comments

The main idea of the approach in this chapter, the elimination of the pressure field, is well known in continuum mechanics. The reduced problem obtained by elimination of the pressure field is easier to study than the original one. Reduced problems in mechanics of incompressible viscous fluid were studied by Ladyzhenskaya (1973), and in mechanics of nonlinearly viscous and viscous-plastic media by Mosolov (1978). In particular, Mosolov (1978) proved the equality of the extremum in the reduced static and kinematic extremum problems of limit analysis.

The approach based on eliminating the pressure field is effective in case certain conditions which guarantee that the reduced problem is equivalent to the original one are established. As far as limit analysis is concerned, the required conditions are those for the equality of the extremums in the reduced and original problems (recall that the original extremum problems of limit analysis determine the safety factor). Various conditions for the equality of the extremums were established by Kamenjarzh (1979), Temam and Strang (1980), and Kamenjarzh (1981).

The equivalence of the reduced and original problems immediately follows from solvability of the pressure field restoration problem. The solvability was established by Ladyzhenskaya and Solonnikov (1976) for the case of vanishing velocity on the boundary of a continuum-occupied domain, see Theorem 2.3. The proof was based on a version of the statement which characterizes properties of a distribution when certain properties of its derivatives are known (Theorem 2.2) and on the compatibility conditions (Theorem 2.1). Edwards (1965) and Lions (1969) extracted Theorem 2.1 from a more abstract proposition, Theorem 17' in the de Rham's book (1955). Ladyzhenskaya and Solonnikov (1976) proved Theorem 2.2. In case the boundary conditions are mixed, that is, they are of the kinematic type on a part of the body boundary and of the static type on the remaining part, the solvability of the pressure field restoration problem (Theorem 2.4) was established by Kamenjarzh (1981). A useful remark made by O.O. Barabanov (private communication) allowed us to weaken the boundary regularity requirements in Theorem 2.4.

Theorems about the pressure field restoration become constructive if complemented with propositions which give some simple approximation conditions for solenoidal vector fields. Ladyzhenskaya and Solonnikov (1976), Maslennikova and Bogovsky (1978), and Bogovsky (1979) studied approximation in the case of vector fields vanishing on the whole of the domain boundary. In particular, Ladyzhenskaya and Solonnikov proved Theorem 4.1. Approximation in the case of vector fields with the zero trace on a part of the domain boundary (Theorems 4.3, 4.4) was established by Kamenjarzh (1981). The approximation problem is adjacent to the problem of constructing in the space $\mathbf{W}_p^1(\Omega)$ a vector field with a given divergence. The latter problem was solved by Ladyzhenskaya and Solonnikov (1976) in case $p = 2$ (Theorem 4.2) and by Bogovsky (1979) in case $p \neq 2$.

The theorems about the pressure field restoration and approximation of solenoidal vector fields hold for stress fields and velocity fields with components in $L_p(\Omega)$ and $W_p^1(\Omega)$ $(1 < p < \infty)$, respectively; see Kamenjarzh (1984). Therefore, the above results are applicable to a wide class of problems of incompressible continuum mechanics.

Additional information about the concepts of distributions theory used in this chapter can be found in books by Gelfund and Shilov (1959) and Schwartz (1950, 1951). Sobolev spaces theory is presented in books by Morrey (1966), Ladyzhenskaya (1973) and Mikhlin (1977). Sobolev's book (1991) is fundamental for this domain. In particular, Theorem B.3 about the trace, a version of the Sobolev embedding theorem, is proved in this book. Spaces of traces are described in the books by Lions and Magenes (1968), and Besov, Il'in and Nikolsky (1975). Theorem B.8 was proved by Volevich and Paneyakh (1965).

CHAPTER VII

# LIMIT STATE

The safety factor $\alpha_l$ characterizes behavior of a rigid-plastic body under the action of load $\mathbf{l}$: the body remains rigid if $\alpha_l > 1$, deforms at an arbitrarily large strain rate if $\alpha_l = 1$ and does not bear the load if $\alpha_l < 1$. These are the main properties of the rigid-plastic problem and this chapter completes their mathematical justification. Existence of a solution to the rigid-plastic problem with a nonzero velocity field is established at $\alpha_l = 1$. At the same time, it turns out that only solutions with zero velocity fields exist when $\alpha_l \geq 1$. We now prove the existence theorem within the framework of the rigid-plastic problem unified formulation using concepts of functional analysis.

The reader not interested in the proof of existence theorem and applications of functional analysis to plasticity may proceed to the next chapter.

## 1. Stress field

In this section, we establish existence of a stress field in the limit state, that is, the existence of an admissible stress field equilibrating the limit load. We start with the original formulation of the rigid-plastic problem and in the form of two extremum problems.

**1.1 Limit state problem.** Consider a rigid perfectly plastic body subjected to load $\mathbf{l}$. To solve the rigid-plastic problem for the load $\mathbf{l}$ means to find a stress field $\boldsymbol{\sigma}$ and a velocity field $\mathbf{v}$ satisfying the relations

$$\boldsymbol{\sigma} \in \Sigma + \mathbf{s}_l, \tag{1.1}$$

$$\mathrm{e} = \mathrm{Def}\,\mathbf{v}, \quad \mathbf{v} \in \mathcal{V}, \tag{1.2}$$

$$\boldsymbol{\sigma} \in C, \quad \langle \mathrm{e}, \boldsymbol{\sigma} - \boldsymbol{\sigma}_* \rangle \geq 0 \quad \text{for every} \quad \boldsymbol{\sigma}_* \in C. \tag{1.3}$$

This is the unified formulation of the rigid-plastic problem, which makes use of the stress fields space $\mathcal{S}$ and velocity fields space $\mathcal{V}$. In this fomulation, $\Sigma$ and $C$ are the sets of self-equilibrated and admissible stress fields in $\mathcal{S}$; $\mathrm{Def}\,\mathbf{v}$ is the strain rate field corresponding to the velocity field $\mathbf{v}$; and $\mathbf{s}_l$ is a fixed stress field equilibrating the given load $\mathbf{l}$. The linear operator Def maps $\mathcal{V}$ into the space $\mathcal{E}$ of strain rate fields. The internal power bilinear form $\langle \cdot, \cdot \rangle$ is a pairing between $\mathcal{E}$ and $\mathcal{S}$, the value $-\langle \mathrm{e}, \boldsymbol{\sigma} \rangle$ being the internal power of the stress field $\boldsymbol{\sigma}$ on the strain rate field e. Relation (1.1) is the equilibrium condition for the stress field $\boldsymbol{\sigma}$ and the load $\mathbf{l}$, where $\Sigma = (\mathrm{Def}\,U)^0 = (\mathrm{Def}\,\mathcal{V})^0$ and $U$ is the subspace of virtual velocity fields. Relations (1.3) are the constitutive maximum principle. The rigid-plastic problem unified formulation is discussed in detail in Section IV.1. We always assume that formulation (1.1) – (1.3) meets requirements (I) – (VI), see Subsection IV.1.4.

The safety factor $\alpha_l$ characterizes the main feature of any solution to (1.1) – (1.3) for the load $\mathbf{l}$. According to the definition of $\alpha_l$, problem (1.1) – (1.3) has no solution if $\alpha_l < 1$: there is no admissible stress field equilibrating load $\mathbf{l}$, so that $\mathbf{l}$ is inadmissible. In case $\alpha_l > 1$, the problem has solutions; however, the velocity field of any solution is zero (Propositions IV.3.3 and IV.3.4), so that the load is safe. The solvability of the rigid-plastic problem has not been studied yet in case $\mathbf{l}$ is a limit load, that is, $\alpha_l = 1$. Existence of solutions with nonzero velocity fields is of the most interest since it results in solutions with arbitrarily large strain rates (Subsection IV.3.5). This is why a nonzero velocity field in a rigid perfectly plastic body is referred to as a failure mechanism. Existence of a failure mechanism explains, within the framework of the rigid-plastic analysis, why the problem loses its solution when the load proceeds from safe loads to inadmissible loads through the limit one.

Let us consider rigid-plastic problem (1.1) – (1.3) for a limit load $\mathbf{l}$. For the above reasons, we restrict ourselves to solutions with nonzero velocity fields and refer to (1.1) – (1.3) with this additional requirement as the *limit state problem*.

Theorem IV.4.2 states that $\boldsymbol{\sigma}$, $\mathbf{v} \neq 0$ are a solution to the limit state problem if and only if the corresponding static and kinematic multipliers of the load $\mathbf{l}$ are equal: $m_s(\boldsymbol{\sigma}) = m_k(\mathbf{v})$. Since every static multiplier is not greater than any kinematic multiplier, the latter equality is equivalent to the three following conditions:

- supremum of static multipliers is attained at the stress field $\boldsymbol{\sigma}$,
- infimum of kinematic multipliers is attained at the velocity field $\mathbf{v}$,
- these two extremums are equal.

Conditions for the equality of the extremums are established in Theorems V.4.2 and V.6.1. Under the assumptions of these theorems, the solvability of the limit state problem is equivalent to the attainability of the extremums in the static and kinematic extremum problems (Subsection V.1.1):

$$m \to \sup; \quad m \geq 0, \text{ there exists } \boldsymbol{\rho} \in \Sigma,\ m(\boldsymbol{\rho} + \mathbf{s}_l) \in C, \tag{1.4}$$

$$\frac{D(\mathrm{Def}\,\mathbf{v})}{\langle\, \mathrm{Def}\,\mathbf{v}, \mathbf{s}_l \,\rangle} \to \inf; \quad \mathbf{v} \in \mathcal{V},\ \langle\, \mathrm{Def}\,\mathbf{v}, \mathbf{s}_l \,\rangle > 0. \tag{1.5}$$

We will separately study the attainability of the extremums in the static and kinematic extremum problems.

**1.2. Stress field.** Let us prove the existence of a stress field in the limit state, that is, the attainability of the supremum in the static extremum problem. We adopt the following additional assumption within the framework of the rigid-plastic problem unified formulation. Let us assume that there is a subspace $\mathcal{E}_0$ in $\mathcal{E}$ such that $\mathcal{S}$ is the conjugate of $\mathcal{E}_0$: $\mathcal{S} = \mathcal{E}_0'$. Let us also assume that, for every $\mathrm{e} \in \mathcal{E}_0$ and every $\boldsymbol{\sigma} \in \mathcal{S}$, $\langle\, \mathrm{e}, \boldsymbol{\sigma} \,\rangle$ equals the value which $\boldsymbol{\sigma}$ (as a functional in $\mathcal{S} = \mathcal{E}_0'$) takes at e: $\langle\, \mathrm{e}, \boldsymbol{\sigma} \,\rangle = \boldsymbol{\sigma}(\mathrm{e})$. We consider weak closure in $\mathcal{S}$ adopting $\mathcal{E}_0'$ as the space paired with $\mathcal{S}$; see the definition of the weak closure in Subsection V.4.2.

THEOREM 1. Suppose 1) the subspace $\mathcal{E}_0$ in $\mathcal{E}$ is a complete normed space; 2) $\mathcal{S}$ is the conjugate of $\mathcal{E}_0$: $\mathcal{S} = \mathcal{E}_0'$; 3) $\operatorname{Def} U \subset \mathcal{E}_0$; 4) the set $C$ of admissible stress fields in $\mathcal{S}$ is weak closed; and 5) $C$ is bounded in $\mathcal{S}$-norm. Then, for limit load $\mathbf{l}$ there exists an admissible stress field $\boldsymbol{\sigma}$ that equilibrates the load $\mathbf{l}$.

PROOF. The safety factor of limit load $\mathbf{l}$ is $\alpha_l = 1$. Therefore, an admissible stress field $\boldsymbol{\sigma}$ equilibrates the load if and only if $\boldsymbol{\sigma} = \boldsymbol{\rho} + \mathbf{s}_l$, where $\boldsymbol{\rho}$ is a self-equilibrated stress field, $\boldsymbol{\rho} \in \Sigma$, and the supremum in (1.4) is attained at $\boldsymbol{\rho}$. This is equivalent to attainability of the infimum in

$$M_C(\boldsymbol{\rho} + \mathbf{s}_l) \to \inf, \quad \boldsymbol{\rho} \in \Sigma$$

(Proposition V.2.1), where $M_C$ is the Minkowski function of the set $C$. Note that considering $\mathcal{E}_0$ as $\mathcal{E}$ in the asssumptions of Theorem V.4.1 satisfies these assumptions. Then Theorem V.4.1 yields attainability of the latter infimum, which finishes the proof.

In the case of bodies with bounded yield surfaces, Theorem 1 immediately implies the existence of the stress field in the limit state. In the case of bodies with cylindrical yield surfaces, we will first obtain the existence of the stress deviator field using Theorem 1, and then, appealing to the pressure field restoration theorem, obtain the existence of the stress field, see Theorem 2.2 below.

## 2. Failure mechanism

In this section, we establish the existence of a failure mechanism, that is, a nonzero velocity field in the rigid-plastic body under the action of a limit load. We also present the final formulation and prove the existence theorem for the limit state problem.

**2.1. Strain rate field.** To prove the failure mechanism existence, we have to establish attainability of the extremum in the kinematic problem (1.5) or, which is the same, in the problem

$$\frac{D(\mathbf{e})}{\langle \mathbf{e}, \mathbf{s}_l \rangle} \to \inf; \quad \mathbf{e} \in \operatorname{Def} \mathcal{V}, \ \langle \mathbf{e}, \mathbf{s}_l \rangle > 0.$$

We start, however, with a different problem:

$$\frac{D(\mathbf{e})}{\langle \mathbf{e}, \mathbf{s}_l \rangle} \to \inf; \quad \mathbf{e} \in \Sigma^0, \ \langle \mathbf{e}, \mathbf{s}_l \rangle > 0,$$

where $\Sigma^0 \supset \operatorname{Def} \mathcal{V}$ according to Remark IV.1.2. The latter infimum is the safety factor and it is attained by (Theorem V.6.1). Let the infimum be attained at $\mathbf{e} \in \mathcal{E}$. If it is possible to interpret $\mathbf{e}$ as a kinematically admissible strain rate field, that is, $\mathbf{e} = \operatorname{Def} \mathbf{v}$ for a certain $\mathbf{v}$ in $\mathcal{V}$, then $\mathbf{e}$ is a minimizer in the kinematic extremum problem, and $\mathbf{v}$ is a failure mechanism. However, it is not obvious that e admits such an interpretation. Actually, we have already encountered an example showing that in case the space $V_0$ of sufficiently regular velocity fields is used as $\mathcal{V}$, there may be no suitable $\mathbf{v}$ in $\mathcal{V}$ at all (Subsection III.6.3). This suggests extending the velocity fields space $V_0$ and the operator Def to ensure the failure mechanism exists. Upon performing this procedure below, it will become apparent that $\Sigma^0$ is the subspace of kinematically admissible strain rate fields, both above extremum problems are the same and the existence of a strain rate field in the limit state is established by Theorem V.6.1.

**2.2. Extension scheme.** We will now consider a general scheme for extending a Banach space and a linear operator and apply it to the limit state problem as was planned in the previous subsection.

Let $X$ be a Banach space and $X'$ be its conjugate, that is, the space of linear continuous functionals on $X$. We denote the value of a functional $\mathbf{x}' \in X'$ at $\mathbf{x} \in X$ by the symbol $\langle \mathbf{x}', \mathbf{x} \rangle$. Let $A$ be a linear operator from $X$ into another Banach space $Y$. We now introduce a linear operator $A^T$ from $Y'$ into $X'$ which maps $\mathbf{y}' \in Y'$ on $A^T\mathbf{y}' \in X'$, $A^T\mathbf{y}'$ being a linear continuous functional whose value at $\mathbf{x} \in X$ is $\langle A^T\mathbf{y}', \mathbf{x} \rangle = \langle \mathbf{y}', A\mathbf{x} \rangle$. We refer to $A^T$ as an *adjoined* operator of $A$. (We use the same symbol $\langle \cdot, \cdot \rangle$ to denote different pairings: between $X$ and $X'$, $Y$ and $Y'$, etc., as long as this does not result in ambiguity.) It is clear that $A^T$ is a linear continuous operator, which allows consideration of the operator conjugate of $A^T$. We denote this operator $(A^T)^T$ by the symbol $A^{TT}$. According to the definition, $A^{TT}$ is an operator from $X''$ into $Y''$; it maps $\mathbf{x}'' \in X''$ on $A^{TT}\mathbf{x}'' \in Y''$ which is a linear continuous functional whose value at $\mathbf{y}' \in Y'$ is $\langle A^{TT}\mathbf{x}'', \mathbf{y}' \rangle = \langle \mathbf{x}'', A^T\mathbf{y}' \rangle$.

Let us show that $A^{TT}$ is an extension of $A$ to all of $X''$. Recall that $X$ is naturally embedded in $X''$, namely, a linear continuous functional $\bar{\mathbf{x}}$ on $X'$ corresponds to $\mathbf{x} \in X$, its value $\langle \bar{\mathbf{x}}, \mathbf{x}' \rangle$ at $\mathbf{x}' \in X'$ being equal $\langle \mathbf{x}', \mathbf{x} \rangle$. It is clear that $\mathbf{x} \rightarrow \bar{\mathbf{x}}$ is a one-to-one mapping. Moreover, the mapping does not alter the norm: $\|\bar{\mathbf{x}}\|_{X''} = \|\mathbf{x}\|_X$. Indeed, using the definition of the norm in the space $X''$ conjugate of $X'$, we obtain:

$$\begin{aligned} \|\bar{\mathbf{x}}\|_{X''} &= \sup \{ \langle \bar{\mathbf{x}}, \mathbf{x}' \rangle : \mathbf{x}' \in X', \ \|\mathbf{x}'\| \leq 1 \} \\ &= \sup \{ \langle \mathbf{x}', \mathbf{x} \rangle : \mathbf{x}' \in X', \ \|\mathbf{x}'\| \leq 1 \} = \|\mathbf{x}\| , \end{aligned}$$

where the latter inequality is established by the Hahn-Banach theorem in its analytical form. Thus, $X$ is embedded in $X''$; therefore, we identify $\bar{\mathbf{x}}$ and $\mathbf{x}$ and consider $X$ as a linear subspace in $X''$. Due to this natural embedding, we can speak about values of the operator $A^{TT}$ on the subspace $X$ in $X''$. Let us show that, at every point in $X$, the operator $A^{TT}$ takes the same value as $A$. Indeed, using subscripts to indicate which spaces are paired by a bilinear form, we obtain from the above definitions the following equalities

$$\begin{aligned} \langle A^{TT}\mathbf{x}, \mathbf{y}' \rangle_{Y'',Y'} &= \langle \mathbf{x}, A^T\mathbf{y}' \rangle_{X'',X'} = \langle A^T\mathbf{y}', \mathbf{x} \rangle_{X',X} = \langle \mathbf{y}', A\mathbf{x} \rangle_{Y',Y} \\ &= \langle A\mathbf{x}, \mathbf{y}' \rangle_{Y'',Y'} \end{aligned}$$

for every $\mathbf{x} \in X$ and every $y' \in Y'$ (the second and the last equalities account for the natural embedding: $X$ in $X''$ and $Y$ in $Y''$). We observe that $A^{TT}\mathbf{x} =$

$A\mathbf{x}$ for every $\mathbf{x} \in X$. Thus, $X''$ *is an extension of the space* $X$, and $A^{TT}$ *is an extension, to all of* $X''$, *of the linear continuous operator* $A$ defined on $X$.

Under certain assumptions about $A$ the operator $A^{TT}$ possesses some important properties described in the following proposition. The formulation of the proposition mentions polar sets; recall that the polar set $L^0$ of a subspace $L$ in a Banach space $B$ is the subspace

$$L^0 = \{\mathbf{x}' \in B' : \langle \mathbf{x}', \mathbf{x} \rangle = 0 \text{ for every } \mathbf{x} \in L\}$$

in $B'$, the latter being conjugate of $B$. In turn, one considers the polar set $(L^0)^0$ for $L^0$, the former being a subspace in $B''$. We denote $(L^0)^0$ by the symbol $L^{00}$ and refer to it as the *bipolar set* of $L$. We also recall that the range $R_A$ of the operator $A$ is the set of all values taken by $A$.

PROPOSITION 1. Suppose $A$ is a one-to-one linear operator from a Banach space $X$ into a Banach space $Y$. If the range $R_A$ of $A$ is closed, then $A^{TT}$ 1) is a one-to-one operator; 2) maps the space $X''$ *on* the bipolar $R_A^{00}$; and 3) when considered as an operator from $X''$ into $R_A^{00}$, $A^{TT}$ possesses a continuous inverse operator.

PROOF. Note that the first statement is valid if $A^T$ maps $Y'$ *on* $X'$. Indeed, according to the definition of $A^{TT}$ the equality $A^{TT}\mathbf{x}_1'' = A^{TT}\mathbf{x}_2''$ means that the relation $\langle \mathbf{x}_1'' - \mathbf{x}_2'', A^T\mathbf{y}' \rangle = 0$ holds for every $\mathbf{y}' \in Y'$. In case the range of $A$ is $X'$, this is the same as

$$\langle \mathbf{x}_1'' - \mathbf{x}_2'', \mathbf{x}' \rangle \quad \text{for every} \quad \mathbf{x}' \in X,$$

and, consequently, $\mathbf{x}_1'' = \mathbf{x}_2''$. Thus, $A^{TT}\mathbf{x}_1'' = A^{TT}\mathbf{x}_2''$ implies $\mathbf{x}_1'' = \mathbf{x}_2''$; $A^{TT}$ is a one-to-one operator. Then, to prove the first statement, it remains to verify that $A^T$ maps $Y'$ *on* $X'$, that is, for every $\mathbf{x}' \in X'$ there is a member $\mathbf{y}'$ in $Y'$ such that $A^T\mathbf{y}' = \mathbf{x}'$. We observe that Banach theorem about the inverse operator is applicable since $R_A$ is closed. Therefore, the operator $A^{-1}$ inverse to $A$ is defined on $R_A \subset Y$, and $A^{-1}$ is continuous: the inequality $\|A^{-1}\mathbf{y}\| \leq c\,\|\mathbf{y}\|$ with a constant $c$ is valid for every $\mathbf{y} \in R_A$. Then, any $\mathbf{x}'$ in $X'$ generates a linear functional $f_{x'}$ on $R_A$:

$$f_{x'}(\mathbf{y}) = \langle \mathbf{x}', A^{-1}\mathbf{y} \rangle, \quad |f_{x'}(\mathbf{y})| \leq \|\mathbf{x}'\|\, c\,\|\mathbf{y}\|, \ \ \mathbf{y} \in R_A. \tag{2.1}$$

The Hahn-Banach theorem extends $f_{x'}$ to all of $Y$ as a linear continuous functional; we denote this functional by $\mathbf{y}'$. Let us show that $A^T\mathbf{y}' = \mathbf{x}'$, where $\mathbf{x}'$ is the very member in $X'$ that generates $\mathbf{y}'$. Indeed, for every $x \in X$:

$$\langle A^T\mathbf{y}', \mathbf{x} \rangle = \langle \mathbf{y}', A\mathbf{x} \rangle = f_{x'}(A\mathbf{x}) = \langle \mathbf{x}', A^{-1}A\mathbf{x} \rangle = \langle \mathbf{x}', \mathbf{x} \rangle,$$

that is, $A^T\mathbf{y}' = \mathbf{x}'$. Thus, $A^T$ maps $Y'$ *on* $X'$, which finishes the proof of the first statement. It is worth mentioning that, by the Hahn-Banach theorem, one can choose an extension $\mathbf{y}'$ of the functional $f_{x'}$ such that the estimation $\|\mathbf{y}'\| \leq c\,\|\mathbf{x}'\|$ is valid.

Let us now show that the bipolar $R_A^{00}$ is the range of $A^{TT}$. Consider any value $A^{TT}\mathbf{x}''$ of the operator $A^{TT}$ and arbitrary $\mathbf{y}' \in R_A^0$. Since $A^T\mathbf{y}' = 0$ for such $\mathbf{y}'$, then

$$\langle A^{TT}\mathbf{x}'', \mathbf{y}' \rangle = \langle \mathbf{x}'', A^T\mathbf{y}' \rangle = 0$$

and, hence, $A^{TT}\mathbf{x}''$ is in $R_A^{00}$. It remains to verify that for any given $\mathbf{y}'' \in R_A^{00}$ there is $\mathbf{x}'' \in X''$ such that $A^{TT}\mathbf{x}'' = \mathbf{y}''$. Let us determine a suitable $\mathbf{x}''$ by defining the value of the linear continuous functional $\mathbf{x}''$ at every $\mathbf{x}' \in X'$. For a given $\mathbf{x}'$, we consider a member $\mathbf{y}' \in Y'$ which satisfies the condition $A^T\mathbf{y}' = \mathbf{x}'$ (we have already established existence of such $\mathbf{y}'$). Then, we consider the number $\langle \mathbf{y}'', \mathbf{y}' \rangle$ which is uniquely determined for the given $\mathbf{y}''$ and $\mathbf{x}'$. Indeed, even if there are different $\mathbf{y}_1'$ and $\mathbf{y}_2'$ for which $A^T\mathbf{y}_1' = A^T\mathbf{y}_2' = \mathbf{x}'$, it is clear that $\mathbf{y}_1' - \mathbf{y}_2' \in R_A^0$ and $\langle \mathbf{y}'', \mathbf{y}_1' \rangle = \langle \mathbf{y}'', \mathbf{y}_2' \rangle$ since $\mathbf{y}'' \in R_A^{00}$. Thus, for the given $\mathbf{y}''$, we constructed a mapping from $X'$ into $\mathbf{R}$, $\mathbf{x}' \to \langle \mathbf{y}'', \mathbf{y}' \rangle$, where $\mathbf{y}' \in Y'$ is chosen so that $A^T\mathbf{y}' = \mathbf{x}'$. This mapping is obviously linear and continuous since $\mathbf{y}'$ can be chosen to satisfy the above-mentioned estimation $\|\mathbf{y}'\| \leq c\,\|\mathbf{x}'\|$. Thus, the mapping is a linear continuous functional on $X'$, that is, there is a certain $\mathbf{x}''$ in $X''$ such that

$$\langle \mathbf{x}'', \mathbf{x}' \rangle = \langle \mathbf{y}'', \mathbf{y}' \rangle, \quad \text{where} \quad \mathbf{x}' = A^T\mathbf{y}', \ \mathbf{y}' \in Y'. \tag{2.2}$$

This means that $A^{TT}\mathbf{x}'' = y''$. Indeed, the equality

$$\langle A^{TT}\mathbf{x}'', \boldsymbol{\eta}' \rangle = \langle \mathbf{x}'', A^T\boldsymbol{\eta}' \rangle$$

is valid for every $\boldsymbol{\eta}' \in Y'$, and the right-hand side of the previous equality is $\langle \mathbf{y}'', \boldsymbol{\eta}' \rangle$ by (2.2). This implies that $A^{TT}\mathbf{x}'' = \mathbf{y}''$, that is $\mathbf{x}''$ satisfying this equality is found for the given $\mathbf{y}'' \in R_A^{00}$, and the second statement of the proposition is proved.

Note that $A^{TT}$ is a one-to-one continuous operator. The range of $A^{TT}$ is the bipolar $R_A^{00}$, that is, a closed set. Then, by the Banach theorem about the inverse operator, $A^{TT}$ possesses a continuous inverse operator on $R_A^{00}$, which finishes the proof.

**2.3. Rigid-plastic problem weak formulation.** We now give a formulation of the rigid-plastic problem which involves an extended space of velocity fields and an extension of the strain rate operator. We make use of the extension scheme presented in the previous subsection.

Let

(a) $V_0$ and $\mathcal{E}_0$ be Banach spaces,

(b) $\mathrm{Def}_0$ be a one-to-one linear continuous operator from $V_0$ into $\mathcal{E}_0$, and its range be closed.

We consider the rigid-plastic problem assuming that

(c) the set $U$ of virtual velocity fields is a linear subspace dense in $V_0$;

(d) $\mathcal{V} = V_0''$ is the space of velocity fields, $\mathcal{L} = V_0'$ is the space of loads (thus, $\mathcal{V} = \mathcal{L}'$);
the external power bilinear form $\langle\langle \cdot, \cdot \rangle\rangle$ is a pairing between $\mathcal{V}$ and $\mathcal{L}$, $\langle\langle \mathbf{v}, \mathbf{l} \rangle\rangle$ being the value of the functional $\mathbf{v} \in \mathcal{V} = \mathcal{L}'$ at $\mathbf{l} \in \mathcal{L}$;

(e) $\mathcal{E} = \mathcal{E}_0'$ is the space of strain rate fields, $\mathcal{S} = \mathcal{E}_0'$ is the space of stress fields (thus, $\mathcal{E} = \mathcal{S}'$);
the internal power bilinear form $\langle \cdot, \cdot \rangle$ is a pairing between $\mathcal{E}$ and $\mathcal{S}$, $\langle \mathbf{e}, \mathbf{s} \rangle$ being the value of the functional $\mathbf{e} \in \mathcal{E} = \mathcal{S}'$ at $\mathbf{s} \in \mathcal{S}$.

(f) $\mathrm{Def} = (\mathrm{Def}_0)^{TT}$.

It is clear that this choice meets requirement (II) in Subsection IV.1.4 and also requirement (III) since Def is a one-to-one operator by Proposition 1. Let us show that requirement (IV) is also satisfied by verifying that the equality

$$\langle \mathrm{Def}_0\, \mathbf{v}, \boldsymbol{\sigma} \rangle = \langle\langle \mathbf{v}, \mathbf{l} \rangle\rangle \quad \text{for every} \quad \mathbf{v} \in U$$

implies

$$\langle \mathrm{Def}\, \mathbf{v}, \boldsymbol{\sigma} \rangle = \langle\langle \mathbf{v}, \mathbf{l} \rangle\rangle \quad \text{for every} \quad \mathbf{v} \in \mathcal{V}.$$

Note that the equality $\langle \mathrm{Def}_0\, \mathbf{v}_0, \boldsymbol{\sigma} \rangle = \langle\langle \mathbf{v}_0, \mathbf{l} \rangle\rangle$ holds for every $\mathbf{v}_0 \in V_0$ since it is valid for every $\mathbf{v}$ in $U$ and $U$ is dense in $V_0$. In other words,

$$\langle \mathbf{v}_0, \mathrm{Def}_0^T\, \boldsymbol{\sigma} \rangle = \langle\langle \mathbf{v}_0, \mathbf{l} \rangle\rangle \quad \text{for every} \quad \mathbf{v}_0 \in V_0. \tag{2.3}$$

Then, every $\mathbf{v}$ in $\mathcal{V} = V_0''$ is a weak limit point of the set $V_0 \subset V_0''$ (we use the standard pairing between $V_0''$ and $V_0'$ to define the weak convergence in $V_0''$). Because of that, (2.3) implies

$$\langle \mathbf{v}, \mathrm{Def}_0^T\, \boldsymbol{\sigma} \rangle = \langle\langle \mathbf{v}, \mathbf{l} \rangle\rangle \quad \text{for every} \quad \mathbf{v} \in V$$

or, which is the same, the required relation

$$\langle (\mathrm{Def}_0)^{TT} \mathbf{v}, \boldsymbol{\sigma} \rangle = \langle\langle \mathbf{v}, \mathbf{l} \rangle\rangle \quad \text{for every} \quad \mathbf{v} \in \mathcal{V}.$$

Thus, if the spaces $V_0$, $\mathcal{E}_0$, $U$, $\mathcal{V}$, $\mathcal{E}$, $\mathcal{S}$ and the operator Def satisfy conditions (a) – (f), the corresponding rigid-plastic problem formulation meets requirements (I) – (IV) in Subsection IV.1.4. To solve this problem means finding stress and velocity fields $\boldsymbol{\sigma}$ and $\mathbf{v}$ satisfying the relations

$$\begin{gathered} \boldsymbol{\sigma} \in \Sigma + \mathbf{s}_l, \\ \mathbf{e} = \mathrm{Def}\, \mathbf{v}, \quad \mathbf{v} \in \mathcal{V}, \\ \boldsymbol{\sigma} \in C, \quad \langle \mathbf{e}, \boldsymbol{\sigma} - \boldsymbol{\sigma}_* \rangle \quad \text{for every} \quad \boldsymbol{\sigma}_* \in C. \end{gathered} \tag{2.4}$$

Here, as usual, $C$ is the set of admissible stress fields and $\mathbf{s}_l$ is a fixed stress field which equilibrates the load $\mathbf{l}$. Although (2.4) is solely the unified formulation of the rigid-plastic problem, however, we specify here that $V_0$, $\mathcal{E}_0$, $U$, $\mathcal{V}$, $\mathcal{E}$, $\mathcal{S}$ and Def satisfy assumptions (a) – (f). In this case, we refer to (2.4) as the *weak formulation* of the rigid-plastic problem and to its solution as the *weak solution.*

It is easily seen that any strong solution to the rigid-plastic problem is a solution in the weak sense. On the other hand, if the velocity field $\mathbf{v}$ in a weak solution $\boldsymbol{\sigma}$, $\mathbf{v}$ is regular, that is, belongs to $V_0$, then $\boldsymbol{\sigma}$, $\mathbf{v}$ is a strong solution to the rigid-plastic problem.

Any proposition established within the framework of the unified formulation of the rigid-plastic problem unified formulation definitely holds within the framework of the weak formulation. In particular, the rigid-plastic problem only possesses solutions with the zero velocity field in the extended space of velocity fields if the load $\mathbf{l}$ is safe, $\alpha_l > 1$; the problem has no solutions if the load is inadmissible, $\alpha_l < 1$. If $\mathbf{l}$ is a limit load, the problem may have no strong solution with a nonzero velocity field (failure mechanism). We will show that a failure mechanism exists within the framework of the weak formulation.

**2.4. Limit state.** We now establish the existence of a weak solution to the limit state problem. In the following proposition we consider the weak closure in the space $\mathcal{S}$ (see Subsection V.4.2) defined through the standard pairing $\langle \cdot, \cdot \rangle$ between $\mathcal{E}_0$ and $\mathcal{S} = \mathcal{E}_0'$: $\langle \mathbf{e}, \boldsymbol{\sigma} \rangle = \boldsymbol{\sigma}(\mathbf{e})$ is the value of the functional $\boldsymbol{\sigma} \in \mathcal{S}$ at $\mathbf{e} \in \mathcal{E}_0'$.

THEOREM 1. Suppose that, within the framework of the rigid-plastic problem unified formulation, 1) the set $C$ of admissible stress fields is bounded in $\mathcal{S}$ and contains the ball $B_r(0)$, $r > 0$, and 2) $C$ is weak closed. Then, for any limit load $\mathbf{l}$ the limit state problem has a solution, that is, there exists a stress field $\boldsymbol{\sigma}$ and a velocity field $\mathbf{v} \neq 0$ satisfying (2.4).

PROOF. We observe that the assumptions of Theorem 1.1 are satisfied, and, therefore, there is a stress field $\boldsymbol{\sigma}$ satisfying the conditions $\boldsymbol{\sigma} \in \Sigma + \mathbf{s}_l$, $\boldsymbol{\sigma} \in C$. Moreover, by Theorem V.6.1 (see also Subsection 2.1), there is a strain rate field $\mathbf{e} \in \Sigma^0$ such that $\langle \mathbf{e}, \mathbf{s}_l \rangle = D(\mathbf{e})$ and $\langle \mathbf{e}, \mathbf{s}_l \rangle > 0$. Then the equality $\langle \mathbf{e}, \boldsymbol{\sigma} \rangle = \langle \mathbf{e}, \mathbf{s}_l \rangle$ is valid since $\boldsymbol{\sigma} - \mathbf{s}_l \in \Sigma$ and $\mathbf{e} \in \Sigma^0$. This allows writing the relation $\langle \mathbf{e}, \mathbf{s}_l \rangle = D(\mathbf{e})$ in the form $\langle \mathbf{e}, \boldsymbol{\sigma} \rangle = D(\mathbf{e})$ or, using the definition of dissipation, in the form

$$\boldsymbol{\sigma} \in C, \quad \langle \mathbf{e}, \boldsymbol{\sigma} - \boldsymbol{\sigma}_* \rangle \geq 0 \quad \text{for every} \quad \boldsymbol{\sigma}_* \in C.$$

It remains to appeal to Proposition 1: $\Sigma^0 = (\operatorname{Def}_0 U)^{00}$ is the range of the one-to-one operator Def. Consequently, for $\mathbf{e} \in \Sigma^0$ there is a velocity field $\mathbf{v} \in \mathcal{V}$ such that $\operatorname{Def} \mathbf{v} = \mathbf{e}$. We note that $\mathbf{v} \neq 0$ since $\mathbf{e} \neq 0$, which finishes the proof.

When considering a certain class of rigid perfectly plastic bodies, one chooses concrete spaces $V_0$, $\mathcal{V}$, $\mathcal{E}_0$, and the operator $\operatorname{Def}_0$. We now discuss, as the main example, this choice in the case of three-dimensional bodies. Consider a body occupying domain $\Omega$ and subjected to the action of the load $\mathbf{l}$. The space $U$ of virtual velocity fields consists of all smooth vector fields which vanish on the fixed part $S_v$ of the body boundary; $\operatorname{Def}_0$ is the regular strain rate operator:

$$(\operatorname{Def}_0 \mathbf{v})_{ij} = \frac{1}{2}\left(\frac{\partial v_i}{\partial x_j} + \frac{\partial v_j}{\partial x_i}\right).$$

We consider bodies of the two types: with bounded and with cylindrical yield surfaces. The sets $C_x$ of admissible stresses are bounded in the first case and are of the form

$$C_x = \{\boldsymbol{\sigma} \in Sym : \boldsymbol{\sigma}^d \in C_x^d\}$$

in the second case. The superscript $d$ as always indicates the deviatoric part of a tensor; $C_x^d$ is a given set in the deviatoric plane $Sym$. We assume that the sets $C_x^d$ are bounded and choose the following stress fields spaces:

$$\mathcal{S} = L_\infty(\Omega; Sym), \qquad \mathcal{S} = L_\infty(\Omega; Sym^d) \times L_2(\Omega; Sym^s)$$

for bodies with bounded and cylindrical yield surfaces, respectively (see Subsections V.5.2 and V.7.2). This choice predetermines the space $\mathcal{E}_0$: to satisfy the requirement (e) in Subsection 2.3, one sets

$$\mathcal{E}_0 = L_1(\Omega; Sym), \qquad \mathcal{E}_0 = L_1(\Omega; Sym^d) \times L_2(\Omega; Sym^s).$$

Consider the closure of $U$ with respect to the norm

$$\|\mathbf{v}\|^2 = \|\,|\mathbf{v}|\,\|^2_{L_1(\Omega)} + \|\,|\mathrm{Def}_0\,\mathbf{v}|\,\|^2_{L_1(\Omega)} \tag{2.5}$$

as the space $V_0$ in the case of bodies with bounded yield surfaces. In the case of bodies with cylindrical yield surfaces, $V_0$ is the closure of $U$ with respect to the norm

$$\|\mathbf{v}\|^2 = \|\,|\mathbf{v}|\,\|^2_{L_1(\Omega)} + \|\,|(\mathrm{Def}_0\,\mathbf{v})^d|\,\|^2_{L_1(\Omega)} + \left\|\frac{1}{3}\,\mathrm{div}\,\mathbf{v}\right\|^2_{L_2(\Omega)}. \tag{2.6}$$

In both cases $\mathrm{Def}_0$ meets requirements (b) in Subsection 2.3. Indeed, $\mathrm{Def}_0$ is a linear continuous operator and it maps $V_0$ into $\mathcal{E}_0$. Let us show that $\mathrm{Def}_0$ is a one-to-one operator, that is, for $\mathbf{v} \in V_0$ the equality $\mathrm{Def}_0\,\mathbf{v} = 0$ implies $\mathbf{v} = 0$. Indeed, the strain rate field $\mathbf{e}$ is zero only for rigid body motion, that is, the velocity field is of the form $\mathbf{v}(\mathbf{r}) = \mathbf{v}_0 + \boldsymbol{\omega}_0 \times \mathbf{r}$, where $\mathbf{v}_0$ and $\boldsymbol{\omega}_0$ do not depend on the position vector $\mathbf{r}$. Since every $\mathbf{v} \in V_0$ satisfies the condition $\mathbf{v}|_{S_v} = 0$, we have $\mathbf{v}_0 = 0$ and $\boldsymbol{\omega}_0 = 0$. Thus, $\mathbf{v} = 0$ if $\mathrm{Def}_0\,\mathbf{v} = 0$, and $\mathrm{Def}_0$ is a one-to-one operator.

The following comment should be added to the above reasoning: 1) we assumed that $S_v$ is of positive area; 2) we stated that $\mathbf{v}|_{S_v} = 0$ by virtue of the trace theorem which is analogous to Theorem VI.B.3 and is valid for $V_0$.

We observe that $\mathrm{Def}_0$ has the closed range in $\mathcal{E}_0$. This follows from the estimation

$$\|\mathbf{v}\|_{V_0(\Omega)} \leq c\,\|\,|\mathrm{Def}_0\,\mathbf{v}|\,\|_{L_1(\Omega)},$$

which, in turn, arises from the well known inequality

$$\|\mathbf{v}\|_{L_1(\Omega)} \leq c\,\|\,|\mathrm{Def}_0\,\mathbf{v}|\,\|_{L_1(\Omega)}.$$

The latter inequality is valid in case $S_v$ has a positive area (or, in case a condition different from $\mathbf{v}|_{S_v} = 0$ forbids rigid motion of the body).

Following the general scheme in Subsection 2.3, we now introduce extended spaces $\mathcal{V} = V_0''$ and $\mathcal{E} = \mathcal{E}_0''$ of velocity and strain rate fields and the extension $\mathrm{Def} = (\mathrm{Def}_0)^{TT}$ of the operator $\mathrm{Def}_0$. We consider $V_0'$ as the space $\mathcal{L}$ of loads. The external power bilinear form $\langle\langle\,\cdot\,,\cdot\,\rangle\rangle$ is defined on $\mathcal{V} \times \mathcal{L}$: $\langle\langle\,\mathbf{v}, \mathbf{l}\,\rangle\rangle$ is the value of the functional $\mathbf{v} \in \mathcal{V}$ at $\mathbf{l} \in \mathcal{L}$. In case $\mathbf{v}$ is a sufficiently regular field and $\mathbf{l}$ is represented by sufficiently regular fields $\mathbf{f}$, $\mathbf{q}$ of body forces and surface tractions on $S_q$, we have

$$\langle\langle \mathbf{v}, \mathbf{l} \rangle\rangle = \int_{\Omega} \mathbf{f}\mathbf{v}\, dV + \int_{S_q} \mathbf{q}\mathbf{v}\, dV.$$

Analogously, the internal power bilinear form $\langle \cdot, \cdot \rangle$ is defined on $\mathcal{E} \times \mathcal{S}$: $\langle \mathbf{e}, \boldsymbol{\sigma} \rangle$ is the value of the functional $\mathbf{e} \in \mathcal{E}$ at $\boldsymbol{\sigma} \in \mathcal{S}$. (It should be recalled that, from the mechanical viewpoint,$-\langle \mathbf{e}, \boldsymbol{\sigma} \rangle$ is the internal power of the stress field $\boldsymbol{\sigma}$ on the strain rate field $\mathbf{e}$.) In case $\mathbf{e}$ is an integrable strain rate field, we have

$$\langle \mathbf{e}, \boldsymbol{\sigma} \rangle = \int_{\Omega} \mathbf{e} \cdot \boldsymbol{\sigma}\, dV.$$

The following proposition establishes the existence of a solution to the limit state problem within the framework of the above weak formulation of the rigid-plastic problem for three-dimensional bodies. We assume, as always, that for every point $x$ of the body the set $C_x$ is convex, closed and contains the origin, and the mapping $x \to C_x$ is measurable.

THEOREM 2. Suppose that 1) in the case of a body with bounded yield surfaces, the collection of all sets $C_x$ is bounded and each of $C_x$ contains the same ball $|\mathbf{s}| \leq r$, $r > 0$, in the stress space, or 2) in the case of a body with cylindrical yield surfaces the collection of all sets $C_x^d$ is bounded, each $C_x^d$ contains the same ball $|\mathbf{s}^d| \leq r$, $r > 0$, in the deviatoric plane, and the body boundary and its free part are sufficiently regular. Then the limit state problem has a solution within the framework of the rigid-plastic problem weak formulation, that is, for any limit load $\mathbf{l}$, there exist a stress field $\boldsymbol{\sigma}$ and a velocity field $\mathbf{v} \neq 0$ satisfying the relations

$$\begin{gathered} \boldsymbol{\sigma} \in \Sigma + \mathbf{s}_l, \\ \mathbf{e} = \operatorname{Def} \mathbf{v}, \quad \mathbf{v} \in \mathcal{V}, \\ \boldsymbol{\sigma} \in C, \quad \langle \mathbf{e}, \boldsymbol{\sigma} - \boldsymbol{\sigma}_* \rangle \geq 0 \quad \text{for every} \quad \boldsymbol{\sigma}_* \in C. \end{gathered} \tag{2.7}$$

PROOF. In the case of a body with bounded yield surfaces, the proposition is a particular case of Theorem 1. In the case of a body with cylindrical yield surfaces, we consider the deviatoric problem:

$$\begin{gathered} \boldsymbol{\tau} \in \Sigma_d + \mathbf{s}_l^d, \\ \mathbf{e} = \operatorname{Def} \mathbf{v}, \quad \mathbf{v} \in \mathcal{V}_d, \\ \boldsymbol{\tau} \in C^d, \quad \langle \mathbf{e}, \boldsymbol{\tau} - \boldsymbol{\tau}_* \rangle \geq 0 \quad \text{for every} \quad \boldsymbol{\tau}_* \in C^d. \end{gathered} \tag{2.8}$$

Comparison of this formulation to the original one shows that (2.8) looks exactly like (2.7) with the space of virtual velocity fields $U$ replaced by its subspace $U_d$ consisting of solenoidal fields, $\mathcal{E}_0$ replaced by its subspace

$$\mathcal{E}_0^d = \{\mathbf{e} \in \mathcal{E}_0 : \mathbf{e}^s = 0\},$$

$C_x$ replaced by $C_x^d = \{\boldsymbol{\sigma} \in C : \boldsymbol{\sigma}^s = 0\}$, $V_0$ by the closure $V_{0d}$ of $U_d$ in norm (2.6), $\mathcal{V}$ by $\mathcal{V}_d = (V_{0d})''$, and $\Sigma$ by $\Sigma_d = (\mathrm{Def}_0\, U_d)^0$. The safety factor of the load $\mathbf{l}$ within the framework of this problem is $\alpha_{ld}$ defined by (IV.1.2). By Theorem VI.1.1, $\alpha_{ld}$ equals the safety factor $\alpha_l$ and, therefore, $\alpha_{ld} = 1$, that is, $\mathbf{l}$ is a limit load for the deviatoric problem (2.8). In this case, Theorem 1 states that (2.8) has solution $\boldsymbol{\tau}$, $\mathbf{v} \neq 0$. Then Theorem VI.2.4 implies that there exists a pressure field $p \in L_2(\Omega)$ such that the stress field $\boldsymbol{\sigma} = \boldsymbol{\tau} - p\mathbf{I}$ equilibrates the load $\mathbf{l}$. It is clear that $\boldsymbol{\sigma}$ and the velocity field $\mathbf{v} \neq 0$ is a solution to the limit state problem, which finishes the proof.

REMARK 1. The regularity assumption in Theorem 2 means that the body boundary and its fixed part satisfy the conditions of the pressure field restoration theorem (Theorem VI.2.4).

## Comments

The solvability of the limit state problem in the case of bodies with bounded yield surfaces was proved by Nayroles (1970), who made use of the spaces $\mathcal{S} = L_\infty(\Omega; Sym)$ and $\mathcal{E} = \mathcal{S}'$. This result holds in the case of bodies with cylindrical yield surfaces; however, it only concerns the deviatoric problems. The existence theorem for the stress field in the limit state was established by Kamenjarzh (1978) and, under somewhat weaker assumptions, by Temam and Strang (1980) and Kamenjarzh (1981). Existence of a failure mechanism in the extended velocity fields space $V_0''$ was proved by Matthies, Strang and Christiansen (1979). The extension scheme was discussed in more detail by Matthies (1979); Subsection 2.2 follows this article.

The range closure of the operator Def is of most importance for the kinematic extremum problem weak formulation and its solvability; it is to ensure this closure that one has to consider the space $V_0''$. (At first sight, the Sobolev space $\mathbf{W}_1^1(\Omega)$ could make an appropriate choice for the velocity fields space; however, $\mathbf{W}_1^1(\Omega)$ does not suit since Korn's inequality is not valid in this space.) Temam and Strang (1978, 1980) studied the space $V_0$ and, in particular, proved the trace theorem.

Failure mechanism is a velocity field at which the infimum in the kinematic extremum problem of limit analysis is attained. Any such field together with any admissible stress field that equilibrates the limit load form a solution to the limit state problem (Subsection 1.1). To ensure the attainability of the infimum, various extensions of the regular velocity fields space and the kinematic functional are used. Mosolov (1967) proposed a general approach to extending a convex extremum problem in a nonreflexive Banach space. Matthies, Strang and Christiansen (1979) considered an extension of the limit analysis kinematic problem with the conjugate space of $C(\bar{\Omega})$ as the space of strain rate fields (recall that $C(\bar{\Omega})$ is the space of continuous functions on $\Omega$). In this case, velocity fields are integrable functions, members of the space $BD(\Omega)$ introduced by Suquet (1978), see also the book by Temam (1983). Within the framework of this formulation, the infimum in the kinematic problem is attained and equals the safety factor, velocity fields being regular functions and not abstract members of $V_0''$. However, using $BD(\Omega)$ as the velocity fields space, one has to narrow down the stress fields space and consider $\mathcal{S} = C(\Omega; Sym)$ instead of $\mathcal{S} = L_\infty(\Omega; Sym)$. This does not look very natural from the mechanical viewpoint since it eliminates, for example, piecewise smooth discontinuous stress fields.

Within the framework of the rigid-plastic problem weak formulation, the velocity field $\mathbf{v}$ is a rather abstract member of $V_0''$. At the same time, there is a procedure which allows better understanding of some features of $\mathbf{v}$, see Kamenjarzh (1985). Namely, since $\mathbf{v}$ is a linear continuous functional on $V_0'$, one can consider a subspace $L$ in $V_0'$ and the restriction of $\mathbf{v}$ to $L$: $\widehat{\mathbf{v}} = \mathbf{v}|_L$. When $L$ is appropriately chosen, $\widehat{\mathbf{v}}$ is a linear continuous functional on $L$, and $\widehat{\mathbf{v}}$ is of a simpler nature than $\mathbf{v}$. In particular, choosing a certain $L$ results in interpretation of $\widehat{\mathbf{v}}$ as a regular function and leads to the rigid-plastic problem formulation which uses $BD(\Omega)$ as the velocity fields space. Analogously, there is a choice of $L$ which results in the extended formulation of the rigid-plastic problem proposed by Seriogin (1983).

The reader can find the Banach and Hahn-Banach theorems and other information about concepts and methods of functional analysis in the books by Kolmogorov and Fomin (1970) and Edwards (1965).

CHAPTER VIII

# DISCONTINUOUS FIELDS IN LIMIT ANALYSIS

Application of discontinuous fields to limit analysis extends the capabilities of its methods. A key to the extension is an explicit formula for the kinematic multiplier of a discontinuous velocity field. We discuss an idea for obtaining this formula in Section 1 and give its rigorous proof in Section 6. The two basic statements of limit analysis remain valid in the case of discontinuous fields: 1) every kinematic multiplier of a load is not less than any of its static multipliers, and 2) the static and kinematic multipliers are equal if and only if they are evaluated for stress and velocity fields which form a solution to the rigid-plastic problem. This time we consider discontinuous solutions, which calls for deriving and using discontinuity relations.

We discuss limit analysis methods dealing with discontinuous fields in Section 2. In Section 4, we generalize the main results to the case of bodies with jump discontinuity of the material properties. Examples of discontinuous fields application to limit analysis are presented in Section 5.

## 1. Kinematic multiplier for discontinuous velocity field.

In this section, we extend the definition of the kinematic multiplier to piecewise regular discontinuous velocity fields and establish a formula for evaluating such a multiplier. The possibility of surface slip, or, in other words, of velocity jump on the body boundary is taken into account. It is shown that, under certain nonrestrictive conditions, the generalized definition does not alter the main property of the kinematic multiplier: every kinematic multiplier of a load is not less than any of its static multipliers. This allows use of discontinuous velocity fields for estimating the safety factor from above.

### 1.1. On definition of kinematic multiplier.

Under conditions pointed out in Chapters V and VI, the safety factor of the load $\mathbf{l} = (\mathbf{f}, \mathbf{q})$ can be calculated as the infimum of kinematic multipliers

$$m_k(\mathbf{v}) = \frac{\int\limits_{\Omega} d_x(\mathrm{Def}\,\mathbf{v})\,dV}{\int\limits_{\Omega} \mathbf{f}\mathbf{v}\,dV + \int\limits_{S_q} \mathbf{q}\mathbf{v}\,dS} \tag{1.1}$$

over the set of regular velocity fields. It should be recalled that the infimum is not necessarily attained (see Subsection III.6.3), and, in case it is not, can be evaluated as the limit of the kinematic multipliers $m_k(\mathbf{v}_n)$ for a certain minimizing sequence of velocity fields $\mathbf{v}_n$, $n = 1, 2, \ldots$. In this connection it looks reasonable to extend the set of velocity fields for which the kinematic multiplier is defined: making use of such a field can replace considering the sequence of regular fields.

In order to be useful, the definition of the kinematic multiplier for a wider class of velocity fields should not alter the main property of the kinematic multiplier (1.1) — the inequality $m_k(\mathbf{v}) \geq m_s(\boldsymbol{\sigma})$, $m_s(\boldsymbol{\sigma})$ being any static multiplier of the load — and should result in a formula which is not much more complicated than (1.1). The definition of the kinematic multiplier within the framework of the unified formulation of the rigid-plastic problem (Subsection IV.4.2) meets the first of these requirements and does not meet the second one. Starting from this definition and operating with generalized velocity fields it is possible to obtain an explicit formula for the kinematic multiplier of the piecewise regular discontinuous velocity field. However, we choose another way to obtain this formula since this way does not call for using the generalized fields. We mollify the jump of a discontinuous velocity field $\mathbf{v}$, that is, construct a sequence of regular fields $\mathbf{v}_\varepsilon$, which approximates $\mathbf{v}$ in a certain sense. The sequence is chosen in such a way that the corresponding sequence of kinematic multipliers $m_k(\mathbf{v}_\varepsilon)$ has a limit when $\varepsilon \to 0$. We find a simple explicit formula for evaluating this limit and take it as the definition of the kinematic multiplier $m_k(\mathbf{v})$. It is clear that passing to the limit does not alter the main property of the kinematic multiplier: $m_k(\mathbf{v}) \geq m_s(\boldsymbol{\sigma})$.

**1.2. Dissipation at discontinuity surface.** We now present an idea of obtaining an explicit formula for the kinematic multiplier of a piecewise regular discontinuous velocity field, starting with the consideration of homogeneous bodies.

Let a rigid perfectly plastic body occupy domain $\Omega$ and $\mathbf{v}$ be a velocity field on $\Omega$ with a discontinuity surface $\gamma$. We assume that $\gamma$ is a sufficiently regular surface which divides $\Omega$ into two parts $\Omega^-$, $\Omega^+$ located on different sides of $\gamma$. It is also assumed that in both $\Omega^-$ and $\Omega^+$ the field $\mathbf{v}$ is, say, continuously differentiable and possesses finite limits $\mathbf{v}^-$ and $\mathbf{v}^+$ when a point tends to $\gamma$ from $\Omega^-$ and $\Omega^+$, respectively. We denote the unit normal to $\gamma$ directed from the side "$-$" to the side "$+$" by $\boldsymbol{\nu}$. The function $[\mathbf{v}]$ is defined on $\gamma$ as $[\mathbf{v}] = \mathbf{v}^+ - \mathbf{v}^-$ and is referred to as *jump* of the field $\mathbf{v}$ on $\gamma$.

Let us now construct a sequence of sufficiently regular fields $\mathbf{v}_\varepsilon$ which converges to the discontinuous field $\mathbf{v}$, and our purpose is to find the limit of kinematic multipliers $m_k(\mathbf{v}_\varepsilon)$ when $\varepsilon \to 0$.

Let us first consider the smoothing field $\mathbf{v}_\varepsilon$ within a thin layer $\gamma_\varepsilon$ surrounding the discontinuity surface $\gamma$. To construct this field, we use a curvilinear coordinate system $x^1, x^2, x^3$ near this surface, the coordinate $x^1$ of a point being the distance from the point to $\gamma$ along the normal to $\gamma$. More precisely, at every point on $\gamma$, we consider the normal $\boldsymbol{\nu}$ and the straight line passing through this point and directed along $\boldsymbol{\nu}$. Every point $x$ near $\gamma$ lies on one of the lines; we denote the point at which this line intersects the surface $\gamma$ by $p(x)$ and the distance between $x$ and $p(x)$ by $\rho(x)$. We set the coordinate $x^1$ of the point $x$ equal to $\rho(x)$ if $x \in \Omega^+$ and equal to $-\rho(x)$ if $x \in \Omega^-$. Then we choose a curvilinear coordinate system $\eta$, $\zeta$ in the surface $\gamma$ and set the coordinates $x^2$ and $x^3$ of the point $x$ equal to $\eta(p(x))$ and $\zeta(p(x))$, respectively.

Thus $x^1, x^2, x^3$ is a coordinate system in a neighborhood of $\gamma$.

Let us consider the layer $\gamma_\varepsilon$ which consists of the points in $\Omega$ satisfying the condition $-\varepsilon/2 < x^1 < \varepsilon/2$. Within this layer we define the following vecto: field $\mathbf{v}_\varepsilon$:

$$\mathbf{v}_\varepsilon(x^1, x^2, x^3) = \frac{1}{2}\left(\mathbf{v}^+(0, x^2, x^3) + \mathbf{v}^-(0, x^2, x^3)\right) + \frac{x^1}{\varepsilon}[\mathbf{v}](0, x^2, x^3). \quad (1.2)$$

The field is continuous on $\gamma$ and linear in $x^1$. The derivative of $\mathbf{v}_\varepsilon$ with respect to $x^1$ is of the order $1/\varepsilon$ while the derivatives with respect to $x^2, x^3$ are of the order 1. Therefore, the components of the strain rate Def $\mathbf{v}_\varepsilon$ are close to

$$e_{11} = \frac{1}{\varepsilon}[v_1], \;\; e_{12} = \frac{1}{2\varepsilon}[v_2], \;\; e_{13} = \frac{1}{2\varepsilon}[v_3], \;\; e_{22} = e_{23} = e_{33} = 0.$$

It is clear that $e_{ij}$ are components of $\mathbf{e}_\nu([\mathbf{v}])/\varepsilon$, where $\mathbf{e}_\nu([\mathbf{v}])$ is the tensor with the components

$$(\mathbf{e}_\nu([\mathbf{v}]))_{ij} = \frac{1}{2}([\mathbf{v}_i]\nu_j + [\mathbf{v}_j]\nu_i).$$

This formula is valid for the components of $\mathbf{e}_\nu([\mathbf{v}])$ with respect to the considered coordinate system $x^1, x^2, x^3$ and also holds in any coordinate system. Then the dissipation in $\gamma_\varepsilon$ has the limit

$$\lim_{\varepsilon\to 0}\int_{\gamma_\varepsilon} d(\mathrm{Def}\,\mathbf{v}_\varepsilon)\,dV = \lim_{\varepsilon\to 0}\int_{\gamma_\varepsilon} d(\frac{1}{\varepsilon}\mathbf{e}_\nu([\mathbf{v}])\,dV = \int_{\gamma} d(\mathbf{e}_\nu([\mathbf{v}])\,dS.$$

We refer to this limit as (total) dissipation at the discontinuity surface $\gamma$ of the velocity field $\mathbf{v}$.

The velocity field $\mathbf{v}$ is regular in $\Omega^+$ and $\Omega^-$, which allows choosing the smoothing field $\mathbf{v}_\varepsilon$ sufficiently close to $\mathbf{v}$ everywhere in $\Omega \setminus \gamma_\varepsilon$. Then the equality

$$\lim_{\varepsilon\to 0}\int_{\Omega\setminus\gamma_\varepsilon} d(\mathrm{Def}\,\mathbf{v}_\varepsilon)\,dV = \int_{\Omega\setminus\gamma} d(\mathrm{Def}\,\mathbf{v})\,dV$$

is valid as well as

$$\lim_{\varepsilon\to 0}\left(\int_{\Omega}\mathbf{f}\mathbf{v}_\varepsilon\,dV + \int_{S_q}\mathbf{q}\mathbf{v}_\varepsilon\,dS\right) = \int_{\Omega}\mathbf{f}\mathbf{v}\,dV + \int_{S_q}\mathbf{q}\mathbf{v}\,dS$$

(the latter presumes that $\mathbf{f}$ and $q$ are sufficiently regular fields). Note that, if the right-hand side is positive, the power on the left-hand side is also positive at least for sufficiently small $\varepsilon > 0$. Therefore, the kinematic multiplier $m_k(\mathbf{v}_\varepsilon)$ is defined. Evaluating $m_k(\mathbf{v}_\varepsilon)$ in terms of definition (1.1) and taking the previous relations into account results in the following limit for $m_k(\mathbf{v}_\varepsilon)$:

$$\lim_{\varepsilon\to 0} m_k(\mathbf{v}_\varepsilon) = \frac{\int_{\Omega\setminus\gamma} d(\mathrm{Def}\,\mathbf{v})\,dV + \int_{\gamma} d(\mathbf{e}_\nu([\mathbf{v}])\,dS}{\int_{\Omega}\mathbf{f}\mathbf{v}\,dV + \int_{S_q}\mathbf{q}\mathbf{v}\,dS}.$$

We take the last expression as the definition of the *kinematic multiplier of a discontinuous velocity field* $\mathbf{v}$. We emphasize that this formula generalizes (1.1) and takes into account the dissipation at the discontinuity surface.

REMARK 1. The derivatives of a piecewise continuously differentiable function $\mathbf{v}$ are defined everywhere with the exception of the discontinuity surfaces of $\mathbf{v}$. The derivatives are regular (piecewise continuous) functions referred to as *local* (or *pointwise*) derivatives of $\mathbf{v}$. The local derivatives do not possess the important property

$$\int\limits_{\Omega} \frac{\partial v_i}{\partial x_j} \varphi \, dV = - \int\limits_{\Omega} v_i \frac{\partial \varphi}{\partial x_j} \, dV$$

(with an arbitrary smooth function $\varphi$ vanishing near $\partial\Omega$). It should be recalled that derivatives of any sufficiently smooth function and the derivatives of any function in the sense of distributions (Appendix VI.A) possess this property. In the sequel, we indicate local derivatives with the subscript $l$; in particular, we denote local derivatives of velocity components by $(\partial v_i/\partial x_j)_l$. If the domain $\Omega$ consists of two domains $\Omega^-$, $\Omega^+$ separated by the surface $\gamma$, and the restrictions $\mathbf{v}_- = \mathbf{v}|_{\Omega^-}$, $\mathbf{v}_+ = \mathbf{v}|_{\Omega^+}$ of $\mathbf{v}$ are sufficiently smooth, then the local derivatives are

$$\left(\frac{\partial \mathbf{v}}{\partial x_i}\right)_l (x) = \begin{cases} \dfrac{\partial \mathbf{v}_-}{\partial x_i}(x) & \text{if } x \in \Omega^-, \\ \dfrac{\partial \mathbf{v}_+}{\partial x_i}(x) & \text{if } x \in \Omega^+. \end{cases}$$

We denote the strain rate tensor composed of the local derivatives $(\partial v_i/\partial x_j)_l$ by the symbol $\mathrm{Def}_l\,\mathbf{v}$. With this notation the above formula for the kinematic multiplier of discontinuous velocity field can be written in the form

$$m_k(\mathbf{v}) = \frac{\int\limits_{\Omega} d(\mathrm{Def}_l\,\mathbf{v})\,dV + \int\limits_{\gamma} d(\mathbf{e}_\nu([\mathbf{v}]))\,dS}{\int\limits_{\Omega} \mathbf{f}\mathbf{v}\,dV + \int\limits_{S_q} \mathbf{q}\mathbf{v}\,dS}. \tag{1.3}$$

The reasoning we used to establish (1.3) calls for some refining, and the rigorous proof of the formula will be given in Section 6. However, Subsection 1.4 presents an accurate formulation of conditions under which (1.3) can be proved.

**1.3. Surface slip.** Under the conditions established in Chapters V and VI, the safety factor can be evaluated as the infimum of the kinematic multipliers over the set of sufficiently regular velocity fields. In this case, there is a sequence $\mathbf{v}_n$, $n = 1, 2, \ldots$, of such fields for which $m_k(\mathbf{v}_n) \to \inf = \alpha_l$ when $n \to \infty$. The minimizing sequence $\mathbf{v}_1, \mathbf{v}_2, \ldots$ does not necessarily converge; however, it may converge to a regular velocity field or to a field with "a jump on the boundary $\partial\Omega$" ($\Omega$ is a domain occupied by the body). The latter means that $\mathbf{v}_1, \mathbf{v}_2, \ldots$ converges to the field $\mathbf{v}$ in $\Omega$, the sequence of the traces $\mathbf{v}_n|_{\partial\Omega}$ converges to the field $\mathbf{v}^e$ on $\partial\Omega$, and $\mathbf{v}^e$ is not equal to $\mathbf{v}|_{\partial\Omega}$. We met this situation in the example in Subsection III.6.3. Thus, it makes sense to consider the pair $\bar{\mathbf{v}} = (\mathbf{v}, \mathbf{v}^e)$, where $\mathbf{v}$ is a field on $\Omega$ and $\mathbf{v}^e$ is a field on $\partial\Omega$. We say that $\bar{\mathbf{v}} = (\mathbf{v}, \mathbf{v}^e)$ is a *field on* $\bar{\Omega}$ *with a surface slip.* We also speak about the *jump of* $\bar{\mathbf{v}}$ *on the body boundary* $\partial\Omega$, which is the function $[\mathbf{v}] = \mathbf{v}^e - \mathbf{v}|_{\partial\Omega}$ on $\partial\Omega$ (more precisely, this is the jump in the direction of the outward normal). In case the part $S_v$ of the body boundary is fixed, we refer to the velocity field $\bar{\mathbf{v}} = (\mathbf{v}, \mathbf{v}^e)$ with the surface slip as kinematically admissible if $\mathbf{v}^e|_{S_v} = 0$. When the body is subjected to the load $\mathbf{l} = (\mathbf{f}, \mathbf{q})$, we consider

$$\int_{\Omega} \mathbf{f}\mathbf{v}\, dV + \int_{S_q} \mathbf{q}\mathbf{v}^e\, dS$$

as the external power of $\mathbf{l}$ on $\bar{\mathbf{v}}$.

We define the kinematic multiplier of a kinematically admissible velocity field $\bar{\mathbf{v}}$ using the same scheme as in Subsection 1.2: we mollify the velocity jump on $\partial\Omega$. We consider a curvilinear coordinate system $x^1, x^2, x^3$ near $\partial\Omega$. The system is analogous to that in Subsection 1.2: $x^1$ is determined by the distance along the normal to $\partial\Omega$, and $x^2$ and $x^3$ are curvilinear coordinates in $\partial\Omega$. In the layer $\partial\Omega_\varepsilon$ which consists of points in $\Omega$ satisfying the condition $-\varepsilon/2 < x^1 < 0$, we consider the velocity field

$$\mathbf{v}_\varepsilon(x^1, x^2, x^3) = \mathbf{v}^e(0, x^2, x^3) + \frac{2x^1}{\varepsilon}[\mathbf{v}](0, x^2, x^3)$$

and find as earlier:

$$\lim_{\varepsilon\to 0} \int_{\partial\Omega_\varepsilon} d(\operatorname{Def} \mathbf{v}_\varepsilon)\, dV = \int_{\partial\Omega} d(\mathbf{e}_\nu([\mathbf{v}]))\, dS.$$

Since $\mathbf{v}$ is sufficiently regular in $\Omega$, we choose the smoothing field $\mathbf{v}_\varepsilon$ sufficiently close to $\mathbf{v}$ so that

$$\lim_{\varepsilon\to 0} \int_{\Omega} d(\operatorname{Def} \mathbf{v}_\varepsilon)\, dV = \int_{\Omega} d(\operatorname{Def} \mathbf{v})\, dV$$

and, under the assumption that $\mathbf{f}$, $\mathbf{q}$ are sufficiently regular, the equality

$$\lim_{\varepsilon\to 0} \left( \int_{\Omega} \mathbf{f}\mathbf{v}_\varepsilon\, dV + \int_{S_q} \mathbf{q}\mathbf{v}_\varepsilon\, dS \right) = \int_{\Omega} \mathbf{f}\mathbf{v}\, dV + \int_{S_q} \mathbf{q}\mathbf{v}^e\, dS$$

is also valid. Then, if the right-hand side is positive, the power on the left-hand side is also positive, at least for sufficiently small $\varepsilon$. Note that $\mathbf{v}_\varepsilon$ satisfies the condition $\mathbf{v}_\varepsilon|_{S_v} = 0$ since $\bar{\mathbf{v}} = (\mathbf{v}, \mathbf{v}^e)$ is kinematically admissible, that is, meets the requirement $\mathbf{v}^e|_{S_v} = 0$. Due to the last two remarks, the kinematic multiplier $m_k(\mathbf{v}_\varepsilon)$ is defined. We evaluate $m_k(\mathbf{v}_\varepsilon)$ using definition (1.1), and find its limit taking into account the previous relations:

$$\lim_{\varepsilon \to 0} m_k(\mathbf{v}_\varepsilon) = \frac{\int\limits_\Omega d(\mathrm{Def}\,\mathbf{v})\,dV + \int\limits_{\partial\Omega} d(\mathbf{e}_\nu([\mathbf{v}])\,dS}{\int\limits_\Omega \mathbf{f}\mathbf{v}\,dV + \int\limits_{S_q} \mathbf{q}\mathbf{v}^e\,dS}.$$

We take the last expression as the definition of the *kinematic multiplier of a velocity field* $\bar{\mathbf{v}} = (\mathbf{v}, \mathbf{v}^e)$ *with surface slip.*

In case not only $\partial\Omega$ is a discontinuity surface of the field $\bar{\mathbf{v}} = (\mathbf{v}, \mathbf{v}^e)$, but so are some surfaces in $\Omega$, a formula for the kinematic multiplier $m_k(\bar{\mathbf{v}})$ can be obtained by adding the dissipation on $\partial\Omega$ that enters the previous formula to the dissipation in (1.3). In the sequel, we include $\partial\Omega$ in the collection $\Gamma$ of discontinuity surfaces and also use the symbol $[\mathbf{v}]$ to denote the jump $\mathbf{v}^e - \mathbf{v}|_{\partial\Omega}$ on $\partial\Omega$. Then the final expression for the kinematic multiplier of a discontinuous velocity field $\bar{\mathbf{v}} = (\mathbf{v}, \mathbf{v}^e)$ is written as

$$m_k(\bar{\mathbf{v}}) = \frac{\int\limits_\Omega d(\mathrm{Def}_l\,\mathbf{v})\,dV + \int\limits_\Gamma d(\mathbf{e}_\nu([\mathbf{v}])\,dS}{\int\limits_\Omega \mathbf{f}\mathbf{v}\,dV + \int\limits_{S_q} \mathbf{q}\mathbf{v}^e\,dS}. \tag{1.4}$$

**1.4. Main property of kinematic multiplier.** The above definition of the kinematic multiplier for a discontinuous velocity field is useful in limit analysis due to the fact that this definition does not alter the main property of kinematic multipliers of regular velocity fields. Namely, every kinematic multiplier $m_k(\bar{\mathbf{v}})$ of a load is not less than any of its static multipliers; therefore, the kinematic multiplier always gives an upper bound for the safety factor: $m_k(\bar{\mathbf{v}}) \geq \alpha_l$. Of course, this statement is to be justified. It is clear that the inequality $m_k(\mathbf{v}_\varepsilon) \geq \alpha_l$, which is always valid for regular velocity fields, implies $m_k(\bar{\mathbf{v}}) \geq \alpha_l$ in case $m_k(\bar{\mathbf{v}}) = \lim m_k(\mathbf{v}_\varepsilon)$ when $\varepsilon \to 0$. An idea of constructing a suitable smoothing sequence of the fields $\mathbf{v}_\varepsilon$ is given in the previous subsections. Let us now formulate conditions which guarantee the existence of such a sequence as well as the validity of the equality

$$m_k(\bar{\mathbf{v}}) = \lim_{\varepsilon \to )} m_k(\mathbf{v}_\varepsilon)$$

and, consequently, of the estimation $m_k(\bar{\mathbf{v}}) \geq \alpha_l$. The rigorous proof of the statement will be given in Section 6.

*Assumptions about material properties.* Consider three-dimensional rigid-plastic bodies with bounded or cylindrical yield surfaces. In the latter case, the sets $C_x$ of admissible stresses in the six-dimensional space $Sym$ are of the form

$$C_x = \{\boldsymbol{\sigma} \in Sym : \boldsymbol{\sigma}^d \in C_x^d\}.$$

Here, the superscript $d$ as always indicates the deviatoric part of a tensor, and $C_x^d$ is a set in the deviatoric plane $Sym^d$; the set $C_x^d$ is given for every point $x$ of the body and specifies the material properties.

We now impose some conditions on the mapping $x \to C_x$, under which the material properties continuously depend on $x$. It is convenient to formulate these conditions in terms of the dissipation

$$d_x(\mathbf{e}) = d(x; \mathbf{e}) = \sup \{\mathbf{e} \cdot \boldsymbol{\sigma}_* : \boldsymbol{\sigma}_* \in C_x\}.$$

The function $d_x$ is defined on $Sym$, the function $d$ on the set $\Omega \times Sym$ ($\Omega$ is the domain occupied by the body). In case the yield surfaces are bounded, $d_x$ is continuous on $Sym$ for every $x \in \Omega$ (Proposition V.6.1). Analogously, $d_x|_{Sym^d}$ is continuous on the deviatoric plane $Sym^d$ if $C_x^d$ is bounded. We say that material properties *depend continuously on the point of the body* if $d$ is a continuous function on $\bar{\Omega} \times Sym$ in the case of bodies with bounded yield surfaces or $d|_{\Omega \times Sym^d}$ is a continuous function on $\bar{\Omega} \times Sym^d$ in the case of bodies with cylindrical yield surfaces. Recall that, for example, the continuity on $\bar{\Omega} \times Sym$ of the function $f$ defined on $\Omega \times Sym$ means that 1) $f$ is continuous on $\Omega \times Sym$, and 2) there is a continuous extension of $f$ to a neighborhood of $\bar{\Omega} \times Sym$ ($\bar{\Omega} = \Omega \cup \partial\Omega$ is the closure of $\Omega$).

In this section we only study bodies whose material properties continuously depend on the point. A more general class of bodies will be considered in Section 4.

*Assumptions about velocity fields.* In this chapter we consider piecewise continuously differentiable velocity and stress fields. We refer to $f$ as *piecewise continuously differentiable in the domain* $\Omega$ if $\Omega$ consists of pair-wise disjoint subdomains $\omega_a$, $a = 1, \ldots, A$, the restriction $f|_{\omega_a}$ being continuously differentiable on $\bar{\omega}_a$ for every $a$; we refer to $\omega_a$, $a = 1, \ldots, A$, as *smoothness domains* (of the function $f$). Collection $\Gamma$ of discontinuity surfaces of $f$ is the union of $\partial\omega_1, \ldots, \partial\omega_A$ or a part of this union. In particular, one can formulate assumptions about discontinuity surfaces in terms of the domains $\omega_a$.

Our main assumption is that all of the smoothness domains are standard. We refer to a domain $\omega$ in $\mathbf{R}^n$ as *standard* if 1) its boundary consists of a finite number of parts of surfaces $F_\kappa(x) = 0$, $\kappa = 1, \ldots, K$, where $F_\kappa$ is a twice continuously differentiable function, and $|\text{grad}\, F_\kappa| \geq a_\kappa$ ($a_\kappa = \text{const} > 0$) when $F_\kappa(x) = 0$; and 2) smooth parts of $\partial\omega$ intersect at nonzero angles pair-wise along a curve or triple-wise at isolated points. The ball, the cylinder, the cube and the tetrahedron are examples of standard domains as well as their sufficiently smooth one-to-one transforms.

We consider, on the closure of the domain $\Omega$, velocity fields with the surface slip. Such field $\bar{\mathbf{v}} = (\mathbf{v}, \mathbf{v}^e)$ is referred to as *piecewise continuously differentiable* if $\mathbf{v}$ is piecewise continuously differentiable in $\Omega$ and $\mathbf{v}^e$ is sufficiently regular (say, piecewise continuously differentiable) on $\partial\Omega$.

We now formulate a proposition establishing an explicit expression for the kinematic multiplier of a discontinuous velocity field $\bar{\mathbf{v}} = (\mathbf{v}, \mathbf{v}^e)$, the expression being obtained as the limit of a sequence of the kinematic multipliers of regular velocity fields. In this case, as we explained earlier, the kinematic multiplier is an upper bound for the safety factor: $m_k(\bar{\mathbf{v}}) \geq \alpha_l$. The formulation of the proposition makes use of local strain rate $\text{Def}_l\, \mathbf{v}$ whose components are composed of the local derivatives $(\partial v_i / \partial x_j)_l$, see Remark 1. Recall that in case $\omega_1, \dots, \omega_A$ are the smoothness domains of the field $\mathbf{v}$, then, using the notation $\mathbf{v}_{(1)} = \mathbf{v}|_{\omega_1}, \dots, \mathbf{v}_{(A)} = \mathbf{v}|_{\omega_A}$, we write

$$\left(\frac{\partial \mathbf{v}}{\partial x_i}\right)_l (x) = \begin{cases} \dfrac{\partial \mathbf{v}_{(1)}}{\partial x_j}(x) & \text{if } x \in \omega_1, \\ \dots \\ \dfrac{\partial \mathbf{v}_{(A)}}{\partial x_j}(x) & \text{if } x \in \omega_A. \end{cases}$$

Also recall that the kinematic admissibility condition for $\bar{\mathbf{v}} = (\mathbf{v}, \mathbf{v}^e)$ is $\mathbf{v}^e|_{S_v} = 0$.

THEOREM 1. Suppose the material properties of a body with bounded or cylindrical yield surfaces continuously depend on the point of the body. Let $\Omega$ be the domain occupied by the body and $\bar{\mathbf{v}} = (\mathbf{v}, \mathbf{v}^e)$ be a piecewise continuously differentiable kinematically admissible velocity field on $\bar{\Omega}$. Suppose the smoothness domains of $\bar{\mathbf{v}}$ are standard and the power of the load $\mathbf{l} = (\mathbf{f}, \mathbf{q})$ on $\bar{\mathbf{v}}$ is positive. Then there is a sequence of regular fields $\mathbf{v}_\varepsilon$ such that their kinematic multipliers possess the following limit:

$$\lim_{\varepsilon \to 0} m_k(\mathbf{v}_\varepsilon) = m_k(\bar{\mathbf{v}}),$$

$$m_k(\bar{\mathbf{v}}) = \frac{\int\limits_\Omega d_x(\mathrm{Def}_l\, \mathbf{v})\, dV + \int\limits_\Gamma d_x(\mathbf{e}_\nu([\mathbf{v}]))\, dS}{\int\limits_\Omega \mathbf{f}\mathbf{v}\, dV + \int\limits_{S_q} \mathbf{q}\mathbf{v}^e\, dS}. \tag{1.5}$$

We postpone proving Theorem 1 until Section 6, where we establish the proposition under somewhat weaker assumptions.

Kinematic multipliers of the regular velocity fields in Theorem 1 are determined, as usual, by (1.1).

Formula (1.5) directly generalizes (1.4) to the case of inhomogeneous bodies whose material properties continuously depend upon the point of the body. The possibility of this generalization is quite understandable: the difference between the functions $d_x$ at different points $x$ lying on the same normal to the discontinuity surface $\gamma$ within the thin layer $\gamma_\varepsilon$ becomes negligible when $\varepsilon \to 0$.

We take (1.5) as the definition of the kinematic multiplier of a discontinuous velocity field. The formula is not more complicated than (1.1) which defines the kinematic multiplier for a regular velocity field. As mentioned above, definition (1.5) does not alter the main property of the kinematic multipliers: *every kinematic multiplier of a load is not less than any of its static multipliers.* This is valid for kinematic multipliers of regular velocity fields and holds for those of discontinuous velocity fields.

## 2. Methods for limit analysis

In the previous section, the definition of the kinematic multiplier was extended from regular to discontinuous velocity fields. The definition does not alter the main property of kinematic multipliers: every kinematic multiplier of a load is not less than any of its static multipliers. In this section, the second basic statement of limit analysis is reestablished in the case of discontinuous velocity fields. This statement gives a criterion for the static and kinematic multipliers equality. Discontinuity relations which complement differential equations of the rigid-plastic problem formulation arise from this criterion. They will be discussed in more detail in the next section.

The results of this and the previous sections allow use of discontinuous fields when solving limit analysis problems by the kinematic or rigid-plastic solutions methods, which extends capabilities of the methods. Both methods are briefly discussed in this section, while some examples of limit analysis making use of discontinuous fields are given in Section 5.

**2.1. Kinematic method.** Theorem 1.1 shows that discontinuous velocity fields can be used when applying the kinematic method to limit analysis. Formula (1.5) defines the kinematic multiplier for a discontinuous velocity field; we are reminded that this multiplier is always an upper bound for the safety factor: $m_k(\bar{\mathbf{v}}) \geq \alpha_l$.

Thus, a wider class of velocity fields is available for evaluating kinematic multipliers, the calculation for a discontinuous field being not more complicated than for a continuous one. This extends capabilities of the kinematic method. First, an appropriate upper bound for the safety factor can sometimes be found as the kinematic multiplier of a simple discontinuous velocity field, for example, the piecewise constant one. Secondly, making use of a discontinuous field may replace considering a sequence of continuous fields. This is of most importance in case the kinematic method allows evaluation of the safety factor and not only its upper bound. This is guaranteed when the conditions of Theorem V.5.1 or VI.1.1 are satisfied. The safety factor equals the infimum of the kinematic multipliers $m_k(\mathbf{v})$ over continuously differentiable velocity fields; however, the infimum is often not attained at such fields. At the same time, the infimum is, possibly, attained at a discontinuous velocity field $\alpha_l = m_k(\bar{\mathbf{v}})$, see example in Subsection 5.4 below. In such cases, it makes sense to look for a discontinuous minimizer rather than for a minimizing sequence of continuous fields.

**2.2. Criterion for static and kinematic multipliers equality.** Now, when a larger stock of velocity fields is available for effective calculation of the kinematic multipliers, new conditions are to be found for the static and kinematic multipliers equality. We will establish a new criterion for the equality assuming that the velocity fields are piecewise continuously differentiable, while in the case of the smooth fields the criterion is given by Theorem III.5.2. The same reason as before stimulates searching for the criterion: the common value of $m_k(\bar{\mathbf{v}})$ and $m_s(\boldsymbol{\sigma})$ is the safety factor, and the criterion for the equality $m_k(\bar{\mathbf{v}}) = m_s(\boldsymbol{\sigma})$ is, actually, a formulation of a problem whose solution results in the value of $\alpha_l$.

To formulate the criterion it is convenient to use the rigid-plastic constitutive relations in the form

$$\boldsymbol{\sigma} \in C_x, \qquad \mathrm{e} \cdot \boldsymbol{\sigma} = d_x(\mathrm{e}), \tag{2.1}$$

where $C_x$ is a given set of admissible stresses at the point $x$ of the body and $d_x$ is the corresponding dissipation

$$d_x(\mathbf{e}) = \sup\{\mathbf{e}\cdot\boldsymbol{\sigma}_* : \boldsymbol{\sigma}_* \in C_x\}. \tag{2.2}$$

Relations (2.1) are equivalent to the constitutive maximum principle (see Subsection I.3.2):

$$\boldsymbol{\sigma} \in C_x, \qquad \mathbf{e}\cdot(\boldsymbol{\sigma}-\boldsymbol{\sigma}_*) \geq 0 \quad \text{for every} \quad \boldsymbol{\sigma} \in C_x. \tag{2.3}$$

The following proposition deals with piecewise continuously differentiable stress and velocity fields $\boldsymbol{\sigma}$ and $\bar{\mathbf{v}}$ in the domain $\Omega$ occupied by the body. We assume that 1) $\Omega$ consists of pair-wise disjoint subdomains $\omega_1, \ldots, \omega_A$ which are the smoothness domains for both fields $\boldsymbol{\sigma}$ and $\bar{\mathbf{v}}$, and 2) each of the domains $\omega_a$ is sufficiently regular to ensure that the integration by parts formula is valid in $\omega_a$. One or both of the fields $\boldsymbol{\sigma}$ and $\bar{\mathbf{v}}$ can be discontinuous on $\partial\omega_a$. We use the symbol $[f]$ to denote the jump of the function $f$ on a certain discontinuity surface and $f^+$, $f^-$ to denote its limit values at the sides of this surface. The symbol $\mathbf{e}_\nu([\mathbf{v}])$ denotes the tensor with the components

$$(\mathbf{e}_\nu([\mathbf{v}]))_{ij} = \frac{1}{2}\left([v_i]\nu_j + [v_j]\nu_i\right)$$

defined at the discontinuity surface of the velocity field $\bar{\mathbf{v}}$; $\boldsymbol{\nu}$ stands for the unit normal to the surface. The subscript $l$ indicates local (point-wise) derivatives of piecewise continuously differentiable functions, see Remark 1.1. The formulation of the following proposition also refers to the continuity of material properties defined in Subsection 1.4 for bodies with bounded as well as cylindrical yield surfaces.

THEOREM 1. Suppose the material properties of a rigid-plastic body continuously depend on the point of the body. Suppose also that 1) $\boldsymbol{\sigma}$ is a piecewise continuously differentiable stress field in the domain $\Omega$ occupied by the body, and $\boldsymbol{\sigma}$ is admissible and equilibrates the load $m\mathbf{l}$, $m > 0$ ($\mathbf{l}$ is a given load); and 2) $\bar{\mathbf{v}} = (\mathbf{v}, \mathbf{v}^e)$ is a piecewise continuously differentiable velocity field on $\bar{\Omega}$, and $\bar{\mathbf{v}}$ is kinematically admissible, that is, $\mathbf{v}^e|_{S_v} = 0$. Then $m_k(\bar{\mathbf{v}}) = m_s(\boldsymbol{\sigma}) = m$ if and only if the constitutive relations

$$\boldsymbol{\sigma} \in C_x, \qquad \mathrm{Def}_l\,\mathbf{v}\cdot\boldsymbol{\sigma} = d_x(\mathrm{Def}_l\,\mathbf{v}), \tag{2.4}$$

are satisfied in the smoothness domains, the discontinuity relations

$$\mathbf{e}_\nu([\mathbf{v}])\cdot\boldsymbol{\sigma} = d_x(\mathbf{e}_\nu([\mathbf{v}])) \tag{2.5}$$

are satisfied on the discontinuity surfaces, and the external power is nonzero:

$$\int_\Omega \mathbf{f}\mathbf{v}\,dV + \int_{S_q} \mathbf{q}\mathbf{v}^e\,dS \neq 0. \tag{2.6}$$

REMARK 1. The expression on the left-hand side in (2.5) makes sense although values of $\boldsymbol{\sigma}$ are not defined on the discontinuity surface. Indeed, by the assumption, $\boldsymbol{\sigma}$ satisfies the equilibrium conditions, one of which requires the equality $[\sigma_{ij}]\nu_j = 0$ to be valid on discontinuity surfaces (Proposition II.1.1). This equality implies that

$$\mathbf{e}_\nu([\mathbf{v}]) \cdot \boldsymbol{\sigma}^- = \mathbf{e}_\nu([\mathbf{v}]) \cdot \boldsymbol{\sigma}^+,$$

due to which the expression $\mathbf{e}_\nu([\mathbf{v}]) \cdot \boldsymbol{\sigma}$ is well defined on the discontinuity surface.

REMARK 2. The collection of discontinuity surfaces may contain the boundary $\partial\Omega$ or its parts. This case is also covered by Theorem 1, which presumes that the dissipation $d_x$ is defined for points $x \in \partial\Omega$. Since Theorem 1 concerns bodies with continuous material properties, $d_x$ for $x \in \partial\Omega$ is just the continuous extension of the dissipation defined for $x \in \Omega$.

PROOF. Let us start with proving the following formula for a stress field $\boldsymbol{\sigma}$ equilibrating the load $m\mathbf{l} = (m\mathbf{f}, m\mathbf{q})$, and for a kinematically admissible velocity field $\bar{\mathbf{v}} = (\mathbf{v}, \mathbf{v}^e)$:

$$\int\limits_\Omega \mathrm{Def}_l\, \mathbf{v} \cdot \boldsymbol{\sigma}\, dV + \int\limits_\Gamma \mathbf{e}_\nu([\mathbf{v}]) \cdot \boldsymbol{\sigma}\, dS = m \left( \int\limits_\Omega \mathbf{f}\mathbf{v}\, dV + \int\limits_{S_q} \mathbf{q}\mathbf{v}^e\, dS \right), \tag{2.7}$$

where $\Gamma$ is the collection of the discontinuity surfaces of the fields $\boldsymbol{\sigma}$ and $\bar{\mathbf{v}}$. Indeed, let $\omega_1, \ldots, \omega_A$ be the smoothness domains of $\boldsymbol{\sigma}$ and $\bar{\mathbf{v}}$. Then the integration by parts formula

$$\int\limits_{\omega_a} \sigma_{ij} \left(\mathrm{Def}_l\, \mathbf{v}\right)_{ij}\, dV = - \int\limits_{\omega_a} \frac{\partial \sigma_{ij}}{\partial x_j} v_i\, dV + \int\limits_{\partial\omega_a} \sigma_{ij}\nu_j v_i\, dS$$

is valid in each of $\omega_a$. We add all of these formulas and transform the sum exactly as we did when deriving (II.1.5). We also make use of the equality $[\sigma_{ij}]\nu_j = 0$ and find that

$$\int\limits_\Omega \boldsymbol{\sigma} \cdot \mathrm{Def}_l\, \mathbf{v}\, dV = - \int\limits_\Omega \left( \frac{\partial \sigma_{ij}}{\partial x_j} \right)_l v_i\, dV - \int\limits_{\Gamma'} \sigma_{ij}\nu_j [v_i]\, dS + \int\limits_{\partial\Omega} \sigma_{ij}\nu_j v_i^-\, dV.$$

Here, $\Gamma' = (\partial\omega_1 \cup \cdots \cup \partial\omega_A) \setminus \partial\Omega$ and $\mathbf{v}^-$ is the trace of $\mathbf{v}$ on $\partial\Omega$. We now appeal to the equilibrium conditions

$$\frac{\partial \sigma_{ij}}{\partial x_j} + m f_i = 0 \quad \text{in } \omega_a,\ a = 1, \ldots, A, \qquad \sigma_{ij}\nu_j|_{S_q} = m q_i,$$

the kinematic admissibility condition $\mathbf{v}^e|_{S_v} = 0$, and the definition of the jump on $\partial\Omega$: $[\mathbf{v}] = \mathbf{v}^e - \mathbf{v}^-$; then the previous formula results in (2.7).

Let us show that $\boldsymbol{\sigma}$, $\bar{\mathbf{v}}$ satisfy (2.4), (2.5) if the equalities $m_k(\bar{\mathbf{v}}) = m_s(\boldsymbol{\sigma}) = m$ are valid. This will prove the first part of the statement as it is not required to verify (2.6): this relation is valid by virtue of the assumption that the kinematic multiplier is defined for the velocity field $\bar{\mathbf{v}}$.

Due to (1.5), the equality $m_k(\bar{\mathbf{v}}) = m$ implies that

$$\int_\Omega d_x(\mathrm{Def}_l\,\mathbf{v})\,dV + \int_\Gamma d_x(\mathbf{e}_\nu([\mathbf{v}]))\,dV = m\left(\int_\Omega \mathbf{f}\mathbf{v}\,dV + \int_{S_q}\mathbf{q}\mathbf{v}^e\,dS\right).$$

Using (2.7), we find that

$$\int_\Omega (d_x(\mathrm{Def}_l\,\mathbf{v}) - \mathrm{Def}_l\,\mathbf{v}\cdot\boldsymbol{\sigma})\,dV + \int_\Gamma (d_x(\mathbf{e}_\nu([\mathbf{v}])) - \mathbf{e}_\nu([\mathbf{v}])\cdot\boldsymbol{\sigma})\,dS = 0.$$

Recall that the restriction $\boldsymbol{\sigma}|_{\omega_a}$ of the stress field on the smoothness domain $\omega_a$ is assumed continuous on $\bar{\omega}_a = \omega_a \cup \partial\omega_a$, $a = 1,\dots,A$, and $\boldsymbol{\sigma}(x)$ is admissible for every $x \in \omega_a$. Therefore, the limit values $\boldsymbol{\sigma}^-$, $\boldsymbol{\sigma}^+$ on both sides of each discontinuity surface are admissible. Then, by definition (2.2) of the dissipation, the previous equality is valid provided (2.4) and (2.5) are satisfied for a.e. point in $\Omega$ and a.e. point on $\Gamma$, respectively. This proves the first statement of the present proposition.

Let us now verify the inverse statement. Suppose 1) an admissible stress field $\boldsymbol{\sigma}$ equilibrates the load $m\mathbf{l}$ and, hence, $m_s(\boldsymbol{\sigma}) = m$; 2) $\bar{\mathbf{v}} = (\mathbf{v}, \mathbf{v}^e)$ is a kinematically admissible velocity field; and 3) conditions (2.4) – (2.6) are satisfied. First, note that the external power (2.6) is positive and, therefore, the kinematic multiplier is defined for the velocity field $\bar{\mathbf{v}}$. Indeed, substituting (2.4) and (2.5) into formula (2.7) we arrive at

$$\int_\Omega d_x(\mathrm{Def}_l\,\mathbf{v})\,dV + \int_\Gamma d_x(\mathbf{e}_\nu([\mathbf{v}]))\,dS = m\left(\int_\Omega \mathbf{f}\mathbf{v}\,dV + \int_{S_q}\mathbf{q}\mathbf{v}^e\,dS\right).$$

Here, the left-hand side is nonnegative, the external power on the right-hand side is nonzero, and $m > 0$ by assumption. Consequently, the left-hand side is positive, kinematic multiplier (1.5) is defined for $\bar{\mathbf{v}}$, and the previous equality shows that $m_k(\bar{\mathbf{v}}) = m$, which finishes the proof.

The criterion established for the static and kinematic multipliers equality is an immediate generalization of the one given by Theorem III.5.2 for the case of regular velocity fields. This generalization results in extension of the rigid-plastic solutions method, which we briefly discuss in the next subsection.

REMARK 3. By simple reasoning like that in Theorem 1, it is possible to establish the inequality $m_k(\bar{\mathbf{v}}) \geq m_s(\boldsymbol{\sigma})$ for piecewise continuously differentiable fields $\bar{\mathbf{v}}$, $\boldsymbol{\sigma}$. However, this statement is much weaker than that of Theorem 1.1: the latter guarantees the inequality $m_k(\bar{\mathbf{v}}) \geq m_s(\boldsymbol{\sigma})$ for piecewise continuously differentiable velocity fields $\bar{\mathbf{v}}$ without assuming any regularity of $\boldsymbol{\sigma}$. It is this more general statement that allows us to conclude that the kinematic multiplier is an upper bound for the safety factor. Indeed, the safety factor is the supremum of static multipliers over all stress fields and not only over the piecewise continuously differentiable ones.

**2.3. Rigid-plastic solutions method.** Under the conditions of Theorem 1, the stress and velocity fields $\boldsymbol{\sigma}$ and $\bar{\mathbf{v}} = (\mathbf{v}, \mathbf{v}^e)$ satisfy certain relations, which will now be divided into three groups.

*In the smoothness domains,* $\boldsymbol{\sigma}$ and $\bar{\mathbf{v}}$ satisfy the relations

$$\begin{gathered} \frac{\partial \sigma_{ij}}{\partial x_j} + m f_i = 0, \\ \boldsymbol{\sigma} \in C_x, \qquad \mathrm{Def}_l\, \mathbf{v} \cdot \boldsymbol{\sigma} = d_x(\mathrm{Def}_l\, \mathbf{v}). \end{gathered} \tag{2.8}$$

The first of these relations is one of the equilibrium conditions; the second is the stress admissibility condition; the third is valid by virtue of Theorem 1. Here, $\mathrm{Def}_l\, \mathbf{v}$ stands for the strain rate field composed of the local derivatives (see Remark 1.1). Recall that the second and third relations in (2.8) are the rigid perfectly plastic constitutive relations.

*On the boundary of the body,* $\boldsymbol{\sigma}$ and $\bar{\mathbf{v}}$ satisfy the relations

$$\sigma_{ij}\nu_j\big|_{S_q} = q_i, \qquad \mathbf{v}^e\big|_{S_v} = 0. \tag{2.9}$$

The first of these relations is one of the equilibrium conditions; the second is the kinematic admissibility condition.

*On discontinuity surfaces,* $\boldsymbol{\sigma}$ and $\bar{\mathbf{v}}$ satisfy the relations

$$[\sigma_{ij}]\nu_j = 0, \tag{2.10}$$

$$\mathbf{e}_\nu([\mathbf{v}]) \cdot \boldsymbol{\sigma} = d_x(\mathbf{e}_\nu([\mathbf{v}])), \tag{2.11}$$

where $\mathbf{e}_\nu([\mathbf{v}])$ is the tensor with the components

$$(\mathbf{e}_\nu([\mathbf{v}]))_{ij} = \frac{1}{2}\left([v_i]\nu_j + [v_j]\nu_i\right).$$

Equality (2.10) is one of the equilibrium conditions (see Proposition II.1.1); (2.11) is valid by virtue of Theorem 1. It should be recalled that the expression $\mathbf{e}_\nu([\mathbf{v}]) \cdot \boldsymbol{\sigma}$ makes sense as it was noted in Remark 1.

The stress and velocity fields $\boldsymbol{\sigma}$ and $\mathbf{v}$ form a strong solution to the rigid-plastic problem with the load $m\mathbf{l}$, as defined in Subsection III.1, if they are sufficiently regular, have no jumps and satisfy (2.8) in $\Omega$ and boundary conditions (2.9). Analogously, piecewise continuously differentiable $\boldsymbol{\sigma}$ and $\bar{\mathbf{v}}$ are referred to as a *discontinuous solution* to this problem if $\boldsymbol{\sigma}$, $\bar{\mathbf{v}}$ satisfy (2.8) in the smoothness domains, boundary conditions (2.9), and relations (2.11) on discontinuity surfaces.

The next proposition reformulates Theorem 1 using this definition of a discontinuous solution.

THEOREM 2. Suppose the material properties of a rigid-plastic body continuously depend on the point of the body. Let $\boldsymbol{\sigma}$ be a piecewise continuously differentiable stress field in the domain $\Omega$ occupied by the body and $\bar{\mathbf{v}} = (\mathbf{v}, \mathbf{v}^e)$ be a piecewise continuously differentiable kinematically admissible velocity field on $\bar{\Omega}$. The static and kinematic multipliers $m_s(\boldsymbol{\sigma})$, $m_k(\bar{\mathbf{v}})$ of the load $\mathbf{l}$ are equal, $m_s(\boldsymbol{\sigma}) = m_k(\bar{\mathbf{v}}) = m > 0$, if and only if $\boldsymbol{\sigma}$ and $\bar{\mathbf{v}}$ are a discontinuous solution to the rigid-plastic problem for the load $m\mathbf{l}$, $m > 0$, and the external power of $\mathbf{l}$ on $\bar{\mathbf{v}}$ is nonzero (or, in case the zero stress field is safe for the body: if and only if $\boldsymbol{\sigma}$ and $\bar{\mathbf{v}}$ are a discontinuous solution to the problem and $\bar{\mathbf{v}} \neq 0$).

The only difference between this theorem and Theorem 1 consists in the statement about possible replacement of the condition

$$\int_{\Omega} \mathbf{f}\mathbf{v}\, dV + \int_{S_q} \mathbf{q}\mathbf{v}^e\, dS \neq 0$$

by the condition $\bar{\mathbf{v}} \neq 0$ in case the zero stress field is safe. More precisely, it remains to show that (2.12) is valid as soon as the velocity field $\bar{\mathbf{v}}$ in a solution $\boldsymbol{\sigma}$, $\bar{\mathbf{v}}$ to rigid-plastic problem (2.8) – (2.11) is nonzero. Indeed, such a velocity field possesses the following property:

$$\begin{aligned} &\mathrm{Def}_l\, \mathbf{v} \neq 0 \quad \text{on a set of positive volume in } \Omega \\ &\qquad\qquad \text{and/or} \\ &\mathbf{e}_\nu([\mathbf{v}]) \neq 0 \quad \text{on a set of positive area in } \Gamma. \end{aligned} \tag{2.13}$$

Indeed, the equality $\mathbf{e}_\nu([\mathbf{v}]) \equiv 0$ implies that $\bar{\mathbf{v}}$ is continuous and $\mathbf{v}|_{S_v} \equiv 0$. In this case the equality $\mathrm{Def}_l\, \mathbf{v} \equiv 0$ implies that $\mathbf{v} \equiv 0$ and $\mathbf{v}^e \equiv 0$, which contradicts the condition $\bar{\mathbf{v}} \neq 0$ in case the zero stress field is safe. Thus, (2.13) is valid. Then, by property 4 in Subsection I.4.2, we have

$$\int_{\Omega} d_x\,(\mathrm{Def}_l\, \mathbf{v})\, dV + \int_{\Gamma} d_x\,(\mathbf{e}_\nu([\mathbf{v}]))\, dS > 0,$$

where the left-hand side equals

$$\int_{\Omega} \boldsymbol{\sigma} \cdot \mathrm{Def}_l\, \mathbf{v}\, dV + \int_{\Gamma} \mathbf{e}_\nu[\mathbf{v}] \cdot \boldsymbol{\sigma}\, dS$$

and also equals

$$m\left(\int_{\Omega} \mathbf{f}\mathbf{v}\, dV + \int_{S_q} \mathbf{q}\mathbf{v}^e\, dS\right)$$

by virtue of (2.7) since $\boldsymbol{\sigma}$, $\bar{\mathbf{v}}$ is a solution to the rigid-plastic problem. Then the previous inequality shows that the external power is nonzero, which finishes the proof.

Theorem 2 is an immediate generalization of Theorem III.5.2 in the case of the piecewise regular solutions to the rigid-plastic problem. According to Theorem 2, discontinuous solutions can be used as well as the continuous ones when applying the rigid-plastic solutions method to limit analysis. The method consists of searching for solutions to the rigid-plastic problem for the loads $m\mathbf{l}$ proportional to the given load $\mathbf{l}$. If such a solution is found for a certain value $m_0$ of the parameter $m$, then $\alpha_l = m_0$ by Theorem 2. This extends capabilities of the rigid-plastic solutions method since discontinuous solutions may exist in case there are no continuous ones.

Some examples of the limit analysis making use of discontinuous fields will be given in Section 5.

## 3. Discontinuity relations

We have just established some relations to be satisfied on discontinuity surfaces of a solution to the rigid-plastic problem. Here, these discontinuity relations are interpreted as a maximum principle for the velocity jump. Some other equivalent formulations of the discontinuity relations (which are sometimes more convenient) are also considered together with their simplest properties.

**3.1. Normality law.** Let us show that, in the case of bodies with smooth yield surfaces, relations (2.11) can be written in the form of the normality law. This formulation is very convenient for solving concrete problems.

REMARK 1. Along with equalities (2.11), we will use the relations $\boldsymbol{\sigma}^- \in C_x$, $\boldsymbol{\sigma}^+ \in C_x$ on the discontinuity surface (the superscripts "−" and "+" indicate limit values of functions at the two sides of the discontinuity surface). We consider piecewise continuously differentiable stress fields $\boldsymbol{\sigma}$ in this chapter, and in this case the admissibility of $\boldsymbol{\sigma}^-$, $\boldsymbol{\sigma}^+$ arises from the admissibility of $\boldsymbol{\sigma}$ in the smoothness domains.

Let the set $C_x$ of admissible stresses be specified by the condition $F_x(\boldsymbol{\sigma}) \leq 0$, where $F_x$ is a convex continuously differentiable yield function. Then the relations

$$\boldsymbol{\sigma} \in C_x, \qquad \mathbf{e} \cdot \boldsymbol{\sigma} = d_x(\mathbf{e})$$

are equivalent to the normality law

$$F_x(\boldsymbol{\sigma}) \leq 0, \quad \mathbf{e} = \dot{\lambda} \frac{\partial F_x}{\partial \boldsymbol{\sigma}}(\boldsymbol{\sigma}), \quad \dot{\lambda} \geq 0, \quad \dot{\lambda} F_x(\boldsymbol{\sigma}) = 0$$

(see Sections I.3 and I.4). Then (2.11) can be written as

$$\frac{1}{2}\left([v_i]x_j + [v_j]x_i\right) = \dot{\lambda}^- \frac{\partial F_x}{\partial \sigma_{ij}}(\boldsymbol{\sigma}^-), \quad \dot{\lambda}^- \geq 0, \quad \dot{\lambda}^- F_x(\boldsymbol{\sigma}^-) = 0 \tag{3.1}$$

or, equivalently, due to Remark 2.1 in the form

$$\frac{1}{2}\left([v_i]x_j + [v_j]x_i\right) = \dot{\lambda}^+ \frac{\partial F_x}{\partial \sigma_{ij}}(\boldsymbol{\sigma}^+), \quad \dot{\lambda}^+ \geq 0, \quad \dot{\lambda}^+ F_x(\boldsymbol{\sigma}^+) = 0. \tag{3.2}$$

Thus, the tensor $\mathbf{e}_\nu([\mathbf{v}])$ is directed along the outward normal to the yield surface $F_x(\boldsymbol{\sigma}) = 0$ at the point $\boldsymbol{\sigma}^-(x)$ as well as at $\boldsymbol{\sigma}^+(x)$, and, hence, the normals at both points have the same direction.

Under the condition $[\sigma_{ij}]\nu_j = 0$, which is one of the equilibrium conditions, each of systems (3.1) and (3.2) is equivalent to (2.11). Thus, $[\sigma_{ij}]\nu_j = 0$ and (3.1) or, equivalently, $[\sigma_{ij}]\nu_j = 0$ and (3.2) form a complete system of the discontinuity relations.

**3.2. Maximum principle.** Let us interpret the discontinuity relations as a maximum principle for the velocity jump.

With the definition of the dissipation, (2.11) can be written in the form

$$\mathbf{e}_\nu([\mathbf{v}]) \cdot (\boldsymbol{\sigma} - \boldsymbol{\sigma}_*) \geq 0 \quad \text{for every} \quad \boldsymbol{\sigma}_* \in C_x. \tag{3.3}$$

Let $\mathbf{t}$ be the traction exerted by the stress $\boldsymbol{\sigma}$ on a discontinuity surface $\gamma$: $\mathbf{t}$ is a vector with the components $t_i = \sigma_{ij}\nu_j$, $\boldsymbol{\nu}$ standing for the unit normal to $\gamma$. Due to the condition $[\sigma_{ij}]\nu_j = 0$, the tractions exerted by stresses $\boldsymbol{\sigma}^+$, $\boldsymbol{\sigma}^-$ on the two sides of $\gamma$ are equal: $\sigma_{ij}^-\nu_j = \sigma_{ij}^+\nu_j = t_i$. Note that the stress admissibility condition restricts possible values of $\mathbf{t}$. In this connection, tractions $\mathbf{t}$ on $\gamma$ are referred to as admissible if there are admissible stresses $\boldsymbol{\sigma}$ that exert this traction: $t_i = \sigma_{ij}\nu_j$, $\boldsymbol{\sigma} \in C_x$. We introduce the *set of admissible tractions* on the surface $\gamma$ at the point $x$:

$$B_{\nu x} = \{\mathbf{b} \in \mathbf{R}^3 : \text{there exists } \mathbf{s} \in C_x,\ s_{ij}\nu_j = b_i\},$$

and we will often use it together with condition $\boldsymbol{\sigma}_- \in C_x$ (or $\boldsymbol{\sigma}_+ \in C_x$) its consequence: $\mathbf{t} \in B_{\nu x}$. This relation and the equalities $\mathbf{e}_\nu([\mathbf{v}]) \cdot \boldsymbol{\sigma} = \sigma_{ij}\nu_j[v_i] = [\mathbf{v}]\mathbf{t}$ allow us to write (3.3) as

$$\mathbf{t} \in B_{\nu x}, \quad [\mathbf{v}](\mathbf{t} - \mathbf{t}_*) \geq 0 \quad \text{for every} \quad \mathbf{t}_* \in B_{\nu x}. \tag{3.4}$$

The velocity jump and the traction appear to be related by the maximum principle analogous to the constitutive maximum principle (I.3.4) which relates the stress and the strain rate in the domains where $\boldsymbol{\sigma}$ and $\mathbf{v}$ are sufficiently smooth.

Now the discontinuity relations (2.10), (2.11) can be formulated as follows: *the tractions on both sides of a discontinuity surface are the same; and the velocity jump and the traction satisfy maximum principle* (3.4).

Consider further simplification of the discontinuity relations in the case of bodies with cylindrical yield surfaces. In this case, the velocity jump $[\mathbf{v}]$ on a discontinuity surface $\gamma$ is tangent to $\gamma$ or, equivalently, satisfies the condition $[\mathbf{v}]\boldsymbol{\nu} = 0$. To verify this equality, note that in case the yield surface is cylindrical the vector $p\boldsymbol{\nu}$ is in $B_{\nu x}$, $p$ being an arbitrary number. Then, if the jump $[\mathbf{v}]$ does not satisfy the condition $[\mathbf{v}]\boldsymbol{\nu} = 0$, neither does it satisfy maximum principle (3.4). Consequently, *the velocity jump in a body with cylindrical yield surfaces is tangent to the discontinuity surface*: $[\mathbf{v}]\boldsymbol{\nu} = 0$.

Due to this property of $[\mathbf{v}]$, only the tangent to $\gamma$ components $\mathbf{t}_t$, $\mathbf{t}_{*t}$ of the tractions $\mathbf{t}$, $\mathbf{t}_*$ enter (3.4), and it can be re-written as

$$[\mathbf{v}](\mathbf{t} - \mathbf{t}_*) \geq 0 \quad \text{for every} \quad \mathbf{t}_* \in B_{\nu x}.$$

To reduce this formulation of the maximum principle, we introduce the set of admissible tangent tractions. Consider the plane orthogonal to $\boldsymbol{\nu}$:

$$\Pi_\nu = \{\mathbf{u} \in \mathbf{R}^3 : \mathbf{u}\boldsymbol{\nu} = 0\}$$

and the *set of admissible tangent tractions* (at the point $x \in \gamma$)

$$A_{\nu x} = \{\mathbf{a} \in \Pi_\nu : \text{there exists } \mathbf{s} \in C_x, \ s_{ij}\nu_j - s_{kj}\nu_k\nu_j\nu_i = a_i\}.$$

It is clear that admissibility of the traction $\mathbf{t}$, that is, the relation $\mathbf{t} \in B_{\nu x}$, implies that $\mathbf{t}_t$ is in $A_{\nu x}$. Conversely, if $\boldsymbol{\tau}$ is in $A_{\nu x}$, that is, $\tau_i = s_{ij}\nu_j - s_{kj}\nu_k\nu_j\nu_i$ for a certain $\mathbf{s} \in C_x$, then the stress $\boldsymbol{\sigma} = \mathbf{s} + p\mathbf{I}$ is admissible for any number $p$, and the corresponding traction $\mathbf{t} = \boldsymbol{\tau} + p\mathbf{I}$ is in $B_{\nu x}$, $\boldsymbol{\tau}$ being its tangent component: $\mathbf{t}_t = \boldsymbol{\tau}$. Therefore, the previous formulation of the maximum principle can be equivalently written as

$$\boldsymbol{\tau} \in A_{\nu x}, \quad [\mathbf{v}] \in \Pi_\nu, \quad [\mathbf{v}](\boldsymbol{\tau} - \boldsymbol{\tau}_*) \geq 0 \quad \text{for every} \quad \boldsymbol{\tau}_* \in A_{\nu x}. \tag{3.5}$$

Then the discontinuity relations (2.10), (2.11) can be formulated as follows: *in the case of a body with cylindrical yield surfaces, the tractions on both sides of a discontinuity surface are the same, the velocity jump is tangent to this surface, and the velocity jump and the tangent tractions satisfy maximum principle* (3.5).

EXAMPLE 1. Let us find the set $A_{\nu x}$ of admissible tangent tractions in case the yield surface is given by the equation $|\boldsymbol{\sigma}^d - \boldsymbol{\rho}| = \sqrt{2}k$. Here, the number $k$ and the symmetric second rank tensor $\boldsymbol{\rho}$ are given, $\boldsymbol{\rho}$ being a deviatoric tensor ($\boldsymbol{\rho} = \boldsymbol{\rho}^d$) with $|\boldsymbol{\rho}| < \sqrt{2}k$. Consider point $x$ on a discontinuity surface. To find the set of admissible tangent tractions at this point, let us use the Cartesian coordinate system with the basis $\mathbf{e}_1 = \boldsymbol{\nu}$, $\mathbf{e}_2$, $\mathbf{e}_3$, where $\boldsymbol{\nu}$ is the unit normal to the discontinuity surface. Any tangent traction $\boldsymbol{\tau}$ has the zero component $\tau_1$ with respect to this basis. By the above definition, the tangent traction $\boldsymbol{\tau}$ is admissible if there exists an admissible stress $\mathbf{s}$ such that $s_{12} = \tau_2$, $s_{13} = \tau_3$ and the inequality

$$(s_{11}^d - \rho_{11})^2 + (s_{22}^d - \rho_{22})^2 + (s_{33}^d - \rho_{33})^2 + \\ + 2(\tau_2 - \rho_{12})^2 + 2(\tau_3 - \rho_{13})^2 + 2(s_{23} - \rho_{23})^2 \leq 2k^2$$

is valid. The inequality $(\tau_2 - \rho_{12})^2 + (\tau_3 - \rho_{13})^2 \leq k^2$ is the criterion for existence of appropriate $\mathbf{s}$, and, therefore,

$$A_{\nu x} = \{\boldsymbol{\tau} = \tau_2\mathbf{e}_2 + \tau_3\mathbf{e}_3 : (\tau_2 - \rho_{12})^2 + (\tau_3 - \rho_{13})^2 \leq k^2\}$$

is the set of admissible tangent tractions.

**3.3. Normality law for velocity jump.** In case the yield surface is smooth, the maximum principle (3.4) or (3.5) can be written in the form of the normality law. Indeed, the boundary of the set $B_{\nu x}$ is smooth due to the smoothness of the yield surface. Let the set $B_{\nu x}$ be specified by the condition

$F_{\nu x}(\mathbf{t}) \le 0$, while $F_{\nu x}(\mathbf{t}) = 0$ is the equation of its boundary. Maximum principle (3.4) is equivalent to the normality law

$$F_{\nu x}(\mathbf{t}) \le 0, \quad [\mathbf{v}] = \dot{\lambda}\frac{\partial F_{\nu x}}{\partial \mathbf{t}}, \quad \dot{\lambda} \ge 0, \quad \dot{\lambda} F_{\nu x}(\mathbf{t}) = 0$$

since $B_{\nu x}$ is convex (see Section I.3). This means that $[\mathbf{v}]$ is directed along the outward normal to the boundary of $B_{\nu x}$. Thus, in case the yield surface is smooth, the discontinuity relations (2.10), (2.11) can be formulated as follows: *the tractions on both sides of a discontinuity surface are the same, and the velocity jump obeys the normality law associated with the set of admissible tractions.*

The discontinuity relations in the case of a body with cylindrical yield surfaces admit an analogous formulation, which can be obtained by replacing $B_{\nu x}$ with the set $A_{\nu x}$ of admissible tangent tractions. Let this set lying in the plane tangent to the discontinuity surface be specified by the inequality $f_{\nu x}(\boldsymbol{\tau}) \le 0$. Then maximum principle (3.5) is equivalent to the normality law

$$f_{\nu x}(\boldsymbol{\tau}) \le 0, \quad [\mathbf{v}] = \dot{\lambda}\frac{\partial f_{\nu x}}{\partial \boldsymbol{\tau}}, \quad \dot{\lambda} \ge 0, \quad \dot{\lambda} f_{\nu x}(\boldsymbol{\tau}) = 0, \tag{3.6}$$

where $\boldsymbol{\tau}$ is the tangent traction on the discontinuity surface.

EXAMPLE 2. *Discontinuity relations in the case of the Mises yield surface.* Consider a body with the Mises stress admissibility condition: $|\boldsymbol{\sigma}^d| \le \sqrt{2}k$ ($k > 0$). To formulate the discontinuity relations at the point $x$ on a discontinuity surface $\gamma$, let us use the Cartesian coordinate system with the basis $\mathbf{e}_1 = \boldsymbol{\nu}$, $\mathbf{e}_2$, $\mathbf{e}_3$, where $\boldsymbol{\nu}$ is the unit normal to $\gamma$ at $x$. The surface traction exerted by the stress $\boldsymbol{\sigma}$ on $\gamma$ is $\mathbf{t} = \sigma_{11}\mathbf{e}_1 + \sigma_{12}\mathbf{e}_2 + \sigma_{13}\mathbf{e}_3$. The set $A_{\nu x}$ of admissible tangent tractions is specified by the condition $\tau_2^2 + \tau_3^2 \le k^2$ (see Example 1). Then, the normality law (3.6) can be written as

$$\tau_2^2 + \tau_3^2 \le k^2, \quad [\mathbf{v}] = \dot{\lambda}\boldsymbol{\tau}, \quad \dot{\lambda} \ge 0, \quad \dot{\lambda}(\tau_2^2 + \tau_3^2 - k^2) = 0.$$

The above discontinuity relations require that the stress $\boldsymbol{\sigma}$ be admissible on both sides of the discontinuity surface and satisfy the following relations:

$$[\sigma_{11}] = 0, \quad [\sigma_{12}] = 0, \quad [\sigma_{13}] = 0,$$
$$[v_1] = 0, \quad [v_2] = \dot{\lambda}\sigma_{12}, \quad [v_3] = \dot{\lambda}\sigma_{13}, \quad \dot{\lambda} \ge 0, \quad \dot{\lambda}(\sigma_{12}^2 + \sigma_{13}^2 - k^2) = 0.$$

Recall that these relations concern stress and velocity components with respect to the Cartesian coordinate system with the basis vector $\mathbf{e}_1$ orthogonal to the discontinuity surface.

**3.4. On the possibility of velocity jump.** Noteworthy are the following simple properties of the discontinuity relations:

- velocity jump cannot occur at some stressed states, as far as strictly convex sets $C_x$ of admissible stress are concerned (or cylindrical sets $C_x$ with strictly convex cross-sections $C_x^d$),
- velocity jump is possible only on surfaces where stresses are continuous.

The following example illustrates the validity of the first of these statements.

EXAMPLE 3. Consider the surface of the velocity jump in a body with the Mises yield criterion: $|\boldsymbol{\sigma}^d| = \sqrt{2}k$, $k > 0$. To show that this imposes certain restrictions on the stressed state, let us use the Cartesian coordinate system with the basis $\mathbf{e}_1 = \boldsymbol{\nu}$, $\mathbf{e}_2 = [\mathbf{v}]/|[\mathbf{v}]|$, $\mathbf{e}_3$ near the point $x$ of the discontinuity surface. One of the basis vectors is orthogonal to the surface; the second one is directed along the velocity jump $[\mathbf{v}]$ (recall that the velocity jump is tangent to the discontinuity surface if the yield surface is cylindrical). Then the component $[v_2]$ of the jump equals $|[\mathbf{v}]|$, and the normality law (see Example 2) implies that

$$|[\mathbf{v}]| = \dot{\lambda}\sigma_{12} \neq 0, \quad \dot{\lambda}\sigma_{13} = 0, \quad \dot{\lambda} \geq 0, \quad \dot{\lambda}(\sigma_{12}^2 - k^2) = 0.$$

These relations can be satisfied only if $\sigma_{12} = k$ since the stress $\boldsymbol{\sigma}$ is admissible, that is, $|\boldsymbol{\sigma}^d| \leq \sqrt{2}k$. Then, the admissibility condition also yields the following relations for the stress components on both sides of the discontinuity surface:

$$\sigma_{12} = k, \quad \sigma_{13} = 0, \quad \sigma_{23} = 0, \quad \sigma_{11} = \sigma_{22} = \sigma_{33} = p, \tag{3.7}$$

where $p$ is an arbitrary number, the same for both sides of the surface due to the condition $[\sigma_{11}] = 0$. Thus, the velocity jump is possible on a surface only in case at every point of the surface there is a Cartesian basis in which (3.7) is valid. It is clear that such basis exists by far not for every $\boldsymbol{\sigma}$. In other words, a velocity jump cannot occur at some stressed states.

Let us now show that the stresses are continuous on a velocity jump surface. First note the following fact.

PROPOSITION 1. Suppose the stress and velocity fields $\boldsymbol{\sigma}$ and $\mathbf{v}$ satisfy discontinuity relations (3.4). Then, if $[\mathbf{v}] \neq 0$, the limit values $\boldsymbol{\sigma}^-$ and $\boldsymbol{\sigma}^+$ on the two sides of the discontinuity surface lie on the boundary of $C_x$, that is, on the yield surface.

PROOF. The stresses $\boldsymbol{\sigma}^-$ and $\boldsymbol{\sigma}^+$ are admissible by Remark 1. Let us show, for example, that $\boldsymbol{\sigma}^-$ stress cannot be in the interior of $C_x$. Indeed, if it were, the corresponding traction $\mathbf{t}$ would be in the interior of $B_{\nu x}$ and the maximum principle (3.4) would imply that $[\mathbf{v}(x)] = 0$. This contradicts the assumption $[\mathbf{v}(x)] \neq 0$, which finishes the proof.

The following proposition applies to the case of strictly convex sets of admissible stresses. The set $A$ in a linear space is referred to as *strictly convex* if the segment joining any two points of $A$ lies in the interior of $A$, possibly with the exception of the end points.

PROPOSITION 2. Let the set $C_x$ of admissible stresses (or its cross-section $C_x^d$ in case $C_x$ is a cylinder) be bounded and strictly convex. Suppose stress and velocity fields $\boldsymbol{\sigma}$ and $\mathbf{v}$ satisfy discontinuity relations (3.4) (or (3.5), respectively). Then 1) the direction of the velocity jump on the discontinuity surface uniquely determines the stresses $\boldsymbol{\sigma}^-$, $\boldsymbol{\sigma}^+$ (or their deviators, respectively), and 2) stresses are continuous on the discontinuity surface.

PROOF. Consider, for example, the case of $C_x$ being a cylinder. Let $\gamma$ be the discontinuity surface with the nonzero velocity jump $[\mathbf{v}]$ and $\boldsymbol{\tau}$ be the tangent traction on $\gamma$. According to (3.5), the maximum

$$\max\{[\mathbf{v}]\boldsymbol{\tau}_* : \boldsymbol{\tau}_* \in A_{\nu x}\} \tag{3.8}$$

is attained at $\boldsymbol{\tau}$ (the attainability arises from the convexity and boundness of $C_x^d$). Moreover, the maximum is attained at the unique point $\boldsymbol{\tau} \in A_{\nu x}$ since $A_{\nu x}$ is strictly convex together with $C_x^d$. Note that the maximum point $\boldsymbol{\tau}$ is uniquely determined by the direction of $[\mathbf{v}]$: it can be easily seen that replacing $[\mathbf{v}]$ by $a[\mathbf{v}]$ with arbitrary $a > 0$ does not alter the maximum point in (3.8). This proves the first statement of the proposition.

Consider now the limit values $\boldsymbol{\sigma}^-$, $\boldsymbol{\sigma}^+$ on the two sides of $\gamma$. According to Proposition 1 they lie on the boundary of $C_x$, and, consequently, their deviators $\boldsymbol{\sigma}^{-d}$, $\boldsymbol{\sigma}^{+d}$ lie on the boundary of $C_x^d$. By virtue of the discontinuity relations the tangent traction $\boldsymbol{\tau}$ corresponds to each of $\boldsymbol{\sigma}^{-d}$, $\boldsymbol{\sigma}^{+d}$, that is, the equality $\sigma_{ij}^d\nu_j - \sigma_{kj}^d\nu_k\nu_j\nu_i = \tau_i$ holds for $\boldsymbol{\sigma}^d = \boldsymbol{\sigma}^{-d}$ and for $\boldsymbol{\sigma}^d = \boldsymbol{\sigma}^{+d}$. Only the unique stress deviator among those lying on the boundary of $C_x^d$ may satisfy this equality. Indeed, if the equality is valid for the deviators $\boldsymbol{\sigma}_1^d$ and $\boldsymbol{\sigma}_2^d$, it also holds, say, for $\boldsymbol{\sigma}^d = (\boldsymbol{\sigma}_1^d + \boldsymbol{\sigma}_2^d)/2$. This point is in the interior of $C_x^d$ as the latter is strictly convex. Then $\boldsymbol{\tau}$ is in the interior of $A_{\nu x}$, and the maximum principle yields $[\mathbf{v}] = 0$. This contradicts the assumption of the proposition, hence, $\boldsymbol{\sigma}^{-d} = \boldsymbol{\sigma}^{+d}$. In this case, the discontinuity relation $[\sigma_{ij}]\nu_j = 0$ implies that $\boldsymbol{\sigma}^- = \boldsymbol{\sigma}^+$, which finishes the proof.

We emphasize that the established properties apply to bodies with continuous material properties. We will see that, in case a jump of material properties occurs on a certain surface, both stress and velocity fields can be discontinuous on this surface.

## 4. Bodies with jump inhomogeneity

So far, we discussed application of discontinuous fields to the limit analysis of bodies with material properties which continually depend on the point of the body. In this section we consider the class of bodies with jump inhomogeneity of material properties. Modifying the reasoning of Section 2, we arrive at a simple explicit formula for the kinematic multiplier of a discontinuous velocity field, thus extending the capabilities of the limit analysis methods.

**4.1. Jump inhomogeneity.** Let us start with the description of the class of bodies we study in this section. Consider three-dimensional rigid perfectly plastic bodies with bounded or cylindrical yield surfaces. In the latter case, the sets $C_x$ of admissible stresses in the six-dimensional space $Sym$ are of the form

$$C_x = \{\boldsymbol{\sigma} \in Sym : \boldsymbol{\sigma}^d \in C_x^d\} \qquad (C_x^d \subset Sym^d).$$

The corresponding dissipation can be written as

$$d_x(\mathbf{e}) = d(x; \mathbf{e}) = \sup\{\boldsymbol{\sigma}_* \cdot \mathbf{e} : \boldsymbol{\sigma}_* \in C_x\}.$$

The function $d_x$ is defined on $Sym$, the function $d$ on the set $\Omega \times Sym$, where $\Omega$ is the domain occupied by the body. We say that material properties of the body *continuously depend on its point* within the domain $\omega \subset \Omega$ if

> $d$ is a continuous function on $\bar{\omega} \times Sym$ in the case of bodies with bounded yield surfaces,
>
> $d|_{\omega \times Sym^d}$ is a continuous function on $\bar{\omega} \times Sym^d$ in the case of bodies with cylindrical yield surfaces.

Recall that the function $f$ defined on $\omega \times Sym$ is continuous on $\bar{\omega} \times Sym$ if $f$ is continuous on $\omega \times Sym$ and, moreover, there is a continuous extension of $f$ to a neighborhood of $\bar{\omega} \times Sym$ ($\bar{\omega}$ is the closure of $\omega$: $\bar{\omega} = \omega \cup \partial\omega$).

We now assume that the domain $\Omega$ occupied by the body consists of pairwise disjoint subdomains $\omega_1, \ldots, \omega_A$ and the material properties continuously depend on the point in each of $\omega_a$. In this case, we say that the material properties *piecewise continuously depend on the point in* $\Omega$.

Let $\gamma$ be a common part of the boundaries of two adjacent subdomains $\omega_a$ and $\omega_b$ ($a \neq b$). Consider the values $d_x(\mathbf{e})$ at a fixed $\mathbf{e}$ when (i) $x$ remains in $\omega_a$ and tends to $x_0 \in \gamma$, and (ii) $x$ remains in $\omega_b$ and tends to $x_0$. Generally speaking, the limits of $d_x(\mathbf{e})$ are different in cases (i) and (ii). If it is so at least for some $\mathbf{e}$, we say that we deal with a body with *jump inhomogeneity*. For example, a composite consisting of a homogeneous material with inclusions of another homogeneous material is a body with jump inhomogeneity. We consider a stress field $\boldsymbol{\sigma}$ in a body with jump inhomogeneity as *admissible* if its values $\boldsymbol{\sigma}(x)$ are admissible in every subdomain where the material properties continuously depend on the point. This means that a "glue" which joints such subdomains is not weaker than the material in the subdomains.

**4.2. Dissipation at a discontinuity surface.** We now establish a formula for the kinematic multiplier of a discontinuous velocity field in a body with jump inhomogeneity. Consider a piecewise continuously differentiable velocity field $\mathbf{v}$ and its discontinuity surface $\gamma$. We assume that there also are surfaces of material properties jump inhomogeneity in the body. In case no part of $\gamma$ lies on those surfaces, the reasoning of Subsection 1.2 needs no revising and formula (1.5) holds for the kinematic multiplier. For this reason, let us consider only the case when the velocity jump and the material properties jump occur on the same surface.

To obtain a formula for the kinematic multiplier, let us construct a regular velocity field $\mathbf{v}_\varepsilon$ which approximates the discontinuous field $\mathbf{v}$. First consider $\mathbf{v}_\varepsilon$ within a thin layer $\gamma_\varepsilon$ which contains $\gamma$ and contracts to it when $\varepsilon \to 0$. It can be seen that the limit

$$\lim_{\varepsilon \to 0} \int_{\gamma_\varepsilon} d_x(\operatorname{Def} \mathbf{v}_\varepsilon)\, dV$$

should be added to the total dissipation in the formula for the kinematic multiplier. Although this was done in Section 1, a new problem should be solved when constructing the approximation $\mathbf{v}_\varepsilon$ to $\mathbf{v}$ in the body with jump inhomogeneity. To make the problem clear, consider first the simplest approximation

(1.2) to the discontinuous field $\mathbf{v}$. Exactly as in Subsection 1.2, this results in the limit of the dissipation

$$\lim_{\varepsilon \to 0} \int_{\gamma_\varepsilon} d_x(\operatorname{Def} \mathbf{v}_\varepsilon)\, dV = \frac{1}{2} \int_{\gamma} \{d_x^-(\mathbf{e}_\nu([\mathbf{v}])) + d_x^+(\mathbf{e}_\nu([\mathbf{v}]))\}\, dS, \tag{4.1}$$

where $\mathbf{e}_\nu([\mathbf{v}])$ is the tensor with the components

$$(\mathbf{e}_\nu([\mathbf{v}]))_{ij} = \frac{1}{2}([v_i]\nu_j + [v_j]\nu_i).$$

The functions $d_x^-$, $d_x^+$ at $x \in \gamma$ represent the dissipation at the two sides of the discontinuity surface. The values $d_x^-(\mathbf{e})$, $d_x^+(\mathbf{e})$ are the limits of $d_y(\mathbf{e})$ when $y$ tends to $x$ remaining on one side of $\gamma$. Example 1 below shows that the limit of the dissipation can be less than (4.1) if a smooth field different from the $\mathbf{v}_\varepsilon$ is used as an approximation to $\mathbf{v}$. This is a crucial point, since the kinematic multiplier is intended to estimate the safety factor from above. Therefore, the approximation to $\mathbf{v}$ should be chosen so as to result in the minimum value of the dissipation limit. (The simplest approximation (1.2) turns out to be suitable in case the material properties continuously depend on the point of the body. For this reason we did not discuss the problem of choosing the approximation before.)

EXAMPLE 1. Consider a rigid-plastic body with jump discontinuity of its material properties on the part $\gamma$ of the plane $x_1 = 0$. Suppose the body is piecewise homogeneous and the sets of admissible stresses are of the form

$$C_x = C^- = \{\boldsymbol{\sigma} \in Sym : |\boldsymbol{\sigma}^d| \le \sqrt{2}k\} \quad \text{if} \quad x_1 < 0,$$
$$C_x = C^+ = \{\boldsymbol{\sigma} \in Sym : |\boldsymbol{\sigma}^d - \boldsymbol{\rho}| \le \sqrt{2}k\} \quad \text{if} \quad x_1 > 0.$$

Let the number $k > 0$ and the tensor $\boldsymbol{\rho}$, which specify the material properties, not depend on $x$, and let the only nonzero components of $\boldsymbol{\rho}$ be $\rho_{13} = \rho_{31} = \rho k$, $0 < \rho < 1$. The yield surfaces are cylindrical, and the dissipation $d_x(\mathbf{e})$ is $+\infty$ if $\mathbf{e}^s \neq 0$. In case $\mathbf{e}^s = 0$, that is, $\mathbf{e} = \mathbf{e}^d$, it can be easily seen that

$$\begin{aligned} d_x(\mathbf{e}) &= d^-(\mathbf{e}) = \sqrt{2}k\,|\mathbf{e}| \qquad \text{if } x_1 < 0, \\ d_x(\mathbf{e}) &= d^+(\mathbf{e}) = \sup\left\{\mathbf{e}\cdot\boldsymbol{\sigma}_* : \boldsymbol{\sigma}_* \in Sym,\ |\boldsymbol{\sigma}_*^d - \boldsymbol{\rho}| \le \sqrt{2}k\right\} \\ &= \sup\left\{\mathbf{e}\cdot(\boldsymbol{\tau}_* + \boldsymbol{\rho}) : \boldsymbol{\tau}_* \in Sym,\ |\boldsymbol{\tau}_*^d| \le \sqrt{2}k\right\} \\ &= \sqrt{2}k\,|\mathbf{e}| + 2k\rho e_{13} \qquad \text{if } x_1 > 0. \end{aligned}$$

Consider a discontinuous velocity field $\mathbf{v}$ with the components

$$v_1 = 0, \qquad v_2 = \begin{cases} u & \text{if } x_1 < 0, \\ -u & \text{if } x_1 > 0, \end{cases} \qquad v_3 = 0,$$

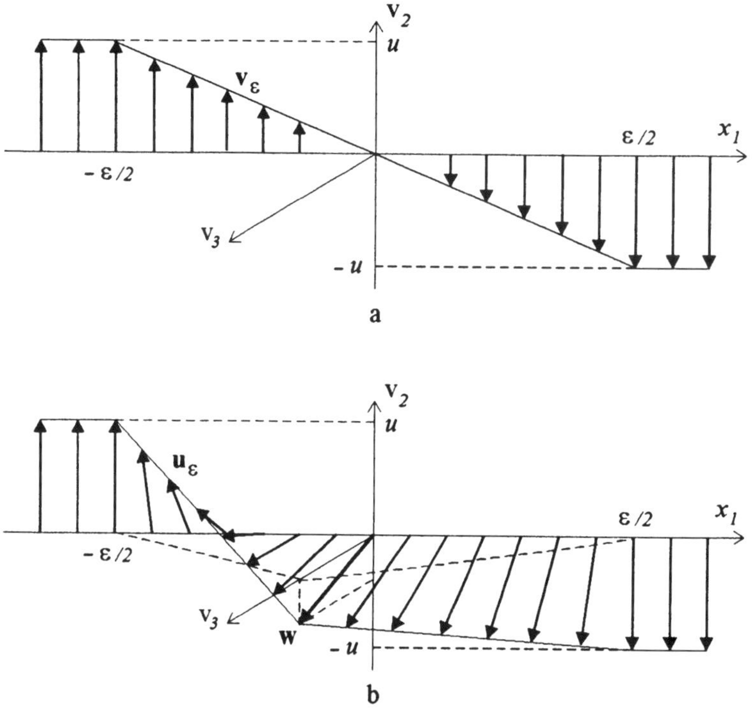


**Figure 4.1**

where $u > 0$ does not depend on $x$. The simplest approximation $\mathbf{v}_\varepsilon$ to $\mathbf{v}$ is determined near $\gamma$ by (1.2). Figure 4.1a shows the diagram of the components $v_{\varepsilon 2}$, $v_{\varepsilon 3}$, the component $v_{\varepsilon 1}$ being identically equal to zero. We evaluate the limit of the dissipation using formula (4.1); the limit is $2kuS_\gamma$ with $S_\gamma$ standing for the area of $\gamma$.

Consider now a different approximation to $\mathbf{v}$. Namely, let $\mathbf{w}$ be a vector with the components $w_1 = 0$, $w_2$, $w_3$. Consider the layer $\gamma_\varepsilon$ specified by the condition $-\varepsilon/2 < x_1 < \varepsilon/2$ and the field $\mathbf{u}_\varepsilon$ in it:

$$\mathbf{u}_\varepsilon = \begin{cases} \mathbf{v}^- + \dfrac{2x_1 + \varepsilon}{\varepsilon}(\mathbf{w} - \mathbf{v}^-) & \text{if } -\dfrac{\varepsilon}{2} \le x_1 \le 0, \\ \mathbf{v}^+ + \dfrac{2x_1 - \varepsilon}{\varepsilon}(-\mathbf{w} + \mathbf{v}^+) & \text{if } \quad 0 \le x_1 \le \dfrac{\varepsilon}{2}. \end{cases} \tag{4.2}$$

It takes the values $\mathbf{v}^-$ at $x_1 = -\varepsilon/2$, $\mathbf{v}^+$ at $x_1 = \varepsilon/2$ and $\mathbf{w}$ at $x_1 = 0$. Figure 4.1b shows the diagram of the components $u_{\varepsilon 2}$, $u_{\varepsilon 3}$, the component $u_{\varepsilon 1}$ being identically equal to zero. Straightforward calculation results in the following formula for the limit of the dissipation:

$$\lim_{\varepsilon \to 0} \int_{\gamma\varepsilon} d_x\,(\mathrm{Def}\,\mathbf{v}_\varepsilon)\,dV = kS_\gamma\left(\sqrt{(w_2 - u)^2 + w_3^2} + \sqrt{(w_2 + u)^2 + w_3^2} - \rho w_3\right),$$

where the $w_2$ and $w_3$ are arbitrary numbers. Now let us choose $w_2$, $w_3$ to make the limit minimum. It can be easily seen that the minimum value equals $kuS\gamma\sqrt{4-\rho^2}$ and is attained at $w_2 = 0$, $w_3 = \rho/\sqrt{4-\rho^2}$. This value is less than $2kuS\gamma$, the latter being the limit of the dissipation for the approximation (1.2). The two values $kuS\gamma\sqrt{4-\rho^2}$ and $2kuS\gamma$ are equal if there is no jump discontinuity of the material properties at $\gamma$, that is, if $\rho = 0$.

Example 1 suggests the following modification of the reasoning in Subsection 1.2. The approximation to the discontinuous velocity field $\mathbf{v}$ in the layer $\gamma_\varepsilon$ surrounding the discontinuity surface $\gamma$ of $\mathbf{v}$ is constructed in the form (4.2), where $\mathbf{v}^-$, $\mathbf{v}^+$, $\mathbf{w}$ depend on $x^2$, $x^3$, and where $x^1$, $x^2$, $x^3$ is the curvilinear coordinate system introduced in Subsection 1.2. The limit of the dissipation is evaluated exactly as in Subsection 1.2, and it equals

$$\lim_{\varepsilon\to 0}\int_{\gamma_\varepsilon} d_x(\operatorname{Def}\mathbf{u}_\varepsilon)\,dV = \int_\gamma \{d_x^-(\mathbf{e}_\nu(\mathbf{w}-\mathbf{v}^-)) + d_x^+(\mathbf{e}_\nu(-\mathbf{w}+\mathbf{v}^+))\}\,dS. \quad (4.3)$$

Then, choosing the vector $\mathbf{w}$ such that the limit is minimum, we take this minimum as the dissipation at the discontinuity surface and add it to the total dissipation in the formula for the kinematic multiplier of the discontinuous velocity field.

Let us consider this construction in more detail, in particular, making it more effective for evaluating of the dissipation at discontinuity surfaces. To be more definite, we restrict ourselves to the (more interesting) case of bodies with cylindrical yield surfaces. Recall that, in this case, $d_x(\mathbf{e}) = +\infty$ if $\mathbf{e}^s \neq 0$. Then, if the velocity jump $[\mathbf{v}]$ does not satisfy the condition $[\mathbf{v}]\boldsymbol{\nu} = 0$ and, consequently, $\operatorname{div}\mathbf{v}_\varepsilon \neq 0$, then $d_x(\operatorname{Def}\mathbf{v}_\varepsilon) = +\infty$ and $m_k(\mathbf{v}_\varepsilon) = +\infty$. Therefore, it is sufficient to consider only the case $[\mathbf{v}]\boldsymbol{\nu} = 0$ and (for the same reason) $\mathbf{w}\boldsymbol{\nu} = \mathbf{v}^+\boldsymbol{\nu} = \mathbf{v}^-\boldsymbol{\nu}$ in (4.3).

Thus, the problem

$$d_x^-(\mathbf{e}_\nu(\mathbf{w}-\mathbf{v}^-)) + d_x^+(\mathbf{e}_\nu(-\mathbf{w}+\mathbf{v}^+)) \to \inf; \quad \mathbf{w}\in\mathbf{R}^3,\ \mathbf{w}\boldsymbol{\nu} = \mathbf{v}^-\boldsymbol{\nu} \quad (4.4)$$

is to be solved to minimize expression (4.3). It is clear that the infimum depends on the parameters $\mathbf{v}^+$, $\mathbf{v}^-$. However, we observe that replacing $\mathbf{v}^+$, $\mathbf{v}^-$ by $\mathbf{v}^+ + \mathbf{a}$, $\mathbf{v}^- + \mathbf{a}$, where $\mathbf{a}$ is any vector satisfying the condition $\mathbf{a}\boldsymbol{\nu} = 0$, does not alter the infimum. In other words, the infimum only depends on $[\mathbf{v}]$. Thus, the number

$$\inf\{d_x^-\,(\mathbf{e}_\nu(\mathbf{w}-\mathbf{v}^-)) + d_x^+\,(\mathbf{e}_\nu(-\mathbf{w}+\mathbf{v}^+)) : \mathbf{w}\in\mathbf{R}^3,\ \mathbf{w}\boldsymbol{\nu} = \mathbf{v}^-\boldsymbol{\nu}\} \quad (4.5)$$

corresponds to the vector

$$[\mathbf{v}] = \mathbf{v}^+ - \mathbf{v}^- \in \Pi_\nu = \{\mathbf{u}\in\mathbf{R}^3 : \mathbf{u}\boldsymbol{\nu} = 0\}.$$

As mentioned before, we will take the integral of (4.5) over the discontinuity surface as the dissipation at this surface. First, we slightly simplify extremum problem (4.5), considering the tangent component $\boldsymbol{\tau}\in\Pi_\nu$ of the traction $\boldsymbol{\tau}$

exerted by the stress $\boldsymbol{\sigma}$ on the discontinuity surface $\gamma$: $\tau_i = \sigma_{ij}\nu_j - \sigma_{kj}\nu_k\nu_j\nu_i$. Obviously, there are various stresses $\boldsymbol{\sigma}$ that result in the same traction $\boldsymbol{\tau}$. If among these stresses there is $\boldsymbol{\sigma}$ admissible on one side of $\gamma$, we say that the tangent traction $\boldsymbol{\tau}$ is admissible on this side of the discontinuity surface. Thus, the sets of tangent tractions admissible on one of the sides of $\gamma$ are

$$A_{\nu x}^- = \{\boldsymbol{\tau} \in \Pi_\nu : \text{there exists } \boldsymbol{\sigma}^- \in C_x^-,\ \sigma_{ij}^-\nu_j - \sigma_{kj}^-\nu_k\nu_j\nu_i = \tau_i\},$$
$$A_{\nu x}^+ = \{\boldsymbol{\tau} \in \Pi_\nu : \text{there exists } \boldsymbol{\sigma}^+ \in C_x^+,\ \sigma_{ij}^+\nu_j - \sigma_{kj}^+\nu_k\nu_j\nu_i = \tau_i\},$$

and $A_{\nu x} = A_{\nu x}^- \cap A_{\nu x}^+$ is the set of tangent tractions admissible on both sides of $\gamma$. We also introduce the function

$$d_{\nu x}(\mathbf{u}) = \sup\{\boldsymbol{\tau}_*\mathbf{u} : \boldsymbol{\tau}_* \in A_{\nu x}\} \tag{4.6}$$

on the plane $\Pi_\nu$. Normally, the set $A_{\nu x}$ can be easily found (see Example 1) and evaluating the function $d_{\nu x}$ creates no problem, as it only requires maximizing of a linear function over the convex set $A_{\nu x}$ in the plane.

The following proposition clarifies the idea of introducing the function $d_{\nu x}$. The proposition is a particular case of a well known duality theorem.

PROPOSITION 1. If the origin is an internal point of $C_x^-$ or/and $C_x^+$, infimum (4.5) is attained and equals $d_{\nu x}([\mathbf{v}])$.

This suggests that the minimizer $\mathbf{w}_0$ in (4.5) be used as the vector $\mathbf{w}$ involved in constructing approximation (4.2) for the discontinuous velocity field $\mathbf{v}$. Then the limit of the dissipation is minimum and equals

$$\lim_{\varepsilon\to 0}\int_{\gamma_\varepsilon} d_x\left(\mathbf{e}_\nu([\mathbf{v}])\right) dV = \int_\gamma d_{\nu x}([\mathbf{v}])\, dS.$$

We refer to $d_{\nu x}([\mathbf{v}])$ as *dissipation at the discontinuity surface* and to its integral as *total dissipation* at this surface.

REMARK 1. The boundary $\partial\Omega$ or its parts can be included in the collection of discontinuity surfaces. In this case we set by definition for the outer side of the surface: $C_x^+ = Sym$, which is the set of admissible stresses for a rigid body. In particular, we consider the fixed part $S_v$ of the body boundary as a border between the body and a rigid body. The function $d_{\nu x}$ is also defined in this case.

REMARK 2. If there is no jump in the material properties on $\gamma$, then $A_{\nu x}^- = A_{\nu x}^+ = A_{\nu x}$ and

$$\begin{aligned} d_{\nu x}([\mathbf{v}]) = \sup\{\mathbf{t}_*[\mathbf{v}] : \mathbf{t}_* \in A_{\nu x}\} = \\ = \sup\{(\sigma_{*ij}\nu_j - \sigma_{*kj}\nu_k\nu_j\nu_i)[v_i] : \boldsymbol{\sigma}_* \in C_x\} = \\ = \sup\{\boldsymbol{\sigma}_* \cdot \mathbf{e}_\nu([\mathbf{v}]) : \boldsymbol{\sigma}_* \in C_x\} = d_x\left(\mathbf{e}_\nu([\mathbf{v}])\right). \end{aligned}$$

The dissipation at the discontinuity surface equals that defined in Subsection 1.2 for bodies with material properties continuously depending on the point.

**4.3. Kinematic multiplier and kinematic method.** Let us now define the kinematic multiplier of a piecewise continuously differentiable velocity field $\bar{\mathbf{v}} = (\mathbf{v}, \mathbf{v}^e)$ in a body with jump inhomogeneity. Suppose that the body occupies domain $\Omega$ consisting of pair-wise disjoint standard (Subsection 1.4) domains $\omega_1, \ldots, \omega_A$, and in each of $\omega_a$ (i) the field $\bar{\mathbf{v}}$ is regular, and (ii) the material properties continuously depend on the point. Approximating $\mathbf{v}$ in thin layers near discontinuity surfaces, as it was described in the previous subsection, results in a sequence of regular fields $\mathbf{v}_\varepsilon$ such that their kinematic multipliers have the limit

$$\lim_{\varepsilon \to 0} m_k(\mathbf{v}_\varepsilon) = m_k(\bar{\mathbf{v}}),$$

$$m_k(\bar{\mathbf{v}}) = \frac{\int\limits_{\Omega} d_x\,(\mathrm{Def}_l\,\mathbf{v})\,dV + \int\limits_{\Gamma} d_{\nu x}([\mathbf{v}])\,dS}{\int\limits_{\Omega} \mathbf{f}\mathbf{v}\,dV + \int\limits_{S_q} \mathbf{q}\mathbf{v}^e\,dS}. \tag{4.7}$$

Here, $\Gamma$ denotes the collection of all discontinuity surfaces of the velocity field; $\Gamma$ is contained in the union of the boundaries $\partial\omega_1, \ldots, \partial\omega_A$. The dissipation at $\Gamma$ is included in the total dissipation in (4.7).

We take (4.7) as the definition of the *kinematic multiplier of piecewise regular velocity field* $\bar{\mathbf{v}}$ *in a body with jump inhomogeneity*. The kinematic multiplier for $\bar{\mathbf{v}} = (\mathbf{v}, \mathbf{v}^e)$ is defined provided the field is kinematically admissible, $\mathbf{v}^e|_{S_v} = 0$, and the power of the load $\mathbf{l} = (\mathbf{f}, \mathbf{q})$ on $\bar{\mathbf{v}}$ (the denominator in (4.7)) is positive. According to Remark 2, the above definition is an immediate generalization of kinematic multiplier definition (1.5) for the case of bodies with jump inhomogeneity.

Definition (4.7) does not alter the main property of kinematic multipliers of regular velocity fields: *every kinematic multiplier* $m_k(\bar{\mathbf{v}})$ *of a load is not less than any of its static multipliers in the case of piecewise regular fields.* Indeed, the inequality $m_k(\mathbf{v}_\varepsilon) \geq m_s(\boldsymbol{\sigma})$ is valid for each of the regular approximations $\mathbf{v}_\varepsilon$ to $\bar{\mathbf{v}}$, and passage to the limit results in $m_k(\bar{\mathbf{v}}) \geq m_s(\boldsymbol{\sigma})$. Thus, any kinematic multiplier is an upper bound for the safety factor: $m_k(\bar{\mathbf{v}}) \geq \alpha_l$. Moreover, $\alpha_l$ is the infimum of $m_k(\bar{\mathbf{v}})$ under the assumptions of Theorems V.5.1 or VI.1.1. As mentioned earlier (Subsection 2.1), application of the discontinuous velocity fields to limit analysis extends the capabilities of the kinematic method.

**4.4. Rigid-plastic solutions method.** Let us obtain a criterion for the equality of the static and kinematic multipliers in the case of bodies with jump inhomogeneity. Consider such a body occupying domain $\Omega$ and subjected to the load $\mathbf{l}$. Suppose $\boldsymbol{\sigma}$ is a piecewise continuously differentiable stress field in $\Omega$ equilibrating the load $m\mathbf{l}$, $m > 0$, which means that $m = m_s(\boldsymbol{\sigma})$ is a static multiplier of $\mathbf{l}$. Let $\bar{\mathbf{v}} = (\mathbf{v}, \mathbf{v}^e)$ be a piecewise regular kinematically admissible velocity field on $\bar{\Omega}$, the admissibility meaning that $\mathbf{v}^e|_{S_v} = 0$. We assume that $\Omega$ consists of pair-wise disjoint standard (Subsection 1.4) subdomains $\omega_1, \ldots, \omega_A$, and in each of $\omega_a$ (i) $\boldsymbol{\sigma}$ and $\mathbf{v}$ are continuously differentiable, and (ii) the material properties continuously depend on the point. The reasoning used in the proofs of Theorems 2.1 and 2.2 is also applicable to bodies with jump inhomogeneity and results in the following *criterion for the equality of the static and kinematic multipliers*: under the aforementioned assumptions, $m_k(\bar{\mathbf{v}}) = m_s(\boldsymbol{\sigma}) = m$ if and only if (i) the relations

$$\begin{gathered}\frac{\partial \sigma_{ij}}{\partial x_j} + m f_i = 0,\\ \boldsymbol{\sigma} \in C_x, \quad \mathrm{Def}_l\, \mathbf{v} \cdot \boldsymbol{\sigma} = d_x\left(\mathrm{Def}_l\, \mathbf{v}\right)\end{gathered} \tag{4.8}$$

are satisfied in the smoothness subdomains $\omega_1, \ldots, \omega_A$, (ii) the boundary conditions

$$\sigma_{ij}\nu_j|_{S_q} = mq_i, \quad \mathbf{v}^e|_{S_v} = 0, \tag{4.9}$$

are satisfied, (iii) the discontinuity relations

$$[\sigma_{ij}]\nu_j = 0, \quad \mathbf{e}_\nu([\mathbf{v}]) \cdot \boldsymbol{\sigma} = d_{\nu x}([\mathbf{v}]) \tag{4.10}$$

are satisfied on the discontinuity surfaces, and (iv) the power of $\mathbf{l} = (\mathbf{f}, \mathbf{q})$ on the velocity field $\bar{\mathbf{v}}$ is nonzero. In case the zero stress field is safe, the latter condition can be replaced by the condition $\bar{\mathbf{v}} \neq 0$. Recall that $d_{\nu x}$ in (4.10) is the dissipation at the discontinuity surface defined by (4.6). The expression $\mathbf{e}_\nu([\mathbf{v}]) \cdot \boldsymbol{\sigma}$ makes sense in spite of a possible discontinuity of $\boldsymbol{\sigma}$ (see Remark 2.1). The components of the strain rate field $\mathrm{Def}_l\, \mathbf{v}$ are composed of the local derivatives $(\partial v_i/\partial x_j)_l$, see Remark 1.1.

We refer to the piecewise continuously differentiable stress and velocity fields $\boldsymbol{\sigma}$ and $\bar{\mathbf{v}} = (\mathbf{v}, \mathbf{v}^e)$ which satisfy (4.8) – (4.10) as the *discontinuous solution* to the rigid-plastic problem with the load $m\mathbf{l}$. Thus, $m_k(\bar{\mathbf{v}}) = m_s(\boldsymbol{\sigma}) = m > 0$ if and only if $\boldsymbol{\sigma}$ and $\bar{\mathbf{v}} \neq 0$ are a discontinuous solution to the rigid-plastic problem with the load $m\mathbf{l}$, $m > 0$. This criterion for the equality of the static and kinematic multipliers results in the extension of the rigid-plastic solutions method (Subsections III.5.3 and 2.3): the discontinuous solutions may exist when the problem has no smooth solutions.

REMARK 3. The definition of the discontinuous solution presumes that (4.10) should be also satisfied in the case when a discontinuity surface lies on $\partial\Omega$ (Subsection 1.3). The dissipation $d_{\nu x}$ on the discontinuity surface in this case is defined in Remark 1.

**4.5. Discontinuity relations.** Let us consider some properties of discontinuity relations (4.10), where both stress and velocity may be discontinuous on the same surface in the case of a body with jump inhomogeneity.

The second of conditions (4.10) can be interpreted as a maximum principle for the velocity jump. We consider this interpretation restricting ourselves to the (more interesting) case of bodies with cylindrical yield surfaces. Then the velocity jump is tangent to the discontinuity surface, that is,

$$[\mathbf{v}] \in \Pi_\nu = \{\mathbf{u} \in \mathbf{R}^3 : \mathbf{u}\boldsymbol{\nu} = 0\}$$

(see Subsection 3.2), where $\boldsymbol{\nu}$ is the normal to the discontinuity surface. The same reasoning as in Subsection 3.2 results in the following form of the second relation in (4.10):

$$\boldsymbol{\tau} \in A_{\nu x}, \quad [\mathbf{v}] \in A_{\nu x}, \quad [\mathbf{v}](\boldsymbol{\tau} - \boldsymbol{\tau}_*) \geq 0 \quad \text{for every} \quad \boldsymbol{\tau}_* \in A_{\nu x} \tag{4.11}$$

(we supplemented the relation with the tangent tractions admissibility condition: $\boldsymbol{\tau} \in A_{\nu x}$).

Thus, *in the case of bodies with cylindrical yield surfaces, the discontinuity relations are as follows: the tractions on both sides of a discontinuity surface are the same, the velocity jump is tangent to this surface, and the velocity jump and the tangent traction satisfy the maximum principle* (4.11).

The form of the discontinuity relations stays the same for bodies with the material properties continuously depending on the point (Subsection 3.2) and bodies with jump inhomogeneity. In the latter case, two different sets $C_x^-$, $C_x^+$ for the two sides of the discontinuity surface are involved in the stress admissibility conditions: $\boldsymbol{\sigma}^- \in C_x^-$, $\boldsymbol{\sigma}^+ \in C_x^+$, while in the former case these sets are the same for both sides, and the conditions read: $\boldsymbol{\sigma}^- \in C_x$, $\boldsymbol{\sigma}^+ \in C_x$. This results in a certain difference between consequences of the discontinuity relations for the two types of bodies. In particular, Proposition 3.1 and 3.2 are not valid in case $C_x^- \neq C_x^+$. Namely, if $C_x^- \neq C_x^+$ at the surface of velocity jump, then 1) not necessarily both stresses $\boldsymbol{\sigma}^-$ and $\boldsymbol{\sigma}^+$ are on the boundaries of the corresponding sets $C_x^-$ and $C_x^+$, that is, on the yield surfaces; 2) the stresses $\boldsymbol{\sigma}^-$, $\boldsymbol{\sigma}^+$ are not uniquely determined by the velocity jump $[\mathbf{v}]$; and 3) the stress and the velocity can be discontinuous at the same surface. The following example illustrates these properties.

EXAMPLE 2. Consider the body described in Example 1. This is a piecewise homogeneous body with the jump inhomogeneity: the part $\gamma$ of the plane $x_1 = 0$ separates the two subdomains in the body, which consist of two different materials. The sets of admissible tangent tractions on two sides of $\gamma$ are

$$A^- = A_{\nu x}^- = \{\boldsymbol{\tau} = \tau_2 \mathbf{e}_2 + \tau_3 \mathbf{e}_3 : \tau_2^2 + \tau_3^2 \leq k^2\},$$
$$A^+ = A_{\nu x}^+ = \{\boldsymbol{\tau} = \tau_2 \mathbf{e}_2 + \tau_3 \mathbf{e}_3 : \tau_2^2 + (\tau_3 - \rho k)^2 \leq k^2\}.$$

The sets are shown in Figure 4.2 as well as $A = A_{\nu x} = A^- \cap A^+$. Consider an admissible tangent traction $\boldsymbol{\tau}$ which is on the yield surface on the "−" side of $\gamma$ and strictly inside the yield surface on the "+" side of $\gamma$:

$$\tau_2^2 + \tau_3^2 = k^2, \quad \tau_2^2 + (\tau_3 - \rho k)^2 < k^2. \tag{4.12}$$

Consider also a velocity jump $[\mathbf{v}]$ directed along the outward normal to the boundary of $A^-$ at its point $\boldsymbol{\tau}$, see Figure 4.2. It is clear that relations (4.11) are satisfied. The remaining discontinuity relations imply that the following relations are valid:

$$\sigma_{11}^- = \sigma_{11}^+, \quad \sigma_{12}^- = \sigma_{12}^+ = \tau_2, \quad \sigma_{13}^- = \sigma_{13}^+ = \tau_3.$$

Besides, the stresses $\boldsymbol{\sigma}^-$, $\boldsymbol{\sigma}^+$ are admissible on both sides of $\gamma$. By virtue of (4.12), the admissibility condition $\boldsymbol{\sigma}^- \in C_x^-$ implies that

$$\sigma_{11}^{-d} = 0, \quad \sigma_{22}^{-d} = 0, \quad \sigma_{33}^{-d} = 0, \quad \sigma_{23}^- = 0.$$

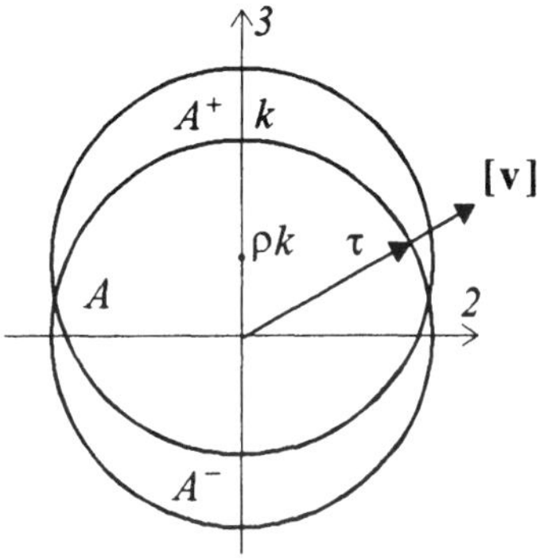


**Figure 4.2**

These equalities together with those given above uniquely determine the stress deviator $\boldsymbol{\sigma}^{-d}$. On the other hand, the admissibility condition $\boldsymbol{\sigma}^{+} \in C_x^+$ means that

$$(\sigma_{11}^{+d})^2 + (\sigma_{22}^{+d})^2 + (\sigma_{33}^{+d})^2 + 2(\sigma_{23}^{+})^2 \leq 2k^2 - 2\tau_2^2 - 2(\tau_3 - k\rho)^2.$$

In contrast to the condition $\boldsymbol{\sigma}^{-} \in C^{-}$, which determined $\boldsymbol{\sigma}^{-d}$, the latter inequality does not determine the stress deviator $\boldsymbol{\sigma}^{+d}$ uniquely (since the right-hand side in the previous formula is positive due to the inequality in (4.12)). In particular, it is clear that 1) the stresses $\boldsymbol{\sigma}^{+}$ may lie strictly inside the yield surface, 2) the deviator $\boldsymbol{\sigma}^{+d}$ is not uniquely determined by the velocity jump $[\mathbf{v}]$, and 3) the components $\sigma_{22}$, $\sigma_{33}$, $\sigma_{23}$ and the velocity $\mathbf{v}$ can be discontinuous on the same surface.

## 5. Examples of limit analysis

In this section some examples illustrating applications of discontinuous fields to limit analysis are presented.

**5.1. Lateral stretching of strip.** Consider an infinitely long strip of a rectangular cross-section. The strip is subjected to a stretching load on its lateral sides, the load being distributed with constant density $p$, see Figure 5.1.

Let us adopt the Mises admissibility condition, $|\boldsymbol{\sigma}^d| \leq \sqrt{2}k$, $k = \text{const} > 0$, for the material of the strip and find the safety factor $\alpha_p$ of the load. To evaluate the safety factor, we apply the rigid-plastic solutions method making use of discontinuous velocity fields. This means looking for a number $m > 0$, an admissible stress field $\boldsymbol{\sigma}$ equilibrating the load $mp$ and a velocity field $\mathbf{v} \neq 0$ which satisfy the normality flow rule

$$\frac{1}{2}\left(\frac{\partial v_i}{\partial x_j} + \frac{\partial v_j}{\partial x_i}\right) = \dot{\lambda}\sigma_{ij}^d, \quad \dot{\lambda} \geq 0, \quad \dot{\lambda}(|\boldsymbol{\sigma}^d| - \sqrt{2}k) = 0$$

in smoothness domains and the discontinuity relations on the discontinuity surfaces. If such $m$, $\boldsymbol{\sigma}$, $\mathbf{v}$ are found, then $m$ is the safety factor of the load, see Theorems 2.1 and 2.2.

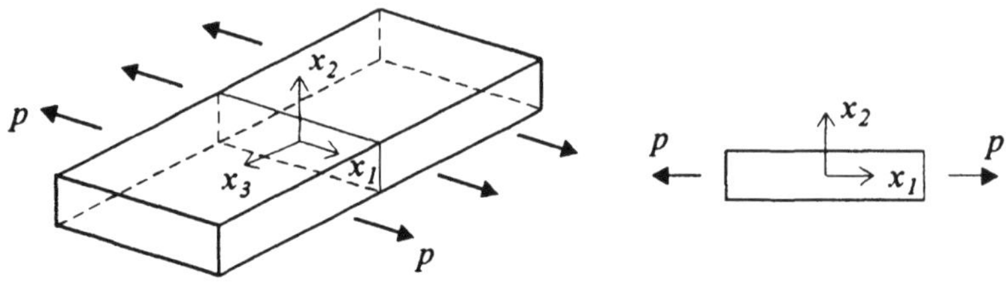


**Figure 5.1**

Due to the symmetry of the problem, the stress and velocity components do not depend on the coordinate $x_3$ (the $x_3$-axis is directed along the axis of the strip), and $v_3 = 0$, $\sigma_{13} = \sigma_{23} = 0$. Let us look for a solution with a homogeneous stress field $\boldsymbol{\sigma}$, that is, $\boldsymbol{\sigma}$ not depending on the point of the body. Then the boundary conditions on the lateral sides of the strip determine the components $\sigma_{11} = mp$, $\sigma_{12} = \sigma_{21} = 0$, $\sigma_{13} = \sigma_{31} = 0$. Similarly, the boundary conditions on the lower and upper sides of the strip imply $\sigma_{22} = 0$, $\sigma_{23} = \sigma_{32} = 0$. Under the plane strain conditions, the normality rule results in $\sigma_{33}^d = 0$ for such a stress field. Then, $\sigma_{33} = \sigma_{11}/2 = mp/2$ and the yield criterion $|\sigma^d| = \sqrt{2}k$ implies $m = 2k/p$. However, this cannot be considered as a final answer until an appropriate velocity field is constructed (or, at least, existence of the velocity field is established).

The normality rule in case of the above determined stress field is reduced to the equations

$$\frac{\partial v_1}{\partial x_2} + \frac{\partial v_2}{\partial x_1} = 0, \quad \frac{\partial v_1}{\partial x_1} + \frac{\partial v_2}{\partial x_2} = 0,$$

which imply that two components $v_1$, $v_2$ of the velocity field satisfy the wave equations

$$\frac{\partial^2 v_1}{\partial x_1^2} - \frac{\partial^2 v_1}{\partial x_2^2} = 0, \quad \frac{\partial^2 v_2}{\partial x_1^2} - \frac{\partial^2 v_2}{\partial x_2^2} = 0.$$

The general solutions of these equations are

$$v_1 = f_1(x_1 - x_2) + g_1(x_1 + x_2), \quad v_2 = f_2(x_1 - x_2) + g_2(x_1 + x_2),$$

where $f_1$, $f_2$, $g_1$, $g_2$ are arbitrary functions. This suggests that the discontinuity surfaces of a discontinuous velocity field are the planes $x_1 - x_2 = c_1$, $x_1 + x_2 = c_2$, where $c_1$ and $c_2$ are constants. The simplest velocity field possessing these properties is a piecewise constant field $\mathbf{v}$ with the single discontinuity surface $x_1 - x_2 = 0$, see Figure 5.2a. In the smoothness domains $x_1 - x_2 < 0$ and $x_1 - x_2 > 0$ this velocity field takes the constant values $\mathbf{v}^-$ and $\mathbf{v}^+$, respectively. The strain rate is zero in both smoothness domains and, therefore, obviously satisfies the normality rule. To show that the above $m$, $\boldsymbol{\sigma}$, $\mathbf{v}$ is the required solution, it remains to verify that $\boldsymbol{\sigma}$, $\mathbf{v}$ satisfy the discontinuity relations on the surface $x_1 - x_2 = 0$.

Here it is convenient to use the discontinuity relations in form (3.1), and they can be written as

$$\frac{1}{2}([v_i]\nu_j + [v_j]\nu_i) = \dot{\lambda}^- \sigma_{ij}^d, \quad \dot{\lambda}^- \geq 0$$

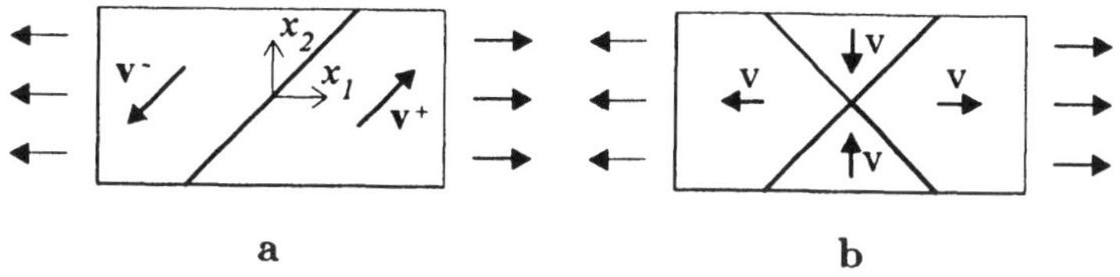


**Figure 5.2**

since the stress $\boldsymbol{\sigma}$ lies on the yield surface. Recall that here $v_3 = 0$, $\nu_1 = -\nu_2 = 1/\sqrt{2}$, $\nu_3 = 0$; and $\sigma_{11} = mp = 2k$ and $\sigma_{33} = k$ are the only nonzero components of the stress. Finally, the normality rule can be written as

$$v_2^+ - v_2^- = v_1^+ - v_1^-, \quad v_1^+ - v_1^- = \dot{\lambda}\sqrt{2}k, \quad \dot{\lambda} \geq 0.$$

If a certain jump $v_1^+ - v_1^- > 0$ is chosen, these relations determine the values of $v_2^+ - v_2^-$ and $\dot{\lambda}$. Then all the discontinuity relations are satisfied, and the velocity field $\mathbf{v}$ is nonzero. Consequently, $m$, $\boldsymbol{\sigma}$, $\mathbf{v}$ are a solution to the above formulated problem and $\alpha_p = m = 2k/p$ by Theorem 2.2.

Note that the problem about lateral stretching of the strip possesses some other solutions, such as those formed by the same stress field and the velocity field shown in Figure 5.2b. The discontinuity surfaces of this velocity field divide the strip into four smoothness domains, the velocity being of the same magnitude in each one of them. The stress and velocity fields satisfy the discontinuity relations, which is easy to verify in the same way as above.

**5.2. Stretching of a strip with a hole.** Consider a body, which under the plane strain conditions is represented by domain $\omega$ in the $(x_1, x_2)$-plane. Let $\omega$ be the strip $|x_2| < h$ with a hole. We assume that the dimension of the hole in $x_2$-direction is $2a$ and also that the hole contour is at the same distances from the upper and lower sides of the strip, see Figure 5.3. Let the strip be subjected to a stretching load at infinity, that is, the stress is to satisfy the conditions

$$\sigma_{11} \to p, \quad \sigma_{12} \to 0, \quad \sigma_{22} \to 0 \quad \text{when} \quad |x_1| \to \infty.$$

Let us adopt the Mises admissibility condition for the material of the strip and find the safety factor $\alpha_p$ of the load. To obtain converging two-sided bounds for $\alpha_p$, we apply the static and kinematic methods of limit analysis.

Let us first establish a kinematic upper bound for $\alpha_p$ using a discontinuous velocity field. To construct the velocity field, consider two points $A_+$ and $A_-$ of the hole contour which are nearest to the upper and lower sides of the strip, respectively. We join $A_+$ with the upper side by the segment $A_+B_+$ and $A^-$ with the lower side by segment $A_-B_-$, both segments forming $\pi/4$ angles with $x_1$-axis. These segments divide the strip into two domains lying to the left and to the right of them, see Figure 5.3. We consider a discontinuous velocity field $\mathbf{v}$ with the components

$$v_1 = v_2 = \begin{cases} u & \text{if } x \in \omega^+, \\ -u & \text{if } x \in \omega^-, \end{cases}$$

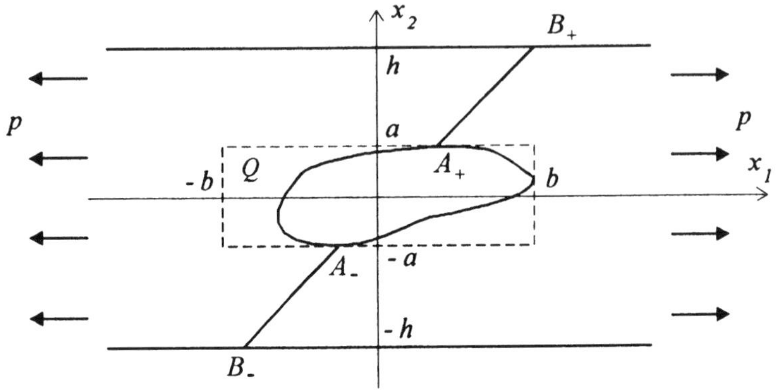


**Figure 5.3**

where $u = \text{const} > 0$. The unit normal $\boldsymbol{\nu}$ to the discontinuity surface $A_+B_+ \cup A_-B_-$ has the components $\nu_1 = 1/\sqrt{2}$, $\nu_2 = -1/\sqrt{2}$ and the tensor $\mathbf{e}_\nu[\mathbf{v}]$ has the components $e_{\nu\,11} = \sqrt{2}u$, $e_{\nu\,12} = 0$, $e_{\nu\,22} = -\sqrt{2}u$. Using (1.3), we find the kinematic multiplier of $\mathbf{v}$:

$$m_k(\mathbf{v}) = \frac{1}{4hpu} \int\limits_{A_+B_+\cup A_-B_-} \sqrt{2}k\,|\mathbf{e}_\nu|\,ds = \frac{2k}{p}\left(1 - \frac{a}{h}\right),$$

which is the upper bound for the safety factor $\alpha_p$.

Let us now establish a static lower bound for $\alpha_p$ using a discontinuous stress field. We will construct the stress field which results in such an estimation in the case of the hole of an arbitrary shape. Namely, consider the rectangle $Q$ centered at the origin, containing the hole and being of the minimum possible height, that is, of the same height $2a$ as the hole. Let the width of $Q$ be $2b$ and let us consider the part $\omega_Q = \omega \setminus Q$ of the domain $\omega$. Now construct the stress field $\boldsymbol{\sigma}$ in $\omega_Q$ satisfying the equilibrium equations $\partial\sigma_{ij}/\partial x_j = 0$ and the boundary conditions

$$\sigma_{ij}\nu_j\big|_{x_2=\pm h} = 0; \quad \sigma_{11} \to mp, \ \sigma_{12} \to 0, \ \sigma_{22} \to 0 \text{ when } |x_1| \to \infty, \quad (5.1)$$

$$\sigma_{ij}\nu_j\big|_{\partial Q} = 0 \quad (5.2)$$

($\boldsymbol{\nu}$ is the unit normal to the boundary). The stress field $\boldsymbol{\sigma}$ is defined in the domain $\omega_Q$, that is, in the strip with the enlarged (rectangular) hole $Q$. If the load $mp$ is applied to $\omega_Q$, it is obviously equilibrated by $\boldsymbol{\sigma}$ in $\omega_Q$. Now, we extend $\boldsymbol{\sigma}$ to all of the $\omega$, assigning the extension $\boldsymbol{\sigma}^c$ zero value $\boldsymbol{\sigma}^c(x) = 0$ at any $x$ in $Q$. Generally speaking, the field $\boldsymbol{\sigma}^c$ is discontinuous on the contour $\partial Q$. At the same time, (5.2) ensures that $\boldsymbol{\sigma}^c$ satisfies the conditions $[\sigma^c_{ij}]\nu_j = 0$ on $\partial Q$, and, consequently, $\boldsymbol{\sigma}^c$ in $\omega$ equilibrates the load $mp$. Since the admissibility of $\boldsymbol{\sigma}^c$ is equivalent to the admissibility of $\boldsymbol{\sigma}$, the static multiplier $m_s(\boldsymbol{\sigma}^c)$ of the load $p$ applied to $\omega$ and the static multiplier $m_s(\boldsymbol{\sigma})$ of the same load applied to $\omega_Q = \omega \setminus Q$ are equal. In particular,

$m_s(\boldsymbol{\sigma}^c) = m_s(\boldsymbol{\sigma})$ is a lower bound for the safety factor $\alpha_p$ in the original problem.

REMARK 1. The above reasoning is well known in limit analysis. It is applicable to a wide class of problems and shows that *adding of material to the free part of the body boundary does not reduce the safety factor.*

Thus, to obtain a lower bound for the safety factor $\alpha_p$ it is sufficient to find a suitable stress field $\boldsymbol{\sigma}$ in the domain $\omega$. Let us construct $\boldsymbol{\sigma}$ using the *stress function* $\Psi$:

$$\sigma_{11} = \frac{\partial^2 \Psi}{\partial^2 x_2}, \quad \sigma_{22} = \frac{\partial^2 \Psi}{\partial^2 x_1}, \quad \sigma_{12} = -\frac{\partial^2 \Psi}{\partial x_1 \partial x_2}.$$

Note that $\boldsymbol{\sigma}$ satisfies the equilibrium equations $\partial \sigma_{ij}/\partial x_j = 0$ no matter how $\Psi$ is chosen. We will now determine the stress function so that the corresponding stress field will result in a suitable estimation for the safety factor. First, let us determine $\Psi$ in the part

$$\omega_Q^+ = \{(x_1, x_2) \in \omega_Q;\ x_1 > 0,\ x_2 > 0\}$$

of $\omega_Q$. Since the stress field should tend to limit (5.1) when $x_1 \to +\infty$, the limit stresses correspond to the stress function

$$\Psi_\infty(x) = \frac{mp}{2}(x_2^2 - h^2).$$

On the other hand, it looks natural to consider constant stresses in the subdomain $0 < x_1 < b$, $0 < x_2 < h$. The only constant stresses which can be the values of a stress field equilibrating the load $mp$ are

$$\sigma_{11} = mp\frac{h}{h-a}, \quad \sigma_{12} = 0, \quad \sigma_{22} = 0.$$

These stresses correspond to a stress function quadratic in $x_2$. To determine the stress function near the side $x_1 = b$, $|x_2| < a$ of the rectangle $Q$, note that the choice $\Psi = \text{const}$ implies that boundary condition (5.2) is satisfied on this side. Thus, we consider the stress function

$$\Psi_0(x) = \begin{cases} \dfrac{mp}{2}\left(\dfrac{a}{h-a}(h - x_2)^2 + x_2^2 - h^2\right) & \text{if } a < x_2 < h, \\ mph(a-h) & \text{if } 0 < x_2 < a \end{cases}$$

near the rectangle $Q$. Finally, we introduce the combination

$$\Psi = a\Psi_0 + (1-a)\Psi_\infty,$$

where $a$ depends on $x_1$ and the multipliers $a$, $1-a$ make $\Psi$ equal to $\Psi_0$ near the rectangle $Q$ and equal to $\Psi_\infty$ at large values of $x_1$. The first and second derivatives of $a$ effect the stresses determined by the stress function

$\Psi$, while we are interested in making the influence negligible. Therefore, we consider a function $\gamma(\xi)$ defined for $-\infty < \xi < +\infty$, which has two continuous derivatives, takes the value 1 if $\xi \leq 0$ and the value 0 if $\xi \geq 1$. Then, we introduce the function $x_1 \to \gamma(\varepsilon(x_1 - b))$, where $\varepsilon > 0$ is a parameter. The derivatives of this function are small if $\varepsilon$ is small enough, and the function takes the value 1 if $0 < x_1 < b$ and the value 1 at sufficient large $x_1$. Therefore, this function can be used as the coefficient $a$ in the above combination of $\Psi_0$, $\Psi_\infty$:

$$\Psi(x) = \gamma(\varepsilon(x_1 - b))\,\Psi_0(x) + \{1 - \gamma(\varepsilon(x_1 - b))\}\Psi_\infty(x).$$

The function $\Psi$ is defined in $\omega_Q^+$ and can be extended to all of $\omega_Q$ as the even function of $x_1$ and even function of $x_2$; we denote the extension by the same symbol $\Psi$.

The function $\Psi$ is continuously differentiable in $\omega_Q$, and its second derivatives are also continuous in $\omega_Q$ with the exception of the segment $b \leq x_1 \leq b + 1/\varepsilon$, $x_2 = a$ and the corresponding symmetric segments (as $\Psi$ is even in $x_1$ and $x_2$). It can be easily verified that the stress field $\boldsymbol{\sigma}$ determined by the stress function $\Psi$ satisfies the condition $[\sigma_{ij}]\nu_j = 0$ at these discontinuity lines. Besides, $\boldsymbol{\sigma}$ satisfies (5.1), (5.2) and, thus, equilibrates the load $mp$ in $\omega_Q$ (the strip with the rectangular hole). It can be easily seen that

$$\max\{\left|\sigma^d(x_1, x_2)\right| : (x_1, x_2) \in \omega_Q\} = \frac{mp}{\sqrt{2}}\frac{h}{h-a} + O(\varepsilon).$$

The first term on the right-hand side is less than $\sqrt{2}k$ if

$$0 < m < \frac{2k}{p}\left(1 - \frac{a}{h}\right).$$

Then, we choose the parameter $\varepsilon$ small enough to ensure the admissibility of $\boldsymbol{\sigma}$, which results in the admissibility of the load $mp$ and the following lower bound for the safety factor:

$$\frac{2k}{p}\left(1 - \frac{a}{h}\right) \leq \alpha_p.$$

Comparing this estimation with the above kinematic upper bound for $\alpha_p$, we arrive at the equality $\alpha_p = 2k(h-a)/hp$.

REMARK 2. If the distance $h - a_1$ between the hole and one of the strip's sides is less than the distance $h - a_2$ between the hole and the other side, the above reasoning results in the following estimation for the safety factor:

$$\frac{2k}{p}\left(1 - \frac{a_1}{h}\right) \leq \alpha_p \leq \frac{2k}{p}\left(1 - \frac{a_1 + a_2}{2h}\right).$$

**5.3. Limit surface for biaxial stretching of the plane with holes.** Consider a body which, under the plane strain condition, is represented by a domain $\Omega$ in the $(x_1, x_2)$-plane. We assume that $\Omega$ is obtained by removing a number of bounded domains from the plane and also that at least one of the removed domains contains a circle of a positive radius $r$.

Let the holes' contours be free and the body be subjected to the load at infinity with the $x_1$-component of the density $p$ and $x_2$-component of the density $q$. Thus, the stresses are to satisfy the conditions

$$\sigma_{11} \to p, \quad \sigma_{12} \to 0, \quad \sigma_{22} \to q \quad \text{when} \quad x_1^2 + x_2^2 \to \infty. \tag{5.3}$$

Let us adopt the Mises admissibility condition $|\sigma^d| \leq \sqrt{2}k$, $k = \text{const} > 0$, for the material of the body and find the limit surface in the $(p, q)$-plane. Recall that the limit surface is the boundary of the safe loads set and consists of the points representing the loads $(p, q)$ for which the safety factor equals 1, see Subsection III.4.3.

First of all, let us show that the limit surfaces are the same for all the domains within the above described class. Let $\Omega$ be one of these domains and consider two circles of the radii $R > 0$ and $r > 0$: the first circle contains all of the holes in $\Omega$; the second lies within one of these holes. Let $\Omega_R$ and $\Omega_r$ be the domains obtained by removing the corresponding circle from the whole plane; then, $\Omega_R \subset \Omega \subset \Omega_r$. We consider the safety factors $\alpha_{(R)}$, $\alpha$ and $\alpha_{(r)}$ of the load $(p, q)$ applied to $\Omega_R$, $\Omega$ and $\Omega_r$, respectively. The relations $\Omega_R \subset \Omega \subset \Omega_r$ imply the inequalities $\alpha_{(R)} \leq \alpha \leq \alpha_{(r)}$ (see Remark 1). On the other hand, the safety factor is dimensionless and, as far as $\Omega_R$ or $\Omega_r$ are concerned, the safety factor cannot depend on the radius of the circle: the other parameteres involved in the formulation of the problem do not contain length in their dimensions. Thus, $\alpha_{(r)} = \alpha_{(R)}$, which together with the above relations $\alpha_{(R)} \leq \alpha \leq \alpha_{(r)}$ proves the equality $\alpha = \alpha_{(R)}$. Since the safety factors of a load applied to $\Omega$ or to $\Omega_R$ are equal, the limit surfaces are the same for $\Omega$ and $\Omega_R$ and, consequently, for all the domains of the class under consideration.

Let us now find this limit surface. It should be noted that any stress field $\boldsymbol{\sigma}$ that equilibrates the load $(p, q)$ satisfies the condition (5.3) and, because of that, $\boldsymbol{\sigma}$ is inadmissible if $|q - p| > 2k$. Thus, the load $(p, q)$ is inadmissible if $|q - p| > 2k$. Let us show that the limit surface is the pair of the straight lines $|q - p| = 2k$ in $(p, q)$-plane or, in other words, the set of safe loads in the strip $|q - p| \leq 2k$. Since the set of safe loads is convex (Proposition III.4.1), it is sufficient to verify that the load $(p, 0)$ is safe when $|p| < 2k$ and the load $(q, q)$ is safe for every $q$: all the loads in the strip $|q - p| < 2k$ are convex combinations of $(p, 0)$- and $(q, q)$-loads.

To verify that the load $(p, 0)$ is safe when $|p| < 2k$, let us consider in the $(x_1, x_2)$-plane the strip $|x_2| < h$ with the hole $|x| < R$. According to Subsection 5.2, the stretching load $p$ is safe for this domain if $|p| < 2k(h-R)/h$. Then, there is a safe stress field $\boldsymbol{\sigma}$ which equilibrates such a safe load. In particular, $\boldsymbol{\sigma}$ satisfies the conditions $\sigma_{12} = 0$, $\sigma_{22} = 0$ on the strip's boundary. We extend $\boldsymbol{\sigma}$ to all of the set $|x_2| > h$, assigning $\sigma_{11}$ the value $p$ in this domain, the components $\sigma_{12} = \sigma_{21}$ and $\sigma_{22}$ being zero. It is clear that the extended stress field 1) equilibrates the load $(p, 0)$ in the domain $\Omega_R$ (the plane with the circular hole), and 2) is safe if $|p| < 2k(h - R)/h$. Here, $h$ can be arbitrarily large, which implies that $(p, 0)$ is a safe load if $|p| < 2k$.

The safety of any load $(q, q)$ arises from the fact that the load $(q', q')$ with $|q'| > |q|$ is admissible. To establish the admissibility of $(q', q')$ it is sufficient to find an admissible stress field $\boldsymbol{\sigma}$ which equilibrates this load. The stress field in the well known analytical solution of the *elastic*-plastic problem about stretching a plane with the circular hole can be taken as $\boldsymbol{\sigma}$.

**5.4. A pipe under internal torsion.** Let us return to the problem discussed in Subsection III.6.3 and construct its solution. Recall that an infinite cylindrical thick-walled pipe is considered with a fixed outer boundary. The pipe is subjected to a torsional load with the constant density $q$ on the interior boundary $S_q$ (see Figure III.6.3). We use the cylindrical coordinate system $\rho$, $\theta$, $z$ with its $z$-axis directed along the pipe axis, $\rho$ and $\theta$ being the polar coordinates in the pipe cross-section. Under the plane strain conditions the pipe is represented by this cross-section, the ring $r_0 < \rho < r_1$.

The problem under consideration has no solution with the regular nonzero velocity field, as shown in Subsection III.3.6. Therefore, we will look for a discontinuous velocity field, more precisely, for a velocity field $\bar{\mathbf{v}} = (\mathbf{v}, \mathbf{v}^e)$ with boundary slip, where $\mathbf{v}$ is a field on $\Omega$ and $\mathbf{v}^e$ is a field on $\partial\Omega$. We choose $\mathbf{v} \equiv 0$ and

$$v^e_\rho|_{S_q} = 0, \quad v^e_\theta|_{S_q} = u = \text{const}$$

(components of vectors and tensors are considered in the Cartesian basis associated with the cylindrical coordinate system, see Subsection III.6.1). Thus, $S_q$ is the only discontinuity surface of the velocity field. The unit outward normal to $S_q$ has the components $\nu_\rho = -1$, $\nu_\theta = 0$; the only nonzero components of the tensor $\mathbf{e} = \mathbf{e}_\nu([\mathbf{v}])$ defined in Subsection 1.3 are $e_{\rho\theta} = \mathbf{e}_{\theta\rho} = -u/2$. Formula (1.4) provides the expression for the kinematic multiplier of the velocity field $\bar{\mathbf{v}} = (\mathbf{v}, \mathbf{v}^e)$:

$$m_k(\bar{\mathbf{v}}) = \int_0^{2\pi} \sqrt{2}k\, |\mathbf{e}_\nu|\, r_0\, d\theta \left( \int_0^{2\pi} q u r_0\, d\theta \right)^{-1} = \frac{k}{q}.$$

This results in the upper bound for the safety factor: $\alpha_q \le k/q$. Recall that this estimation was obtained in Subsection III.6.3 with the help of a sequence of regular velocity fields.

Actually, the above velocity field $\bar{\mathbf{v}} = (\mathbf{v}, \mathbf{v}^e)$ together with the stress field $\boldsymbol{\sigma}$ introduced in Subsection III.6.3 forms a discontinuous solution to the rigid-plastic problem under consideration when the load is limit, $q = k$. Indeed, $\boldsymbol{\sigma}$ is an admissible stress field and equilibrates this load as shown in Subsection III.6.3. The velocity field $\bar{\mathbf{v}} = (\mathbf{v}, \mathbf{v}^e)$ is kinematically admissible. The corresponding strain rate field is zero in $\Omega$ and, therefore, satisfies the normality flow rule in $\Omega$. It remains to verify that relations (3.1) are satisfied on the

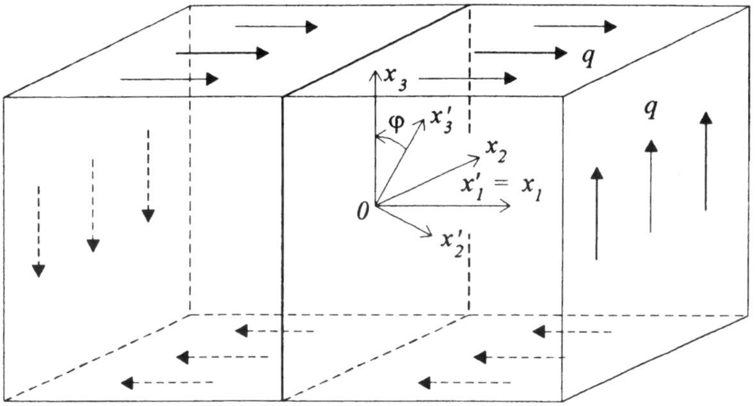


**Figure 5.4**

discontinuity surface $S_q$. Note that the only nonzero components of the tensors $\mathbf{e} = \mathbf{e}_\nu([\mathbf{v}])$ and $\boldsymbol{\sigma}$ are $e_{\rho\theta} = e_{\theta\rho} = -u/2$ and $\sigma_{\rho\theta} = \sigma_{\theta\rho} = -k$. Therefore, relations (3.1), which can be written as

$$\mathbf{e}_\nu([\mathbf{v}]) = \dot{\lambda}\boldsymbol{\sigma}^d, \quad \dot{\lambda} \geq 0$$

in the case of Mises yield surface, are satisfied on $S_q$ with any $u > 0$ (and $\dot{\lambda} = u/2k$ in these relations).

**5.5. Shear of a parallelepiped with jump inhomogeneity.** Consider a rigid perfectly plastic body occupying the domain $-a < x_1 < a$, $-b < x_2 < b$, $-c < x_3 < c$, which is a rectangular parallelepiped. Let the sides $x_2 = \pm b$ of the parallelepiped be free while the rest of the sides are subjected to the tangent load with the density $q = \text{const}$, see Figure 5.4. Thus, the stress should satisfy the following boundary conditions:

$$\begin{aligned} \sigma_{21} = \sigma_{22} = \sigma_{23} = 0 \quad &\text{at } x_2 = \pm b, \\ \sigma_{11} = \sigma_{12} = 0, \quad \sigma_{13} = q \quad &\text{at } x_1 = \pm a, \\ \sigma_{13} = q, \quad \sigma_{23} = \sigma_{33} = 0 \quad &\text{at } x_3 = \pm c. \end{aligned} \tag{5.4}$$

Let the material properties of the body be piecewise homogeneous with the jump discontinuity on the plane $x_1 = 0$. Namely, let the sets of admissible stresses be of the form

$$C_x = C_x^- = \{\boldsymbol{\sigma} \in Sym : |\boldsymbol{\sigma}^d| \leq \sqrt{2}k\} \quad \text{if } x_1 < 0,$$
$$C_x = C_x^+ = \{\boldsymbol{\sigma} \in Sym : |\boldsymbol{\sigma}^d - \boldsymbol{\rho}| \leq \sqrt{2}k\} \quad \text{if } x_1 > 0,$$

where the number $k > 0$ and the tensor $\boldsymbol{\rho}$ are constant. The tensor $\boldsymbol{\rho}$ is given by its components in the Cartesian coordinate system $x'_1 = x_1$, $x'_2$, $x'_3$ with the same origin and $x'_1$-axis as the origin and $x_1$-axis of the system $x_1$, $x_2$, $x_3$:

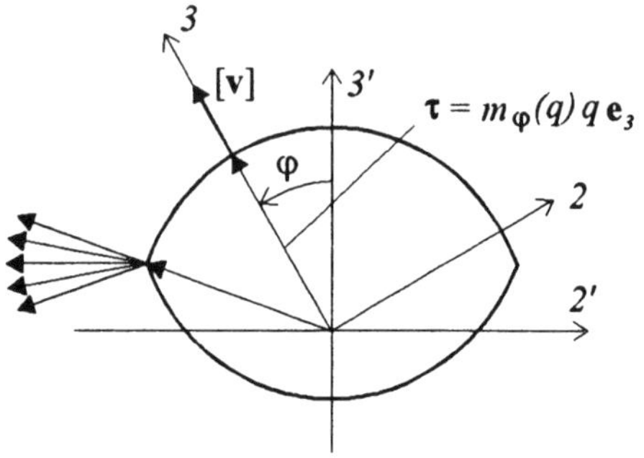


**Figure 5.5**

all the components of $\boldsymbol{\rho}$ are zero with the exception of $\rho'_{13} = \rho'_{31} = \rho k$, where $\rho$ is a given number, $0 < \rho < 1$. Thus, the parallelepiped can be thought of as being cut out of the body considered in Examples 4.1, 4.2. Two of the parallelepiped sides are parallel to and equidistant from the plane of the material properties jump discontinuity. The positions of the rest of the sides are determined by the angle $\varphi$ through which the axes $x'_2$, $x'_3$ should be rotated about the $x'_1 = x_1$-axis to coincide with the axes $x_2$, $x_3$. Thus, the constants $k$ and $\rho$ determine the properties of the rigid perfectly plastic material, while the response of the body to load (5.4) also depends upon the angle $\varphi$. The latter describes the orientation of the load with respect to the direction of the $x'_3$-axis characterizing the anisotropy of the body.

Let us find the safety factor of the load as a function of the angle $\varphi$ applying the rigid-plastic solutions method and looking for a solution with a velocity field discontinuous on the surface of the material properties jump. In other words, we will look for a number $m > 0$, an admissible stress field $\boldsymbol{\sigma}$ equilibrating the load $mq$, and velocity field $\mathbf{v} \neq 0$, which satisfy the normality flow rule at $x_1 < 0$ and $x_1 > 0$ and also discontinuity relations (4.11) on the plane $x_1 = 0$. If such $m$, $\boldsymbol{\sigma}$ and $\mathbf{v}$ are found, $m$ is the safety factor of the load $q$ according to Theorem 2.2.

To construct the solution, consider a homogeneous stress field $\mathbf{s}$ with the only nonzero components $s_{13} = s_{31} = q$. The stress field $m\mathbf{s}$ equilibrates the load $mq$ and is admissible if and only if the tangent traction $\boldsymbol{\tau} = ms_{13}\mathbf{e}_3 = mq\mathbf{e}_3$ is admissible on both sides of the jump surface $x_1 = 0$. Recall the set $A_{\nu x}$ of admissible tangent tractions $\boldsymbol{\tau}$ from Example 4.2 shown in Figure 5.5 (the axes $x_2$, $x_3$ characterizing the anisotropy of the body in Example 4.2 are replaced here with $x'_2$, $x'_3$). The boundary of $A_{\nu x}$ determines, for every angle $\varphi$, the maximum of the numbers $m > 0$ for which the tangent traction $\boldsymbol{\tau} = mq\mathbf{e}_3$ is admissible (and, hence, the stress field $m\mathbf{s}$ is admissible too). We denote this number by the symbol $m_\varphi(q)$.

Let us show that there is a velocity field $\mathbf{v} \neq 0$ such that $\mathbf{v}$ and the stress field $\boldsymbol{\sigma} = m_\varphi(q)\mathbf{s}$ form a solution to the rigid-plastic problem with the load $m_\varphi(q)q$. This implies that $m_\varphi(q)$ is the safety factor of the load $q$ when the direction of the load forms the angle $\varphi$ with the direction of the $x'_3$-axis (the latter characterizes the anisotropy of the body).

Let the velocity field take a constant value at $x_1 < 0$ and another constant value at $x_1 > 0$. The strain rate field is zero in both smoothness domains, and therefore, together with $\boldsymbol{\sigma}$ satisfies the normality flow rule in these domains. To show that $\boldsymbol{\sigma}$, $\mathbf{v}$ form a solution to the rigid-plastic problem, it remains to verify that they satisfy discontinuity relations (4.11) or, which is the same, maximum principle (4.12). The latter means that the velocity jump is directed along the outward normal to the boundary of the admissible tangent tractions set $A_{\nu x}$ at the point $\boldsymbol{\tau} = m_\varphi(q)q\mathbf{e}_3$ if the boundary is smooth at this point. If $\boldsymbol{\tau} = m_\varphi(q)q\mathbf{e}_3$ is a corner point of the boundary, then the maximum principle means (Section I.3) that the velocity jump $[\mathbf{v}]$ is any of the vectors $\dot{\lambda}_1\mathbf{n}_1+\dot{\lambda}_2\mathbf{n}_2$, where $\dot{\lambda}_1 \geq 0$, $\dot{\lambda}_2 \geq 0$ and $\mathbf{n}_1$, $\mathbf{n}_2$ stand for the outward normals to the two smooth parts of the boundary which intersect at the point, see Figure 5.5. In both cases $\boldsymbol{\sigma}$, $\mathbf{v} \neq 0$ form a solution to the rigid-plastic problem with the load $m_\varphi(q)q$, and, consequently, the number $m_\varphi(q)$ is the safety factor of the load $q$.

## 6. Derivation of the formula for kinematic multiplier

We have already seen that the explicit formula for the kinematic multiplier of a discontinuous velocity field plays an important role in limit analysis. However, we only described the idea of deriving this formula and proving the main property of the kinematic multiplier in Section 1. This section implements the idea and presents a proof of the formula.

**6.1. Formula for kinematic multiplier.** Let us formulate the proposition about the kinematic multiplier to be proved in this section. The proposition deals with piecewise regular velocity fields here, and we start with explaining what we mean by "piecewise regular field". Consider a vector field $\mathbf{v}$ defined on a Lipschitz domain $\Omega$. We say that $\mathbf{v}$ is *piecewise regular in* $\Omega$ if $\Omega$ is divided into pair-wise disjoint standard (Subsection 1.4) subdomains $\omega_1, \ldots, \omega_A$ such that the restriction $\mathbf{v}_a = \mathbf{v}|_{\omega_a}$ to each of the subdomains is in $\mathbf{W}_p^1(\omega_a)$, $p > 1$, and the trace $\mathbf{v}_a|_{\partial\omega_a}$ is continuously differentiable on each of the smooth parts of $\partial\omega_a$. (It is worth noting that the partition $\omega_1, \ldots, \omega_A$ depends on $\mathbf{v}$.) Consider also a vector field $\mathbf{v}^e$ on $\partial\Omega$, the field being continuously differentiable on each of the smooth parts of $\partial\Omega$. The trace $\mathbf{v}|_{\partial\Omega}$ is, generally speaking, not equal to $\mathbf{v}^e$, and, in case it is not, we consider the pair $\bar{\mathbf{v}} = (\mathbf{v}, \mathbf{v}^e)$ and refer to it as a *field with surface slip* (Subsection 1.4). If $\Omega$, $\mathbf{v}$, $\mathbf{v}^e$ satisfy the aforementioned conditions, we refer to $\bar{\mathbf{v}} = (\mathbf{v}, \mathbf{v}^e)$ as a *piecewise regular field on* $\bar{\Omega}$ and to $\omega_1, \ldots, \omega_A$ as *regularity domains* of this field.

Recall that the boundary of any standard domain consists of a finite number of smooth parts. As far as the regularity domains $\omega_1, \ldots, \omega_A$ of a piecewise regular field $\bar{\mathbf{v}}$ are concerned, we assume that each smooth part of $\partial\omega_a$ either separates $\omega_a$ and one of the other regularity domains or lies on $\partial\Omega$. (If the original partition does not meet this requirement, there is a suitable refinement of the partition which does.)

Let $\bar{\mathbf{v}} = (\mathbf{v}, \mathbf{v}^e)$ be a piecewise regular vector field on $\bar{\Omega}$, and $\omega^+$ and $\omega^-$ be two of its regularity domains separated by the surface $\gamma$. Consider the traces $\mathbf{v}^+$ and $\mathbf{v}^-$ on $\gamma$ of the restrictions $\mathbf{v}|_{\omega^+}$ and $\mathbf{v}|_{\omega^-}$, respectively, and their difference $[\mathbf{v}] = \mathbf{v}^+ - \mathbf{v}^-$, which we refer to as a *jump* of the field $\bar{\mathbf{v}}$ or $\mathbf{v}$ on $\gamma$ (in the direction from $\omega^-$ to $\omega^+$). Analogously, we refer to the difference $[\mathbf{v}] = \mathbf{v}^e - \mathbf{v}|_{\partial\Omega}$ as a jump of $\bar{\mathbf{v}}$ on the boundary $\partial\Omega$. It is clear that the collection of all discontinuity surfaces of $\bar{\mathbf{v}}$ is contained in $\Gamma = \partial\omega_1 \cup \cdots \partial\omega_A$.

In this section, we derive the formula for the kinematic multiplier of piecewise regular velocity field $\bar{\mathbf{v}} = (\mathbf{v}, \mathbf{v}^e)$ satisfying the aforegoing assumptions. The field $\bar{\mathbf{v}}$ is assumed to be kinematically admissible, that is, satisfying the condition $\mathbf{v}^e|_{S_v} = 0$. We also assume that the load for which we define the kinematic multiplier is sufficiently regular; namely, under the condition $\mathbf{v}|_{\omega_a} \in \mathbf{W}^1_p(\omega_a)$, $p > 1$, $a = 1, \ldots, A$, we consider loads $\mathbf{l} = (\mathbf{f}, \mathbf{q})$ with

$$\mathbf{f} \in L_r(\Omega; \mathbf{R}^3), \quad \text{where } r > \max\{2, \frac{p}{p-1}\}, \quad \mathbf{q} \in L_1(S_q; \mathbf{R}^3).$$

Here, as usual, the boundary $\partial\Omega$ is divided into two pairs $S_v$ and $S_q$, the former being fixed, the latter being the free part of $\partial\Omega$.

The above assumptions are mostly of a technical nature and can be weakened.

Recall that $\mathbf{e}_\nu([\mathbf{v}])$ denotes the tensor with the components

$$(\mathbf{e}_\nu([\mathbf{v}]))_{ij} = \frac{1}{2}([\mathbf{v}_i]\nu_j + [\mathbf{v}_j]\nu_i),$$

where $\boldsymbol{\nu}$ is the unit normal to $\Gamma$. The notation $(\partial v_i/\partial x_j)_l$ is used for the local derivatives of $\mathbf{v}$ (see Subsections 1.1 and 1.4). We denote the strain rate with the components composed of the local derivatives by the symbol $\mathrm{Def}_l\,\mathbf{v}$:

$$(\mathrm{Def}_l\,\mathbf{v})_{ij} = \frac{1}{2}\left(\left(\frac{\partial v_i}{\partial x_j}\right)_l + \left(\frac{\partial v_j}{\partial x_i}\right)_l\right).$$

THEOREM 1. Let a rigid perfectly plastic body with bounded or cylindrical yield surfaces occupy bounded domain $\Omega$, and the material properties of the body continuously depend on the point. Let $\bar{\mathbf{v}} = (\mathbf{v}, \mathbf{v}^e)$ be a piecewise regular kinematically admissible velocity field on $\bar{\Omega}$ and the external power of a load $\mathbf{l} = (\mathbf{f}, \mathbf{q})$ on $\bar{\mathbf{v}}$ be positive:

$$\int_\Omega \mathbf{f}\mathbf{v}\,dV + \int_{S_q} \mathbf{q}\mathbf{v}^e\,dS > 0.$$

Then, for some $s > 1$, there is a sequence of fields $\mathbf{v}_\varepsilon \in \mathbf{W}^1_s(\Omega)$, such that the kinematic multipliers $m_k(\mathbf{v}_\varepsilon)$ of the load $\mathbf{l}$ have the limit

$$\lim_{\varepsilon\to 0} m_k(\mathbf{v}_\varepsilon) = m_k(\bar{\mathbf{v}})$$

$$m_k(\bar{\mathbf{v}}) = \frac{\int_\Omega d_x(\mathrm{Def}_l\,\mathbf{v})\,dV + \int_\Gamma d_x\,(\mathbf{e}_\nu([\mathbf{v}]))\,dS}{\int_\Omega \mathbf{f}\mathbf{v}\,dV + \int_{S_q} \mathbf{q}\mathbf{v}^e\,dS}.$$

The proof of this theorem consists mostly in constructing the fields $\mathbf{v}_\varepsilon$, which can be done by certain regular procedures, the first of them smoothing the jump.

**6.2. Smoothing the jump.** A discontinuous function can be smoothed by volume averaging. Let us discuss some properties of this operation and construct certain approximations to the derivatives of the averaged function.

Consider function $f$ on $\mathbf{R}^n$ integrable over every bounded domain. The average value of $f$ defined for every $x \in \mathbf{R}^n$ is the integral of $f$ over a cube centered at $x$ divided by the volume of the cube:

$$f_h(x) = \int q(x-\xi) f(\xi)\, dV_\xi,$$

$$q(x) = \begin{cases} \dfrac{1}{h^n} & \text{if} \quad |x_1| < \dfrac{h}{2}, \dots, |x_n| < \dfrac{h}{2}, \\ 0 & \text{if} \quad |x_i| \geq \dfrac{h}{2} \quad \text{for some } i, \ 1 \leq i \leq n \end{cases}$$

(here and in the sequel we mean integration over all of $\mathbf{R}^n$ if no integration domain is indicated). The function $x \to f_h(x)$ is referred to as the *average* of $f$. Due to the equality $q(x-\xi) = q(\xi - x)$, the definition of the average immediately implies that

$$\int f_h g\, dV = \int f g_h\, dV. \tag{6.1}$$

We only consider averaging of finite functions, that is, functions vanishing everywhere in the exterior of a certain bounded set. The following well known proposition is valid for such functions.

THEOREM 2. If $f$ is a finite function in $L_p(\mathbf{R}^n)$, $p \geq 1$, the sequence of its averages $f_h$ approximates $f$ in $L_p(\mathbf{R}^n)$

$$\|f - f_h\|_{L_p(\mathbf{R}^n)} \to 0 \quad \text{when} \quad h \to 0.$$

For a continuously differentiable function $f$, the definition of averaging immediately implies the equality

$$\frac{\partial f_h}{\partial x_i} = \left(\frac{\partial f}{\partial x_i}\right)_h. \tag{6.2}$$

Applying Theorem 2 to $\partial f/\partial x_i$, we obtain from (6.2) that, in case $f$ is continuously differentiable, the derivatives $\partial f_h/\partial x_i$ and $\partial f/\partial x_i$ are infinitely close when $h \to 0$. Let us now consider a piecewise continuously differentiable discontinuous function $f$ and establish an approximation for the derivatives $\partial f_h/\partial x_i$ of the average $f_h$. It is clear *a priori* that $\partial f_h/\partial x_i$ and the local derivative $(\partial f/\partial x_i)_l$ are close everywhere in the exterior of the thin layer surrounding the discontinuity surface of $f$. Within this layer, the derivative $\partial f_h/\partial x_i$ is "large": the function $f_h$ varies continuously but steeply. The variation replaces the jump of $f$ across the discontinuity surface. Briefly, $\partial f_h/\partial x_i$ is close to the derivative $\partial f/\partial x_i$ in the sense of distributions and not to the local derivative $(\partial f/\partial x_i)_l$.

We start constructing the approximation to $\partial f_h/\partial x_i$ in case the function $f$ vanishes everywhere in the exterior of the negative orthant

$$\mathbf{R}^n_- = \{x \in \mathbf{R}^n : x_1 < 0, \dots, x_n < 0\}$$

and is sufficiently regular in $\mathbf{R}^n_-$:

$$f(x) = \begin{cases} f^-(x) & \text{if } x \in \mathbf{R}^n_-, \\ 0 & \text{if } x \notin \mathbf{R}^n_-, \end{cases} \qquad f^- \in W^1_p(\mathbf{R}^n_-),\ p > 1.$$

Recall that every function in the Sobolev space has the trace on the domain boundary (Theorem VI.B.3) which allows introducing the functions $f_i^-$ on the planes $x_i = 0$, $i = 1, \dots, n$:

$$f_i^-(x) = \begin{cases} f^-\big|_{\partial\mathbf{R}^n_-}(x) & \text{if } x_i = 0,\ x \in \partial\mathbf{R}^n_-, \\ 0 & \text{if } x_i = 0,\ x \notin \partial\mathbf{R}^n_-. \end{cases}$$

In the following proposition, the symbol $p_i$ denotes the projection of $x$ on the plane $x_i = 0$. We also make use of the functions $\chi_{(h)i}$, $i = 1, \dots, n$:

$$\chi_{(h)i}(x) = \begin{cases} 1 & \text{if } |x_i| < h/2, \\ 0 & \text{if } |x_i| \geq h/2. \end{cases}$$

PROPOSITION 1. Suppose that 1) finite function $f$ vanishes everywhere in the exterior of the negative orthant $\mathbf{R}^n_-$, 2) the function $f^- = f|_{\mathbf{R}^n_-}$ is in $W^1_p(\mathbf{R}^n_-)$, and 3) the trace $f^-|_{\partial\mathbf{R}^n_-}$ is continuously differentiable on each of the faces of $\mathbf{R}^n_-$. Then the averages $f_h$ are in $W^1_p(\mathbf{R}^n)$ and converge to $f$ in $L_p(\mathbf{R}^n)$ when $h \to 0$, while their derivatives are approximated in $L_s(\mathbf{R}^n)$, $1 \leq s < \min\{p, 2\}$, by the functions $g_{(h)j}$:

$$\begin{gathered} g_{(h)j}(x) = \left(\frac{\partial f}{\partial x_j}\right)_l - \frac{1}{h}\chi_{(h)j}(x) f_j^-(p_j x) \quad \text{(no summation over } j\text{)}, \\ \left\| \frac{\partial f_h}{\partial x_j} - g_{(h)j} \right\|_{L_s(\mathbf{R}^n)} \to 0 \quad \text{when} \quad h \to 0. \end{gathered} \tag{6.3}$$

PROOF. By Theorem 2 the averages $f_h \in L_p(\mathbf{R}^n)$ converge to $f$ in $L_p(\mathbf{R}^n)$ when $h \to 0$. Then consider, for example, the derivative $\partial f_h/\partial x_1$. Recall that $\partial f_h/\partial x_1 = F_1$ if the function $F_1$ satisfies the equality

$$\int F_1 \varphi \, dV = -\int f_h \frac{\partial \varphi}{\partial x_1} \, dV$$

for every test function $\varphi$ (Appendix VI.A). Using the definition of averaging and (6.1), (6.2), it is easy to verify that the following function possesses the above property:

$$\frac{\partial f_h}{\partial x_1}(x) = F_1(x) = \left(\frac{\partial f}{\partial x_1}\right)_{lh}(x) - \frac{1}{h}\chi_{(h)1}(x)(f_1^-)_h(p_1 x).$$

Here, the average $(f_1^-)_h$ of $f_1^-$ is considered in $\mathbf{R}^{n-1}$ (the plane $x_1 = 0$ in $\mathbf{R}^n$). By Theorem 2, $\partial f_h/\partial x_1$ is in $L_p(\mathbf{R}^n)$; analogously, $\partial f_h/\partial x_2, \dots, \partial f_h/\partial x_n$ are also in $L_p(\mathbf{R}^n)$; and consequently $f_h$ is in $W_p^1(\mathbf{R}^n)$.

Let us show, for example, that functions (6.3) with $j = 1$ approximate the derivative $\partial f_h/\partial x_1$. Indeed, the first term on the right-hand side in the previous formula converges in $L_p(\mathbf{R}^n)$ to $(\partial f/\partial x_1)_l$, and it remains to verify that, for a suitable $s$,

$$\left\| \frac{1}{h}\chi_{(h)1}(f_1^-)_h \circ p_1 - \frac{1}{h}\chi_{(h)1} f_1^- \circ p_1 \right\|_{L_s(\mathbf{R}^n)} \to 0 \quad \text{when} \quad h \to 0.$$

According to the definition of the function $\chi_{(h)1}$, the expression on the right-hand side can be written down as

$$\left\| \frac{1}{h}\chi_{(h)1}(f_1^-)_h \circ p_1 - \frac{1}{h}\chi_{(h)1} f_1^- \circ p_1 \right\|_{L_s(\mathbf{R}^n)} = \frac{1}{h} h^{\frac{1}{s}} \left\|(f_1^-)_h - f_1^-\right\|_{L_s(\mathbf{R}^{n-1})}. \qquad (6.4)$$

To estimate this value, consider the difference

$$(f_1^-)_h(x') - f_1^-(x') = \frac{1}{h^{n-1}} \int\limits_{Q_h(0)} (f_1^-(x'+\xi') - f_1^-(x'))\, dV_{\xi'}, \qquad (6.5)$$

$$x' = (x_2, \dots, x_n), \quad Q_h(0) = \{\xi' \in \mathbf{R}^{n-1} : |\xi_2| < h/2, \dots, |\xi_n| < h/2\}.$$

Note that $f_1^-(x') = 0$ if $x_i > 0$ at least at some $i$ of $2, \dots, n$. Besides, there is a number $R > 0$ such that $f_1^-(x') = 0$ when $|x| > R$ (since $f_1^-$ is a finite function). Therefore, the difference $f_1^-(x' + \xi') - f_1^-(x')$ with $\xi' \in Q_h(0)$, for sufficiently small $h$, may take nonzero values only on the set, which for convenience we represent as the union of the following two sets:

$$\Omega_1 = \{x' \in \mathbf{R}^{n-1} : x_2 < -h/2, \dots, x_n < -h/2,\ |x'| < R+1\},$$
$$\Omega_2 = \{x \in \mathbf{R}^{n-1}: |x_i| < h/2 \text{ at least for some } i = 2, \dots, n,\ |x'| < R+1\}.$$

Since $f_1^-$ is continuously differentiable in $\Omega_1$ at $x_2 < 0, \dots, x_n < 0$, the inequality

$$\left|f_1^-(x'+\xi') - f_1^-(x')\right| < m_1 h\sqrt{n-1}/2 \quad \text{at} \quad \xi' \in Q_h(0)$$

is valid, where the constant $m_1$ estimates $\left|\operatorname{grad} f_1^-\right|$ from above. In $\Omega_2$, the inequality

$$\left|f_1^-(x'+\xi') - f_1^-(x')\right| \le 2m_0 \quad \text{at} \quad \xi' \in Q_h(0)$$

is valid, where the constant $m_0$ estimates $\left|f_1^-\right|$ from above. Due to (6.5), this implies

$$\left|(f_1^-)_h(x') - f_1^-(x')\right| < m_1 h\sqrt{n-1}/2 \quad \text{at} \quad x' \in \Omega_1,$$
$$\left|(f_1^-)_h(x') - f_1^-(x')\right| \le 2m_0 \quad \text{at} \quad x' \in \Omega_2.$$

These relations together with the inequalities

$$\operatorname{mes}\Omega_1 < c_1(n,R), \quad \operatorname{mes}\Omega_2 < c_2(n,R)h$$

result in the estimation

$$\|(f_1^-)_h - f_1^-\|_{L_s(\mathbf{R}^n)} \le c(n,R,m_0,m_1)h^{\frac{1}{s}},$$

which implies that, for $1 \le s < 2$, (6.4) tends to zero when $h \to 0$, which finishes the proof.

**6.3. Smoothing the jump on standard domain boundary.** In the previous subsection, function $f$ was considered with the jump discontinuity on the negative orthant boundary, $f$ being sufficiently regular in the orthant and zero in its exterior. It turned out that the averages of $f_h$ smooth the jump and approximate $f$ when $h \to 0$. As to the derivatives $\partial f_h/\partial x_i$, say, $\partial f_h/\partial x_1$ only slightly differs from the local derivative $(\partial f/\partial x_1)_l$ everywhere with the exception of the thin layer $|x_1| < h/2$. Within this layer, $\partial f_h/\partial x_1$ is close to the function which takes the constant value $[f]/h$ along each straight line orthogonal to the plane $x_1 = 0$.

Actually, the similar proposition is valid for function $u$ with the jump discontinuity on the boundary $\partial\omega$ of a standard domain $\omega$, the function being sufficiently regular in $\omega$ and zero in its exterior:

$$u(x) = \begin{cases} u^-(x) & \text{if } x \in \omega, \\ 0 & \text{if } x \notin \omega, \end{cases} \qquad u^- \in W_p^1(\omega), \quad p > 1.$$

Let us construct a sequence of functions $u_{(h)}$ smoothing and approximating $u$. We will show that the derivatives $\partial u_{(h)}/\partial x_j$ only slightly differ from $(\partial u/\partial x_j)_l$ everywhere with exception of the thin layer surrounding the surface $\partial\omega$. If the latter consists of the smooth parts

$$S_\kappa = \{x \in \partial\omega : F_\kappa(x) = 0\} \quad (\kappa = 1, \ldots, K)$$

(see the definition of standard domain in Subsection 1.4), the layer consists of points satisfying the conditions $|F_\kappa(x)| < h/2$, $\kappa = 1, \ldots, K$. Within this layer, near the surface $F_\kappa(x) = 0$, the derivative $\partial u_{(h)}/\partial x_j$ is close to the function which takes the constant value

$$\frac{1}{h}[u]\frac{\partial F_\kappa}{\partial x_j}$$

along each curve orthogonal to the collection of the surfaces $F_\kappa(x) = \text{const}$.

To formulate this statement more accurately we introduce the function $u_\kappa^-$ on the surface $F_\kappa(x) = 0$ $(\kappa = 1, \ldots, K)$:

$$u_\kappa^-(x) = \begin{cases} u^-\big|_{\partial\omega}(x) & \text{if } F_\kappa(x) = 0, \ x \in S_\kappa, \\ 0 & \text{if } F_\kappa(x) = 0, \ x \notin S_\kappa, \end{cases}$$

and the function $\chi_{(h)}(F_\kappa;\cdot)$:

$$\chi_{(h)}(F_\kappa;x)=\begin{cases}1 & \text{if} \quad |F_\kappa(x)|<h/2,\\ 0 & \text{if} \quad |F_\kappa(x)|>h/2.\end{cases}$$

Let us consider a curve which passes through the point $x$ and is orthogonal to the collection of the surfaces $F_\kappa$ = const, and let $n_\kappa(x)$ be the point at which the curve intersects the surface $F_\kappa = 0$. The function $n_\kappa$ is defined and continuously differentiable on the set consisting of points $x$ which satisfy the condition $|F_\kappa(x)| < \delta$, where $\delta > 0$ is a fixed sufficiently small number. Then, the following expression makes sense on this set:

$$\chi_{(h)}(F_\kappa;x)\psi(n_\kappa(x))$$

where $0 < h/2 < \delta$ and $\psi$ is a function on the surface $F_\kappa = 0$. We will also use expressions of this form for any $x$, assigning the zero value to such an expression in case $|F_\kappa(x)| > h/2$ (the multiplier $\chi_{(h)}(F_\kappa;x)$ reminds about that). For brevity, we restrict ourselves to the case of the three-dimensional space.

PROPOSITION 2. Suppose that 1) a finite function $u$ vanishes everywhere in the exterior of a standard domain $\omega$; 2) the function $u^- = u|_\omega$ is in $W_p^1(\omega)$, $p > 1$; and 3) the trace $u^-|_{\partial\omega}$ is continuously differentiable on each of the smooth parts $S_\kappa$ of the boundary $\partial\omega$. Then there is a sequence of functions $u_{(h)} \in W_p^1(\mathbf{R}^3)$ which converges to $u$ in $L_s(\mathbf{R}^3)$ when $h \to 0$, while their derivatives $\partial u_{(h)}/\partial x_j$ are approximated in $L_s(\mathbf{R}^3)$, $1 \le s < \min\{p, 2\}$, by the functions

$$g_{(h)j}=\left(\frac{\partial u}{\partial x_j}\right)_l+G_{(h)j},$$

$$G_{(h)j}(x)=-\frac{1}{h}\sum_{\kappa=1}^{K}\chi_{(h)}(F_\kappa;x)u_\kappa^-(n_\kappa(x))\frac{\partial F_\kappa}{\partial x_j}(n_\kappa(x)) \tag{6.6}$$

(no summation over $j$),

$$\left\|\frac{\partial u_{(h)}}{\partial x_j}-g_{(h)j}\right\|_{L_s(\mathbf{R}^3)}\to 0 \quad \text{when} \quad h\to 0.$$

PROOF. Using local coordinate systems, let us reduce the proof to appealing to Proposition 1. To make clear how the coordinate systems are chosen, consider, for example, a point $x_0 \in \partial\omega$ at which three smooth parts $F_{\kappa_1}(x) = 0$, $F_{\kappa_2}(x) = 0$, $F_{\kappa_3}(x) = 0$ of $\partial\omega$ intersect. Then, using the functions $\phi_1 = F_{\kappa_1}$, $\phi_2 = F_{\kappa_2}$, $\phi_3 = F_{\kappa_3}$, we introduce a curvilinear coordinate system $y_i = \phi_i(x)$, $i = 1, 2, 3$, in a certain neighborhood $\mathcal{O}$ of $x_0$. The coordinate functions $\phi_i(x)$ map $\mathcal{O}$ on a certain neighborhood $\widetilde{\mathcal{O}}$ of the origin, the domain $\omega \cap \mathcal{O}$ on $\widetilde{\mathcal{O}} \cap \mathbf{R}^3_-$, and the surface $\partial\omega \cap \mathcal{O}$ on a part of the orthant's boundary (see Figure 6.1). Next consider the case of only two (not three as before) smooth parts $F_{\kappa_1}(x) = 0$ and $F_{\kappa_2}(x) = 0$ of $\partial\omega$ intersecting at the point $x_0$. Then the local coordinate system is introduced through the functions $\phi_1 = F_{\kappa_1}$, $\phi_2 = F_{\kappa_2}$

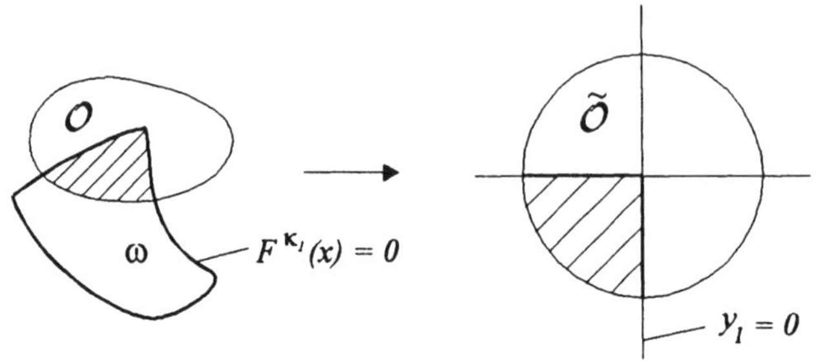


**Figure 6.1**

and a suitable additional function $\phi_3$. Without loss of generality, we assume in this case that the distance between $\widetilde{\mathcal{O}}$ and the plane $y_3$ is positive. Thus we cover the closure $\bar{\omega} = \omega \cup \partial\omega$ by a number of domains $\mathcal{O}_1, \ldots, \mathcal{O}_M$ with a local coordinate system $y = \Phi_\mu = (\phi_{\mu 1}, \phi_{\mu 2}, \phi_{\mu 3})$, defined in $\mathcal{O}_\mu$, $\mu = 1, \ldots, M$. All coordinate functions are supposed to be twice continuously differentiable and possess the same properties as in the above examples.

Let $\alpha_1, \ldots, \alpha_M$ be smooth functions such that $\alpha_\mu$ vanishes everywhere in the exterior of a certain closed set $\operatorname{supp}\alpha_\mu \subset \mathcal{O}_\mu$, $\mu = 1, \ldots, M$, and the sum $\alpha_1 + \cdots + \alpha_M$ takes the value 1 everywhere in $\omega$. Using these functions, we write $u$ as $u = \alpha_1 u + \cdots + \alpha_M u$. We will smooth each of the summands $f_\mu = \alpha_\mu u$ and use the sum of the obtained smoothed functions as an approximation to $u$.

Let us first introduce the functions $\widetilde{f}_\mu$:

$$\widetilde{f}_\mu(y) = \begin{cases} f_\mu(\Phi_\mu^{-1}(y)) & \text{if} \quad y \in \widetilde{\mathcal{O}}_\mu, \\ 0 & \text{if} \quad y \notin \widetilde{\mathcal{O}}_\mu. \end{cases}$$

It is clear that they satisfy the conditions of Proposition 1. Therefore, their averages $\widetilde{f}_{\mu h}$ converge to $\widetilde{f}_\mu$ when $h \to 0$, and their derivatives are approximated by the functions of the form (6.3). Using the inverse change of variables, we define the functions

$$f_{\mu(h)}(x) = \begin{cases} \widetilde{f}_{\mu h}(\Phi_\mu(x)) & \text{if} \quad x \in \mathcal{O}_\mu, \\ 0 & \text{if} \quad x \notin \mathcal{O}_\mu. \end{cases}$$

It is clear that the sum $u_{(h)} = f_{1(h)} + \cdots + f_{M(h)}$ converges to $u$ in $L_p(\Omega)$ when $h \to 0$.

To finish the proof, it remains to verify that the derivatives $\partial u_{(h)}/\partial x_j$ are approximated by the functions $g_{(h)j}$, see (6.6). Let us verify this statement for any of the functions $f_{\mu(h)}$. To simplify the notation, we omit the subscript $\mu$ in what follows; in particular, we write $f_{(h)}$ instead of $f_{\mu(h)}$. It is clear that

$$\frac{\partial f_{(h)}(x)}{\partial x_j} = \begin{cases} \dfrac{\partial \widetilde{f}_h}{\partial y_i}(\Phi(x))\dfrac{\partial \phi_i}{\partial x_j} & \text{if} \quad x \in \mathcal{O}, \\ 0 & \text{if} \quad x \notin \mathcal{O}. \end{cases}$$

According to Proposition 1, there is a function of the form (6.3) which approximates the derivative $\partial \widetilde{f}_h / \partial y_i$. Therefore, the function

$$\left( \frac{\partial \alpha u}{\partial x_j} \right)_l + r_{(h)i} \frac{\partial \phi_i}{\partial x_j} \tag{6.7}$$

approximates the derivative $\partial f_{(h)} / \partial x_j$, where, for example, the function $r_{(h)1}$ can be written down as follows:

$$r_{(h)1}(x) = 0 \quad \text{if } x \notin \mathcal{O},$$
$$r_{(h)1}(x) = \frac{1}{h} \chi_h(F_{\kappa_1}; x)\, \alpha(P_1(x))\, u^-_{\kappa_1}(P_1(x)) \quad \text{if } x \in \mathcal{O}$$

(we use the notation $P_1(x) = \Phi^{-1}(p_1(\Phi(x)))$, $\phi_1 = F_{\kappa_1}$, and, without loss of generality, assume that $\widetilde{\mathcal{O}}$ is a ball with its center in the plane $y_1 = 0$, which implies $P_1(x) \in \mathcal{O}$). An estimation analogous to that in the proof of Proposition 1 replaces $P_1(x)$ by $n_{\kappa_1}(x)$ keeping (6.7) close to $\partial f_{(h)} / \partial x_j$. Similarly, $(\partial \phi_i / \partial x_j)(x)$ in (6.7) can be replaced by $(\partial \phi_i / \partial x_j)(n_\kappa(x))$ since $|x - n_\kappa(x)| < kh$ when $|F_\kappa(x)| < h/2$ with the constant $k$ which does not depend on $h$.

Thus, the derivative $\partial f_{(h)} / \partial x_j$ is approximated in $L_s(\mathbf{R}^3)$ by the function $\widehat{g}_{(h)j}$,

$$\widehat{g}_{(h)j}(x) = \left( \frac{\partial \alpha u}{\partial x_j} \right)_l (x) - \frac{1}{h} \sum_i \chi_{(h)}(F_{\kappa_i}; x) \left( \alpha u^-_{\kappa_i} \frac{\partial F_{\kappa_i}}{\partial x_j} \right) (n_{\kappa_i}(x)),$$

with summation over the values of $i$ for which the component $\phi_i$ of the coordinate mapping $\Phi$ is the function $F_{\kappa_i}$. Without loss of generality, we assume that, if the function $F_\kappa$ is not one of the components $\phi_1$, $\phi_2$, $\phi_3$, then $\chi_{(h)}$ or $\alpha$ vanishes for sufficiently small $h$. Therefore, the previous expression can be considered as a sum over all values of the index $\kappa$ and written as

$$\left( \frac{\partial \alpha u}{\partial x_j} \right)_l (x) - \frac{1}{h} \sum_{\kappa=1}^{K} \chi_{(h)}(F_\kappa; x) \left( \alpha u^-_\kappa \frac{\partial F_\kappa}{\partial x_j} \right) (n_\kappa(x)). \tag{6.8}$$

Finally, recall that the subscript $\mu$ was omitted in the notations, and $\alpha$, actually, stands for the function $\alpha_\mu$ associated with the domain $\mathcal{O}_\mu$ and (6.8) approximates the derivative of $f_{\mu(h)}$. Restoring the subscript $\mu$ in (6.8) and summing functions (6.8) over $\mu = 1, \ldots, M$, we observe that the sum approximates $\partial u_{(h)} / \partial x_j$, which finishes the proof.

**6.4. Smoothing with a given trace.** An approximation to a piecewise smooth discontinuous function $u$ was constructed in the previous subsection. The approximation smooths the jump of $u$, which allows using it for deriving a formula for the kinematic multiplier of a discontinuous velocity field. In this connection, the following additional requirement is to be imposed on the approximation: the smoothing function should have a certain given trace on the discontinuity surface of $u$. Such approximation will be constructed by homothetic transformation.

Consider the homothety in the space $\mathbf{R}^n$ with its center at the origin and the coefficient $\lambda > 0$: $x \to \lambda x$. This transformation induces a correspondence $f \to f^\lambda$ between functions on $\mathbf{R}^n$:

$$f^\lambda(x) = f\left(\frac{x}{\lambda}\right). \tag{6.9}$$

The following well known proposition states that $f^\lambda$ converges to $f$.

THEOREM 3. Suppose $f \in L_p(\mathbf{R}^n)$, $p \geq 1$, is a finite function. Then the sequence of the functions $f^\lambda$, $0 < \lambda < 1$ approximates $f$:

$$\left\|f - f^\lambda\right\|_{L_p(\mathbf{R}^n)} \to 0 \quad \text{when} \quad \lambda \to 1.$$

This property allows solving of the above problem of constructing a smoothing function for $u$ with the given trace on the discontinuity surface of $u$. Such a smoothing function is easily constructed if $u$ vanishes in the exterior of a star-shaped domain $\omega$, whose boundary is the discontinuity surface of $u$. Without loss of generality, we assume $\omega$ to be star-shaped with respect to the origin, that is, the domain $\lambda\omega$ lies in the interior of $\omega$ for every $\lambda$, $0 < \lambda < 1$.

The required smoothing function is constructed in the following proposition, which makes use of function (6.6). Note that the latter function depends on the collection of parameters $F = (F_1, \dots, F_K)$, that is, on the functions which specify the smooth parts $S_\kappa$, $\kappa = 1, \dots, K$, of $\partial\omega$, see Subsection 6.3. Function (6.6) also depends on the parameters $u_1^-, \dots, u_K^-$ which are functions on the surfaces $F_1(x) = 0, \dots, F_K(x) = 0$, respectively. Actually, $u_1^-, \dots, u_K^-$ are expressed through one function given on $\partial\omega$. In Subsection 6.3, this function was the trace $u^-|_{\partial\omega}$. In the sequel we consider an arbitrary function $v$ on $\partial\omega$ instead of $u^-|_{\partial\omega}$, and $v$ generates the functions

$$u_\kappa^-(v; x) = \begin{cases} v(x) & \text{if} \quad F_\kappa(x) = 0, \ x \in S_\kappa, \\ 0 & \text{if} \quad F_\kappa(x) = 0, \ x \notin S_\kappa \end{cases}$$

on $S_\kappa$, $\kappa = 1, \dots, K$. The collection $u_1^-, \dots, u_K^-$ participates in determining function (6.6), and we consider (6.6) as depending on the parameter $v$ which is the function given on $\partial\omega$ and generating $u_1^-, \dots, u_K^-$. Sometimes, we will use a vector-valued function $\mathbf{v}$ as this parameter. It is convenient to explicitly indicate the parameters $F = (F_1, \dots, F_K)$ and $v$ in the expression for function (6.6); thus, we write

$$G_{(h)j}(F, v; x) = -\frac{1}{h}\sum_{\kappa=1}^{K} \chi_{(h)}(F_\kappa; x) u_\kappa^-(v, n_\kappa(x)) \frac{\partial F_\kappa}{\partial x_j}(n_\kappa(x)).$$

Recall that $\left(G_{(h)j}(F, v; \cdot)\right)^\lambda$ is determined by formula (6.9) and $\|\cdot\|_{\Omega,r}$ stands for the norm in the space $L_r(\Omega; \mathbf{R}^3)$ of vector fields (see Example V.A.5).

PROPOSITION 3. Suppose $\omega$ is a star-shaped (with respect to origin), standard domain in $\mathbf{R}^3$, $\mathbf{u}$ is a vector field on $\omega$, $\mathbf{u} \in \mathbf{W}^1_p(\omega)$, $p > 1$, and the trace $\mathbf{u}^- = \mathbf{u}|_{\partial\omega}$ is continuously differentiable on each of the smooth parts $S_1, \ldots, S_K$ of $\partial\omega$. Let $\mathbf{u}^0$ be a vector field on $\partial\omega$ continuously differentiable on each of $S_1, \ldots, S_K$. Then, for every $\varepsilon > 0$ and every $1 < s < \min\{p, 2\}$, there are numbers $\lambda$ and $h$, $1 - \varepsilon < \lambda < 1$, $0 < h < \varepsilon$, and a vector field $\mathbf{u}_\varepsilon \in \mathbf{W}^1_s(\omega)$ such that

$$\|\mathbf{u}_\varepsilon - \mathbf{u}\|_{\omega,s} < \varepsilon, \qquad \mathbf{u}_\varepsilon|_{\partial\omega} = \mathbf{u}^0,$$
$$\left\| \frac{\partial \mathbf{u}_\varepsilon}{\partial x_j} - \frac{\partial \mathbf{u}}{\partial x_j} - \frac{1}{\lambda}\left(G_{(h)j}(F, \mathbf{u}^- - \mathbf{u}^0; \cdot)\right)^\lambda \right\|_{\omega,s} < \varepsilon. \tag{6.10}$$

In case $\mathbf{u}$ is solenoidal, $\operatorname{div} \mathbf{u} = 0$, and $\mathbf{u}^- \boldsymbol{\nu} = \mathbf{u}^0 \boldsymbol{\nu}$ on $\partial\omega$ ($\boldsymbol{\nu}$ is the normal to $\partial\omega$), the above field $\mathbf{u}_\varepsilon$ can also be solenoidal.

PROOF. Let us construct the solenoidal smoothing field $\mathbf{u}_\varepsilon$ in case $\operatorname{div} \mathbf{u} = 0$ in $\omega$ and $\mathbf{u}^0 \boldsymbol{\nu} = \mathbf{u}^- \boldsymbol{\nu}$ on $\partial\omega$. A simplified version of the reasoning below is also applicable to the case of nonsolenoidal fields.

First, let us show that the statement is valid in case $\mathbf{u}^0$ is the trace $\mathbf{u}^0 = \mathbf{u}^s|_{\partial\omega}$ of a solenoidal field $\mathbf{u}^s \in \mathbf{W}^1_r(\omega)$, where $1 < r < \min\{p, 2\}$. In this case, we consider the field $\mathbf{w} = \mathbf{u} - \mathbf{u}^s$ and extend it to all of $\mathbf{R}^3$, setting the extension equal to zero in the exterior of $\omega$. We denote the extension by the same symbol $\mathbf{w}$. We also consider its local derivatives $(\partial \mathbf{w}/\partial x_j)_l$ (see Remark 1.1) and corresponding fields (6.9), that is, $\mathbf{w}^\lambda$ and

$$\left(\left(\frac{\partial \mathbf{w}}{\partial x_j}\right)_l\right)^\lambda$$

$(0 < \lambda < 1)$, the following obvious relation

$$\left(\left(\frac{\partial \mathbf{w}}{\partial x_j}\right)_l\right)^\lambda = \lambda \left(\frac{\partial \mathbf{w}^\lambda}{\partial x_j}\right)_l$$

being valid. Let us apply Theorem 3 to $\mathbf{w}$ and its local derivatives, taking into account the previous formula. Theorem 3 implies that, for every $\varepsilon > 0$, there is a number $\bar{\lambda}(\varepsilon)$ such that

$$\|\mathbf{w} - \mathbf{w}^\lambda\|_{\mathbf{R}^3,r} < \frac{\varepsilon}{2}, \quad \left\| \left(\frac{\partial \mathbf{w}}{\partial x_j}\right)_l - \left(\frac{\partial \mathbf{w}^\lambda}{\partial x_j}\right)_l \right\|_{\mathbf{R}^3,r} < \frac{\varepsilon}{2} \text{ when } \bar{\lambda}(\varepsilon) < \lambda < 1.$$

The field $\mathbf{w}^\lambda$ satisfies the conditions of Proposition 2 in the domain $\lambda\omega$; let $(\mathbf{w}^\lambda)_{(h)}$ be the smoothing field for $\mathbf{w}^\lambda$ constructed in this proposition. For brevity, we will denote $(\mathbf{w}^\lambda)_{(h)}$ by the symbol $\mathbf{w}_{\lambda,h}$. By virtue of Proposition 2, there is a number $\bar{h}(\varepsilon, \lambda)$ such that, for $0 < h < \bar{h}(\varepsilon, \lambda)$, we have

$$\|\mathbf{w}^\lambda - \mathbf{w}_{\lambda,h}\|_{\mathbf{R}^3,r} < \frac{\varepsilon}{2}, \qquad \mathbf{w}_{\lambda,h}|_{\partial\omega} = 0,$$
$$\left\| \frac{\partial \mathbf{w}_{\lambda,h}}{\partial x_j} - \left(\frac{\partial \mathbf{w}^\lambda}{\partial x_j}\right)_l - \frac{1}{\lambda}\left(G_{(h)j}(F, \mathbf{u}^- - \mathbf{u}^0; \cdot)\right)^\lambda \right\|_{\mathbf{R}^3,s} < \frac{\varepsilon}{2}$$

(we took into account that $G_{(h)j}(F^\lambda, \mathbf{v}^\lambda; \cdot) = \left(G_{(h)j}(F, \mathbf{v}; \cdot)\right)^\lambda / \lambda$). Thus, choosing for a given $\varepsilon$ a number $\lambda$, $\bar\lambda(\varepsilon) < \lambda < 1$, and then a number $h$, $0 < h < \bar h(\varepsilon, \lambda)$, we observe that due to the previous relation the field $\mathbf{w}_{\lambda,h}$ satisfies the conditions

$$\|\mathbf{w} - \mathbf{w}_{\lambda,h}\|_{\mathbf{R}^3,r} < \varepsilon, \qquad \mathbf{w}_{\lambda,h}|_{\partial\omega} = 0,$$

$$\left\| \frac{\partial \mathbf{w}_{\lambda,h}}{\partial x_j} - \left(\frac{\partial \mathbf{w}}{\partial x_j}\right)_l - \frac{1}{\lambda}\left(G_{(h)j}(F, \mathbf{u}^- - \mathbf{u}^0; \cdot)\right)^\lambda \right\|_{\mathbf{R}^3,s} < \varepsilon.$$

Let us show that the latter inequality implies the following estimation:

$$\|\operatorname{div} \mathbf{w}_{\lambda,h}\|_{\mathbf{R}^3,s} < 3\varepsilon.$$

It is sufficient to verify that

$$\left(\frac{\partial w_j}{\partial x_j}\right)_l + \frac{1}{\lambda}\left(G_{(h)j}(F, u^{-j} - u^{0j}; \cdot)\right)^\lambda = 0 \tag{6.11}$$

(with summation over the repeated index). Recall that here $\mathbf{w} = \mathbf{u} - \mathbf{u}^s$ and the fields $\mathbf{u}$, $\mathbf{u}^s$ are solenoidal in $\omega$. Therefore $(\partial w_j / \partial x_j)_l = 0$. It remains to show that $G_{(h)j}(F, u^{-j} - u^{0j}; x) = 0$. We establish this equality using the formula which defines $G_{(h)j}$, where we only have to take into account that the normal $\boldsymbol{\nu}$ to the surface $F_\kappa(x) = 0$ is directed along the gradient $\operatorname{grad} F_\kappa$ and that $(\mathbf{u}^- - \mathbf{u}^0)\boldsymbol{\nu} = 0$ by assumption. Then, equality (6.11) is valid and hence $\|\operatorname{div} \mathbf{w}_{\lambda,h}\|_{\mathbf{R}^3,s} < 3\varepsilon$.

Then Theorem VI.4.2 (more precisely, its extension to the case $s \neq 2$; see Comments in Chapter VI) implies that there is a field $\mathbf{v}$ in $\mathbf{W}_s^1(\omega)$ such that

$$\mathbf{v}|_{\partial\omega=0}, \quad \operatorname{div} \mathbf{v} = -\operatorname{div} \mathbf{w}_{\lambda,h}, \quad \|\mathbf{v}\|_{\mathbf{W}_s^1(\omega)} \le c\,\|\operatorname{div} \mathbf{w}_{\lambda,h}\|_{\mathbf{R}^3,s}.$$

It is clear that the field $\mathbf{w}_{\lambda,h} + \mathbf{v}$ is solenoidal. Finally, we observe that the field $\mathbf{u}_\varepsilon = \mathbf{w}_{\lambda,h} + \mathbf{v} + \mathbf{u}^s$ possesses all of the required properties (6.10) (with $\varepsilon$ on the right-hand side replaced by $C\varepsilon$, which does not matter since the constant $C$ does not depend on $\mathbf{u}$, $\mathbf{u}^0$, $\varepsilon$).

To finish the proof, it remains to verify that the field $\mathbf{u}^0$ given on $\partial\omega$ can be considered as the trace of a solenoidal field $\mathbf{u}^s \in \mathbf{W}_r^1(\omega)$, as it was previously assumed. First, we observe that there is an extension $\mathbf{u}^c \in \mathbf{W}_r^1(\omega)$ of $\mathbf{u}^0$ to all of $\omega$: $\mathbf{u}^c|_{\partial\omega} = \mathbf{u}^0$ (the extension is not necessarily solenoidal). Indeed, using the local coordinate systems as in the proof of Proposition 2 reduces construction of the extension to solving the similar problem in the orthant $\mathbf{R}^3_-$. It is easy to solve this problem since its input, that is, a field given on $\partial\mathbf{R}^3_-$, is continuously differentiable on each of the sides of $\mathbf{R}^3_-$. Finally, we modify $\mathbf{u}^c$ by adding to it a field $\mathbf{v}^c$ which possesses the following properties:

$$\mathbf{v}^c \in \mathbf{W}_r^1(\omega), \quad \operatorname{div} \mathbf{v}^c = -\operatorname{div} \mathbf{u}^c, \quad \mathbf{v}^c|_{\partial\omega} = 0;$$

such a field exists by virtue of Theorem VI.4.2 since

$$\int_\omega \operatorname{div} \mathbf{u}^c\, dV = \int_{\partial\omega} \mathbf{u}^0 \boldsymbol{\nu}\, dS = \int_{\partial\omega} \mathbf{u}^- \boldsymbol{\nu}\, dS = \int_\omega \operatorname{div} \mathbf{u}\, dV = 0.$$

The proposition is proved.

**6.5. Derivation of the formula for kinematic multiplier.** The above established proposition allows us to prove Theorem 1, that is, to construct for a given piecewise regular velocity field $\bar{\mathbf{v}} = (\mathbf{v}, \mathbf{v}^e)$ on $\bar{\Omega}$ a sequence of fields $\mathbf{v}_\varepsilon$ approximating $\bar{\mathbf{v}}$ and find the limit of their kinematic multipliers.

Consider the regularity domains of the field $\mathbf{v}$. Let $S_{a\kappa}$, $\kappa = 1, \ldots, K_a$, be the smooth parts which form the boundary $\partial\omega_a$. These are the surfaces on which a jump of $\mathbf{v}$ may occur. We denote the restriction of $\mathbf{v}$ on the domain $\omega_a$ by the symbol $\mathbf{u}_a$ and the trace of $\mathbf{u}_a$ on $\partial\omega_a$ by $\mathbf{u}_a^-$:

$$\mathbf{u}_a = \mathbf{v}\big|_{\omega_a}, \qquad \mathbf{u}_a^- = \mathbf{u}_a\big|_{\partial\omega_a}.$$

By assumption, the following function is defined on the surfaces $S_{a\kappa}$ and continuously differentiable on each of them:

$$\mathbf{u}_a^0(x) = \begin{cases} \mathbf{u}_a^-(x) + \dfrac{1}{2}[\mathbf{v}](x) & \text{if } x \in S_{a\kappa} \subset \Omega, \\ & \text{or } x \in S_{a\kappa} \cap S_v, \\ \mathbf{v}^e(x) & \text{if } x \in S_{a\kappa} \cap S_q. \end{cases} \tag{6.12}$$

Since $\omega_a$ is a Lipschitz domain, we assume without loss of generality that it is star-shaped (one easily achieves this by refining the original partition $\omega_1, \ldots, \omega_A$ if necessary). Then the domain $\omega_a$, the field $\mathbf{u}_a$ on it and the field $\mathbf{u}^0$ on $\partial\omega_a$ satisfy the conditions of Proposition 3. Consequently, for every $\varepsilon > 0$ there is a field $\mathbf{u}_{a,\varepsilon} \in \mathbf{W}_s^1(\omega_a)$ possessing properties (6.10). This allows introducing the field $\mathbf{v}_\varepsilon$ on $\Omega$:

$$\mathbf{v}_\varepsilon(x) = \begin{cases} \mathbf{u}_{1,\varepsilon}(x) & \text{if } x \in \omega_1, \\ \cdots & \\ \mathbf{u}_{A,\varepsilon}(x) & \text{if } x \in \omega_A. \end{cases}$$

The fields $\mathbf{u}_{a,\varepsilon}$ were chosen in such a way that

$$\mathbf{u}_{a,\varepsilon} \in \mathbf{W}_s^1(\omega_a), \quad \mathbf{u}_{a,\varepsilon}\big|_{\partial\omega_a} = \mathbf{u}_a^0 \qquad (a = 1, \ldots, A)$$

and that their traces on the surface $\gamma$ separating two domains $\omega_a$ and $\omega_b$ are equal: $\mathbf{u}_{a,\varepsilon}\big|_\gamma = \mathbf{u}_{b,\varepsilon}\big|_\gamma$. Therefore, $\mathbf{v}_\varepsilon$ is in $\mathbf{W}_s^1(\Omega)$ (Theorem VI.B.4). Proposition 3 and kinematic admissibility of $\bar{\mathbf{v}} = (\mathbf{v}, \mathbf{v}^e)$ imply that the following relations are valid:

$$\|\mathbf{v}_\varepsilon - \mathbf{v}\|_{L_s(\Omega;\mathbf{R}^3)} < C\varepsilon, \quad \mathbf{v}_\varepsilon\big|_{S_v} = 0, \quad \mathbf{v}_\varepsilon\big|_{S_q} = \mathbf{v}^e$$

(the constant $C$ does not depend on $\mathbf{v}$ and $\varepsilon$).

Thus, $\mathbf{v}_\varepsilon$ is sufficiently regular and kinematically admissible. Besides, the previous relations imply that the external power

$$P(\mathbf{v}_\varepsilon) = \int_\Omega \mathbf{f}\mathbf{v}_\varepsilon \, dV + \int_{S_q} \mathbf{q}\mathbf{v}_\varepsilon \, dS$$

has the limit

$$P(\bar{\mathbf{v}}) = \int_{\Omega} \mathbf{f}\mathbf{v}\, dV + \int_{S_q} \mathbf{q}\mathbf{v}^e\, dS$$

when $\varepsilon \to 0$.

Then, by virtue of the condition $P(\bar{\mathbf{v}}) > 0$, the inequality $P(\mathbf{v}_\varepsilon) > 0$ is valid for sufficiently small $\varepsilon > 0$. Consequently, the kinematic multiplier is defined:

$$m_k(\mathbf{v}_\varepsilon) = \frac{1}{P(\mathbf{v}_\varepsilon)} \int_{\Omega} d_x(\operatorname{Def}\mathbf{v}_\varepsilon)\, dV.$$

Let us find the limit of the kinematic multipliers

$$m_k(\bar{\mathbf{v}}) = \lim_{\varepsilon\to 0} m_k(\mathbf{v}_\varepsilon) = \frac{1}{P(\mathbf{v}_\varepsilon)} \sum_{a=1}^{A} \lim_{\varepsilon\to 0} D_{\omega_a}(\operatorname{Def}\mathbf{u}_{a,\varepsilon}), \tag{6.13}$$

where we made use of the notation

$$D_\omega(\mathbf{e}) = \int_{\omega} d_x(\mathbf{e})\, dV.$$

Let us consider bodies with cylindrical yield surfaces (a simplified version of the following reasoning is applicable to the case of bodies with bounded yield surfaces). If the velocity field $\bar{\mathbf{v}} = (\mathbf{v}, \mathbf{v}^e)$ in a body with cylindrical yield surfaces does not satisfy the conditions

$$\operatorname{div}\mathbf{v} = 0 \quad \text{in} \quad \omega_a, \qquad [\mathbf{v}]\boldsymbol{\nu} = 0 \quad \text{on} \quad \partial\omega_a \qquad (a = 1, \dots, A),$$

the statement of Theorem 1 is trivial: the kinematic multiplier $m_k(\bar{\mathbf{v}})$ as it was defined in the formulation of the theorem is $+\infty$. Therefore, we assume that these conditions are satisfied in what follows. Then, due to Proposition 3, the field $\mathbf{v}_\varepsilon$ is solenoidal: $\operatorname{div}\mathbf{v}_\varepsilon = 0$.

Let $\omega$ be one of the domains $\omega_1, \dots, \omega_A$. For brevity, we will also omit the subscript $a$ in the notations $\mathbf{u}_a$, $\mathbf{u}_{a,\varepsilon}$, etc. Let us find the limit of the dissipation $D_\omega(\operatorname{Def}\mathbf{v}_\varepsilon)$ when $\varepsilon \to 0$. Without loss of generality, we assume that $\omega$ is star-shaped with respect to the origin. Then, due to approximation (6.11), the field $\operatorname{Def}\mathbf{u}_\varepsilon$ can be written as

$$\begin{gathered} \operatorname{Def}\mathbf{u}_\varepsilon = \mathbf{e} + \boldsymbol{\eta}, \qquad \|\boldsymbol{\eta}\|_{L_s(\omega;Sym)} < \varepsilon, \\ e_{ij} = (\operatorname{Def}\mathbf{u})_{ij} + \frac{1}{2\lambda}\left(G_{(h)j}(F, u_i^- - u_i^0; \cdot) + G_{(h)i}(F, u_j^- - u_j^0; \cdot)\right)^\lambda \end{gathered} \tag{6.14}$$

with certain $\lambda$ and $h$, satisfying the conditions $1 - \varepsilon < \lambda < 1$, $0 < h < \varepsilon$. Here, $\mathbf{e}^s = \boldsymbol{\eta}^s = 0$ since

$$\operatorname{div}\mathbf{u}_\varepsilon = 0, \quad \operatorname{div}\mathbf{u} = 0, \quad G_{(h)j}(F, u_j^- - u_j^0; \cdot) = 0.$$

The latter equality arises from (6.6), see the proof of Proposition 3. The following remark simplifies evaluating the limit. Since the material properties continuously depend on the point of the body, there is a constant $d_*$ such that $d_x(\mathbf{e}) \leq d_* \,|\,\mathbf{e}\,|$ for every $\mathbf{e} \in Sym^d$ and every $x$ in $\omega$. This inequality and convexity of the dissipation imply the estimation

$$|D_\omega(\mathbf{e}+\boldsymbol{\eta}) - D_\omega(\mathbf{e})| \leq c\,||\boldsymbol{\eta}||_{L_1(\omega;Sym)} \tag{6.15}$$

for any two fields $\mathbf{e}$, $\boldsymbol{\eta}$ in $L_1(\omega; Sym^d)$, which, together with (6.14), results in the equality

$$\lim_{\varepsilon\to 0} D_\omega(\operatorname{Def}\mathbf{u}_\varepsilon) = \lim_{\varepsilon\to 0} D_\omega(\mathbf{e}). \tag{6.16}$$

We now write formula (6.14) for $\mathbf{e}$ as $\mathbf{e} = \operatorname{Def}\mathbf{u} + \mathbf{e}_*$, where $\mathbf{e}_*$ is expressed through the functions $G_{(h)i}$, $G_{(h)j}$. Because of the multiplier $\chi_{(h)}(F_\kappa;\cdot)$ in formula (6.6) for $G_{(h)j}$, the field $\mathbf{e}_*$ may take nonzero values in the domain $\omega$ only on the set

$$\omega_* = \bigcup_{\kappa=1}^{K} \omega_{\kappa,\lambda}, \quad \omega_{\kappa,\lambda} = \left\{ x \in \omega :\ \left|(F_\kappa)^\lambda(x)\right| < \frac{h}{2} \right\}.$$

Therefore, the equality

$$D_\omega(\mathbf{e}) = D_{\omega_*}(\operatorname{Def}\mathbf{u} + \mathbf{e}_*) + D_{\omega\setminus\omega_*}(\operatorname{Def}\mathbf{u})$$

is valid. Since $\operatorname{mes}\omega_* \to 0$ when $\varepsilon \to 0$, we use the estimation similar to (6.15):

$$|D_{\omega_*}(\operatorname{Def}\mathbf{u} + \mathbf{e}_*) - D_{\omega_*}(\mathbf{e}_*)| \leq c\,||\operatorname{Def}\mathbf{u}||_{L_1(\omega_*;Sym)}\,,$$

and find from the previous equation that

$$\lim_{\varepsilon\to 0} D_\omega(\mathbf{e}) = \lim_{\varepsilon\to 0} D_{\omega_*}(\mathbf{e}_*) + D_\omega(\operatorname{Def}\mathbf{u}). \tag{6.17}$$

Consider the first term in more detail. According to formula (6.6) for the function $G_{(h)j}$, we write the field $\mathbf{e}_*$ as

$$\mathbf{e}_* = \frac{1}{\lambda}\sum_{\kappa=1}^{K} \mathbf{e}^\lambda_{(\kappa)},$$

with

$$\mathbf{e}^\lambda_{(\kappa)}(x) = \mathbf{e}_{(\kappa)}\left(\frac{x}{\lambda}\right), \qquad \mathbf{e}_{(\kappa)}(x) = \frac{1}{h}\,\chi_{(h)}(F_\kappa;x)\,\boldsymbol{\varepsilon}_{(\kappa)}(n_\kappa(x)),$$

where we use the notation

$$\varepsilon_{(\kappa)ij} = \frac{1}{2}\left((u_i^- - u_i^0)\frac{\partial F_\kappa}{\partial x_j} + (u_j^- - u_j^0)\frac{\partial F_\kappa}{\partial x_i}\right).$$

The field $\mathbf{e}^{\lambda}_{(\kappa)}$ may take nonzero values in the domain $\omega$ only on the set $\omega_{\kappa,\lambda}$. We denote the union of all intersections $\omega_{\kappa_1,\lambda} \cap \omega_{\kappa_2,\lambda}$, $\kappa_1 \neq \kappa_2$, by the symbol $\omega_{**}$. Only one term in the sum $\mathbf{e}_*$ may be nonzero in the exterior of this set; therefore,

$$D_{\omega_*}(\mathbf{e}_*) = \sum_{\kappa=1}^{K} D_{\omega_{\kappa,\lambda}\setminus\omega_{**}}(\mathbf{e}^{\lambda}_{(\kappa)}) + D_{\omega_{**}}(\mathbf{e}_*).$$

The estimation $\operatorname{mes}\omega_{**} < c_1 h^2$ is valid since different surfaces $F_\kappa(x) = 0$ by assumption intersect at nonzero angles, $c_1$ being constant. On the other hand, the inequality

$$\left|\mathbf{e}^{\lambda}_{(\kappa)}(x)\right| < \frac{c_2}{h}, \qquad x \in \omega,$$

is valid with a certain constant $c_2$. Therefore, $\lim_{\varepsilon\to 0} D_{\omega_{**}}(\mathbf{e}_*) = 0$ and

$$\lim_{\varepsilon\to 0} D_{\omega_*}(\mathbf{e}_*) = \sum_{\kappa=1}^{K} \lim_{\varepsilon\to 0} D_{\omega_{\kappa,\lambda}}(\mathbf{e}^{\lambda}_{(\kappa)}). \tag{6.18}$$

Let us now evaluate the limit

$$\lim_{\varepsilon\to 0} D_{\omega_{\kappa,\lambda}}(\mathbf{e}^{\lambda}_{(\kappa)}) = \lim_{\varepsilon\to 0} \int_{\omega_{\kappa,\lambda}} d_x\left(\mathbf{e}_{(\kappa)}\left(\frac{x}{\lambda}\right)\right)\, dV.$$

It is convenient to introduce new variables $\xi = x/\lambda$ and consider the domain

$$\omega'_{\kappa,\lambda} = \left\{\xi \in \frac{1}{\lambda}\omega : |F_\kappa(\xi)| < \frac{h}{2}\right\}.$$

The Jacobian of this change of variables tends to 1 since $\lambda \to 1$ when $\varepsilon \to 0$. Because of that, the limit under consideration equals

$$\lim_{\varepsilon\to 0} \int_{\omega'_{\kappa,\lambda}} d_{\lambda\xi}(\mathbf{e}_{(\kappa)}(\xi))\, dV = \lim_{\varepsilon\to 0} \int_{\omega'_{\kappa,\lambda}} \frac{1}{h}\, d_{\lambda\xi}(\boldsymbol{\varepsilon}_{(\kappa)}(n_\kappa(\xi)))\, dV,$$

where $\boldsymbol{\varepsilon}_{(\kappa)}$ is the tensor with the components $\varepsilon_{(\kappa)ij}$ defined above. Here we may replace $d_{\lambda\xi}$ by $d_{n_\kappa(\xi)}$ since

$$\max\{|\lambda\xi - n_\kappa(\xi)\}| : \xi \in \omega'_{\kappa,\lambda}\} \to 0 \quad \text{when} \quad \varepsilon \to 0,$$

the values $\boldsymbol{\varepsilon}_{(\kappa)}(n_\kappa(\xi))$ are bounded, and the function $d(x; \mathbf{e})$ is uniformly continuous on $\bar{\Omega} \times Sym^d$. Thus,

$$\lim_{\varepsilon\to 0} D_{\omega_{\kappa,\lambda}}(\mathbf{e}^{\lambda}_{(\kappa)}) = \lim_{\varepsilon\to 0} \int_{\omega'_{\kappa,\lambda}} \frac{1}{h}\, d_{n_\kappa(\xi)}(\boldsymbol{\varepsilon}_{(\kappa)}(n_\kappa(\xi)))\, dV = \int_{S_\kappa} \frac{d_x(\boldsymbol{\varepsilon}_{(\kappa)}(x))}{|\operatorname{grad} F_\kappa(x)|}\, dS$$

$$= \int_{S_\kappa} d_x\left(\frac{\boldsymbol{\varepsilon}_{(\kappa)}(x)}{|\operatorname{grad} F_\kappa(x)|}\right) dS = \int_{S_\kappa} d_x\left(\mathbf{e}_\nu(\mathbf{u}^- - \mathbf{u}^0)\right) dS.$$

Using this equality and the chain of relations (6.18), (6.17), (6.16) we find

$$\lim_{\varepsilon \to 0} D_{\omega_a}(\operatorname{Def} \mathbf{u}_{a,\varepsilon}) = \int_{\partial \omega_a} d_x \left(\mathbf{e}_\nu(\mathbf{u}_a^- - \mathbf{u}_a^0)\right) dS + \int_{\omega_a} d_x(\operatorname{Def} \mathbf{u}_a)\, dV,$$

where we restored the index $a$ which numerates the domains $\omega_a$.

To obtain the formula for the kinematic multiplier $m_k(\bar{\mathbf{v}})$, let us sum these expressions over $a = 1, \ldots, A$ and observe that integrating over each of the surfaces separating two regularity domains enters the sum twice. Namely, let $\gamma$ be the common part of the boundaries of the domains $\omega_a$, $\omega_b$, which we denote $\omega^+$, $\omega^-$. Then, the above mentioned sum includes the two terms

$$\int_\gamma d_x \left(\mathbf{e}_{\nu^-}(\mathbf{u}^- - \mathbf{u}^0)\right) dS, \quad \int_\gamma d_x \left(\mathbf{e}_{\nu^+}(\mathbf{u}^+ - \mathbf{u}^0)\right) dS.$$

Since

$$\mathbf{u}^0 = \frac{1}{2}(\mathbf{u}^- + \mathbf{u}^+), \quad \mathbf{e}_{\nu^+}(\mathbf{u}^+ - \mathbf{u}^0) = \mathbf{e}_{\nu^-}(\mathbf{u}^- - \mathbf{u}^0) = \mathbf{e}_\nu([\mathbf{v}]),$$

the sum of the two integrals over $\gamma$ equals

$$\int_\gamma d_x(\mathbf{e}_\nu([\mathbf{v}]))\, dS.$$

In view of the definition of the function $\mathbf{u}_0$ and the jump $[\mathbf{v}]$ on the boundary $\partial\Omega$, the integrals over $\partial\omega_a \cap \partial\Omega$ may be written in exactly the same form. Substituting these expressions for the limit dissipation in (6.13), we arrive at the formula for the kinematic multiplier given in the formulation of the theorem, which finishes the proof.

REMARK 1. It is easy to extend the proof to the case of bodies with jump inhomogeneity by choosing the appropriate $\mathbf{u}^0$ in (6.12). Namely, the minimizer $\mathbf{w}^0(x)$ in extremum problem (4.4) should be found at every point $x$ of the surface where the jump of both material properties and velocity occurs. The vector $\mathbf{w}^0(x)$ should be taken as the value of $\mathbf{u}^0$ at $x$. Then constructing the smoothing sequence and passing to the limit results in formula (4.7) for the kinematic multiplier of a discontinuous velocity field.

## Comments

The formula for the kinematic multiplier of a discontinuous velocity field was proposed by Drucker, Prager and Greenberg (1951). To justify its application to establishing the upper bounds for the safety factor, $m_k(\bar{\mathbf{v}}) \geq \alpha_l$, it is necessary to prove the inequality $m_k(\bar{\mathbf{v}}) \geq m_s(\boldsymbol{\sigma})$ for every static multiplier $m_s(\boldsymbol{\sigma})$. It is easy to verify that this inequality is valid in case the stress field $\boldsymbol{\sigma}$ is piecewise continuously differentiable. However, this does not imply that $m_k(\bar{\mathbf{v}}) \geq \alpha_l$. Indeed, $\alpha_l$ is the supremum $\sup m_s(\boldsymbol{\sigma})$ taken over all stress fields, not necessarily piecewise continuous differentiable.

The inequality $m_k(\bar{\mathbf{v}}) \geq m_s(\boldsymbol{\sigma})$ for the kinematic multiplier of a discontinuous velocity field and every static multiplier was proved, under different assumptions, by Temam and Strang (1978), Seriogin (1983), and Kamenjarzh (1985). These results concern homogeneous bodies and are easily extendable to the case of bodies whose material properties continuously depend on the point of the body. In the case of bodies with jump inhomogeneity the formula for the kinematic multiplier of a discontinuous velocity field and the inequality $m_k(\bar{\mathbf{v}}) \geq \alpha_l$ were obtained by Kamenjarzh (1986). Chapter VIII basically follows the reasoning of the latter paper (which is applicable, in particular, to the case of continuous material properties).

Note that the formula for the kinematic multiplier is derived in case the strain rate field is only integrable and not necessarily bounded in the regularity domains. In particular, the strain rate $\mathbf{e}(x)$ can tend to infinity when $x$ approaches the boundary of a regularity domain, that is, a discontinuity surface or the body boundary. For example, the Prandtle solution to the problem of plastic layer compression between two rigid plates possesses this property. Durban and Fleck (1992) also made use of a strain rate field with singularities; Aleksandrov (1992) studied possible types of singularities.

As to the author's knowledge, Hill (1961) was the first who considered discontinuity relations in limit analysis and established some of their properties (Propositions 3.1 and 3.2). Ivlev and Bykovtsev (1971), Drugan and Rice (1984), and Telega (1985) also proposed different versions of discontinuity relations. These relations are considered as a necessary complement of the normality flow rule since the latter makes no sense at discontinuity surfaces. A different approach which consists of obtaining the discontinuity relations as conditions for the equality of the static and kinematic multipliers is adopted in Chapter VIII. In case the body is homogeneous, these relations are exactly Hill's discontinuity relations. Note that not only are the discontinuity relations involved in the formulation of sufficient conditions for the static and kinematic multiplier equality, but they are also necessary conditions for this equality. Therefore, Hill's relations turn out to be the only version of discontinuity relations, which possesses the following property: if the discontinuity relations (together with the normality flow rule) are adopted as constitutive relations, the stress and velocity field in the body subjected to a limit load have equal static and kinematic multipliers. In other words, Hill's relations are (in the case of material properties continuously depending on the point of the body) the only version of the discontinuity relations resulting in a definition of the rigid-plastic problem discontinuous solution which does not alter formulation of the rigid-plastic solutions method (Theorem 2.2). Recall that, in the case of smooth solutions, the normality flow rule similarly plays the double role of the constitutive relation and the part of the criterion for the static and kinematic multipliers equality (see Comments to Chapter III).

The discontinuity relations, introduced in case both velocity field and material properties of the body are discontinuous on the same surface, generalize Hill's relations. These relations were obtained by Kamenjarzh (1986) under the assumption that the stress field $\boldsymbol{\sigma}$ in a body with jump inhomogeneity is admissible if $\sigma$ is admissible in each of the body parts where the material properties continuously depend on the point. In other words, the body is considered as "glued" to several parts whose material properties continuously depend on the point of the body, while the "glue" is not weaker than the materials of the body parts. It is clear that this is by far not the only situation of interest, and another discontinuity relations may be obtained under different assumptions about "gluing".

Relation between piecewise regular discontinuous solutions to the rigid-plastic problem and its weak solutions (Section VIII.2) was discussed by Kamenjarzh (1986). It turned out that, for any piecewise regular solution, there exists a weak solution with velocity field $\mathbf{v}$ whose restriction $\tilde{\mathbf{v}}$ (see Comments to Chapter VII) is exactly the velocity field in the piecewise regular solution. In some cases, there also exists a piecewise regular solution corresponding to a weak solution of the rigid-plastic problem. Namely, if the restriction $\tilde{\mathbf{v}}$ of the velocity field $\mathbf{v}$ of the weak solution is piecewise continuously differentiable, then $\tilde{\mathbf{v}}$ together with the stress field $\sigma$ of this solution form a piecewise regular solution to the rigid-plastic problem (as defined in this chapter). In particular, $\tilde{\mathbf{v}}$ and $\sigma$ satisfy the discontinuity relations.

Justifying the formula for the kinematic multiplier of a discontinuous velocity field (Section 6) made use of several well known procedures. The main one, averaging, was proposed by V.A. Steklov; the reader can find more detailed information about this and other useful tools in the book by Mikhlin (1977).

Proposition 4.1 is a particular case of Theorem 3.4.1 in the book by Ioffe and Tikhomirov (1979).

CHAPTER IX

# NUMERICAL METHODS FOR LIMIT ANALYSIS

This chapter presents some numerical methods for calculating the safety factor, which is the main problem of limit analysis. Under the above established conditions, the safety factor can be evaluated as the infimum of kinematic multipliers. We discretize this extremum problem, replacing it by a sequence of extremum problems in finite-dimensional spaces. Solving these problems results in approximations to the safety factor, whose convergence will be shown here. Concrete methods for the discretization (finite element method) and for solving of the discretized extremum problems (separating plane method) are also described in this chapter.

## 1. Approximations for the kinematic extremum problem

The space of velocity fields over which the kinematic multiplier should be minimized in order to evaluate the safety factor is not finite-dimensional. However, this space can be approximated by a sequence of its finite-dimensional subspaces, and it is not a very complicated problem to find the minimum of the kinematic multiplier over each of them. These minimums are upper bounds for the safety factor and we will show that their sequence converges to the safety factor if the finite-dimensional approximations for the space of velocity fields are chosen properly. A concrete method for constructing these approximations is presented in the next section.

**1.1. Formulation of the problem.** We consider the problem of evaluating the safety factor of the load $\mathbf{l}$. Let $\Omega$ be the domain occupied by the body, $\mathbf{f}$ be the density of body force in $\Omega$ and $\mathbf{q}$ be the density of surface tractions given on the part $S_q$ of the body boundary. The remaining part $S_v$ of the boundary is fixed. For every point $x$ of the body, the set $C_x$ of admissible stresses is given in the (six-dimensional) space $Sym$, the boundary of $C_x$ being the yield surface. We consider bodies with bounded or cylindrical yield surfaces. In the latter case, $C_x$ is of the form

$$C_x = \{\boldsymbol{\sigma} \in Sym : \boldsymbol{\sigma}^d \in C_x^d\},$$

where $C_x^d$ is a given set in the deviatoric plane $Sym^d$.

Under the hypothesis of Theorem V.5.1 or Theorem VI.1.1 the safety factor can be evaluated as the infimum in the kinematic extremum problem

$$\alpha_l = \inf \left\{ \frac{\int\limits_\Omega d_x(\mathbf{e})\, dV}{\int\limits_\Omega \mathbf{e} \cdot \mathbf{s}_l \, dV} : \mathbf{e} \in \widehat{E},\ \int\limits_\Omega \mathbf{e} \cdot \mathbf{s}_l \, dV > 0 \right\}. \tag{1.1}$$

Here as usual $\mathbf{s}_l$ is a fixed stress field equilibrating the load $\mathbf{l} = (\mathbf{f}, \mathbf{q})$;

$$d_x(\mathbf{e}) = \sup\{\mathbf{e}\cdot\boldsymbol{\sigma}_* : \boldsymbol{\sigma}_* \in C_x\}$$

is the dissipation; and $\widehat{E}$ is the subspace of kinematically admissible strain rate fields. By virtue of Theorems V.5.1 and VI.1.1, it is sufficient to consider $\widehat{E}$ consisting of only smooth strain rate fields, that is, $\widehat{E} = \operatorname{Def} U$ in the case of bodies with bounded yield surfaces and $\widehat{E} = \operatorname{Def} U_d$ in the case of bodies with cylindrical yield surfaces. Here, $\operatorname{Def} V$ stands for the set of strain rate fields corresponding to all velocity fields in the set $V$; Def is the strain rate operator:

$$(\operatorname{Def}\mathbf{v})_{ij} = \frac{1}{2}\left(\frac{\partial v_i}{\partial x_j} + \frac{\partial v_j}{\partial x_i}\right).$$

Also recall that the set $U$ of virtual velocity fields consists of infinitely differentiable vector fields vanishing on $S_v$; $U_d$ is the subspace of solenoidal fields:

$$U_d = \{\mathbf{v} \in U : \operatorname{div}\mathbf{v} = 0\}.$$

The above-mentioned choice $\widehat{E} = \operatorname{Def} U$ ($\widehat{E} = \operatorname{Def} U_d$) in (1.1) is sufficient to evaluate the safety factor. However, it is not the best one from the computational viewpoint: the finite element method, for example, makes use of piecewise smooth fields rather than smooth ones. The following proposition gives a more convenient formulation of the kinematic extremum problem involving the Sobolev space $\mathbf{W}_2^1(\Omega)$ of vector fields (see Appendix VI.B, in particular, Example VI.B.1).

PROPOSITION 1. Suppose the hypotheses of Theorem V.5.1 or Theorem VI.1.1 are satisfied. Then the safety factor of the load $\mathbf{l}$ can be evaluated using the extremum problem

$$\alpha_l = \inf\left\{\int_\Omega d_x(\operatorname{Def}\mathbf{v})\,dV : \mathbf{v}\in V,\ \int_\Omega \operatorname{Def}\mathbf{v}\cdot\mathbf{s}_l\,dV = 1\right\}, \qquad (1.2)$$

where $V$ is the closure in $\mathbf{W}_2^1(\Omega)$ of the set $U$ or $U_d$ (for bodies with bounded or cylindrical yield surfaces, respectively).

PROOF. Let us first show that the safety factor can be evaluated as the infimum in (1.1) with $\widehat{E} = \operatorname{Def} V$. For example, in the case of bodies with bounded yield surfaces, this is what Theorem V.5.1 states under the assumption that the sets $\operatorname{Def} V$ and $\operatorname{Def} U$ have the same closure in the space $L_1(\Omega; Sym)$. Actually, $\operatorname{Def} V$ and $\operatorname{Def} U$ possess an even stronger property: $[\operatorname{Def} U]_2 = \operatorname{Def} V$, where $[\,\cdot\,]_2$ stands for the closure in $L_2(\Omega; Sym)$. Indeed, for every $\mathbf{v} \in V$ there is a sequence of smooth fields $\mathbf{u}_n \in U$, $n = 1, 2, \ldots$, which converges to $\mathbf{v}$ in $\mathbf{W}_2^1(\Omega)$ when $n \to \infty$. Then $\operatorname{Def}\mathbf{u}_n \to \operatorname{Def}\mathbf{v}$ in $L_2(\Omega; Sym)$ and, consequently, $\operatorname{Def}\mathbf{v}$ is in $[\operatorname{Def} U]_2$, that is, $\operatorname{Def} V \subset [\operatorname{Def} U]_2$. Let us now verify that $\operatorname{Def} V \supset [\operatorname{Def} U]_2$ as well. Consider $\mathbf{e} \in [\operatorname{Def} U]_2$; then there is a sequence of vector fields $\mathbf{u}_n \in U$ such that $\operatorname{Def}\mathbf{u}_n \to \mathbf{e}$ in $L_2(\Omega; Sym)$

when $n \to \infty$. This implies that $\operatorname{Def} \mathbf{u}_n - \operatorname{Def} \mathbf{u}_m \to 0$ in $L_2(\Omega; Sym)$ when $m, n \to \infty$.

We now will make use of Korn's inequality:

$$\|\mathbf{u}\|_{\mathbf{W}_2^1(\Omega)} \leq c \, \|\operatorname{Def} \mathbf{u}\|_{L_2(\Omega; Sym)}$$

(here, the constant $c$ does not depend on $\mathbf{u}$). It is valid for any domain $\Omega$ with a not too irregular boundary and for every $\mathbf{u} \in \mathbf{W}_2^1(\Omega)$, which satisfies the condition $\mathbf{u}|_{S_v} = 0$, the part $S_v$ of the boundary $\partial\Omega$ being of positive area. In case $S_v = \emptyset$, the inequality is valid for every $\mathbf{u} \in \mathbf{W}_2^1(\Omega)$ which satisfies the conditions

$$\int_\Omega \mathbf{u} \, dV = 0, \qquad \int_\Omega \mathbf{r} \times \mathbf{u} \, dV = 0$$

(these conditions guarantee that the zero strain rate field $\operatorname{Def} \mathbf{u} = 0$ corresponds only to the zero velocity field: $\mathbf{u} = 0$).

Due to Korn's inequality, the relation

$$\|\mathbf{u}_n - \mathbf{u}_m\|_{\mathbf{W}_2^1(\Omega)} \to 0 \quad \text{when} \quad n, m \to \infty$$

is valid and implies that the sequence $\mathbf{u}_1, \mathbf{u}_2, \ldots$ converges in $\mathbf{W}_2^1(\Omega)$. By the definition of $V$, the limit of this sequence is in $V$: $\mathbf{u}_n \to \mathbf{v} \in V$. Then, $\operatorname{Def} \mathbf{u}_n \to \operatorname{Def} \mathbf{v}$ in $L_2(\Omega; Sym)$ when $n \to \infty$. Recall that the sequence $\mathbf{u}_1, \mathbf{u}_2, \ldots$ was chosen so that $\operatorname{Def} \mathbf{u}_n \to \mathbf{e}$; thus, it turns out that $\mathbf{e} = \operatorname{Def} \mathbf{v} \in \operatorname{Def} V$, which finishes verifying the relation $[\operatorname{Def} U]_2 = \operatorname{Def} V$.

The analogous statement is also valid for bodies with bounded yield surfaces. Thus, in cases of bodies with both bounded and cylindrical yield surfaces, equality (1.1) is valid with $\widehat{E} = \operatorname{Def} V$.

Let us now verify that (1.1) is equivalent to the problem

$$\int_\Omega d_x(\mathbf{e}) \, dV \to \inf; \quad \mathbf{e} \in \widehat{E}, \; \int_\Omega \mathbf{e} \cdot \mathbf{s}_l \, dV = 1. \tag{1.3}$$

Indeed, let $\mathbf{e}_1, \mathbf{e}_2, \ldots$ be a minimizing sequence in (1.1). Then the fields

$$\mathbf{e}_{*n} = \mathbf{e}_n \left( \int_\Omega \mathbf{e}_n \cdot \mathbf{s}_l \, dV \right)^{-1}, \quad n = 1, 2, \ldots,$$

satisfy the restrictions of problem (1.3). Note that the equality

$$\int_\Omega d_x(\mathbf{e}_n) \, dV \left( \int_\Omega \mathbf{e}_n \cdot \mathbf{s}_l \, dV \right)^{-1} = \int_\Omega d_x(\mathbf{e}_{*n}) \, dV$$

is valid due to the first degree homogeneity of the dissipation. Consequently, infimum (1.3) is not greater than infimum (1.1). On the other hand, if $\{\mathbf{e}_{0n}\}$,

$n = 1, 2, \ldots$, is a minimizing sequence in problem (1.3), each of the fields $\mathbf{e}_{0n}$ satisfies the restrictions of problem (1.1), and, consequently, infimum (1.1) is not greater than infimum (1.3). Thus, the infimums are equal and

$$\alpha_l = \inf\left\{ \int_\Omega d_x(\mathbf{e})\, dV : \mathbf{e} \in \widehat{E},\ \int_\Omega \mathbf{e}\cdot \mathbf{s}_l\, dV = 1 \right\} \quad (\widehat{E} = \operatorname{Def} V),$$

which finishes the proof.

Note that, by definition, $V$ is a closed subspace in $\mathbf{W}_2^1(\Omega)$ and, therefore, $V$ is a complete normed space with the norm $\|\mathbf{v}\|_V = \|\mathbf{v}\|_{\mathbf{W}_2^1(\Omega)}$.

**1.2. Approximations.** According to Proposition 1, the safety factor $\alpha_l$ can be evaluated using extremum problem (1.2), which requires minimization of a function over the space $V$ which is not finite-dimensional. However, this problem can be approximated by a sequence of discrete problems, that is, the problems requiring minimization over a finite-dimensional space. The discrete problems can be solved easily, and we will now construct them and show that the sequence of their infimums converges to infimum (1.2), the safety factor $\alpha_l$.

Let $V_h$ be a subspace in $V$. Consider the extremum problem in $V$ analogous to (1.2):

$$\alpha_{l(h)} = \inf\left\{ \int_\Omega d_x(\operatorname{Def}(\mathbf{v}))\, dV : \mathbf{v} \in V_h,\ \int_\Omega \operatorname{Def}\mathbf{v}\cdot \mathbf{s}_l\, dV = 1 \right\}.$$

The inequality $\alpha_{l(h)} \geq \alpha_l$ is obviously valid since $V_h \subset V$. The difference between $\alpha_{l(h)}$ and $\alpha_l$ can be large. However, considering not only a single subspace $V_h$ but a collection of subspaces, it is easy to ensure the convergence of the corresponding infimums $\alpha_{l(h)}$ to $\alpha_l$. To obtain the convergence, the collection of $V_h$ should be chosen in such a way that every $\mathbf{v} \in V$ (or, equivalently, every member of the set $\widetilde{V}$ dense in $V$) is approximated by a sequence of $\mathbf{v}_h \in V_h$ when $h \to 0$.

More precisely, let $\widetilde{V}$ be a dense subset in $V$ and $h > 0$ be a parameter with its values in a neighborhood of zero. We also assume that $h$ may take the values which form a sequence converging to zero. Let $\{V_h\}$ be a collection of subspaces in $V$ such that for every $\widetilde{\mathbf{v}} \in \widetilde{V}$ there is a sequence of $\mathbf{v}_h \in V_h$ converging to $\widetilde{\mathbf{v}}$: $\|\mathbf{v}_h - \widetilde{\mathbf{v}}\|_V \to 0$ when $h \to 0$. In this case, we say that the collection of the subspaces $V_h$ forms an *internal approximation* for the space $V$.

THEOREM 1. Suppose that 1) the assumptions of Theorem V.5.1 or Theorem VI.1.1 are satisfied and $V$ is the closure of the set of virtual velocity fields $U$ or $U_d$ (for bodies with bounded or cylindrical yield surfaces, respectively) with respect to the norm $\|\cdot\|_V = \|\cdot\|_{\mathbf{W}_2^1(\Omega)}$, 2) the collection of the subspaces $V_h$ forms an internal approximation for $V$. Then the sequence of the infimums

$$\alpha_{l(h)} = \inf\left\{ \int_\Omega d_x(\operatorname{Def}\mathbf{v})\, dV : \mathbf{v} \in V_h,\ \int_\Omega \operatorname{Def}\mathbf{v}\cdot \mathbf{s}_l\, dV = 1 \right\} \tag{1.4}$$

converges to the safety factor: $\alpha_{l(h)} \to \alpha_l$ when $h \to 0$, each of the infimums being an upper bound for the safety factor: $\alpha_{l(h)} \geq \alpha_l$.

PROOF. The inequality $\alpha_{l(h)} \geq \alpha_l$ is obviously valid since $V_h \subset V$. To prove the convergence $\alpha_{l(h)} \to \alpha_l$, note first that the extremum problem determining the safety factor can be written as

$$\alpha_l = \inf \left\{ \int_\Omega d_x(\mathrm{Def}\, \widetilde{\mathbf{v}})\, dV : \widetilde{\mathbf{v}} \in \widetilde{V},\ \int_\Omega \mathrm{Def}\, \widetilde{\mathbf{v}} \cdot \mathbf{s}_l \, dV = 1 \right\},$$

where $\widetilde{V} = U$ or $\widetilde{V} = U_d$ (for bodies with bounded or cylindrical yield surfaces, respectively). Indeed, the above equality is valid since Def $\widetilde{V}$ can be taken as $\widehat{E}$ in (1.1) due to Theorem V.5.1 or Theorem VI.1.1 and the condition

$$\int_\Omega \mathrm{Def}\, \widetilde{\mathbf{v}} \cdot \mathbf{s}_l \, dV > 0$$

can be replaced by

$$\int_\Omega \mathrm{Def}\, \widetilde{\mathbf{v}} \cdot \mathbf{s}_l \, dV = 1,$$

exactly as in the proof of Proposition 1. Then, by the definition of the infimum, for every $\varepsilon > 0$ there is a field $\widetilde{\mathbf{v}}_\varepsilon \in \widetilde{V}$ such that

$$\int_\Omega d_x(\mathrm{Def}\, \widetilde{\mathbf{v}}_\varepsilon)\, dV \leq \alpha_l + \varepsilon, \qquad \int_\Omega \mathrm{Def}\, \widetilde{\mathbf{v}}_\varepsilon \cdot \mathbf{s}_l \, dV = 1. \tag{1.5}$$

Note that the theorem will be proved if for sufficiently small $h$ it is shown that there is a field $\mathbf{u}_{\varepsilon,h}$ in $V_h$ such that

$$\int_\Omega d_x(\mathrm{Def}\, \mathbf{u}_{\varepsilon,h})\, dV \leq \int_\Omega d_x(\mathrm{Def}\, \widetilde{\mathbf{v}}_\varepsilon)\, dV + R\varepsilon, \quad \int_\Omega \mathrm{Def}\, \mathbf{u}_{\varepsilon,h} \cdot \mathbf{s}_l \, dV = 1, \tag{1.6}$$

where $R$ is a positive constant depending only on the set $C$ of admissible stress fields. Indeed, in this case, $\mathbf{u}_{\varepsilon,h}$ satisfies the restrictions in extremum problem (1.4), and, hence, the following estimation is valid

$$\alpha_{l(h)} \leq \int_\Omega d_x(\mathrm{Def}\, \mathbf{u}_{\varepsilon,h})\, dV \leq \alpha_l + \varepsilon + R\varepsilon,$$

when $h$ is sufficiently small (the second inequality arises from (1.5), (1.6)). This estimation together with the inequality $\alpha_{l(h)} \geq \alpha_l$ shows that $\alpha_{l(h)} \to \alpha_l$ when $h \to 0$.

To finish the proof, it remains to verify that there is a suitable approximation to $\widetilde{\mathbf{v}}_\varepsilon$, that is, $\mathbf{u}_{\varepsilon,h}$ satisfying (1.6). Recall that the subspaces $V_h$ form an internal approximation for $V$; therefore for the fixed $\varepsilon > 0$ there is $\mathbf{v}_{\varepsilon,h} \in V_h$ such that $\|\mathbf{v}_{\varepsilon,h} - \widetilde{\mathbf{v}}_\varepsilon\|_V \to 0$ when $h \to 0$, which in particular implies that

$$\|\mathrm{Def}\,\mathbf{v}_{\varepsilon,h} - \mathrm{Def}\,\widetilde{\mathbf{v}}_\varepsilon\|_{L_2(\Omega;Sym)} \to 0, \quad \int_\Omega \mathrm{Def}\,\mathbf{v}_{\varepsilon,h}\cdot\mathbf{s}_l\,dV \to \int_\Omega \mathrm{Def}\,\widetilde{\mathbf{v}}_\varepsilon\cdot\mathbf{s}_l\,dV = 1$$

$$\text{when } h \to 0.$$

Now consider the sequence of the fields

$$\mathbf{u}_{\varepsilon,h} = \mathbf{v}_{\varepsilon,h}\left(\int_\Omega \mathrm{Def}\,\mathbf{v}_{\varepsilon,h}\cdot\mathbf{s}_l\,dV\right)^{-1} \in V_h.$$

By virtue of the previous relations, it is clear that

$$\begin{gathered} \|\mathrm{Def}\,\mathbf{u}_{\varepsilon,h} - \mathrm{Def}\,\widetilde{\mathbf{v}}_\varepsilon\|_{L_2(\Omega;Sym)} \to 0 \quad \text{when } h \to 0, \\ \int_\Omega \mathrm{Def}\,\mathbf{u}_{\varepsilon,h}\cdot\mathbf{s}_l\,dV = 1 \end{gathered} \tag{1.7}$$

Now, it can be shown that $\mathbf{u}_{\varepsilon,h}$ satisfies inequality (1.6). Recall that the relation $d_x(\mathbf{e}_1) \le d_x(\mathbf{e}_2) + d_x(\mathbf{e}_1 - \mathbf{e}_2)$ is valid for every $\mathbf{e}_1$ and $\mathbf{e}_2$ in $Sym$ due to the convexity and homogeneity of the dissipation. Besides, in the case of bodies with bounded yield surfaces, the estimation $d_x(\mathbf{e}) \le R\,|\mathbf{e}|$ is valid for every $\mathbf{e}$ in $Sym$ (this arises from the assumptions of Theorem V.5.1, which are satisfied here). Analogously, in the case of bodies with cylindrical yield surfaces, the same estimation is valid for every $\mathbf{e} \in Sym^d$ (see the assumptions of Theorem VI.1.1, which are satisfied here). Applying these estimations to $\mathbf{e}_1(x) = \mathrm{Def}\,\mathbf{u}_{\varepsilon,h}(x)$, $\mathbf{e}_2(x) = \mathrm{Def}\,\widetilde{\mathbf{v}}_\varepsilon(x)$, we find

$$\int_\Omega d_x(\mathrm{Def}\,\mathbf{u}_{\varepsilon,h})\,dV \le \int_\Omega d_x(\mathrm{Def}\,\widetilde{\mathbf{v}}_\varepsilon)\,dV + R\int_\Omega |\mathrm{Def}(\mathbf{u}_{\varepsilon,h} - \widetilde{\mathbf{v}}_\varepsilon)|\,dV.$$

Finally, we observe that, due to (1.7), the second integral on the right-hand side is less than $\varepsilon$ for sufficiently small $h$. This results in inequality (1.6), which finishes the proof.

Thus, in case a collection of subspaces $V_h \in V$ forms an internal approximation for $V$, solutions of extremum problems (1.4) converge to the safety factor $\alpha_l$ when $h \to 0$. The next section presents a method for constructing such a collection, while Section 3 describes a method for solving discrete extremum problems.

REMARK 1. The extremum in kinematic problem (1.2) is taken over the set of regular velocity fields and is not necessarily attained even if the inputs of the problem (the body boundary and the load) are smooth. However, this is no obstacle for constructing a minimizing sequence consisting of regular velocity fields. Actually, solutions to discrete extremum problems (1.4) form such a sequence. The sequence does not necessarily converge to a regular velocity field but the sequence of the corresponding kinematic multipliers does converge to the safety factor $\alpha_l$. Recall that the safety factor $\alpha_l$ (rather than the velocity field at which $\alpha_l$ is attained) is of the most practical interest.

## 2. Discretization: finite element method

In this section, the finite element method is presented for constructing an internal approximation for the functional space $V$. The approximation is given by a collection of finite-dimensional subspaces in $V$. Using this method, we discretize the kinematic extremum problem of limit analysis, which provides the rational basis for calculating of the safety factor. Limit analysis is just one of numerous domains where the finite element method is used.

**2.1. Idea of the method.** The finite element method suggests the following scheme for constructing an internal approximation for the space $V$ of functions on the domain $\Omega$.

- Divide the domain $\Omega$ into a number of standard subdomains, *finite elements*.
  For example, tetrahedral finite elements are often used if $\Omega$ is a polyhedron. The partition is normally characterized by a parameter $h$ which is the maximum dimension of the finite elements.
- Select a set of points, *nodes*, associated with the partition.
  For example, the vertices of the tetrahedrons can be chosen as the nodes in the case of a tetrahedral partition.
- Define *basis functions* associated with each of the nodes and introduce a finite-dimensional space $V_h$ consisting of all linear combinations of the basis functions.
  Normally, a basis function is piecewise linear or piecewise quadratic (or, more generally, piecewise polynomial) and vanishes in the exterior of the finite elements containing the node with which the function is associated.
- Specify a procedure for constructing of *interpolant* $v_h \in V_h$ for any $\widetilde{v}$ in $\widetilde{V}$, the latter being a subset dense in $V$; the interpolant approximates $\widetilde{v}$ in a certain sense.
  Normally, the coefficients in the decomposition of the interpolant $v_h$ with respect to the basis functions are determined by the values of $\widetilde{v}$ and a few of its derivatives at the corresponding nodes.
- Establish the convergence: $v_h \to \widetilde{v}$ when $h \to 0$.

The last step presumes that $V_h \subset V$ at least for sufficiently small $h$ and means that an *internal approximation* for the space $V$ is constructed.

We now illustrate the above described scheme by constructing a finite element internal approximation for a space of functions on the internal $I = (0, 1)$. We divide $(0, 1)$ into $N$ equal intervals of the length $h = 1/N$ and consider these intervals as finite elements. Let us take their ends

$$x_0 = 0,\ x_1 = h,\ \ldots,\ x_{N-1} = (N-1)h,\ x_N = 1$$

as the nodes. We define the following basis function $\varphi_k$ associated with the $k$-th node: $\varphi_k$ is linear on each of the finite elements, takes the value 1 at the $k$-th node and vanishes at the rest of the nodes, see Figure 2.1.

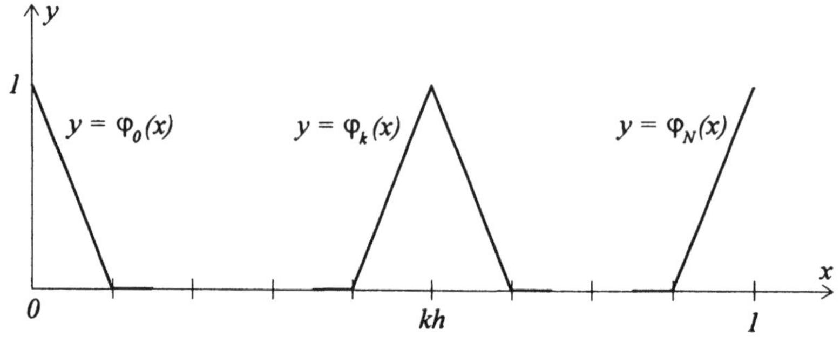


**Figure 2.1**

We introduce the space $V_h$ consisting of all linear combinations $\alpha_0\varphi_0 + \cdots + \alpha_N\varphi_N$ of the basis functions ($\alpha_0, \cdots, \alpha_N$ are arbitrary numbers). For a smooth function $\widetilde{v}$ on $I$ we define its interpolant $v_h \in V_h$:

$$v_h = \widetilde{v}(0)\varphi_0 + \widetilde{v}(x_1)\varphi_1 + \cdots + \widetilde{v}(x_N)\varphi_N.$$

It is clear that the interpolant takes the same values as $\widetilde{v}$ at all nodes: $v_h(kh) = \widetilde{v}(kh)$ since $\varphi_k(kh) = 1$ and $\varphi_k(lh) = 0$ if $l \neq k$.

The chosen finite elements and interpolation procedure allow construction of an internal approximation for the Sobolev space $V = W_2^1(I)$. Recall that this space consists of square integrable functions $u$ which have square integrable derivatives $u'$, and the norm in $V = W_2^1(I)$ is

$$\|u\|_V = \left(\int_0^1 [u^2 + (u')^2]\, dx\right)^{\frac{1}{2}}$$

(see Appendix VI.B for more details). The example of constructing an internal approximation for $W_2^1(I)$ is of certain interest in connection with limit analysis since Theorem 1.1 suggests using the velocity fields space with the Sobolev norm for evaluating of the safety factor.

PROPOSITION 1. The collection of the above described spaces $V_h$ forms an internal approximation for the space $V = W_2^1(I)$.

PROOF. All of the above defined functions $v_h$ are continuous and piecewise linear on $I$, and, therefore, form a subspace $V_h$ in $V$.

The subspace $\widetilde{V}$ of smooth, for example, infinitely differentiable functions on $I$ is dense in $V = W_2^1(I)$ by Theorem VI.B.2. Consider any $\widetilde{v}$ in $\widetilde{V}$ and its interpolant $v_h = \widetilde{v}(0)\varphi_0 + \widetilde{v}(x_1)\varphi_1 + \cdots + \widetilde{v}(x_N)\varphi_N$ in $V_h$. Let us show that $\|v_h - \widetilde{v}\|_V \to 0$ when $h \to 0$, which means that the collection of the spaces $V_h$ is an internal approximation for $V$. We estimate the difference $v_h - \widetilde{v}$ on each of the finite elements. The function $v_h$ is linear on the segment $x_k \leq x \leq x_{k+1}$, and takes on the values

$$v_h(x) = \widetilde{v}(x_k) + \frac{\widetilde{v}(x_{k+1}) - \widetilde{v}(x_k)}{h}(x - x_k) = \widetilde{v}(x_k) + \widetilde{v}'(\xi_k)(x - x_k),$$

where $\xi_k$ is a certain point in $[x_k, x_{k+1}]$. On this segment, the function $\widetilde{v}$ can be written in the form

$$\widetilde{v}(x) = \widetilde{v}(x_k) + \widetilde{v}'(x_k)(x - x_k) + \frac{1}{2}\widetilde{v}(\eta(x))(x - x_k)^2,$$

where $\eta(x)$ is a certain point in $[x_k, x_{k+1}]$. Then, the following estimations are valid at $x \in [x_k, x_{k+1}]$:

$$|\widetilde{v}(x) - v_h(x)| \leq \frac{3}{2}c_k h^2, \quad |\widetilde{v}'(x) - v_h'(x)| \leq c_k h,$$

where $c_k$ is the maximum of $|\widetilde{v}''(x)|$ over $[x_k, x_{k+1}]$. These estimations hold if $c_k$ is replaced by $C$, $C$ being the maximum of $|\widetilde{v}''(x)|$ over $[0,1]$. This immediately implies the convergence $v_h \to \widetilde{v}$ in $V = W_2^1(I)$ when $h \to 0$, which finishes the proof.

**2.2. Approximation for velocity fields space.** For evaluating the safety factor in the case of a body with bounded yield surfaces, one can use the velocity fields space $V$ in which the set $U$ of virtual velocity fields is dense (Theorem 1.1). If the body occupies the domain $\Omega$ and $S_v$ is the fixed part of the body boundary, the set $U$ consists of smooth (infinitely differentiable) vector fields on $\Omega$ vanishing on $S_v$. To describe the finite element approximation for $V$, let us start with approximating a space of number-valued functions and then use this approximation for every component of $\mathbf{v} \in V$. For simplicity, we restrict ourselves to the case of a polyhedral domain $\Omega$; more precisely, we assume that $\Omega$ consists of a finite number of parts of planes. (Domains with holes and cavities can satisfy this assumption.)

Consider a *triangulation* of $\Omega$, that is, split $\Omega$ into a finite number of tetrahedrons, which may have common parts of their boundaries and no common internal points. We restrict ourselves to triangulations which satisfy the following condition: a side of each tetrahedron either is a side of another tetrahedron or lies on $\partial\Omega$. Thus, if the vertex $A$ of a tetrahedron lies on a side of another tetrahedron, $A$ is also a vertex of the latter tetrahedron. We choose all vertices $x_{(1)}, \ldots, x_{(N)}$ of the tetrahedrons as the *nodes* of the triangulation.

We now define the *basis function* associated with node $x_{(k)}$, which is, generally speaking, a vertex of several tetrahedrons. On each of these tetrahedrons, consider a linear function which takes the value 1 in the vertex $x_{(k)}$ and 0 at each of the rest of the vertices of the tetrahedron. Note that a linear function on a tetrahedron is uniquely determined by the values the function takes at the tetrahedron vertices. Analogously, a linear function on a side of the tetrahedron is uniquely determined by the values the function takes at the tetrahedron vertices which belong to this side. In particular, if two of the tetrahedrons have a common side, the linear functions defined in these tetrahedrons take the same values on this side. Thus, a continuous function is defined on the union of the tetrahedrons which contain the vertex $x_{(k)}$: the function is identical to the above introduced linear function on each of the tetrahedrons. Note that this function vanishes on those sides of the tetrahedrons that do not contain $x_{(k)}$. Therefore, extending the function as zero to all of $\Omega$, we obtain a continuous function on $\Omega$. Let us denote this extended function by the symbol $\varphi_k$ and consider $\varphi_k$ as the basis function associated with the node $x_{(k)}$. Briefly, the basis function $\varphi_k$ is a piecewise linear function which is linear on each of the tetrahedrons of the triangulation, takes the value 1 at the node $x_{(k)}$ and vanishes at all other nodes.

All linear combinations $\alpha_1\varphi_1+\cdots+\alpha_N\varphi_N$ of the basis functions ($\alpha_1,\ldots,\alpha_N$ are arbitrary numbers) form a finite-dimensional space we denote by the symbol $W_h$. Here, $h$ is the maximum diameter of the tetrahedrons which form the triangulation (we understand the diameter of a set as the maximum distance between its points). We denote the triangulation by the symbol $\mathcal{T}_h$.

We now define an *interpolant* $w_h \in W_h$ for a smooth (for example, infinitely differentiable) function $\widetilde{w}$:

$$w_h = \widetilde{w}(x_{(1)})\varphi_1 + \cdots + \widetilde{w}(x_{(N)})\varphi_N.$$

The interpolant takes the same values as $\widetilde{w}$ at all of the nodes $x_{(k)}$, $k = 1,\ldots,N$.

Note that every function in $W_h$ is continuous and piecewise linear; therefore, it belongs to $W_2^1(\Omega)$. Actually, we have constructed an internal approximation for $W_2^1(\Omega)$. More precisely, the following well known proposition is valid when the triangulations $\mathcal{T}_h$ satisfy the additional requirement: there is a constant $C$ such that, for sufficiently small $h$ and for every tetrahedron in $\mathcal{T}_h$, the ratio of its diameter to the maximum diameter over all balls contained in the tetrahedron is less than $C$. We refer to a collection of triangulations $\mathcal{T}_h$ which satisfy this condition as regular. It can be easily seen that the regularity means that the tetrahedrons do not flatten. The following well known approximation theorem is valid.

THEOREM 1. Suppose $\Omega$ is a polyhedral domain, $\mathcal{T}_h$ and $W_h$ are the above defined triangulation and space of piecewise linear continuous functions on $\Omega$. If the collection of the triangulations is regular, the collection of the spaces $W_h$ forms an internal approximation for $W_2^1(\Omega)$: for every smooth function $\widetilde{w}$ on $\Omega$, its interpolants $w_h = \widetilde{w}(x_{(1)})\varphi_1 + \cdots + \widetilde{w}(x_{(N)})\varphi_N$ converge to $\widetilde{w}$ in $W_2^1(\Omega)$ when $h \to 0$.

Using this approximation, it is easy to construct the required internal approximation for the velocity fields space $V$. To show this, we should first verify that, if a smooth function $\widetilde{w}$ vanishes on the surface $S_v \subset \partial\Omega$, its interpolants $w_h$ also vanish on $S_v$. Indeed, let $\Omega$ be a polyhedron (as assumed before) and the contour of $S_v$ consists of a finite number of segments. Then we choose a collection of triangulations $\mathcal{T}_h$ such that, for sufficiently small $h$, $S_v$ consists of a number of sides of some tetrahedrons in $\mathcal{T}_h$. We restrict ourselves to such triangulations and observe in this case that $\widetilde{w}$ vanishes at all nodes of these triangulations that lie on $S_v$. Therefore, the corresponding coefficients in the expression $w_h = \widetilde{w}(x_{(1)})\varphi_1 + \cdots + \widetilde{w}(x_{(N)})\varphi_N$ vanish. In addition, the basis functions associated with the rest of the nodes (that is, the nodes not in $S_v$) vanish on $S_v$. Thus, $w_h|_{S_v} = 0$, that is, $w_h$ belongs to the subspace

$$W_{(0)h} = \{u \in W_h : \ u|_{S_v} = 0\}.$$

Under the foregoing assumptions about $\Omega$ and $S_v$, the collection of the finite-dimensional spaces $V_h = (W_{(0)h})^3$ forms an internal approximation for the velocity fields space $V$. Recall that $V$ is the closure of the set $U$ with respect to the norm $\|\cdot\|_V = \|\cdot\|_{\mathbf{W}_2^1(\Omega)}$, $U$ consisting of smooth vector fields which vanish on $S_v$. We observe that $V_h$ is a subspace in $V$ (see Theorem VI.3.1). For every $\tilde{\mathbf{v}} \in U$, we consider its components $v_i$ and their interpolants $v_{ih}$ in $W_{(0)h}$. It is clear that the vector fields $\mathbf{v}_h$ with the components $(v_{1h}, v_{2h}, v_{3h})$ belong to $V_h$ and $\mathbf{v}_h \to \mathbf{v}$ in $V$. Thus, the collection of the spaces $V_h$ forms an internal approximation for $V$.

REMARK 1. Let $\varphi_1, \ldots, \varphi_M$ be the (scalar) basis functions associated with those nodes $x_{(1)}, \ldots, x_{(M)}$ of the triangulation that do not lie in $S_v$. Then the vector fields

$$\varphi_1\mathbf{e}_1,\ \varphi_1\mathbf{e}_2,\ \varphi_1\mathbf{e}_3,\ \ldots,\ \varphi_M\mathbf{e}_1,\ \varphi_M\mathbf{e}_2,\ \varphi_M\mathbf{e}_3$$

form a basis in $V_h$.

REMARK 2. For the space $V$ of velocity fields on a two-dimensional domain $\Omega$, it is possible to construct the internal approximation similar to the above one in the three-dimensional case. One only has to replace the tetrahedrons with triangles.

**2.3. Approximation for solenoidal velocity fields space.** For evaluating the safety factor in the case of a body with cylindrical yield surfaces, one can use the velocity fields space $V$ in which the set $U_d$ of smooth solenoidal vector fields is dense (Theorem 1.1). The spaces $V_h$ introduced in the previous subsection do not form an internal approximation for $V$: in contrast to $V$ the spaces $V_h$ contain not only solenoidal fields. Let us now describe a suitable internal approximation for $V$.

For simplicity, we consider the plane strain state, assuming that only two components $v_1$, $v_2$ of any velocity field $\mathbf{v}$ are nonzero, and that they do not depend on the coordinate $x_3$. Such a state gets realized, for example, under appropriate conditions in a body which is a cylinder with its axis parallel to the $x_3$-axis. The cylinder is represented by a two-dimensional domain $\Omega$, the cross-section of the cylinder by the $(x_1, x_2)$-plane. We restrict ourselves to the case when $\Omega$ is a polygon.

*Using stream function.* The main idea of constructing the internal approximation for the space $V$ of solenoidal velocity fields consists of expressing each of these fields through corresponding stream function and approximating the stream function. Namely, let $\tilde{\mathbf{v}}$ be an arbitrary smooth solenoidal field on $\Omega$. Since $\tilde{\mathbf{v}}$ is solenoidal, that is,

$$\frac{\partial \tilde{v}_1}{\partial x_1} + \frac{\partial \tilde{v}_2}{\partial x_2} = 0,$$

there is a smooth function $\tilde{\phi}$ on $\Omega$ such that

$$\tilde{v}_1 = -\frac{\partial \tilde{\phi}}{\partial x_2} \quad \tilde{v}_2 = \frac{\partial \tilde{\phi}}{\partial x_1}.$$

We refer to such $\widetilde{\phi}$ as a *stream function*. Suppose now that there is a sequence of interpolants $\phi_h$ such that not only $\phi_h$ approximates $\widetilde{\phi}$, but also the sequence of the derivatives $\partial\phi_h/\partial x_i$ approximates the derivative $\partial\widetilde{\phi}/\partial x_i$, $1 = 1, 2$. In this case, the vector field $\mathbf{v}_h$ with the components

$$v_{h1} = -\frac{\partial\phi_h}{\partial x_2}, \quad v_{h2} = \frac{\partial\phi_h}{\partial x_1}$$

is solenoidal and the sequence of the fields $\mathbf{v}_h$ approximates $\widetilde{\mathbf{v}}$ when $h \to 0$. This results in the internal approximation for the space $V$, and the approximation satisfies the assumptions of Theorem 1.1 if $\mathbf{v}_h$ converges to $\widetilde{\mathbf{v}}$ in $\mathbf{W}_2^1(\Omega)$. In terms of stream functions, the latter condition means that the stream functions $\phi_h$ and their first and second derivatives should approximate $\widetilde{\phi}$ and its corresponding derivatives. This calls for a bit more sophisticated structure than we used in the previous subsection: piecewise linear interpolants considered there do not possess the second derivatives. Let us now present a suitable internal approximation for the space of stream functions.

*Triangulation and Fraeijs de Veubeke – Sander quadrangle.* Let us make use of quadrangle finite elements with piecewise polynomial basis functions of power 3. We split $\Omega$ into a finite number of convex quadrangles which may have common parts of their boundaries and no common internal points. We only consider triangulations which satisfy the following condition: a side of each quadrangle either is a side of another quadrangle or lies on $\partial\Omega$. Thus, if the vertex $A$ of a quadrangle lies on a side of another one, then $A$ is also a vertex of the latter quadrangle. Midpoints of all sides of the quadrangles possess the same property. We choose all vertices of the quadrangles and all midpoints of their sides as the nodes of the triangulation.

We now define the basis functions associated with the vertices $\mathbf{a}_1$, $\mathbf{a}_2$, $\mathbf{a}_3$, $\mathbf{a}_4$ of the quadrangle $Q$ and with the midpoints $\mathbf{b}_1$, $\mathbf{b}_2$, $\mathbf{b}_3$, $\mathbf{b}_4$ of its sides, see Figure 2.2. Let us first define these functions only on $Q$. Three polynomials $P_1$, $P_2$, $P_3$ of power 3 in the coordinates $x, y$ represent each of the basis functions on $Q$: the function is $P_1$ in the triangle $\mathbf{a}_1\mathbf{a}_2\mathbf{a}_0$, $P_1 + P_2$ in the triangle $\mathbf{a}_2\mathbf{a}_3\mathbf{a}_0$, $P_1 + P_2 + P_3$ in the triangle $\mathbf{a}_3\mathbf{a}_4\mathbf{a}_0$, and $P_1 + P_3$ in the triangle $\mathbf{a}_4\mathbf{a}_1\mathbf{a}_0$ ($\mathbf{a}_0$ is the intersection of the quadrangle diagonals). Each of the three polynomials is determined by ten coefficients; however, several obvious restrictions are imposed on these thirty parameters. Namely, the basis function has to possess second derivatives; therefore, both functions and their first derivatives should be continuous on the diagonals of $Q$. This means that the polynomial $P_2$ should vanish on the diagonal $\mathbf{a}_2\mathbf{a}_4$ as well as the derivative of $P_2$ in the direction of the normal to the diagonal. The polynomial $P_3$ should possess the analogous properties on the diagonal $\mathbf{a}_1\mathbf{a}_3$. These conditions result in fourteen linear equations for the coefficients of the polynomials $P_1, P_2, P_3$. These restrictions leave sixteen parameters whose values are to be chosen to determine the basis function $\varphi$. We will use the following sixteen parameters to determine $\varphi$:

$$\varphi(\mathbf{a}_i), \quad \frac{\partial\varphi}{\partial x}(\mathbf{a}_i), \quad \frac{\partial\varphi}{\partial y}(\mathbf{a}_i), \quad \frac{\partial\varphi}{\partial n}(\mathbf{b}_i), \quad i = 1, 2, 3, 4, \tag{2.1}$$

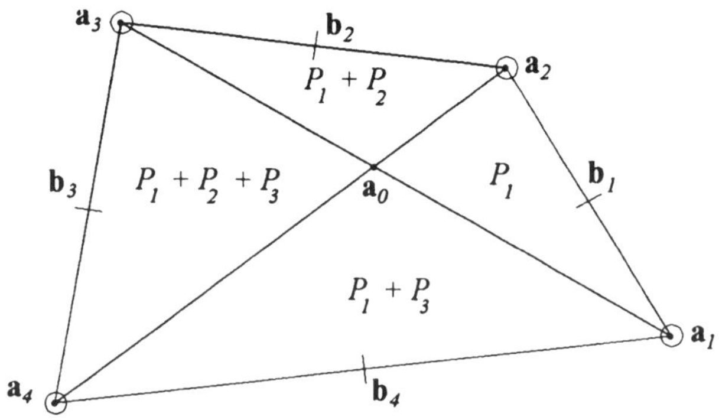


**Figure 2.2**

where $\partial/\partial n$ stands for the derivative in the direction normal to the quadrangle sides. We refer to (2.1) as *degrees of freedom* of the basis function $\varphi$.

Sixteen basis functions are considered in the quadrangle $Q$; each of them is determined by selecting one of the degrees of freedom equal to 1, while the rest are zero. Thus, three basis functions are associated with the vertex $\mathbf{a}_i$ of $Q$, the only nonzero degrees of freedom being either the value of the function at $\mathbf{a}_i$ or the value of its derivative with respect to $x$ or to $y$. We denote these functions by the symbols $\psi_{\mathbf{a}_i}$, $\psi_{\mathbf{a}_i,1}$, $\psi_{\mathbf{a}_i,2}$, respectively; we will also use the following brief notations: $\psi_i = \psi_{\mathbf{a}_i}$, $\psi_{i1} = \psi_{\mathbf{a}_i,1}$, $\psi_{i2} = \psi_{\mathbf{a}_i,2}$. One basis function is associated with the midpoint of each of the quadrangle sides. We denote the basis function associated with the midpoint $\mathbf{b}_j$ by the symbol $\chi_{\mathbf{b}_j}$ or, briefly, by $\chi_j$. The only nonzero degree of freedom for this function is the normal derivative $\partial\chi_{\mathbf{b}_j}/\partial n$. It can be easily seen that the above conditions uniquely determine the basis functions on $Q$. The quadrangle together with the basis functions defined on it is referred to as the *Fraeijs de Veubeke – Sander finite element.*

EXAMPLE 2. *Basis functions on rectangle.* Let us find the basis functions on the rectangle $[x_0 - a, x_0 + a] \times [y_0 - b, y_0 + b]$, $a > 0$, $b > 0$. It is easy to express them through the basis functions on the square $[-1, 1] \times [-1, 1]$. Indeed, basis function $\varphi$ on the square and the corresponding basis function $\Phi$ on the rectangle are related by the formula

$$\Phi(x, y) = \varphi\left(\frac{x - x_0}{a}, \frac{y - y_0}{b}\right),$$

the only nonzero degree of freedom being the value of the function at a vertex, or by the formula

$$\Phi(x, y) = a\varphi\left(\frac{x - x_0}{a}, \frac{y - y_0}{b}\right),$$

the only nonzero degree of freedom being the value of the derivative with respect to $x$ at a vertex or at the midpoint of a vertical side, or by the formula

$$\Phi(x, y) = b\varphi\left(\frac{x - x_0}{a}, \frac{y - y_0}{b}\right),$$

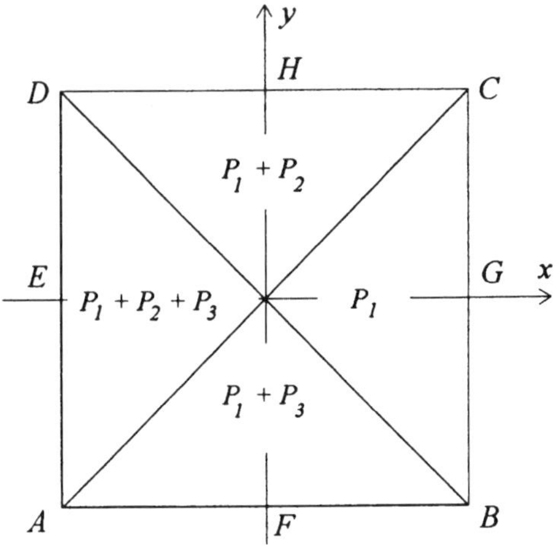


**Figure 2.3**

the only nonzero degree of freedom being the value of the derivative with respect to $y$ at a vertex or at the midpoint of a horizontal side.

It remains to find the basis function on the square $[-1,1]\times[-1,1]$. Due to the symmetry of the conditions which determine the basis functions, the following relations are obviously valid (see Figure 2.3):

$$\psi_B(x,y)=\psi_A(-x,y),\quad \psi_C(x,y)=\psi_A(-x,-y),$$

$$\psi_D(x,y)=\psi_A(x,-y),$$

$$\psi_{B,1}(x,y)=-\psi_{A,1}(-x,y),\quad \psi_{C,1}(x,y)=-\psi_{A,1}(-x,-y),$$

$$\psi_{D,1}(x,y)=\psi_{A,1}(x,-y),$$

$$\psi_{B,2}(x,y)=\psi_{A,2}(-x,y),\quad \psi_{C,2}(x,y)=-\psi_{A,2}(-x,-y),$$

$$\psi_{D,2}(x,y)=-\psi_{A,2}(x,-y),$$

$$\chi_G(x,y)=-\chi_E(-x,y),\quad \chi_H(x,y)=-\chi_F(x,-y),$$

$$\psi_{A,2}(x,y)=\psi_{A,1}(y,x),\quad \chi_F(x,y)=\chi_E(y,x).$$

Thus, all of the basis functions on the square are expressed through the three functions $\psi_A$, $\psi_{A,1}$, $\chi_E$. Straightforward calculation of these functions is simplified by the symmetry relations, for example, $\psi_A(x,y)=\psi_A(y,x)$. We present the final results of the calculation in the form of the expressions for the polynomials $P_1$, $P_2$, $P_3$. Figure 2.3 shows which sums of the polynomials determine the basis function in each of the four triangles constituting the square. The polynomials $P_1$, $P_2$, $P_3$ are as follows:

$$P_1=\frac{1}{8}x^3-\frac{3}{8}x^2y+\frac{3}{4}xy-\frac{3}{8}x-\frac{3}{8}y+\frac{1}{4},\quad P_2=\frac{1}{8}(y-x)^3,\quad P_3=\frac{1}{8}(x+y)^3$$

for the basis function $\psi_A$,

$$P_1 = -\frac{1}{12}x^3 + \frac{1}{4}x^2 + xy - \frac{5}{4}x + \frac{1}{12}, \quad P_2 = \frac{1}{4}(x-y)^2\left(\frac{1}{3}x + \frac{2}{3}y - 1\right),$$
$$P_3 = \frac{1}{4}(x+y)^2\left(\frac{1}{3}x - \frac{2}{3}y - 1\right)$$

for the basis function $\chi_E$, and

$$P_1 = \frac{5}{48}x^3 - \frac{1}{8}x^2y - \frac{9}{16}x^2 + \frac{1}{4}xy - \frac{11}{16}x - \frac{1}{8}y - \frac{17}{48},$$
$$P_2 = \frac{1}{4}(x-y)^2\left(-\frac{5}{12}x - \frac{1}{3}y + \frac{3}{4}\right), \quad P_3 = \frac{1}{4}(x+y)^2\left(\frac{1}{12}x + \frac{1}{3}y + \frac{1}{4}\right)$$

for the basis function $\psi_{A,1}$.

Return now to the basis functions on the quadrangle $Q$. Noteworthy is one of their properties which allows extending them to all of the domain $\Omega$ and arises from the following simple proposition.

PROPOSITION 2. Suppose polynomial $p$ of power 3 in the variables $x, y$ satisfy the conditions

$$p(\mathbf{a}) = 0, \quad \frac{\partial p}{\partial x}(\mathbf{a}) = 0, \quad p(\mathbf{b}) = 0, \quad \frac{\partial p}{\partial x}(\mathbf{b}) = 0,$$
$$\frac{\partial p}{\partial y}(\mathbf{a}) = 0, \quad \frac{\partial p}{\partial n}\left(\frac{\mathbf{a}+\mathbf{b}}{2}\right) = 0, \quad \frac{\partial p}{\partial y}(\mathbf{b}) = 0,$$

where $\mathbf{a} \neq \mathbf{b}$ and $\partial/\partial n$ is the derivative in the direction normal to the segment $[\mathbf{a}, \mathbf{b}]$. Then the polynomial $p$ vanishes on $[\mathbf{a}, \mathbf{b}]$ together with its first derivatives.

PROOF. Without loss of generality, $\mathbf{a} = (0,0)$, $\mathbf{b} = (2c, 0)$ can be considered as the ends of the segment. Then the assumptions of the proposition can be written as

$$p(0,0) = 0, \quad \frac{\partial p}{\partial x}(0,0) = 0, \quad p(0,2c) = 0, \quad \frac{\partial p}{\partial x}(0,2c) = 0,$$
$$\frac{\partial p}{\partial y}(0,0) = 0, \quad \frac{\partial p}{\partial y}(0,c) = 0, \quad \frac{\partial p}{\partial y}(0,2c) = 0.$$

The first four relations imply that there are no terms containing $y$ in the power 0 in $p$. Analogously, the next three relations imply that neither are there terms with $y$ in the power 1. Thus, $p(x,y) = y^2(\alpha x + \beta y + \gamma)$ which proves the proposition.

COROLLARY. Each of the sixteen basis functions of the above quadrangle finite element vanishes together with its first derivatives on those sides of the quadrangle that do not contain the node the function is associated with.

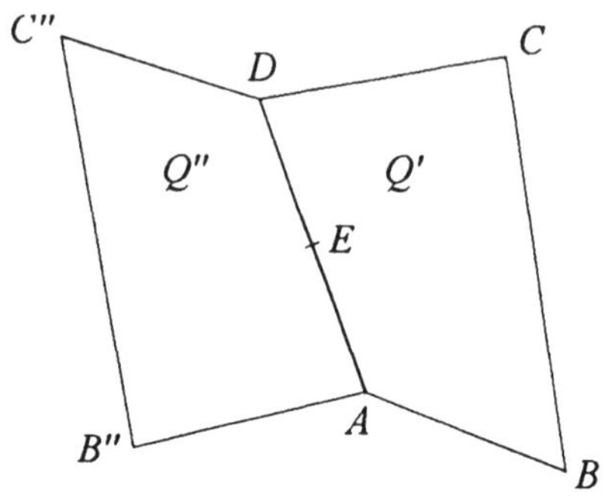


**Figure 2.4**

Indeed, let $p$ be the basis function associated with the node $A$ of the quadrangle. It is sufficient to observe that $p$ satisfies the conditions of Proposition 2, where the segment $[\mathbf{a}, \mathbf{b}]$ is the side of the quadrangle which does not contain $A$.

*Basis functions.* At this point, the basis functions associated with a node of triangulation are defined only on the quadrangles which contain this node. We now extend these functions to all of $\Omega$. Consider, for example, the definition of the function $\psi_A$ associated with the node $A$, which is, generally, a vertex of several quadrangles. Let $AD$ be a common side of the quadrangles $Q'$ and $Q''$, see Figure 2.4. The basis functions $\psi'_A$ and $\psi''_A$ are defined on $Q'$ and $Q''$, respectively, the only nonzero degrees of freedom for these functions being their values at $A$. Each of the functions is a polynomial of power 3 near the segment $AD$. By the definitions of $\psi'_A$ and $\psi''_A$, their difference satisfies the assumptions of Proposition 2 (at the segment $AD$). Therefore, the function

$$\psi_A(x,y) = \begin{cases} \psi'_A(x,y) & \text{if} \quad (x,y) \in Q', \\ \psi''_A(x,y) & \text{if} \quad (x,y) \in Q'' \end{cases}$$

is continuously differentiable on $Q' \cup Q''$. If there are quadrangles adjacent to $Q'$ or $Q''$ at the sides $AB$ or $AB''$, the basis functions defined on these quadrangles are continuously differentiable extensions of the corresponding function defined on $Q' \cup Q''$. Thus, the basis function $\psi_A$ is, actually, defined on the union $Q_A$ of all the quadrangles which contain the vertex $A$. We observe that this function vanishes together with its first derivatives on those sides of the quadrangles that do not contain $A$ (Proposition 2). To extend $\psi_A$ to all of $\Omega$, we set it equal to 0 everywhere in the exterior of $Q_A$. The the extension is continuously differentiable on $\Omega$; on each of the quadrangles containing $A$ the extension equals the basis function originally defined on this quadrangle. We keep denoting the extension by the symbol $\psi_A$ and take $\psi_A$ as the basis function on $\Omega$. This function is associated with the node $A$ and corresponds to the only nonzero degree of freedom: the value of $\psi_A$ at $A$. The basis functions $\psi_{A,1}$, $\psi_{A,2}$, $\chi_E$ and others are analogously defined on $\Omega$.

*Approximation for Sobolev space.* As mentioned before, evaluating the safety factor requires approximation of the stream function together with its first and second derivatives, more precisely, in the Sobolev space $W_2^2(\Omega)$ (see Remark VI.B.1). Thus, we need an internal approximation for $W_2^2(\Omega)$. To construct this approximation, we will use the above introduced triangulation of $\Omega$ and consider all linear combinations

$$\alpha_i\psi_i + \alpha_{i1}\psi_{i1} + \alpha_{i2}\psi_{i2} + \beta_j\chi_j$$

of the basis functions ($\alpha_i$, $\alpha_{i1}$, $\alpha_{i2}$, $\beta_j$ are arbitrary numbers). These combinations form a finite-dimensional space which we denote by the symbol $W_h$. Here, $h$ is the maximum of the diameters of the quadrangles that form the triangulation. We denote the triangulation by the symbol $\mathcal{T}_h$.

Let us define an *interpolant* $w_h \in W_h$ for a smooth function $\widetilde{w}$ on $\Omega$:

$$w_h = \widetilde{w}(\mathbf{a}_i)\psi_i + \frac{\partial\widetilde{w}}{\partial x}(\mathbf{a}_i)\psi_{i1} + \frac{\partial\widetilde{w}}{\partial y}(\mathbf{a}_i)\psi_{i2} + \frac{\partial\widetilde{w}}{\partial n}(\mathbf{b}_j)\chi_j.$$

Here, the subscript $i$ is a number of a vertex, $j$ is a number of a midpoint, and summation, as usual, runs over the repeated indices. It is clear that, at the vertices $\mathbf{a}_i$, $w_h$ and its first derivatives take the same values as $\widetilde{w}$ and its corresponding derivatives. Analogously, at each of the midpoints $\mathbf{b}_j$, the derivative of $w_h$ in the direction normal to the side takes the same value as the corresponding derivative of $\widetilde{w}$.

Note that all of the functions $w_h$ are continuously differentiable; they are polynomials of power 3 on each of the triangles in $\mathcal{T}_h$ (recall that every quadrangle in $\mathcal{T}_h$ is split into four triangles). Therefore, $W_h$ is in $W_2^2(\Omega)$. Moreover, the collection of the subspaces $W_h$ forms an internal approximation for $W_2^2(\Omega)$ at least if the collection of the triangulations $\mathcal{T}_h$ satisfies the following additional condition: the ratio of the lengths of segments into which the quadrangle diagonal is divided by the diagonals intersection 1) is separated from 0 by a constant $c_1 > 0$, and 2) is bounded from above by a constant $c_2$ ($c_1$ and $c_2$ are the same for all quadrangles of all triangulations $\mathcal{T}_h$ in the collection under consideration when $h$ is sufficiently small). We refer to a collection of triangulations $\mathcal{T}_h$ which satisfy this condition as *regular*. The following well known approximation theorem is valid.

THEOREM 2. Suppose $\Omega$ is a polygon, $\mathcal{T}_h$ and $W_h$ are its triangulation and the above defined space of piecewise polynomial functions of power 3. If the collection of triangulations $\mathcal{T}_h$ is regular, the collection of the spaces $W_h$ forms an internal approximation for the space $W_2^2(\Omega)$: for every smooth function $\widetilde{w}$ on $\Omega$ its interpolants

$$w_h = \widetilde{w}(\mathbf{a}_i)\psi_i + \frac{\partial\widetilde{w}}{\partial x}(\mathbf{a}_i)\psi_{i1} + \frac{\partial\widetilde{w}}{\partial y}(\mathbf{a}_i)\psi_{i2} + \frac{\partial\widetilde{w}}{\partial n}(\mathbf{b}_j)\chi_j$$

converge to $\widetilde{w}$ in $W_2^2(\Omega)$ when $h \to 0$.

In the sequel, we consider stream functions in $W_2^2(\Omega)$ and approximate them with the above defined interpolants, whose convergence is established by Theorem 2.

*Approximation for solenoidal velocity fields space.* Using the above internal approximation for $W_2^2(\Omega)$, it is easy to construct an internal approximation for the space $V$ of solenoidal velocity fields. Indeed, let $\widetilde{\mathbf{v}}$ be a smooth field in $V$ (the set of such fields is dense in $V$) and $\widetilde{\phi}$ is the corresponding stream function. Consider the sequence of the interpolants $\phi_h$, which converges to $\widetilde{\phi}$ in $W_2^2(\Omega)$ when $h \to 0$. Note that the fields $\mathbf{v}_h$ with the components

$$v_{1h} = -\frac{\partial \phi_h}{\partial x_2}, \quad v_{2h} = -\frac{\partial \phi_h}{\partial x_1}$$

are obviously solenoidal and their sequence converges to $\widetilde{\mathbf{v}}$ in $\mathbf{W}_2^1(\Omega)$. It remains to verify that the fields $\mathbf{v}_h$ have zero traces on $S_v$ and, consequently, are in $V$ due to Theorem VI.4.3 or VI.4.4.

Indeed, let $\Omega$ be a polygon (as assumed before) and, for a certain partition of $\Omega$ into convex quadrangles, the fixed part $S_v$ of its boundary consists of a finite number of the quadrangles sides. Then we choose a regular collection of triangulations $\mathcal{T}_h$ such that $\mathcal{T}_h$ possesses the previous property for sufficiently small $h$. We restrict ourselves to such triangulations and show that in this case $\mathbf{v}|_{S_v} = 0$. Note that this condition is equivalent to the equality $\operatorname{grad} \phi|_{S_v} = 0$ for the corresponding stream function. Thus, it is to be verified that, if a stream function $\widetilde{\phi}$ satisfies the condition $\operatorname{grad} \widetilde{\phi}|_{S_v} = 0$, then the interpolant $\phi_h$ possesses the analogous property: $\operatorname{grad} \phi_h|_{S_v} = 0$. To show this, we divide the summands in the expression for the interpolant

$$\phi_h = \widetilde{\phi}(\mathbf{a}_i)\psi_i + \frac{\partial \widetilde{\phi}}{\partial x}(\mathbf{a}_i)\psi_{i1} + \frac{\partial \widetilde{\phi}}{\partial y}(\mathbf{a}_i)\psi_{i2} + \frac{\partial \widetilde{w}}{\partial n}(\mathbf{b}_j)\chi_j \tag{2.2}$$

into three groups: (i) the terms which include the basis functions $\psi_{i1}$, $\psi_{i2}$ and $\chi_j$ associated with the nodes which lie in $S_v$; (ii) the terms which include the basis functions $\psi_i$ associated with the nodes which lie in $S_v$; and (iii) the terms which include all of the remaining basis functions, that is, those associated with the nodes which are not in $S_v$. Due to the condition $\operatorname{grad} \widetilde{\phi}|_{S_v} = 0$, the coefficients in (2.2) which correspond to the basis functions of the first group equal zero. Consequently, there are no such terms in (2.2). Also note that the gradients of the functions in the third group vanish at $S_v$ (Corollary of Proposition 2). Taking into account the last two remarks, we obtain from (2.2) that

$$\operatorname{grad} \phi_h = \widetilde{\phi}(\mathbf{a}_i) \operatorname{grad} \psi_i \quad \text{on} \quad S_v, \tag{2.3}$$

where the sum is taken over the values of the index $i$ corresponding to the nodes in $S_v$. Consider (2.3) on any of the segments which form $S_v$; let $\mathbf{a}_1$ and $\mathbf{a}_2$ be the ends of the segment. Only two terms enter the sum on the right-hand side:

$$\widetilde{\phi}(\mathbf{a}_1) \operatorname{grad} \psi_1 + \widetilde{\phi}(\mathbf{a}_2) \operatorname{grad} \psi_2$$

(gradients of the rest of the functions $\psi_i$ vanish at $[\mathbf{a}_1, \mathbf{a}_2]$ due to the definition and Corollary of Proposition 2). Then the condition $\operatorname{grad} \widetilde{\phi}|_{S_v} = 0$ implies the equality $\widetilde{\phi}(\mathbf{a}_1) = \widetilde{\phi}(\mathbf{a}_2) = c$ and the polynomial $p = \widetilde{\phi}(\mathbf{a}_1)\psi_1 + \widetilde{\phi}(\mathbf{a}_2)\psi_2 - c$

satisfies the assumptions of Proposition 2 on the segment $[\mathbf{a}_1, \mathbf{a}_2]$. By this proposition, $\operatorname{grad} p = 0$ on $[\mathbf{a}_1, \mathbf{a}_2]$, which implies that $\operatorname{grad} \phi_h = 0$ on $[\mathbf{a}_1, \mathbf{a}_2]$, and this is exactly what had to be verified.

Thus, under the foregoing assumptions about $\Omega$ and $S_v$, the collection of finite-dimensional spaces

$$V_h = \{\mathbf{v} = v_1\mathbf{e}_1 + v_2\mathbf{e}_2 : v_1 = -\frac{\partial \phi}{\partial x_2},\ v_2 = \frac{\partial \phi}{\partial x_1},\ \phi \in W_h,\ \operatorname{grad} \phi|_{S_v} = 0\}$$

forms an internal approximation for the space $V$ of solenoidal velocity fields.

REMARK 3. To make use of an internal approximation for solving concrete problems, it is necessary to choose a certain basis in each of the finite-dimensional spaces that form the approximation. Let us specify a suitable basis in the finite-dimensional space $V_h$ involved in the internal approximation for the space of solenoidal velocity fields.

Every field $\mathbf{v}$ in $V_h$ is of the form

$$\mathbf{v} = -\frac{\partial \phi}{\partial x_2}\mathbf{e}_1 + \frac{\partial \phi}{\partial x_1}\mathbf{e}_2, \tag{2.4}$$

where $\phi = \alpha_i \psi_i + \alpha_{i1}\psi_{i1} + \alpha_{i2}\psi_{i2} + \beta_j \chi_j$. The coefficients $\alpha_i$, $\alpha_{i1}$, $\alpha_{i2}$, $\beta_j$ are not arbitrary here: the condition $\mathbf{v}|_{S_v} = 0$ (or, equivalently, $\operatorname{grad} \phi|_{S_v} = 0$) restricts their values. Let us list these relations using the above subdivision of the basis functions into three groups. This will result in constructing a basis in $V_h$. The coefficients of the basis functions in group (i) equal the values of the corresponding derivatives of $\phi$. By virtue of the condition $\operatorname{grad} \phi|_{S_v} = 0$ this immediately implies that all of these coefficients are zero. Coefficients of the basis functions in group (iii) are arbitrary, as the condition $\operatorname{grad} \phi|_{S_v} = 0$ imposes no restrictions on their values since the derivatives of these functions vanish on $S_v$. It remains to consider the coefficients of the basis functions in group (ii). Let $\mathbf{a}_p$ and $\mathbf{a}_q$ be two vertices of the triangulation quadrangles which can be joined by a polygonal line lying in $S_v$ and consisting of the triangulation quadrangles' sides. Briefly, $\mathbf{a}_p$ and $\mathbf{a}_q$ belong to a connected component of $S_v$. Then the condition $\operatorname{grad} \phi|_{S_v} = 0$ implies the equality $\phi(\mathbf{a}_p) = \phi(\mathbf{a}_q)$. Note that $\alpha_p = \phi(\mathbf{a}_p)$ and $\alpha_q = \phi(\mathbf{a}_q)$; therefore, $\alpha_p = \alpha_q$. Thus the coefficients of the basis functions $\psi_i$ associated with the nodes which belong to a connected component of $S_v$ are equal. In other words, only the sum of these functions multiplied by an arbitrary number enters the decomposition of $\phi$. We will denote this sum by the symbol $\Psi^{(\kappa)}$ where $\kappa$ is the number of the connected component. It is now clear that the basis in $V_h$ consists of vector fields (2.4) with $\phi$ ranging over group (ii) and all of the functions $\Psi^{(\kappa)}$.

**2.4. Discretized problem of limit analysis.** If a collection of spaces $V_h$ forms an internal approximation for the velocity fields space $V$, the safety factor can be evaluated as the limit of the infimums

$$\alpha_{l(h)} = \inf\left\{\int_\Omega d_x(\operatorname{Def}\mathbf{v})\,dV : \mathbf{v} \in V_h,\ \int_\Omega \operatorname{Def}\mathbf{v} \cdot \mathbf{s}_l\,dV = 1\right\} \tag{2.5}$$

(Theorem 1.1). Let us consider in more detail the extremum problem which determines $\alpha_{l(h)}$, assuming that the spaces $V_h$ are finite-dimensional. Let $\boldsymbol{\varphi}_{(1)}, \cdots, \boldsymbol{\varphi}_{(M)}$ be vector fields which form a basis in $V_h$ (see the examples in Remarks 1 and 3), so that every field $\mathbf{v} \in V_h$ can be written as

$$\mathbf{v} = v_1 \boldsymbol{\varphi}_{(1)} + \cdots + v_M \boldsymbol{\varphi}_{(M)}.$$

We identify this field with the member $(v_1, \ldots, v_M)$ in $\mathbf{R}^M$, which in particular determines the strain rate components:

$$(\operatorname{Def} \mathbf{v})_{ij} = v_1 \frac{1}{2}\left(\frac{\partial \varphi_{(1)i}}{\partial x_j} + \frac{\partial \varphi_{(1)j}}{\partial x_i}\right) + \cdots + v_M \frac{1}{2}\left(\frac{\partial \varphi_{(M)i}}{\partial x_j} + \frac{\partial \varphi_{(M)j}}{\partial x_i}\right),$$

($(\varphi_{(\mu)})_i$ is the $i$-th component of the basis field $\boldsymbol{\varphi}_{(\mu)}$) as well as the value of the function

$$F(v_1, \ldots, v_M) = \int_\Omega d_x(\operatorname{Def} \mathbf{v})\, dV, \tag{2.6}$$

which is minimized in problem (2.5), and the value of the linear function

$$\int_\Omega \operatorname{Def} \mathbf{v} \cdot \mathbf{s}_l \, dV = v_1 \int_\Omega s_{l\,ij} \frac{\partial \varphi_{(1)i}}{\partial x_j} dV + \cdots + v_M \int_\Omega s_{l\,ij} \frac{\partial \varphi_{(M)i}}{\partial x_j} dV$$

involved in the formulation of the restriction in (2.5). The values of the integrals on the right-hand side are known; we denote them by $a_\mu$, $\mu = 1, \ldots, M$, and write the restriction in (2.5) as $a_1 v_1 + \cdots + a_M v_M = 1$. Thus, (2.5) is reduced to the minimization problem for the function $F$ over the linear manifold in $\mathbf{R}^M$:

$$\alpha_{l(h)} = \inf\{F(v_1, \ldots, v_M) : (v_1, \ldots, v_M) \in \mathbf{R}^M,\ a_1 v_1 + \cdots + a_M v_M = 1\}.$$

To write this formulation in a more convenient form, we choose $(v_1^0, \ldots, v_M^0)$ in $\mathbf{R}^M$ such that $a_1 v_1^0 + \cdots + a_M v_M^0 = 1$ and introduce a new variable:

$$\mathbf{u} = (u_1, \ldots, u_M) = (v_1 - v^0, \ldots, v_M - v_M^0).$$

Then the strain rate can be written as $\operatorname{Def} \mathbf{v} = \mathbf{e_u} + \boldsymbol{\varepsilon}$, where $e_{\mathbf{u}}$ linearly depends on $\mathbf{u}$ and has the following components:

$$(\mathbf{e_u})_{ij} = u_1 \frac{1}{2}\left(\frac{\partial \varphi_{(1)i}}{\partial x_j} + \frac{\partial \varphi_{(1)j}}{\partial x_i}\right) + \cdots + u_M \frac{1}{2}\left(\frac{\partial \varphi_{(M)i}}{\partial x_j} + \frac{\partial \varphi_{(M)j}}{\partial x_i}\right),$$

while $\boldsymbol{\varepsilon}$ does not depend on $\mathbf{u}$ and has the components:

$$\varepsilon_{ij} = v_1^0 \frac{1}{2}\left(\frac{\partial \varphi_{(1)i}}{\partial x_j} + \frac{\partial \varphi_{(1)j}}{\partial x_i}\right) + \cdots + v_M^0 \frac{1}{2}\left(\frac{\partial \varphi_{(M)i}}{\partial x_j} + \frac{\partial \varphi_{(M)j}}{\partial x_i}\right).$$

We also introduce the function

$$f(\mathbf{u}) = F(u_1 + v_1^0, \dots, u_M + v_M^0) = \int_\Omega d_x(\mathbf{e_u} + \boldsymbol{\varepsilon})\, dV$$

and write the discretized problem of limit analysis as

$$\alpha_{l(h)} = \inf \left\{ f(\mathbf{u}) : \mathbf{u} \in \mathbf{R}^M,\ \mathbf{au} = 0 \right\},$$

where $\mathbf{a}$ stands for $(a_1, \dots, a_M)$.

Note that the function $f$ is convex. Indeed, the dissipation $d_x$ is convex, and the intergrand in (2.6) is the superposition of $d_x$ and $\mathbf{e_u}$, the latter being linear in $\mathbf{u}$; therefore, the intergrand is convex in $\mathbf{u}$, which immediately implies the convexity of $f$. On the other hand, the function $f$ is not differentiable everywhere, which is a consequence of the corresponding property of the dissipation: $d_x$ is not differentiable at least at the origin (as a first degree homogeneous function). Even if the origin is the only point at which $d_x$ is not differentiable, the function $f$ may be nondifferentiable on a wide set. The following example illustrates this statement.

EXAMPLE 3. Consider the function $f$ in the simplest case of the plane strain rate state of a body with the yield surface $|\boldsymbol{\sigma}| = k$, $k = \text{const} > 0$. The corresponding dissipation is $d(\mathbf{e}) = k\,|\mathbf{e}|$. Under the plane strain conditions only two components $v_1$, $v_2$ of a velocity field can be nonzero; they do not depend on $x_3$ and are defined on the domain $\Omega \subset \mathbf{R}^2$. We restrict ourselves to the case of $\Omega$ being a polygon and discretize the kinematic extremum problem of limit analysis by the finite element method as in Subsection 2.2. Namely, we split $\Omega$ into triangles, consider their vertices as the nodes, and associate a (scalar) basis function with each node. The basis function $\varphi_A$ associated with the node $A$ is continuous, linear on every finite element, and takes the value 1 at $A$ and the value 0 at all other nodes. Consider a certain triangulation $\mathcal{T}_h$ and the corresponding finite-dimensional space $V_h$ of velocity fields on $\Omega$ with the basis described in Remark 1. Then the above introduced function $f$ involved in the discretized kinematic extremum problem of limit analysis can be written as:

$$f(\mathbf{u}) = k \sum_\alpha \int_{T_\alpha} |\mathbf{e_u} + \boldsymbol{\varepsilon}|\, dS.$$

Here the sum is taken over all of the triangles $T_\alpha$ of the triangulation. Note that the basis functions are linear on $T_\alpha$, and, therefore, the fields $\mathbf{e_u}$ and $\boldsymbol{\varepsilon}$ are constant on $T_\alpha$, which implies that

$$f(\mathbf{u}) = k \sum_\alpha (|\mathbf{e_u} + \boldsymbol{\varepsilon}|) S_\alpha$$

($S_\alpha$ stands for the area of the triangle $T_\alpha$). Consider one of the summands on the right-hand side; let it be the term corresponding to the triangular finite element with the vertices $A$, $B$, $C$.

The basis functions $\varphi_A$, $\varphi_B$, $\varphi_C$ are associated with these vertices, and on the triangle $ABC$ every velocity field in $V_h$ is of the form

$$(u_1\varphi_A + u_3\varphi_B + u_5\varphi_C)\mathbf{e}_1 + (u_2\varphi_A + u_4\varphi_B + u_6\varphi_C)\mathbf{e}_2.$$

The corresponding strain rate $\mathbf{e}_u$ is easily evaluated, which results in the following formula on $ABC$ for the summand under consideration:

$$kS_{ABC}\left\{\left(u_1\frac{\partial\varphi_A}{\partial x_1} + u_3\frac{\partial\varphi_B}{\partial x_1} + u_5\frac{\partial\varphi_C}{\partial x_1} + \varepsilon_{11}\right)^2 + \right.$$
$$+\left(u_2\frac{\partial\varphi_A}{\partial x_2} + u_4\frac{\partial\varphi_B}{\partial x_2} + u_6\frac{\partial\varphi_C}{\partial x_2} + \varepsilon_{22}\right)^2 +$$
$$\left.+\frac{1}{2}\left(u_1\frac{\partial\varphi_A}{\partial x_2}+u_3\frac{\partial\varphi_B}{\partial x_2}+u_5\frac{\partial\varphi_C}{\partial x_2}+u_2\frac{\partial\varphi_A}{\partial x_1}+u_4\frac{\partial\varphi_B}{\partial x_1}+u_6\frac{\partial\varphi_C}{\partial x_1}+2\varepsilon_{12}\right)^2\right\}^{\frac{1}{2}}.$$

Here, the derivatives of the basis functions are constant on $ABC$ and $u_1, \ldots, u_6$ are independent variables. Observe that function (2.7) is nondifferentiable with respect to $u_1, \ldots, u_6$ at any point where the three expressions in parentheses vanish. It turns out that the nondifferentiability of the dissipation at the origin implies the nondifferentiability of $f$ on some manifold.

Thus, the discretized kinematic extremum problem of limit analysis consists of minimizing a convex nondifferentiable function on a finite-dimensional space.

REMARK 4. Although we discretized the limit analysis problem in a polyhedral domain, the finite element method also allows discretizing in the case of the domain $\Omega$ with a curved boundary. If the shape of $\Omega$ is sufficiently simple, finite elements can be used with suitable curved boundaries which coincide with parts of $\partial\Omega$. Normally, to construct such elements, a simple mapping is considered: the elements are the transforms of tetrahedral or other polyhedral finite elements.

Another disretization procedure can be based on the fact that the set $\mathcal{U}$ or $\mathcal{U}_d$ of smooth functions which vanish *near* $S_v$ is dense in $V$ for bodies with bounded or cylindrical yield surfaces, respectively (see Remarks VI.3.1 and VI.4.2). Let us describe the idea of the discretization in case the boundary $\partial\Omega$ is fixed: $\partial\Omega = S_v$. Let $\Omega_h \subset \Omega$ be an expanding sequence of polyhedral domains, the distance between $\partial\Omega_h$ and $\partial\Omega$ tending to 0 when $h \to 0$. Consider a triangulation of $\Omega_h$, the corresponding basis functions and the space of their linear combinations which have the zero trace on $\partial\Omega$. Let $\mathbf{v}$ be a field in this space; we extend it to all of $\Omega$ as zero in $\Omega \setminus \Omega_h$. It can be easily seen that the collection of the spaces $V_h$ forms an internal approximation for $V$. Indeed, consider the set $\mathcal{U}$ or $\mathcal{U}_d$ in the case of bodies with bounded or cylindrical yield surfaces, respectively. This set is dense in $V$, and any field $\tilde{\mathbf{v}}$ in this set vanishes near $S_v$. When $h$ is sufficiently small, $\tilde{\mathbf{v}}$ vanishes everywhere in the exterior of $\Omega_h$. Therefore, if $\mathbf{v}_h$ is an interpolant for $\tilde{\mathbf{v}}$ in the domain $\Omega_h$, then, extending $\mathbf{v}_h$ as zero to all of $\Omega$, we obtain an interpolant for $\tilde{\mathbf{v}}$ in $\Omega$.

An internal approximation for $V$ can be similarly constructed in the case of a polyhedral domain $\Omega$ when the contour of $S_v$ is curvilinear and in the more general case when $S_v$ is a curved surface while $S_q$ is piecewise plane.

## 3. Minimization: separating plane method

In the previous sections evaluating the safety factor was reduced to solving a sequence of discrete extremum problems. Each of these problems consists of minimizing a convex nondifferentiable function over a finite-dimensional space. This section presents an iterative method for the minimization; its iteration step consists of solving a standard linear programming problem, computing one value of the minimized function, and finding one of its subgradients. An algorithm for finding the subgradient is also described in this section. The separating plane method presented here is applicable to solving of a wide class of extremum problems.

**3.1. Subgradients.** To formulate the method for evaluating the infimum

$$\inf\left\{f(\mathbf{x}) : \mathbf{x} \in \mathbf{R}^N\right\}$$

of a convex function $f$, we first consider a useful generalization of the concept of the gradient to the case of nondifferentiable convex functions. Let $f$ be such a function on $\mathbf{R}^N$ and its value $f(\mathbf{x}_0)$ at $\mathbf{x}_0$ be finite. Consider the graph of this function in the space $\mathbf{R} \times \mathbf{R}^N$, that is, the set consisting of the points $(f(\mathbf{x}), \mathbf{x})$, $\mathbf{x} \in \mathbf{R}^N$.

Consider now a plane in $\mathbf{R} \times \mathbf{R}^N$ passing through the point $(f(\mathbf{x}_0), \mathbf{x}_0)$ of the graph. The plane is described by the equation

$$y = \mathbf{g}(\mathbf{x} - \mathbf{x}_0) + f(\mathbf{x}_0), \quad \mathbf{x} \in \mathbf{R}^N,\ y \in \mathbf{R},$$

where $\mathbf{g}$ is a certain member in $\mathbf{R}^N$. If the whole plane lies below the graph of $f$, that is,

$$f(\mathbf{x}) \geq \mathbf{g}(\mathbf{x} - \mathbf{x}_0) + f(\mathbf{x}_0) \quad \text{for every} \quad \mathbf{x} \in \mathbf{R}^N,$$

we refer to $\mathbf{g}$ as a *subgradient* of $f$ at the point $\mathbf{x}_0$. In case $f$ is differentiable at $\mathbf{x}_0$, $f$ possesses only one subgradient at the point, namely, $\mathbf{g} = \operatorname{grad} f(\mathbf{x}_0)$. If $f$ is nondifferentiable at $x_0$, its subgradient at $\mathbf{x}_0$ is non-unique; see Figure 3.1.

EXAMPLE 1. Consider the function $f(\mathbf{x}) = |\mathbf{x}|$ on $\mathbf{R}^N$. Its subgradients $\mathbf{g}$ at the point $\mathbf{x}_0$ are determined by the condition

$$|\mathbf{x}| \geq \mathbf{g}\mathbf{x} \quad \text{for every} \quad \mathbf{x} \in \mathbf{R}^N,$$

that is, they form the ball in $\mathbf{R}^N$ of the radius 1 centered at the origin.

Consider now the same structure as in the definition of the subgradient starting this time with the point $(f(\mathbf{x}_0) - \varepsilon, \mathbf{x}_0)$, $\varepsilon > 0$, located below the point $(f(\mathbf{x}_0), \mathbf{x}_0)$ considered earlier. A plane in $\mathbf{R} \times \mathbf{R}^N$ passing through $(f(\mathbf{x}_0) - \varepsilon, \mathbf{x}_0)$ is described by the equation

$$y = \mathbf{g}(\mathbf{x} - \mathbf{x}_0) + f(\mathbf{x}_0) - \varepsilon, \quad \mathbf{x} \in \mathbf{R}^N,\ y \in \mathbf{R},$$

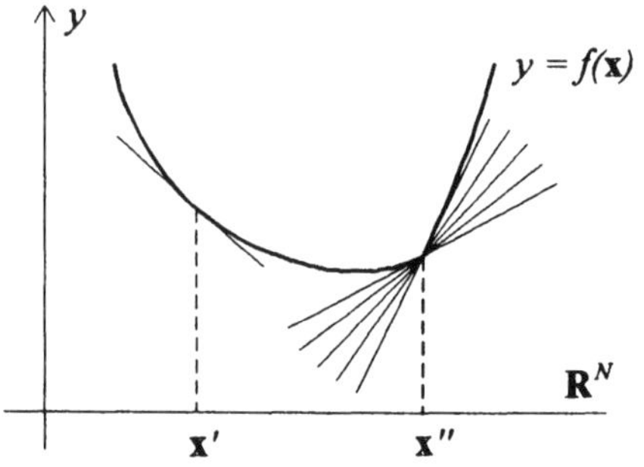


**Figure 3.1**

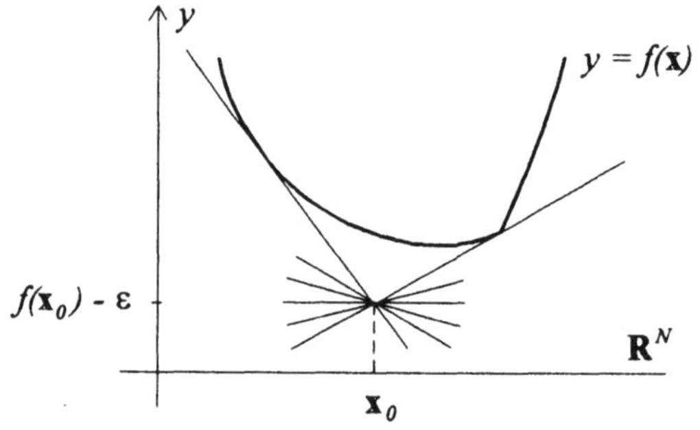


**Figure 3.2**

with a fixed $\mathbf{g} \in \mathbf{R}^N$. If the whole plane is located below the graph of $f$, that is,

$$f(\mathbf{x}) \geq \mathbf{g}(\mathbf{x} - \mathbf{x}_0) + f(\mathbf{x}_0) - \varepsilon \quad \text{for every} \quad \mathbf{x} \in \mathbf{R}^N, \tag{3.1}$$

we refer to $\mathbf{g}$ as an *$\varepsilon$-subgradient* of $f$ at the point $\mathbf{x}_0$, see Figure 3.2.

We denote the set of all $\varepsilon$-subgradients of $f$ at $\mathbf{x}_0$ by the symbol $\partial_\varepsilon f(\mathbf{x}_0)$ and refer to it as an *$\varepsilon$-subdifferential* of $f$ at the point $\mathbf{x}_0$,

$$\partial_\varepsilon f(\mathbf{x}_0) = \{\mathbf{g} \in \mathbf{R}^N : f(\mathbf{x}) \geq \mathbf{g}(\mathbf{x} - \mathbf{x}_0) + f(\mathbf{x}_0) - \varepsilon \text{ for every } \mathbf{x} \in \mathbf{R}^N\}.$$

If $\varepsilon \geq 0$ is a parameter while $\mathbf{x}_0$ is fixed, this expression induces a set-valued mapping which maps the number $\varepsilon$ on the set $\partial_\varepsilon f(\mathbf{x}_0)$ in $\mathbf{R}^N$.

**3.2. Infimum and $\varepsilon$-subdifferentials.** The infimum

$$m = \inf \{f(\mathbf{x}) : \mathbf{x} \in \mathbf{R}^N\}$$

is related to certain properties of the mapping $\varepsilon \longrightarrow \partial_\varepsilon f(0)$, where $\varepsilon$ is a non-negative number and the $\varepsilon$-subdifferential $\partial_\varepsilon f(0)$ is considered at the origin in $\mathbf{R}^N$ (or at any other fixed point in $\mathbf{R}^N$). This relation allows formulating a method for minimizing a nondifferentiable convex function. Without loss of generality we consider in the sequel only functions $f$ which take the value 0 at the origin: $f(0) = 0$. Any function which takes a finite value at least at one

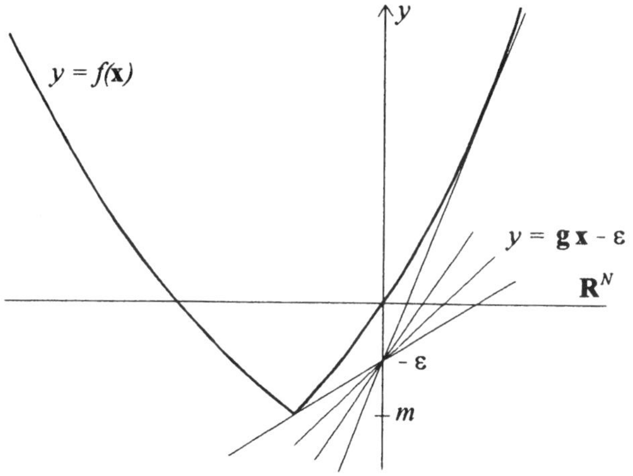


**Figure 3.3**

point can be reduced to the one satisfying this requirement: it is sufficient to add a constant to the original function and shift its argument. In the case of such a function, condition (3.1) determining the $\varepsilon$-subgradient can be written as

$$f(\mathbf{x}) \geq \mathbf{g}\mathbf{x} - \varepsilon \quad \text{for every} \quad \mathbf{x} \in \mathbf{R}^N. \tag{3.2}$$

Let us find the relation of (3.2) to the value $m$ of the infimum, restricting ourselves to the case of the finite infimum. Note that $m$ is non-positive: $m \leq f(0) = 0$. It is clear that at $-\varepsilon \leq m$ and $\mathbf{g} = 0$ condition (3.2) is satisfied, which means that 0 is $\varepsilon$-subgradient of $f$ at the origin when $-\varepsilon \leq m$. On the other hand, Figure 3.3 shows that, in case $-\varepsilon > m$ (recall that $\varepsilon$ is non-negative here), the "horizontal" plane in $\mathbf{R} \times \mathbf{R}^N$ passing through the point $(-\varepsilon, 0) \in \mathbf{R} \times \mathbf{R}^N$ intersects the graph of $f$. This means that $\mathbf{g} = 0$ is not an $\varepsilon$-subgradient of $f$ at 0. Thus,

$$0 \in \partial_\varepsilon f(0) \text{ if } -\varepsilon \leq m, \qquad 0 \notin \partial_\varepsilon f(0), \text{ if } m < -\varepsilon \leq 0.$$

More precisely, the following well known statement of convex analysis is valid.

THEOREM 1. Let $f$ be a convex function on $\mathbf{R}^N$ taking only finite values on a certain neighborhood of the origin and the zero value at the origin, $f(0) = 0$. Then

$$\inf \left\{ f(\mathbf{x}) : \mathbf{x} \in \mathbf{R}^N \right\} = -\inf \left\{ \varepsilon \geq 0 : 0 \in \partial_\varepsilon f(0) \right\}. \tag{3.3}$$

A minimization method based on this theorem is described in the next subsection.

**3.3. Separating plane method.** To formulate an iterative method for solving the nondifferentiable convex optimization problem, we interpret this problem geometrically, using Theorem 1.

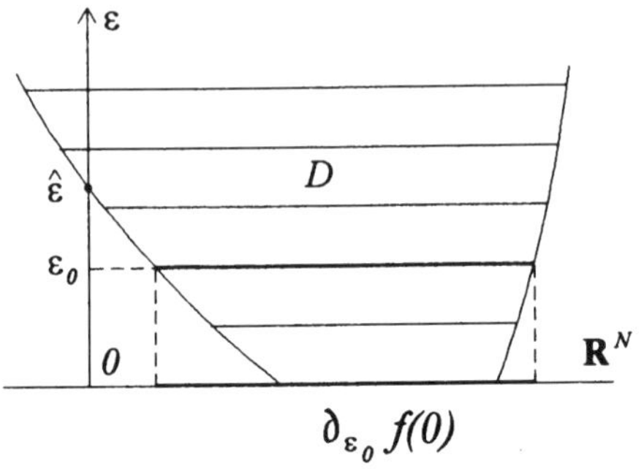


**Figure 3.4**

Consider the set

$$D = \{(\varepsilon, \mathbf{g}) \in \mathbf{R}_+ \times \mathbf{R}^N : \mathbf{g} \in \partial_\varepsilon f(0)\}$$

in $\mathbf{R}^{N+1}$ ($\mathbf{R}_+$ is the set of non-negative numbers), see Figure 3.4. Using this notation, we write the condition $0 \in \partial_\varepsilon f(0)$ as $(\varepsilon, 0) \in D$ and equality (3.3) as

$$\inf \{f(\mathbf{x}) : \mathbf{x} \in \mathbf{R}^N\} = -\inf \{\varepsilon \geq 0 : (\varepsilon, 0) \in D\}.$$

Thus, the unknown infimum of $f$ equals $-\widehat{\varepsilon}$, where $\widehat{\varepsilon}$ is the infimum on the right-hand side. In other words, $\widehat{\varepsilon}$ is the value of the parameter $\varepsilon$ at which the $\varepsilon$-axis intersects the boundary of the set $D$, see Figure 3.4. Note that $D$ is convex, which immediately follows from its definition.

The idea of the separating plane method consists of constructing a contracting sequence of the sets

$$D_k^+ \supset D, \quad D_k^+ \supset D_{k+1}^+, \quad k = 1, 2, \ldots.$$

For every $D_k^+$, one determines the number $\varepsilon_k$ possessing the property analogous to that of $\widehat{\varepsilon}$: $\varepsilon_k$ is the value of the parameter $\varepsilon$ at which the $\varepsilon$-axis intersects the boundary of $D_k^+$:

$$\varepsilon_k = \inf \{\varepsilon \geq 0 : (\varepsilon, 0) \in D_k^+\}.$$

It is clear that $\varepsilon_k$, $k = 1, 2, \ldots$, are the lower bounds for $\widehat{\varepsilon}$. We observe that they increase (more precisely, do not decrease) with $k$: $\varepsilon_{k+1} \geq \varepsilon_k$ since $D_{k+1}^+ \subset D_k^+$. The sequence $D_1^+, D_2^+, \ldots$ should be constructed so as to ensure the convergence of $\varepsilon_k$ to $\widehat{\varepsilon}$, which we are looking for. We refer to $D_k^+$ as to *external approximation* to $D$.

To find an appropriate sequence of shrinking external approximations, it is convenient to construct it together with an expanding sequence of *internal approximations*:

$$D_k^- \subset D, \quad D_k^- \subset D_{k+1}^-, \quad k = 1, 2, \ldots.$$

The internal approximations can be constructed in the form of convex polyhedrons $D_k^- = \text{conv}\{\mathbf{p}_1, \ldots, \mathbf{p}_k\}$ with the vertices $(\mathbf{p}_1, \ldots, \mathbf{p}_k)$ and the external approximations in the form of the intersection of several subspaces in $\mathbf{R}^{N+1}$:

$$D_k^+ = \{\mathbf{z} \in \mathbf{R}^{N+1} : \mathbf{b}_1\mathbf{z} + \beta_1 \geq 0, \ldots, \mathbf{b}_k\mathbf{z} + \beta_k \geq 0\}.$$

We now describe a procedure for constructing the internal and external approximations and the sequence of the numbers $\varepsilon_k$, which converges to the unknown $\widehat{\varepsilon}$.

*Choice of initial approximations.* We will use the subspace $D_1^+ = \mathbf{R}_+ \times \mathbf{R}^N$ as the initial external approximation $D_1^+$ and the set consisting of one point $\mathbf{p}_1 = (\eta_1, 0) \in \mathbf{R}_+ \times \mathbf{R}^N$ with $\eta_1 \geq \widehat{\varepsilon}$ as the initial internal approximation $D_1^-$. The inequality $\eta_1 \geq \widehat{\varepsilon}$ means that $-\eta_1$ is the lower bound for the unknown infimum:

$$-\eta_1 \leq \inf\left\{f(\mathbf{x}) : \mathbf{x} \in \mathbf{R}^N\right\}.$$

Normally, there is no problem in finding such a lower bound.

*Iteration step.* Suppose the $k$-th approximations $D_k^+$, $D_k^-$ and

$$\varepsilon_k = \inf\left\{\varepsilon \geq 0 : (\varepsilon, 0) \in D_k^+\right\}$$

are found. To find the $(k+1)$-th approximations we solve the following problems (algorithms for solving these problems and the iteration convergence are considered in the next subsection).

*Problem A.* For the given point $\mathbf{q}_k = (\varepsilon_k, 0) \in \mathbf{R}_+ \times \mathbf{R}^N$ and the set $D_k^- = \text{conv}\{\mathbf{p}_1, \ldots, \mathbf{p}_k\}$, find $\mathbf{n}_k$ in $\mathbf{R}^{N+1}$ satisfying the condition

$$\mathbf{n}_k\mathbf{q}_k \leq \mathbf{n}_k\mathbf{z} \quad \text{for every} \quad \mathbf{z} \in D_k^-. \tag{3.4}$$

If the solution $\mathbf{n}_k$ is known, plane $\Pi$ orthogonal to $\mathbf{n}_k$ and separating the point $\mathbf{q}_k$ and the set $D_k^-$ can be chosen in $\mathbf{R}^{N+1}$, see Figure 3.5. This means that $\mathbf{q}_k$ and $D_k^-$ lie on different sides of $\Pi$. (The plane $\Pi$ is described by the equation $\mathbf{n}_k\mathbf{z} + \gamma = 0$, $\mathbf{z} \in \mathbf{R}^{N+1}$, where $\gamma$ can be chosen as $\gamma = -\inf\left\{\mathbf{n}_k\mathbf{z} : \mathbf{z} \in D_k^-\right\}$. It is clear that the set $D_k^-$ lies in the half-space $\mathbf{n}_k\mathbf{z} + \gamma \geq 0$, while $\mathbf{q}_k$ lies in the half-space $\mathbf{n}_k\mathbf{z} + \gamma \leq 0$ since $\mathbf{n}_k\mathbf{q}_k - \inf\left\{\mathbf{n}_k\mathbf{z} : \mathbf{z} \in D_k^-\right\} \leq 0$ due to (3.4). The number $\gamma$ is of no importance for the following consideration.)

It turns out that a certain translation moves the plane $\Pi$ to a location $\Pi'$ such that $D$ lies in one of the two half-spaces bounded by $\Pi'$. We use this half-space to construct the next external approximation $D_{k+1}^+$: we add the half-space to the collection of those specifying $D_k^+$ and define $D_{k+1}^+$ as the intersection of all these half-spaces. The closer $\Pi'$ is to the boundary of $D$, the better the external approximation $D_{k+1}^+$. The best choice is the plane $\Pi'$ passing through the point $\mathbf{d}_k$ which belongs to the boundary of $D$; see Figure 3.5, where the boundary is shown as the dotted line.

Moreover, the point $\mathbf{d}_k$ can be used to construct the next internal approximation $D_{k+1}^-$: we add $\mathbf{d}_k$ to the vertices of the polyhedron $D_k^-$ and define $D_{k+1}^-$ as the convex hull of all these points.

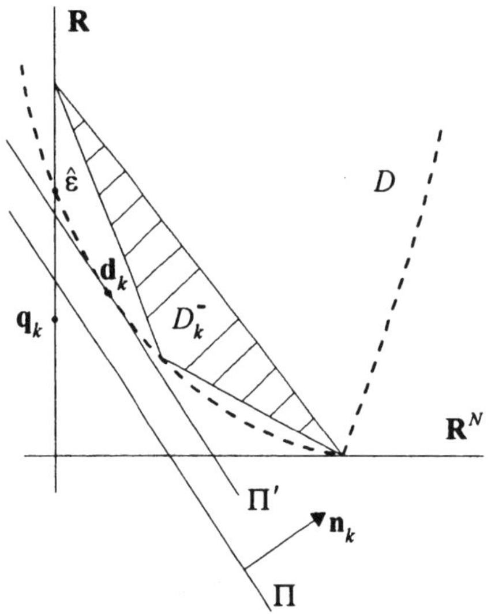


**Figure 3.5**

At first sight, there is no easy way to find $\mathbf{d}_k$ since the set $D$ is not described constructively. However, the following problem can be effectively solved, and its solution determines $\mathbf{d}_k$ and $\mathbf{n}_k$.

*Problem B.* For the solution $\mathbf{n}_k$ to Problem A, find the point $\mathbf{d}_k \in D$ at which the infimum $\inf\{\mathbf{n}_k\mathbf{z} : \mathbf{z} \in D\}$ is attained:

$$\inf\{\mathbf{n}_k\mathbf{z} : \mathbf{z} \in D\} = \mathbf{n}_k\mathbf{d}_k.$$

If $\mathbf{d}_k$ is a solution to Problem B, the equation of the plane $\Pi$ can be written as $\mathbf{n}_k\mathbf{z} - \mathbf{n}_k\mathbf{d}_k = 0$, $\mathbf{z} \in \mathbf{R}^{N+1}$. It is clear that the set $D$ lies on one side of this plane since

$$\mathbf{n}_k\mathbf{z} - \mathbf{n}_k\mathbf{d}_k \geq 0 \quad \text{for every} \quad \mathbf{z} \in D. \tag{3.5}$$

*Modifying approximations.* The convex polyhedron $D_k^-$ is given by the collection $\{\mathbf{p}_1, \ldots, \mathbf{p}_k\}$ of its vertices and lies in $D$. Also recall that $\mathbf{d}_k$ is in $D$ and $D$ is convex; therefore, we can set $\mathbf{p}_{k+1} = \mathbf{d}_k$, consider the convex hull of the points $\{\mathbf{p}_1, \ldots, \mathbf{p}_k, \mathbf{p}_{k+1}\}$ which lies in $D$, and define the next internal approximation as

$$D_{k+1}^- = \text{conv}\{\mathbf{p}_1, \ldots, \mathbf{p}_k, \mathbf{p}_{k+1}\}.$$

The external approximation $D_k^+$ is given by a system of inequalities, and we complement this system with the inequality $\mathbf{b}_{k+1}\mathbf{z} + \beta_{k+1} \geq 0$, where $\mathbf{b}_{k+1} = \mathbf{n}_k$ and $\beta_{k+1} = -\mathbf{n}_k\mathbf{d}_k$. According to (3.5), $D$ lies in the half-space $\mathbf{b}_{k+1}\mathbf{z} + \beta_{k+1} \geq 0$, and the intersection of $D_k^+$ with this half-space can be taken as the next external approximation:

$$D_{k+1}^+ = D_k^+ \cap \{\mathbf{z} \in \mathbf{R}^{N+1} : \mathbf{b}_{k+1}\mathbf{z} + \beta_{k+1} \geq 0\}.$$

The external approximation $D_{k+1}^{+}$ determines the $(k+1)$-th approximation to $\widehat{\varepsilon}$:

$$\varepsilon_{k+1} = \inf\left\{\varepsilon \geq 0 : (\varepsilon, 0) \in D_{k+1}^{+}\right\}.$$

Let us show that it is easy to find this value if solutions to Problems A and B are known. Indeed, let $\mathbf{n}_k = (\xi, \mathbf{x}_*) \in \mathbf{R} \times \mathbf{R}^N$ be a solution to Problem B with $\xi > 0$ (we will show that $\xi$ is positive). Then the condition $(\varepsilon, 0) \in D_{k+1}^{+}$ can be written as

$$(\varepsilon, 0) \in D_k^{+} \quad \text{and} \quad \xi\varepsilon - \mathbf{n}_k \mathbf{d}_k \geq 0,$$

which due to the definition of $\varepsilon_{k+1}$ implies that

$$\varepsilon_{k+1} = \inf\left\{\varepsilon : (\varepsilon, 0) \in D_k^{+}, \xi\varepsilon - \mathbf{n}_k\mathbf{d}_k \geq 0\right\} = \max\{\varepsilon_k, \frac{1}{\xi}\mathbf{n}_k\mathbf{d}_k\}. \tag{3.6}$$

REMARK 1. Note that constructing the subsequent approximations does not require accumulating the information about previous external approximations $D_k^{+}$. This is one of the advantages of the separating plane method.

**3.4. Algorithm and convergence of iterations.** Let us first describe algorithms for solving Problems A and B.

Problem A has, generally speaking, a set of solutions. One of the solutions is determined by solving the linear programming problem formulated below. Its formulation involves the polyhedron

$$P = \{\mathbf{z} \in \mathbf{R}^{N+1} : \mathbf{a}_1\mathbf{z} + \alpha_1 \geq 0, \ldots, \mathbf{a}_M\mathbf{z} + \alpha_M \geq 0\},$$

where $\mathbf{a}_1, \ldots \mathbf{a}_M$ in $\mathbf{R}^{N+1}$ and the numbers $\alpha_1, \ldots, \alpha_M$ are fixed and chosen so that the polyhedron is bounded and contains the origin strictly inside. We will use the condition $\mathbf{z} \in P$ as a restriction in the formulation of the above-mentioned linear programming problem. Besides, we introduce several restrictions related to the internal approximation $D_k^{-} = \operatorname{conv}\{\mathbf{p}_1, \ldots, \mathbf{p}_k\}$ and the approximation $\varepsilon_k$ to $\widehat{\varepsilon}$ (the $k$-th approximations are considered as known). We denote the point $(\varepsilon_k, 0) \in \mathbf{R} \times \mathbf{R}^N$ by $\mathbf{q}_k$ and impose the restrictions

$$(\mathbf{p}_1 - \mathbf{q}_k)\mathbf{z} - w \geq 0, \ \ldots, \ (\mathbf{p}_k - \mathbf{q}_k)\mathbf{z} - w \geq 0$$

on the variables $\mathbf{z} \in \mathbf{R}^{N+1}$ and $w \in \mathbf{R}$. These restrictions together with $\mathbf{z} \in P$ specify a certain domain, on which we consider a simple function $f(\mathbf{z}, w) = w$ and the extremum problem of its maximizing:

$$\begin{gathered} w \longrightarrow \sup, \\ \mathbf{a}_1\mathbf{z} + \alpha_1 \geq 0, \ \ldots, \ \mathbf{a}_M\mathbf{z} + \alpha_M \geq 0, \\ (\mathbf{p}_1 - \mathbf{q}_k)\mathbf{z} - w \geq 0, \ \ldots, \ (\mathbf{p}_k - \mathbf{q}_k)\mathbf{z} - w \geq 0. \end{gathered} \tag{3.7}$$

The function and the restrictions are linear; (3.7) is a standard linear programming problem. It can be proved that its solution results in a solution to Problem A; more precisely, the following proposition is valid.

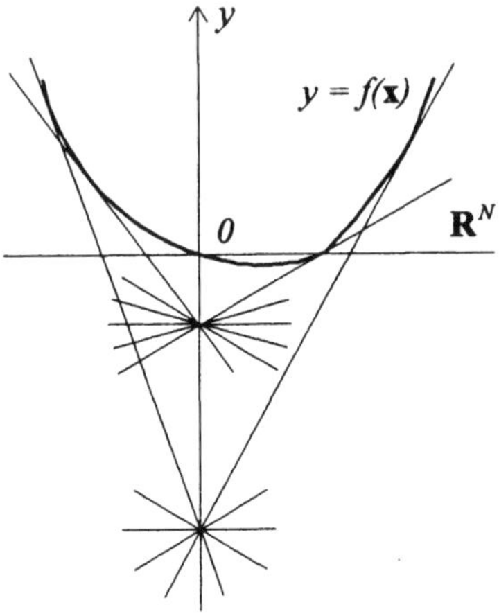


**Figure 3.6**

THEOREM 2. Suppose the point $\mathbf{z} = (\xi, \mathbf{x}_*) \in \mathbf{R} \times \mathbf{R}^N$ and the number $w = w_0$ solve problem (3.7). Then $\mathbf{n}_k = (\xi, \mathbf{x}_*)$ is a solution to Problem A, $\xi$ being positive.

Problem (3.7) can be solved, for example, by the well known simplex method of linear programming.

Let us now describe the algorithm for solving Problem B. The following proposition reduces Problem B to computing the value of the function $f$ and its subgradient at one point.

THEOREM 3. Suppose the point $(\xi, \mathbf{x}_*) \in \mathbf{R} \times \mathbf{R}^N$ and the number $w = w_0$ solve problem (3.7). Then the infimum in Problem B is attained at $\mathbf{z} = \mathbf{d}_k = (\eta, \mathbf{g}_*) \in \mathbf{R} \times \mathbf{R}^N$, where $\mathbf{g}_*$ is any of the subgradients of $f$ at the point $-\mathbf{x}_*/\xi$ and

$$\eta = -\frac{1}{\xi}\,\mathbf{g}_*\mathbf{x}_* - f\left(-\frac{1}{\xi}\,\mathbf{x}_*\right).$$

The proof is based on the fact that the complete information about the function $f$ can be extracted from the mapping $\varepsilon \to \partial_\varepsilon f(0)$: the graph of $f$ is the envelope of the planes $y = \mathbf{g}\mathbf{x} - \varepsilon$ constructed for every $\varepsilon \geq 0$ and $\mathbf{g} \in \partial_\varepsilon f(0)$. Figure 3.6 illustrates this statement (recall that $f(0) = 0$ by assumption); see also the definition of $\varepsilon$-subdifferential in Subsection 3.1. In other words, the following formula is valid:

$$f(\mathbf{x}) = \sup\left\{\mathbf{g}\mathbf{x} - \varepsilon : \varepsilon \in \mathbf{R}_+,\ \mathbf{g} \in \partial_\varepsilon f(0)\right\}.$$

It allows us to evaluate the minimum in Problem B. Namely, this problem can be written as

$$\inf\left\{\mathbf{n}_k\mathbf{z} : \mathbf{z} \in D\right\} = \inf\left\{\xi\varepsilon + \mathbf{x}_*\mathbf{g} : \varepsilon \in \mathbf{R},\ \mathbf{g} \in \mathbf{R}^N,\ (\varepsilon, \mathbf{g}) \in D\right\} =$$
$$= -\xi \sup\left\{-\varepsilon - \frac{1}{\xi}\,\mathbf{x}_*\mathbf{g} : \varepsilon \in \mathbf{R}_+,\ \mathbf{g} \in \partial_\varepsilon f(0)\right\}. \quad (3.8)$$

Here, the supremum equals $f(-\mathbf{x}_*/\xi)$ by the previous formula, and it is easy to find $\varepsilon$, $\mathbf{g}$ at which the extremum is attained. Namely, consider

$$\varepsilon = \eta = -\frac{1}{\xi}\mathbf{g}_*\mathbf{x}_* - f\left(-\frac{1}{\xi}\mathbf{x}_*\right) \quad \text{and} \quad \mathbf{g} = \mathbf{g}_*,$$

where $\mathbf{g}_*$ is any of the subgradients of $f$ at the point $-\mathbf{x}_*/\xi$. The definitions of the subgradient and $\varepsilon$-subdifferential immediately imply that $\eta \geq 0$ and $\mathbf{g}_* \in \partial_\eta f(0)$, that is, $\varepsilon = 0$ and $\mathbf{g} = \mathbf{g}_*$ satisfy the restrictions of extremum problem (3.8). Finally, we find the value of the function maximized in (3.8):

$$-\varepsilon - \frac{1}{\xi}\mathbf{x}_*\mathbf{g} = f\left(-\frac{1}{\xi}\mathbf{x}_*\right).$$

As we saw earlier, this is the value of the supremum in problem (3.8). Thus, the maximum in (3.8) or, equivalently, the minimum in Problem B is attained at $\varepsilon = \eta$, $\mathbf{g} = \mathbf{g}_*$, which is exactly what Theorem 3 states.

Thus Theorems 2 and 3 provide us with the procedures for solving Problems A and B to accomplish the iteration step of the separating plane method.

It only remains to simplify expression (3.6) for the $(k+1)$-th approximation to $\widehat{\varepsilon}$. Recall that $\mathbf{n}_k = (\xi, \mathbf{x}_*)$ in this expression is a solution to Problem A, while $\mathbf{d}_k = (\eta, \mathbf{g}_*)$ solves Problem B. Then the value $\mathbf{n}_k\mathbf{d}_k$ in (3.6) equals

$$\mathbf{n}_k\mathbf{d}_k = \xi\left(-\frac{1}{\xi}\mathbf{g}_*\mathbf{x}_* - f\left(-\frac{1}{\xi}\mathbf{x}_*\right)\right) + \mathbf{x}_*\mathbf{g}_* = -\xi f\left(-\frac{1}{\xi}\mathbf{x}_*\right),$$

where the result of Theorem 3 is taken into account. This reduces formula (3.6) for evaluating of $\varepsilon_{k+1}$ to

$$\varepsilon_{k+1} = \max\{\varepsilon_k, -f(-\mathbf{x}_*/\xi\}.$$

Finally, let us describe the separating plane algorithm for minimizing a convex nondifferentiable function $f$ on $\mathbf{R}^N$ (recall that $f$ is assumed to satisfy the nonrestrictive condition $f(0) = 0$). The algorithm allows calculation of the successive approximations $-\varepsilon_k$, $k = 1, 2, \ldots$, to the unknown infimum

$$-\widehat{\varepsilon} = \inf\left\{f(\mathbf{x}) : \mathbf{x} \in \mathbf{R}^N\right\},$$

each of the approximations $-\varepsilon_k$ being an upper bound for $-\widehat{\varepsilon}$.

Note that the algorithm requires finding a subgradient of the minimized function, which can be done in the way described in the next subsection.

*Separating plane algorithm*

0.1. The algorithm formulation involves the polyhedron

$$P = \{\mathbf{z} \in \mathbf{R}^{N+1} : \mathbf{a}_1\mathbf{z} + \alpha_1 \geq 0, \ldots, \mathbf{a}_M\mathbf{z} + \alpha_M \geq 0\}$$

where $\mathbf{a}_1, \ldots, \mathbf{a}_M \in \mathbf{R}^{N+1}$ and the numbers $\alpha_1, \ldots, \alpha_M$ are chosen so that $P$ is convex and contains the origin strictly inside.

0.2. The lower bound for the infimum

$$-\eta_1 \le \inf\left\{f(\mathbf{x}) : \mathbf{x} \in \mathbf{R}^N\right\}$$

is assumed to be known; it determines the initial approximation.

1. *Initial approximation.* Set $\varepsilon_1 = 0$, $\mathbf{p}_1 = (\eta_1, 0) \in \mathbf{R} \times \mathbf{R}^N$.
2. *Iteration step* ($k = 1, 2, \ldots$).

2.1. Set $\mathbf{q} = (\varepsilon_k, 0) \in \mathbf{R} \times \mathbf{R}^N$. Solve the linear programming problem: find $\mathbf{n} \in \mathbf{R}^{N+1}$ such that at $\mathbf{z} = \mathbf{n}$ and a certain number $w = w_0$ the following extremum is attained:

$$\begin{gathered} w \longrightarrow \sup, \\ \mathbf{a}_1\mathbf{z} + \alpha_1 \ge 0, \;\ldots,\; \mathbf{a}_M\mathbf{z} + \alpha_M \ge 0, \\ (\mathbf{p}_1 - \mathbf{q})\mathbf{z} - w \ge 0, \;\ldots,\; (\mathbf{p}_k - \mathbf{q})\mathbf{z} - w \ge 0. \end{gathered}$$

Write $\mathbf{n}$ in the form $(\xi, \mathbf{x}_*)$, where $\xi$ is a number and $\mathbf{x}_*$ belongs to $\mathbf{R}^N$.

2.2. Compute the value $f(-\mathbf{x}_*/\xi)$.

2.3. Find subgradient $\mathbf{g}_* \in \mathbf{R}^N$ of $f$ at the point $-\mathbf{x}_*/\xi$.

2.4. Set

$$\begin{aligned} \varepsilon_{k+1} &= \max\left\{\varepsilon_k, \, -f\left(-\frac{1}{\xi}\mathbf{x}_*\right)\right\}, \\ \mathbf{p}_{k+1} &= \left(-\frac{1}{\xi}\,\mathbf{g}_*\mathbf{x}_* - f\left(-\frac{1}{\xi}\mathbf{x}_*\right), \; \mathbf{g}_*\right) \in \mathbf{R}^{N+1}. \end{aligned}$$

3. *Conditional transfer.* Check whether the preliminary adopted criterion for finishing the calculations is satisfied; if it is, stop computing; if it is not, go to the next iteration step.

The following convergence theorem is valid.

THEOREM 4. Let $f$ be a convex finite function on $\mathbf{R}^N$ and $f(0) = 0$. Then the sequence of the approximations $-\varepsilon_k$ determined by the separating plane algorithm converges to the unknown infimum $-\widehat{\varepsilon}$:

$$\lim_{k\to\infty}(-\varepsilon_k) = -\widehat{\varepsilon} = \inf\left\{f(\mathbf{x}) : \mathbf{x} \in \mathbf{R}^N\right\}.$$

REMARK 2. The separating plane algorithm was formulated above in the case of minimization over all of $\mathbf{R}^N$. However, solving a more general minimization problem

$$\inf\left\{f(\mathbf{x}) : \mathbf{x} \in L\right\}, \quad \text{where } L \text{ is a subspace in } \mathbf{R}^N, \tag{3.9}$$

is often required. For example, the discretized problem of limit analysis (Subsection 2.4) is of the form (3.9): the kinematic variables $\mathbf{u} = (u_1, \ldots, u_M)$ are subjected to the restriction $\mathbf{au} = 0$, that is, minimization is considered over a hyperplane in $\mathbf{R}^N$.

Problem (3.9) is reduced to the problem of minimization over the whole space if one solves the system of equations specifying the subspace $L$. This is a linear system $A\mathbf{x} = 0$, $\mathbf{x} \in \mathbf{R}^N$, and it allows us to express the variables $\mathbf{x}$ through independent variables $\mathbf{y} = (y_1, \ldots, y_n)$: $\mathbf{x} = B\mathbf{y}$. Then (3.9) can be written as

$$\inf \{F(\mathbf{y}) : \mathbf{y} \in \mathbf{R}^n\}, \quad \text{where} \quad F(\mathbf{y}) = f(B\mathbf{y}).$$

It is clear that $F$ possesses the same properties as $f$: it is convex, finite, and $F(0) = 0$. Thus the latter problem can be solved using the separating plane algorithm.

The following modification of the above algorithm avoids introducing the new independent variables:

(i) after accomplishing 2.1 of the algorithm, in its remaining part replace $\mathbf{x}_*$ by its projection $\mathrm{pr}_L \mathbf{x}_*$ on the subspace $L$,

(ii) after accomplishing 2.3, which results in the subgradient $\mathbf{g}_*$ (this time at the point $\mathrm{pr}_L \mathbf{x}_*$), in the remaining part of the algorithm, replace $\mathbf{g}_*$ by its projection $\mathrm{pr}_L \mathbf{g}_*$ on the subspace $L$.

REMARK 3. The most important element in the method under consideration is constructing the separating plane (solving Problem A). The above algorithm suggests that this plane be determined by solving linear programming problem (3.8). Procedures different from this may also be applied to constructing the separating plane. However, to ensure the convergence $\varepsilon_k \to \hat{\varepsilon}$, the procedure for determining the separating plane has to meet the following *uniform separation* requirement. Without loss of generality, we will formulate this requirement in case the point to be separated from a convex set is the origin. Let $M_k$, $k = 1, 2, \ldots$, be convex, closed sets in $\mathbf{R}^{N+1}$ and the origin not belong to any of them. Suppose a certain procedure determines planes separating $M_k$ and the origin, that is, the procedure results in $\mathbf{n}_k \in \mathbf{R}^{N+1}$ such that $0 \leq \mathbf{n}_k \mathbf{z}$ for every $\mathbf{z} \in M_k$. The uniform separation requirement deals with the property of this solution in case the origin is not in the neighborhood $M_k + \lambda B_1(0)$ of $M_k$ ($\lambda$ is a positive number and $B_1(0)$ is the ball in $\mathbf{R}^{N+1}$ of the radius 1 centered at 0). The requirement consists of the existence of a positive number $\lambda' > 0$ such that the solution $\mathbf{n}_k$ separates the origin not only from $M_k$ but also from its neighborhood $M_k + \lambda' B_1(0)$, that is, $0 \leq \mathbf{n}_k \mathbf{z}$ for every $\mathbf{z} \in M_k + \lambda' B_1(0)$.

Note that the procedure for constructing the separating plane in the above algorithm by solving (3.8) meets the uniform separation requirement.

**3.5. Finding a subgradient.** One of the steps in the separating plane algorithm consists of determining a subgradient of the function to be minimized. We will now describe a way for solving this problem. To be more definite, we restrict ourselves to the discretized problem of limit analysis, although the method is also applicable to constructing subgradients of various functions arising from the discretization of integral functionals.

The discretized kinematic problem of limit analysis introduced in Subsection 2.4 determines the approximations to the safety factor:

$$\alpha_{l(h)} = \inf\left\{f(\mathbf{u}) : \mathbf{u} \in \mathbf{R}^M, \ \mathbf{au} = 0\right\}.$$

Here, the variable $\mathbf{u}$ represents velocity field, the function $f$ is determined as

$$f(\mathbf{u}) = \int_\Omega d_x(\mathbf{e_u} + \boldsymbol{\varepsilon})\, dV,$$

the symmetric second rank tensor field $\boldsymbol{\varepsilon}$ and $\mathbf{a} \in \mathbf{R}^M$ are given, and the components of the strain rate $\mathbf{e_u}$ are linearly expressed through $\mathbf{u}$:

$$(\mathbf{e_u})_{ij} = u_1 \frac{1}{2}\left(\frac{\partial \varphi_{(1)i}}{\partial x_j} + \frac{\partial \varphi_{(1)j}}{\partial x_i}\right) + \cdots + u_M \frac{1}{2}\left(\frac{\partial \varphi_{(M)i}}{\partial x_j} + \frac{\partial \varphi_{(M)j}}{\partial x_i}\right).$$

Recall that $\varphi_{(\mu)k}$ are the components of the basis function $\boldsymbol{\varphi}_\mu$ of the finite element triangulation.

The following proposition reduces constructing a subgradient of $f$ to finding subgradients of the dissipation $d_x$, which can be easily done.

PROPOSITION 1. Suppose subgradient $\boldsymbol{\tau}(x) \in Sym$ of the dissipation $d_x$ at the point $\mathbf{e_u}(x) + \boldsymbol{\varepsilon}(x)$ is known for every $x$ in $\Omega$, the field $\boldsymbol{\tau}$ being integrable in $\Omega$. Then $\mathbf{g} = (g_1, \ldots, g_M) \in \mathbf{R}^M$ with the components

$$g_\mu = \int_\Omega \tau_{ij} \frac{\partial \varphi_{(\mu)i}}{\partial x_j}\, dV, \qquad \mu = 1, \ldots, M,$$

is a subgradient of the function $f$ at the point $\mathbf{u}$.

PROOF. According to the definition of subgradient, we have to verify that $\mathbf{g}$ satisfies the condition

$$f(\mathbf{u}') - f(\mathbf{u}) \geq \mathbf{g}(\mathbf{u}' - \mathbf{u}) \quad \text{for every} \quad \mathbf{u}' \in \mathbf{R}^M$$

or, in more detail, that

$$\int_\Omega \{d_x(\mathbf{e_{u'}} + \boldsymbol{\varepsilon}) - d_x(\mathbf{e_u} + \boldsymbol{\varepsilon})\}\, dV$$
$$\geq (u_1' - u_1)\int_\Omega \tau_{ij} \frac{\partial \varphi_{(1)i}}{\partial x_j}\, dV + \cdots + (u_M' - u_M)\int_\Omega \tau_{ij} \frac{\partial \varphi_{(M)i}}{\partial x_j}\, dV$$
$$\text{for every} \quad \mathbf{u}' \in \mathbf{R}^M.$$

It is clear that the expression on the right-hand side equals

$$\int_\Omega \{(\mathbf{e_{u'}} + \boldsymbol{\varepsilon}) - (\mathbf{e_u} + \boldsymbol{\varepsilon})\} \cdot \boldsymbol{\tau}\, dV,$$

and therefore the inequality

$$d_x(\mathbf{e}_{\mathbf{u}'}+\boldsymbol{\varepsilon})-d_x(\mathbf{e}_{\mathbf{u}}+\boldsymbol{\varepsilon})\geq\{(\mathbf{e}_{\mathbf{u}'}+\boldsymbol{\varepsilon})-(\mathbf{e}_{\mathbf{u}}+\boldsymbol{\varepsilon})\}\cdot\boldsymbol{\tau}\quad\text{for every}\quad \mathbf{e}_{\mathbf{u}'}\in Sym$$

guarantees that the previous condition is satisfied. The latter inequality is valid since $\boldsymbol{\tau}(x)$ is a subgradient of $d_x$ at the point $\mathbf{e}_{\mathbf{u}}(x)+\boldsymbol{\varepsilon}(x)$, which finishes the proof.

REMARK 4. The proof of the above proposition does not make use of the homogeneity of $d_x$. Therefore, the proposition is valid not only for the dissipation but also for any convex function, and it allows construction of subgradients of various functions arising from the discretization of integral functionals.

Proposition 1 expresses a subgradient of $f$ through the subgradient $\boldsymbol{\tau}(x)$, $x \in \Omega$, of the dissipation $d_x$ under the additional requirement: the function $\boldsymbol{\tau}(x)$ is to be integrable. Normally, it is not a complicated task to find subgradients of $d_x$ and to choose a suitable subgradient at each of the points $\mathbf{e}_{\mathbf{u}}(x) + \boldsymbol{\varepsilon}(x)$, $x \in \Omega$, to ensure the integrability of $\boldsymbol{\tau}$. Actually, $d_x$ and $\mathbf{e}_{\mathbf{u}}(x) + \boldsymbol{\varepsilon}(x)$ are rather simple functions, the latter being piecewise constant or piecewise linear if one uses the discretization described in Subsection 2.2 or 2.3, respectively.

EXAMPLE 2. Let us find a subgradient of the dissipation $d_x$ in the case of the Mises yield surface $|\sigma^d| = \sqrt{2}k_x$, $k_x > 0$, that is,

$$d_x(\mathbf{e}) = \begin{cases} \sqrt{2}k_x\,|\mathbf{e}| & \text{if } \mathbf{e}^s = 0, \\ +\infty & \text{if } \mathbf{e}^s \neq 0 \end{cases}$$

(see Example I.4.1). If $\mathbf{e}^s \neq 0$, the set of subgradients of $d_x$ at $\mathbf{e}$ is apparently empty. At the same time, the limit analysis kinematic extremum problem discretized in Subsection 2.3 only makes use of strain rate fields $\mathbf{e}$ satisfying the condition $\mathbf{e}^s = 0$. In this case, it is easy to find all subgradients of $d_x$ at $\mathbf{e}$, that is, those $\boldsymbol{\tau}$ in $Sym$ which satisfy the condition

$$d_x(\mathbf{e}') - d_x(\mathbf{e}) \geq \boldsymbol{\tau}\cdot(\mathbf{e}'-\mathbf{e}) \quad \text{for every} \quad \mathbf{e}' \in Sym.$$

Note that this inequality imposes no restriction on $\boldsymbol{\tau}$ when $\mathbf{e}'$ with $\mathbf{e}'^s \neq 0$ is considered: $d_x(\mathbf{e}') = +\infty$ in this case. This allows re-writing the above condition determining the subgradients as

$$d_x(\mathbf{e}') - d_x(\mathbf{e}) \geq \boldsymbol{\tau}\cdot(\mathbf{e}'-\mathbf{e}) \quad \text{for every} \quad \mathbf{e}' \in Sym^d$$

or, equivalently (since $\mathbf{e}^s = 0$ and $\mathbf{e}'^s = 0$), as

$$d_x(\mathbf{e}'^d) - d_x(\mathbf{e}^d) \geq \boldsymbol{\tau}^d\cdot(\mathbf{e}'-\mathbf{e}) \quad \text{for every} \quad \mathbf{e}' \in Sym. \tag{3.10}$$

This condition means exactly that $\boldsymbol{\tau}^d$ is a subgradient of the function

$$d_x^{(d)}(\mathbf{e}) = d_x(\mathbf{e}^d)$$

at the point e. In the case of the Mises yield surface, $d_x^{(d)}(\mathbf{e}) = \sqrt{2}k_x\,|\mathbf{e}^d|$, this function is differentiable everywhere with the exception of the point $\mathbf{e} = 0$. Therefore, $d_x^{(d)}$ possesses the unique subgradient

$$\frac{\partial d^{(d)}}{\partial \mathbf{e}}(\mathbf{e}) = \frac{\partial}{\partial \mathbf{e}}(\sqrt{2}k_x\,|\mathbf{e}^d|) = \sqrt{2}k_x\frac{\mathbf{e}^d}{|\mathbf{e}^d|} \tag{3.11}$$

if $\mathbf{e} \neq 0$. At $\mathbf{e} = 0$, relation (3.10) takes the form

$$\sqrt{2}k_x|\mathbf{e}'^d| \geq \boldsymbol{\tau}^d \cdot \mathbf{e}' \quad \text{for every} \quad \mathbf{e}' \in Sym.$$

It is clear that $\boldsymbol{\tau}$ satisfies this condition if and only if $|\boldsymbol{\tau}^d| \leq \sqrt{2}k_x$. Therefore, the set of subgradients at $\mathbf{e} = 0$ is the cylinder $|\boldsymbol{\tau}^d| \leq \sqrt{2}k_x$ (cf. Example 1).

Now, in the case of the Mises yield surface, it is clear how to find the subgradient when solving the discretized kinematic extremum problem by the separating plane algorithm. On every finite element where $\mathbf{e_u} + \boldsymbol{\varepsilon}$ is nonzero, the subgradient is determined by formula (3.11). On a finite element where $\mathbf{e_u} + \boldsymbol{\varepsilon} = 0$, any admissible stress $\boldsymbol{\tau}$ is a subgradient.

## Comments

The finite element method is successfully applied to solving a wide class of problems. The books by Strang and Fix (1973) and Ciarlet (1978) contain its detailed description, in particular, the proof of the approximation theorem (Theorem 2.1) for the Sobolev space $W_2^1(\Omega)$.

The finite element approximation for the stream function (Subsection 2.3) was proposed by Mercier (1982). The approximation makes use of the quadrangle finite element with piecewise polynomial basis functions. Sander (1964) and Fraeijs de Veubeke (1968) introduced this finite element; Ciavaldini and Nedelek (1974) studied it in detail, and in particular, established the approximation theorem for the Sobolev space $W_2^2(\Omega)$ (Theorem 2.2). The basic functions for the square finite element were calculated by T.S.Galkina per the author's request.

In this chapter, we only make use of the continuous velocity field when discretizing the limit analysis kinematic extremum problem. It is also possible to use piecewise continuous fields with the jump discontinuity on the finite element boundaries. This allows reduction of the power of the basis function polynomials. For example, van Rij and Hodge (1978) considered piecewise constant velocity fields; the corresponding strain rate was zero on each of the finite elements and the dissipation was produced only by the slip at the elements' boundaries. Seriogin (1987) studied approximation to solenoidal vector fields by interpolants with the jump discontinuity and proved the convergence of the corresponding approximations to the safety factor.

The separating plane method is described in detail in the book by Nurminsky (1991), where the reader can find proofs of the propositions presented in Section 3.

CHAPTER X

# SHAKEDOWN THEORY

Residual stresses normally remain in a body unloaded after plastic deformation. The residual stresses can neutralize those caused by subsequent loading and this can substantially restrain the plastic flow. In particular, it may occur that the plastic work be bounded for all loading histories within a given set of loads. This type of body response is referred to as shakedown (or adaptation) to the given set of loads. It is often required to find out whether the body shakes down to a certain set of loads. In particular, this is an important question to be answered when estimating the response of a structure to a cyclic loading.

This chapter deals with a theory which answers this question. The first two sections present formulation of the elastic-plastic problem and examples illustrating various types of body response to variable loadings. Then we establish conditions for shakedown and nonshakedown; Section 3 presents formulations of the results.

The concept of the shakedown safety factor naturally arises from the shakedown conditions. The safety factor is determined by a static extremum problem which turns out to be a direct generalization of that of limit analysis. Therefore, the subsequent consideration follows the scheme used in limit analysis theory: 1) the dual kinematic extremum problem is constructed, 2) conditions are established for the equality of the extremums in the static and kinematic problems extremums, 3) the extremum problems are reduced in the case of bodies with cylindrical yield surfaces, and 4) the kinematic problem is discretized and convergence of finite element approximations to the safety factor is established.

## 1. Elastic-plastic problem

In this section, we present constitutive relations and formulate the quasistatic problem for the elastic perfectly plastic body. First, we consider the strong formulation of the problem for three-dimensional bodies. Then we introduce a unified formulation of the problem for elastic perfectly plastic systems of all types (beams, plates, frames, etc.) with no smoothness involved.

**1.1. Elastic perfectly plastic body.** Consider a body whose deformation $\boldsymbol{\varepsilon}$ is the sum of the elastic and plastic components $\boldsymbol{\varepsilon}^e$ and $\boldsymbol{\varepsilon}^p$: $\boldsymbol{\varepsilon} = \boldsymbol{\varepsilon}^e + \boldsymbol{\varepsilon}^p$. We assume that the elastic strain $\boldsymbol{\varepsilon}^e$ is related to the stress $\boldsymbol{\sigma}$ by the *linear elasticity law*: $\varepsilon^e_{ij} = A_{ijkl}\sigma_{kl}$ or, briefly, $\boldsymbol{\varepsilon}^e = \mathbf{A}\boldsymbol{\sigma}$, see Subsection I.A.3. Here, the elasticity modules tensor $\mathbf{A}$ is symmetric, $A_{ijkl} = A_{jikl} = A_{ijlk} = A_{klij}$, and positive definite: $\boldsymbol{\sigma} \cdot \mathbf{A}\boldsymbol{\sigma} > 0$ for every symmetric second rank tensor $\boldsymbol{\sigma}$. We assume that $\mathbf{A}$ does not vary with time, but may depend on the point of the body. We adopt the same constitutive relations for the plastic strain as in the rigid perfectly plastic model. Namely, let the set of admissible stresses $C_x$ be given for every point $x$ of the body. As always, it is a convex, closed set in the space *Sym* of symmetric second rank tensors, and $C_x$ contains the origin. The *constitutive maximum principle* (Subsection I.3.2)

$$\boldsymbol{\sigma}(x) \in C_x, \qquad \mathbf{e}^p \cdot (\boldsymbol{\sigma}(x) - \boldsymbol{\sigma}_*) \geq 0 \quad \text{for every} \quad \boldsymbol{\sigma}_* \in C_x$$

relates the plastic strain rate $\mathbf{e}^p = \dot{\boldsymbol{\varepsilon}}^p$ to the stress $\boldsymbol{\sigma}$ (as usual, the dot denotes the timederivative). It is convenient to differentiate the elasticity law with respect to the time and write the constitutive relations as

$$\mathbf{e} = \mathbf{e}^e + \mathbf{e}^p,$$
$$\mathbf{e}^e = \mathbf{A}\dot{\boldsymbol{\sigma}},$$
$$\boldsymbol{\sigma}(x) \in C_x, \qquad \mathbf{e}^p \cdot (\boldsymbol{\sigma}(x) - \boldsymbol{\sigma}_*) \geq 0 \quad \text{for every} \quad \boldsymbol{\sigma}_* \in C_x.$$

Here, $\mathbf{e}$, $\mathbf{e}^e$ and $\mathbf{e}^p$ are the rates of the total, elastic and plastic strains.

In the case of a smooth yield surface, the yield function $F_x$, which specifies the set $C_x$ with the inequality $F_x(\boldsymbol{\sigma}) \leq 0$, is continuously differentiable. Then the constitutive maximum principle is reduced to the normality flow rule (Subsection I.3.1):

$$\mathbf{e} = \mathbf{A}\dot{\boldsymbol{\sigma}} + \dot{\lambda}\,\frac{\partial F_x}{\partial \boldsymbol{\sigma}}, \quad F_x(\boldsymbol{\sigma}) \leq 0, \quad \dot{\lambda} \geq 0, \quad \dot{\lambda}\, F_x(\boldsymbol{\sigma}) = 0.$$

In the case of an arbitrary yield surface eliminating $\boldsymbol{\varepsilon}^e$ and $\boldsymbol{\varepsilon}^p$ re-writes the constitutive maximum principle as

$$\boldsymbol{\sigma}(x) \in C_x, \quad (\mathbf{e} - \mathbf{A}\dot{\boldsymbol{\sigma}}) \cdot (\boldsymbol{\sigma}(x) - \boldsymbol{\sigma}_*) \geq 0 \text{ for every } \boldsymbol{\sigma}_* \in C_x. \tag{1.1}$$

A body with the strain rate $\mathbf{e}$ and stress $\boldsymbol{\sigma}$ related by the constitutive maximum principle (1.1) is referred to as *elastic perfectly plastic*.

**1.2. Elastic-plastic problem: strong formulation.** Consider an elastic perfectly plastic body which occupies domain $\Omega$. The body is subjected to body forces with the volume density $\mathbf{f}$ given in $\Omega$ and to surface tractions with the density $\mathbf{q}$ given on the part $S_q$ of the body boundary. The remaining part $S_v$ of the boundary is fixed: every kinematically admissible velocity field satisfies the condition $\mathbf{v}|_{S_v} = 0$. The displacement $\mathbf{u}$ and stress $\boldsymbol{\sigma}$ are given at the initial moment: $\mathbf{u}|_{t=0} = \mathbf{u}_0$, $\boldsymbol{\sigma}|_{t=0} = \boldsymbol{\sigma}_0$ ($\mathbf{u}_0$, $\boldsymbol{\sigma}_0$ are given functions of the point of the body).

We assume displacements and strains to be small under the action of the variable load $\mathbf{l} = (\mathbf{f}, \mathbf{q})$. Hence, the geometrically linear theory is applicable. It is also assumed that the load varies slowly enough to result only in small accelerations that can be neglected. Then the stress field $\boldsymbol{\sigma}$ equilibrates the load $\mathbf{l}(t)$ at any moment $t$, that is, the equilibrium conditions

$$\frac{\partial\sigma_{ij}}{\partial x_j} + f_i = 0, \qquad \sigma_{ij}\nu_j|_{S_q} = q_i$$

are satisfied ($\boldsymbol{\nu}$ is the unit outward normal).

We presume here that the stress field is sufficiently regular, say, continuously differentiable. If the velocity field $\mathbf{v}$ and the corresponding strain rate field $\mathbf{e}$ are also regular, they satisfy the kinematic relations

$$e_{ij} = \frac{1}{2}\left(\frac{\partial v_i}{\partial x_j} + \frac{\partial v_j}{\partial x_i}\right), \quad \mathbf{v}|_{S_v} = 0, \quad \dot{\mathbf{u}} = \mathbf{v}.$$

Besides, $\boldsymbol{\sigma}$ and $\mathbf{v}$ satisfy constitutive relations (1.1). Thus, we arrive at the following formulation of the quasistatic problem for the elastic perfectly plastic body:

$$\frac{\partial\sigma_{ij}}{\partial x_j} + f_i = 0, \qquad \sigma_{ij}\nu_j|_{S_q} = q_i, \tag{1.2}$$

$$e_{ij} = \frac{1}{2}\left(\frac{\partial v_i}{\partial x_j} + \frac{\partial v_j}{\partial x_i}\right), \qquad \mathbf{v}|_{S_v} = 0, \tag{1.3}$$

$$\boldsymbol{\sigma}(x) \in C_x, \quad (\mathbf{e} - \mathbf{A}\dot{\boldsymbol{\sigma}})\cdot(\boldsymbol{\sigma}(x) - \boldsymbol{\sigma}_*) \geq 0 \text{ for every } \boldsymbol{\sigma}_* \in C_x, \tag{1.4}$$

$$\dot{\mathbf{u}} = \mathbf{v}, \quad \mathbf{u}|_{t=0} = u_0, \quad \boldsymbol{\sigma}|_{t=0} = \boldsymbol{\sigma}_0. \tag{1.5}$$

Nonzero boundary values may also be assigned to the velocity field on $S_v$; however, we restrict ourselves to the case $\mathbf{v}|_{S_v} = 0$ and consider the shakedown theory under this assumption.

It is easy to get rid of the stress field regularity assumption in the above formulation of the elastic-plastic problem by replacing equilibrium conditions (1.2) with the virtual work principle, see Subsection II.1.3. Namely, let $\mathcal{S}$ be a linear space of stress fields on $\Omega$ (it is not necessary to specify $\mathcal{S}$ now; we just assume that all of the fields in $\mathcal{S}$ are integrable). According to the virtual work principle the stress field $\boldsymbol{\sigma}$ in $\mathcal{S}$ equilibrates the load $\mathbf{l} = (\mathbf{f}, \mathbf{q})$ if and only if

$$\int_\Omega \operatorname{Def}\mathbf{v}\cdot\boldsymbol{\sigma}\,dV = \int_\Omega \mathbf{f}\mathbf{v}\,dV + \int_{S_q}\mathbf{q}\mathbf{v}\,dS \quad \text{for every} \quad \mathbf{v} \in U, \tag{1.6}$$

where $\operatorname{Def}\mathbf{v}$ is the strain rate field

$$(\operatorname{Def}\mathbf{v})_{ij} = \frac{1}{2}\left(\frac{\partial v_i}{\partial x_j} + \frac{\partial v_j}{\partial x_i}\right) \tag{1.7}$$

and $U$ is the set of virtual velocity fields consisting of all smooth vector fields on $\Omega$ vanishing on $S_v$. Thus, we replace equilibrium conditions (1.2) in formulation (1.2) – (1.5) of the elastic-plastic problem by (1.6). We keep assuming that the velocity field $\mathbf{v}$ in this formulation is sufficiently regular and that is why formulation (1.3) – (1.6) is referred to as *strong*.

**1.3. A way to generalize formulation: examples.** In the previous subsection, the elastic-plastic problem for a three-dimensional body was formulated under the velocity field regularity assumption. Formulation of the problem for various continual and discrete systems (plates, beams, trusses, etc.) is analogous to it, and our immediate goal is to give a general unified formulation of the problem to cover all these systems. The general formulation is also convenient since it does not involve the velocity field regularity assumption. Therefore, developing the shakedown theory within the framework of the unified formulation we will obtain results valid for elastic perfectly plastic systems of all types.

Let us start with two examples of setting up the elastic-plastic problem showing the way to the unified formulation.

EXAMPLE 1. *Truss.* Consider an elastic perfectly plastic truss consisting of $m$ rods. Some of the truss hinges are not fixed; let $k$ be their number. The truss is subjected to the load $\mathbf{l} = (\mathbf{f}_1, \ldots, \mathbf{f}_k)$, where $\mathbf{f}_\alpha$ stands for the force applied to the $\alpha$-th (unfixed) hinge. We denote the velocity of the $\alpha$-th hinge by $\mathbf{v}_\alpha$, the axial force in the $i$-th rod by $N_i$ and the elongation of this rod by $e_i$. Thus, the velocity $\mathbf{v} = (\mathbf{v}_1, \ldots, \mathbf{v}_k)$ and the load $\mathbf{l} = (\mathbf{f}_1, \ldots, \mathbf{f}_k)$ are represented by the members $\mathbf{v}$ and $\mathbf{l}$ of the spaces $\mathcal{V} = \mathbf{R}^{3k}$ and $\mathcal{L} = \mathbf{R}^{3k}$. Analogously, the stress field $\boldsymbol{\sigma}$ and the strain rate field $\mathbf{e}$ are represented by the members $\boldsymbol{\sigma} = (N_1, \ldots, N_m)$ and $\mathbf{e} = (e_1, \ldots, e_m)$ of the spaces $\mathcal{S} = \mathbf{R}^m$ and $\mathcal{E} = \mathbf{R}^m$. The strain rate field $\mathbf{e}$ corresponding to the velocity field $\mathbf{v}$ is determined by formula (IV.1.2), see also Section I.6.

The external power (the power of the load $\mathbf{l}$ on the velocity field $\mathbf{v}$) is

$$\langle\langle \mathbf{v}, \mathbf{l} \rangle\rangle = \mathbf{f}_1\mathbf{v}_1 + \cdots + \mathbf{f}_k\mathbf{v}_k,$$

while the internal power is

$$-\langle \mathbf{e}, \boldsymbol{\sigma} \rangle = -N_1 e_1 - \cdots - N_m e_m.$$

The virtual work principle formulates the equilibrium conditions for the truss in the following form:

$$\langle \operatorname{Def} \mathbf{v}, \boldsymbol{\sigma} \rangle = \langle\langle \mathbf{v}, \mathbf{l} \rangle\rangle \quad \text{for every} \quad \mathbf{v} \in U,$$

where $U$ is the set of virtual velocity fields. In the case of a truss, $U$ is nothing but the velocity fields space $\mathcal{V}$. The equilibrium conditions can also be written in the brief form: $\boldsymbol{\sigma} \in \Sigma + \mathbf{s}_l$, where $\Sigma = (\operatorname{Def} U)^0$ is the set of self-equilibrated stress fields and $\mathbf{s}_l$ is a fixed stress field equilibrating the given load, see Subsection IV.1.1.

The elastic-plastic constitutive relations for the $j$-th rod are of the form

$$\varepsilon_j = \varepsilon_j^e + \varepsilon_j^p, \quad \varepsilon_j^e = a_j N_j,$$
$$N_j \in C_j, \qquad \dot{\varepsilon}_j^p (N_j - N_*) \geq 0 \quad \text{for every} \quad N_* \in C_j$$

(no summation over $j$), where $\varepsilon_j$, $\varepsilon_j^e$ and $\varepsilon_j^p$ are the total, elastic and plastic elongations of the rod, $a_j$ is its elasticity coefficient, and

$$C_j = \{N \in \mathbf{R} : -Y_j^- \leq N \leq Y_j^+\}$$

is the set of admissible stresses, $Y_j^+$, $Y_j^-$ being the yield stresses in tension and compression. Recall that the collection of the constitutive maximum principles written for every rod of the truss is equivalent to the total principle (Example IV.1.1):

$$\boldsymbol{\sigma} \in C, \qquad \langle \mathbf{e}^p, \boldsymbol{\sigma} - \boldsymbol{\sigma}_* \rangle \geq 0 \quad \text{for every} \quad \boldsymbol{\sigma}_* \in C,$$
$$C = \{(N_1, \ldots, N_m) \in \mathcal{S} : N_i \in C_i,\ i = 1, \ldots, m\},$$

where $\langle \cdot, \cdot \rangle$ is the scalar multiplication in $\mathbf{R}^m$. We express the plastic strain rate $\mathbf{e}^p$ through $\mathbf{e}$ and $\boldsymbol{\sigma}$: $\mathbf{e}^p = \mathbf{e} - \mathbf{A}\dot{\boldsymbol{\sigma}}$, where $\mathbf{A}$ is the diagonal matrix with the components $a_{11} = a_1, \cdots, a_{mm} = a_m$.

Summarizing the above-mentioned relations we arrive to the following formulation of the elastic-plastic problem with the given initial conditions at $t = 0$ and the given loading $\mathbf{l}(t)$ at $t \geq 0$:

$$\boldsymbol{\sigma} \in \Sigma + \mathbf{s}_l,$$
$$\mathbf{e} = \operatorname{Def} \mathbf{v}, \quad \mathbf{v} \in \mathcal{V},$$
$$\boldsymbol{\sigma} \in C, \quad \langle \mathbf{e} - \mathbf{A}\dot{\boldsymbol{\sigma}}, \boldsymbol{\sigma} - \boldsymbol{\sigma}_* \rangle \geq 0 \quad \text{for every} \quad \boldsymbol{\sigma}_* \in C,$$
$$\dot{\mathbf{u}} = \mathbf{v}, \quad \mathbf{u}|_{t=0} = \mathbf{u}_0, \quad \boldsymbol{\sigma}|_{t=0} = \boldsymbol{\sigma}_0.$$

Recall that the unified formulation of the rigid-plastic problem satisfies conditions (I) – (VI) in Subsection IV.1.4 and note that the above formulation of the elastic-plastic problem also meets requirements (I) – (V).

EXAMPLE 2. *Three-dimensional body (strong formulation of the elastic-plastic problem).* Let us show that the strong formulation (1.3) – (1.6) of the elastic-plastic problem can be written in the same form as the problem for the truss in the previous example. To be more definite, we restrict ourselves to a body with bounded yield surfaces. The stress fields space $\mathcal{S}$ consists of bounded integrable fields in this case, while the set of all integrable fields (of symmetric second rank tenors) is considered as the space $\mathcal{E}$ of strain rate fields. The internal power is defined for every stress field $\boldsymbol{\sigma}$ in $\mathcal{S}$ and every strain rate field $\mathbf{e}$ in $\mathcal{E}$:

$$-\langle \mathbf{e}, \boldsymbol{\sigma} \rangle = -\int_\Omega \mathbf{e} \cdot \boldsymbol{\sigma}\, dV$$

($\Omega$ is the domain occupied by the body). Let $\mathcal{V} = V_0$ be the space of sufficiently regular kinematically admissible velocity fields (recall that they vanish on $S_v$). The external power of the load $\mathbf{l} = (\mathbf{f}, \mathbf{q})$ is defined for every $\mathbf{v} \in \mathcal{V}$:

$$\langle\langle \mathbf{v}, \mathbf{l} \rangle\rangle = \int_\Omega \mathbf{f}\mathbf{v}\, dV + \int_{S_q} \mathbf{q}\mathbf{v}\, dS,$$

the load $\mathbf{l}$ being a member of a certain linear space $\mathcal{L}$ of loads. The virtual work principle (1.6) can now be written as

$$\langle \operatorname{Def} \mathbf{v}, \boldsymbol{\sigma} \rangle = \langle\langle \mathbf{v}, \mathbf{l} \rangle\rangle \quad \text{for every} \quad \mathbf{v} \in U$$

or, equivalently, as $\boldsymbol{\sigma} \in \Sigma + \mathbf{s}_l$, where $\mathbf{s}_l$ is a fixed stress field equilibrating the load $\mathbf{l}$ and $\Sigma$ is the set of self-equilibrated stress fields:

$$\Sigma = \{\boldsymbol{\sigma} \in \mathcal{S} : \langle \operatorname{Def} \mathbf{v}, \boldsymbol{\sigma} \rangle = 0 \text{ for every } \mathbf{v} \in U\} = (\operatorname{Def} U)^0$$

(Subsection IV.1.1). Let us write kinematic relations (1.3) in the form $\mathbf{e} = \operatorname{Def} \mathbf{v}$, $\mathbf{v} \in \mathcal{V}$. The point-wise formulation of constitutive relations (1.4) is equivalent to their integral formulation (Subsection IV.1.3 and Section IV.4):

$$\boldsymbol{\sigma} \in C, \qquad \langle \mathbf{e}^p, \boldsymbol{\sigma} - \boldsymbol{\sigma}_* \rangle \geq 0 \quad \text{for every} \quad \boldsymbol{\sigma}_* \in C,$$

where $C$ is the set of admissible stress fields, and $\mathbf{e}^p = \mathbf{e} - \mathbf{A}\dot{\boldsymbol{\sigma}}$ is the plastic strain rate.

Finally, the strong formulation of the elastic-plastic problem with the given initial conditions at $t = 0$ and the given loading $\mathbf{l}(t)$ can be written down as

$$\begin{gathered} \boldsymbol{\sigma} \in \Sigma + \mathbf{s}_l, \\ \mathbf{e} = \operatorname{Def} \mathbf{v}, \quad \mathbf{v} \in \mathcal{V}, \\ \boldsymbol{\sigma} \in C, \quad \langle \mathbf{e} - \mathbf{A}\dot{\boldsymbol{\sigma}}, \boldsymbol{\sigma} - \boldsymbol{\sigma}_* \rangle \geq 0 \text{ for every } \boldsymbol{\sigma}_* \in C, \\ \dot{\mathbf{u}} = \mathbf{v}, \quad \mathbf{u}|_{t=0} = \mathbf{u}_0, \quad \boldsymbol{\sigma}|_{t=0} = \boldsymbol{\sigma}_0. \end{gathered}$$

This formulation is completely analogous to that of the elastic-plastic problem formulation for trusses (see Example 1).

Let us mention some additional features of the above formulation. They arise from the elastic properties of the body or, in other words, from properties of the elastic modules tensor $\mathbf{A}$.

Recall that $\mathbf{A}$ is always assumed to be symmetric and positive definite. We also assume that $\mathbf{A}$ is bounded: there are positive constants $a_-$, $a_+$ such that, for every point $x$ of the body and every symmetric tensor $\mathbf{s}$, the following equalities are valid:

$$a_- |\,\mathbf{s}\,|^2 \leq A_{ijkl}(x) s_{ij} s_{kl} \leq a_+ |\,\mathbf{s}\,|^2. \tag{1.8}$$

Note that

$$\frac{1}{2}\langle \boldsymbol{\varepsilon}^e, \boldsymbol{\sigma} \rangle = \frac{1}{2}\langle \mathbf{A}\boldsymbol{\sigma}, \boldsymbol{\sigma} \rangle = \frac{1}{2}\int_\Omega A_{ijkl}(x)\sigma_{ij}(x)\sigma_{kl}(x)\, dV$$

is the elastic energy.

The bilinear form

$$\langle \boldsymbol{\varepsilon}, \boldsymbol{\sigma} \rangle = \int_\Omega \boldsymbol{\varepsilon}(x) \cdot \boldsymbol{\sigma}(x)\, dV$$

is defined not only for every $\boldsymbol{\varepsilon}$ in $\mathcal{E}$ (which is integrable) and every $\boldsymbol{\sigma}$ in $\mathcal{S}$ (which is bounded and integrable), but also for any symmetric second rank tensor fields $\boldsymbol{\varepsilon}$ and $\boldsymbol{\sigma}$ with the bounded integrals

$$\int_\Omega |\,\boldsymbol{\varepsilon}\,|^2\, dV, \qquad \int_\Omega |\,\mathbf{s}\,|^2\, dV.$$

Such fields form a linear space $\mathcal{H} = L_2(\Omega; Sym)$ with the scalar multiplication determined by the bilinear form $\langle \boldsymbol{\varepsilon}, \mathbf{s} \rangle$ and the corresponding norm

$$\| \mathbf{s} \|_{\mathcal{H}}^2 = \langle \mathbf{s}, \mathbf{s} \rangle = \int_{\Omega} | \mathbf{s} |^2 \, dV$$

(see Example V.A.4). It is clear that $\mathcal{E}$ contains $\mathcal{H}$ and $\mathcal{H}$ contains $\mathcal{S}$: $\mathcal{E} \supset \mathcal{H} \supset \mathcal{S}$.

The elastic modules tensor generates a linear operator on $\mathcal{H}$ which maps a field $\mathbf{s} \in \mathcal{H}$ on the field $\mathbf{As} \in \mathcal{H}$:

$$(\mathbf{As}(x))_{ij} = A_{ijkl}(x) s_{kl}(x).$$

Note that for any two fields $\boldsymbol{\varepsilon}$ and $\boldsymbol{\sigma}$ in $\mathcal{H}$, the equality $\langle \mathbf{A}\boldsymbol{\varepsilon}, \mathbf{s} \rangle = \langle \boldsymbol{\varepsilon}, \mathbf{As} \rangle$ is valid due to the symmetry of the elastic modules tensor.

The quadratic form $a(\mathbf{s}) = \frac{1}{2}\langle \mathbf{As}, \mathbf{s} \rangle$ is also defined on $\mathcal{H}$; we refer to it as *elastic energy*, and this is the meaning of $a(\mathbf{s})$ when $\mathbf{s}$ is the stress field in an elastic-plastic body. The elastic energy is positive definite due to (1.8), and the two-sided estimation for its values

$$c_-\langle \mathbf{s}, \mathbf{s} \rangle \leq a(\mathbf{s}) \leq c_+\langle \mathbf{s}, \mathbf{s} \rangle$$

is valid with certain constants $c_-, c_+ > 0$.

Thus, there are

1) the linear space $\mathcal{H}$, $\mathcal{S} \subset \mathcal{H} \subset \mathcal{E}$, with the bilinear form $\langle \cdot, \cdot \rangle$ defining scalar multiplication in $\mathcal{H}$,

2) the linear operator $\mathbf{A}$ which maps $\mathcal{H}$ into $\mathcal{H}$, $(\mathbf{As})_{ij} = A_{ijkl}s_{kl}$, and is symmetric: $\langle \mathbf{A}\boldsymbol{\varepsilon}, \mathbf{s} \rangle = \langle \boldsymbol{\varepsilon}, \mathbf{As} \rangle$,

3) the elastic energy quadratic form $a(\mathbf{s}) = \frac{1}{2}\langle \mathbf{As}, \mathbf{s} \rangle$ on $\mathcal{H}$ with the two-sided estimation

$$c_-\langle \mathbf{s}, \mathbf{s} \rangle \leq a(\mathbf{s}) \leq c_+\langle \mathbf{s}, \mathbf{s} \rangle, \qquad c_- = \text{const} > 0, \quad c_+ = \text{const} > 0.$$

These properties of the elastic-plastic problem formulation are very important for the shakedown theory.

**1.4. General formulation of the problem.** We now give a general, unified formulation of the elastic-plastic problem. This formulation covers all types of elastic perfectly plastic systems: solids, beams, trusses, etc. The unified formulation only keeps the most general common properties of concrete formulations like those in Examples 1 and 2. At the same time, these properties are all one needs to develop the shakedown theory. In this connection it is also important that the general formulation does not make use of any smoothness assumptions. Therefore, the results obtained within the framework of this formulation hold in case no smoothness is guaranteed for a solution to the elastic-plastic problem.

We keep using three-dimensional terminology to avoid introducing new terms. We take the term body to mean not only three-dimensional bodies but also beams, shells and other systems, the term stress to mean all kinds of internal forces such as cutting force or moments. We speak about fields although we consider not only continua but also discrete systems, whose state is described by functions of a discrete variable.

Consider a body whose stressed state and location at any moment $t \geq 0$ are described by the stress field $\boldsymbol{\sigma}_t$ and displacement field $\mathbf{u}_t$. We denote their time derivatives by $\dot{\boldsymbol{\sigma}}_t = \partial \boldsymbol{\sigma}_t / \partial t$. In the sequel, we will omit the subscript $t$, so that $\boldsymbol{\sigma}$ will stand for the stress field and $\dot{\mathbf{u}} = \mathbf{v}$ for the velocity field. For example, in the case of a three-dimensional body, $\boldsymbol{\sigma}$ is a tensor-valued function defined on the domain $\Omega$ occupied by the body, and this function also depends on the parameter $t$. Thus it should be remembered that loads, stress, velocity and other variables depend on time although the parameter $t$ is not involved in their notation.

Let the body undergo small strains and displacements under the action of the load $\mathbf{l}$. The load is assumed to vary slowly enough to result only in small accelerations that can be neglected. In other words, we adopt the quasistatic approximation: the stress field $\boldsymbol{\sigma}$ equilibrates the given load $\mathbf{l}$ at every moment $t$. The general formulation of the quasistatic elastic-plastic problem is as follows.

An elastic perfectly plastic body and loads are characterized by:

(I) the linear space $\mathcal{V}$ of kinematically admissible displacement and velocity fields $\mathbf{u}$ and $\mathbf{v} = \dot{\mathbf{u}}$ with the subspace $U$ of virtual velocity fields;
the linear space $\mathcal{L}$ of loads;
the external power bilinear form $\langle\langle \cdot, \cdot \rangle\rangle$ defined on $\mathcal{V} \times \mathcal{L}$; its value $\langle\langle \mathbf{v}, \mathbf{l} \rangle\rangle$ is the power of the load $\mathbf{l}$ on the velocity field $\mathbf{v}$;

(II) the linear space $\mathcal{S}$ of stress fields;
the linear space $\mathcal{E}$ of strain rate fields;
the internal power bilinear form $\langle\langle \cdot, \cdot \rangle\rangle$; the value $-\langle \mathbf{e}, \boldsymbol{\sigma} \rangle$ is the power of the stress field $\boldsymbol{\sigma}$ on the strain rate field $\mathbf{e}$;
if $\langle \mathbf{e}, \boldsymbol{\sigma}_0 \rangle = 0$ for every $\mathbf{e} \in \mathcal{E}$, then $\boldsymbol{\sigma}_0 = 0$;
if $\langle \mathbf{e}_0, \boldsymbol{\sigma} \rangle = 0$ for every $\boldsymbol{\sigma} \in \mathcal{S}$, then $\mathbf{e}_0 = 0$.

(III) the linear operator Def on $\mathcal{V}$ with values in $\mathcal{E}$; its value $\mathrm{Def}\,\mathbf{v}$ is the strain rate field corresponding to the velocity field $\mathbf{v}$;
the only velocity field with the zero strain rate field is zero;

(IV) the virtual work principle: a stress field $\boldsymbol{\sigma}$ equilibrates the load $\mathbf{l}$ if and only if

$$\langle \mathrm{Def}\,\mathbf{v}, \boldsymbol{\sigma} \rangle = \langle\langle \mathbf{v}, \mathbf{l} \rangle\rangle \quad \text{for every} \quad \mathbf{v} \in U;$$

if $\boldsymbol{\sigma}$ satisfies the virtual work principle, the equality $\langle \mathrm{Def}\,\mathbf{v}, \boldsymbol{\sigma} \rangle = \langle\langle \mathbf{v}, \mathbf{l} \rangle\rangle$ also holds for every $\mathbf{v} \in \mathcal{V}$;

(V) the linear operator $\mathbf{A}$ on $\mathcal{S}$ with values in $\mathcal{E}$, its value $\boldsymbol{\varepsilon}^e = \mathbf{A}\boldsymbol{\sigma}$ being the elastic strain field corresponding to the stress field $\boldsymbol{\sigma}$.

(VI) the linear space $\mathcal{H}$ of stress rate fields $\dot{\boldsymbol{\sigma}}$;
$\mathcal{S}$ is embedded in $\mathcal{H}$, $\mathcal{H}$ is embedded in $\mathcal{E}$: $\mathcal{S} \subset \mathcal{H} \subset \mathcal{E}$;

the bilinear form $\langle \cdot, \cdot \rangle$ is also defined on $\mathcal{H}$ and determines scalar multiplication and the corresponding norm $\|\mathbf{s}\|_{\mathcal{H}} = \sqrt{\langle \mathbf{s}, \mathbf{s} \rangle}$ in $\mathcal{H}$;
the operator $\mathbf{A}$ is also defined on $\mathcal{H}$ and takes its values in $\mathcal{H}$;
$\mathbf{A}$ is symmetric: $\langle \mathbf{A}\boldsymbol{\varepsilon}, \mathbf{s} \rangle = \langle \boldsymbol{\varepsilon}, \mathbf{A}\mathbf{s} \rangle$ for every $\boldsymbol{\varepsilon}$ and $\mathbf{s}$ in $\mathcal{H}$;
the quadratic form $a(\mathbf{s}) = \frac{1}{2}\langle \mathbf{s}, \mathbf{s} \rangle$ is also defined on $\mathcal{H}$, its value $a(\mathbf{s})$ being the elastic energy corresponding to the stress field $\mathbf{s}$;
the two-sided estimation $c_-\langle \mathbf{s}, \mathbf{s} \rangle \leq a(\mathbf{s}) \leq c_+\langle \mathbf{s}, \mathbf{s} \rangle$ is valid with positive constants $c_-$ and $c_+$;

(VII) the set $C$ of admissible stress fields in $\mathcal{S}$;
$C$ is convex and contains the origin;

(VIII) the constitutive maximum principle: the stress and plastic strain rate fields $\boldsymbol{\sigma}$ and $\mathbf{e} - \mathbf{A}\dot{\boldsymbol{\sigma}}$ (where $\mathbf{e}$ is the total strain rate) satisfy the constitutive relations if and only if

$$\boldsymbol{\sigma} \in C, \quad \langle \mathbf{e} - \mathbf{A}\dot{\boldsymbol{\sigma}}, \boldsymbol{\sigma} - \boldsymbol{\sigma}_* \rangle \geq 0 \quad \text{for every} \quad \boldsymbol{\sigma}_* \in C.$$

*Solution to the elastic-plastic problem* for the given initial displacement and stress fields $\mathbf{u}_0$ and $\boldsymbol{\sigma}_0$ and loading $\mathbf{l}$ is the stress and velocity fields $\boldsymbol{\sigma}$ and $\mathbf{v}$ satisfying the relations:

$$\begin{gathered}
\boldsymbol{\sigma} \in \Sigma + \mathbf{s}_l, \\
\mathbf{e} = \operatorname{Def} \mathbf{v}, \quad \mathbf{v} \in \mathcal{V}, \\
\boldsymbol{\sigma} \in C, \quad \langle \mathbf{e} - \mathbf{A}\dot{\boldsymbol{\sigma}}, \boldsymbol{\sigma} - \boldsymbol{\sigma}_* \rangle \geq 0 \quad \text{for every} \quad \boldsymbol{\sigma}_* \in C, \\
\dot{\mathbf{u}} = \mathbf{v}, \quad \mathbf{u}|_{t=0} = \mathbf{u}_0, \quad \boldsymbol{\sigma}|_{t=0} = \boldsymbol{\sigma}_0.
\end{gathered} \tag{1.9}$$

Here, $\mathbf{s}_l$ is a (time dependent) stress field equilibrating the given load $\mathbf{l}$ and $\Sigma = (\operatorname{Def} U)^0$ is the subspace of self-equilibrated stress fields. The solution $\boldsymbol{\sigma}$, $\mathbf{v}$ also depends on $t$. The formulation of the equilibrium conditions in (1.9) is a form of the virtual work principle, see Subsection IV.I.I.

REMARK 1. Due to (IV) the definition of the self-equilibrated stress fields set $\Sigma = (\operatorname{Def} U)^0$ can also be written as $\Sigma = (\operatorname{Def} \mathcal{V})^0$. In particular, the equality $\langle \mathbf{e}, \boldsymbol{\sigma} \rangle = 0$ holds for every kinematically admissible strain rate field $\mathbf{e} = \operatorname{Def} \mathbf{v}$, $\mathbf{v} \in \mathcal{V}$ and every self-equilibrated stress field $\boldsymbol{\sigma} \in \Sigma$.

REMARK 2. Requirements (I) – (VIII) contain some hypotheses about solutions to elastic-plastic problems. For example, we assumed that displacement and velocity fields belong to the same space $\mathcal{V}$ while stress fields and stress rate fields are in $\mathcal{S}$ and $\mathcal{H}$, respectively. The assumptions are not arbitrary, but every solution to the elastic-plastic problem satisfies them as it follows from the existence theorem (not considered in this book). Thus, any statement established within the framework of general formulation (1.9) is valid for all solutions to the elastic-plastic problems.

REMARK 3. The loading $\mathbf{l}$ and the initial stress field $\boldsymbol{\sigma}_0$ in elastic-plastic problem formulation (1.9) are to be consistent. In fact, at $t = 0$ the load $\mathbf{l}(0)$ is equilibrated by the stress field $\boldsymbol{\sigma}_0$ and, therefore, is uniquely determined as soon as the latter is given.

The elastic-plastic problem formulations for a three-dimensional body and a truss (Examples 1 and 2) are particular cases of the unified formulation. Let us now consider another example of a concrete formulation.

EXAMPLE 3. *Discrete system.* A discrete system admits a description making use of finite-dimensional spaces of velocity, strain rate and stress fields (see Subsection IV.2.2). Here, we temporarily leave the above terminology convention and simply speak about velocities, strain rates, etc., and not about fields.

The column-matrices

$$\mathbf{u} = (u_1, \ldots, u_n)^T, \quad \mathbf{v} = (v_1, \ldots, v_n)^T, \quad \mathbf{l} = (l_1, \ldots, l_n)^T,$$
$$\boldsymbol{\sigma} = (\sigma_1, \ldots, \sigma_m)^T, \quad \mathbf{e} = (e_1, \ldots, e_m)^T$$

represent displacements, velocities, loads, strain rates and stresses (the subscript $T$ indicates transposition). The external and interval powers are

$$\langle\langle \mathbf{v}, \mathbf{l} \rangle\rangle = \mathbf{l}^T \mathbf{v} \quad \text{and} \quad -\langle \mathbf{e}, \boldsymbol{\sigma} \rangle = -\mathbf{e}^T \boldsymbol{\sigma}.$$

The operator Def which maps velocity $\mathbf{v}$ on the corresponding strain rate Def $\mathbf{v}$ is represented by a $m \times n$-matrix; we denote the latter by $D$. The equilibrium conditions are given by the virtual work principle and written as $\boldsymbol{\sigma} \in \Sigma + \mathbf{s}_l$, where $\Sigma$ stands for the subspace of self-equilibrated stresses (the equilibrium conditions can be also written as $D^T \boldsymbol{\sigma} + \mathbf{l} = 0$, see Subsection IV.2.2).

The plastic strain rate in the discrete system is $\mathbf{e} - A\dot{\boldsymbol{\sigma}}$, where $A$ is the $m \times m$-matrix of elasticity modules. We adopt the constitutive maximum principle for the plastic strain rates, the set $C$ of admissible stresses being given in the stress space $\mathcal{S} = \mathbf{R}^m$.

We supplement the above relations with the initial conditions and arrive at the following formulation of the elastic-plastic problem for the discrete system:

$$D^T \boldsymbol{\sigma} + \mathbf{l} = 0,$$
$$\mathbf{e} = D\mathbf{v}, \quad \mathbf{v} \in \mathcal{V},$$
$$\boldsymbol{\sigma} \in C, \quad (\mathbf{e} - A\dot{\boldsymbol{\sigma}})^T (\boldsymbol{\sigma} - \boldsymbol{\sigma}_*) \geq 0 \ \text{ for every } \ \boldsymbol{\sigma}_* \in C,$$
$$\dot{\mathbf{u}} = \mathbf{v}, \quad \mathbf{u}|_{t=0} = \mathbf{u}_0, \quad \boldsymbol{\sigma}|_{t=0} = \boldsymbol{\sigma}_0.$$

It is clear that the formulation meets requirements (I) – (VIII) and is a realization of the unified formulation of the elastic-plastic problem.

**1.5. Formulation in stresses.** It is easy to eliminate the kinematic variables from the unified formulation of the elastic-plastic problem. Indeed, let us use in the constitutive maximum principle only admissible stress fields $\boldsymbol{\sigma}_*$ which satisfy the additional condition $\boldsymbol{\sigma}_* \in \Sigma + \mathbf{s}_l$, that is, equilibrate the load $\mathbf{l}$. Then the difference $\boldsymbol{\sigma} - \boldsymbol{\sigma}_*$ is a self-equilibrated stress field, $\boldsymbol{\sigma} - \boldsymbol{\sigma}_* \in \Sigma$, and the equality $\langle \mathbf{e}, \boldsymbol{\sigma} - \boldsymbol{\sigma}_* \rangle = 0$ is valid according to Remark 1. It turns out that the kinematic variables are no longer involved in the constitutive maximum principle when $\boldsymbol{\sigma}_*$ is considered in $C \cap (\Sigma + \mathbf{s}_l)$. Then we extract from (1.9) the relations concerning the stress field:

$$\begin{gathered} \boldsymbol{\sigma} \in \Sigma + \mathbf{s}_l, \quad \boldsymbol{\sigma} \in C, \quad \boldsymbol{\sigma}|_{t=0} = \boldsymbol{\sigma}_0, \\ \langle -\mathbf{A}\dot{\boldsymbol{\sigma}}, \boldsymbol{\sigma} - \boldsymbol{\sigma}_* \rangle \geq 0 \ \text{ for every } \ \boldsymbol{\sigma}_* \in C \cap (\Sigma + \mathbf{s}_l) \end{gathered} \tag{1.10}$$

and, thus, obtain the general formulation of the elastic-plastic problem *in stresses*.

It is clear that the stress field of any solution to elastic-plastic problem (1.9) also solves (1.10). At the same time, the original problem (1.9) and the problem in stresses (1.10) are more closely related. Namely, we will now show that the problem in stresses can only have a unique solution $\boldsymbol{\sigma}$. Consequently, $\boldsymbol{\sigma}$ is nothing but the stress field of the solution to the original problem (1.9) (if both problems are solvable).

Let us verify the uniqueness of the solution to the elastic-plastic problem in stresses. Suppose $\boldsymbol{\sigma}_1$ and $\boldsymbol{\sigma}_2$ are two solutions to (1.10) and consider the admissible stress field $\boldsymbol{\sigma}_* = \boldsymbol{\sigma}_2$ in the constitutive maximum principle for the solution $\boldsymbol{\sigma}_1$ and, analogously, $\boldsymbol{\sigma}_* = \boldsymbol{\sigma}_1$ for the solution $\boldsymbol{\sigma}_1$. This results in the inequalities

$$\langle -\mathbf{A}\dot{\boldsymbol{\sigma}}_1, \boldsymbol{\sigma}_1 - \boldsymbol{\sigma}_2 \rangle \geq 0, \qquad \langle -\mathbf{A}\dot{\boldsymbol{\sigma}}_2, \boldsymbol{\sigma}_2 - \boldsymbol{\sigma}_1 \rangle \geq 0,$$

which together with symmetry of the operator $\mathbf{A}$ result in the estimation for the time derivative

$$\begin{aligned} \frac{d}{dt}\langle \mathbf{A}(\boldsymbol{\sigma}_1 - \boldsymbol{\sigma}_2), \boldsymbol{\sigma}_1 - \boldsymbol{\sigma}_2 \rangle &= \langle \mathbf{A}(\dot{\boldsymbol{\sigma}}_1 - \dot{\boldsymbol{\sigma}}_2), \boldsymbol{\sigma}_1 - \boldsymbol{\sigma}_2 \rangle + \langle \mathbf{A}(\boldsymbol{\sigma}_1 - \boldsymbol{\sigma}_2), \dot{\boldsymbol{\sigma}}_1 - \dot{\boldsymbol{\sigma}}_2 \rangle \\ &= \langle \mathbf{A}\dot{\boldsymbol{\sigma}}_1, \boldsymbol{\sigma}_1 - \boldsymbol{\sigma}_2 \rangle + \langle \mathbf{A}\dot{\boldsymbol{\sigma}}_2, \boldsymbol{\sigma}_2 - \boldsymbol{\sigma}_1 \rangle \\ &\leq 0. \end{aligned}$$

Recall that both $\boldsymbol{\sigma}_1$ and $\boldsymbol{\sigma}_2$ satisfy the same initial condition: $\boldsymbol{\sigma}_1|_{t=0} = \boldsymbol{\sigma}_2|_{t=0} = \boldsymbol{\sigma}_0$, and therefore $(\boldsymbol{\sigma}_1 - \boldsymbol{\sigma}_2)|_{t=0} = 0$. Then the previous inequality and the positive definiteness of the quadratic form $\langle \mathbf{s}, \mathbf{A}\mathbf{s} \rangle$ result in the equality $\boldsymbol{\sigma}_1 = \boldsymbol{\sigma}_2$ at every $t \geq 0$. Thus, the elastic-plastic problem in stresses can have only one solution. This implies that the solution to original elastic-plastic problem (1.9) is also unique in stresses.

REMARK 4. There are examples showing that the solution to problem (1.9) is nonunique in displacements and velocities even for very simple discrete systems subjected to safe loads. Therefore, formulation (1.10) of the problem in stresses is of utmost importance for the elastic-plastic analysis.

## 2. Elastic-plastic body under variable load: examples

Nonzero stresses normally remain in a body unloaded after plastic deformation; they are referred to as residual. The residual stresses can neutralize the stresses caused by the subsequent loading. In some cases they even completely prevent plastic deformation, although the same loading, when applied to the body free of residual stresses, results in plastic flow. In other cases the

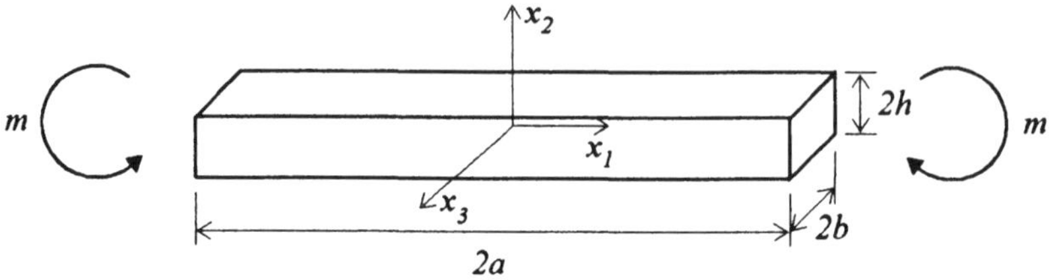


**Figure 2.1**

neutralizing effect of the residual stresses substantially restrains the plastic flow. In both cases we say that the body shakes down to the loading.

Examples of the typical response of elastic-plastic bodies to variable loadings are presented in this section to illustrate the body behavior when shakedown occurs and when it does not. We will precisely define the concept of shakedown in the next section and present the shakedown theory in subsequent sections. The goal of the theory is developing methods to answer the question of whether a given body shakes down to a given class of loadings or not.

**2.1. Residual stresses and shakedown.** Let us first consider an example showing formation of residual stresses and clarifying the way they prevent the subsequent plastic flow or, in other words, result in shakedown.

Consider a beam of rectangular cross-section subjected to the bending moments applied at the ends of the beam. We use the Cartesian coordinate system $x_1$, $x_2$, $x_3$ with its origin in the beam's center of symmetry and the basis vector $\mathbf{e}_1$ directed along the beam's axis, the other basis vectors $\mathbf{e}_2$, $\mathbf{e}_3$ being directed along the symmetry axes of the cross-section (see Figure 2.1).

Let the lateral surface of the beam be free and the load at each of its ends have the zero resultant and the total moments $m\mathbf{e}_3$ and $-m\mathbf{e}_3$ at the left and right ends, respectively. Since no load is applied to the lateral surface of the beam, the stress $\boldsymbol{\sigma}$ satisfies the boundary conditions

$$\sigma_{12} = \sigma_{22} = \sigma_{23} = 0 \text{ at } x_2 = \pm h, \quad \sigma_{13} = \sigma_{23} = \sigma_{33} = 0 \text{ at } x_3 = \pm b.$$

In case the beam is thin, that is, $b$, $h \ll a$, these conditions allow us to neglect the components $\sigma_{12}$, $\sigma_{13}$, $\sigma_{22}$, $\sigma_{23}$, $\sigma_{33}$ in the part of the beam sufficiently distant from its ends. We only focus on the behavior of this part of the beam and assume $\sigma_{11}$ to be the only nonzero stress component in it. Then the equilibrium equations are reduced to the relation $\partial\sigma_{11}/\partial x_1 = 0$, which means that $\sigma_{11}$ is a function of the coordinates $x_2$, $x_3$ alone. We also assume that far enough from the ends of the thin beam the stress is determined by the resultant and total moment of the loads applied to the ends, that is, does not depend on details of the load distribution. Then, the equilibrium boundary conditions at the ends of the beam are replaced by the integral equalities

$$\int_S \sigma_{11}\, dS = 0, \qquad \int_S (x_2\mathbf{e}_2 + x_3\mathbf{e}_3) \times \sigma_{11}\mathbf{e}_1\, dS = -m\mathbf{e}_3, \tag{2.1}$$

where $S$ stands for the cross-section of the beam.

We adopt the elastic-plastic constitutive relations for the material of the beam: Hooke's law for elastic strain and the normality flow rule associated with the Mises yield surface. Let us consider the response of the beam 1) to the initial loading starting at the unstressed and unstrained configuration: the moment $m$ increases from 0 up to $M$; 2) to the subsequent unloading: $m$ decreases from $M$ to 0; and 3) to the repeated loading – unloading cycles.

*Under the initial loading*, the behavior of the beam is purely elastic as long as the moment $m$ is sufficiently small. The elastic response of the beam to the loading is described by the following equations for the stress $\sigma_{11} = \sigma(x_2, x_3)$ and displacement $\mathbf{u}(x_1, x_2, x_3)$:

$$\frac{\partial u_1}{\partial x_1} = \frac{1}{\mathrm{E}}\sigma, \qquad \frac{\partial u_2}{\partial x_3} + \frac{\partial u_3}{\partial x_2} = 0, \tag{2.2}$$

$$\frac{\partial u_1}{\partial x_2} + \frac{\partial u_2}{\partial x_1} = 0, \qquad \frac{\partial u_2}{\partial x_2} = -\nu\frac{\partial u_1}{\partial x_1}, \tag{2.3}$$

$$\frac{\partial u_1}{\partial x_3} + \frac{\partial u_3}{\partial x_1} = 0, \qquad \frac{\partial u_3}{\partial x_3} = -\nu\frac{\partial u_1}{\partial x_1}. \tag{2.4}$$

These equations represent Hooke's law in case $\sigma_{11} = \sigma(x_2, x_3)$ is the only nonzero stress component (E is the Young's modules, $\nu$ is the Poisson's ratio). The first equation in (2.2) results in $\partial^2 u_1/\partial x_1^2 = 0$ since $\partial\sigma/\partial x_1 = 0$. Then (2.3) implies $\partial^2 u_1/\partial x_2^2 = 0$, and (2.4) implies $\partial^2 u_1/\partial x_3^2 = 0$. Consequently,

$$u_1 = \alpha_{12}x_1x_2 + \alpha_{13}x_1x_3 + \alpha_{23}x_2x_3 + \beta_1x_1 + \beta_2x_2 + \beta_3x_3 + \gamma,$$

where the coefficients $\alpha_{ij}$, $\beta_i$, $\gamma$ do not depend on $x_1$, $x_2$, $x_3$.

We restrict ourselves to the case of the displacement field being symmetric with respect to the $(x_1, x_2)$-plane, which is guaranteed, for example, when the given load possesses such symmetry. Then $\alpha_{13}$, $\alpha_{23}$ and $\beta_3$ vanish. Note also that eliminating rigid motion results in vanishing of $\gamma$, and the expressions for $u_1$ and $\sigma$ can be written as:

$$u_1 = \alpha x_1x_2 + \beta_1x_1 + \beta_2x_2, \qquad \sigma = \mathrm{E}(\alpha x_2 + \beta_1) \tag{2.5}$$

(here and in what follows, the symbol $\alpha$ is used instead of $\alpha_{12}$). The coefficients $\alpha_1$, $\beta_1$ are determined by conditions (2.1):

$$\alpha = \alpha(m) = m/J\mathrm{E}, \qquad \beta_1 = 0,$$

where $J = 4bh^3/3$. Thus, the stress is determined by the formula $\sigma = mx_2/J$ until the magnitude $|\sigma| = m|x_2|/J$ reaches the yield stress $\sigma_Y = \sqrt{3}k$ somewhere in the beam. It can be easily seen that this happens at both the lower and upper sides of the beam when the moment $m$ attains the value $m = m_Y = \sigma_Y J/h = 4\sigma_Y h^2 b/3$.

When the moment increases beyond this value, two plastic zones propagate from the surfaces $x_2 = \pm h$ inside the beam. Within the elastic zone located

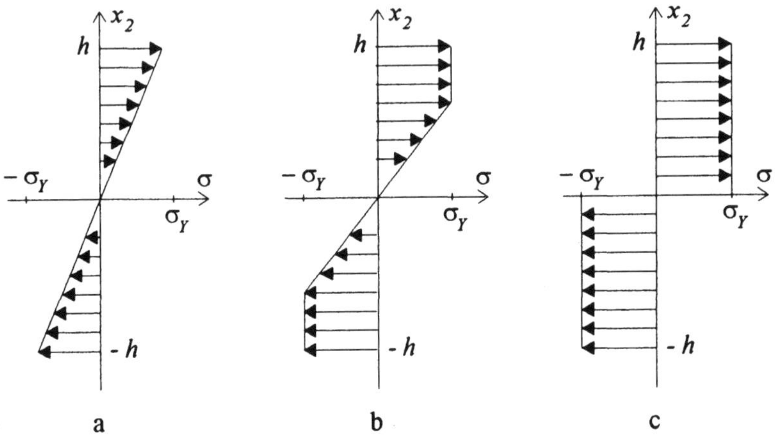


**Figure 2.2**

between the two plastic ones, equations (2.2) – (2.4) hold and, consequently, formulas (2.5) remain valid. In particular, the stress $\sigma$ depends linearly on $x_2$ within the elastic zone and reaches the yield stress on its boundaries. Within the plastic zones $|\sigma|$ equals the yield stress $\sigma_Y$, that is, $\sigma = \pm\sigma_Y$. It is clear that the stress is tensing within the upper plastic zone, which makes $\sigma = \sigma_Y$ in it, while $\sigma$ is compressing and equals $-\sigma_Y$ within the lower zone. The equalities $\sigma = \sigma_Y$, $\sigma = -\sigma_Y$ in the plastic zones and linearity of the stress distribution in the elastic zone reduce the first of conditions (2.1) to the equality $\beta_1 = 0$. Therefore, the stress distribution in the beam cross-section is given by the formula

$$\sigma(x_2) = \begin{cases} \sigma_Y & \text{if} \quad \sigma_Y/A \le x_2 \le h, \\ Ax_2 & \text{if} \quad -\sigma_Y/A \le x_2 \le \sigma_Y/A, \\ -\sigma_Y & \text{if} \quad -h \le x_2 \le -\sigma_Y/A. \end{cases}$$

Here, the coefficient $A$ is to be determined, $h - \sigma_Y/A$ being the thickness of each of the plastic zones. The second of conditions (2.1) determines the coefficient $A$ through the given moment $m$:

$$A = \frac{\sigma_Y}{h}\left(\frac{2b\sigma_Y h^2}{3(2\sigma_Y h^2 b - m)}\right)^{\frac{1}{2}} = \frac{\sigma_Y}{h}\left(\frac{m_Y}{3m_Y - 2m}\right)^{\frac{1}{2}} \quad \text{when } m > m_Y.$$

Figures 2.2a and 2.2b show the stress distributions at $m < m_Y$ and $m > m_Y$, respectively.

The above formula for the stress distribution is valid until the moment reaches the limit value $m_{\lim}$, which corresponds to the case of the beam cross-section completely covered by the plastic zones, see Figure 2.2c. We assume that the maximum moment $M$ is less than $m_{\lim}$; thus, the above formula describes the response of the beam to the whole of the initial loading, that is, increasing of the moment $m$ from 0 up to $M$. In particular, the stress

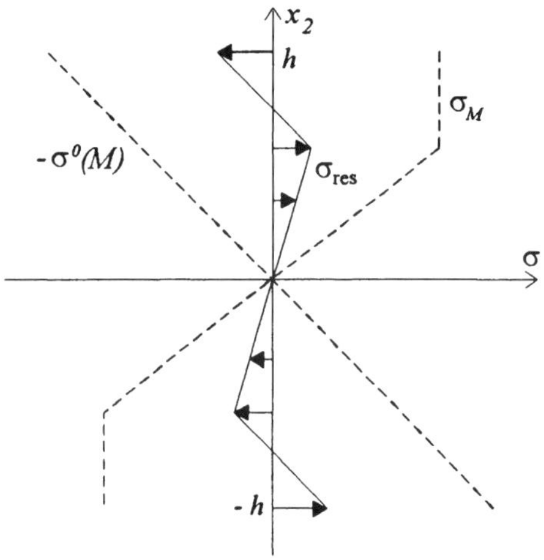


**Figure 2.3**

and velocity fields are determined at $m = M$, that is, at the end of the first loading; we will denote them by $\boldsymbol{\sigma}_M$ and $\mathbf{v}_M$, respectively.

Let us now describe the process in the beam under *unloading*, that is, decreasing of the moment from $M$ to 0, which we will consider as applying additional load to the beam: the negative moment $-\Delta m$. It is convenient to introduce an auxiliary elastic problem for a beam, namely, for the purely elastic beam of the same shape and under the same conditions as the elastic-plastic beam under consideration. Let $\boldsymbol{\sigma}^0(m')$ and $\mathbf{u}^0(m')$ be the stress and displacement fields in the elastic beam under the action of the moment $m'$. It is clear that the stress and displacement fields $\boldsymbol{\sigma}_M - \boldsymbol{\sigma}^0(m')$ and $\mathbf{u}_M - \mathbf{u}^0(m')$ with $m'$ ranging between 0 and $\Delta m$ solve the problem about unloading the elastic-plastic beam when the moment decreases from $M$ to $M - \Delta m$ *if the stresses* $\boldsymbol{\sigma}_M - \boldsymbol{\sigma}^0(m')$ *are admissible for every* $m' \in [0, \Delta m]$. Indeed, the following estimation is valid when $M < m_{\rm lim} = 2\sigma_Y h^2 b$:

$$|\boldsymbol{\sigma}^0(m')| = \frac{m'}{J}|x_2| < \frac{m_{\rm lim}}{J}|x_2| = \frac{3\sigma_Y |x_2|}{2h}$$

and implies the inequality $|\boldsymbol{\sigma}_M - \boldsymbol{\sigma}^0(m')| < \sigma_Y$ for $m' \in [0, M)$, that is, the stress field $\boldsymbol{\sigma}_M - \boldsymbol{\sigma}^0(m')$ is safe for such $m'$ and is admissible for $0 \le m' \le M$. Consequently, $\boldsymbol{\sigma}_M - \boldsymbol{\sigma}^0(m')$, $\mathbf{u}_M - \mathbf{u}^0(m')$ form a solution to the unloading problem for the moment $m = M - m'$, where $m'$ increases from 0 up to $M$.

It is worth noting that 1) the plastic strain does not vary during the unloading, and 2) the residual stress remains in the beam after the unloading, that is, at $m = 0$. Figure 2.3 shows the residual stress distribution.

Let us now consider the *second loading cycle*. At the beginning there are the residual stress and displacement fields

$$\boldsymbol{\sigma}_{\rm res} = \boldsymbol{\sigma}_M - \boldsymbol{\sigma}^0(M) \quad \text{and} \quad \mathbf{u}_{\rm res} = \mathbf{u}_M - \mathbf{u}^0(M).$$

Note that superimposing the elastic stress field $\boldsymbol{\sigma}^0(m)$ on $\boldsymbol{\sigma}_{\text{res}}$ results in the safe stress field

$$\boldsymbol{\sigma}_{\text{res}} + \boldsymbol{\sigma}^0(m) = \boldsymbol{\sigma}_M - \boldsymbol{\sigma}^0(M) + \boldsymbol{\sigma}^0(m) = \boldsymbol{\sigma}_M - \boldsymbol{\sigma}^0(M - m)$$

for every $m$, $0 \leq m < M$. Safety of this field has already been established: it is the actual stress field in the beam at $m' = M - m$ during the previous unloading. Therefore, the stress and displacement fields

$$\boldsymbol{\sigma}_{\text{res}} + \boldsymbol{\sigma}^0(m) = \boldsymbol{\sigma}_M - \boldsymbol{\sigma}^0(M - m), \quad \mathbf{u}_{\text{res}} + \mathbf{u}^0(m) = \mathbf{u}_M - \mathbf{u}^0(M - m)$$

form a solution to the elastic-plastic problem for the second loading. At $m \leq M$ the solution returns to the state reached at the end of the first loading: $\boldsymbol{\sigma} = \boldsymbol{\sigma}_M$, $\mathbf{u} = \mathbf{u}_M$, and the plastic strain remains unchanged during the second loading. Therefore, the response of the beam to the subsequent (second) unloading replicates that of the first unloading.

It is clear that under the third and *subsequent loading cycles* the stressed-strained state of the beam varies cyclically following the pattern of the second cycle.

It should be emphasized that this cyclic response with the unvarying plastic strain is made possible by the residual stress field remaining in the beam after the first loading cycle. The residual stresses neutralize those caused by the subsequent loading, and the actual stresses are safe.

REMARK 1. For brevity we did not consider the displacement field of the above solution. It is easy to find them analytically, at least, if the material of the beam is incompressible.

**2.2. Nonshakedown at bounded plastic strain.** The elastic-plastic body does not necessarily shake down to a variable loading. Even some of the simplest cyclic loadings may result in nonstop plastic deformation in contrast to the example in the previous subsection. Let us consider an example showing the possibility of unlimited accumulation of the plastic work (which is nonshakedown), although the plastic strain remains bounded.

Consider the structure consisting of three rods of the same length. The rods are parallel, one end of each rod being fixed and the other ends connected with a rigid beam; see Figure 2.4a. The structure is symmetric with respect to the axis of rod 1, in particular, rods 2 and 3 are identical. The cross-section area of rods 2 and 3 is $S/2$, that of rod 1 is $S$. A uniformly distributed load parallel to the rods is applied to the rigid beam, its resultant being $p$. The strain, which is the same for all the rods, is denoted by $\varepsilon$. Stresses in rods 2 and 3 are equal and denoted by $\sigma''$, while $\sigma'$ stands for the stress in rod 1. The equilibrium condition for the structure is written as $(\sigma' + \sigma'')S = p$.

We adopt the elastic perfectly plastic constitutive relations for the rods, assuming the Young's modules and the yield stress of rods 2 and 3 to be greater than those of rod 1. Figure 2.4b shows the strain-stress diagrams for the rods under uniaxial tension.

Let us consider the response of the structure 1) to the initial loading starting at the unstrained and unstressed configuration: the load $p$ increases from 0

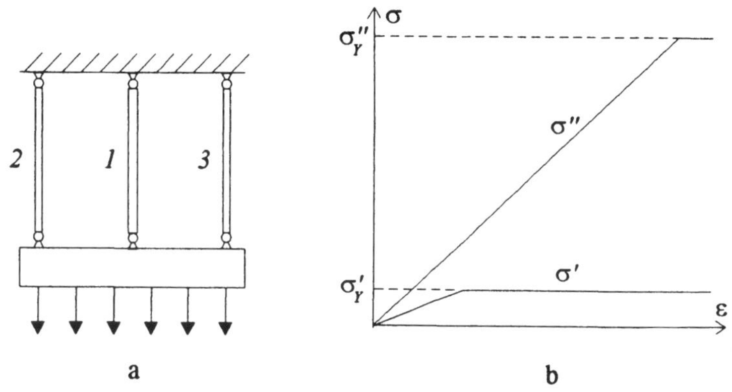


**Figure 2.4**

up to $P$; 2) to the unloading: $p$ decreases from $P$ to 0; and 3) to the repeated loading – unloading cycles.

Let the ratio of the yield stress $\sigma_Y''$ of rods 2, 3 to the yield stress $\sigma_Y'$ of rod 1 be sufficiently large to ensure that the yield stress is not reached in rods 2 and 3 during the initial loading, while rod 1 undergoes plastic deformation at the end of the loading. The state of rod 1 at the end of the initial loading corresponds to the point $A'$ in Figure 2.5, the state of rods 2, 3 to the point $A''$.

When the unloading starts and the load $p$ decreases, the point corresponding to the current state of rods 2, 3 moves down along the elastic part of the stress-strain diagram. The stress $\sigma''$ in these rods is related to the strain $\varepsilon$ by Hooke's law, $\sigma'' = \mathrm{E}''\varepsilon$ (E'' stands for the Young's modules of rods 2, 3). The point corresponding to the current state of rod 1 moves first along the unloading part $A'B'$ of the stress-strain diagram, then along its elastic compression part $B'C'$ and, finally, along the compression yield plateau $C'O'$. The process in rod 1 is described by the following relation between the stress $\sigma'$ and the strain $\varepsilon$:

$$\sigma' = \begin{cases} \sigma_Y' + \mathrm{E}'(\varepsilon - \varepsilon_A) & \text{if} \quad \varepsilon_A - 2\sigma_Y'/\mathrm{E}' \le \varepsilon \le \varepsilon_A, \\ -\sigma_Y' & \text{if} \quad \varepsilon \le \varepsilon_A - 2\sigma_Y'/\mathrm{E}', \end{cases} \tag{2.6}$$

where $\varepsilon_A$ stands for the maximum strain reached at the end of the initial loading (E' is the Young's modules of rod 1, and the yield stresses at tension and compression are assumed equal).

Note that at a certain value $p = p_B > 0$ of the load the stress $\sigma'$ in rod 1 vanishes, while there is a tensile (positive) stress $\sigma''$ in rods 2, 3; points $B'$ and $B''$ in Figure 2.5 correspond to the states of rod 1 and rods 2, 3, respectively. Two different types of structure response to further unloading are possible depending on whether the inequality

$$\mathrm{E}''(\varepsilon_A - 2\frac{\sigma_Y'}{\mathrm{E}'}) > \sigma_Y' \tag{2.7}$$

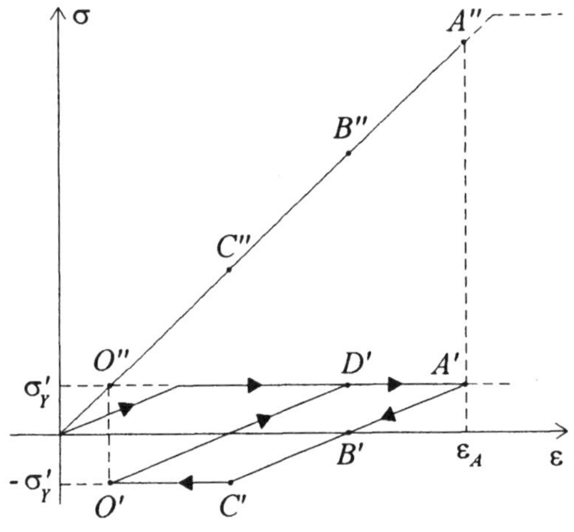


**Figure 2.5**

is valid or not. We will focus on the case when it is valid. Then the response of the structure to the unloading with $p$ decreasing from $p_B$ to

$$p_C = [\mathrm{E}''(\varepsilon_A - 2\frac{\sigma'_Y}{\mathrm{E}'}) - \sigma'_Y]S > 0$$

is described by the equilibrium condition $(\sigma'+\sigma'')S = p$, Hooke's law $\sigma'' = \mathrm{E}''\varepsilon$ and (2.6), and the yield stress in compression is reached in rod 1 at $p = p_C$. Points $C'$ and $C''$ in Figure 2.5 correspond to the states of the rods at $p = p_C$. Note that $p_C > 0$ and the structure is not unloaded at $p = p_C$, the strain $\varepsilon$ being equal to $\varepsilon_C = \varepsilon_A - 2\sigma'_Y/\mathrm{E}'$.

Under further decreasing of the load $p$ from $p_C$ to 0, rod 1 undergoes plastic compression and the stress in it does not vary: $\sigma' = -\sigma'_Y$. Then, at the end of the unloading, when $p = 0$, the equilibrium condition $\sigma' + \sigma'' = 0$ implies $\sigma'' = -\sigma' = \sigma'_Y$. Thus, the residual stresses $\sigma'_{\mathrm{res}} = -\sigma'_Y$ and $\sigma''_{\mathrm{res}} = \sigma'_Y$ remain in rod 1 and rods 2, 3 after unloading. Points $O'$ and $O''$ in Figure 2.5 correspond to the states of the rods after the unloading, and the additional plastic strain $\Delta_{C'O'}\varepsilon^p$ was accumulated in rod 1 when it deformed along the yield plateau $C'O'$ during the unloading. The strain $\varepsilon_O$ in the unloaded state is determined by Hooke's law: $\varepsilon = \varepsilon_O = \sigma''_{\mathrm{res}}/\mathrm{E}'' = \sigma'_Y/\mathrm{E}''$. Then the additional plastic strain accumulated during unloading is

$$\Delta_{C'O'}\varepsilon^p = \varepsilon_O - \varepsilon_C = \frac{\sigma'_Y}{\mathrm{E}''} + 2\frac{\sigma'_Y}{\mathrm{E}'} - \varepsilon_A.$$

Consider now the *second loading cycle*. It can be easily seen that rods 2 and 3 deform elastically during the whole cycle. The point depicting their current stressed-strained states moves along the segment $O''A''$ in Figure 2.5. Rod 1 undergoes elastic deformation (along the path $O'D'$ in Figure 2.5) and then plastic deformation (along the yield plateau $D'A'$). The plastic strain

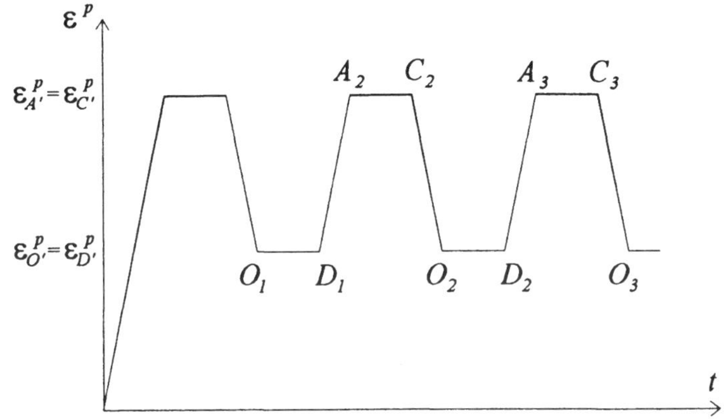


**Figure 2.6**

$\Delta_{D'A'}\varepsilon^p = -\Delta_{C'O'}\varepsilon^p$ is accumulated in rod 1 during this process. When the load $p$ attains its maximum value $P$, the stresses in the rods take the same values as at $p = P$ during the first loading cycle. The stressed-strained state of rod 1 corresponds to the point $A'$ in Figure 2.5, those of rods 2 and 3 to the point $A''$. The response of the rods to the subsequent unloading replicates that to the first unloading. The corresponding stress-strain paths are $A'B'C'O'$ and $A''O''$ for rod 1 and for rods 2, 3, respectively. The plastic strain $\Delta_{C'O'}\varepsilon^p$ is accumulated in rod 1 during the unloading. The total plastic strain accumulated in the second loading cycle is $\Delta_{D'A'}\varepsilon^p + \Delta_{C'O'}\varepsilon^p=0$, that is, compression during the unloading neutralizes the elongation accumulated under the loading. The second loading cycle ends at the same stressed-strained states as the first one.

It is clear that under the third and *subsequent loading cycles* the stressed-strained state of the rods vary cyclically following the pattern of the second cycle.

Consider now the evolution of the plastic strain $\varepsilon^p$ in rod 1. Figure 2.6 shows $\varepsilon^p$ versus "time" $t$ (recall that elastic-plastic response is rate insensitive; "time" means any parameter monotonically increasing with the time). The evolution of $\varepsilon^p$ is cyclic starting from the second loading cycle. Each of the cycles starts at the stressed-strained state of rod 1 depicted by the point $O'$ in Figure 2.5. During the loading, $\varepsilon^p$ remains constant until the stress $\sigma'$ attains the value $\sigma'_Y$; the segments $O'D'$ in Figure 2.5 and $O_1D_1$ in Figure 2.6 correspond to this process. Then, $\varepsilon^p$ increases at the constant stress $\sigma' = \sigma'_Y$ (the segments $D'A'$ and $D_1A_2$ in the figures). During the unloading, $\varepsilon^p$ remains constant until $\sigma'$ attains the value $-\sigma'_Y$ (the segments $A'C'$ and $A_2C_2$ in the figures); then, $\varepsilon^p$ decreases at $\sigma' = -\sigma'_Y$ (the segments $C'O'$ and $C_2O_2$). The cycle ends at the stressed-strained state where it started.

Thus, no additional plastic strain is accumulated in the second and subsequent loading cycles. However, plastic deformation occurs and produces the plastic work during each of the cycles. The plastic work per cycle can be easily evaluated since the rod undergoes plastic deformation at the constant stress $\sigma' = \sigma'_Y$ or $\sigma' = -\sigma'_Y$:

$$\int_{t_0}^{t_0+\tau} e^p \sigma' \, dt = \sigma'_Y \Delta_{D'A'} \, ep^p + (-\sigma'_Y)\Delta_{C'O'}\varepsilon^p = 2\sigma'_Y \Delta_{D'A'}\varepsilon^p > 0$$

(here, $e^p$ is the rate of plastic stretching and $\tau$ is the cycle time). Then the accumulated plastic work

$$W^p(T) = \int_0^T e^p \sigma' \, dt$$

infinitely increases with $T$. In such a case we say that the body does not shake down with respect to the loading.

The unbounded growth of the plastic work $W^p(T)$ is dangerous itself even in case the strain remains small. The problem is that microdamage increases with $W^p(T)$, which results in so called material fatigue and the fatigue failure of the body.

REMARK 2. Recall that the above reasoning established nonshakedown in case inequality (2.7) is valid. Let us show that the structure shakes down otherwise, that is, the plastic work is bounded if

$$\mathrm{E}'' \left( \varepsilon_A - 2\frac{\sigma'_Y}{\mathrm{E}'} \right) < \sigma_Y. \tag{2.8}$$

Moreover, we will show that the structure undergoes the plastic deformation only during the first loading cycle.

Let the points $A'$ and $A''$ in Figure 2.7 correspond to the stressed-strained state of rod 1 and rods 2, 3, respectively, at the end of the first loading. Let us consider the subsequent unloading and verify that rod 1 does not undergo plastic deformation during the unloading (in contrast to the previously considered case (2.7)). This immediately arises from the fact that the stress remains below the yield limit. Indeed, due to inequality (2.8) there is such $\rho$, $0 < \rho < \sigma'_Y$, that

$$\mathrm{E}'' \left( \varepsilon_A + 2\frac{-\rho - \sigma'_Y}{\mathrm{E}'} \right) = \rho. \tag{2.9}$$

Then the stress $\sigma' = -\rho$ in rod 1, the strain

$$\varepsilon = \varepsilon_A + \frac{\sigma' - \sigma'_Y}{\mathrm{E}'} = \varepsilon_A + \frac{-\rho - \sigma'_Y}{\mathrm{E}'},$$

determined by (2.6), and the stress $\sigma'' = \mathrm{E}''\varepsilon$ in rods 2, 3 corresponds to the end of unloading. Indeed, equality (2.9) means that the stresses $\sigma'$ and $\sigma''$ satisfy the equilibrium condition $\sigma' + \sigma'' = 0$ at the zero load $p = 0$.

Under the repeated loading cycles the stressed-strained state of the structure varies cyclically. The point corresponding to the state of rod 1 cyclically

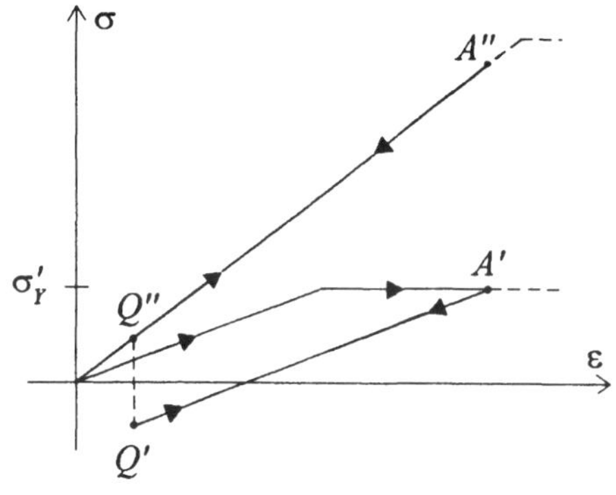


**Figure 2.7**

moves along the segment in Figure 2.7: $Q' \to A' \to Q' \to A' \to \cdots$. Similarly, the point corresponding to the state of rods 2, 3 cyclically moves along the segment $Q''A''$. The structure undergoes no additional plastic deformation during the second and subsequent loading cycles and, therefore, shakes down to the cyclic loading.

It is worth noting that nonshakedown and shakedown conditions (2.7) and (2.8) are of rather simple forms: $P > P_0$ and $P < P_0$, where $P$ is the maximum load during the cycle and $P_0$ is a certain critical value. Indeed, let the yield stress be attained during the first loading; this is the case if and only if $P > S(1 + \mathrm{E}''/\mathrm{E}')\sigma'_Y$. Then it is clear that $\sigma' = \sigma'_Y$ at $p = P$ and the strain $\varepsilon_A$ corresponding to the maximum load can be easily found: the equilibrium equation $(\sigma' + \sigma'')S = p$ at $p = P$, the equality $\sigma' = \sigma'_Y$ and Hooke's law $\sigma'' = \mathrm{E}''\varepsilon_A$ result in

$$\varepsilon_A = \frac{1}{\mathrm{E}''}\left(\frac{P}{S} - \sigma'_Y\right).$$

Then, inequality (2.7) reduces to $P > 2S(1 + \mathrm{E}''/\mathrm{E}')\sigma'_Y$ and the structure does not shake down under this condition. In case $P < 2S(1 + \mathrm{E}''/\mathrm{E}')\sigma'_Y$, the structure either shakes down to the loading or does not undergo plastic deformation at all if $P \le S(1 + \mathrm{E}''/\mathrm{E}')\sigma'_Y$.

**2.3. Nonshakedown at unbounded plastic strain.** The stress response of an elastic perfectly plastic body to a cyclic loading is often cyclic, which, however, does not mean that the plastic strain also varies cyclically, thus remaining bounded. It is possible that it is the plastic strain rate that varies cyclically, and the increment of the plastic strain during each of the loading cycles is not zero as well as the increment of the plastic work per cycle. This increment is the same during each of the loading cycles if the strain rate and stress vary cyclically; therefore, the plastic work grows infinitely with the number of cycles and the body does not shake down. The following example illustrates this type of elastic-plastic response.

Consider the structure consisting of three rods of the same length and with the same material properties. The rods are parallel; one end of each rod is

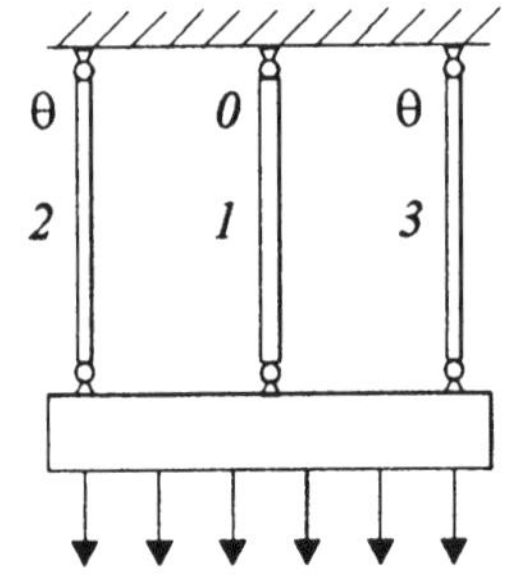


**Figure 2.8**

fixed, the other ends being connected with a rigid beam as is shown in Figure 2.8. The structure is symmetric with respect to the axis of rod 1. The cross-section area of rod 1 is $S$; each of rods 2, 3 is of the cross-section area $S/2$.

The structure is subjected to a permanent load and a variable *thermal loading*. The load is parallel to the rods and uniformly distributed along the beam with the resultant magnitude $p_0 = \text{const}$. As to the thermal loading, it determines the temperatures of the rods: they are zero at the initial moment, then the temperature of rod 1 remains zero while the temperatures of rods 2, 3 are equal and vary within the interval $[0, \theta_{\max}]$.

The stresses in rods 2, 3 are obviously equal and we denote them by $\sigma''$. The stress in rod 1 is denoted by $\sigma'$. The condition for equilibrium of the structure is of the form $\sigma' + \sigma'' = p_0/S$. Strains of all the three rods are the same.

We adopt elastic perfectly plastic constitutive relations for the material of the rods and assume that their yield stresses do not depend on the temperature. The only effect of the temperature $\theta$ is the temperature strain $\varepsilon^\theta = \alpha\theta$, where the constant $\alpha$ is the thermal expansion coefficient. Thus, the material properties are described by the following constitutive relations:

$$\begin{gathered} \varepsilon = \varepsilon^e + \varepsilon^p + \varepsilon^\theta, \quad \varepsilon^e = \frac{1}{\mathrm{E}}\sigma, \quad \varepsilon_\theta = \alpha\theta, \\ |\sigma| \le \sigma_Y, \\ \dot{\varepsilon}^p = 0 \text{ at } |\sigma| < \sigma_Y, \quad \dot{\varepsilon}^p \ge 0 \text{ at } \sigma = \sigma_Y, \quad \dot{\varepsilon}^p \le 0 \text{ at } \sigma = -\sigma_Y, \end{gathered} \tag{2.10}$$

where E is the Young's modules, $\sigma_Y$ is the yield stress.

We assume that the value $\sigma_0 = p_0/2S$ characterizing the load is less than the yield stress $\sigma_Y$ and the plastic strains of the rods are zero at the initial moment $t = 0$. Then the stresses in the three rods are the same at $t = 0$: $\sigma_0' = \sigma_0'' = p_0/2S = \sigma_0$, while the strain is $\varepsilon_0 = \sigma_0/\mathrm{E}$. Let us consider the response of the structure 1) to the initial thermal loading: increasing of the temperature $\theta$ of rods 2, 3 from 0 up to $\theta_{\max}$; 2) to subsequent decreasing of $\theta$ from $\theta_{\max}$ to 0; and 3) to repeated cycles of the temperature.

During the *initial thermal loading* the stresses in the rods do not reach the yield stress as long as their temperature $\theta$ is sufficiently small. In this case,

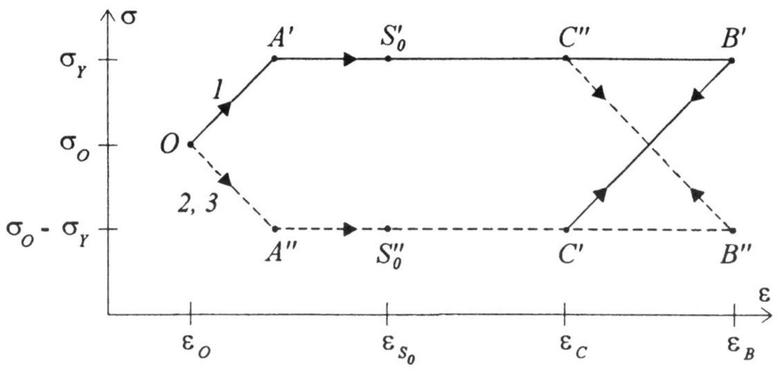


**Figure 2.9**

the constitutive relations and equilibrium condition

$$\sigma' = \mathrm{E}\varepsilon, \quad \sigma'' = \mathrm{E}(\varepsilon - \alpha\theta), \quad \sigma' + \sigma'' = 2\sigma_0 \quad (\sigma_0 = p_0/2S)$$

determine the stressses and strain

$$\sigma' = \sigma_0 + \frac{\alpha \mathrm{E}}{2}\theta, \quad \sigma'' = \sigma_0 - \frac{\alpha \mathrm{E}}{2}\theta, \quad \varepsilon = \varepsilon_0 + \frac{\alpha}{2}\theta \quad (\varepsilon_0 = \frac{\sigma_0}{\mathrm{E}}).$$

Figure 2.9 shows the variation of the stress and strain in rod 1 (the solid line) and in rods 2, 3 (the dotted line). The segments $OA$ and $OA'$ of the diagrams correspond to this (elastic) part of the process. The purely elastic deformation of rod 1 ends at the temperature $\theta_A = 2(\sigma_Y - \sigma_0)/\alpha\mathrm{E}$, when the stress $\sigma'$ reaches the yield stress value $\sigma_Y$ ($\theta_{\max}$ is assumed to be greater than $2(\sigma_Y - \sigma_0)/\alpha\mathrm{E}$, which ensures attainability of the yield stress). Point $A'$ in Figure 2.9 corresponds to the state of rod 1 at $\theta = \theta_A$, point $A''$ to the state of rods 2, 3.

Under the subsequent increasing of $\theta$ from $\theta_A$ up to $\theta_{\max}$, the plastic strain $\varepsilon^{p\prime}$ is accumulated in rod 1 at $\sigma' = \sigma_Y$, and the temperature strain $\varepsilon_\theta = \alpha\theta$ increases in rods 2, 3 at $\sigma'' = 2\sigma_0 - \sigma_Y$. Note that $\sigma''$ is less than the yield stress since $\sigma_0 < \sigma_Y$ by the above assumption. The points corresponding to the stressed-strained states of rod 1 and rods 2, 3 move along the segments $A'B'$ and $A''B''$ in Figure 2.9, respectively. The strain reaches the value

$$\varepsilon_B = \frac{\sigma_Y}{\mathrm{E}} + \varepsilon_B^{p\prime} = \frac{2\sigma_0 - \sigma_Y}{\mathrm{E}} + \alpha\theta_{\max} \tag{2.11}$$

at $\theta = \theta_{\max}$.

After the maximum value $\theta_{\max}$ has been reached, the temperature of rods 2, 3 decreases from $\theta_{\max}$ to 0. At the beginning of this process, the stresses in rods 2, 3 remain close to the values they had at $\theta = \theta_{\max}$. This makes $\sigma''$ less than $\sigma_Y$ and also shows that rod 1 undergoes unloading. Indeed, if the stress $\sigma'$ is equal to the yield stress $\sigma_Y$, thus remaining constant, the equilibrium equation implies that neither does the stress $\sigma''$ vary in rods 2, 3. This results in the equality $\varepsilon = \varepsilon_0 + \alpha\theta/2$, which means, in particular, that the strain rate $\dot{\varepsilon} = \alpha\dot{\theta}$ is negative when the temperature decreases. At the same time, constitutive relations (2.10) for rod 1 (whose temperature is constant and the stress is also constant by the above assumption $\sigma' = \sigma_Y$) result in $\dot{\varepsilon} = \dot{\varepsilon}^p \geq 0$. This contradiction shows that the assumption $\sigma' = \sigma_Y$ is wrong. Thus, $\sigma' < \sigma_Y$ and rod 1 undergoes unloading. Then constitutive relations (2.10) and the equilibrium condition describing the process take the form

$$\varepsilon - \varepsilon_B = \frac{1}{\mathrm{E}}(\sigma' - \sigma_Y), \quad \sigma'' = \mathrm{E}(\varepsilon - \alpha\theta), \quad \sigma' + \sigma'' = 2\sigma_0. \tag{2.12}$$

These relations admit two different types of the structure response depending on whether the inequality

$$\theta_{\max} > 4\frac{\sigma_Y - \sigma_0}{\alpha \mathrm{E}} \tag{2.13}$$

is valid or not. Let us now focus on the case when (2.13) is valid. Then, due to (2.12) the stresses in rods 2, 3 reach the yield stress at the temperature $\theta_C$ and the strain $\varepsilon_C$:

$$\theta_C = \theta_{\max} - 4\frac{\sigma_Y - \sigma_0}{\alpha \mathrm{E}} > 0, \qquad \varepsilon_C = \varepsilon_B - 2\frac{\sigma_Y - \sigma_0}{\mathrm{E}}. \tag{2.14}$$

When the temperature $\theta$ decreases from $\theta_{\max}$ to $\theta_C$, the points corresponding to the current stressed-strained states of rod 1 and rods 2, 3 move along the segments $B'C'$ and $B''C''$ in Figure 2.9, respectively. Under the subsequent decreasing of the temperature from $\theta_C$ to 0, the stress $\sigma''$ remains equal to the yield stress $\sigma_Y$, which makes the stress $\sigma'$ in rod 1 constant and equal to $2\sigma_0 - \sigma_Y$. This implies that neither does the strain $\varepsilon$ vary, which can be easily explained: the plastic stretching of rods 2, 3 is neutralized by their thermal compressing. The point $C''$ in Figure 2.9 depicts the stressed-strained state of rods 2, 3 during this process; the point $C'$ corresponds to the process in rod 1. The first cycle ends when the temperature $\theta$ of rods 2, 3 attains 0.

Let us now consider the *second cycle* of the thermal loading. During its first part, the temperature increases from 0 up to $\theta_{\max}$. At the beginning of this process the stress in rod 1 remains close to the value it had at $\theta = 0$, thus being less than the yield stress. Note that rods 2, 3 undergo unloading (the assumption that $\sigma'' = \sigma_Y$ results in the contradiction). Then constitutive relations (2.10) and the equilibrium equation describing the process as long as $\sigma' < \sigma_Y$ can be written as

$$\varepsilon - \varepsilon_C = \frac{1}{\mathrm{E}}[\sigma' - (2\sigma_0 - \sigma_Y)], \quad \varepsilon - \varepsilon_C = \frac{1}{\mathrm{E}}(\sigma' - \sigma_Y) + \alpha\theta, \quad \sigma' + \sigma'' = 2\sigma_0.$$

Taking into account (2.14) and (2.11) for the strains $\varepsilon_C$ and $\varepsilon_B$, we observe that the latter equations give (in the parametric form) the same relations between $\sigma'$, $\sigma''$ and $\varepsilon$ as equations (2.12) does. Therefore, the segments $C'B'$ and $C''B''$ in Figure 2.9 again correspond to the process under consideration. However, this time the points depicting the current states of the rods move in the opposite direction: from $C'$ to $B'$ and from $C''$ to $B''$. The above system of equations also implies that $\sigma'$ attains the yield stress at the temperature $\theta = 4(\sigma_Y - \sigma_0)/\alpha\mathrm{E}$ (which is less than $\theta_{\max}$ by the above assumption). At this temperature, the stressed-strained states are depicted by the points $B'$ and $B''$ for rod 1 and rods 2, 3, respectively. The temperature and the stresses in this state are

$$\theta = \theta_1 = 4\frac{\sigma_Y - \sigma_0}{\alpha\mathrm{E}}, \quad \sigma' = \sigma_Y, \quad \sigma'' = 2\sigma_0 - \sigma_Y, \tag{2.15}$$

the strain and the plastic strains being

$$\varepsilon = \varepsilon_B = \frac{2\sigma_0 - \sigma_Y}{\mathrm{E}} + \alpha\theta_{\max}, \quad \varepsilon^{p\prime} = \varepsilon - \frac{\sigma}{\mathrm{E}}, \quad \varepsilon^{p\prime\prime} = \varepsilon - \frac{2\sigma_0 - \sigma_Y}{\mathrm{E}} - \alpha\theta_1 .$$

Let us denote this state by $S_1$.

Note that there was the moment $t_0$ during the first loading cycle when the temperature $\theta$ of rods 2, 3 took the value $\theta_1$. This occurred during the temperature increasing from $\theta_A = 2(\sigma_Y - \sigma_0)/\alpha\mathrm{E}$ to $\theta_B = \theta_{\max}$ at the constant stresses $\sigma' = \sigma_Y$, $\sigma'' = 2\sigma_0 - \sigma_Y$; the segments $A'B'$, $A''B''$ in Figure 2.9 correspond to that part of the process. Thus, at $t = t_0$ the stresses and temperature took the same values (2.15) as in the state $S_1$; however, the strain at $t = t_0$ was less than that in the state $S_1$. Let us denote the state of the structure at $t = t_0$ by $S_0$, the stresses and strains of the rods in this state being depicted by the points $S_0'$ and $S_0''$ in Figure 2.9. Note that the plastic strain $\Delta\varepsilon^p = \varepsilon_B - \varepsilon_{S_0} > 0$ (the same in all three rods) was accumulated during the process between $S_0$ and $S_1$, and this is the only difference between the two states.

The latter remark suggests subdividing the whole cyclic process into certain parts more convenient for further discussion. Namely, let us reconsider the process starting from the moment $t = t_0$ when the temperature $\theta$ reached the value $\theta_1$ for the first time. The state of the structure was $S_0$ at that moment. Then the temperature increased up to $\theta_{\max}$, decreased from this value to 0, and again increased up to $\theta_{\max}$, which resulted in the state $S_1$. The subsequent process consists of repeating this cycle of thermal loading: $\theta_1 \to \theta_{\max} \to 0 \to \theta_1 \to \cdots$. Due to constitutive relations (2.10), the stress, and the elastic and temperature strains do not depend on the initial plastic strain. Neither do the *increments* of the plastic total strains depend on the initial plastic strain. Therefore, the second and *subsequent cycles* $\theta_1 \to \theta_{\max} \to 0 \to \theta_1$ of the thermal loading result in the states $S_2, S_3, \ldots$ of the structure with values (2.15) of the temperature and stresses and the additional plastic strain $\Delta\varepsilon^p = \varepsilon_B - \varepsilon_{S_0} > 0$ accumulated during each of the cycles. Thus, the total strain of the structure infinitely increases with the number of loading cycles. Similarly, the plastic work increases during each loading cycle, the increment being the same in every cycle. Thus, the total plastic work

$$\int_0^T e^p \sigma \, dt$$

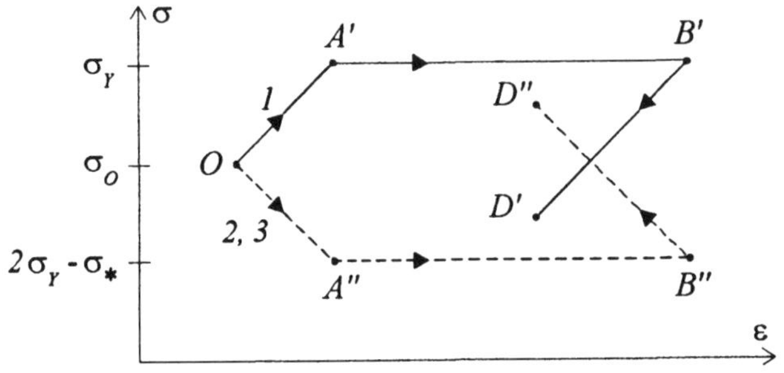


**Figure 2.10**

grows infinitely when $T \to \infty$; the structure does not shake down to the loading.

REMARK 3. Recall that nonshakedown was established by the above reasoning in case inequality (2.13) is valid. Let us show that otherwise the structure shakes down, that is, the plastic work is bounded if

$$\theta_{\max} < 4\frac{\sigma_Y - \sigma_0}{\alpha E}. \tag{2.16}$$

Indeed, equations (2.12) together with expression (2.11) for $\varepsilon_B$ result in the relation

$$\sigma'' = 2\sigma_0 - \sigma_Y + \frac{1}{2}\alpha E\theta_{\max} < \sigma_Y \quad \text{at} \quad \theta = 0,$$

where the inequality arises from (2.16).

Thus, the stress $\sigma''$ does not attain the yield limit $\sigma_Y$ when the temperature decreases from $\theta_{\max}$ to 0 during the first loading cycle. Then the states of the rods at the end of the first cycle $0 \to \theta_{\max} \to 0$ are depicted by the points $D'$, $D''$ in Figure 2.10. During the subsequent loading cycles, the process is described by equations (2.12) and no plastic deformation occurs. The points corresponding to the current stressed-strained states of the rods move along the segments $D' \to B' \to D' \to B' \to \cdots$ and $D'' \to B'' \to D'' \to B'' \to \cdots$ in Figure 2.10. The structure shakes down to the loading.

It is worth noting that nonshakedown and shakedown conditions (2.13) and (2.16) are of rather simple forms: $\theta_{\max} > \theta_0$ and $\theta_{\max} < \theta_0$. Thus, there is the critical value $\theta_0$ of the maximum temperature $\theta_{\max}$ which separates the cases of shakedown and nonshakedown.

**2.4. Shakedown at nonstop plastic flow.** Sometimes, the stress response of an elastic perfectly plastic body to a cyclic loading becomes cyclic after several loading cycles. This was the case in all of the above considered examples. However, this is not the only possible type of response. The following example shows that the stress variation may *tend* to a cyclic pattern without reaching it. Shakedown is quite possible in this case: the plastic work remains bounded when the number $m$ of the cycles tends to infinity, although the plastic deformation does not stop. The intensity of the plastic

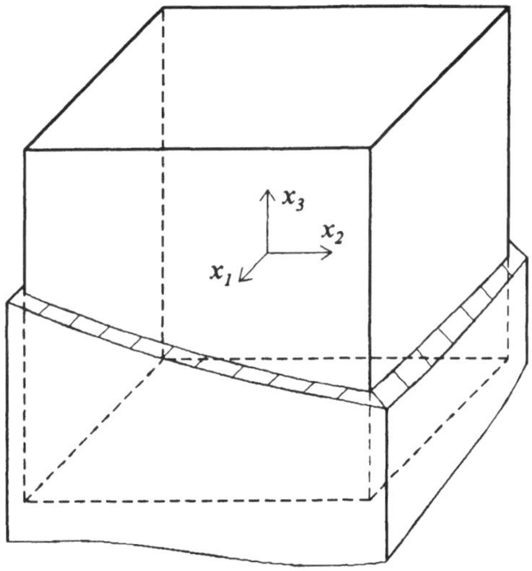


**Figure 2.11**

deformation decays due to a favorable distribution of residual stresses, although the latter do not reach this distribution and only tend to it when $m \to \infty$.

Consider a cube made of elastic perfectly plastic material, the length of the cube edge being 1. The body is placed inside the rigid pipe of a square cross-section, see Figure 2.11. It is assumed that there is no friction between the body and the walls of the pipe. Let us write the corresponding boundary conditions using the Cartesian coordinate system with the $x_3$-axis directed along the pipe axis and the $x_2$- and $x_3$-axis orthogonal to the cube sides. The conditions are written as

$$\begin{gathered} u_1|_{x_1=\pm\frac{1}{2}} = u_2|_{x_2=\pm\frac{1}{2}} = 0, \\ \sigma_{12}|_{x_1=\pm\frac{1}{2}} = \sigma_{13}|_{x_1=\pm\frac{1}{2}} = \sigma_{12}|_{x_2=\pm\frac{1}{2}} = \sigma_{23}|_{x_2=\pm\frac{1}{2}} = 0. \end{gathered} \tag{2.17}$$

The upper and lower sides of the cube are subjected to the load which is uniformly distributed with the densities

$$\begin{gathered} q_1 = 0, \quad q_2 = 0, \quad q_3 = q \quad \text{at} \quad x_3 = \frac{1}{2}, \\ q_1 = 0, \quad q_2 = 0, \quad q_3 = -q \quad \text{at} \quad x_3 = -\frac{1}{2}. \end{gathered}$$

This means that the stress field should also satisfy the boundary conditions

$$\sigma_{13}|_{x_3=\pm\frac{1}{2}} = \sigma_{23}|_{x_2=\pm\frac{1}{2}} = 0.$$

Consider the behavior of the cube under the slowly varying load $q$. We will find a solution to the quasistatic problem, the stress field being uniform, that is, taking the same value at every point of the body. Boundary conditions (2.7) immediately determine the components $\sigma_{12} = \sigma_{13} = \sigma_{23} = 0$ of this stress field. Thus, three numbers $\sigma_{11}$, $\sigma_{22}$, $\sigma_{33}$ completely characterize the current stressed state of the body. These components will be denoted by $\sigma_1$, $\sigma_2$, $\sigma_3$, and $\boldsymbol{\sigma} = (\sigma_1, \sigma_2, \sigma_3) \in \mathbf{R}^3$ will be referred to as stress. Note that the equilibrium condition for the body is of the form $\sigma_3 = q$.

The strain components

$$\varepsilon_{ij} = \frac{1}{2}\left(\frac{\partial u_i}{\partial x_j} + \frac{\partial u_j}{\partial x_i}\right)$$

are the same at every point of the cube if the strained state of the body is uniform. Together with boundary conditions (2.17), this immediately results in the following formulas for the displacement field: $u_1 = u_2 = 0$, $u_3 = wx_3 + c$, where $w$ and $c$ do not depend on the coordinates $x_1$, $x_2$, $x_3$. Eliminating the rigid body motion results in $c = 0$, due to which the number $w$ completely characterizes the displacement field $\mathbf{u}$ with the components $u_1 = u_2 = 0$, $u_3 = wx_3$. The corresponding strain has the components

$$\varepsilon_{11} = \varepsilon_{22} = \varepsilon_{12} = \varepsilon_{13} = \varepsilon_{23} = 0, \quad \varepsilon_{33} = w.$$

We refer to $\boldsymbol{\varepsilon} = (\varepsilon_1, \varepsilon_2, \varepsilon_3) \in \mathbf{R}^3$ with $\varepsilon_1 = \varepsilon_2 = 0$, $\varepsilon_3 = w$ as the strain corresponding to the displacement field $\mathbf{u}$.

The strain of the elastic-plastic cube is the sum of the elastic and plastic strains: $\boldsymbol{\varepsilon} = \boldsymbol{\varepsilon}^e + \boldsymbol{\varepsilon}^p$,

$$\boldsymbol{\varepsilon} = \boldsymbol{\varepsilon}^e + \boldsymbol{\varepsilon}^p, \quad \boldsymbol{\varepsilon}^e = (\varepsilon_1^e, \varepsilon_2^e, \varepsilon_3^e), \quad \boldsymbol{\varepsilon}^p = (\varepsilon_1^p, \varepsilon_2^p, \varepsilon_3^p).$$

We adopt Hooke's law for the elastic strain: $\boldsymbol{\varepsilon}^e = \boldsymbol{\sigma}/\mathrm{E}$ (for simplicity we set the Poisson's ratio equal to zero). Then the constitutive maximum principle (1.1) is written as

$$\boldsymbol{\sigma} \in C, \quad (\dot{\boldsymbol{\varepsilon}} - \frac{1}{\mathrm{E}}\dot{\boldsymbol{\sigma}})(\boldsymbol{\sigma} - \boldsymbol{\sigma}_*) \geq 0 \quad \text{for every} \quad \boldsymbol{\sigma}_* \in C,$$

where $C$ is the set of admissible stresses in $\mathbf{R}^3$. Note that under the above conditions the cube may be considered as a discrete elastic-plastic system, and the quasistatic problem for it has the following formulation:

$$\sigma_3 = q, \quad \boldsymbol{\sigma} \in C,$$

$$-\frac{1}{\mathrm{E}}\dot{\sigma}_1(\sigma_1 - \sigma_{*1}) - \frac{1}{\mathrm{E}}\dot{\sigma}_2(\sigma_2 - \sigma_{*2}) + \left(\dot{w} - \frac{1}{\mathrm{E}}\dot{\sigma}_3\right)(\sigma_3 - \sigma_{*3}) \geq 0$$
$$\text{for every } \boldsymbol{\sigma}_* \in C,$$

$$w|_{t=0} = 0, \quad \boldsymbol{\sigma}|_{t=0} = 0,$$

where $w$ is the unknown displacement, $\boldsymbol{\sigma} = (\sigma_1, \sigma_2, \sigma_3)$ is the unknown stress, and the load $q = q(t)$ is given for $t \geq 0$ (with $q(0) = 0$).

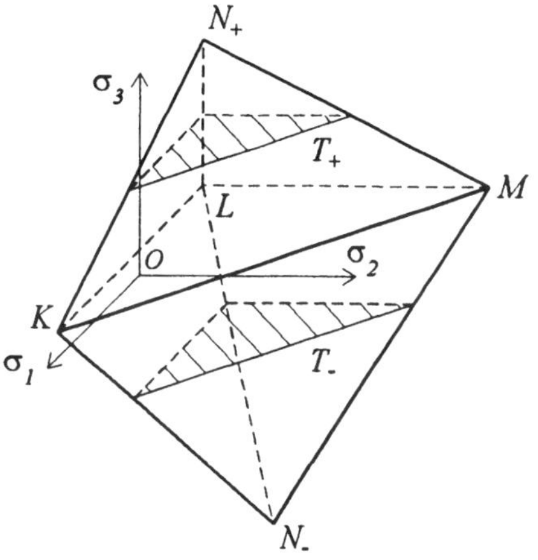


**Figure 2.12**

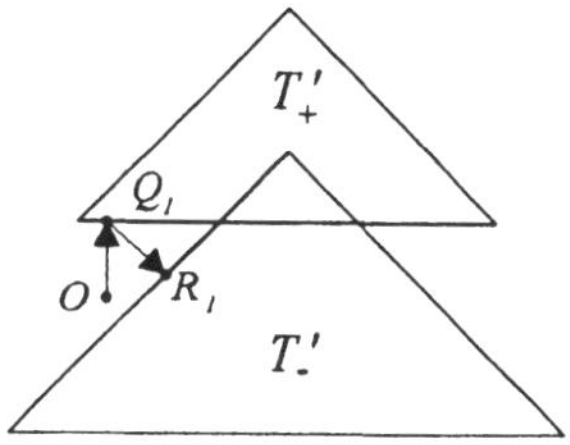


**Figure 2.13**

To make the example as simple as possible, let us choose the following set $C$ of admissible stresses in $\mathbf{R}^3$. Consider the polyhedron $KLMN_+N_-$ consisting of two tetrahedrons with the common side $KLM$, see Figure 2.12. The triangle $KLM$ lies in the plane $\sigma_3 = 0$ and is isosceles, with the right angle $A$. The vertices $N_+$ and $N_-$ are located on different sides of the plane $\sigma_3 = 0$; the projection of $N_+$ on this plane being $L$ and the projection of $N_-$ being the middle point of the segment $KM$. The $\sigma_3$-axis intersects the sides $KN_+M$ and $KN_-L$ of the tetrahedron.

We will consider the response of the body to the load $q$ varying between the values $-q_0$ and $q_0$, where $q_0$ is chosen so that the points $(0, 0, q_0)$ and $(0, 0, -q_0)$ belong to the exterior of $C$ and the planes $\sigma_3 = q_0$, $\sigma_3 = -q_0$ intersect $C$. The sections are triangles which we denote $T_+$ and $T_-$, respectively; we assume that their projections on the plane $\sigma_3 = 0$ have a nonempty intersection (see Figure 2.13).

Let us consider the response of the body 1) to the initial loading starting at the unstressed and unstrained configuration: the load increases from 0 up to $q_0$; 2) to the unloading followed by the loading in the opposite direction: $q$ decreases from $q_0$ to 0 and then to $-q_0$; 3) to the subsequent unloading from $-q_0$ to 0; and 4) to the repeated loading cycles.

Note that, due to the stress admissibility and the equilibrium condition $\sigma_3 = 0$, the point $\boldsymbol{\sigma} = (\sigma_1, \sigma_2, \sigma_3)$ corresponding to the current stress belongs to $C$ and is located between the cross-sections $T_+$ and $T_-$. At the initial moment, this point is 0.

During the *initial loading*, the point $\boldsymbol{\sigma} = (\sigma_1, \sigma_2, \sigma_3)$ belongs to the elasticity domain as long as the load $q$ is sufficiently small. Then the constitutive maximum principle implies that $\dot{\sigma}_1 = 0$ and $\dot{\sigma}_2 = 0$. Thus, the projection of $\boldsymbol{\sigma} = (\sigma_1, \sigma_2, \sigma_3)$ on the plane $\sigma_3 = 0$ is 0 as long as the load is sufficiently small. At a certain value $q_1$ of the load $q$, the stress $\boldsymbol{\sigma}$ reaches the yield surface (the side $KN_+M$ of the polyhedron in Figure 2.12). Under the subsequent increasing of the load $q$ the stress remains on the yield surface. Let us find the projection of the stress path on the plane $\sigma_3 = 0$. The velocity of the projection of the point $\boldsymbol{\sigma} = (\sigma_1, \sigma_2, \sigma_3)$ on the plane $\sigma_3 = 0$ is $(\dot{\sigma}_1, \dot{\sigma}_2, 0)$, and it can be easily found with the constitutive maximum principle. Note that $\boldsymbol{\sigma}$ belongs to the boundary of the cross-section of $C$ by the plane $\sigma_3 = q$, which is triangle $T_q$ similar to the triangle $KLM$. Any point in $T_q$ can be taken as $\boldsymbol{\sigma}_*$ in the constitutive maximum principle, and this implies that $(\dot{\sigma}_1, \dot{\sigma}_2, 0)$ is directed along the inward normal to the boundary of $T_q$ at the point $\boldsymbol{\sigma}$. We restrict ourselves to the case when $\boldsymbol{\sigma}$ does not reach the edge $KN_+$ of the yield surface, thus staying within the side $KN_+M$. Then the above-mentioned normal is the normal to the side of the triangle $T_q$ parallel to $KM$. The first component of the normal is nonpositive, $\dot{\sigma}_1 < 0$, and the second one is zero, $\dot{\sigma}_2 = 0$, which means that the stress path projection on the plane is a segment orthogonal to $KM$. The projection of $\boldsymbol{\sigma}$ moves along this segment until $\boldsymbol{\sigma}$ reaches $T_+$, the latter being the section of $C$ formed by the stresses which equilibrate the maximum load $q = q_0$. Consequently, $\boldsymbol{\sigma}$ reaches $T_+$ at the end of the initial loading. Thus, the projection of the stress path during the initial loading is the segment $OQ_1$ in the plane $\sigma_3 = 0$. The segment is orthogonal to the hypotenuse of the triangle $T'_+$, the projection of $T_+$, see Figure 2.13.

Let us now consider the response of the body to the *unloading and the subsequent loading in the opposite direction*, that is, to decreasing of the load $q$ from $q_0$ to $-q_0$. At the beginning of this process, the stress leaves the yield surface and enters the elasticity domain, and the constitutive maximum principle results in the relations $\dot{\sigma}_1 = 0$, $\dot{\sigma}_2 = 0$. Therefore, the point $Q_1$ is the projection of $\boldsymbol{\sigma}$ until the stress reaches again the yield surface at a certain (negative) value of the load $q$. Under further decreasing of $q$, $\boldsymbol{\sigma}$ stays on the yield surface, more precisely, on its side $KN_-L$, see Figure 2.12. Similarly to the above considered case, the projection of $\dot{\boldsymbol{\sigma}}$ on the plane $\sigma_3 = 0$ has the direction of the inward normal to the boundary of the triangle $T_q$ $(q < 0)$. We restrict ourselves to the case when $\boldsymbol{\sigma}$ stays within the side $KN_-M$ of the yield surface. Then, the above-mentioned normal is the normal to the side of $T_q$ parallel to $KL$, and the projection of the stress path on the plane $\sigma_3 = 0$ is the segment $Q_1R_1$ orthogonal to $KL$, see Figure 2.13. The point $R_1$ corresponds to the stress at the load $q = -q_0$.

Under the subsequent *increase of the load $q$ from $-q_0$ to* 0, which is unloading, the stress $\boldsymbol{\sigma}$ leaves the yield surface, enters the elasticity domain and stays inside it. The point $R_1$ is the projection of $\boldsymbol{\sigma}$ on the plane $\sigma_3$, in particular, corresponding to the stress at $q = 0$, that is, the residual stress at the end of the first loading cycle.

Under the *second loading cycle*, the body response is completely analogous to that during the first cycle; the only difference between them consists of replacing the initial (zero) stress with the one corresponding to the point $R_1$. The projection of the stress path on the plane $\sigma_3$ is the broken line $R_1Q_2R_2$, see Figure 2.14. The point $R_2$ corresponds to the residual stress at the end of the second loading cycle. The body behavior under the *subsequent loading cycles* follows the same pattern. In particular, the projection of the stress path on the plane $\sigma_3 = 0$ is the broken line $R_1Q_2R_2Q_3R_3\ldots$ (Figure 2.14).

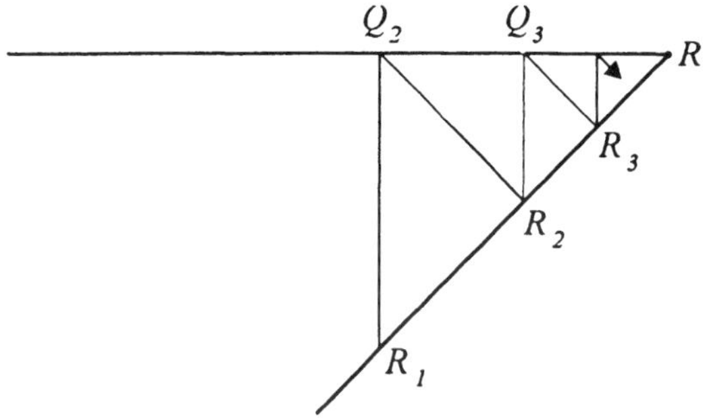


**Figure 2.14**

Let us show that the body shakes down to the loading, that is, the plastic work

$$W_p(T) = \int_0^T \mathbf{e}^p \boldsymbol{\sigma}\, dt$$

remains bounded by a certain constant no matter how large the number of loading cycles. Note that the plastic strain rate $\mathbf{e}^p$ in the above formula is

$$\mathbf{e}^p = \dot{\boldsymbol{\varepsilon}} - \dot{\boldsymbol{\varepsilon}}^e = (0, 0, \dot{w}) - \frac{1}{\mathrm{E}}(\dot{\sigma}_1, \dot{\sigma}_2, \dot{\sigma}_3). \tag{2.18}$$

To show the boundness of the plastic work, let us first consider the part of the loading cycle during which the stress $\boldsymbol{\sigma}$ stays on the side $KN_+M$ of the yield surface. In this case, the constitutive maximum principle is equivalent to the normality flow rule

$$e_1^p = \dot{\lambda} n_1^+, \quad e_2^p = \dot{\lambda} n_2^+, \quad e_3^p = \dot{\lambda} n_3^+, \quad \dot{\lambda} \geq 0, \tag{2.19}$$

where $n_1^+$, $n_2^+$, $n_3^+$ are the components of the unit outward normal to the side $KN_+M$ of $C$. Note that the plane $KN_+M$ is described by the equation $n_1^+\sigma_1 + n_2^+\sigma_2 + n_3^+\sigma_3 = c_+$ with a certain positive constant $c_+$. Then the normality flow rule implies that $\mathbf{e}^p\boldsymbol{\sigma} = \dot{\lambda} c_+$, and relations (2.18), (2.19) result in the expression $\dot{\lambda} = -\dot{\sigma}_1/\ \mathrm{E}n_1^+$ and in the formula

$$\mathbf{e}^p\boldsymbol{\sigma} = -c_+ \frac{\dot{\sigma}_1}{\mathrm{E}n_1^+} \qquad (n_1^+ > 0,\ \dot{\sigma}_1 < 0).$$

Analogously, in case the stress belongs to the side $KN_-L$, whose outward unit normal has the components $n_1^- = 0$, $n_2^-$, $n_3^-$, the following formula

$$\mathbf{e}^p\boldsymbol{\sigma} = \dot{\lambda} c_- = -c_- \frac{\dot{\sigma}_2}{\mathrm{E}n_2^-} = c_- \frac{\dot{\sigma}_2}{\mathrm{E}|n_2^-|} \qquad (n_2^- < 0,\ \dot{\sigma}_2 > 0)$$

is valid with a certain positive constant $c_-$. Consequently, the increment of the plastic work during the loading cycle, when the projection of the stresses moves along the broken line $R_m Q_{m+1} R_{m+1}$ $(m = 1, 2, \dots)$, equals

$$w_m = \int_0^T \dot{\mathrm{e}}^p \sigma \, dt = \frac{c_+}{\mathrm{E} n_1^+} \frac{|R_m Q_{m+1}|}{\sqrt{2}} + \frac{c_-}{\mathrm{E}|n_2^-|} |Q_{m+1} R_{m+1}|.$$

Here, $|R_m Q_{m+1}|$ and $|Q_{m+1} R_{m+1}|$ are the lengths of the corresponding segments. Note that the broken line $R_m Q_{m+1} R_{m+1}$ is half as long as the analogous path during the previous cycle; therefore, the series $w_1 + w_2 + \cdots$ converges. Its sum is an upper bound for the plastic work $W_p(T)$; consequently, the body shakes down to the cyclic loading.

It should be emphasized that none of the finite numbers of loading cycles results in the residual stress which remains unchanged by the subsequent loading. The residual stress at the end of the $m$-th cycle is depicted by the point $R_m$ in Figure 2.14, and these points tend to the limit $R$, the latter being the residual stress favorable for stopping the plastic deformation. Although the limit is not attained, the evolution of the residual stresses towards $R$ results in fading of the plastic deformation and in shakedown. At the same time, it can be easily seen that the plastic deformation does not stop; this follows from the fact that the residual stress varies during each cycle, which is impossible in case the plastic strain remains constant. Thus, the example shows that shakedown may occur at unstopping plastic flow.

REMARK 4. Note that normally the plastic strain remains bounded if the plastic work is bounded. Indeed, for example, the set $C$ of admissible stresses in the above considered system contains the ball $B_r(0)$ of the radius $r > 0$ with its center at the origin. Then the constitutive maximum principle results in the inequality

$$\dot{\mathrm{e}}^p \boldsymbol{\sigma} = \sup\{\dot{\mathrm{e}}^p \boldsymbol{\sigma}_* : \boldsymbol{\sigma}_* \in C\} \geq \sup\{\dot{\mathrm{e}}^p \boldsymbol{\sigma}_* : \boldsymbol{\sigma}_* \in B_r(0)\} = r|\dot{\mathrm{e}}^p|,$$

which implies the estimation

$$|\boldsymbol{\epsilon}^p(T)| \leq \int_0^T |\dot{\mathrm{e}}^p| \, dt \leq \frac{1}{r} \int_0^T \sigma \dot{\mathrm{e}}^p \, dt = \frac{1}{r} W_p(T).$$

## 3. Conditions for shakedown. Safety factor

Several examples of shakedown and nonshakedown of elastic-plastic bodies to variable loadings were considered in the previous section. Here, the precise meaning is prescribed to these concepts. We also formulate the main statements about shakedown and nonshakedown conditions (they will be proved in the next section). These conditions naturally induce the concept of the shakedown safety factor. It is determined by a static extremum problem, the latter being a direct generalization of the limit analysis static extremum problem.

**3.1. Definitions of shakedown and nonshakedown.** We consider an elastic perfectly plastic body with the displacement and stress field $\mathbf{u}_0$ and $\boldsymbol{\sigma}_0$ given at the initial moment $t = 0$. The body is subjected to a quasistatic loading: the load $\mathbf{l}(t)$ is given at $t \geq 0$. The following system of relations describes the response of the body to this loading:

$$\boldsymbol{\sigma} \in \Sigma + \mathbf{s}_l, \tag{3.1}$$

$$\mathbf{e} = \operatorname{Def} \mathbf{v}, \quad \mathbf{v} \in \mathcal{V}, \tag{3.2}$$

$$\boldsymbol{\sigma} \in C, \quad \langle \mathbf{e} - \mathbf{A}\dot{\boldsymbol{\sigma}}, \boldsymbol{\sigma} - \boldsymbol{\sigma}_* \rangle \geq 0 \ \text{ for every } \ \boldsymbol{\sigma}_* \in C, \tag{3.3}$$

$$\dot{\mathbf{u}} = \mathbf{v}, \quad \mathbf{u}|_{t=0} = \mathbf{u}_0, \quad \boldsymbol{\sigma}|_{t=0} = \boldsymbol{\sigma}_0 \tag{3.4}$$

(see Subsection 1.4). Here, $\mathbf{u}$, $\mathbf{v}$, $\mathbf{e}$ and $\boldsymbol{\sigma}$ are the displacement, velocity, strain rate and stress fields, respectively; $\Sigma$ and $C$ are the sets of self-equilibrated and admissible stress fields.

The *plastic work*

$$W_p(T) = \int_0^T \langle \mathbf{e}^p, \boldsymbol{\sigma} \rangle \, dt$$

is an important characteristic of a solution to (3.1) – (3.4). Recall that $\mathbf{e}^p = \mathbf{e} - \mathbf{A}\dot{\boldsymbol{\sigma}}$ is the plastic strain rate and $\langle \mathbf{e}^p, \boldsymbol{\sigma} \rangle$ is the dissipation. The latter is nonnegative and, therefore, the plastic work does not decrease with $T$. Engineering experience relates the plastic work to microdamage of the body. The accumulated microdamage results in failure even if the body does not undergo large deformation. This suggests consideration of the following two possibilities of the body response to a given loading: either the plastic work is bounded (there exists a number $c$ such that $W_p(T) \leq c$ for every $T \geq 0$) or $W_p(T)$ increases infinitely with $T$. We say that the body *shakes down* to the loading in the first case, and *does not shake down* in the second case.

It is often required to answer the question of whether the body shakes down to a certain loading or not. Frequently the loading is cyclic, say, in the case of cyclically working devices. In many cases, analysis is needed of the body response with respect not to a concrete loading but to all loading within a certain class. Consider, for example, the beam shown in Figure 3.1a; it is subjected to the forces $q_1$ and $q_2$ at points 1 and 2. The forces vary independently within the limits $0 \leq q_1 \leq q_{1\,\max}$, $0 \leq q_2 \leq q_{2\,\max}$; the load $\mathbf{l} = (q_1, q_2)$ may take any value within the rectangle $L$ (Figure 3.1b), and the beam response to all loading within $L$ is to be estimated. In connection with such problems the following generalization of the above introduced shakedown concept is useful.

Consider an elastic perfectly plastic body which can be subjected to any loading within the given set L of *possible loads.* If there is a number $c$ such that for all loadings within $L$ and for all moments $T \geq 0$ the plastic work is bounded, $W_p(T) \leq c$, we say that the body *shakes down* to loadings within $L$. In case the plastic work is not bounded, that is, for any number $N$ there is a loading within $L$ and a moment $T \geq 0$ such that $W_p(T) > N$, we say that

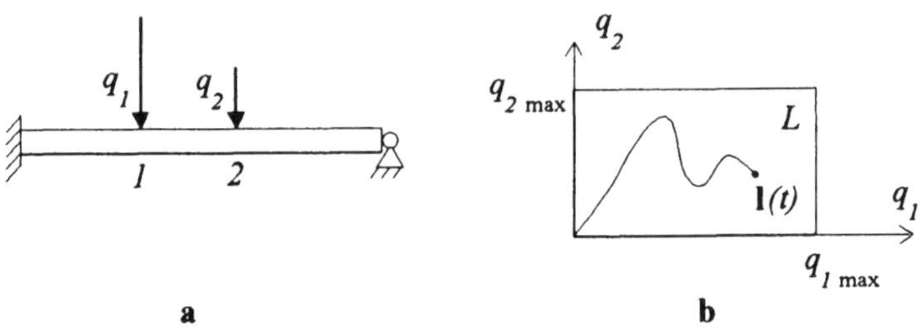


**Figure 3.1**

the body *does not shake down* to loadings within $L$. Briefly, we say the body shakes down to $L$ or does not shake down to $L$, respectively.

We will mostly consider shakedown to a set $L$ of possible loads (and not to a particular loading). Some examples of typical sets of possible loads and formulations of shakedown problems are given in Section 5 below.

**3.2. Elastic reference body.** Our nearest objective is to formulate shakedown conditions. In this connection it is convenient to make use of a certain auxiliary purely elastic body, in particular, to compare the stresses in the given elastic-plastic body with those in the auxiliary body.

Let $\mathcal{B}$ be an elastic perfectly plastic body occupying domain $\Omega$. The body is subjected to a quasistatic loading by body forces with the density $\mathbf{f}$ given in $\Omega$ and surface tractions with the density $\mathbf{q}$ given on the part $S_q$ of the body boundary. The remaining part $S_v$ of the boundary is fixed. Consider also a purely elastic body $\mathcal{B}^0$ which occupies the same domain $\Omega$ and is subjected to the same loading $\mathbf{l} = (\mathbf{f}, \mathbf{q})$. We adopt the linear elasticity law $\varepsilon_{ij}^0 = A_{ijkl}\sigma_{kl}^0$ as the constitutive relation for $\mathcal{B}^0$, the elasticity modules tensor $\mathbf{A}$ being the same as that of the elastic-plastic body $\mathcal{B}$ ($\boldsymbol{\varepsilon}^0$ and $\boldsymbol{\sigma}^0$ stand for stress and strain in $\mathcal{B}^0$). The state of the body $\mathcal{B}^0$ at any moment $t \geq 0$ can be determined by solving the following elastic problem:

$$
\begin{gathered}
\frac{\partial \sigma_{ij}^0}{\partial x_j} + f_i = 0, \quad \sigma_{ij}^0 \nu_j\big|_{S_q} = q_i, \\
\varepsilon_{ij}^0 = \frac{1}{2}\left(\frac{\partial u_i^0}{\partial x_j} + \frac{\partial u_j^0}{\partial x_i}\right), \quad u_i^0\big|_{S_v} = 0, \\
\varepsilon_{ij}^0 = A_{ijkl}\sigma_{kl}^0.
\end{gathered}
\tag{3.5}
$$

Here, $\mathbf{u}^0$, $\boldsymbol{\varepsilon}^0$ and $\boldsymbol{\sigma}^0$ are the displacement, strain and stress fields. We refer to $\mathcal{B}^0$ as an *elastic reference body* (for the elastic-plastic body $\mathcal{B}$).

The elastic problem for the body $\mathcal{B}^0$ is much simpler than the elastic-plastic problem for $\mathcal{B}$; in particular, solutions of (3.5) possess better regularity properties. We assume that, for any of the loadings under consideration, the fields $\mathbf{u}^0$, $\boldsymbol{\varepsilon}^0$ and $\boldsymbol{\sigma}^0$ belong to the spaces $\mathcal{V}$, $\mathcal{E}$ and $\mathcal{S}$ involved in the *strong* formulation of the elastic-plastic problem for $\mathcal{B}$. With these spaces and the notations of Subsection 1.4, elastic problem (3.5) can be re-written in the brief

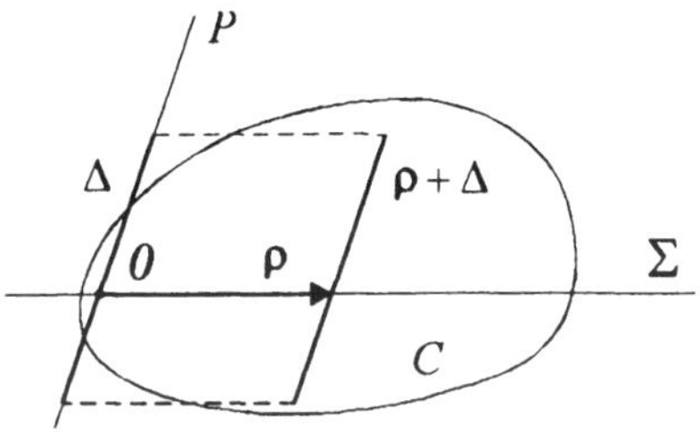


**Figure 3.2**

form

$$\sigma^0 \in \Sigma + s_l, \quad \varepsilon^0 = \mathrm{Def}\, \mathbf{u}^0, \quad \mathbf{u} \in \mathcal{V}, \quad \varepsilon = \mathbf{A}\sigma^0. \tag{3.6}$$

In the general case, unified formulation (3.1) – (3.4) of the elastic-plastic problem is considered for the body $\mathcal{B}$ (which is not necessarily three-dimensional). We also consider elastic problem (3.6) with $\Sigma$, $\mathcal{V}$, Def and $\mathbf{A}$ the same as in (3.1) – (3.4) and refer to this problem as the *elastic problem for the reference body* $\mathcal{B}^0$.

In the shakedown theory it is convenient to represent the loads applied to the body $\mathcal{B}$ by the corresponding stress fields in the elastic reference body $\mathcal{B}^0$. In order to do this, consider any possible load $\mathbf{l}$ and the (unique) solution to elastic problem (3.6) for this load. Let $\sigma^0$ be the stress field of this solution. It is easily seen that the mapping $\mathbf{l} \to \sigma^0$ is a one-to-one correspondence of the possible loads and the stress fields in $\mathcal{B}^0$. Indeed, if $\mathbf{l}_1$, $\mathbf{l}_2$ are two different possible loads and $\sigma_1^0$, $\sigma_2^0$ are the corresponding stress fields, the equality $\sigma_1^0 = \sigma_2^0$ is not valid (otherwise, there would be the zero stress field $\sigma_1^0 - \sigma_2^0$ in the elastic body under the action of the nonzero load $\mathbf{l}_1 - \mathbf{l}_2$). The one-to-one correspondence $\mathbf{l} \leftrightarrow \sigma^0$ allows us to represent the loads $\mathbf{l} \in L$ by the stress fields in the reference body $\mathcal{B}^0$. Note that the elastic stress fields $\sigma^0$ are members of the very space $\mathcal{S}$ where the sets of admissible and self-equilibrated stress fields $C$ and $\Sigma$ lie, and this is why the above representation of the loads is convenient for considering shakedown problems. In particular, a certain set $\Delta_L$ of the elastic stress fields corresponds to the given set of possible loads $L$ and represents it in $\mathcal{S}$; in what follows we omit the subscript $L$ and write $\Delta$ instead of $\Delta_L$.

The correspondence $\mathbf{l} \leftrightarrow \sigma^0$ allows to speak about the loading $\sigma^0$ with values $\sigma^0(t)$ in $\mathcal{S}$ (instead of the loading $\mathbf{l}$ with values $\mathbf{l}(t)$ in $\mathcal{L}$), and we will use this terminology.

**3.3. Shakedown conditions and safety factor: main results.** Let us formulate shakedown conditions in terms of the set of possible loads $\Delta$, the set of admissible stress fields $C$ and the subspace of self-equilibrated stress fields $\Sigma$. Figure 3.2 shows these sets in the stress fields space $\mathcal{S}$. Note that any load $\sigma^0$ in $\Delta$ (that is, any elastic stress field) is a solution to problem (3.6), thus satisfying the condition $\mathbf{A}\sigma^0 = \mathrm{Def}\, \mathbf{u}^0$. This implies that the set $\Delta$ of possible loads is always contained in the subspace $P \subset \mathcal{S}$ consisting of the stress fields that can be written in the form $\sigma^0 = \mathbf{A}^{-1}\, \mathrm{Def}\, \mathbf{u}^0$.

It should be emphasized that discussing the shakedown problem makes sense provided every load in $\Delta$ is admissible; otherwise, the elastic-plastic problem has no solutions for some loads in $\Delta$. Therefore, all of the loads in $\Delta$ are assumed to be admissible, which means that each of the loads can be equilibrated by an admissible stress field. In other words, for every $\boldsymbol{\tau}$ in $\Delta$ there is a self-equilibrated stress field $\boldsymbol{\rho}_\tau$ such that the stress field $\boldsymbol{\rho}_\tau + \boldsymbol{\tau}$ is admissible, $\boldsymbol{\rho}_\tau + \boldsymbol{\tau} \in C$.

The shakedown condition is much stronger: there should exist a self-equilibrated stress field $\boldsymbol{\rho}$ such that $\boldsymbol{\rho} + \boldsymbol{\tau}$ is admissible for every $\boldsymbol{\tau}$ in $\Delta$ or, which is the same, translating by $\boldsymbol{\rho}$ places the set $\Delta$ of possible loads in $C$: $(\boldsymbol{\rho}+\Delta) \subset C$. Strictly speaking, the shakedown condition is even stronger: $\boldsymbol{\rho}+\Delta$ should be contained not only in $C$, but in the narrower set $C/m$, $m > 1$. The shakedown condition reads:

> elastic perfectly plastic body shakes down to the set $\Delta$ of possible loads if there are self-equilibrated stress field $\boldsymbol{\rho}$ and number $m > 1$ such that $m(\boldsymbol{\rho} + \Delta) \subset C$.

This condition is sufficient for shakedown and it is almost necessary since the nonshakedown condition reads:

> elastic perfectly plastic body does not shake down to the set $\Delta$ of possible loads if there is no self-equilibrated stress field $\boldsymbol{\rho}$ such that $(\boldsymbol{\rho} + \Delta) \subset C$.

This condition also admits a simple geometric interpretation: the body does not shake down if there is no translation by $\boldsymbol{\rho} \in \Sigma$, placing the set $\Delta$ of possible loads in $C$.

The shakedown and nonshakedown conditions naturally induce the following definition. The number

$$s_\Delta = \sup\{m \geq 0 : \text{ there exists } \boldsymbol{\rho} \in \Sigma,\ m(\boldsymbol{\rho} + \Delta) \subset C\} \tag{3.7}$$

is referred to as the (shakedown) *safety factor* with respect to the set $\Delta$ of possible loads. With this concept, the shakedown and nonshakedown conditions can be formulated as follows:

> the strict inequality $s_\Delta > 1$ is a sufficient condition for shakedown of the body to the set $\Delta$ of possible loads; the inequality $s_\Delta \geq 1$ is a necessary condition for the shakedown.

It should be emphasized that the above conditions only concern the value of the safety factor $s_\Delta$ or, which is the same, the shape and location of the sets $\Delta$, $C$, $\Sigma$.

The definition of the safety factor $s_\Delta$ is analogous to that of the safety factor $\alpha_l$ in the limit analysis theory, see (III.2.3). In particular, if the set $L$ of possible loads contains only one member $\mathbf{l}$ (or, equivalently, $\Delta$ consists of the only member $\boldsymbol{\sigma}^0$), the definitions of both safety factors determine the same number $s_\Delta = \alpha_l$. The above sufficient condition for shakedown is analogous to the load $\mathbf{l}$ safety condition $\alpha_l > 1$, the necessary condition for shakedown to the load $\mathbf{l}$ admissibility condition $\alpha_l \geq 1$. We will see that the analogy extends much further: to the methods for evaluating the safety factors $s_\Delta$ and $\alpha_l$.

The necessary condition for shakedown, $s_\Delta \geq 1$, is valid under some minor additional assumptions about the material properties. The sufficient condition for shakedown is always valid; more precisely, the following main statement is valid within the framework of the unified formulation of the elastic-plastic problem.

THEOREM 1. Let $\Delta$ be a set of possible loads for a given elastic perfectly plastic body with the set $C$ of admissible stress fields. If there exist a self-equilibrated stress field $\boldsymbol{\rho}$ and a number $m > 1$ such that $m(\boldsymbol{\rho} + \Delta) \subset C$ (or, equivalently, $s_\Delta > 1$), the body shakes down to the set $\Delta$ of possible loads.

The theorem will be proved in Section 4.

Let us now accurately formulate the necessary conditions for shakedown. As was mentioned earlier, this calls for some additional assumptions about material properties. We will consider two classes of elastic perfectly plastic bodies: with bounded and cylindrical yield surfaces. In the first case, the sets $C_x$ of admissible stresses are bounded in a certain finite-dimensional stress space $\mathbf{S}$. In the second case, to be more definite, we restrict ourselves to three-dimensional bodies; then the sets $C_x$ lie in the space $Sym$ of symmetric second rank tensors, and we only consider the cylinders $C_x$ with their axes directed along the unity tensor $\mathbf{I}$. In this case, the cylinder $C_x$ is completely determined by its cross-section $C_x^d$, that is, the intersection of $C_x$ with the deviatoric plane $Sym^d$, and

$$C_x = \{\boldsymbol{\sigma} \in Sym : \boldsymbol{\sigma}^d \in C_x^d\}.$$

In the case of bodies with cylindrical yield surfaces, we will often use the set $\Delta^d$ consisting of the deviatoric fields $\boldsymbol{\tau}^d$ that correspond to the fields $\boldsymbol{\tau}$ in $\Delta$.

The following proposition makes use of assumptions analogous to those of limit analysis theorems. In particular, they concern the domain $\Omega$ occupied by the body and the free part $S_q$ of its boundary (where the surface tractions are given). We will assume that $\partial\Omega$ is sufficiently regular and $S_q$ is its sufficiently regular part; these assumptions are to ensure that the pressure field restoration problem is solvable, see Remark VI.1.1 and Theorems VI.2.3 and VI.2.4. We will also assume that $\Delta$ (or $\Delta^d$) is bounded, which means that there is a number $c$ estimating $|\boldsymbol{\tau}(x)|$ from above: $|\boldsymbol{\tau}(x)| < c$ for every $\boldsymbol{\tau} \in \Delta$ and a.e. $x \in \Omega$.

THEOREM 2. Suppose 1) there is a ball $|\mathbf{s}| \leq r$, $r > 0$, which lies inside the yield surface for every point of the body; 2) for every point of the body the yield surface lies in the same ball $|\mathbf{s}| \leq R$, $R > 0$, or, in the case of a body with cylindrical yield surfaces, in the same cylinder $|\mathbf{s}^d| \leq R$, $R > 0$; 3) the set $\Delta$ of possible loads is bounded or, in the case of a body with cylindrical yield surfaces, the set $\Delta^d$ is bounded; and 4) in the case of a body with cylindrical yield surfaces, the body boundary is sufficiently regular and the free surface $S_q$ is its sufficiently regular part. Then, if there is no self-equilibrated stress field $\boldsymbol{\rho} \in \Sigma$ such that $(\boldsymbol{\rho} + \Delta) \subset C$ (in particular, if $s_\Delta < 1$), the body does not shake down to the set $\Delta$ of possible loads.

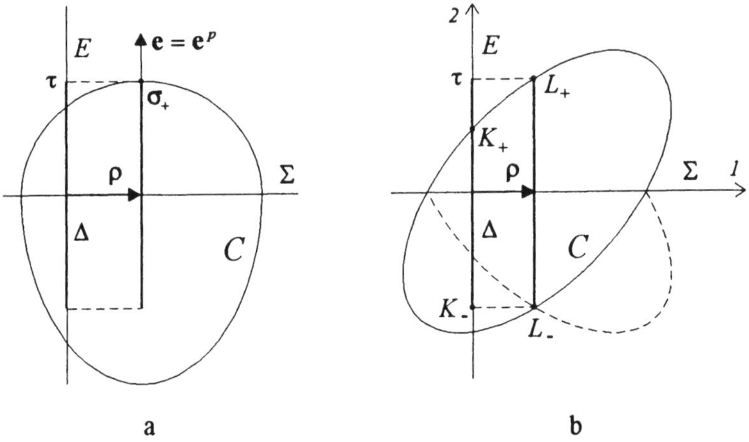


**Figure 3.3**

This proposition is valid within the framework of the unified formulation of the elastic-plastic problem; we will prove it in the next section.

Theorems 1 and 2 say nothing about shakedown or nonshakedown in case $s_\Delta = 1$, and the following example shows that both shakedown and nonshakedown are possible at $s_\Delta = 1$.

EXAMPLE 1. Consider a simple discrete system within the framework of the unified formulation of the elastic-plastic problem. Let the spaces of stress and strain rate field be two-dimensional: $\mathcal{S} = \mathcal{E} = \mathbf{R}^2$, the space $\mathcal{V}$ of velocity fields be one-dimensional, the spaces $\Sigma$ and $E$ of self-equilibrated stress fields and kinematically admissible strain rate fields be one-dimensional orthogonal subspaces in $\mathcal{S} = \mathcal{E}$, and the elasticity law be of the form $\boldsymbol{\varepsilon}^e = \boldsymbol{\sigma}/\mathrm{E}$ ($\mathrm{E}$ = const $> 0$). Figures 3.3a and 3.3b illustrate the shakedown analysis problems for two of such systems, with the only difference between them consisting of the shapes of the admissible stress fields sets. Let us find out whether the systems shake down to the set $\Delta = [-\tau, \tau]$ of possible loads.

Note first that the safety factor is $s_\Delta = 1$ for both systems. The equality immediately arises from the above geometric interpretation of the shakedown conditions. Indeed the translation of the segment $\Delta$ by a certain $\boldsymbol{\rho} \in \Sigma$ places $\Delta$ in $C$ and places $m\Delta$, where $0 < m < 1$, strictly inside $C$. On the other hand, no such translation can place the set $m\Delta$ with $m > 1$ in $C$. The latter statement is obvious in the case (a); in the case (b), it becomes obvious if one considers the mirror reflection of $C$ with respect to $\Sigma$ (the boundary of this set is depicted by the dotted line in Figure 3.3b). Thus, $s_\Delta = 1$ for each of the two systems.

Let us now show that in the case (a) the system does not shake down to $\Delta$, while in the case (b) it does. Indeed, consider in the case (a) the loading during which the load increases from 0 up to the end point $\boldsymbol{\tau}$ of the segment $\Delta$ and then remains constant. There is a unique admissible stress field $\boldsymbol{\sigma}_+ \in C$ equilibrating the load $\boldsymbol{\tau}$, see Figure 3.3a. Hence, under the action of the load

$\boldsymbol{\tau}$, the stress field does not vary and equals $\boldsymbol{\sigma}_+$. Therefore, the elastic strain rate is zero and the total strain rate equals the plastic strain rate: $\mathbf{e} = \mathbf{e}^p$. Then equations (3.1) – (3.4) which describe the behavior of the body are reduced to the following relations for the velocity $\mathbf{v}$ and strain rate $\mathbf{e}$:

$$\mathbf{e} = \operatorname{Def} \mathbf{v}, \quad \mathbf{v} \in \mathcal{V}, \qquad \langle \mathbf{e}, \boldsymbol{\sigma}_+ - \boldsymbol{\sigma}_* \rangle \geq 0 \quad \text{for every} \quad \boldsymbol{\sigma}_* \in C.$$

This means that $\mathbf{e}$ should be kinematically admissible and together with the stress $\boldsymbol{\sigma}_+$ should satisfy the normality flow rule. Solution $\mathbf{e} \neq 0$ to this problem is the failure mechanism corresponding to the limit load $\boldsymbol{\tau}$ (Subsection III.3.3). It is clear that $\mathbf{e}$ directed along the outward normal to the boundary of $C$ at the point $\boldsymbol{\sigma}_+$ (see Figure 3.3a) satisfies the above relations. These relations do not restrict the magnitude of $\mathbf{e} = \mathbf{e}^p$, and, therefore, the rate of plastic work $\mathbf{e}\boldsymbol{\sigma}_+$ is unbounded. Thus, the body does not shake down to the set $\Delta = [-\tau, \tau]$, and this happens at $s_\Delta = 1$.

Let us now show that in the case (b) the body shakes down to $\Delta = [-\tau, \tau]$, and this also happens at $s_\Delta = 1$. Consider an arbitrary loading within the segment $\Delta$ and note that the body undergoes only elastic deformation as long as the load varies within the segment $K_-K_+$ in Figure 3.3b. If the *increasing* load reaches beyond $K_-K_+$, the corresponding stress, admissible and equilibrating the load, is depicted by a point of the curve $K_+L_+$. Let us show that, if after such loading the load *decreases*, the stress moves inside the yield surface. In other words, the point depicting the stress in Figure 3.3b cannot move along the curve $K_+L_+$ in the direction from $L_+$ to $K_+$. Indeed, the plastic strain rate satisfies the normality flow rule

$$e_1^p = \dot{\lambda} n_1, \quad e_2^p = \dot{\lambda} n_2, \quad \dot{\lambda} \geq 0,$$

where $n_1$, $n_2$ are the components of the unit outward normal to the yield surface (all components are considered with respect to the Cartesian coordinate system whose 1- and 2-axis are the one-dimensional subspaces $\Sigma$ and $E$, respectively). Then the kinematic admissibility of $\mathbf{e}$ means that $e_1 = 0$, which according to the equality $\mathbf{e}^p = \mathbf{e} - \dot{\boldsymbol{\sigma}}/\mathrm{E}$ implies $e_1^p = -\dot{\sigma}_1/\mathrm{E}$. Together with the normality rule, this results in

$$\dot{\lambda} = -\dot{\sigma}_1/\mathrm{E}n_1. \tag{3.8}$$

Recall that $n_1$ is negative on $K_+L_+$; therefore, the point depicting the stress in Figure 3.3b cannot move along $K_+L_+$ at $\dot{\sigma}_1 < 0$, that is, in the direction from $L_+$ to $K_+$: this would break the condition $\dot{\lambda} \geq 0$. Thus, if the load decreases after the stress has reached the part $K_+L_+$ of the yield surface, the system undergoes only elastic deformation. This means that $\dot{\boldsymbol{\sigma}} = \mathrm{E}\mathbf{e}$, where $\mathbf{e}$ belongs to the subspace $E$ of kinematically admissible strain rate fields; consequently, $\dot{\sigma}_1 = 0$. In other words, the point depicting the current stress in Figure 3.3b moves along the segment parallel to the axis 2. Thus, if the unloading starts from the state $L_+$, the stress can reach again the yield surface only at the point $L_-$. It is easily seen that $\dot{\lambda} = 0$ and plastic flow is impossible at this state. Indeed, $L_-$ corresponds to the minimum value $-\tau$ of the load, which implies $\dot{\sigma}_1 \geq 0$ at $L_-$. Note also that $n_1 > 0$ at $L_-$; then the assumption that $\dot{\lambda} > 0$ contradicts equality (3.8), and this is why $\dot{\mathbf{e}}^p = 0$ at $L_-$. Under the subsequent loading, when the load increases starting from the minimum value $-\tau$, the process replicates the above pattern. The only difference is that the point on the curve $K_+L_+$ corresponding to the maximum load of the previous loading plays now the role that $K_+$ did earlier.

Thus, under an arbitrary loading within the segment $\Delta$, 1) the only reachable stresses are those depicted by points of the segments $K_-K_+$, $L_-L_+$, $K_-L_-$ and the curve $K_+L_+$, and 2) plastic deformation is possible only during the time intervals when the current stress corresponds to points in $K_+L_+$ and $\dot{\sigma}_1 \geq 0$. Therefore, to evaluate the plastic work, one actually integrates only over these intervals:

$$W_p(T) = \int_0^T \dot{\mathbf{e}}^p \boldsymbol{\sigma}\, dt = \int_I \dot{\lambda}(n_1\sigma_1 + n_2\sigma_2)\, dt = \int_I \frac{n_1\sigma_1 + n_2\sigma_2}{\mathrm{E}|n_1|}\dot{\sigma}_1\, dt,$$

$$I = \{t \geq 0 : \boldsymbol{\sigma}(t) \in K_+L_+,\ \dot{\boldsymbol{\sigma}}(t) \geq 0\},$$

where expression (3.8) for $\dot{\lambda}$ is taken into account. It is clear that

$$\frac{n_1\sigma_1 + n_2\sigma_2}{\mathrm{E}|n_1|}$$

is bounded on $K_+L_+$; let $\varepsilon_*$ be its upper bound. Then the following estimation is obviously valid for the plastic work:

$$W_p(T) \leq \varepsilon_* \int_I \dot{\sigma}_1\, dt \leq \varepsilon_*\sigma_{+1},$$

where $\boldsymbol{\sigma}_+$ is the stress at $L_+$. Thus, in the case (b) the body shakes down to the set $\Delta$, and this happens at $s_\Delta = 1$.

## 4. Shakedown and nonshakedown theorems

Theorems on shakedown and nonshakedown conditions will be proved in this section. Together they mean that the inequality $s_\Delta \geq 1$ is a necessary condition for shakedown of the body to the set $\Delta$ of possible loads, while the inequality $s_\Delta > 1$ is a sufficient condition for shakedown. We will establish both statements within the framework of the unified formulation of the elastic-plastic problem. Some assumptions of the nonshakedown theorem will be reduced for three-dimensional bodies in Subsection 4.5, which will result in the above formulated Theorem 3.2.

**4.1. Shakedown theorem.** We consider elastic perfectly plastic body $\mathcal{B}$ subjected to quasistatic loadings. The following system describes the process if the load $\mathbf{l}(t)$ is given at every moment $t \geq 0$ and the initial displacement and stress fields $\mathbf{u}_0$ and $\boldsymbol{\sigma}_0$ are given at $t = 0$:

$$\boldsymbol{\sigma} \in \Sigma + \mathbf{s}_l, \tag{4.1}$$

$$\mathbf{e} = \operatorname{Def} \mathbf{v}, \quad \mathbf{v} \in \mathcal{V}, \tag{4.2}$$

$$\boldsymbol{\sigma} \in C, \quad \langle \mathbf{e} - \mathbf{A}\dot{\boldsymbol{\sigma}}, \boldsymbol{\sigma} - \boldsymbol{\sigma}_* \rangle \geq 0 \ \text{ for every } \ \boldsymbol{\sigma}_* \in C, \tag{4.3}$$

$$\dot{\mathbf{u}} = \mathbf{v}, \quad \mathbf{u}|_{t=0} = \mathbf{u}_0, \quad \boldsymbol{\sigma}|_{t=0} = \boldsymbol{\sigma}_0. \tag{4.4}$$

Here $\mathbf{u}$, $\mathbf{v}$, $\mathbf{e}$ and $\boldsymbol{\sigma}$ are the displacement, velocity, strain rate and stress fields; problem (4.1) – (4.4) was discussed in detail in Section 1.

We assume that the body $\mathcal{B}$ may be subjected to any loading with the load $\mathbf{l}(t)$ varying within the given set $L$ of possible loads. Recall that there is a one-to-one correspondence $\mathbf{l} \leftrightarrow \boldsymbol{\sigma}^0$ between the loads and the stress fields in the elastic reference body $\mathcal{B}^0$ (Subsection 3.2). This allows representing the loads by elastic stress fields and the set $L$ of possible loads by the corresponding set $\Delta$ in the stress fields space $\mathcal{S}$; $\Delta$ is referred to as the set of possible loads. The following proposition establishes shakedown conditions within the framework of the unified formulation of the elastic-plastic problem.

THEOREM 1 (*Melan*). Let $\mathcal{B}$ be an elastic perfectly plastic body with the set $C$ of admissible stresses and $\Delta$ be a set of possible loads in the stress fields space. If there are a self-equilibrated stress field $\boldsymbol{\rho}$ and a number $m > 1$ such that $m(\boldsymbol{\rho} + \Delta) \subset C$ or, which is the same, $s_\Delta > 1$, then the body shakes down to the set $\Delta$ of possible loads.

PROOF. Let $\boldsymbol{\sigma}^0$ be a loading with values $\boldsymbol{\sigma}^0(t)$ in $\Delta$, and let $\boldsymbol{\sigma}$, $\mathbf{e}$ and $\mathbf{e}^p = \mathbf{e} - \mathbf{A}\dot{\boldsymbol{\sigma}}$ be the stress, strain rate and plastic strain rate fields of the solution to (4.1) – (4.4). Consider the stress field $\boldsymbol{\sigma}_* = m(\boldsymbol{\rho} + \boldsymbol{\sigma}^0)$ admissible by the assumption of the theorem. With this $\boldsymbol{\sigma}_*$, the inequality of constitutive maximum principle (4.3) results in the following estimation for the dissipation $\langle \mathbf{e}^p, \boldsymbol{\sigma} \rangle$:

$$\langle \mathbf{e}^p, \boldsymbol{\sigma} \rangle \geq \langle \mathbf{e}^p, m\boldsymbol{\rho} + m\boldsymbol{\sigma}^0 \rangle.$$

We re-write it as

$$m \langle \mathbf{e}^p, \boldsymbol{\sigma} - \boldsymbol{\sigma}^0 - \boldsymbol{\rho} \rangle \geq (m-1) \langle \mathbf{e}^p, \boldsymbol{\sigma} \rangle$$

or, which is the same, as

$$\langle \mathbf{e}^p, \boldsymbol{\sigma} \rangle \leq \frac{m}{m-1} \langle \mathbf{e}^p, \boldsymbol{\sigma} - \boldsymbol{\sigma}^0 - \boldsymbol{\rho} \rangle. \tag{4.5}$$

The expression on the right-hand side is

$$\begin{aligned} \langle \mathbf{e}^p, \boldsymbol{\sigma} - \boldsymbol{\sigma}^0 - \boldsymbol{\rho} \rangle &= \langle \mathbf{e} - \mathbf{A}\dot{\boldsymbol{\sigma}}, \boldsymbol{\sigma} - \boldsymbol{\sigma}^0 - \boldsymbol{\rho} \rangle \\ &= \langle \mathbf{e} - \mathbf{A}\dot{\boldsymbol{\sigma}}^0, \boldsymbol{\sigma} - \boldsymbol{\sigma}^0 - \boldsymbol{\rho} \rangle - \langle \mathbf{A}(\dot{\boldsymbol{\sigma}} - \dot{\boldsymbol{\sigma}}^0), \boldsymbol{\sigma} - \boldsymbol{\sigma}^0 - \boldsymbol{\rho} \rangle. \end{aligned}$$

Note that $\mathbf{e} - \mathbf{A}\dot{\boldsymbol{\sigma}}^0$ in the first term is a kinematically admissible strain rate field since it is equal to $\operatorname{Def}(\mathbf{v} - \dot{\mathbf{u}}^0)$, where $\mathbf{v}$ and $\dot{\mathbf{u}}^0$ stand for the velocity fields of the solutions to elastic-plastic problem (4.1) – (4.4) and elastic problem (3.6), respectively. Then, according to Remark 1.1, the equality

$$\langle \mathbf{e} - \mathbf{A}\dot{\boldsymbol{\sigma}}^0, \boldsymbol{\sigma} - \boldsymbol{\sigma}^0 - \boldsymbol{\rho} \rangle = 0$$

is valid as $\boldsymbol{\sigma} - \boldsymbol{\sigma}^0 - \boldsymbol{\rho}$ is a self-equilibrated stress field. Thus, the first term on the right-hand side of the above formula is zero, while the second one can be written as

$$-\langle \mathbf{A}(\dot{\boldsymbol{\sigma}} - \dot{\boldsymbol{\sigma}}^0), \boldsymbol{\sigma} - \boldsymbol{\sigma}^0 - \boldsymbol{\rho} \rangle = -\frac{1}{2}\frac{d}{dt}\langle \mathbf{A}(\boldsymbol{\sigma} - \boldsymbol{\sigma}^0 - \boldsymbol{\rho}), \boldsymbol{\sigma} - \boldsymbol{\sigma}^0 - \boldsymbol{\rho} \rangle,$$

where we took into account that $\boldsymbol{\rho}$ does not depend on $t$ and the operator $\mathbf{A}$ is symmetric.

Finally, inequality (4.5) can be written as

$$\langle \mathbf{e}^p, \boldsymbol{\sigma} \rangle \leq -\frac{m}{2(m-1)}\frac{d}{dt}\langle \mathbf{A}(\boldsymbol{\sigma} - \boldsymbol{\sigma}^0 - \boldsymbol{\rho}), \boldsymbol{\sigma} - \boldsymbol{\sigma}^0 - \boldsymbol{\rho} \rangle.$$

This yields the following estimation for the plastic work:

$$W_p(T) = \int_0^T \langle \mathbf{e}^p, \boldsymbol{\sigma} \rangle\, dt \leq \frac{m}{2(m-1)} \left[ \langle \mathbf{A}(\boldsymbol{\sigma}_0 - \boldsymbol{\sigma}_0^0 - \boldsymbol{\rho}), \boldsymbol{\sigma}_0 - \boldsymbol{\sigma}_0^0 - \boldsymbol{\rho} \rangle - \right.$$
$$\left. - \langle \mathbf{A}(\boldsymbol{\sigma}_T - \boldsymbol{\sigma}_T^0 - \boldsymbol{\rho}), \boldsymbol{\sigma}_T - \boldsymbol{\sigma}_T^0 - \boldsymbol{\rho} \rangle \right],$$

where $\boldsymbol{\sigma}_0$, $\boldsymbol{\sigma}_0^0$ and $\boldsymbol{\sigma}_T$, $\boldsymbol{\sigma}_T^0$ are the stress fields in the elastic-plastic body $\mathcal{B}$ and elastic reference body $\mathcal{B}^0$ at $t = 0$ and $t = T$. Since the quadratic form $a(\mathbf{s}) = \frac{1}{2}\langle \mathbf{A}\mathbf{s}, \mathbf{s} \rangle$ is positive definite, the previous inequality implies that

$$W_p(T) \leq \frac{m}{2(m-1)} \langle \mathbf{A}(\boldsymbol{\sigma}_0 - \boldsymbol{\sigma}_0^0 - \boldsymbol{\rho}), \boldsymbol{\sigma}_0 - \boldsymbol{\sigma}_0^0 - \boldsymbol{\rho} \rangle. \tag{4.6}$$

Note that the right-hand side in (4.6) depends neither on the loading nor on the moment $T$. Hence, the plastic work is bounded and the body shakes down to the set $\Delta$ of possible loads. The theorem is proved.

It should be emphasized that inequality (4.6) estimates the plastic work for any loading within $\Delta$. The upper bound, the right-hand side in (4.6), depends on the sets $C$ and $\Delta$, on the number $m$ and the self-equilibrated stress field $\boldsymbol{\rho}$ (implicitly); it also depends on the operator $\mathbf{A}$ which gives the elastic properties of the body and on initial stress field $\boldsymbol{\sigma}_0$. The field $\boldsymbol{\sigma}_0^0$ of initial stresses in the reference body $\mathcal{B}^0$ also enters the expression for the upper bound; however it is not an independent parameter, as $\boldsymbol{\sigma}_0^0$ is completely determined by $\boldsymbol{\sigma}_0$. Indeed, it is the stress field which solves the elastic problem for the body $\mathcal{B}^0$ under the action of the load $\mathbf{l}(0)$, the latter being completely determined by the given stress field $\boldsymbol{\sigma}_0$ (see Remark 1.3).

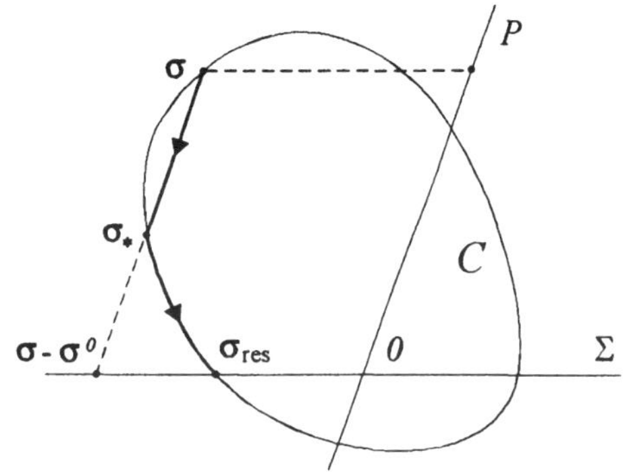


**Figure 4.1**

**4.2. Lower bound for plastic work.** Elastic perfectly plastic body $\mathcal{B}$ does not shake down to the set $\Delta$ of possible loads if the plastic work $W_p$ is unbounded for loadings within $\Delta$. To detect nonshakedown, it is sufficient to find a sequence of lower bounds for $W_p$ containing arbitrarily large members. We will construct such lower bounds in terms of the difference $\mathbf{s} = \boldsymbol{\sigma} - \boldsymbol{\sigma}^0$ of the stress fields $\boldsymbol{\sigma}$ and $\boldsymbol{\sigma}^0$ in the elastic-plastic body $\mathcal{B}$ and the corresponding elastic reference body $\mathcal{B}^0$; we refer to $\boldsymbol{\sigma} - \boldsymbol{\sigma}^0$ as a *"residual"* stress field.

REMARK 1. In some cases, the field $\mathbf{s} = \boldsymbol{\sigma} - \boldsymbol{\sigma}^0$ is equal to the actual residual stress field $\boldsymbol{\sigma}_{\text{res}}$, the latter being the stress field that remains in the body after unloading which starts from the current state (see examples in Section 2). However, generally, the equality $\boldsymbol{\sigma} - \boldsymbol{\sigma}^0 = \boldsymbol{\sigma}_{\text{res}}$ is not valid, as the plastic flow may occur in the body during unloading and effect the residual stress, while the "residual" stress $\mathbf{s} = \boldsymbol{\sigma} - \boldsymbol{\sigma}^0$ is completely determined by the stress fields in $\mathcal{B}$ and $\mathcal{B}^0$ at the current moment. Figure 4.1 illustrates the possibility of unequality of $\boldsymbol{\sigma} - \boldsymbol{\sigma}^0$ and $\boldsymbol{\sigma}_{\text{res}}$. The sets of admissible and self-equilibrated stress fields $C$ and $\Sigma$ are shown in the figure as well as the stress fields $\boldsymbol{\sigma}$ and $\boldsymbol{\sigma}^0$ at a certain moment. To compare $\boldsymbol{\sigma}_{\text{res}}$ and $\boldsymbol{\sigma} - \boldsymbol{\sigma}^0$, consider unloading which starts at the current state. No plastic deformation occurs at the beginning of the unloading, and the stress increment is a solution to the elastic problem corresponding to the increment of the load. Therefore, there is a part of the stress path parallel to the subspace $P$ of elastic stress fields in the reference body $\mathcal{B}^0$ (Subsection 3.2). This is the segment $\boldsymbol{\sigma}\boldsymbol{\sigma}_*$ in Figure 4.1, and the elastic process is no longer possible after the stress has reached the value $\boldsymbol{\sigma}_*$. Under further unloading, plastic deformation begins again, and the stress varies along the path $\boldsymbol{\sigma}_*\boldsymbol{\sigma}_{\text{res}}$. The process ends at the zero load (the unloading is completed) and results in the residual stress $\boldsymbol{\sigma}_{\text{res}} \neq \boldsymbol{\sigma} - \boldsymbol{\sigma}^0$.

Let us now construct a lower bound for the plastic work during the time interval $t_1 \leq t \leq t_2$. We will express it through the elastic energy:

$$a(\mathbf{s}_2 - \mathbf{s}_1) = \frac{1}{2} \langle \mathbf{A}(\mathbf{s}_2 - \mathbf{s}_1), \mathbf{s}_2 - \mathbf{s}_1 \rangle,$$

where $\mathbf{s}_1$ and $\mathbf{s}_2$ are the "residual" stress fields at the moments $t_1$ and $t_2$. The plastic work will be estimated within the framework of the unified formulation of the elastic-plastic problem (4.1) – (4.4) under the assumptions that the space $\mathcal{S}$ of stress fields is a normed space and the internal power bilinear form $\langle \mathbf{e}, \mathbf{s} \rangle$ is continuous in $\mathbf{s} \in \mathcal{S}$ with respect to this norm. The assumptions are not restrictive and are satisfied for a wide class of elastic-plastic bodies, and so is the second assumption of the following proposition.

PROPOSITION 1. Let $\mathcal{B}$ be an elastic perfectly plastic body. Suppose that the space $\mathcal{S}$ of stress fields in $\mathcal{B}$ is a complete normed linear space and, for every $\mathbf{e} \in \mathcal{E}$, the function $\mathbf{s} \to \langle \mathbf{e}, \mathbf{s} \rangle$ is continuous on $\mathcal{S}$. Let the set $C$ of admissible stress fields in $\mathcal{B}$ contain the ball $B_r(0)$ of the radius $r > 0$ centered at 0. If the body is subjected to the loading $\boldsymbol{\sigma}^0$, the following estimation is valid for the plastic work $W_p$ during the time interval $t_1 \le t \le t_2$:

$$W_p \ge \frac{r}{R} a(\mathbf{s}_2 - \mathbf{s}_1),$$

where $\mathbf{s}_1$ and $\mathbf{s}_2$ are the fields of "residual" stresses at $t_1$ and $t_2$ and $R$ is a number greater than $\|\mathbf{s}_1\|$, $\|\mathbf{s}_2\|$.

PROOF. To estimate the dissipation, we use the inequality of constitutive maximum principle (4.3) with $\boldsymbol{\sigma}_* = r(\mathbf{s}_1 - \mathbf{s}_2)/2R$ (this stress field is admissible according to the assumptions of the proposition):

$$\langle \mathbf{e}^p, \boldsymbol{\sigma} \rangle \ge \frac{r}{2R} \langle \mathbf{e}^p, \mathbf{s}_1 - \mathbf{s}_2 \rangle. \tag{4.7}$$

Note that the plastic strain rate field can be written as

$$\mathbf{e}^p = \mathbf{e} - \mathbf{A}\dot{\boldsymbol{\sigma}} = \mathbf{e} - \mathbf{A}\dot{\boldsymbol{\sigma}}^0 - \mathbf{A}\dot{\mathbf{s}} = \operatorname{Def} \mathbf{v} - \operatorname{Def} \dot{\mathbf{u}}^0 - \mathbf{A}\dot{\mathbf{s}},$$

where $\mathbf{v}$ and $\dot{\mathbf{u}}^0$ are the velocity fields of the solutions to the elastic-plastic problem for $\mathcal{B}$ and the elastic problem for the corresponding reference body $\mathcal{B}^0$, respectively. According to this remark we re-write the right-hand side in (4.7) as

$$\langle \mathbf{e}^p, \mathbf{s}_1 - \mathbf{s}_2 \rangle = \langle \operatorname{Def}(\mathbf{v} - \dot{\mathbf{u}}^0), \mathbf{s}_1 - \mathbf{s}_2 \rangle - \langle \mathbf{A}\dot{\mathbf{s}}, \mathbf{s}_1 - \mathbf{s}_2 \rangle.$$

The first term on the right-hand side is zero according to Remark 1.1 ($\mathbf{s}_1 - \mathbf{s}_2$ is self-equilibrated together with $\mathbf{s}_1$ and $\mathbf{s}_2$). The second term obviously equals

$$\frac{d}{dt} \langle \mathbf{A}\mathbf{s}, \mathbf{s}_2 - \mathbf{s}_1 \rangle,$$

which re-writes (4.7) as

$$\langle \mathbf{e}^p, \boldsymbol{\sigma} \rangle \ge \frac{r}{2R} \frac{d}{dt} \langle \mathbf{A}\mathbf{s}, \mathbf{s}_2 - \mathbf{s}_1 \rangle.$$

Integrating this inequality over the time interval $[t_1, t_2]$ results in the estimation

$$W_p = \int_{t_1}^{t_2} \langle \mathbf{e}^p, \boldsymbol{\sigma} \rangle \, dt \ge \frac{r}{2R} \langle \mathbf{A}(\mathbf{s}_2 - \mathbf{s}_1), \mathbf{s}_2 - \mathbf{s}_1 \rangle,$$

which finishes the proof.

**4.3. Damaging cyclic loading.** As was mentioned earlier, the plastic work results in microdamage accumulation, which, in turn, causes failure of the body if $W_p$ infinitely increases. Therefore, we refer to the loading, under which $W_p(T) \to \infty$ when $T \to \infty$, as *damaging*. Let us now establish a condition under which the cyclic loading is damaging. We will show later that in case $s_\Delta < 1$ it is easy to construct a possible cyclic loading satisfying this condition, which makes it damaging. This is the way we will prove the nonshakedown theorem.

First, consider the case when the response of the body to the cyclic loading is also cyclic. More precisely, this means that the stress and plastic strain rate fields starting from a certain moment vary as periodic functions of time and the process is only considered after this moment. Let us show that, if the "residual" stress $\mathbf{s} = \boldsymbol{\sigma} - \boldsymbol{\sigma}^0$ varies with time, the cyclic loading is damaging. Indeed, let $t_1$ and $t_2$ be two moments within one loading cycle and the corresponding "residual" stresses $\mathbf{s}_1$ and $\mathbf{s}_2$ are different. According to Proposition 1, the plastic work during the time interval $[t_1, t_2]$ is positive, and the plastic work during the whole cycle is not less than this value. Therefore, $W_p$ grows infinitely with the number of cycles since the increment of the plastic work is the same during each cycle. In other words, the body does not shake down to the loading. An example of the cyclic response to a cyclic loading has already been encountered in Subsection 2.2, and the body did not shake down to the loading in that case.

The stress response of an elastic perfectly plastic body to a cyclic loading is not necessarily cyclic, see Subsection 2.4. At the same time, it can be shown that the response tends to a cyclic pattern when the number of cycles tends to infinity, and this fact allows establishing the nonshakedown theorem. However, it is not easy to prove the basic statement about convergence of the response to the cyclic pattern. In addition, the question of whether shakedown occurs is not logically connected with that about approaching the cyclic response. Therefore, we will leave the latter question undiscussed and get rid of the cyclic response assumption in the above reasoning. This is the purpose of the following proposition. The proposition makes use of the norm $||\mathbf{s}||_{\mathcal{H}} = (\langle \mathbf{s}, \mathbf{s} \rangle)^{1/2}$ induced by the bilinear form $\langle \cdot, \cdot \rangle$; see (VI) in Subsection 1.2.

PROPOSITION 2. Suppose the assumptions of Proposition 1 are satisfied and $\boldsymbol{\sigma}^0$ is a cyclic loading of the period $\bar{t}$. Let $t_1, \dots, t_N$ be certain moments during the first loading cycle, and $\mathbf{s}_{k,1}, \dots, \mathbf{s}_{k,N}$ be the "residual" stress fields at the moments $t_1 + (k-1)\bar{t}, \dots, t_N + (k-1)\bar{t}$ of the $k$-th loading cycle. If a) the collection of the fields $\mathbf{s}_{k,1}, \dots, \mathbf{s}_{k,N}$, $k = 1, 2, \dots$, is bounded, that is,

$$||\mathbf{s}_{k,1}||_{\mathcal{S}} \le R, \; \dots, \; ||\mathbf{s}_{k,N}||_{\mathcal{S}} \le R, \quad k = 1, 2, \dots, \quad R = \text{const},$$

and b) the sequence of the numbers

$$b_k = ||\mathbf{s}_{k,2} - \mathbf{s}_{k,1}||^2_{\mathcal{H}} + ||\mathbf{s}_{k,3} - \mathbf{s}_{k,2}||^2_{\mathcal{H}} + \dots + ||\mathbf{s}_{k,N} - \mathbf{s}_{k,N-1}||^2_{\mathcal{H}}$$

does not converge to zero when $k \to \infty$, then $\boldsymbol{\sigma}^0$ is a damaging loading.

PROOF. By property (VI) of the elastic-plastic problem unified formulation (Subsection 1.4), the inequality

$$a(\mathbf{s}_{k,2}-\mathbf{s}_{k,1})+a(\mathbf{s}_{k,3}-\mathbf{s}_{k,2})+\cdots+a(\mathbf{s}_{k,N}-\mathbf{s}_{k,N-1})\geq c_{-}b_k$$

is valid. Then Proposition 1 implies the following lower bound for the plastic work $w_k$ during the $k$-th cycle: $w_k \geq c_{-}b_k r/R$. Due to assumption (b), there is $\delta > 0$ such that the sequence $b_1, b_2, \ldots$ has an unbounded number of members which are greater than $\delta$. Therefore, the sum $b_1 + \cdots + b_M$ grows infinitely when $M \to \infty$. Then $w_1 + \cdots + w_M \to \infty$ when $M \to \infty$ and, therefore, $W_p(T) \to \infty$ when $T \to \infty$, which finishes the proof.

We will show later that the assumptions of Proposition 2 are satisfied if $s_\Delta < 1$, and this will result in the nonshakedown condition.

**4.4. Nonshakedown theorem.** Let us discuss the idea and prove the nonshakedown theorem formulated in Subsection 3.3. The theorem states that, under some nonrestrictive conditions, the elastic perfectly plastic body does not shake down to the set $\Delta$ of possible loads if no translation by $\boldsymbol{\rho} \in \Sigma$ places $\Delta$ in the set $C$ of admissible stress fields. In particular, this is the case when $s_\Delta < 1$.

We start with re-formulating this nonshakedown condition. Recall that, in shakedown problems, all possible loads belonging to the set $\Delta$ are assumed admissible, that is, for any $\boldsymbol{\tau}$ in $\Delta$ there is a self-equilibrated stress field $\boldsymbol{\rho}_{\boldsymbol{\tau}}$ in $\Sigma$ such that the stress field $\boldsymbol{\rho}_{\boldsymbol{\tau}} + \boldsymbol{\tau}$ is admissible: $\boldsymbol{\rho}_{\boldsymbol{\tau}} + \boldsymbol{\tau} \in C$. Consider for a fixed $\boldsymbol{\tau} \in \Delta$ all $\boldsymbol{\rho}$ in $\Sigma$ which can be taken as $\boldsymbol{\rho}_{\boldsymbol{\tau}}$, that is, for which $\boldsymbol{\rho} + \boldsymbol{\tau}$ is an admissible stress field. It can be easily seen that the set of these $\boldsymbol{\rho}$ is written as

$$R_{\boldsymbol{\tau}} = (\Sigma + \boldsymbol{\tau}) \cap C - \boldsymbol{\tau} = \Sigma \cap (C - \boldsymbol{\tau}),$$

see Figure 4.2a.

Consider now all of the sets $R_{\boldsymbol{\tau}}$, $\boldsymbol{\tau} \in \Delta$. In case they have a common member $\boldsymbol{\rho}$, the translation by $\boldsymbol{\rho}$ places the set $\Delta$ inside $C$: $\Delta + \boldsymbol{\rho} \subset C$ or, which is the same, $\boldsymbol{\tau} + \boldsymbol{\rho} \in C$ for every $\boldsymbol{\tau}$ in $\Delta$. Therefore, if no translation $\boldsymbol{\rho} \in \Sigma$ places $\Delta$ inside $C$, the intersection of all sets $R_{\boldsymbol{\tau}}$, $\boldsymbol{\tau} \in \Delta$, is empty (and vice versa); see Figure 4.2b. In other words, there is the following obvious re-formulation of the above nonshakedown condition: the intersection of all sets $R_{\boldsymbol{\tau}}$, $\boldsymbol{\tau} \in \Delta$, is empty. As will be shown later, under this condition and some nonrestrictive assumptions, there exist an integer $N \geq 2$ and such $\boldsymbol{\tau}_1, \ldots, \boldsymbol{\tau}_N$ in $\Delta$ that the intersection of the sets $R_{\boldsymbol{\tau}_1}, \ldots, R_{\boldsymbol{\tau}_N}$ is empty (not only the above-mentioned intersection of all sets $R_{\boldsymbol{\tau}}$, $\boldsymbol{\tau} \in \Delta$), see Figure 4.2b. We will now use this statement to sketch the idea of the nonshakedown theorem. It consists in considering a cyclic loading which includes the above loads $\boldsymbol{\tau}_1, \ldots, \boldsymbol{\tau}_N$. We will show that this loading is damaging, the body does not shake down to it and, consequently, to the set $\Delta$ of possible loads.

Consider a cyclic loading taking the values $\boldsymbol{\tau}_1, \ldots, \boldsymbol{\tau}_N$ at the moments $t_1, \ldots, t_N$ during the first cycle, $\bar{t}$ being the period of the loading. The loading also takes the values $\boldsymbol{\tau}_1, \ldots, \boldsymbol{\tau}_N$ at the moments $t_1 + (k-1)\bar{t}, \ldots, t_N + (k-1)\bar{t}$ during the $k$-th cycle, $k = 1, 2, \ldots$. Let $\mathbf{s}_{k,1}, \ldots, \mathbf{s}_{k,N}$ be the corresponding

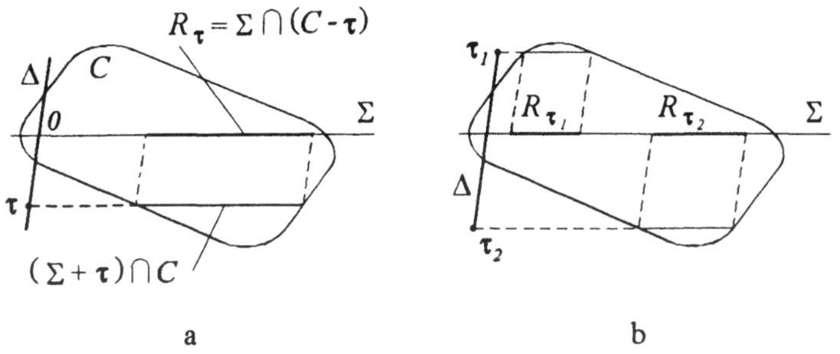


**Figure 4.2**

"residual" stress fields at these moments (recall that the "residual" stress $\mathbf{s}$ is the difference $\mathbf{s} = \boldsymbol{\sigma} - \boldsymbol{\sigma}^0$ of the stresses in the elastic-plastic body $\mathcal{B}$ and the reference body $\mathcal{B}^0$). To show that the loading under consideration is damaging, it is sufficient to verify that (a) the collection of the stress fields $\mathbf{s}_{k,1}, \ldots, \mathbf{s}_{k,N}$, $k = 1, 2, \ldots$, is bounded and (b) the sequence of the numbers

$$b_k = ||\mathbf{s}_{k,2} - \mathbf{s}_{k,1}||_{\mathcal{H}}^2 + ||\mathbf{s}_{k,3} - \mathbf{s}_{k,2}||_{\mathcal{H}}^2 + \cdots + ||\mathbf{s}_{k,N} - \mathbf{s}_{k,N-1}||_{\mathcal{H}}^2$$

does not converge to zero when $k \to \infty$ (see Proposition 2).

We now assume that the sets $R_{\tau_1}, \ldots, R_{\tau_N}$ are bounded and restrict ourselves to discrete systems, in which case the stress fields space $\mathcal{S}$ is finite-dimensional (the general case will be discussed later). As $\mathbf{s}_{k,1} \in R_{\tau_1}, \ldots$, $\mathbf{s}_{k,N} \in R_{\tau_N}$, the first of these assumptions implies that condition (a) is satisfied.

To show that (b) is also satisfied, let us denote the point $(\mathbf{s}_{k,1}, \ldots, \mathbf{s}_{k,N})$ in $\mathcal{S} \times \cdots \times \mathcal{S} = \mathcal{S}^N$ by $\bar{\mathbf{s}}_k$ and consider the sequence of the points $\bar{\mathbf{s}}_k$, $k = 1, 2, \ldots$. It has a limit point (as any bounded sequence in a finite-dimensional space does); we denote it by $\bar{\mathbf{p}} = (\mathbf{p}_1, \ldots, \mathbf{p}_N)$ and, assuming that the sets $R_{\tau_1}, \ldots, R_{\tau_N}$ are closed, arrive at the relations $\mathbf{p}_1 \in R_{\tau_1}, \ldots, \mathbf{p}_N \in R_{\tau_N}$. Suppose now that the sequence $\bar{\mathbf{s}}_1, \bar{\mathbf{s}}_2, \ldots$ does not possess property (b), that is,

$$||\mathbf{s}_{k,2} - \mathbf{s}_{k,1}||_{\mathcal{H}} \to 0, \quad \ldots, \quad ||\mathbf{s}_{k,N} - \mathbf{s}_{k,N-1}||_{\mathcal{H}} \to 0 \quad \text{when} \quad k \to \infty.$$

Note, that in the case of a discrete system, the (finite-dimensional) spaces $\mathcal{S}$ and $\mathcal{H}$ consist of the same members and convergence of a sequence in one of their norms implies its convergence in the other one. Therefore, the limit point $\bar{\mathbf{p}} = (\mathbf{p}_1, \cdots, \mathbf{p}_N)$ of the sequence $\bar{\mathbf{s}}_1, \bar{\mathbf{s}}_2, \ldots$ in $\mathcal{S}^N$ is also its limit point in $\mathcal{H}^N$. Then the assumption that $b_k \to 0$ results in the equalities $\mathbf{p}_1 = \cdots = \mathbf{p}_N$. Together with the relations $\mathbf{p}_1 \in R_{\tau_1}, \ldots, \mathbf{p}_N \in R_{\tau_N}$, this implies that the sets $R_{\tau_1}, \ldots, R_{\tau_N}$ have a common element, which contradicts the above relation $R_{\tau_1} \cap \cdots \cap R_{\tau_N} = \emptyset$. Thus, the assumption $b_k \to 0$ is wrong, the cyclic loading under consideration is damaging by Proposition 2, and the body does not shake down to the set $\Delta$ of possible loads.

In the general case when $\mathcal{S}$ is not finite-dimensional, the above reasoning calls for certain refining. The modifying basically consists of replacing the convergence by the weak convergence (Subsection V.4.2). This, in particular, overcomes the difficulty connected with the fact the spaces $\mathcal{S}$ and $\mathcal{H}$ differ in the general case.

The following nonshakedown theorem is valid within the framework of the unified formulation of the elastic-plastic problem (we will reduce its assumptions in the next subsection). We presume in this proposition that the weak convergence in $\mathcal{S}$ is defined with respect to the pairing between $\mathcal{S}$ and $\mathcal{E}_0$.

THEOREM 2. Suppose that for an elastic perfectly plastic body 1) the stress fields space $\mathcal{S}$ is a complete normed linear space and for every strain rate field $\mathbf{e} \in \mathcal{E}$ the linear function $\mathbf{s} \to \langle \mathbf{e}, \mathbf{s} \rangle$ is continuous on $\mathcal{S}$, 2) $\mathcal{S}$ is the conjugate of the complete normed linear space $\mathcal{E}_0$: $\mathcal{S}' = \mathcal{E}_0'$, 3) $\operatorname{Def} U \subset \mathcal{E}_0$ and $\mathcal{S} \subset \mathcal{E}_0$, 4) the set $C$ of admissible stress fields is weak closed and contains the ball $B_r(0)$ of the radius $r > 0$ with its center at zero, and 5) the set $\Sigma \cap (C - \Delta)$ is bounded in the norm of $\mathcal{S}$. Then, if there is no self-equilibrated stress field $\rho \in \Sigma$ such that $\rho + \Delta \subset C$ (in particular, if $s_\Delta < 1$ ), the body does not shake down to set $\Delta$ of possible loads.

PROOF. To prove the theorem, we will consider the case when there is no $\rho$ in $\Sigma$ such that $(\rho + \Delta) \subset C$, and construct a damaging cyclic loading. In order to do this, we start with introducing the set $R_{\boldsymbol{\tau}} = \Sigma \cap (C - \boldsymbol{\tau})$ for every $\boldsymbol{\tau}$ in $\Delta$ and showing that there is an integer $N \geq 2$ and $\boldsymbol{\tau}_1, \ldots, \boldsymbol{\tau}_N \in \Delta$ such that the intersection $R_{\boldsymbol{\tau}_1} \cap \cdots \cap R_{\boldsymbol{\tau}_N}$ is empty. Indeed, let us assume this is wrong, that is, any finite subcollection in the collection of the sets $R_{\boldsymbol{\tau}}$, $\boldsymbol{\tau} \in \Delta$, has a nonempty intersection. Note that all of the sets $R_{\boldsymbol{\tau}}$ are contained in a certain ball $B$ in $\mathcal{S}$ since $\Sigma \cap (C - \Delta)$ is bounded. Then by the well known compactness theorem for a ball in conjugate space, a collection of subsets in the ball has a nonempty intersection if any finite subcollection in this collection has a nonempty intersection. Thus, under the above assumption, there is $\rho_0$ that belongs to every $R_{\boldsymbol{\tau}}$, $\boldsymbol{\tau} \in \Delta$. Then $\rho \in \Sigma$ and $\rho_0 + \Delta \subset C$, which contradicts the condition specifying the case under consideration. Hence, the above assumption is wrong, and for a certain integer $N \geq 2$ there are $\boldsymbol{\tau}_1, \ldots, \boldsymbol{\tau}_N$ in $\Delta$ such that $R_{\boldsymbol{\tau}_1} \cap \cdots \cap R_{\boldsymbol{\tau}_N} = \emptyset$.

Consider now a cyclic loading taking the values $\boldsymbol{\tau}_1, \ldots, \boldsymbol{\tau}_N$ at the moments $t_1, \ldots, t_N$ during the first cycle. Let $\bar{t}$ be the period of the loading; then the loading also takes the values $\boldsymbol{\tau}_1, \ldots, \boldsymbol{\tau}_N$ at the moments

$$t_1 + (k-1)\bar{t}, \ldots, t_N + (k-1)\bar{t}$$

during the $k$-th cycle, $k = 1, 2, \ldots$. Let $\mathbf{s}_{k,1}, \ldots, \mathbf{s}_{k,N}$ be the corresponding "residual" stress fields $\mathbf{s} = \boldsymbol{\sigma} - \boldsymbol{\sigma}^0$ at these moments. These stress fields obviously belong to $R_{\boldsymbol{\tau}_1}, \ldots, R_{\boldsymbol{\tau}_N}$, respectively, and hence all of them belong to the ball $B$. Then the theorem about a ball in a conjugate space implies that the sequence of the points $\bar{\mathbf{s}}_k = (\mathbf{s}_{k,1}, \ldots, \mathbf{s}_{k,N})$, $k = 1, 2, \ldots$, in the space $\mathcal{S}^N = (\mathcal{E}_0')^N$ has a weak limit point; we denote it $\bar{\mathbf{p}} = (\mathbf{p}_1, \ldots, \mathbf{p}_N)$. Note that the set $C$ and the subspace $\Sigma = (\operatorname{Def} U)^0$ are weak closed: the former by the assumption (4), the latter as a polar set. Then each of the sets $R_{\boldsymbol{\tau}}$, $\boldsymbol{\tau} \in \Delta$, is also weak closed, which implies that $\mathbf{p}_1 \in R_{\tau_1}, \ldots, \mathbf{p}_N \in R_{\tau_N}$. Recall that there are different points among $\mathbf{p}_1, \ldots, \mathbf{p}_N$ since $R_{\tau_1} \cap \cdots \cap R_{\tau_N} = \emptyset$.

To show that the above introduced cyclic loading is damaging, we appeal to Proposition 2. Its first assumption is satisfied since the "residual" stress fields $\mathbf{s}_{k,1}, \ldots, \mathbf{s}_{k,N}$, $k = 1, 2, \ldots$, are contained in the ball $B \subset \mathcal{S}$. It only remains to verify that assumption (b) of Proposition 2 is satisfied, that is, the sequence of the numbers

$$b_k = \|\mathbf{s}_{k,2} - \mathbf{s}_{k,1}\|_{\mathcal{H}}^2 + \|\mathbf{s}_{k,3} - \mathbf{s}_{k,2}\|_{\mathcal{H}}^2 + \cdots + \|\mathbf{s}_{k,N} - \mathbf{s}_{k,N-1}\|_{\mathcal{H}}^2$$

does not converge to zero when $k \to \infty$. Indeed, the assumption that $b_k \to 0$ when $k \to \infty$ implies the equalities $\mathbf{p}_1 = \cdots = \mathbf{p}_N$ for the weak limit point $\bar{\mathbf{p}} = (\mathbf{p}_1, \ldots, \mathbf{p}_N)$ of the sequence $\bar{\mathbf{s}}_k = (\mathbf{s}_{k,1}, \ldots, \mathbf{s}_{k,N})$, $k = 1, 2, \ldots$. Let us verify, for example, that $\mathbf{p}_1 = \mathbf{p}_2$. Consider the difference $\mathbf{p}_2 - \mathbf{p}_1$, which we will denote by $\mathbf{x}_0$; it is clear that $\mathbf{x}_0$ is a weak limit point of the sequence of the points $\mathbf{x}_k = \mathbf{s}_{k,2} - \mathbf{s}_{k,1}$, $k = 1, 2, \ldots$. By the definition of the weak limit point, for every $\varepsilon > 0$ and every $\mathbf{e}$ in $\mathcal{E}_0$ there is $\mathbf{x}_k$ satisfying the inequality $|\langle \mathbf{x}_k - \mathbf{x}_0, \mathbf{e} \rangle| < \varepsilon$, the number $k$ being arbitrarily large. The relation $\mathcal{S} \subset \mathcal{E}_0$ allows us to take $\mathbf{x}_0 = \mathbf{p}_2 - \mathbf{p}_1 \in \mathcal{S}$ as $\mathbf{e}$, in which case the previous inequality implies that $|\langle \mathbf{x}_0, \mathbf{x}_0 \rangle| < \varepsilon + |\langle \mathbf{x}_k, \mathbf{x}_0 \rangle|$. Recall that the form $\langle \cdot, \cdot \rangle$ is the scalar multiplication in the space $\mathcal{H}$ (see (VI) in Subsection 1.4). Therefore, the last inequality results in

$$\|\mathbf{x}_0\|_{\mathcal{H}}^2 \leq \varepsilon + \|\mathbf{x}_k\|_{\mathcal{H}} \, \|\mathbf{x}_0\|_{\mathcal{H}} .$$

By the above assumption, $\|\mathbf{x}_k\|_{\mathcal{H}} \to 0$ when $k \to \infty$, and, apart from that, $\varepsilon$ is an arbitrarily small positive number. Together with the previous inequality, this yields $\mathbf{x}_0 = 0$, that is, $\mathbf{p}_2 = \mathbf{p}_1$. Analogously, $\mathbf{p}_1 = \mathbf{p}_2 = \cdots = \mathbf{p}_N$, which contradicts the above established fact that there are different points among $\mathbf{p}_1, \ldots, \mathbf{p}_N$. Thus, the above assumption ($b_k \to 0$ when $k \to \infty$) is wrong, which means that assumption (b) of Proposition 2 is satisfied. Then this proposition is applicable to the cyclic loading under consideration and implies that the loading is damaging and the body does not shake down to the set $\Delta$ of possible loads. The theorem is proved.

**4.5. Reduction of nonshakedown theorem assumptions.** Assumptions of Theorem 2 can be reduced in the case of bodies with a local description of material properties (Subsection IV.1.5). Let us show that this results in the nonshakedown theorem given in Subsection 3.3, which is rather simple and easily applicable to concrete problems.

We consider two types of elastic perfectly plastic bodies: with bounded and cylindrical yield surfaces. In the first case, the sets $C_x$ of admissible stresses are in a certain finite-dimensional space $\mathbf{S}$. In the second case, to be more definite, we restrict ourselves to the case of three-dimensional bodies. Then $\mathbf{S} = Sym$ and the sets of admissible stresses are of the form

$$C_x = \{\boldsymbol{\sigma} \in Sym : \boldsymbol{\sigma}^d \in C_x^d\}.$$

We will make use of the same spaces $\mathcal{S}$ and $\mathcal{E}_0$ as in limit analysis (Sections V.5 and V.7). Namely, for a body with bounded yield surfaces that occupies domain $\Omega$, $\mathcal{E}_0$ is the space of all integrable fields on $\Omega$ with values in $\mathbf{S}$, and $\mathcal{S}$ is the space of all essentially bounded integrable fields on $\Omega$ with values in $\mathbf{S}$. For three-dimensional bodies with cylindrical yield surfaces, analogous spaces of fields with values in $Sym^d$ are considered as the spaces of strain rate deviator and stress deviator fields, and the space of square integrable fields is considered as the space of spheric parts of strain rate and stress fields. The internal power is determined for every stress field s in $\mathcal{S}$ and every strain rate field e in $\mathcal{E}$:

$$-\langle \mathbf{e}, \mathbf{s} \rangle = -\int_{\Omega} \mathbf{e}(x) \cdot \mathbf{s}(x).$$

The above choice of the spaces obviously satisfies assumptions (1) – (3) of Theorem 2 and also guarantees that assumption (4) of Theorem 2 is satisfied: the set $C$ is weak closed if the mapping $x \to C_x$ is measurable (Subsection V.5.1). Recall that all plasticity problems of practical interest possess the latter property.

Assumption (5) of Theorem 2 is satisfied if the ball $|\mathbf{s}| \leq r$, $r > 0$, of the stress space $\mathbf{S}$ lies inside the yield surface for every point of the body. Indeed, in this case the ball $B_r(0)$ of the stress fields space $\mathcal{S}$ is contained in $C$.

It remains to discuss the last of the assumptions of Theorem 2: the boundness of the set $\Sigma \cap (C - \Delta)$. It is obviously satisfied for bodies with bounded yield surfaces in case both sets $\Delta$ and $C$ are bounded: this immediately implies that $C - \Delta$ is bounded too. As to bodies with cylindrical yield surfaces, the set $\Sigma \cap (C - \Delta)$ is bounded, for example, if (a) the sets

$$C^d = \{\boldsymbol{\sigma}^d; \boldsymbol{\sigma} \in C\}, \quad \Delta^d = \{\boldsymbol{\tau}^d : \boldsymbol{\tau} \in \Delta\}$$

are bounded and (b) the boundary of the body is sufficiently regular and the free surface is its sufficiently regular part. The latter assumption is to guarantee solvability of the pressure field restoration problem, see Theorem VI.2.4. It is clear that the set $\Sigma \cap (C - \Delta)$ is bounded under these conditions. Indeed, if the stress field s belongs to this set, then, according to (a), the estimation $\|\mathbf{s}^d\|_{\mathcal{S}^d} < c_1$ is valid for its deviator $\mathbf{s}^d$, where the constant $c_1$ does depend on s. Then, by Theorem VI.2.4, $\|\mathbf{s}\|_{\mathcal{S}}$ is also bounded by a constant which does not depend on s, that is, $\Sigma \cap (C - \Delta)$ is bounded.

Thus, the assumptions of Theorem 2 are reduced and we arrive at Theorem 3.2. Recall that we use the stress fields spaces $\mathcal{S} = L_\infty(\Omega; \mathbf{S})$ and $\mathcal{S} = L_\infty(\Omega; Sym^d) \times L_2(\Omega; Sym^s)$ for bodies with bounded and cylindrical yield surfaces, respectively. Then, for example, boundness of the set $\Delta$ means that there is a number $c$ such that the inequality $|\boldsymbol{\tau}(x)| < c$ is valid for every $\boldsymbol{\tau} \in \Delta$ almost everywhere in the domain occupied by the body.

## 5. Problems of shakedown analysis

Shakedown and nonshakedown are two absolutely different types of the elastic perfectly plastic body response to a given collection of loadings. Nonshakedown is the failure of the body caused by unbounded accumulation of plastic strain or microdamage; shakedown is safety against these effects. Examples of shakedown and nonshakedown were considered in Section 2, and the concepts were precisely defined in Subsection 3.1. Here we consider various formulations of shakedown analysis problems. All of them are reduced to the problem of evaluating the safety factor defined in Subsection 3.3.

**5.1. Shakedown to a set of loads.** Consider an elastic perfectly plastic body $\mathcal{B}$ which can be subjected to variable loadings within a given set $\Delta$ of possible loads.

EXAMPLE 1. Let the body $\mathcal{B}$ occupy domain $\Omega$ in a three-dimensional space. The body is subjected to body forces with the density $\mathbf{f}$ given in $\Omega$ and to surface tractions with the density $\mathbf{q}$ given on the part $S_q$ of the body boundary. There are many problems of practical interest in which the load $\mathbf{l} = (\mathbf{f}, \mathbf{q})$ varies with time in such a way that at any moment $t$ it can be decomposed with respect to basis loads $\mathbf{l}_1 = (\mathbf{f}_1, \mathbf{q}_1)$, ..., $\mathbf{l}_N = (\mathbf{f}_N, \mathbf{q}_N)$ which do not depend on $t$:

$$\mathbf{l}(t) = \alpha_1(t)\mathbf{l}_1 + \cdots + \alpha_N(t)\mathbf{l}_N$$

or, in more detail,

$$\mathbf{f}(t) = \alpha_1(t)\mathbf{f}_1 + \cdots + \alpha_N(t)\mathbf{f}_N \quad \text{in} \quad \Omega,$$
$$\mathbf{q}(t) = \alpha_1(t)\mathbf{q}_1 + \cdots + \alpha_N(t)\mathbf{q}_N \quad \text{on} \quad S_q.$$

It should be emphasized that here only the coefficients $\alpha_i$ may depend on time. Briefly speaking, the load varies in the finite-dimensional space $\mathcal{L} = \mathbf{R}^N$. Note that possible fields of the surface tractions are defined on the same part $S_q$ of the body boundary, and this is what allows consideration of linear combinations of the loads.

In the case under consideration, the set $L$ of possible loads can be identified as the subset of possible values of the parameters $\alpha_1, \ldots, \alpha_N$. For example, $L$ is often specified by the conditions $a_1 \le \alpha_1 \le b_1, \ldots, a_N \le \alpha_N \le b_N$, where $a_1, b_1, \ldots, a_N, b_N$ are given numbers.

We will represent possible loads by the corresponding elastic stress fields in the reference body $\mathcal{B}^0$, see Subsection 3.2. These fields form the set $\Delta$ in the stress fields space $\mathcal{S}$, and we refer to $\Delta$ as a set of possible loads (it replaces the set $L$ referred to the same way). In case $L$ is specified by the above inequalities, $\Delta$ is the parallelepiped

$$\Delta = \{\boldsymbol{\tau} \in \mathcal{S} : \boldsymbol{\tau} = \alpha_1\boldsymbol{\tau}_1 + \cdots + \alpha_N\boldsymbol{\tau}_N,\ a_1 \le \alpha_1 \le b_1, \ldots, a_N \le \alpha_N \le b_N\},$$

where $\boldsymbol{\tau}_1, \ldots, \boldsymbol{\tau}_N$ are the elastic stress fields in $\mathcal{B}^0$ corresponding to the basis loads $\mathbf{l}_1, \ldots, \mathbf{l}_N$.

EXAMPLE 2. A convex polyhedron is the form of the possible loads set slightly more general than that in the previous example. Let $\mathbf{l}_1, \ldots, \mathbf{l}_m$ be time-independent loads which can be applied to the body $\mathcal{B}$, and a loading can take any of the values

$$\mathbf{l} = \alpha_1\mathbf{l}_1 + \cdots + \alpha_N\mathbf{l}_m, \quad 0 \le \alpha_1, \ldots \alpha_m \le 1, \quad \alpha_1 + \cdots + \alpha_m = 1.$$

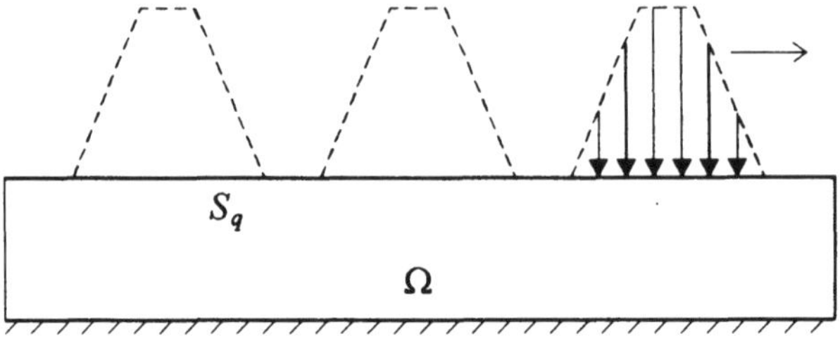


**Figure 5.1**

These loads form a convex polyhedron in the loads space $\mathcal{L}$, the vertices of the polyhedron being $\mathbf{l}_1, \ldots, \mathbf{l}_m$. We denote this polyhedron by the symbol $\text{conv}\{\mathbf{l}_1, \ldots, \mathbf{l}_m\}$. The loads $\mathbf{l}_1, \ldots, \mathbf{l}_m$ are fixed, and only the parameters $\alpha_1, \ldots, \alpha_m$ may vary during a loading. The set $\Delta$ of possible loads (elastic stress fields in the reference body $\mathcal{B}^0$) is a polyhedron too: $\Delta = \text{conv}\{\boldsymbol{\tau}_1, \ldots \boldsymbol{\tau}_m\}$. Here, $\boldsymbol{\tau}_i$ are the elastic stress fields in $\mathcal{B}^0$ corresponding to the loads $\mathbf{l}_i$.

As far as the polyhedral sets $\Delta$ of possible loads are concerned, it is sufficient to restrict the shakedown analysis to the case $\Delta$ is convex (see Proposition 1 below).

EXAMPLE 3. Consider now an example of a possible loads set which substantially differs from the above ones. Let a body occupy the domain whose boundary has a plane part $S_q$ and the possible loading consists of moving a certain load along $S_q$, see Figure 5.1. The load is given by the function $\mathbf{Q}$: the surface tractions density at the moment $t$ is $\mathbf{q}(x, t) = \mathbf{Q}(x_1 - ut)$, $u$ being the speed of the load motion.

It is clear that all of the loads $\mathbf{q}$ during the time interval $t_1 \leq t \leq t_2$ cannot be decomposed with respect to a finite number of time-independent basis loads. Thus, the loads space $\mathcal{L}$ is not finite-dimensional in this case.

The safety factor $s_\Delta$ characterizes the type of the elastic perfectly plastic body $\mathcal{B}$ response to all possible loadings within the given set $\Delta$. According to Theorems 3.1 and 3.2, the body shakes down to the set $\Delta$ of possible loads if $s_\Delta > 1$ and does not shake down if $s_\Delta < 1$. Thus, to find out whether the shakedown occurs or not, one may evaluate the safety factor

$$s_\Delta = \sup \{m \geq 0 : \text{ there exists } \boldsymbol{\rho} \in \Sigma,\ m(\boldsymbol{\rho} + \Delta) \subset C\}, \qquad (5.1)$$

where $\Sigma$ and $C$ are the sets of self-equilibrated and admissible stress fields, respectively. In many cases, it is sufficient to obtain an appropriate estimation for the safety factor. If, for example, a lower bound $s_\Delta \geq m_*$ with $m_* > 1$ is established for the safety factor, the body shakes down to $\Delta$. Analogously, the estimation $s_\Delta \leq m^*$ with $m_* < 1$ shows that the body does not shake down.

Evaluating the safety factor $s_\Delta$ is the main problem of shakedown analysis; we refer to this problem as *standard*.

Let us now show that, as far as polyhedral sets $\Delta$ of possible loads are concerned, it is sufficient to restrict the shakedown analysis to the case when $\Delta$ is convex. In order to do this, we will use the following structure. Let $A$ be a set in a linear space $X$. The intersection of all convex subsets in $X$ containing $A$ is referred to as a *convex hull* of $A$; we denote the hull by the symbol $\operatorname{conv} A$. The convex hull also admits the following description. Let $\mathbf{x}_1, \ldots, \mathbf{x}_m$ be members of $X$; any $\mathbf{x}$ in $X$ that can be written as

$$x = \alpha_1 x_1 + \cdots + \alpha_m x_m, \quad 0 \le \alpha_1, \ldots \alpha_m \le 1, \quad \alpha_1 + \cdots + \alpha_m = 1,$$

is referred to as a *convex combination* of $x_1, \ldots, x_m$. It should be emphasized that the number $m$ of the summands in this formula is not fixed. *The convex hull of $A$ is the set of all convex combinations of members of $A$.* Indeed, any convex set containing $A$ also contains all convex combination of its members. On the other hand, convex combinations of members of $A$ form a convex set; thus, this set is the "smallest" convex set containing $A$, that is, the convex hull of $A$.

The following proposition shows that the set $\Delta$ of possible loads in the shakedown analysis problem can be replaced by $\operatorname{conv} \Delta$ and vice versa.

PROPOSITION 1. For an elastic perfectly plastic body, the shakedown safety factors with respect to the sets of possible loads $\Delta$ and $\Delta' = \operatorname{conv} \Delta$ are equal: $s_\Delta = s_{\Delta'}$.

PROOF. The inequality $s_\Delta \ge s_{\Delta'}$ is obviously valid since the set $\Delta' = \operatorname{conv} \Delta$ contains $\Delta$. Let us show that the inequality $s_\Delta \le s_{\Delta'}$ is also valid, and, hence, $s_\Delta = s_{\Delta'}$. To establish the inequality $s_\Delta \le s_{\Delta'}$, we will verify that, for every number $m$ satisfying the conditions $0 \le m < s_\Delta$, the inequality $m \le s_{\Delta'}$ is valid. Indeed, if $0 \le m < s_\Delta$, by definition (5.1) there are a number $m_0$, $m \le m_0 \le s_\Delta$, and a self-equilibrated stress field $\rho_0$ in $\Sigma$ for which $m_0(\rho_0 + \Delta) \subset C$. In particular, for any collection $\boldsymbol{\tau}_1 \in \Delta, \ldots, \boldsymbol{\tau}_n \in \Delta$

$$m_0(\rho_0 + \boldsymbol{\tau}_1) \in C, \ \ldots, \ m_0(\rho_0 + \boldsymbol{\tau}_n) \in C.$$

Then, if numbers $\alpha_1, \ldots, \alpha_n \ge 0$ satisfy the condition $\alpha_1 + \cdots + \alpha_n = 1$, convexity of C implies that

$$\alpha_1 m_0(\rho_0 + \boldsymbol{\tau}_1) + \cdots + \alpha_n m_0(\rho_0 + \boldsymbol{\tau}_n) = m_0(\rho_0 + \alpha_1 \boldsymbol{\tau}_1 + \cdots + \alpha_n \boldsymbol{\tau}_n) \in C$$

or, briefly, $m_0(\rho_0 + \boldsymbol{\tau}) \in C$ for every $\boldsymbol{\tau} = \alpha_1 \boldsymbol{\tau}_1 + \cdots + \alpha_n \boldsymbol{\tau}_n$ in $\operatorname{conv} \Delta$. Thus, the inequalities $s_{\Delta'} \ge m_0 \ge m$ are valid, which finishes the proof.

EXAMPLE 4. Let the loads space $\mathcal{L}$ in the shakedown analysis problem be a normed space with the norm $|\cdot|$. Consider the ball $|\mathbf{l}| \le R$, the spheric layer $r \le |\mathbf{l}| \le R$ and the sphere $|\mathbf{l}| = R$ as the sets of possible loads. According to Proposition 1, the shakedown safety factors with respect to these three sets are equal.

**5.2. One-parametric problems of shakedown analysis.** According to Theorems 4.1 and 4.2, an elastic perfectly plastic body shakes down to the

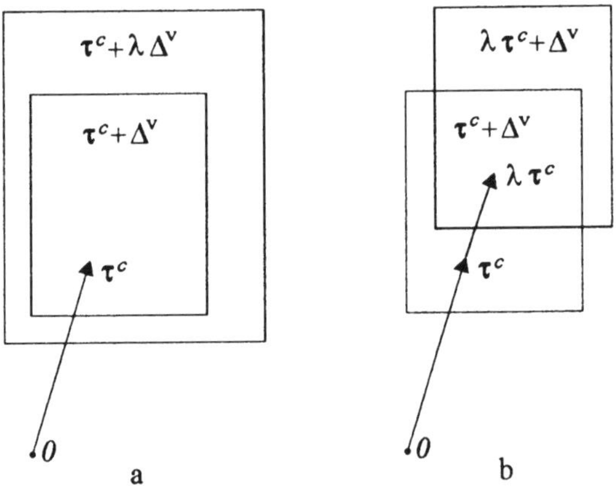


**Figure 5.2**

set $\Delta$ of possible loads if $s_\Delta > 1$ and does not shake down if $s_\Delta < 1$. There is a simple relation between the safety factors with respect to $\Delta$ and to any homothetic set $\lambda\Delta$, $\lambda > 0$:

$$\begin{aligned} s_{\lambda\Delta} &= \sup\{m \geq 0 : \text{ there exists } \rho \in \Sigma,\ m(\rho + \lambda\Delta) \subset C\} \\ &= \sup\{m \geq 0 : \text{ there exists } \rho' \in \Sigma,\ \lambda m(\rho' + \Delta) \subset C\} \\ &= \sup\left\{\frac{m'}{\lambda} \geq 0 : \text{ there exists } \rho' \in \Sigma,\ m'(\rho' + \Delta) \subset C\right\} \\ &= \frac{1}{\lambda} s_\Delta . \end{aligned}$$

Then by Theorems 4.1 and 4.2 the body shakes down to the set $\lambda\Delta$ of possible loads if $0 \leq \lambda < s_\Delta$ and does not shake down to it if $\lambda > s_\Delta$. Thus, calculating the safety factor $s_\Delta$ solves the shakedown problem with respect to all sets of possible loads in the one-parametric collection $\lambda\Delta$, $\lambda \geq 0$.

Some other one-parametric collections of sets of possible loads are also important in connection with the shakedown problems of practical interest.

EXAMPLE 5. Consider an elastic perfectly plastic body $\mathcal{B}$ subjected to a permanent load, say, its weight and also to variable loading within a certain set $\Delta^v$ of loads. As usual, we represent the loads by the elastic stress fields in the reference body $\mathcal{B}^0$. In particular, let $\boldsymbol{\tau}^c$ represent the given permanent load. Apart from the question about the shakedown to the set $\boldsymbol{\tau}^c + \Delta^v$ of possible loads, the following question often arises: how many times the limits of the variable load may be increased without switching from shakedown to nonshakedown. In other words, what are the values of the multiplier $\lambda \geq 0$ at which the body shakes down to the set $\boldsymbol{\tau}^c + \lambda\Delta^v$ of possible loads (see Figure 5.2a)?

EXAMPLE 6. Let $\boldsymbol{\tau}^c$ be the permanent load and $\Delta^v$ be the set of possible values of the variable load as in the previous example. The following question often arises when $\Delta^v$ is fixed: how many times the permanent component of the load may be increased without switching from shakedown to nonshakedown. This means that the values of the multiplier $\lambda \geq 0$ are to be found for which the body shakes down to the set $\lambda \boldsymbol{\tau}^c + \Delta^v$ of possible loads, see Figure 5.2b.

EXAMPLE 7. Consider the following set of possible loads:

$$\Delta = \{\alpha_1 \boldsymbol{\tau}_1 + \alpha_2 \boldsymbol{\tau}_2 \in \mathcal{S} : a_1 \leq \alpha_1 \leq b_1, \; a_2 \leq \alpha_2 \leq b_2\}$$

($a_1 < b_1$, $a_2 < b_2$). In case the limits $a_1$,$b_1$ for the first component of the load are fixed, while the limits $a_2$, $b_2$ for the second component may be $\lambda \geq 0$ times increased, the following one-parametric collection of the sets of possible loads enters the picture:

$$\Delta_\lambda = \{\alpha_1 \boldsymbol{\tau}_1 + \alpha_2 \boldsymbol{\tau}_2 \in \mathcal{S} : a_1 \leq \alpha_1 \leq b_1, \; \lambda a_2 \leq \alpha_2 \leq \lambda b_2\}.$$

Then the question arises about the values of the multiplier $\lambda \geq 0$ for which the body shakes down to the set $\Delta_\lambda$ of possible loads. Analogously, in case the load is of the form

$$\boldsymbol{\tau} = \alpha_1 \boldsymbol{\tau}_1 + \cdots + \alpha_N \boldsymbol{\tau}_N, \quad a_1 \leq \alpha_1 \leq b_1, \ldots a_N \leq \alpha_N \leq b_N$$

(Example 1), the limits for some components of the load may be fixed, while the limits for the rest of its components may be $\lambda$ times increased. Then the question arises about the values of the multiplier $\lambda \geq 0$ for which the body shakes down to the set

$$\Delta_\lambda = \{\alpha_1 \boldsymbol{\tau}_1 + \cdots + \alpha_N \boldsymbol{\tau}_N \in \mathcal{S} : a_1 \leq \alpha_1 \leq b_1, \ldots a_n \leq \alpha_n \leq b_n, \\ \lambda a_{n+1} \leq \alpha_1 \leq \lambda b_{n+1}, \ldots, \lambda a_N \leq \alpha_N \leq \lambda b_N\}$$

of possible loads.

Note that the formulations of the shakedown analysis problems in the above examples are of the same form. Namely, for the given sets of loads $\Delta^0$ and $\Delta^1$, the values of the parameter $\lambda \geq 0$ are to be found for which the body shakes down to the set $\Delta_\lambda = \Delta^0 + \lambda \Delta^1$ of possible loads. We refer to this problem as a *one-parametric problem* of shakedown analysis. In particular, the sets $\Delta^0$, $\Delta^1$ in Examples 5 and 6 are $\Delta^0 = \{\boldsymbol{\tau}^c\}$, $\Delta^1 = \Delta^v$ and $\Delta^0 = \Delta^v$, $\Delta^1 = \{\boldsymbol{\tau}^c\}$, respectively, while

$$\Delta^0 = \{\alpha_1 \boldsymbol{\tau}_1 + \cdots + \alpha_N \boldsymbol{\tau}_N \in \mathcal{S} : a_1 \leq \alpha_1 \leq b_1, \ldots a_n \leq \alpha_n \leq b_n\},$$
$$\Delta^1 = \{\alpha_{n+1} \boldsymbol{\tau}_{n+1} + \cdots + \alpha_N \boldsymbol{\tau}_N \in \mathcal{S} : \\ a_{n+1} \leq \alpha_{n+1} \leq b_{n+1}, \ldots, a_N \leq \alpha_N \leq b_N\}$$

in Example 7.

In what follows we will assume that for a certain value $\lambda_0 \geq 0$ of the parameter $\lambda$ the safety factor with respect to the set $\Delta_{\lambda_0} = \Delta^0 + \lambda_0 \Delta^1$ is greater than 1. Without loss of generality, only $\lambda_0 = 0$ can be considered. Indeed, if the above condition is satisfied for $\lambda = \lambda_0 > 0$, we replace $\lambda$ by the parameter $\bar{\lambda} = \lambda - \lambda_0$ and introduce the one-parametric problem of shakedown analysis with the following sets of possible loads:

$$\bar{\Delta}_{\bar{\lambda}} = \bar{\Delta}^0 + \bar{\lambda}\bar{\Delta}^1, \quad \text{where} \quad \bar{\Delta}^0 = \Delta^0 + \lambda_0\Delta^1, \quad \bar{\Delta}^1 = \Delta^1.$$

In particular, at $\bar{\lambda} = 0$, the set $\bar{\Delta}_{\bar{\lambda}} = \bar{\Delta}_0$ is exactly $\Delta^0 + \lambda_0\Delta^1$, and the safety factor with respect to the latter set is greater than 1. It is clear that the shakedown analysis problem with respect to the sets $\bar{\Delta}_{\bar{\lambda}}$ at $\bar{\lambda} \geq 0$ is equivalent to the original problem with the sets $\Delta_\lambda = \Delta^0 + \lambda\Delta^1$ of possible loads considered for $\lambda \geq \lambda_0$. Analogously, the original shakedown problem at $0 \leq \lambda \leq \lambda_0$ can be replaced by the one-parametric problem of shakedown analysis with the parameter $\bar{\lambda} = \lambda_0 - \lambda \geq 0$ and the following sets of possible loads:

$$\bar{\Delta}_{\bar{\lambda}} = \bar{\Delta}^0 + \bar{\lambda}\bar{\Delta}^1, \quad \text{where} \quad \bar{\Delta}^0 = \Delta^0 + \lambda_0\Delta^1, \quad \bar{\Delta}^1 = -\Delta^1.$$

According to this remark we restrict ourselves to the case $\lambda_0 = 0$ and assume that $s_{\Delta^0} > 1$.

A particular case of this problem (at $\Delta^0 = \{0\}$) is the one-parametric problem of shakedown analysis with the sets $\lambda\Delta$ of possible loads. It was considered in the beginning of Subsection 5.2 and reduced to the standard problem of calculating the safety factor $s_\Delta$. We will now show that, in the case of the one-parametric problem of shakedown analysis with the sets $\Delta^0 + \lambda\Delta^1$ of possible loads, there is a number $s(\Delta^0, \Delta^1)$ which plays the same role as $s_\Delta$ did in the above-mentioned particular case. It is remarkable that the number $s(\Delta^0, \Delta^1)$ can be interpreted as the safety factor $s_{\Delta^1}$ with respect to the set $\Delta^1$ in the standard shakedown analysis problem for a certain auxiliary fictitious body $\widehat{\mathcal{B}}$. The following proposition indicates, in particular, how the sets $\widehat{C}_x$ of admissible stresses are to be chosen for the auxiliary body $\widehat{\mathcal{B}}$.

PROPOSITION 2. Suppose an elastic perfectly plastic body $\mathcal{B}$ and the sets $\Delta^0 + \lambda\Delta^1$ of possible loads for every $\lambda \geq 0$ satisfy the conditions of Theorem 4.2, and $s_{\Delta^0}$ is greater than 1. Then 1) there is a number $s(\Delta^0, \Delta^1) > 0$ such that the body shakes down to the set $\Delta^0 + \lambda\Delta^1$ of possible loads if $0 \leq \lambda < s(\Delta^0, \Delta^1)$ and does not shake down if $\lambda > s(\Delta^0, \Delta^1)$, and 2) the number $s(\Delta^0, \Delta^1)$ equals the safety factor $s_{\Delta^1}$ for the elastic perfectly plastic body $\widehat{\mathcal{B}}$ which occupies the same domain as $\mathcal{B}$ and has the sets of admissible stresses of the form

$$\widehat{C}_x = \bigcap_{\tau \in \Delta^0} (C_x - \tau(x)).$$

PROOF. Let us show that the first statement is valid with

$$s(\Delta^0, \Delta^1) = \sup\left\{m \geq 0 : \text{there exists } \rho \in \Sigma,\ \rho + \Delta^0 + m\Delta^1 \subset C\right\}. \tag{5.2}$$

Proving this fact is easily reduced to verifying the following implication: if the inequality $s_{\Delta_{\lambda^*}} \geq 1$ is valid for a certain value $\lambda^* > 0$ of the parameter $\lambda$, then the strict inequality $s_{\Delta_\lambda} > 1$ holds for every $0 \leq \lambda < \lambda^*$ (recall that

$\Delta_\lambda = \Delta^0 + \lambda\Delta^1$). Indeed, let this implication be valid. Consider $\lambda$ satisfying the condition $0 \le \lambda < s(\Delta^0, \Delta^1)$; by the definition of supremum it follows from (5.2) that there are number $\lambda^* > \lambda$ and self-equilibrated stress field $\rho_*$ such that $\rho_* + \Delta^0 + \lambda^*\Delta^1 \subset C$ and, consequently, $s_{\Delta_{\lambda^*}} \ge 1$. Due to the above assumption, this means that $s_{\Delta_\lambda} > 1$, and the body shakes down to the set $\Delta_\lambda$ of possible loads (Theorem 3.1). In case $\lambda > s(\Delta^0, \Delta^1)$, formula (5.2) implies that no $\rho$ in $\Sigma$ satisfies the condition $\rho + \Delta^0 + \lambda\Delta^1 \subset C$, and the body does not shake down to $\Delta_\lambda$ (Theorem 4.2).

To finish the proof of the first statement, it remains to verify that $s_{\Delta_{\lambda^*}} \ge 1$ implies $s_{\Delta_\lambda} > 1$ for every $\lambda$, $0 \le \lambda < \lambda_*$. To show this, note that, by definition (5.1) of the safety factor, if the inequality $s_{\Delta_{\lambda^*}} \ge 1$ is valid, then for every $\varepsilon > 0$ there are the number $m_\varepsilon > 1 - \varepsilon$ and the stress field $\rho_\varepsilon$ such that $m_\varepsilon(\rho_\varepsilon + \Delta_{\lambda^*}) \subset C$. Recall that $\Delta_{\lambda^*} = \Delta^0 + \lambda^*\Delta^1$, which allows us to write the previous relation as

$$\rho_\varepsilon + \Delta^0 + \lambda^*\Delta^1 \subset \frac{C}{m_\varepsilon}$$

or, equivalently,

$$\alpha\rho_\varepsilon + \alpha\Delta^0 + \lambda\Delta^1 \subset \alpha\frac{C}{m_\varepsilon}, \quad \text{where} \quad \alpha \equiv \frac{\lambda}{\lambda_*}, \quad 0 \le \alpha < 1.$$

On the other hand, the condition $s_{\Delta^0} > 1$ means that there are the number $m^0 > 1$ and the self-equilibrated stress field $\rho^0$ for which $\rho^0 + \Delta^0 \subset C/m^0$. Together with the previous formula, this implies that

$$\begin{aligned} \alpha\rho_\varepsilon + (1-\alpha)\rho^0 + \Delta^0 + \lambda\Delta^1 &= \alpha\rho_\varepsilon + \alpha\Delta^0 + \lambda\Delta^1 + (1-\alpha)(\rho^0 + \Delta^0) \\ &\subset \alpha\frac{C}{m_\varepsilon} + (1-\alpha)\frac{C}{m^0}. \end{aligned}$$

The left-hand side of this relation is of the form $\rho + \Delta_\lambda$, where $\rho = \alpha\rho_\varepsilon + (1-\alpha)\rho^0$ is a self-equilibrated field. If, in addition, the relation

$$\alpha\frac{C}{m_\varepsilon} + (1-\alpha)\frac{C}{m^0} \subset \frac{C}{m} \tag{5.3}$$

is valid for a certain $m > 1$, then $s_{\Delta_\lambda} > 1$ by the definition of the safety factor.

Let us now show that (5.3) is valid for a certain $m > 1$ if $\varepsilon > 0$ is sufficiently small; this will finish the proof of the first statement. Consider number $\beta$, $0 < \beta < 1$, and write the expression on the left-hand side in (5.3) as

$$\begin{gathered} \alpha\frac{C}{m_\varepsilon} + (1-\alpha)\frac{C}{m^0} = \beta\frac{C}{p} + (1-\beta)\frac{C}{q}, \\ p = \frac{m_\varepsilon\beta}{\alpha}, \quad q = \frac{m^0(1-\beta)}{1-\alpha}. \end{gathered} \tag{5.4}$$

Let us denote the minimum of $p$ and $q$ by $m$; the relations

$$\frac{C}{p} \subset \frac{C}{m}, \qquad \frac{C}{q} \subset \frac{C}{m}$$

are valid since $C$ is convex and contains 0. Then (5.4) implies that

$$\alpha \frac{C}{m_\varepsilon} + (1-\alpha)\frac{C}{m^0} \subset \beta \frac{C}{m} + (1-\beta)\frac{C}{m} = \frac{C}{m}.$$

It remains to verify that, for sufficiently small $\varepsilon > 0$, the number $\beta$ can be chosen so that $p > 1$, $q > 1$ and, consequently, $m > 1$. The inequalities $p > 1$, $q > 1$ are obviously equivalent to the relations

$$\frac{\alpha}{m_\varepsilon} < \beta < 1 - \frac{1-\alpha}{m^0},$$

and appropriate $\beta$ exists if

$$\frac{\alpha}{m_\varepsilon} < 1 - \frac{1-\alpha}{m^0}.$$

The number $m_\varepsilon$ can be chosen sufficiently close to 1 for sufficiently small $\varepsilon$; therefore, the previous condition is satisfied if

$$\alpha < 1 - \frac{1-\alpha}{m_0},$$

which is valid since $\alpha < 1$, $m_0 > 1$. This finish the proof of the first statement.

Let us now show that extremum problem (5.2) is reduced to evaluating the safety factor for a certain fictitious body $\widehat{\mathcal{B}}$. Recall that the relation $A+B \subset C$ for sets $A$, $B$, $C$ in a linear space means that the sum of any $\mathbf{a} \in A$ and $\mathbf{b} \in B$ belongs to $C$, that is, $\mathbf{a} \in C - \mathbf{b}$. It is clear that the relation $A + B \subset C$ is equivalent to

$$A \subset \bigcap_{\mathbf{b}\in B} (C - \mathbf{b}).$$

In particular, the relation $\boldsymbol{\rho}+\Delta^0+m\Delta^1 \subset C$ is valid if and only if

$$\boldsymbol{\rho} + m\Delta^1 \subset \widehat{C}, \quad \text{where } \widehat{C} = \bigcap_{\boldsymbol{\tau}\in\Delta^0} (C - \boldsymbol{\tau}).$$

Then, definition (5.2) of the $s(\Delta^0, \Delta^1)$ can be re-written as

$$\begin{aligned} s(\Delta^0, \Delta^1) &= \sup\left\{m \geq 0 : \text{there exists } \boldsymbol{\rho} \in \Sigma,\ \boldsymbol{\rho} + m\Delta^1 \subset \widehat{C}\right\} \\ &= \sup\left\{m \geq 0 : \text{there exists } \boldsymbol{\rho} \in \Sigma,\ m(\boldsymbol{\rho} + \Delta^1) \subset \widehat{C}\right\}. \end{aligned} \tag{5.5}$$

This is the standard problem of evaluating the safety factor for the body $\widehat{\mathcal{B}}$ with the set $\widehat{C}$ of admissible stress fields.

Finally, let us show that, if the body $\mathcal{B}$ occupies domain $\Omega$ and the set $C$ is of the form

$$C = \{\boldsymbol{\sigma} \in \mathcal{S} : \boldsymbol{\sigma}(x) \in C_x \text{ for a.e. } x \in \Omega\},$$

the set $\widehat{C}$ admits the similar description:

$$\widehat{C} = \{\boldsymbol{\sigma} \in \mathcal{S} : \boldsymbol{\sigma}(x) \in \widehat{C}_x \text{ for a.e. } x \in \Omega\}$$

with the sets $\widehat{C}_x$ defined as

$$\widehat{C}_x = \bigcap_{\boldsymbol{\tau} \in \Delta^0} (C_x - \boldsymbol{\tau}(x)).$$

In other words, it should be verified that the relations

$$\boldsymbol{\sigma} \in \bigcap_{\boldsymbol{\tau} \in \Delta^0} (C - \boldsymbol{\tau}) \quad \text{and} \quad \boldsymbol{\sigma}(x) \in \bigcap_{\boldsymbol{\tau} \in \Delta^0} (C_x - \boldsymbol{\tau}(x))\} \text{ for a.e. } x \in \Omega$$

are equivalent. It is clear that the latter relation can be written as

$$\boldsymbol{\sigma}(x) + \boldsymbol{\tau}(x) \in C_x \text{ for a.e. } x \in \Omega \text{ and for every } \boldsymbol{\tau} \in \Delta^0,$$

which is the same as

$$\boldsymbol{\sigma} + \boldsymbol{\tau} \in C \text{ for every } \boldsymbol{\tau} \in \Delta^0$$

or, finally, $\boldsymbol{\sigma}$ belongs to each of the sets $C - \boldsymbol{\tau}$, $\boldsymbol{\tau} \in \Delta^0$, as is required. The proposition is proved.

EXAMPLE 8. Let us construct the sets $\widehat{C}_x$ for the auxiliary body $\widehat{\mathcal{B}}$ in case the set $\Delta^0$ of possible loads is a convex polyhedron: $\Delta^0 = \text{conv}\{\boldsymbol{\tau}_1, \ldots, \boldsymbol{\tau}_n\}$. According to Proposition 2, $\widehat{C}_x$ is the intersection of all sets of the form $C_x - \boldsymbol{\tau}(x)$, where $\boldsymbol{\tau}$ is in $\Delta^0$. We will now show that $\widehat{C}_x$ is the intersection of only $n$ sets:

$$\widehat{C}_x = (C_x - \boldsymbol{\tau}_1(x)) \cap \cdots \cap (C_x - \boldsymbol{\tau}_n(x)),$$

where $\boldsymbol{\tau}_1, \ldots, \boldsymbol{\tau}_n$ are the vertices of the polyhedron $\Delta^0$.

Let us denote the set on the right-hand side in the previous formula by $\widetilde{C}_x$. It is clear that $\widehat{C}_x \subset \widetilde{C}_x$; we will now verify that also $\widehat{C}_x \supset \widetilde{C}_x$ and, hence, $\widehat{C}_x = \widetilde{C}_x$. Consider $\boldsymbol{\sigma}$ in $\widetilde{C}_x$; then there are members $\boldsymbol{\sigma}_1, \ldots \boldsymbol{\sigma}_n \in C_x$ such that

$$\boldsymbol{\sigma} = \boldsymbol{\sigma}_1 - \boldsymbol{\tau}_1(x) = \cdots = \boldsymbol{\sigma}_n - \boldsymbol{\tau}_n(x).$$

Also consider any $\boldsymbol{\tau}$ in $\Delta^0 = \text{conv}\{\boldsymbol{\tau}_1, \ldots, \boldsymbol{\tau}_n\}$; it is of the form

$$\boldsymbol{\tau} = \alpha_1 \boldsymbol{\tau}_1 + \cdots + \alpha_n \boldsymbol{\tau}_n, \quad \alpha_1, \ldots, \alpha_n \geq 0, \quad \alpha_1 + \cdots + \alpha_n = 1. \tag{5.6}$$

Making use of the equality $\alpha_1 + \cdots + \alpha_n = 1$ we write $\boldsymbol{\sigma}$ as

$$\boldsymbol{\sigma} = \alpha_1(\boldsymbol{\sigma}_1 - \boldsymbol{\tau}_1(x)) + \cdots + \alpha_n(\boldsymbol{\sigma}_n - \boldsymbol{\tau}_n(x)) = \alpha_1 \boldsymbol{\sigma}_1 + \cdots + \alpha_n \boldsymbol{\sigma}_n - \boldsymbol{\tau}(x).$$

Note that $\alpha_1\boldsymbol{\sigma}_1+\cdots+\alpha_n\boldsymbol{\sigma}_n$ belongs to $C_x$ together with $\boldsymbol{\sigma}_1,\ldots,\boldsymbol{\sigma}_n$. Thus, $\boldsymbol{\sigma}$ belongs to $C_x-\boldsymbol{\tau}(x)$, where $\boldsymbol{\tau}$ is any member of $\Delta^0$; therefore, $\boldsymbol{\sigma}$ also belongs to the intersection of all such sets, which is $\widehat{C}_x$.

REMARK 1. The conditions of Proposition 2 are easily verifiable for concrete bodies. In particular, the sets $\Delta^0+\lambda\Delta^1$ satisfy condition (5) of Theorem 4.2, in the case of a body with cylindrical yield surfaces if the sets $(\Delta^0)^d$, $(\Delta^1)^d$ are bounded (see Subsection 4.5).

REMARK 2. The sets $\widehat{C}_x$ possess the regular properties of the set of admissible stresses. They are convex and closed as the intersection of convex and closed sets $C_x-\boldsymbol{\tau}(x)$. Without loss of generality, it can also be assumed that $0\in\widehat{C}_x$. Indeed, the restriction

$$\text{there exists } \boldsymbol{\rho}\in\Sigma,\ m(\boldsymbol{\rho}+\Delta^1)\subset\widehat{C}$$

in extremum problem (5.5) determining the safety factor can be equivalently re-written as

$$\text{there exists } \bar{\boldsymbol{\rho}}\in\Sigma,\ m(\bar{\boldsymbol{\rho}}+\Delta^1)\subset\widehat{C}-\boldsymbol{\rho}^0,$$

where $\boldsymbol{\rho}^0$ is any fixed self-equilibrated stress field. Therefore, for the shakedown analysis purpose, the set $\widehat{C}_x$ can be replaced by $\widehat{C}_x-\boldsymbol{\rho}^0(x)$. Note that the inequality $s_{\Delta^0}>1$ ensures that the field $\boldsymbol{\rho}^0$ can be chosen so that $\boldsymbol{\rho}^0+\Delta^0\subset C$, that is, $\boldsymbol{\rho}^0(x)\in C_x-\boldsymbol{\tau}(x)$ for every $\boldsymbol{\tau}\in\Delta^0$ and a.e. $x\in\Omega$. Then $\boldsymbol{\rho}^0(x)\in\widehat{C}_x$ and the set $\widehat{C}_x-\boldsymbol{\rho}^0(x)$ replacing $\widehat{C}_x$ contains the origin.

**5.3. Shakedown under thermomechanical loading.** Many structures work under variable thermal conditions, which should be taken into account when considering the shakedown problems. We will now incorporate one of the main thermal effects, thermal extension, in the shakedown analysis. The simplest way to describe thermal expansion is by replacing Hooke's law with the more general relation

$$\varepsilon_{ij}=-\frac{\nu}{\mathrm{E}}\sigma_{kk}\delta_{ij}+\frac{1+\nu}{\mathrm{E}}\sigma_{ij}+\alpha\theta\delta_{ij}, \tag{5.7}$$

where $\theta$ is the temperature and $\alpha$ is the thermal expansion coefficient. Another important thermal effect is thermal softening, that is, decreasing of the yield stress with the temperature. In some cases of practical interest, softening and other thermal effects are negligible with respect to thermal expansion, and we will show that under this assumption the shakedown analysis problem is reduced to the standard problem of evaluating the safety factor $s_\Delta$. The latter is to be determined with respect to the set of possible loads appropriately modified to take thermal expansion into account.

Consider the body $\mathcal{B}$ whose total strain $\boldsymbol{\varepsilon}$ consists of the elastic, plastic and temperature components: $\boldsymbol{\varepsilon}^e$, $\boldsymbol{\varepsilon}^p$, $\boldsymbol{\varepsilon}^\theta$. The constitutive relations we adopt for the elastic and plastic strains are the same as in the case of the elastic perfectly plastic body (Subsection 1.1). The temperature strain $\boldsymbol{\varepsilon}^\theta$ is assumed to be determined by the temperature, for example, $\varepsilon^\theta_{ij}=\alpha\theta\delta_{ij}$, likewise (5.7), the temperature field $\theta$ being given. We refer to the body $\mathcal{B}$ with the above material properties as *thermoelastic-plastic*. The general formulation of the quasistatic problem for $\mathcal{B}$ is analogous to (1.9) and is written as

$$\begin{gathered}\boldsymbol{\sigma} \in \Sigma + \mathbf{s}_l, \\ \mathbf{e} = \operatorname{Def} \mathbf{v}, \quad \mathbf{v} \in \mathcal{V}, \\ \boldsymbol{\sigma} \in C, \quad \langle \mathbf{e} - \dot{\boldsymbol{\varepsilon}}^\theta - \mathbf{A}\dot{\boldsymbol{\sigma}}, \boldsymbol{\sigma} - \boldsymbol{\sigma}_* \rangle \geq 0 \text{ for every } \boldsymbol{\sigma}_* \in C, \\ \dot{\mathbf{u}} = \mathbf{v}, \quad \mathbf{u}|_{t=0} = \mathbf{u}_0, \quad \boldsymbol{\sigma}|_{t=0} = \boldsymbol{\sigma}_0.\end{gathered} \tag{5.8}$$

Here, $\mathbf{s}_l$ is any stress field equilibrating the given mechanical load $\mathbf{l}$, and $\boldsymbol{\varepsilon}^\theta$ is the *given* field of temperature strain representing the *thermal load*. The pair $(\mathbf{l}(t), \boldsymbol{\varepsilon}^\theta(t))$ represents the *load* in the case of thermoelastic-plastic body $\mathcal{B}$. If the load is given for every moment $t \geq 0$, we say that (thermo-mechanical) loading is given.

The definitions of shakedown and nonshakedown (boundness and unboundness of the plastic work, respectively) remain unchanged for thermoelastic-plastic bodies. The only difference is that the load is now given not only by $\mathbf{l}$ but also by $\boldsymbol{\varepsilon}^\theta$. However, introducing a thermoelastic reference body makes the difference insignificant. Namely, consider, for the thermoelastic-plastic body $\mathcal{B}$, the body $\mathcal{B}^0$ with the displacement, stress and strain fields $\mathbf{u}^0$, $\boldsymbol{\sigma}^0$ and $\boldsymbol{\varepsilon}^0$ determined by the following relations:

$$\boldsymbol{\sigma}^0 \in \Sigma + \mathbf{s}_l, \quad \mathbf{e}^0 = \operatorname{Def} \mathbf{u}^0, \quad \mathbf{u}^0 \in \mathcal{V}, \quad \boldsymbol{\varepsilon}^0 = \mathbf{A}\boldsymbol{\sigma}^0 + \boldsymbol{\varepsilon}^\theta, \tag{5.9}$$

where $(\mathbf{l}, \boldsymbol{\varepsilon}^\theta)$ is the given load and $\Sigma$, $\mathcal{V}$, Def, $\mathbf{A}$ are the same as in (5.8). The body $\mathcal{B}^0$ is called a *thermoelastic reference body*. It is assumed that (5.9) has a unique solution for the given load $(\mathbf{l}, \boldsymbol{\varepsilon}^\theta)$, and we will use the stress field $\boldsymbol{\sigma}^0$ of this solution to represent the load. The mapping $(\mathbf{l}, \boldsymbol{\varepsilon}^\theta) \to \boldsymbol{\sigma}^0$ is not a one-to-one correspondence in contrast to the case when thermal effects are neglected. This, however, does not call for any changes in the reasoning that results in shakedown conditions.

Consider the thermoelastic-plastic body $\mathcal{B}$ which can be subjected to any loading $(\mathbf{l}, \boldsymbol{\varepsilon}^\theta)$ within the limits of the given set $L$ of possible loads. Let $\Delta$ be the set of the stress fields in the thermoelastic reference body $\mathcal{B}^0$ that corresponds to possible loads in $L$. Note that $\Delta$ is a subset in the same stress fields space as in the case of purely mechanical loading. Therefore, the safety factor $s_\Delta$ is determined by formula (5.1) no matter where $\Delta$ originated. The following proposition relates shakedown of the body $\mathcal{B}$ to the value of $s_\Delta$.

PROPOSITION 3. Let $L$ be the set of possible loads $(\mathbf{l}, \boldsymbol{\varepsilon}^\theta)$ for a thermoelastic-plastic body $\mathcal{B}$, $\Delta$ be the set of corresponding stress fields in the thermoelastic reference body $\mathcal{B}^0$, and $s_\Delta$ be the safety factor (5.1). Then the body $\mathcal{B}$ shakes down to the set of possible loads $L$ if $s_\Delta > 1$ and does not shake down if $s_\Delta < 1$ and the assumptions of Theorem 4.2 are satisfied.

PROOF. In case $s_\Delta > 1$, the plastic work

$$W_p(T) = \int_0^T \langle \mathbf{e}^p, \boldsymbol{\sigma} \rangle \, dt$$

can be evaluated the same way as in Theorem 4.1; this only requires that the strain rate $\mathbf{e}$ be replaced by $\mathbf{e} - \dot{\boldsymbol{\varepsilon}}^{\theta}$. Upper bound (4.6) remains valid, and the body shakes down to the set of possible loads $L$.

In case $s_{\Delta} < 1$, the present statement can be proved with the same lower bound for plastic work as in Proposition 4.1. The only difference consists of determining the stress field $\boldsymbol{\sigma}^0$ as the solution to thermoelastic problem (5.9) rather than to elastic problem (3.6). Proposition 4.2 and Theorem 4.2 based on it are applicable without any changes to the case under consideration. This implies that $\mathcal{B}$ does not shake down to the set $L$ if $s_{\Delta} < 1$, which finishes the proof.

Thus, the shakedown analysis problem for a thermoelastic-plastic body is reduced to the standard problem of evaluating the safety factor $s_{\Delta}$, where $\Delta$ is the set of stress fields in the thermoelastic reference body $\mathcal{B}^0$, and the thermal load is taken into account when determining $\Delta$.

## 6. Extremum problems of shakedown analysis

We have already seen that the shakedown analysis problems are reduced to evaluation of the safety factor (Subsection 3.3 and Section 5). The safety factor is determined as the supremum in a certain static extremum problem, and any value of the function to be maximized in this problem is a lower bound for the safety factor. The dual kinematic extremum problem is constructed in this section, any value of the function to be minimized in this problem being an upper bound for the safety factor. Under certain conditions these estimations may converge to the safety factor, in which case solving the kinematic extremum problem is a way to compute the safety factor.

Constructing the dual problem follows the same scheme as in limit analysis theory, and the function to be minimized in the kinematic problem is analogous to the kinematic multiplier in limit analysis. However, there is no simple explicit formula for this function like that for the kinematic multiplier. We will overcome this difficulty in the next section in the case of a polyhedral set of possible loads, which is of most importance for applications of the shakedown theory.

**6.1. Static extremum problem.** The main problem of the shakedown analysis of elastic perfectly plastic body $\mathcal{B}$ is evaluating of the safety factor $s_{\Delta}$ with respect to the set $\Delta$ of possible loads. The safety factor characterizes the type of the body response to loadings within $\Delta$: the body shakes down to the set $\Delta$ of possible loads if $s_{\Delta} > 1$ and does not shake down if $s_{\Delta} < 1$. The safety factor $s_{\Delta}$ is determined as the supremum

$$s_{\Delta} = \sup \{ m \geq 0 : \text{there exists } \boldsymbol{\rho} \in \Sigma,\ m(\boldsymbol{\rho} + \Delta) \subset C \}, \tag{6.1}$$

where $\Sigma$ and $C$ are the sets of self-equilibrated and admissible stress fields, respectively. Recall also that $\Delta$ consists of elastic stress fields in the reference body $\mathcal{B}^0$, each of the fields representing one of the possible loads (Subsection 3.2). Problem (6.1) is referred to as the *static extremum problem of the shakedown analysis* and is a direct generalization of the static extremum problem of limit analysis (V.2.3). We will now construct the dual extremum problem of (6.1) following the same scheme as in the limit analysis theory. The dual problem allows us to obtain upper bounds for the safety factor and, in many cases, also its value.

The first step in constructing the dual problem of (6.1) is re-writing the latter in the standard form. In order to do this, we make use of the Minkowski function

$$M_C(\mathbf{s}) = \inf\left\{\mu > 0 : \frac{1}{\mu}\mathbf{s} \in C\right\}$$

(see Subsection V.2.1). Recall that this function is convex and positive homogeneous of degree 1. If the set $C$ is convex, contains 0 and is closed along rays (which conditions are always satisfied for elastic-plastic bodies), then

$$\text{the relation } \boldsymbol{\sigma} \in C \text{ is equivalent to } M_C(\boldsymbol{\sigma}) \leq 1 \tag{6.2}$$

(Subsection V.2.1). Using these properties of the Minkowski function, we write (6.1) as

$$\begin{aligned}\frac{1}{s_\Delta} &= \inf\left\{\frac{1}{m} > 0 : \text{there exists } \boldsymbol{\rho} \in \Sigma,\ m(\rho + \Delta) \subset C\right\} \\ &= \inf\left\{\mu > 0 : \text{there exists } \boldsymbol{\rho} \in \Sigma,\ \frac{1}{\mu}(\rho + \Delta) \subset C\right\} \\ &= \inf\left\{\mu > 0 : \text{there exists } \boldsymbol{\rho} \in \Sigma,\ \frac{1}{\mu}(\rho + \boldsymbol{\tau}) \in C \text{ for every } \boldsymbol{\tau} \in \Delta\right\}.\end{aligned} \tag{6.3}$$

Due to (6.2) and homogeneity of the Minkowski function, the condition

$$\frac{1}{\mu}(\boldsymbol{\rho} + \boldsymbol{\tau}) \in C \quad \text{for every} \quad \boldsymbol{\tau} \in \Delta$$

can be written equivalently as

$$M_C(\boldsymbol{\rho} + \boldsymbol{\tau}) \leq \mu \quad \text{for every} \quad \boldsymbol{\tau} \in \Delta$$

or, which is the same, as the inequality

$$\sup\left\{M_C(\boldsymbol{\rho} + \boldsymbol{\tau}) : \boldsymbol{\tau} \in \Delta\right\} \leq \mu.$$

We introduce the function

$$F(\mathbf{s}) = \sup\left\{M_C(\mathbf{s} + \boldsymbol{\tau}) : \boldsymbol{\tau} \in \Delta\right\}, \tag{6.4}$$

which allows writing the previous inequality as $F(\boldsymbol{\rho}) \leq \mu$. Recall that this is the restriction in extremum problem (6.3), and the latter takes the form

$$\frac{1}{s_\Delta} = \inf\left\{\mu > 0 : \text{there exists } \boldsymbol{\rho} \in \Sigma,\ F(\boldsymbol{\rho}) \leq \mu\right\}$$

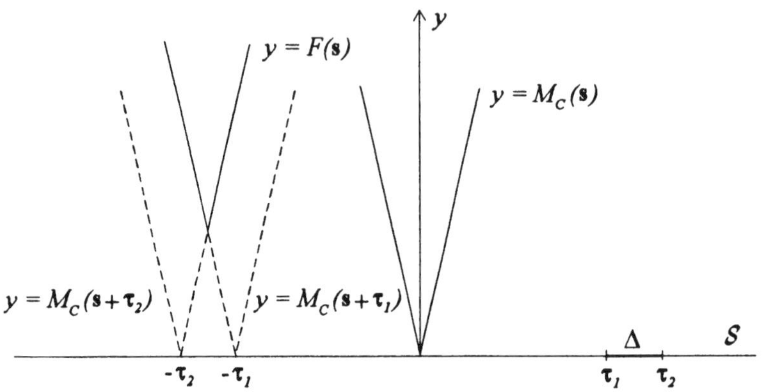


**Figure 6.1**

or, equivalently,

$$\frac{1}{s_\Delta} = \inf\{F(\rho) : \rho \in \Sigma\}. \tag{6.5}$$

Note that $F$ is the supremum

$$F(\mathbf{s}) = \sup\{f_\tau(\mathbf{s}) : \boldsymbol{\tau} \in \Delta\} \tag{6.6}$$

of the collection of the convex functions $f_\tau$, $\boldsymbol{\tau} \in \Delta$, where

$$f_\tau(\mathbf{s}) = M_C(\mathbf{s} + \boldsymbol{\tau}).$$

Therefore, $F$ is a convex function, and static extremum problem (6.5) is standard, that is, consists of minimizing a convex function over a subspace in a linear space:

$$\begin{gathered} F(\boldsymbol{\rho}) \to \inf; \quad \boldsymbol{\rho} \in \Sigma \\ (F(\mathbf{s}) = \sup\{M_C(\mathbf{s} + \boldsymbol{\tau}) : \boldsymbol{\tau} \in \Delta\}). \end{gathered} \tag{6.7}$$

The infimum in this problem is equal to $1/s_\Delta$, and computing of $F(\boldsymbol{\rho})$ for any self-equilibrated stress field $\boldsymbol{\rho} \in \Sigma$ results in an upper bound for $1/s_\Delta$ (which gives a lower bound for the safety factor). Note that evaluating $F(\boldsymbol{\rho})$ itself calls for solving extremum problem (6.6), which is not easy to do. However, this problem can be substantially reduced in the case of a polyhedral set of possible loads, the most important for applications.

EXAMPLE 1. Let the set of possible loads be the convex polyhedron with the vertices $\boldsymbol{\tau}_1, \ldots, \boldsymbol{\tau}_m$: $\Delta = \text{conv}\{\boldsymbol{\tau}_1, \ldots, \boldsymbol{\tau}_m\}$. The value $F(\mathbf{s})$ of the function to be minimized in the static extremum problem is the supremum of the function $\phi(\boldsymbol{\tau}) = M_C(\mathbf{s} + \boldsymbol{\tau})$ over $\boldsymbol{\tau} \in \Delta$, while $\mathbf{s}$ is fixed. Since $\phi$ is convex, this supremum is attained in one of the vertices of the polyhedron and, consequently,

$$F(\mathbf{s}) = \max\{M_C(\mathbf{s} + \boldsymbol{\tau}_1), \ldots, M_C(\mathbf{s} + \boldsymbol{\tau}_m)\}.$$

Thus, to evaluate $F(\mathbf{s})$, one has to find values of the Minkowski function in $m$ points and choose the greatest of them, see Figure 6.1.

**6.2. Kinematic extremum problem.** Let us now formulate the extremum problem dual of the static extremum problem determining the safety factor. The dual problem allows estimating the safety factor from above. Under certain conditions, the extremum in the dual problem equals $s_\Delta$, in which case solving the problem is a way to find not only estimations for the safety factor but also its value.

To construct the dual kinematic problem, we apply the general scheme of Section V.3 starting with the static problem in its standard form (6.7). Recall that the latter is considered within the framework of the unified formulation of the elastic-plastic problem and meets the requirements (I) – (VIII) of Subsection 1.4. In particular, the value $\langle \mathbf{e}, \mathbf{s} \rangle$ of the internal power bilinear form is defined for every $\mathbf{e}$ in the strain rate fields space $\mathcal{E}$ and every $\mathbf{s}$ in the stress fields space $\mathcal{S}$. The bilinear form is a pairing between $\mathcal{S}$ and $\mathcal{E}$ and allows us to introduce the Fenhel transformation (Subsection V.3.1) needed for constructing the dual problem as is suggested by the general scheme.

Consider the Fenhel transformation of function (6.4) involved in the static extremum problem:

$$G(\mathbf{e}) = F^*(\mathbf{e}) = \sup \{\langle \mathbf{e}, \boldsymbol{\sigma} \rangle - F(\boldsymbol{\sigma}) : \boldsymbol{\sigma} \in \mathcal{S}\}. \tag{6.8}$$

The definition of the Fenhel transformation immediately results in the inequality

$$F(\boldsymbol{\sigma}) \geq \langle \mathbf{e}, \boldsymbol{\sigma} \rangle - G(\mathbf{e}) \quad \text{for every} \quad \boldsymbol{\sigma} \in \mathcal{S} \quad \text{and every} \quad \mathbf{e} \in \mathcal{E}.$$

Restricting ourselves to self-equilibrated stress fields $\boldsymbol{\sigma} \in \Sigma$ and kinematically admissible strain rate fields $\mathbf{e} \in E = \operatorname{Def} \mathcal{V}$, we obtain $\langle \mathbf{e}, \boldsymbol{\sigma} \rangle = 0$ (see Remark 1.1), and the previous relation implies that

$$\inf \{F(\boldsymbol{\rho}) : \boldsymbol{\rho} \in \Sigma\} \geq -G(\mathbf{e}) \quad \text{for every} \quad \mathbf{e} \in E$$

or, by virtue of (6.5),

$$\frac{1}{s_\Delta} \geq -G(\mathbf{e}) \quad \text{for every} \quad \mathbf{e} \in E.$$

Thus, evaluating $-G(\mathbf{e})$ for any kinematically admissible strain rate field $\mathbf{e}$ results in the lower bound for $1/s_\Delta$ (that is, gives an upper bound for the safety factor). The best estimation obtainable this way is

$$\frac{1}{s_\Delta} \geq \sup \{-G(\mathbf{e}) : \mathbf{e} \in E\}.$$

Thus, we arrived at the following *kinematic extremum problem* of the shakedown analysis:

$$\begin{gathered} G(\mathbf{e}) \to \sup; \quad \mathbf{e} \in E \\ (G = F^*, \quad E = \operatorname{Def} \mathcal{V}). \end{gathered} \tag{6.9}$$

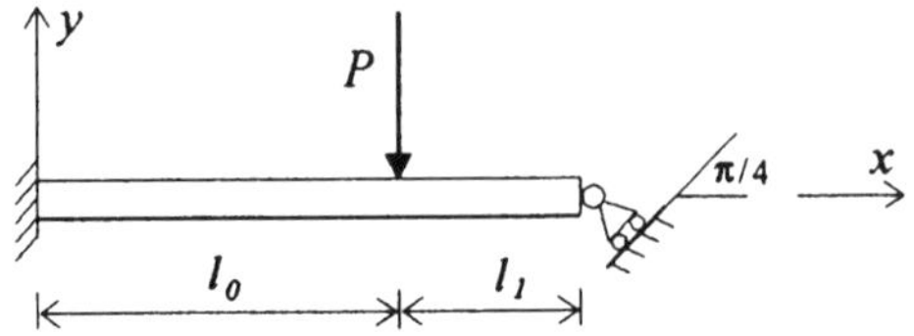


**Figure 6.2**

Problem (6.9) is analogous to the limit analysis kinematic extremum problem (V.2.5), and this analogy is complete in some important cases considered below in Section 7. In general, for the function $G$ involved in (6.9), there is no simple explicit formula like that for the kinematic multiplier $m_k$ in limit analysis. To evaluate $G(\mathbf{e})$, extremum problem (6.8) is to be solved, which is not easy to do in the case of continual elastic-plastic systems. However, in case the set $\Delta$ of possible loads is a polyhedron, modifying of problem (6.9) results in a simple formula for evaluating the function to be minimized; the modification will be considered in the next section. Note also that evaluating of $G(\mathbf{e})$ is not very complicated in the case of discrete systems.

EXAMPLE 2. Let us consider an example of deriving explicit formulas for the functions $F$ and $G$ involved in the static and kinematic extremum problems of the shakedown analysis in the case of a discrete system. Consider a beam working in in-plane bending under the action of the concentrated load $P$. Figure 6.2 shows the geometric parameters and constraints, the latter imposing the following restrictions on the velocity components $u$, $v$ along the $x$- and $y$-axis, respectively, bending moment $M$, cutting and axial forces $Q$ and $N$:

$$\begin{aligned} &u = v = 0, \quad v' = 0 \quad \text{at} \quad x = 0, \\ &u - v = 0, \quad M = 0, \quad N + Q = 0 \quad \text{at} \quad x = l_1 + l_2 \end{aligned} \tag{6.10}$$

(the prime denotes the derivative with respect to $x$).

The space of stress fields in the beam is finite-dimensional. Indeed, since there is no distributed load, the equilibrium equations (Subsection I.6.2) with boundary conditions (6.10) imply that the axial force is constant,

$$N(x) = \text{const} = N,$$

the moment linearly depends on $x$,

$$M(x) = \begin{cases} M - \dfrac{M + Nl_1}{l_0} x & \text{if } 0 \le x \le l_0, \\ N(x - l_0 - l_1) & \text{if } l_0 \le x \le l_0 + l_1, \end{cases} \qquad M = \text{const},$$

and the cutting force is piecewise constant,

$$Q(x) = \begin{cases} Q & \text{if } 0 \le x \le l_0, \\ -N & \text{if } l_0 \le x \le l_0 + l_1, \end{cases} \qquad Q = \text{const},$$

where $N$ is the above axial force. Thus, the parameters $M$, $N$, $Q$ represent each stress field in the beam; the stress fields space is three-dimensional: $\mathcal{S} = \mathbf{R}^3$. Note that the equilibrium conditions are reduced to the equation $-N - Q = P$.

In what follows, we consider the beam of a rectangular cross-section of the width $b$ and height $h$. Let $\sigma_Y$ be the yield stress of the material of the beam and the inequality

$$\frac{|M(x)|}{M_Y} + \frac{N^2(x)}{N_Y^2} \leq 1 \qquad (M_Y = \frac{1}{4}bh^2\sigma_Y, \quad N_Y = bh\sigma_Y) \tag{6.11}$$

be the stress admissibility condition. This condition is normally adopted for a rectangular cross-section beam working in in-plane bending. We introduce dimensionless parameters $m = M/M_Y$, $n = N/N_Y$, $q = Q/N_Y$, their space being the stress fields space $\mathcal{S} = \mathbf{R}^3$ for the beam under consideration. Expressing $M(x)$ and $Q(x)$ through $m$, $n$ and using (6.11), we find that the set of admissible stress fields in the beam is

$$C = \{(m, n, q) \in \mathbf{R}^3 : |m| + n^2 \leq 1, \ |n| \leq a\}$$
$$(a = \sqrt{1 + \frac{\alpha^2}{4}} - \frac{\alpha}{2}, \quad \alpha = \frac{l_1 N_Y}{M_Y}).$$

Consider the shakedown analysis problem for the beam when the load $P$ may vary within the limits $P^- \leq P \leq P^+$. The load $P$ is represented by the stress field in the elastic reference body (reference beam), see Subsection 3.2. The elastic stress field, in its turn, is identified with the point in the space $\mathcal{S} = \mathbf{R}^3$; let $m^0(P)$, $n^0(P)$, $q^0(P)$ be the corresponding values of the parameters $m$, $n$, $q$. The points $(m^0(P), n^0(P), q^0(P))$ form a segment in $\mathcal{S}$, as the load $P$ may vary within the limits $P^- \leq P \leq P^+$. We denote the segment of possible loads by the symbol $\Delta$ and its ends by $(m^-, n^-, q^-)$ and $(m^+, n^+, q^+)$.

Let us now find the function $F$ involved in the static extremum problem determining the safety factor $s_\Delta$. According to Example 1, $F$ is given by the formula

$$F(m, n, q) = \max\{M_C(m+m^-, n+n^-, q+q^-), \ M_C(m+m^+, n+n^+, q+q^+)\}.$$

It is easy to evaluate the Minkowski function $M_C$, which does not depend on $q$ since $C$ is a cylinder with the axis parallel to the $q$-axis in $\mathcal{S} = \mathbf{R}^3$. The definition of the Minkowski function immediately implies that

$$M_C(m, n) = \begin{cases} \dfrac{|n|}{a} & \text{if } \dfrac{|n|}{|m|} \geq \dfrac{a}{1 - a^2}, \\[2ex] \dfrac{2n^2}{\sqrt{m^2 + 4n^2} - |m|} & \text{if } \dfrac{|n|}{|m|} < \dfrac{a}{1 - a^2}, \end{cases}$$

where $a$ is the parameter involved in the above expression for $C$.

According to this formula, the function $F$ does not depend upon $q$, which describes it the following way. There is a curve $\Gamma$ dividing the $(m,n)$-plane into two parts so that

$$F(m,n,q) = M_C(m+m^-, n+n^-, q+q^-) > M_C(m+m^+, n+n^+, q+q^+)$$
$$\text{if the point } (m,n) \text{ lies above } \Gamma,$$

while

$$F(m,n,q) = M_C(m+m^+, n+n^+, q+q^+) > M_C(m+m^-, n+n^-, q+q^-)$$
$$\text{if the point } (m,n) \text{ lies below } \Gamma.$$

To describe $\Gamma$, we note that $n = km$ on the segment $\Delta$, where the constant $k$ is expressed through the geometric parameters of the beam, and restrict ourselves to the case $a/(1-a)^2 \le k \le 1/2a$. It is convenient to use the following new coordinates in the $(m,n)$-plane:

$$\mu = m + c, \quad \nu = n + kc \quad \left(c = \frac{m^- + m^+}{2}\right),$$

and it can be easily seen that $\Gamma$ consists of the segment

$$-\frac{1-a^2}{a}kd + d \le \mu \le \frac{1-a^2}{a}kd - d, \quad \nu = 0 \quad \left(d = \frac{m^+ - m^-}{2}\right)$$

and the curves:

$$\mu = -d - \frac{\nu - kd}{a} + a\left(\frac{\nu + kd}{\nu - kd}\right)^2, \quad \nu \le 0,$$

and

$$\mu = d - \frac{\nu + kd}{a} + a\left(\frac{\nu - kd}{\nu + kd}\right)^2, \quad \nu \ge 0,$$

which are parts of hyperbolas. Thus, the curve $\Gamma$ is completely determined, which together with the formula for $M_C(m,n)$ allows easy evaluation of the function $F$ involved in the static extremum problem.

Let us now evaluate the function $G$ involved in the kinematic extremum problem. Recall that $G$ is the Fenhel transformation of $F$; the following remark (valid in the general case of any polyhedral set $\Delta$) makes evaluating it simpler: the space $\mathcal{S}$ is subdivided into several subsets, so that on each of them $F$ is the Minkowski function with a shifted argument (see Example 1) and, consequently, is linear along rays. For the beam under consideration, $F$ does not depend upon $q$, which makes the above-mentioned subsets cylindrical. Therefore, it is sufficient to consider their cross-sections by the plane orthogonal to the $q$-axis, which are the sets in the $(m,n)$-plane separated by the curve $\Gamma$;

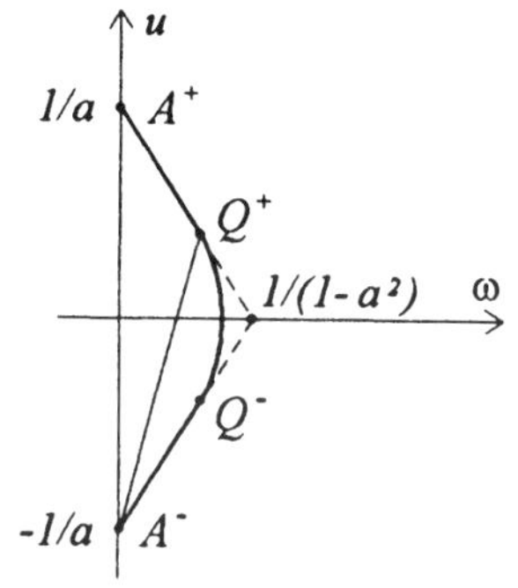


**Figure 6.3**

$F$ is linear along rays within each of them. This reduces evaluation of the Fenhel transformation of $F$,

$$G(\omega, u) = \sup \left\{ m\omega + nu - F(m,n) : (m,n) \in \mathbf{R}^2 \right\},$$

to solving of an extremum problem on the curve $\Gamma$. Indeed, the function $m\omega - nu - F(m,n)$ is linear along the rays, and its supremum over each of them either is attained at the point where the ray intersects $\Gamma$ or is $+\infty$. Note also that it is sufficient to determine $G(\omega, u)$ only for $\omega \geq 0$ since $G(-\omega, -u) = G(\omega, u)$. Straightforward calculation results in the following expressions for $G(\omega, u)$ in the subdomains of the half-plane $\omega \geq 0$ shown in Figure 6.3. In the triangle with the vertices

$$A^- \left(0, -\frac{1}{a}\right), \quad A^+ \left(0, \frac{1}{a}\right), \quad Q^+ \left(\frac{1}{1+a^2}, \frac{2a}{1+a^2}\right)$$

$G$ is given by the formula

$$G(\omega, u) = (k/a - ak - 1)\omega d - c(\omega + ku) - kd/a.$$

In the curvilinear triangle bounded by the arc $Q^-Q^+$ of the ellipse

$$4\left(\omega - \frac{1}{2}\right)^2 + u^2 = 1, \quad \omega \geq \frac{2a}{1+a^2}$$

and the segments $A^-Q^-$ and $Q^+A^+$, the values $G(\omega, u)$ are written as

$$G(\omega, u) = 4ak(1 - R)\omega d - (\omega - ku)d - c(\omega + ku),$$

where

$$R = (1 - 1/a^2 + (1 + au)/(a^2\omega))^{1/2}.$$

In the exterior of the curve $A^-Q^-Q^+A^+A^-$ in the half-plane $\omega \geq 0$ the function $G$ takes the value $+\infty$.

**6.3. Conditions for equality of the extremums – I.** Let us discuss the possibility of equality of the extremums in the static and kinematic problems (6.7) and (6.9). When the extremums are equal, the safety factor $s_\Delta$ can be evaluated not only by the static but also by the kinematic method. We give two types of conditions which guarantee the equality of the extremums: bodies with bounded yield surfaces are considered here, bodies with cylindrical yield surfaces in the next subsection.

In the following proposition,

for bodies with bounded yield surfaces, we use the spaces

$$\mathcal{S} = L_\infty(\Omega; \mathbf{S}), \quad \mathcal{E} = L_1(\Omega; \mathbf{S})$$

and the pairing

$$\langle \mathbf{e}, \boldsymbol{\sigma} \rangle = \boldsymbol{\sigma}(\mathbf{e}) = \int_\Omega \mathbf{e}(x) \cdot \boldsymbol{\sigma}(x)\, dV$$

between them;

the mapping $x \to C_x$ is measurable, as usual.

Here, $\mathbf{S}$ is the finite-dimensional space of stresses, for example, $\mathbf{S} = Sym$ in the case of three-dimensional bodies. The choice of the spaces was discussed in Section 5.2; their definitions can be found in Appendix V.A.

THEOREM 1. Suppose the foregoing assumptions are satisfied and for every point of the body the yield surface lies in the same ball of the stress space. Suppose also that the set $\Delta$ of possible loads is bounded in the stress fields space $\mathcal{S}$. Then

1) the safety factor can be evaluated by the kinematic as well as by the static method:

$$\frac{1}{s_\Delta} = \inf\{F(\boldsymbol{\rho}) : \boldsymbol{\rho} \in \Sigma\} = \sup\{-G(\mathbf{e}) : \mathbf{e} \in E\}, \tag{6.12}$$

2) the extremum in the kinematic problem remains the same if the set $E$ in its formulation is replaced by any subspace $\widehat{E}$ in $\mathcal{E}$ satisfying the condition $\widehat{E}^0 = \Sigma$; in particular, $E$ can be replaced by $\operatorname{Def} U$ or by any linear subspace in $\mathcal{E}$ whose closure is the same as the closure of $\operatorname{Def} U$.

PROOF. To prove the first statement, we will apply Theorem V.3.1 to the problem

$$-G(\mathbf{e}) \to \sup; \quad \mathbf{e} \in E.$$

Let us verify that the conditions of the theorem are satisfied. Since the function $G = F^*$ is convex according to Proposition V.3.1, it remains to verify that $G$ is continuous on a certain neighborhood of 0. To show this, note that the set of admissible stress fields is bounded, that is, there is a number $R$ such that $\|\boldsymbol{\sigma}\| \leq R$ for every $\boldsymbol{\sigma}$ in $C$, and, consequently, $M_C(\boldsymbol{\sigma}) \geq \|\boldsymbol{\sigma}\|/R$ (Subsection V.2.1). This inequality results in

$$F(\boldsymbol{\sigma}) = \sup\{M_c(\boldsymbol{\sigma}+\boldsymbol{\tau}) : \boldsymbol{\tau} \in \Delta\} \geq \sup\left\{\frac{1}{R}\|\boldsymbol{\sigma}+\boldsymbol{\tau}\| : \boldsymbol{\tau} \in \Delta\right\}$$

$$\geq \frac{1}{R}\|\boldsymbol{\sigma}\| - \frac{1}{R}\sup\{\|\boldsymbol{\tau}\| : \boldsymbol{\tau} \in \Delta\},$$

which, in turn, implies the estimation

$$G(\mathrm{e}) = \sup \{\langle \mathrm{e}, \boldsymbol{\sigma} \rangle - F(\boldsymbol{\sigma}) : \boldsymbol{\sigma} \in \mathcal{S}\}$$
$$\leq \sup \left\{ \langle \mathrm{e}, \boldsymbol{\sigma} \rangle - \frac{1}{R} \|\boldsymbol{\sigma}\| : \boldsymbol{\sigma} \in \mathcal{S} \right\} + \frac{1}{R} \sup \{\|\boldsymbol{\tau}\| : \boldsymbol{\tau} \in \Delta\}.$$

It is clear that the first term on the right-hand side is zero when $\|\mathrm{e}\| < 1/R$, while the second one is bounded due to boundness of $\Delta$. Then $G$ is continuous at 0 according to Proposition V.6.1. Thus, the kinematic extremum problem satisfies the assumptions of Theorem V.3.1, and the equality

$$\sup \{-G(\mathrm{e}) : \mathrm{e} \in E\} = \inf \{G^*(\rho) : \rho \in \Sigma\}$$

is valid by this theorem (we took into account that $E^0 = \Sigma$).

To prove (6.12), it remains to verify that $G^* = F$ or, which is the same, $F^{**} = F$. Theorem V.4.1 states that for the nonnegative function $F$ the latter equality is valid if $F$ is weak closed. Note that according to (6.6) $F$ is the supremum of the collection of the functions $f_{\boldsymbol{\tau}}$, $\boldsymbol{\tau} \in \Delta$, where $f_{\boldsymbol{\tau}}(\mathrm{s}) = M_C(\mathrm{s} + \boldsymbol{\tau})$. Each of these functions is weak closed (see Subsection V.4.2 and Proposition V.5.1), that is, the set $\{\mathrm{s} \in \mathcal{S} : f_{\boldsymbol{\tau}}(\mathrm{s}) \leq a\}$ is weak closed whatever the number $a$ is. Then the set

$$\{\mathrm{s} \in \mathcal{S} : F(\mathrm{s}) \leq a\} = \{\mathrm{s} \in \mathcal{S} : f_{\boldsymbol{\tau}}(\mathrm{s}) \leq a \ \text{ for every } \ \boldsymbol{\tau} \in \Delta\}$$
$$= \bigcap_{\boldsymbol{\tau} \in \Delta} \{\mathrm{s} \in \mathcal{S} : f_{\boldsymbol{\tau}}(\mathrm{s}) \leq a\}$$

is weak closed as the intersection of weak closed sets. This means that the function $F$ is weak closed and, consequently, $F^{**} = F$, which finishes the proof of equality (6.12).

We observe that the above reasoning only made use of the following two properties of the linear subspace $E$: $E^0 = \Sigma$ and $0 \in E$. Any subspace possesses the latter property, and it is clear that any subspace $\widehat{E}$, whose closure is the same as the closure of $\operatorname{Def} U$, also possesses the former one; see the proof of Theorem V.4.2. Therefore, equality (6.12) holds if $E$ is replaced by such $\widehat{E}$, which finishes the proof.

**6.4. Conditions for equality of the extremums – II.** Let us establish conditions for equality of the extremums in the static and kinematic extremum problems of the shakedown theory in the case of three-dimensional bodies with cylindrical yield surfaces. Both problems are considered in their reduced formulations, which are obtainable the same way as in limit analysis theory (Section VI.1). Namely, the spaces $\mathcal{S}$ and $\mathcal{E}$ of stress and strain rate fields in the original formulation are replaced by their subspaces

$$\mathcal{S}^d = \{\boldsymbol{\sigma} \in \mathcal{S} : \boldsymbol{\sigma}^s = 0\}, \quad \mathcal{E}^d = \{\mathrm{e} \in \mathcal{E} : \mathrm{e}^s = 0\},$$

respectively. Analogously, the space $\mathcal{V}$ of velocity fields and the set $U$ of virtual velocity fields are replaced by

$$\mathcal{V}_d = \{\mathbf{v} \in \mathcal{V} : \operatorname{div} \mathbf{v} = 0\}, \quad U_d = \{\mathbf{v} \in U : \operatorname{div} \mathbf{v} = 0\},$$

respectively, where div v stands for $(\mathrm{Def}\,\mathbf{v})^s$. The set $\Sigma = (\mathrm{Def}\,\mathbf{v})^0$ of self-equilibrated stress fields is replaced by the set of weak self-equilibrated deviatoric stress fields:

$$\Sigma_d = (\mathrm{Def}\,U_d)^0 = \{\boldsymbol{\tau} \in \mathcal{S}^d : \int_\Omega \boldsymbol{\tau} \cdot \mathrm{Def}\,\mathbf{v}\,dV = 0 \ \text{ for every } \ \mathbf{v} \in U_d\}.$$

We assume that the spaces $\mathcal{V}$ and $\mathcal{E}$ are such that the set of kinematically admissible strain rate fields $E_d = \mathrm{Def}\,\mathcal{V}_d$ is contained in the closure $[\mathrm{Def}\,U_d]$ of the set $\mathrm{Def}\,U_d$ in the space $\mathcal{E}^d$ (for example, it is sufficient to take $\mathcal{V}_d = U_d$). Then the relations

$$\mathrm{Def}\,U_d \subset \mathrm{Def}\,\mathcal{V}_d \subset [\mathrm{Def}\,U_d]$$

result in

$$\Sigma_d = (\mathrm{Def}\,U_d)^0 = (\mathrm{Def}\,\mathcal{V}_d)^0. \tag{6.13}$$

Finally, to formulate the reduced static problem, let us consider the set of admissible deviatoric stress fields

$$C^d = \{\boldsymbol{\sigma} \in \mathcal{S}^d : \boldsymbol{\sigma}(x) \in C_x^d \ \text{ for a.e. } \mathbf{x} \in \Omega\}$$

($\Omega$ is the domain occupied by the body) and the set $\Delta^d = \{\boldsymbol{\tau}^d : \boldsymbol{\tau} \in \Delta\}$, where $\Delta$ is the given set of possible loads in the shakedown analysis problem. The reduced, deviatoric static problem is analogous to (6.7) and is formulated as follows:

$$\begin{gathered} F_d(\boldsymbol{\rho}) \to \inf; \quad \boldsymbol{\rho} \in \Sigma_d \\ \left(F_d(\mathbf{s}) = \sup\left\{M_C(\mathbf{s} + \boldsymbol{\tau}) : \boldsymbol{\tau} \in \Delta^d\right\}\right), \end{gathered} \tag{6.14}$$

where the Minkowski function is considered in the space $\mathcal{S}^d$.

The reduced kinematic problem is analogous to (6.9):

$$\begin{gathered} -G_d(\mathbf{e}) \to \sup; \quad \mathbf{e} \in E_d \\ (G_d = F_d^*, \quad E_d = \mathrm{Def}\,\mathcal{V}_d), \end{gathered} \tag{6.15}$$

where *the space $\mathcal{V}$ consists of sufficiently regular velocity fields.*

The following proposition establishes conditions for equality of the extremums in the reduced static and kinematic problems. In this proposition,

for bodies with cylindrical yield surfaces, we use the spaces

$$\mathcal{S} = \mathcal{S}^d \times \mathcal{S}^s, \ \mathcal{S}^d = L_\infty(\Omega; Sym^d), \ \mathcal{S}^s = L_2(\Omega; Sym^s),$$
$$\mathcal{E} = \mathcal{E}^d \times \mathcal{E}^s, \ \mathcal{E}^d = L_1(\Omega; Sym^d), \ \mathcal{E}^s = L_2(\Omega; Sym^s)$$

and the pairing

$$\langle \mathbf{e}, \boldsymbol{\sigma} \rangle = \int_\Omega \mathbf{e}(x) \cdot \boldsymbol{\sigma}(x)\,dV$$

between them;

the mapping $x \to C_x$ is measurable, as usual.

The choice of the spaces was discussed in Subsections V.7.2 and VI.1.2; their definitions can be found in Appendix V.A.

THEOREM 2. Suppose the foregoing assumptions are satisfied and the body occupies domain $\Omega$ with a sufficiently regular boundary, the free surface being its sufficiently regular part. Suppose also that for every point of the body the cylindrical yield surface lies inside the same cylinder $|\mathbf{s}^d| \leq R$, $R > 0$. Then, if the given set $\Delta$ of possible loads is bounded in $\mathcal{S}$,

1) the safety factor can be evaluated by both the static and kinematic methods using the reduced extremum problems:

$$\frac{1}{s_\Delta} = \inf\{F_d(\rho) : \rho \in \Sigma_d\} = \sup\{-G_d(\mathrm{e}) : \mathrm{e} \in E_d\}, \tag{6.16}$$

2) the extremum in the reduced kinematic problem remains the same if the set $E_d$ of kinematically admissible strain rate fields is replaced by any subspace $\widehat{E}_d$ in $\mathcal{E}^d$ satisfying the condition $(\widehat{E}_d)^0 = \Sigma_d$; in particular, $E_d$ can be replaced by $\operatorname{Def} U_d$ or by any linear subspace in $\mathcal{E}^d$ whose closure is the same as the closure of $\operatorname{Def} U_d$.

PROOF. Let us denote the extremum in the reduced static problem (6.14) by the symbol $1/s_{\Delta d}$ and prove that $s_{\Delta d} = s_\Delta$, that is, the first of equalities (6.16) is valid. Note that the inequality $1/s_{\Delta d} \leq 1/s_\Delta$ is always valid. Indeed, if $\{\rho_n\}$, $n = 1, 2, \ldots$, is a minimizing sequence in problem (6.7), that is,

$$\lim_{n\to\infty} F(\rho_n) = \frac{1}{s_\Delta}, \qquad \rho_n \in \Sigma \quad (n = 1, 2, \ldots),$$

then $\rho_n^d$ satisfies the restriction in problem (6.14): $\rho_n^d \in \Sigma_d$. Hence, the estimation

$$\frac{1}{s_{\Delta d}} \leq \lim_{n\to\infty} F_d(\rho_n^d)$$

is valid. With the equality $F_d(\rho_n^d) = F(\rho_n)$ arising from the obvious relation $M_C(\sigma) = M_{C^d}(\sigma^d)$, it can be written as

$$\frac{1}{s_{\Delta d}} \leq \lim_{n\to\infty} F(\rho_n) = \frac{1}{s_\Delta},$$

which proves the above inequality.

To prove the equality $s_{\Delta d} = s_\Delta$, it remains to show that $1/s_{\Delta d} \geq 1/s_\Delta$. It can be easily verified by considering a minimizing sequence $\{\rho_n'\}$, $n = 1, 2, \ldots$, in problem (6.14). Note that the following relations are valid:

$$\frac{1}{s_{\Delta d}} = \lim_{n\to\infty} F_d(\rho_n'), \quad \rho_n' \in \Sigma_d \quad (n = 1, 2, \ldots),$$

and consider the solution to the pressure field restoration problem for $\rho_n'$. The problem is solvable by virtue of Theorem VI.2.4, whose conditions are satisfied since $\partial\Omega$ and the free surface $S_q$ are sufficiently regular. Let $p_n$ be the solution, that is, the pressure field such that the stress field $\rho_n = \rho_n' - p_n\mathbf{I}$ is self-equilibrated, thus satisfying the restriction in problem (6.7). This results in the estimation

$$\frac{1}{s_\Delta} \le \lim_{n\to\infty} F(\rho_n),$$

which using the equality $F(\rho_n) = F_d(\rho'_n)$ can be written as

$$\frac{1}{s_\Delta} \le \lim_{n\to\infty} F_d(\rho'_n) = \frac{1}{s_{\Delta d}},$$

which finishes the proof of the first equality in (6.16).

Let us now show that the second equality in (6.16) is also valid, that is, the extremums in the reduced static and kinematic problems (6.14) and (6.15) are equal. Note that these problems are of the same forms as (6.7) and (6.9), where $\mathcal{S}$, $\mathcal{E}$, $\mathcal{V}$, $U$, $C$, $\Sigma$, $E$ and $\Delta$ are replaced by $\mathcal{S}^d$, $\mathcal{E}^d$, $\mathcal{V}_d$, $U_d$, $C^d$, $\Sigma_d$, $E_d$ and $\Delta^d$, respectively. This choice meets requirements (I) – (VIII) of Subsection 1.4 on the formulation of the elastic-plastic problem. In particular, (6.13) implies that $\Sigma_d = (E_d)^0$, which plays the role here the relation $\Sigma = (E)^0$ did in the original formulation of the elastic-plastic problem. Note also that, for all points of the body, the sets $C_x^d$ lie in the same ball of the space $Sym^d$. Thus, the reduced problems satisfy the assumptions of Theorem 1, which implies that the extremums in these problems are equal, and the equality holds in case $E_d$ is replaced by any subspace $\widehat{E}_d \subset \mathcal{E}^d$ which has the same closure as Def $U_d$. The theorem is proved.

Theorems 1 and 2 are direct generalizations of Theorems V.5.1 and VI.1.1 of the limit analysis theory.

REMARK 1. It is clear that the statements analogous to Theorems 1 and 2 are valid for discrete systems, that is, in case the spaces $\mathcal{S}$ and $\mathcal{E}$ are finite-dimensional.

## 7. Kinematic method for safety factor evaluation

This section deals with evaluation of the shakedown safety factor in the case of the polyhedral set of possible loads, which is of most importance from the practical viewpoint. The kinematic extremum problem determining the safety factor is modified so that the function to be minimized in it is given by a simple explicit formula. The formula is a direct generalization of the one determining the kinematic multiplier in the limit analysis problem. Subsection 1 presents the main results; the rest of the section deals with their justifications and finite element formulation of the safety factor calculation problem.

**7.1. Formulation of the method.** Consider an elastic perfectly plastic body $\mathcal{B}$ with the set $C_x$ of admissible stresses given for every point $x$ of the body, $d_x$ being the corresponding dissipation. Suppose that the body can be subjected to variable loadings within the polyhedral set $\Delta$ of possible loads. Without loss of generality, we restrict ourselves to the case when $\Delta$ is convex (see Proposition 5.1). Let $\tau_1, \ldots, \tau_m$ be the vertices of $\Delta$, that is, $\Delta = \operatorname{conv}\{\tau_1, \ldots, \tau_m\}$. As usual, the loads are represented by the corresponding elastic stress fields in the reference body $\mathcal{B}^0$ (see Subsection 3.2). It turns out that, under some nonrestrictive conditions,

> the safety factor can be evaluated as the extremum in the following modified kinematic extremum problem:
>
> $$s_\Delta = \inf \left\{ \frac{\int_\Omega (d_x(\mathbf{e}_1) + \cdots + d_x(\mathbf{e}_m))\, dV}{\int_\Omega (\mathbf{e}_1 \cdot \boldsymbol{\tau}_1 + \cdots + \mathbf{e}_m \cdot \boldsymbol{\tau}_m)\, dV} : \mathbf{e}_1, \ldots, \mathbf{e}_m \in \mathcal{E}, \right.$$
>
> $$\left. \mathbf{e}_1 + \cdots + \mathbf{e}_m \in E, \quad \int_\Omega (\mathbf{e}_1 \cdot \boldsymbol{\tau}_1 + \cdots + \mathbf{e}_m \cdot \boldsymbol{\tau}_m)\, dV > 0 \right\},$$
>
> where $\mathcal{E}$ is the space of regular strain rate fields, and $E$ is its subspace consisting of kinematically admissible strain rate fields (in the case of bodies with cylindrical yield surfaces, $\mathcal{E}$ contains only deviatoric fields).

This modified problem is a direct generalization of the limit analysis kinematic problem, and propositions analogous to Theorems V.5.1 and VI.1.1 are valid for it. Namely, to evaluate the safety factor, it is sufficient to consider the minimization only over the space of smooth velocity fields. Other options for the velocity fields space are also available; in particular, we will specify the choice most convenient for applying the finite element method to the shakedown analysis.

Note that the function to be minimized in the modified kinematic problem is given by a simple explicit formula. However, the number of its arguments equals the number of vertices of the polyhedron $\Delta$ and is greater than the number of arguments in the original formulation, each of the arguments being a field in both cases.

It is easy to get rid of the restriction $\mathbf{e}_1 + \cdots + \mathbf{e}_m \in E$ in the formulation of the modified problem. Indeed, this restriction means that the sum $\mathbf{e}_1 + \cdots + \mathbf{e}_m$ is kinematically admissible, that is, equals Def $\mathbf{v}$, where $\mathbf{v}$ is a kinematically admissible velocity field. Therefore, the variables $\mathbf{e}_1, \ldots, \mathbf{e}_m$ can be replaced by $\mathbf{v}, \mathbf{e}_1, \ldots, \mathbf{e}_{m-1}$ and the terms $d_x(\mathbf{e}_m)$, $\mathbf{e}_m \cdot \boldsymbol{\tau}$ in the formula for the goal function by

$$d_x(\mathbf{e}_m) = d_x(\operatorname{Def} \mathbf{v} - \mathbf{e}_1 - \cdots - \mathbf{e}_{m-1}),$$
$$\mathbf{e}_m \cdot \boldsymbol{\tau} = (\operatorname{Def} \mathbf{v} - \mathbf{e}_1 - \cdots - \mathbf{e}_{m-1}) \cdot \boldsymbol{\tau}.$$

This makes the minimization problem unrestricted.

The modified kinematic method is applicable to calculating the safety factor under the following conditions.

> In the case of the polyhedral set $\Delta$ of possible loads, the safety factor $s_\Delta$ can be evaluated as the infimum in the modified kinematic extremum problem
>
> (I) for a body with bounded yield surfaces, if for every point of the body the yield surface lies in the same ball of the stress space;

(II) for a body with cylindrical yield surfaces, if there is a cylinder, $|\sigma^d| \le r$, $r > 0$, which lies inside the yield surface for every point of the body, the body boundary is sufficiently regular and the free surface is its sufficiently regular part.

Here, as always, the free surface $S_q$ is the part of the body boundary on which the surface tractions are given; it complements the fixed part $S_v$ of the boundary. Regularity of the boundary and its free part guarantees the solvability of the pressure field restoration problem; for example, this is the case if the assumptions of Theorems VI.2.3 or VI.2.4 are satisfied.

**7.2. Modified kinematic problem.** In order to construct the modified kinematic problem, we will modify the static problem determining the safety factor:

$$\begin{gathered} F(\boldsymbol{\rho}) \to \inf; \quad \boldsymbol{\rho} \in \Sigma \\ (F(\mathbf{s}) = \sup\{M_C(\mathbf{s}+\boldsymbol{\tau}) : \boldsymbol{\tau} \in \Delta\}) \end{gathered} \tag{7.1}$$

and apply the procedure described in Subsection V.3 to the modified static problem.

Let the set of possible loads in the shakedown analysis problem be the convex polyhedron $\Delta = \operatorname{conv}\{\boldsymbol{\tau}_1, \ldots, \boldsymbol{\tau}_m\}$ with the vertices $\boldsymbol{\tau}_1, \ldots, \boldsymbol{\tau}_m$. Then the function $F$ is written as

$$F(\boldsymbol{\sigma}) = \max\{M_C(\boldsymbol{\sigma}+\boldsymbol{\tau}_1), \ldots, M_C(\boldsymbol{\sigma}+\boldsymbol{\tau}_m)\},$$

see Example 6.1. It can be easily seen that this function is related to the Minkowski function of the set $C^m = C \times \cdots \times C$ in the space $\mathcal{S}^m$. Indeed, according to the definition and property (III) of the Minkowski function (Subsection V.3.2), we obtain at $\mathbf{s} = (\mathbf{s}_1, \ldots, \mathbf{s}_m) \in \mathcal{S}^m$ :

$$\begin{aligned} M_{C^m}(\mathbf{s}) &= \inf\left\{\mu > 0 : \frac{1}{\mu}\mathbf{s}_1 \in C, \ldots, \frac{1}{\mu}\mathbf{s}_m \in C\right\} \\ &= \inf\{\mu > 0 : M_C(\mathbf{s}_1) \le \mu, \ldots, M_C(\mathbf{s}_m) \le \mu\} \\ &= \max\{M_C(\mathbf{s}_1), \ldots, M_C(\mathbf{s}_m)\}, \end{aligned}$$

and, therefore, $F(\boldsymbol{\sigma})$ can be written as

$$F(\boldsymbol{\sigma}) = M_{C^m}(\boldsymbol{\sigma}+\boldsymbol{\tau}_1, \ldots, \boldsymbol{\sigma}+\boldsymbol{\tau}_m).$$

For more convenience we introduce the element $\boldsymbol{\tau} = (\boldsymbol{\tau}_1, \ldots, \boldsymbol{\tau}_m)$ to represent in $\mathcal{S}^m$ the given set of possible loads $\Delta = \operatorname{conv}\{\boldsymbol{\tau}_1, \ldots, \boldsymbol{\tau}_m\}$ . We also consider for any $\boldsymbol{\sigma} \in \mathcal{S}$ the element $(\boldsymbol{\sigma}, \ldots, \boldsymbol{\sigma})$ in $\mathcal{S}^m$, each of its components being $\boldsymbol{\sigma}$; we denote $(\boldsymbol{\sigma}, \ldots, \boldsymbol{\sigma})$ by $\bar{\boldsymbol{\sigma}}$. Then $F(\boldsymbol{\sigma})$ can be written as

$$F(\boldsymbol{\sigma}) = M_{C^m}(\bar{\boldsymbol{\sigma}}+\boldsymbol{\tau}),$$

and extremum problem (7.1) as

$$\frac{1}{s_\Delta} = \inf\left\{M_{C^m}(\bar{\boldsymbol{\rho}}+\boldsymbol{\tau}) : \bar{\boldsymbol{\rho}} \in \Sigma_{(m)}\right\}, \tag{7.2}$$

where $\Sigma_{(m)}$ is the subspace in $\mathcal{S}^m$:

$$\Sigma_{(m)} = \{\bar{\mathbf{r}} \in \mathcal{S}^m : \bar{\mathbf{r}} = (\mathbf{r}, \ldots, \mathbf{r}), \ \ \mathbf{r} \in \Sigma\}. \tag{7.3}$$

Let us apply to (7.2) the general scheme of constructing the dual extremum problem. This calls for making use of a linear space paired with $\mathcal{S}^m$ by a certain bilinear form. Recall that the space $\mathcal{S}$ of strain rate fields is paired with the space $\mathcal{E}$ of strain rate fields by the internal power bilinear form $\langle \cdot, \cdot \rangle$. This suggests considering the space $\mathcal{E}^m = \mathcal{E} \times \cdots \times \mathcal{E}$ and the following pairing $\langle \cdot, \cdot \rangle_{(m)}$ between $\mathcal{E}^m$ and $\mathcal{S}^m$:

$$\langle \mathbf{e}, \boldsymbol{\sigma} \rangle_{(m)} = \langle \mathbf{e}_1, \boldsymbol{\sigma}_1 \rangle + \cdots + \langle \mathbf{e}_m, \boldsymbol{\sigma}_m \rangle,$$
$$\text{for every } \ \mathbf{e} = (\mathbf{e}_1, \ldots, \mathbf{e}_m) \in \mathcal{E}^m \ \text{ and every } \ \boldsymbol{\sigma} = (\boldsymbol{\sigma}_1, \ldots, \boldsymbol{\sigma}_m) \in \mathcal{S}^m.$$

The pairing allows us to introduce the Fenhel transformation $G_{(m)}$ of the function to be minimized in problem (7.2):

$$G_{(m)}(\mathbf{e}) = \sup \left\{ \langle \mathbf{e}, \mathbf{s} \rangle_{(m)} - M_{C^m}(\mathbf{s} + \boldsymbol{\tau}) : \ \mathbf{s} \in \mathcal{S}^m \right\}.$$

Then the inequality

$$M_{C^m}(\mathbf{s} + \boldsymbol{\tau}) \geq \langle \mathbf{e}, \mathbf{s} \rangle_{(m)} - G_{(m)}(\mathbf{e}) \tag{7.4}$$

is valid for every $\mathbf{s} \in \mathcal{S}^m$ and every $\mathbf{e} \in \mathcal{E}^m$.

Consider now the subspace

$$E_{(m)} = \{\mathbf{e} = (\mathbf{e}_1, \ldots, \mathbf{e}_m) \in \mathcal{E}^m : \ \mathbf{e}_1 + \cdots + \mathbf{e}_m \in E\}$$

in $\mathcal{E}^m$, where $E$ is the set of kinematically admissible strain rate fields. For every $\mathbf{e} = (\mathbf{e}_1, \ldots, \mathbf{e}_m)$ in $E_{(m)}$ and every $\bar{\boldsymbol{\rho}} = (\boldsymbol{\rho}, \ldots, \boldsymbol{\rho})$ in $\Sigma_{(m)}$, the equality

$$\langle \bar{\boldsymbol{\rho}}, \mathbf{e} \rangle_{(m)} = \langle \boldsymbol{\rho}, \mathbf{e}_1 \rangle + \cdots + \langle \boldsymbol{\rho}, \mathbf{e}_m \rangle = \langle \boldsymbol{\rho}, \mathbf{e}_1 + \cdots + \mathbf{e}_m \rangle = 0$$

is valid according to Remark 1.1. Then, restricting (7.4) to such $\bar{\boldsymbol{\rho}}$ and $\mathbf{e}$, we obtain the relation

$$\inf \left\{ M_{C^m}(\bar{\boldsymbol{\rho}} + \boldsymbol{\tau}) : \ \bar{\boldsymbol{\rho}} \in \Sigma_{(m)} \right\} \geq -G_{(m)}(\mathbf{e}) \quad \text{for every} \quad \mathbf{e} \in E_{(m)}$$

or

$$\frac{1}{s_\Delta} \geq -G_{(m)}(\mathbf{e}) \quad \text{for every} \quad \mathbf{e} \in E_m. \tag{7.5}$$

Thus, we arrive at the modified kinematic extremum problem

$$-G_{(m)}(\mathbf{e}) \to \sup; \quad \mathbf{e} \in E_{(m)} \tag{7.6}$$

whose extremum is the best kinematic lower bound for $1/s_\Delta$ (that is, gives the best kinematic upper bound for the safety factor). We will now derive an explicit formula for $G_{(m)}$ and establish conditions which guarantee that the best kinematic upper bound of the type under consideration equals the safety factor.

**7.3. Formula for the safety factor upper bound.** To obtain an explicit formula for the function $G_{(m)}$ involved in the modified kinematic problem, let us find the Fenhel transformation of the function $\mathbf{s} \to M_{C^m}(\mathbf{s}+\boldsymbol{\tau})$ which is maximized in the static problem. Here, $\boldsymbol{\tau} = (\boldsymbol{\tau}_1, \dots, \boldsymbol{\tau}_m) \in \mathcal{S}^m$ is fixed and represents the set of possible loads $\Delta = \operatorname{conv}\{\boldsymbol{\tau}_1, \dots, \boldsymbol{\tau}_m\}$. Recall that the Fenhel transformation of the analogous function $\boldsymbol{\sigma} \to M_C(\boldsymbol{\sigma}+\mathbf{s}_l)$ has already been evaluated in Subsection V.6.1. Replacing $C$ by $C^m$ and $\mathbf{s}_l$ by $\boldsymbol{\tau}$ in that formula, we obtain the following expression for the Fenhel transformation $G_{(m)}$:

$$G_{(m)}(\mathbf{e}) = -\langle \mathbf{e}, \boldsymbol{\tau} \rangle_{(m)} + M^*_{C^m}(\mathbf{e}),$$

where

$$M^*_{C^m}(\mathbf{e}) = \begin{cases} 0 & \text{if } D_{(m)}(\mathbf{e}) \le 1, \\ +\infty & \text{if } D_{(m)}(\mathbf{e}) > 1. \end{cases}$$

Then the modified kinematic problem (7.6) can be written as

$$\langle \mathbf{e}, \boldsymbol{\tau} \rangle_{(m)} - M^*_{C^m}(\mathbf{e}) \to \sup; \quad \mathbf{e} \in E_{(m)}$$

or, which is the same, as

$$\langle \mathbf{e}, \boldsymbol{\tau} \rangle_{(m)} \to \sup; \quad \mathbf{e} \in E_{(m)}, \quad D_{(m)}(\mathbf{e}) \le 1.$$

According to this, estimation (7.5) takes the form

$$\frac{1}{s_\Delta} \ge \langle \mathbf{e}, \boldsymbol{\tau} \rangle_{(m)} \quad \text{for every} \quad \mathbf{e} \in E_{(m)} \text{ with } D_{(m)}(\mathbf{e}) \le 1. \tag{7.7}$$

Note that $\mathbf{s}_*$ in the above definition of $D_{(m)}$ is the collection of $\mathbf{s}_{*1}, \dots, \mathbf{s}_{*m}$ in $\mathcal{S}$, and these variables are independent, which resullts in the equality

$$D_{(m)}(\mathbf{e}) = D(\mathbf{e}_1) + \dots + D(\mathbf{e}_m),$$

where

$$D(\mathbf{e}_i) = \sup\{\langle \mathbf{e}_i, \boldsymbol{\sigma}_* \rangle : \boldsymbol{\sigma}_* \in C\}.$$

Using this formula for $D_{(m)}(\mathbf{e})$ and taking into account the definition of $E_{(m)}$, we arrive at the following formulation of the modified kinematic problem:

$$\begin{gathered} \langle \mathbf{e}_1, \boldsymbol{\tau}_1 \rangle + \dots + \langle \mathbf{e}_m, \boldsymbol{\tau}_m \rangle \to \sup; \\ \mathbf{e}_1, \dots, \mathbf{e}_m \in \mathcal{E}, \quad \mathbf{e}_1 + \dots + \mathbf{e}_m \in E, \quad D(\mathbf{e}_1) + \dots + +D(\mathbf{e}_m) \le 1, \end{gathered} \tag{7.8}$$

while estimation (7.7) can be written as

$$\frac{1}{s_\Delta} \ge \langle \mathbf{e}_1, \boldsymbol{\tau}_1 \rangle + \dots + \langle \mathbf{e}_m, \boldsymbol{\tau}_m \rangle$$
$$\text{for every } \mathbf{e}_1, \dots, \mathbf{e}_m \in \mathcal{E}, \ \mathbf{e}_1 + \dots + \mathbf{e}_m \in E, \ D(\mathbf{e}_1) + \dots + D(\mathbf{e}_m) \le 1.$$

Simple reasoning analogous to that in Proposition V.2.2 allows us to equivalently reformulate the modified kinematic problem as

$$\frac{D(\mathbf{e}_1) + \cdots + D(\mathbf{e}_m)}{\langle \mathbf{e}_1, \boldsymbol{\tau}_1 \rangle + \cdots + \langle \mathbf{e}_m, \boldsymbol{\tau}_m \rangle} \to \inf; \tag{7.9}$$
$$\mathbf{e}_1, \ldots, \mathbf{e}_m \in \mathcal{E}, \ \mathbf{e}_1 + \cdots + \mathbf{e}_m \in E, \ \langle \mathbf{e}_1, \boldsymbol{\tau}_1 \rangle + \cdots + \langle \mathbf{e}_m, \boldsymbol{\tau}_m \rangle > 0,$$

and the estimation for the safety factor as

$$s_\Delta \le \frac{D(\mathbf{e}_1) + \cdots + D(\mathbf{e}_m)}{\langle \mathbf{e}_1, \boldsymbol{\tau}_1 \rangle + \cdots + \langle \mathbf{e}_m, \boldsymbol{\tau}_m \rangle}$$
$$\text{for every } \ \mathbf{e}_1, \ldots, \mathbf{e}_m \in \mathcal{E}, \ \mathbf{e}_1 + \cdots + \mathbf{e}_m \in E, \ \langle \mathbf{e}_1, \boldsymbol{\tau}_1 \rangle + \cdots + \langle \mathbf{e}_m, \boldsymbol{\tau}_m \rangle > 0.$$

Finally, recall that

$$\langle \mathbf{e}_1, \boldsymbol{\tau}_1 \rangle + \cdots + \langle \mathbf{e}_m, \boldsymbol{\tau}_m \rangle = \int_\Omega (\mathbf{e}_1 \cdot \boldsymbol{\tau}_1 + \cdots + \mathbf{e}_m \cdot \boldsymbol{\tau}_m) \, dV$$

in case $\mathbf{e}_1, \ldots \mathbf{e}_m$ are integrable fields, and the dissipation $D(\mathbf{e}_i)$ can be evaluated by simply integrating $d_x(\mathbf{e}_i)$ (see Section IV.5). Then the previous estimation can be written as

$$s_\Delta \le \frac{\int_\Omega d_x(\mathbf{e}_1) + \cdots + d_x(\mathbf{e}_m) \, dV}{\int_\Omega (\mathbf{e}_1 \cdot \boldsymbol{\tau}_1 + \cdots + \mathbf{e}_m \cdot \boldsymbol{\tau}_m) \, dV},$$

where $\mathbf{e}_1, \ldots, \mathbf{e}_m$ are integrable fields in $\mathcal{E}$, satisfying the conditions

$$\mathbf{e}_1 + \cdots + \mathbf{e}_m \in E, \qquad \langle \mathbf{e}_1, \boldsymbol{\tau}_1 \rangle + \cdots + \langle \mathbf{e}_m, \boldsymbol{\tau}_m \rangle > 0.$$

This is obviously a direct generalization of the estimation for the safety factor $\alpha_l$ in limit analysis:

$$\alpha_l \le \frac{\int_\Omega d_x(\mathbf{e}) \, dV}{\int_\Omega \mathbf{e} \cdot \mathbf{s}_l \, dV},$$

where $\mathbf{e} \in E$ is integrable and satisfies the condition $\langle \mathbf{e}, \mathbf{s}_l \rangle > 0$.

**7.4. Possibility of safety factor evaluation.** Let us establish conditions under which the extremum in modified kinematic problem (7.8) is the safety factor and not only its upper bound. The following proposition concerns bodies with bounded yield surfaces, the case of cylindrical yield surfaces will be considered later.

THEOREM 1. Suppose the assumptions of Theorem 6.1 are satisfied (in particular, the same spaces of stress and strain rate fields are considered) and the set of possible loads in the shakedown analysis problem is the polyhedron $\Delta = \operatorname{conv}\{\boldsymbol{\tau}_1, \ldots, \boldsymbol{\tau}_m\}$. Then

1) the safety factor can be evaluated as the extremum in the modified kinematic problem:

$$s_\Delta = \inf \left\{ \frac{\int\limits_\Omega d_x(\mathbf{e}_1) + \cdots + d_x(\mathbf{e}_m)\, dV}{\int\limits_\Omega (\mathbf{e}_1 \cdot \boldsymbol{\tau}_1 + \cdots + \mathbf{e}_m \cdot \boldsymbol{\tau}_m)\, dV} : \mathbf{e}_1, \ldots, \mathbf{e}_m \in \mathcal{E}, \right.$$

$$\left. \mathbf{e}_1 + \cdots + \mathbf{e}_m \in E, \quad \int\limits_\Omega \mathbf{e}_1 \cdot \boldsymbol{\tau}_1 + \cdots + \mathbf{e}_m \cdot \boldsymbol{\tau}_m\, dV > 0 \right\},$$

2) the extremum remains the same if the space $\mathcal{E}$ is replaced by any linear subspace $\widehat{\mathcal{E}}$ dense in it and the set $E$ of kinematically admissible strain rate fields is replaced by any subspace $\widehat{E}$ in $\widehat{\mathcal{E}}$ satisfying the condition $\widehat{E}^0 = \Sigma$; in particular, $E$ can be replaced by $\operatorname{Def} U$ or by any linear subspace in $\mathcal{E}$ whose closure is the same as the closure of $\operatorname{Def} U$.

PROOF. The equality to be proved is equivalent to

$$\frac{1}{s_\Delta} = \sup \left\{ \frac{\int\limits_\Omega (\mathbf{e}_1 \cdot \boldsymbol{\tau}_1 + \cdots + \mathbf{e}_m \cdot \boldsymbol{\tau}_m)\, dV}{\int\limits_\Omega d_x(\mathbf{e}_1) + \cdots + d_x(\mathbf{e}_m)\, dV} : \mathbf{e}_1, \ldots, \mathbf{e}_m \in \mathcal{E}, \right.$$

$$\left. \mathbf{e}_1 + \cdots + \mathbf{e}_m \in E \right\},$$

or, which is the same, to

$$\frac{1}{s_\Delta} = \sup \left\{ \langle \mathbf{e}, \boldsymbol{\tau} \rangle_{(m)} : \mathbf{e} \in E_{(m)},\ D_{(m)}(\mathbf{e}) \le 1 \right\}$$

(see the previous subsection). Recall also that according to (7.2) the equality

$$\frac{1}{s_\Delta} = \inf \left\{ M_{C^m}(\bar{\boldsymbol{\rho}} + \boldsymbol{\tau}) : \bar{\boldsymbol{\rho}} \in \Sigma_{(m)} \right\}$$

is valid. Thus the equality of the extremums

$$\inf \left\{ M_{C^m}(\bar{\boldsymbol{\rho}} + \boldsymbol{\tau}) : \bar{\boldsymbol{\rho}} \in \Sigma_{(m)} \right\} = \sup \left\{ \langle \mathbf{e}, \boldsymbol{\tau} \rangle_{(m)} : \mathbf{e} \in E_{(m)},\ D_{(m)}(\mathbf{e}) \le 1 \right\} \quad (7.10)$$

is what should be verified in the present theorem.

We are reminded that the extremum problems analogous to (7.10) were considered in Theorem V.5.1. It can be easily seen that replacing $\mathcal{S}$, $\mathcal{E}$, $C$, $\Sigma$, $E$ in formulations of those problems by $\mathcal{S}^m$, $\mathcal{E}^m$, $C^m$, $\Sigma_{(m)}$, $E_{(m)}$ and $\mathbf{s}_l \in \mathcal{S}$ by $\boldsymbol{\tau} \in \mathcal{S}^m$ results in the problems involved in (7.10). We will now verify that these problems satisfy the assumptions of Theorem V.5.1 and apply the theorem to them. In the case under consideration, the main assumption of the theorem takes the form of boundness of the set $C_x^m = C_x \times \cdots \times C_x$ and is obviously satisfied. The other assumptions are: measurability of the mapping $x \to C_x$ and validity of the equality $\Sigma = E^0$. Measurability of $x \to C_x^m$ corresponds to the former assumption, and it is clear that the mapping possesses this property. The latter of the assumptions takes the form $\Sigma_{(m)} = (E_{(m)})^0$; let us verify that it is also satisfied. We will show that the more general relation $\Sigma_{(m)} = (\widehat{E}_{(m)})^0$ is valid with

$$\widehat{E}_{(m)} = \{\mathbf{e} \in \widehat{\mathcal{E}}^m : \mathbf{e}_1 + \cdots + \mathbf{e}_m \in \widehat{E}\}. \tag{7.11}$$

Note that the closures of $\widehat{E}$ and Def $U$ are the same according to the assumption of the present proposition, and this implies that $\widehat{E}^0 = \Sigma$ (see the proof of Theorem V.4.2). This results in the relation $\Sigma_{(m)} \subset \left(\widehat{E}_{(m)}\right)^0$. Indeed, (7.3), (7.11) and the relation $\widehat{E}^0 = \Sigma$ imply the equality

$$\langle \mathbf{e}, \boldsymbol{\rho} \rangle_{(m)} = \langle \mathbf{e}_1 + \cdots + \mathbf{e}_m, \boldsymbol{\rho} \rangle = 0$$

for every $\bar{\boldsymbol{\rho}} = (\boldsymbol{\rho}, \ldots, \boldsymbol{\rho})$ in $\Sigma_{(m)}$ and every $\mathbf{e} = (\mathbf{e}_1, \ldots, \mathbf{e}_m)$ in $\widehat{E}_{(m)}$; consequently, $\Sigma_{(m)} \subset (\widehat{E}_{(m)})^0$. We will now also show that $\Sigma_{(m)} \supset (\widehat{E}_{(m)})^0$. Let us consider any $\boldsymbol{\sigma} = (\boldsymbol{\sigma}_1, \ldots, \boldsymbol{\sigma}_m)$ in $(\widehat{E}_{(m)})^0$ and verify that $\boldsymbol{\sigma} \in \Sigma_{(m)}$, that is, $\boldsymbol{\sigma}_1 = \cdots = \boldsymbol{\sigma}_m \in \Sigma$. Let $\mathbf{e}_1$ be an arbitrary member of $\widehat{E}$, in which case $\mathbf{e} = (\mathbf{e}_1, 0, \ldots, 0)$ belongs to $\widehat{E}_{(m)}$ and, therefore, $\langle \mathbf{e}_1, \boldsymbol{\sigma}_1 \rangle = \langle \mathbf{e}, \boldsymbol{\sigma} \rangle_{(m)} = 0$ since $\boldsymbol{\sigma}$ is in $(\widehat{E}_{(m)})^0$. This means that $\boldsymbol{\sigma}_1$ is in $(\widehat{E})^0 = \Sigma$ as the equality $\langle \mathbf{e}_1, \boldsymbol{\sigma}_1 \rangle = 0$ holds for every $\mathbf{e}_1 \in \widehat{E}$. Analogously, $\boldsymbol{\sigma}_2, \ldots, \boldsymbol{\sigma}_m$ belong to $\Sigma$, and it remains to verify that $\boldsymbol{\sigma}_1 = \cdots = \boldsymbol{\sigma}_m$. Let us show, for example, that $\boldsymbol{\sigma}_1 = \boldsymbol{\sigma}_2$. Consider arbitrary $\mathbf{e}'$ in $\widehat{E}$ and arbitrary $\mathbf{e}''$ in $\widehat{\mathcal{E}}$; then $\mathbf{e} = (\mathbf{e}' + \mathbf{e}'', \mathbf{e}' - \mathbf{e}'', 0, \ldots, 0)$ belongs to $\widehat{E}_{(m)}$ and, consequently,

$$\langle \mathbf{e}' + \mathbf{e}'', \boldsymbol{\sigma}_1 \rangle + \langle \mathbf{e}' - \mathbf{e}'', \boldsymbol{\sigma}_2 \rangle = \langle \mathbf{e}, \boldsymbol{\sigma} \rangle_{(m)} = 0 \tag{7.12}$$

for $\boldsymbol{\sigma}$ in $(\widehat{E}_{(m)})^0$. Here, $\langle \mathbf{e}', \boldsymbol{\sigma}_1 \rangle = 0$ since $\boldsymbol{\sigma}_1 \in \Sigma$, $\mathbf{e}' \in \widehat{E}$ and $(\widehat{E})^0 = \Sigma$, and, analogously, $\langle \mathbf{e}', \boldsymbol{\sigma}_2 \rangle = 0$. Then (7.12) is reduced to the equality $\langle \mathbf{e}'', \boldsymbol{\sigma}_1 - \boldsymbol{\sigma}_2 \rangle = 0$, which results in $\boldsymbol{\sigma}_1 = \boldsymbol{\sigma}_2$ as $\mathbf{e}''$ is an arbitrary member of the set $\widehat{\mathcal{E}}$ dense in $\mathcal{E}$. This finishes the proof of the relation $\boldsymbol{\sigma} \in \Sigma_{(m)}$ and, consequently, the proof of $\Sigma_{(m)} = (\widehat{E}_{(m)})^0$. In particular, the latter relation is valid if $\widehat{\mathcal{E}} = \mathcal{E}$, $\widehat{E} = E$.

Thus, the extremum problems in (7.10) satisfy the assumptions of Theorem V.5.1, where $\mathcal{S}$, $\mathcal{E}$, $C$, $\Sigma$, $E$ and $\mathbf{s}_l$ are replaced by $\mathcal{S}^m$, $\mathcal{E}^m$, $C^m$, $\Sigma_{(m)}$, $E_{(m)}$ and $\boldsymbol{\tau}$, respectively. By virtue of this theorem, statement 1 is valid. The above relation $(\widehat{E}_{(m)})^0 = \Sigma_{(m)}$ also allows us to conclude from the same theorem that $E_{(m)}$ in the kinematic problem can be replaced by $\widehat{E}_{(m)}$ without changing the extremum. The theorem is proved.

The following proposition shows that the safety factor can be evaluated as the extremum in the modified kinematic problem also in the case of bodies with cylindrical yield surfaces. The proposition makes use of the spaces $\mathcal{S}^d$ and $\mathcal{E}^d$ of deviatoric stress and strain rate fields, the spaces

$$\mathcal{V}_d = \{\mathbf{v} \in \mathcal{V} : \operatorname{div} \mathbf{v} = 0\}, \quad U_d = \{\mathbf{v} \in U : \operatorname{div} \mathbf{v} = 0\}$$

of solenoidal velocity fields, the subspace $\Sigma_d = (\operatorname{Def} U_d)^0 \subset \mathcal{S}^d$ of weak self-equilibrated deviatoric stress fields and the subspace $E_d = \operatorname{Def} \mathcal{V}_d \subset \mathcal{E}^d$ of kinematically admissible deviatoric strain rate fields, see Subsection 6.4.

THEOREM 2. Suppose the assumptions of Theorem 6.2 are satisfied (in particular, the same spaces of stress and strain rate fields are considered) and the set of possible loads in the shakedown analysis problem is the polyhedron $\Delta = \operatorname{conv}\{\boldsymbol{\tau}_1, \ldots, \boldsymbol{\tau}_m\}$. Then

1) the safety factor can be evaluated as the extremum in the modified kinematic problem:

$$s_\Delta = \inf \left\{ \frac{\int_\Omega d_x(\mathbf{e}_1) + \cdots + d_x(\mathbf{e}_m)\, dV}{\int_\Omega (\mathbf{e}_1 \cdot \boldsymbol{\tau}_1^d + \cdots + \mathbf{e}_m \cdot \boldsymbol{\tau}_m^d)\, dV} : \mathbf{e}_1, \ldots, \mathbf{e}_m \in \mathcal{E}^d, \right.$$
$$\left. \mathbf{e}_1 + \cdots + \mathbf{e}_m \in E_d, \quad \int_\Omega \mathbf{e}_1 \cdot \boldsymbol{\tau}_1^d + \cdots + \mathbf{e}_m \cdot \boldsymbol{\tau}_m^d\, dV > 0 \right\},$$

2) the extremum remains the same if the space $\mathcal{E}^d$ is replaced by any linear subspace $\widehat{\mathcal{E}}^d$ dense in it and the set $E_d$ of kinematically admissible strain rate fields is replaced by any subspace $\widehat{E}^d$ in $\widehat{\mathcal{E}}^d$ satisfying the condition $(\widehat{E}^d)^0 = \Sigma_d$; in particular $E_d$ can be replaced by $\operatorname{Def} U_d$ or by any linear subspace in $\mathcal{E}^d$ whose closure is the same as the closure of $\operatorname{Def} U_d$.

PROOF. According to Theorem 6.2, the safety factor $s_\Delta$ can be determined as the extremum in the deviatoric static problem

$$\frac{1}{s_\Delta} = \inf \{F_d(\boldsymbol{\rho}) : \boldsymbol{\rho} \in \Sigma_d\},$$
$$F_d(\boldsymbol{\rho}) = \max\{M_{C^d}(\boldsymbol{\rho} + \boldsymbol{\tau}_1^d), \ldots, M_{C^d}(\boldsymbol{\rho} + \boldsymbol{\tau}_m^d)\}.$$

This problem is of the form (7.1), with $\mathcal{S}$, $C$, $\Delta$ replaced by $\mathcal{S}^d$, $C^d$, $\Delta^d$ and $\Sigma$ replaced by $\Sigma_d$ (the latter change presumes that the set of solenoidal virtual velocity fields $U_d$ is used instead of $U$). We also replace $\mathcal{E}$, $\mathcal{V}$ and $E = \operatorname{Def} \mathcal{V}$ by $\mathcal{E}^d$, $\mathcal{V}_d$ and $E_d = \operatorname{Def} \mathcal{V}_d$. Then the same reasoning as in Subsection 7.2 results in transforming the problem under consideration to the form analogous to (7.2) and constructing the corresponding modified kinematic problem. The latter is analogous to (7.6) and can also be written in the form analogous to (7.8). Theorem 1 is applicable to this modified problem and, with the above changes, turns out to be the present theorem, which finishes the proof.

**7.5. Finite element method.** Let us show that the modified kinematic problem can be discretized by the finite element method and the extremums of the discrete problems converge to the safety factor. We start with the following proposition which allows re-writing the kinematic problem in a more convenient form.

PROPOSITION 1. Suppose the assumptions of Theorem 1 or Theorem 2 are satisfied. Then the safety factor can be evaluated as the extremum

$$s_\Delta = \inf\left\{ \int_\Omega d_x(\mathbf{e}_1) + \cdots + d_x(\mathbf{e}_m)\, dV : \mathbf{e}_1, \ldots, \mathbf{e}_m \in \mathcal{E}, \right.$$

$$\left. \mathbf{e}_1 + \cdots + \mathbf{e}_m \in \operatorname{Def} V, \quad \int_\Omega \mathbf{e}_1 \cdot \boldsymbol{\tau}_1 + \cdots + \mathbf{e}_m \cdot \boldsymbol{\tau}_m \, dV = 1 \right\}, \quad (7.13)$$

where $V$ is the closure in $\mathbf{W}_2^1(\Omega)$ of the set $U$ or $U_d$ in the case of bodies with bounded or cylindrical yield surfaces, respectively (and $\mathcal{E}$ is replaced by $\mathcal{E}^d$ in the latter case).

The statement is completely analogous to Proposition IX.1.1 and can be proved the same way.

To discretize problem (7.13), consider an internal approximation of the space $V$, the latter being the subspace in $\mathbf{W}_2^1(\Omega)$ and a complete normed space with the same Sobolev norm. The definition of the approximation is given in Subsection IX.1.2, while some concrete approximations of velocity fields spaces are considered in Subsections IX.2.2 and IX.2.3. Apart from the approximation for the velocity fields space, that for the space $\mathcal{E}$ of strain rate fields is also needed here, and these approximations should be *consistent*: collections of the subspaces $V_h \subset V$ and $\mathcal{E}_h \subset \mathcal{E}$, which form the internal approximations for $V$ and $\mathcal{E}$, are to satisfy the condition $\operatorname{Def} V_h \subset \mathcal{E}_h$. The following examples show that the consistency condition can be easily satisfied.

EXAMPLE 1. In the case of a body with bounded yield surfaces, consider for the space $V$ of velocity fields the internal approximation formed by the subspaces $V_h$ described in Subsection IX.2.2. Note that the set $\operatorname{Def} V_h$ consists of the fields that take constant values on everyone of the tetrahedrons of the triangulation $\mathcal{T}_h$. This allows us to introduce the following approximation for the space $\mathcal{E}$ satisfying the condition $\operatorname{Def} V_h \subset \mathcal{E}_h$. For every tetrahedron $T$ of the triangulation $\mathcal{T}_h$ we define the function $f_T$ that takes the value 1 on $T$ and vanishes in its exterior. Then, we choose a certain Cartesian coordinate system and introduce the symmetric second order tensor field $\mathbf{a}$ with the only nonzero component $a_{11} = f_T$. We also introduce analogous field $\mathbf{b}$ with the only nonzero components $b_{12} = b_{21} = f_T$ and another four fields with the only nonzero components with the indices (i) 22, (ii) 33, (iii) 13 and 31, (iv) 23 and 32, respectively, each of these components being equal to $f_T$. Let $\mathcal{E}_h \subset \mathcal{E}$ be the finite-dimensional subspace consisting of all linear combinations of the above tensor fields defined for every tetrahedron of the triangulation. It can be easily seen that the collection of the subspaces $\mathcal{E}_h$ is an internal approximation for $\mathcal{E}$ and the condition $\operatorname{Def} V_h \subset \mathcal{E}_h$ is satisfied.

EXAMPLE 2. In the case of a body with cylindrical yield surfaces, consider for the space $V$ of solenoidal velocity fields the internal approximation formed by the subspaces $V_h$ described in Subsection IX.2.3. Note that the set $\operatorname{Def} V_h$ consists of second rank symmetric tensor fields which are 1) compatible, that is, are strain rate fields corresponding to kinematically admissible velocity fields; 2) deviatoric; 3) piecewise continuous; and 4) linear on every triangle in $\mathcal{T}_h$. To construct an appropriate internal approximation for the space $\mathcal{E} = \mathcal{E}^d$ of deviatoric strain rate fields, we make use of the triangulation $\mathcal{T}_h$ involved in the approximation for $V$. We consider the set $\mathcal{E}_h$ of second rank symmetric tensor fields which are deviatoric, piecewise continuous, and linear on every triangle in $\mathcal{T}_h$. It is clear that $\mathcal{E}_h$ is a finite-dimensional subspace in $\mathcal{E} = \mathcal{E}^d$ and the collection of these subspaces forms an internal approximation for $\mathcal{E}$, the condition $\operatorname{Def} V_h \subset \mathcal{E}_h$ being satisfied.

To discretize extremum problem (7.13), we simply replace the spaces $V$, $\mathcal{E}$ in its formulation by their finite-dimensional approximations $V_h$, $\mathcal{E}_h$. This results in the following extremum problem determining the approximation $s_{\Delta h}$ to the safety factor:

$$s_{\Delta h} = \inf \left\{ \int_\Omega d_x(\mathbf{e}_1) + \cdots + d_x(\mathbf{e}_m)\, dV : \mathbf{e}_1, \ldots, \mathbf{e}_m \in \mathcal{E}_h, \right.$$
$$\left. \mathbf{e}_1 + \cdots + \mathbf{e}_m \in \operatorname{Def} V_h, \quad \int_\Omega \mathbf{e}_1 \cdot \boldsymbol{\tau}_1 + \cdots + \mathbf{e}_m \cdot \boldsymbol{\tau}_m \, dV = 1 \right\}. \quad (7.14)$$

THEOREM 3. Suppose 1) conditions of Theorem 1 or Theorem 2 are satisfied and $V$ is the closure of the set of virtual velocity fields $U$ or $U_d$ (for bodies with bounded and cylindrical yield surfaces, respectively) in the norm $\|\cdot\|_V = \|\cdot\|_{\mathbf{W}_2^1(\Omega)}$; and 2) the collections of the subspaces $V_h$ in $V$ and $\mathcal{E}_h$ in $\mathcal{E}$ form internal approximations for the corresponding spaces $V$ and $\mathcal{E}$, satisfying the condition $\operatorname{Def} V_h \subset \mathcal{E}_h$. Then the sequence of the infimums (7.14) converges to the safety factor: $s_{\Delta h} \to s_\Delta$ when $h \to \infty$, each of the infimums being an upper bound for the safety factor: $s_\Delta \le s_{\Delta h}$.

The proof of the theorem is completely analogous to that of Theorem IX.1.1.

Theorem 3 reduces evaluation of the safety factor to solving the sequence of finite-dimensional extremum problems. Each of them consists of minimizing a convex nondifferentiable function over a finite-dimensional space and can be solved, for example, by the separating plane method.

## Comments

The shakedown theorem, the first main statement of the theory, was established by Melan (1938) and improved by Koiter (1960), who introduced the multiplier, allowing estimation of the plastic work from above. The reasoning that resulted in these statements remains valid in the case of the unified formulation of the elastic-plastic problem (Theorem 4.1). The second main statement of the theory, the nonshakedown theorem, seemed so obvious that no attempts to prove it were made for a long time. Only two decades ago, Débordes and Nayroles (1976) proved the nonshakedown theorem in the case of discrete systems. Their reasoning turned out to be partially applicable to the shakedown theory within the framework of the unified formulation of the elastic-plastic problem. However, proving Theorem 4.2 in this case calls for additional reasoning since the spaces of stress and strain rate fields for a continual system, generally, are different.

The shakedown and nonshakedown theorems are established on the basis of the static approach: they treat the problem in terms of possible loads, self-equilibrated and admissible stress fields. The static approach in the shakedown theory directly generalizes that of limit analysis. As to the kinematic approach, Koiter (1960) proved the kinematic theorem of the shakedown theory, which plays the same role as the kinematic upper bound theorem does in limit analysis. However, these two propositions are by far not analogous: in the case of limit analysis the upper bounds are determined by time-independent velocity fields, while in the case of shakedown analysis by the histories of plastic strain fields.

Several approaches were developed to eliminate the time variable from the kinematic problem of the shakedown theory. Chiras (1969) and Corradi and Zavelani (1974), in the case of bodies with piecewise linear yield surfaces, proposed a kinematic method for shakedown analysis analogous to that of limit analysis. Gokhfeld and Cherniavsky (1980) considered the shakedown analysis problem in two cases: with the safety factor determining (1) the value of the permanent load, and (2) the limits of variation of the variable load. In the first case, the kinematic problem of the shakedown analysis was reduced to that of limit analysis of an auxiliary fictitious body. In the second case, the time variable was also eliminated; however evaluating the safety factor was reduced to a nonstandard problem inverse to the limit analysis problem. Polizzotto *et al.* (1991) replaced the integration over the time interval in the kinematic extremum problem by the integration over the set of basis loads, which resulted in the kinematic method similar to that of limit analysis. Kamenjarzh and Weichert (1992) made use of the convex analysis approach and constructed the kinematic extremum problem dual of the static one arising from Melan's theorem and determining the safety factor. The kinematic problem only involves time-independent kinematic fields and is analogous to the limit analysis kinematic problem. This result originally obtained for spheric yield surface was extended by Kamenjarzh and Merzljakov (1994a, 1994b) to cover a wide class of yield surfaces. The presentation of the kinematic approach in Sections 6, 7 mostly follows these two articles.

Shakedown theory is dealt with in the books by Gokhfeld (1970), Martin (1975), Gokhfeld and Cherniavsky (1980), König (1987) and reviews by König and Maier (1981), Polizzotto (1982), and Maier and Lloyd Smith (1986). Generalization of the main statements of shakedown theory takes into account some important effects. Prager (1957) and Gokhfeld (1970) considered the thermal loading, Ceradini (1969, 1980), and Corradi and Maier (1974) the dynamic effects. Within the framework of more general constitutive relations including hardening, creep and damage, shakedown was considered in the works by Neal (1950), Maier (1970, 1973), Martin (1975), Ponter (1975),

Mandel (1976), Gokhfeld and Cherniavsky (1980), König (1982, 1987), Hachemi and Weichert (1992), Nayroles and Weichert (1993), and others. Maier (1973), Weichert (1983, 1986), Gross-Weege (1990), and Saczuk and Stumpf (1990) studied effects of geometric nonlinearity. Shakedown analysis methods allow estimating of not only the plastic work but also displacement, plastic strain, etc. (Polizzotto (1978, 1982)).

A lot of shakedown analysis problems are solved in the above-mentioned and other works, many of the problems being rather complicated and of substantial practical interest. Examples in Section 2 are not of this type. They just illustrate characteristic features of the elastic-plastic response, and the structures for these examples were selected to be as simple as possible. The example in Subsection 2.1 is taken from the book by König (1987), the idea of the example in Subsection 2.3 from the book by Kachanov (1971), and the example in Subsection 2.4 was constructed together with B.S. Safronov.

The main propositions of the shakedown theory were considered here within the framework of the unified formulation of the elastic-plastic problem. Therefore, they cover a wide class of continual and discrete systems and avoid smoothness assumptions concerning solutions to the elastic-plastic problem. The latter remark is important since the known theorems guarantee solvability of the problem in rather wide classes of generalized velocity and strain rate fields (Duvault and Lions (1972), Johnson (1976), and Matthies (1979)).

Note also that the solution to the elastic-plastic problem is unique in stress and is, generally speaking, nonunique in velocity. The nonuniqueness is not a peculiarity of the limit state but may also occur at a safe load. A simple example of the velocity and strain rate nonuniqueness at a safe load is given by White and Hodge (1980).

# BIBLIOGRAPHY


**Aleksandrov, S. E. (1992)**, Discontinuous velocity fields due to arbitrary strains in an ideal rigid-plastic body, *Sov. Phys. Doklady*, **37**, 6, 283.

**Annin, B. D., Bytev, V. O. and Senashov, S. I. (1985)**, *Group properties of elastic-plastic equations*, Nauka, Novosibirsk (in Russian).

**Annin, B. D. and Cherepanov, G. P. (1983)**, *Elastic-plastic problem*, Nauka, Novosibirsk (in Russian).

**Avitzur, B. (1980)**, *Metal forming: the application of limit analysis*, New York, Basel.

**Barabanov, O. O. (1989)**, On limit surface loads in plasticity, *Prikladnaya matematika i mechanika*, **53**, 5, 824 (in Russian).

**Berdichevsky, V. L. (1983)**, *Variational principles in continuum mechanics*, Nauka, Moscow (in Russian).

**Besov, O. V., Il'in, V. O. and Nikolsky, S. M. (1975)**, *Integral representations of functions and embedding theorems*, Nauka, Moscow (in Russian).

**Bishop, J. F. W. and Hill, R. (1951)**, A theory of plastic distortion of a polycrystalline aggregate under combined stresses, *Phil. Mag.*, **42**, 7, 414.

**Bogovsky, M. E. (1979)**, Solution to first boundary value problem for mass conservation equation for incompressible medium, *Doklady Akad. Nauk SSSR*, **248**, 5, 1037.

**Bolotin, V.V., Goldenblat, I. I. and Smirnov, A. F. (1972)**, *Structural mechanics. State of the art and prospects*, 2nd edition, Stroyizdat, Moscow (in Russian).

**Bouchitté, G. (1985)**, Homogénéisation sur $BV(\Omega)$ de fonctionnelles intégrales à croissance linéaire. Application à un problème d'analyse limite en plasticité, *C. R. Acad. Sci. Paris*, Sér. 1, **301**, 17, 785.

**Bouchitté, G. and Suquet, P. (1987)**, Charges limites, plasicité et homogénéisation: la cas d'un bord chargé, *C. R. Acad. Sci. Paris*, Sér 1, **305**, 441.

**Ceradini, G. (1969)**, Sul l'adattomento dei corpi elasto-plasici sogetti ad azioni dinamiche, *Giornale del Genio Civile*, 415, 239.

**Ceradini, G. (1980)**, Dynamic shakedown in elastic-plastic bodies, *J. Engng. Mech. Div. Proc. ASCE*, **106**, EM3, 481.

**Charnes, A. and Greenberg, H. J. (1951)**, Plastic collapse and linear programming, *Bull. Amer. Math. Soc.*, **57**, 6, 480.

**Charnes, A., Lemke, C. and Zienkievicz, O. (1959)**, Virtual work, linear programming and plastic limit analysis, *Proc. Roy. Soc.*, Ser. A, **251**, 1264, 110.

**Chiras, A. A. (1969)**, *Linear programming methods for analysis of one-dimensional elastic-plastic systems*, Stroyizdat, Leningrad (in Russian).

**Chiras, A. A. (1971)**, *Theory of optimization in limit analysis of deformable bodies*, Mintas, Vilnius (in Russian).

**Christiansen, E. (1980)**, Limit analysis for plastic plates, *SIAM J. Math. Anal.*, **11**, 3, 514.

**Ciarlet, P. G. (1978)**, *The finite element method for elliptic problems*, North Holland, Amsterdam.

**Ciavaldini, J. F. and Nedelec, J. C. (1974)**, Sur l'élément de Fraeijs de Veubeke et Sander, *Rev. Fransaise Automat. Informat. Recherche Opérationnelle, Sér. Rouge, Anal. Numér.*, R-2, 29.

**Corradi, L. and Zavelani, A. (1974)**, A linear programming approach to shakedown analysis of structures, *Comput. Meth. Appl. Mech. and Eng.*, **3**, 1, 73.

**Corradi, L. and Maier, G. (1974)**, Dynamic inadaptation theorem for elastic perfectly plastic continua, *J. Mech. Phys. Solids*, **22**, 401.

**Débordes, O. (1976)**, Dualité des théorèmes statique et cinématique dans la théorie de l'adaptation des milieux continus élastoplastiques, *C. R. Acad. Sci. Paris*, Sér. A, **282**, 535.

**Débordes, O. and Nayroles, B. (1976)**, Sur la théorie et le calcul à l'adaptation des structures élastoplastiques, *J. Mécanique*, **15**, 4, 1.

**Demengel, F. and Tang Qi (1986)**, Homogénéisation en plasticité, *C. R. Acad. Sci. Paris*, Sér. 1, **303**, 8, 339.

**Dorn, W. S. and Greenberg, H. J. (1957)**, Linear programming and plastic limit analysis of structures, *Quart. Appl. Math.*, **15**, 2, 155.

**Drucker, D. C. (1951)**, A more fundamental approach to plastic stress-strain relations, *Proc. 1st U.S. Nat. Congr. Appl. Mech.*, 487.

**Drucker, D. C. (1967)**, *Introduction to mechanics of deformable solids*, McGraw - Hill, New York.

**Drucker, D. C., Prager, W. and Greenberg, H. J. (1951)**, Extended limit design theorems for continuous media, *Quart. Appl. Math.*, **9**, 381.

**Drugan, W. H. and Rice, J. R. (1984)**, Restrictions on quasi-statically moving surfaces of strong discontinuity in elastic-plastic solids, *Mech. Mater. Behav. Daniel C. Drucker Aniv. Vol.*, Amsterdam, 59.

**Durban, D. and Fleck, N. A. (1992)**, Singular plastic fields in steady penetration of a rigid cone, *Trans. ASME, J. Appl. Mech.*, **59**, 4, 706.

**Duvault, G. and Lions, J.-L. (1972)**, *Les inéquations en mécanique et en physique*, Dunod, Paris.

**Edwards, R. E. (1965)**, *Functional analysis*, Holt, Rinehart and Winston, New York.

**Ekeland, I. and Temam, R. (1976)**, *Convex analysis and variational problems*, North Holland, Amsterdam.

**Feinberg, S. M. (1948)**, Limit stress state principle, *Prikladnaya matematika i mechanika*, **12**, 1, 63 (in Russian).

**Fenchel, W. (1949)**, On conjugate convex functions, *Canad. J. Math.*, **1**, 73.

**Fenchel, W. (1951)**, *Convex cones, sets and functions*, Princeton Univ., Prinston, NJ.

**Fraeijs de Veubeke, B. (1968)**, A conforming finite element for plate bending, *Intern. J. Solids Struct.*, **4**, 95.

**Frederick, C. O. and Armstrong, P. J. (1966)**, Convergent internal stresses and steady cyclic states of stress, *J. Strain Anal.*, **1**, 2, 154.

**Freidental, A. M. and Geiringer, H. (1958)**, The mathematical theories of inelastic continuum, *Handbuch der Physik*, **VI**, Springer, Berlin.

**Frémond, M. and Friâa, A. (1982)**, Les méthodes statique et cinématique en calcul à la rupture et en analyse limite, *J. Mécanique*, **21**, 5, 881.

**Frietas, J. A. T. (1985)**, An efficient simplex method for the limit analysis of structures, *Computers and Structures*, **21**, 1255.

**Galin, L. A. (1984)**, *Elastic-plastic problem*, Nauka, Moscow (in Russian).

**Gelfund, I. M. and Shilov, G. E. (1959)**, *Generalized functions and operations on them*, 2nd edition, Fizmatgiz, Moscow (in Russian).

**Germain, P. (1973)**, *Cours de méchanique des milieux continus. Théorie générale*, Masson, Paris.

**Gokhfeld, D. A. (1970)**, *Bearing capacity of structures under thermal loading*, Mashinostroenie, Moscow (in Russian).

**Gokhfeld, D. A. and Cherniavsky, O. F. (1980)**, *Limit Analysis of Structures at Thermal Cycling*, Sijthoff & Noordhoff, The Netherlands.

**Goldstein, E. G. (1967)**, Dual problems in convex and convex fractional programming in functional spaces, *Doklady Akad. Nauk SSSR*, **172**, 5, 1007 (in Russian).

**Goldstein, E. G. (1968)**, Dual problems in convex and convex fractional programming, *Issledovaniya po matematicheskomu programmirovaniyu*, Moscow (in Russian).

**Goldstein, E. G. (1971)**, *Duality theory in mathematical programming*, Nauka, Moscow (in Russian).

**Gross-Weege, J. (1990)**, A unified formulation of statical shakedown criteria for geometrically nonlinear problems, *Intern. J. Plasticity*, **6**, 433.

**Gudramovich, V. S., Gerasimov, V. P. and Demenkov, A. F. (1990)**, *Limit analysis of structural members*, Naukova dumka, Kiev (in Russian).

**Gvozdev, A. A. (1938)**, Determining failure load for systems undergoing plastic deformation, *Trudy konferentsii po plasticheckim deformatsiyam*, Izdat. AN SSSR, 19 (in Russian).

**Gvozdev, A. A. (1948)**, On limit equilibrium, *Inzhenernyii sbornik*, **1**, 32 (in Russian).

**Gvozdev, A. A. (1949)**, *Evaluating of structural bearing capacity by limit equilibrium method. Part I. Essentials and justification of the method*, Stroyizdat, Moscow (in Russian).

**Hachemi, A. and Weichert, D. (1992)**, An extension of the static shakedown theorem to a certain class of inelastic materials with damage, *Arch. Mech.*, **44**, 491.

**Halphen, B. and Nguen, Q. S. (1975)**, Sur la matériaux standards généralisés, *J. Mécanique*, **14**, 1, 36.

**Hill, R., (1948)**, A variational principle of maximum plastic work in classical theory of plasticity, *Quart. J. Mech. Appl. Math.*, **1**, 1, 18.

**Hill, R., (1950)**, *Mathematical theory of plasticity*, Oxford.

**Hill, R., (1961)**, Discontinuity relations in mechanics of solids, Sneddon, I. N., Hill, R. (Eds.), *Progress in solid mechanics*, **II**, North Holland, Amsterdam, 247.

**Hodge, P. G. (1981)**, *Plastic analysis of structures*, R.E. Krieger Publ. Comp., Malabar, Florida.

**Hodge, P. G. and Belytschko, T. (1970)**, Plane strain limit analysis by finite elements, *J. Eng. Mech. Div. Proc. ASCE*, **96**, 6, 931.

**Ioffe, A. D. and Tikhomirov, V. M. (1979)**, *Theory of extremal problems*, North Holland, Amsterdam, New York.

**Ivlev, D. D. (1966)**, *Theory of perfect plasticity*, Nauka, Moscow (in Russian).

**Ivlev, D. D. (1967)**, On dissipation in theory of plastic media, *Doklady Akad. Nauk SSSR*, **176**, 5, 1037 (in Russian).

**Ivlev, D. D. and Ershov, L. V. (1978)**, *Perturbation method in theory of elastic-plastic solids*, Nauka, Moscow (in Russian).

**Jikov, V. V. (1986)**, Homogenization of functionals in calculus of variations and elasticity, *Izvestiya AN SSSR, Matematika*, **50**, 4, 155 (in Russian).

**Jikov, V. V., Kozlov, S. M. and Oleinik, O.A. (1994)**, *Homogenization of differential operators and integral functionals*, Springer, Berlin, Heidelberg, New York.

**Johnson, C. (1976)**, Existence theorems for plasticity problems, *J. Meth. Pures et Appl.*, **55**, 431.

**Kachanov, L. M. (1971)**, *Foundations of the theory of plasticity*, North Holland, Amsterdam.

**Kamenjarzh, J. (1978)**, Plastic body subjected to limit load: stress field, *Sov. Phys. Dokl.*, **23**, 7, 518.

**Kamenjarzh, J. (1979)**, On dual problems in theory of limit load for perfectly plastic bodies, *Sov. Phys. Dokl.*, **24**, 3, 177.

**Kamenjarzh, J. (1981)**, Stresses in incompressible media. Equivalence of certain formulations of plasticity problems, *Sov. Phys. Dokl.*, **26**, 11, 1051.

**Kamenjarzh, J. (1984)**, Integral estimate of pressure in incompressible medium, *PMM Journal Appl. Math. Mech.*, **48**, 1, 81.

**Kamenjarzh, J. (1985)**, On discontinuous solutions in limit load theory, *Sov. Phys. Dokl.*, **30**, 1, 42.

**Kamenjarzh, J. (1986)**, Conditions at discontinuity surfaces in rigid-plastic analysis, *Sov. Phys. Dokl.*, **31**, 1, 38.

**Kamenjarzh, J. and Weicher, D. (1992)**, On kinematic upper bound for the safety factor in shakedown theory, *Intern. J. Plasticity*, **8**, 827.

**Kamenjarzh, J. and Merzljakov, A. (1985)**, On theory of limit load, *PMM Journal Appl. Math. Mech.*, **49**, 4, 515.

**Kamenjarzh, J. and Merzljakov, A. (1994a)**, On kinematic method in shakedown theory: I. Duality of extremum problems, *Intern. J. Plasticity*, **10**, 363.

**Kamenjarzh, J. and Merzljakov, A. (1994b)**, On kinematic method in shakedown theory: II. Modified kinematic method, *Intern. J. Plasticity*, **10**, 381.

**Kazinczy, G. (1914)**, Experiments with clamped beams, *Betonszemle*, 4, p. 68, 5, p. 83, 6, p. 101 (in Hungarian).

**Khan, A. S. and Sujian Huang (1995)**, *Continuum theory of plasticity*, John Wiley & Sons, New York.

**Kist, N. C. (1917)**, Does a stress analysis based on Hook's law lead to a satisfactory design of steel bridges and buildings? – *Inaugural lecture*, Delft.

**Klyushnikov, V. D. (1979)**, *Mathematical theory of plasticity*, Izdat. MGU, Moscow (in Russian).

**Kohn, R. and Temam, R. (1982)**, Principles variationnels duaux et théoreme de l'énergie dans le modéle de plasticité de Hencky, *C. R. Acad. Sci. Paris*, **294**, I-205.

**Koiter, W. T. (1960)**, General theorems for elastic-plastic solids, Sneddon, I. N. and Hill, R. (Eds.), *Progress in Solid Mechanics*, **I**, North Holland, Amsterdam.

**Kolmogorov, A. N. and Fomin, S. V. (1970)**, *Introductory real analysis*, Prentice-Hall, Englewood Cliffs, N.J.

**König, J. A. (1982)**, On some recent developments in the shakedown theory, *Adv. Mech.*, **5**, 235.

**König, J. A. (1987)**, *Shakedown of elastic-plastic structures*, PWN – Polish Scientific, Warsaw, and Elsevier, Amsterdam.

**König, J. A. and Maier, G. (1981)**, Shakedown analysis of elastic-plastic structures: a review of recent developments, *Nucl. Engng. Design*, **6**, 81.

**Koopman, D. and Lance, R. (1965)**, On linear programming and plastic limit analysis, *J. Mech. Phys. Solids*, **3**, 2, 77.

**Kukudzhanov, V. N., Lyubimov, V. M. and Myshev, V. D. (1984)**, Method for determining of lower bounds for limit load, *Chislennye metody v meh. deform. tv. tela*, Moscow, 138 (in Russian).

**Ladyzhenskaya, O. A. (1973a)**, *Mathematical problems in dynamics of incompressible fluid*, 2nd edition, Nauka, Moscow (in Russian).

**Ladyzhenskaya, O. A. (1973b)**, *Boundary value problems in mathematical physics*, Nauka, Moscow (in Russian).

**Ladyzhenskaya, O. A. and Solonnikov, V. A. (1976)**, On some problems of vector analysis and generalized formulations of problems for Navier–Stokes equations, *Zapiski nauchnyh seminarov LOMI AN SSSR*, **59**, 9, 81.

**Lagrange, J.-L. (1787)**, *Mechnique analitique*, Paris.

**Lance, R. H. (1967)**, On automatic construction of velocity fields in plastic limit analysis, *Acta Mech.*, **3**, 1, 22.

**Lions, J.-L. (1969)**, *Quelques méthodes de résolution des problèms aux limites non linèaires*, Dunod, Paris.

**Lions, J.-L. and Magenes, E. (1968)**, *Problèmes aux limites non homogènes et applications*, Dunod, Paris.

**Maier, G. (1970)**, A matrix structural theory of piecewise linear plasticity with interacting yield planes, *Meccanica*, **5**, 55.

**Maier, G. (1973)**, A shakedown matrix theory allowing for workhardening second order geometrical effects, *Mech. Plast. Solids*, **I**, Noordhoff, Leyden, 417.

**Maier, G. and Lloyd Smith, D. (1986)**, Update to mathematical programming applications to engineering plastic analysis, Steele, C. R. and Springer, G. S. (Eds.), *Applied Mechanics Update*, New York, 377.

**Mandel, J. (1976)**, Adaptation d'une structure plastique ecrouissable et approximations, *Mech. Res. Comm.*, **3**, 483.

**Markov, A. A. (1947)**, On variational principles in plasticity, *Prikladnaya matematika i mehanika*, **11**, 3, 339.

**Martin, J. B. (1975)**, *Plasticity: fundamentals and general results*, MIT Press, Cambridge, MA.

**Masslennikova, V. N. and Bogovsky, M. E. (1978)**, On density of solenoidal vector fields, *Sibirskiy matem. jurnal*, **19**, 5, 1092.

**Matthies, H. (1979)**, Existence theorems in thermoplasticity, *J. Mécanique*, **18**, 4, 695.

**Matthies, H., Strang, G. and Christiansen, E. (1979)**, The saddle point of a differential program, Glowinski, R., Rodin, E. Y. and Zienkiewicz O. C. (Eds.), *Energy methods in finite element analysis*, John Willey & Sons.

**Melan, E. (1938)**, Theorie statisch uberstimmer tragwerke aus ideal plastischem baustoff, *Sitzungbericht der Academie der Wissenschaften, Wien*, **IIA**, 145, 195.

**Mercier, B. (1977)**, Une méthode pour résourde le probléme des charges limites, *J. Mécanique*, **16**, 3, 467.

**Mercier, B. (1982)**, Numerical methods for problems of viscousplasticity, Lions, J.-L. and Marchuk, G. I. (Eds.), *Numerical methods in applied mathematics*, Nauka, Novosibirsk, 63 (in Russian).

**Mikhlin, S. G. (1977)**, *Linear equations in partial derivatives*, Vysshaya shkola, Moscow (in Russian).

**Mises, R. (1913)**, Mechanik der festen körper im plastisch deformablen zustand, *Göttinger Nuchr., Math. Phys. Kl.*, 582.

**Mises, R. (1928)**, Mechanik der plastischen formänderung von kristallen, *ZAMM*, **8**, 161.

**Moreau, J. J. (1962)**, Fonctions convexes en dualité, *Séminaire de Math.*, Faculte des sciences de Montpellier.

**Moreau, J. J. (1964)**, Sur la fonction polaire d'une fonction semi-continue supérieurement, *C. R. Acad. Sci. Paris*, **258**, 1128.

**Moreau, J. J. (1966)** Fonctionnelles convexes, *Séminaire equations aux dérivées partielles*, Collège de France.

**Morrey, C. B. (1966)**, *Multiple integrals in the calculus of variation*, Springer, Berlin.

**Mosolov, P. P. (1967)**, On minimum of integral functional, *Izvestiya AN SSSR, Ser. Matem.*, **31**, 6, 1289.

**Mosolov, P. P. (1978)**, On some mathematical problems in theory of incompressible viscousplastic media, *Prikladnaya matematika i mechanika*, **42**, 4, 737 (in Russian).

**Mosolov, P. P. and Myasnikov, V. P. (1971)**, *Variational methods in theory of rigid-viscousplastic flow*, Izdatelstvo MGU, Moscow (in Russian).

**Mosolov, P. P. and Myasnikov, V. P. (1977)**, Asymptotic theory of rigid-plastic shells, *Prikladnaya matematika i mechanika*, **41**, 3, 538 (in Russian).

**Mosolov, P. P. and Myasnikov, V. P. (1981)**, *Mechanics of viscousplastic media*, Nauka, Moscow (in Russian).

**Nadai, A. L. (1949)**, *Theory of flow and fracture of solids*, revised edition, McGraw, New York.

**Nayroles, B. (1970)**, Essai de théorie fonctionnele des structures rigides plastiques parfaites, *J. Mécanique*, **9**, 491.

**Nayroles, B. (1971)**, Quelques applications variationnelles de la théorie des fonctions duales a la mécanique des solides, *J. Mécanique*, **10**, 2, 491.

**Nayroles, B. and Weichert, D. (1993)**, La notion de sanctuaire d'élasticité et d'adaptation des structures, *C. R. Sci. Paris*, Sér. II, **316**, 1493.

**Neal, B. G. (1950)**, Plastic collapse and shake-down theorems for structures of strain-hardening materials, *J. Aeronaut. Sci.*, **17**, 297.

**Neal, B. G. (1963)**, *The plastic methods of structural analysis*, 2nd edition, John Wiley & Sons, New York.

**Nurminsky, E. A. (1991)**, *Numerical methods for convex optimization*, Nauka, Moscow (in Russian).

**Ovechkin, A. M. (1961)**, *Analysis of reinforced concrete axially symmetric structures*, Gosstroyizdat, Moscow (in Russian).

**Panagiatopoulos, P. D. (1985)**, *Inequality problems in mechanics and applications*, Birkhäuser, Boston, MA.

**Panarelie, J. E. and Hodge, P. G. (1963)**, Interaction of pressure and load and twisting moment for a rigid-plastic circular tube, *J. Appl. Mech.*, 30, 4, 396.

**Polizzotto, C. (1978)**, A unified approach to quasi-static shakedown problems for elastic-plastic solids with piecewise linear yield surfaces, *Meccanica*, **13**, 2, 109.

**Polizzotto, C. (1982)**, A unified treatment of shakedown theory and related bounding technique, *Solid Mech. Arch.*, **7**, 1, 19.

**Polizzotto, C., Borino, G., Caddemi, S. and Fuschi, P. (1991)**, Shakedown problems for material models with internal variables, *Eur. J. Mech., A /Solids* 10, *6, 621.*

**Ponter, A. R. S. (1975)**, *A general shakedown theorem for elastic-plastic bodies with hardening, 3rd Int. Conf. Struct. Mech. Reactor Technol.*,

*London 1975*, **5**, *Amsterdam e.a., L5.2/1.*

**Prager, W. (1974)**, *Limit analysis: the development of a concept, Sawczuk, A. (Ed.), Problems of Plasticity, Noordhoff Intern. Publ., Leyden, 3.*

**Prager, W. (1974)**, *Introduction to structural optimization, Springer.*

**Prager, W. and Hodge, P. G. (1968)**, *Theory of perfect plastic solids, Dover Publications, New York.*

**Prandtl, L. (1923)**, *Anwedungsbeispiele zu einem Henckyschen satz uber das plastische gleichgeweieht, Zeits. ang. math. mech.*, **6**, *401.*

**Protsenko, A. M. (1982)**, *Theory of elastic perfectly plastic systems, Nauka, Moscow (in Russian).*

**Rabotnov, U. N. (1979)**, *Mechanics of deformable solids, Nauka, Moscow (in Russian).*

**Reitman, M. I. and Shapiro, G. S. (1964)**, *Optimal design theory in structural mechanics, elasticity and plasticity, Mechanics. Elasticity and plasticity, Izdat. VINITY, Moscow (in Russian).*

**de Rham, G. (1955)**, *Variétes diffèrentiables, Hermann, Paris.*

**van Rij, H. M. (1979)**, *A finite element model for plane strain plasticity with velocity discontinuities, Trans. 5th Int. Conf. Struct. Mech. React. Technol.*, **M**, *Berlin, Amsterdam e.a., 1.*

**van Rij, H. M. and Hodge, P. G. (1978)**, *A slip model for finite-element plasticity, J. Appl. Mech.*, **45**, *527.*

**Rockafellar, R. T. (1967)**, *Duality and stability in extremum problems involving convex functions, Pacific J. Math.*, **21**, *167.*

**Rockafellar, R. T. (1968)**, *Integrals which are convex functionals, Pacific J. Math.*, **24**, *525.*

**Rockafellar, R. T. (1969)**, *Convex functions and duality in optimization problems and dynamics, Lecture Notes in Oper. Res. and Math. Ec.*, **II**, *Springer, Berlin.*

**Rockafellar, R. T. (1970)**, *Convex Analysis, Princeton Univ. Press, Princeton, NJ.*

**Rzhanitsin, A. R. (1954)**, *Analysis of structures allowing for plastic material properties, 2nd edition, Stroyizdat (in Russian).*

**Rzhanitsin, A. R. (1983)**, *Limit equilibrium of plates and shells, Nauka, Moscow (in Russian).*

**Saczuk, J. and Stumpf, H. (1990)**, *On statical shakedown theorems for nonlinear problems, Ruhr-Universität Bochum, Mitteilungen Inst. für Mech.*, **47**, *1.*

**de Saint Vénant, B. (1870)**, *Mémoire sur l'établissement des équations différentielles des mouvements intérieurs opérés dans les corps ductiles au delà des limites où l'élasticité pourrait les ramener à leur premier état, C. R. Acad. Sci. Paris*, **70**, *473.*

**Salençon, J. (1974)**, *Application of the theory of plasticity in soil mechanics, John Wiley & Sons, New York.*

**Sander, G. (1964)**, *Bornes supérieures et inférieures dans l'analyse matricielle des plaques en flexion - torsion, Bull. Soc. Roy. Sci. Liège*, **33**, *456.*

**Save, M. A. (1985)**, *Limit analysis design: an up-to-date subject of engineering plasticity, Plasticity today: modelling, methods, and applications, Elsevier Publ., London, New York, 767.*

**Save, M. A. and Massonet, C. E. (1972)**, *Plastic analysis and design of plastic plates, shells and disks, North Holland, Amsterdam.*

**Schwartz, L. (1950)**, *Théorie des distributions,* **I**, *Hermann, Paris.*

**Schwartz, L. (1951)**, *Théorie des distributions,* **II**, *Hermann, Paris.*

**Sedov, L. I. (1994)**, *Continuum mechanics, 5th edition, Nauka, Moscow (in Russian).*

**Seriogin, G. A. (1983)**, *Extension of variational formulation of rigid-plastic problem including velocity fields with slip discontinuities, Prikladnaya matematika i mechanika,* **47**, *6, 1030 (in Russian).*

**Seriogin, G. A. (1987)**, *On a variational-difference scheme for limit analysis problem, Zhurnal vychisl. matem. i matem. fiziki,* **27**, *1, 83 (in Russian).*

**Sobolev, S. L. (1991)**, *Some applications of functional analysis to mathematical physics, AMS, Providence, Rhode Island.*

**Sokolovsky, V. V. (1969)**, *Theory of plasticity, 3rd edition, Vysshaya shkola, Moscow (in Russian).*

**Strang, G. and Fix, G. J. (1973)**, *An analysis of the finite element method, Prentice-Hall.*

**Suquet, P. (1978)**, *Sur un nouveau cadre fonctionnel pour les equations de la plasicité, C. R. Acad. Sci. Paris, Sér. A,* **286**, *23, 1129.*

**Suquet, P. (1983)**, *Analyse limite et homogénéisation, C. R. Acad. Sci. Paris, Sér. II,* **296**, *18, 1355.*

**Tarnovsky, I. Y. and Pozdeev, A. A. (1963)**, *Theory of metal forming (variational methods for stress and strain analysis), Metallurgizdat, Moscow (in Russian).*

**Telega, J. J. (1985)**, *Variational methods and extremum principles for some non-classical problems in plasticity: unilateral boundary conditions, friction, discontinuities, Variational Methods in Engineering, Proc. 2nd Ind. Conf., Southampton, Berlin, 8-27.*

**Temam, R. (1983)**, *Problèmes mathématiques en plasticité, Gauthier - Villars, Paris.*

**Temam, R. and Strang, G. (1978)**, *Existence de solutions relaxées pour les equations de la plasicité: étude d'un espace fonctionnel, C. R. Acad. Sci. Paris, Sér. A,* **287**, *515.*

**Temam, R. and Strang, G. (1980)**, *Duality and relaxation in the variational problems of plasticity, J. Mécanique,* **19**, *3, 494.*

**Thomsen, E. G., Yang, C. T. and Kobayashi, S. (1965)**, *Mechanics of plastic deformation in metal processing, Macmillan Company, New York.*

**Tresca, H. (1864)**, *Mémoire sur l'écoulement des corps solides soumis à de fortes pressions, C. R. Acad. Sci. Paris,* **59**, *754.*

**Volevich, L. R. and Paneyakh, B. P. (1965)**, *Spaces of generalized functions and embedding theorems, Uspehi matem. nauk,* **20**, *1, 3 (in Russian).*

**Weichert, D. (1983)**, *Shakedown at finite displacements: a note on Melan's theorem, Mech. Res. Comm.*, **11**, *121.*

**Weichert, D. (1986)**, *On the influence of geometrical nonlinearities on the shakedown of elastic-plastic structures, Intern. J. Plasticity*, **2**, *135.*

**White, D. L. and Hodge, P. G. (1980)**, *Computation of non-unique solutions of elastic-plastic trusses, Computers and Structures*, **12**, *769.*

**Zadoian, M. A. (1992)**, *Spatial problems in plasticity, Nauka, Moscow (in Russian).*

**Ziegler, H. (1963)**, *Some extremum principles in irreversible thermodynamics with application to continuum mechanics, Sneddon, I. N., Hill, R. (Eds.), Progress in solids mechanics*, **IV**, *North Holland, Amsterdam.*

## INDEX


approximation
 external 338
 internal 316, 320, 338
 to solenoidal vector field 222, 225, 228
averaging 295
axial force 27
Banach space 193
basis functions 319, 321
 degrees of freedom 325
beam 25, 28, 54, 177, 414
bearing capacity 24, 72, 114
boundary conditions 28, 35, 47, 54
 integral 52
 mixed 51
 surface slip 51
bending moment 27
composite body 124
constitutive maximum principle 12, 350
 integral formulation 101, 102, 120
constitutive minimum principle 43
cutting force 27
cyclic loading 361, 365, 370, 377, 394
 damaging 393
deformation 36
discontinuity relations 47, 264, 267, 271, 282
discrete system 109, 358
dissipation 15, 17, 116, 121, 126
 at discontinuity surface 256, 275, 279
distribution 209, 233
dual extremum problems 157, 158, 195
duality theorem 158, 195
elastic energy 354, 357
elastic modules tensor 350, 354
elastic-plastic body 2, 350
elastic-plastic problem 350
 discrete system 358
 general formulation 355, 357
 strong formulation 350, 353
  in stresses 358
elastic reference body 382
elasticity
 domain 3, 5
 law 37, 349, 354
 limit 2
epigraph 38
$\varepsilon$-subdifferential 336
equilibrium conditions 34, 47, 96
 generalized 50, 52, 53
 weak 202
extremum problem 41, 157
failure mechanism 70, 71, 90, 115, 245, 250, 252
Fenhel transformation 154, 160, 162
finite element 319, 321, 324
finite element method 319
 basis functions 319, 321, 325
 interpolant 319, 322, 328
 triangulation 319, 321, 324, 325
function
 convex 38, 174
 essentially bounded 192
 homogeneous 17
 measurable 140
functional 127, 194
hinge 31
infimum 40

integral functional 127, 130
interpolant 319, 322, 328
jump 46, 49
  of velocity field 256, 259, 272
kinematic extremum problem 144, 164, 175, 180, 182, 314
  approximations
  deviatoric 203
  discretization 333
  dual of the static problem 172
  in shakedown analysis 413, 420, 430
  reduced 204, 206
  standard form 151
kinematic method 81, 84, 89, 119, 145, 164, 171, 175, 180, 182, 206, 263, 280, 314
  in shakedown analysis 418, 421, 423, 424, 430
kinematic multiplier 80, 115, 258, 260, 262, 280, 294
kinematic relations 98
Kirchhoff hypothesis 25
Korn's inequality 315
Kuhn-Tucker theorem 14, 42
limit analysis 114
  discretized problem 331, 333
  examples 87, 88, 283, 285, 289, 290, 291
  in presence of a permanent load 76
  main problem 73
limit multiplier 144, 145, 164, 171, 175, 180, 182, 206
  kinematic 144
  static 144
limit load 23, 70, 111, 114
limit state 244
  stress 245, 250, 252
  failure mechanism 250, 252
limit stress state principle 72
limit surface 76, 289
linear space 136
  conjugate 194
  normed 190
  subspace 137
load 19
  admissible 24, 64, 66, 67, 111, 114
  inadmissible 24, 64, 66, 67, 111, 114
  safe 68, 74, 113
loading 4
  cyclic 361, 365, 370, 377, 394
  proportional 73
  quasistatic 19
  rigid punch 52
  thermal 370, 408
material properties
  continuous 261
  discontinuous 261, 275, 276, 291
  local description 105
maximum 40
measure 139
Melan theorem 389
minimum 39
Minkowsky function 147
nonshakedown 364, 382
  conditions 384, 385, 396
  examples 364,369, 386
nonshakedown theorem 396
normed space 190
  internal approximation 320
normality flow rule 8, 10, 13
pairing 138, 153
pipe 87, 88, 290
plane with holes 289
plastic body 4
  constitutive relations 8, 12
  perfectly plastic body 6
  elastic-plastic body 350
  rigid-plastic body 9, 14
  thermoelastic-plastic body 408
plastic failure 70, 72, 114
plastic work 381
  lower bound 391, 392
  upper bound 390
polar set 138
power
  external 46, 51, 53, 96, 356
  internal 46, 356
pressure field restoration 204, 210, 212
rate insensitivity 3, 5

residual stresses 359, 363, 366, 378, 391
rigid-plastic problem 20, 23, 24, 34, 105
discrete system 109
general formulation 103
strong formulation 63, 104
weak formulation 249
rigid-plastic solution method 84, 87, 119, 268, 280
safety criterion 69, 113
safety factor 66, 111, 113
lower bound 78, 115
upper bound 117
separating plane 38, 339
separating plane method 338
algorithm 343
convergence 344
external approximation 338
internal approximation 338
set
absorbing 112
bipolar 247
convex 37
measurable 139
polar 138
shakedown 349, 360, 381, 408, 409
conditions 384, 385
examples 360, 368, 374, 386
extremum problems 374, 410, 413, 418, 421
one-parametric problems 401, 403, 404
safety factor 384, 400, 410, 418, 421, 423, 428, 430
under thermomechanical loading 408, 409
shakedown safety factor 384, 395, 410, 418, 421, 423, 428, 430
upper bound 426
shakedown theorem 389
Sobolev space 235, 236, 238
static extremum problem 144, 164, 175, 180, 182
dual of the kinematic problem 159
in shakedown analysis 410, 420
reduced 202, 206
standard form 149
static method 79, 88, 119, 154, 164, 171, 175, 180, 182, 205, 206
in shakedown analysis 418, 421
static multiplier 78, 80, 115
strain 1, 36
elastic 1, 349
plastic 5, 349
residual 2, 4
temperature 408
strain rate 36, 62, 97, 98 175, 178, 205, 356
stream function 324
stress 6, 46, 166, 202, 275
admissible 7, 14, 60, 166
strip 283
with a hole 285
subgradient 335, 346
supremum 40
surface slip 51, 259
thermal expansion 408
thermal loading 370, 408
traction 35
admissible 270
truss 31, 56, 165, 175, 352
vector field
approximation 217, 221
piecewise regular 261
solenoidal 222, 225, 228
with a given divergence 224
unloading 4
velocity fields 62, 96, 98, 248, 316, 356
discontinuous 256, 261
kinematically admissible 62
virtual 46, 51, 96
virtual work principle 46, 52, 53, 62, 96, 356
yield function 7, 29
yield plateau 3, 8
yield stress 2
yield surface 6, 8, 11, 177

Printed and bound by CPI Group (UK) Ltd, Croydon, CR0 4YY
17/10/2024
01775690-0017